Medicinal Chemistry

An Introduction

Gareth Thomas

University of Portsmouth

John Wiley & Sons, Ltd

Chichester • New York • Weinheim • Brisbane • Singapore • Toronto

Published by John Wiley & Sons, Ltd, The Atrium,
Southern Gate, Chichester
West Sussex PO19 8SQ, England
Telephone (+44) 1243 779777

Email (for orders and customer service enquiries): cs-books@wiley.co.uk
Visit our Home Page www.wileyeurope.com or www.wiley.com

Reprinted with corrections November 2001, November 2002, November 2003

Other Wiley Editorial Offices

John Wiley & Sons Inc., 111 River Street, Hoboken, NJ 07030, USA

Jossey-Bass, 989 Market Street, San Francisco, CA 94103-1741, USA

Wiley-VCH Verlag GmbH, Boschstr. 12, D-69469 Weinheim, Germany

John Wiley & Sons Australia Ltd, 33 Park Road, Milton, Queensland 4064, Australia

John Wiley & Sons (Asia) Pte Ltd, 2 Clementi Loop #02-01, Jin Xing Distripark, Singapore 129809

John Wiley & Sons (Canada) Ltd, 22 Worcester Road, Etobicoke, Ontario M9W 1L1

Library of Congress Cataloging-in-Publication Data

Thomas, Garry, Dr.
Medicinal chemistry : an introduction / Garry Thomas.
p. ; cm.
Includes bibliographical references and index.
ISBN 0–471–98807–3 (cloth : alk. paper) – ISBN 0–471–48935–2 (pbk. : alk. paper)
1. Pharmaceutical chemistry. I. Title.
[DNLM: 1. Chemistry, Pharmaceutical. 2. Drug Design. 3. Drug Evaluation.
4. Phamacokinetics QV 744 T4567m 2000]
RS403 .T447 2000
615′.19 – dc21 00-043439

British Library Cataloguing in Publication Data
A catalogue record for this book is available from the British Library

ISBN 0–471–988073 (Hardback)
0–471–489352 (Paperback)

Typeset in 9.75 on 13 pt Times by Best-set Typesetter Ltd, Hong Kong
Printed and bound in Great Britain by Biddles Ltd, Guildford and King's Lynn
This book is printed on acid-free paper responsibly manufactured from sustainable forestry, in which at least two trees are planted for each one used for paper production.

Contents

Preface

This book is written for second, and subsequent, year undergraduates studying for degrees in medicinal chemistry, pharmaceutical chemistry, pharmacy, pharmacology and other related degrees. It assumes that the reader has a knowledge of chemistry at level one of a university life sciences degree. The text discusses the chemical principles used for drug discovery and design with relevant physiology and biology introduced as required. Readers do not need any previous knowledge of biological subjects.

Chapter 1 is intended to give an overview of the subject and also includes some topics of peripheral interest to medicinal chemists that are not discussed further in the text. Chapter 2 discusses the approaches used to discover and design drugs. The remaining chapters cover the major areas that have a direct bearing on the discovery and design of drugs. These chapters are arranged, as far as is possible, in a logical succession.

The approach to medicinal chemistry is kept as simple as possible. Each chapter has a summary of its contents in which the key words are printed in bold type. The text is also supported by a set of questions at the end of each chapter. Answers, sometimes in the form of references to sections of the book, are listed separately. A list of recommended further reading, classified according to subject, is also included.

Acknowledgements

I wish to thank all my colleagues, past and present, without whose help this book would have not been written. In particular, I would like to thank Dr J. Tsibouklis and Dr J. Wong, both of the University of Portsmouth, Dr Christopher G. Frost, University of Bath, Dr Andrew W. Lloyd and Dr A. Christy Hunter, both of the University of Brighton, Dr Anita R. Maguire, University College Cork, and Dr Alison Rodger, University of Warwick, for reading and commenting upon the complete manuscript, Dr S. Arkle for access to his lecture notes, Dr J. Brown for acting as a living pharmacology dictionary, Miss J. Chapman for sterling work at the photocopier, Dr P. Cox for the molecular model diagrams and Mr A. Barrow and Dr D. Brimage for the library searches they conducted. I wish also to thank the following friends and colleagues for proofreading chapters and supplying information; Dr L. Adams, Dr C. Alexander, Dr E. Allen, Dr L. Banting, Dr D. Brown, Dr S. Campbell, Dr B. Carpenter, Mr P. Clark, Mr M. Colville, Dr P. Howard, Dr A. Hunt, Mrs W. Jones, Dr T. Mason, Dr T. Nevell, Dr M. Norris, Dr J. Smart, Professor D. Thurston, Dr G. White and Mr S. Wills. They will all be relieved to know that next time they see me I will not be carrying the ubiquitous brown envelope! Thanks are also due to Dr Erland and Stevens of Davidson College for helpful comments and suggestions on the first printing.

Finally, I would like to thank my wife for the cover design and the sketches included in the text. Her support through the years when I was writing the text made its completion possible.

Abbreviations

A	Adenine
ACE	Angiotensin-converting enzyme
ADP	Adenosine diphosphate
Ala	Alanine
AMP	Adenosine monophosphate
Arg	Arginine
Asp	Aspartate
ATP	Adenosine triphosphate
AUC	Area under the curve
C	Cytosine
CaM	Calmodulin
cAMP	Cyclic adenosine monophosphate
CNS	Central nervous system
CYP-450	Cytochrome P-450 family
Cys	Cysteine
DAG	Diacylglycerol
DCC	Dicyclohexylcarbodiimide
DDD	Dichlorodiphenyldichloroethene
DDE	Dichlorodiphenyldichloroethane
DDT	Dichlorodiphenyltrichloroethane
d.e.	diastereoisometric excess
DNA	Deoxyribonucleic acid
e.e.	enantiomeric excess
FAD	Flavin adenine dinucleotide
FH_2	Dihydrofolic acid
FMO	Flavin monooxygenases
Fmoc	9-Fluorenylmethoxychloroformyl group
G	Guanine
GABA	γ-Aminobutyric acid
GC	Guanylyl cyclase
GDP	Guanosine diphosphate
Glu	Glutamate
Gly	Glycine
cGMP	Cyclic gnanosine monophosphate
5′-GMP	Guanosine 5′-monophosphate
GSH	Glutathione
GTP	Guanosine triphosphate

HbS	Sickel cell haemoglobin
hGH	Human growth hormone
His	Histidine
HIV	Human immunodeficiency disease
HMP	Hydroxymethylphenoxy-resin
hnRNA	heterogeneous nuclear RNA
I	Inosine
IDDM	Insulin dependent diabetes mellitus
IL-1	Interleukin 1
Ile	Isoleucine
IP_3	Inositol-1,4,5-triphosphate
Leu	Leucine
Lys	Lysine
mACh	Muscarinic cholinergic receptor
$MeFH_4$	Methylenetetrahydrofolic acid
MESNA	Sodium 2-mercaptoethanesulphonate
Met	Methionine
Moz	4-Methoxybenzyloxychloroformyl group
MR	Molar refractivity
mRNA	messenger RNA
nACh	Nicotinic cholinergic receptor
NAD^+	Nicotinamide adenine dinucleotide (oxidised form)
NADH	Nicotinamide adenine dinucleotide (reduced form)
$NADP^+$	Nicotinamide dinucleotide phosphate (oxidised form)
NADPH	Nicotinamide dinucleotide phosphate (reduced form)
NAG	β-*N*-Acetylglucosamine
NAM	β-*N*-Acetylmuramic acid
NAME	N^G-Nitro-L-arginine methyl ester
NMDA	*N*-Methyl-D-aspartic acid
NMMA	N^G-Monomethyl-L-arginine
NOS	Nitric oxide synthase
NVOC	6-Nitroveratryloxycarbonyl group
ONs	Sequence defined oligonucleotides
P-450	Cytochrome P-450 oxidases
PABA	*p*-Aminobenzoic acid
PAPS	3′-Phosphoadenosine-5′-phosphosulphate
PCR	Polymerase chain reaction primer
PEG	Polyethyleneglycol
PG	Prostaglandin
PIP_2	Phosphatidylinositol biphosphate
pre-mRNA	Premessenger RNA
Pro	Proline
ptRNA	Primary transcript RNA

QSAR	Quantitative structural-activity relationships
RNA	Ribonucleic acid
S	Svedberg units
SAM	*S*-Adenosylmethionine
SAR	Structural-activity relationships
Ser	Serine
SIN-1	3-Morpholino-sydnomine
T	Thymine
TdRP	Deoxythymidylic acid
Thr	Threonine
tRNA	transfer RNA
Try	Tyrosine
U	Uracil
UDP	Uridine diphosphate
UDPGA	Uridine diphosphate glucuronic acid
UdRP	Deoxyuridylic acid
Val	Valine

1 Introduction

1.1 Introduction

The primary objective of medicinal chemistry is the design and discovery of new compounds that are suitable for use as drugs. This process requires a **team effort**. It not only involves chemists but also workers from a wide range of disciplines such as biology, biochemistry, pharmacology, mathematics, computing and medicine, amongst others.

The discovery of a new drug not only requires its design and synthesis but also the development of testing methods and procedures, which are needed to establish how a substance operates in the body and its suitability for use as a drug. However, the role of the testing methods used in medicinal chemistry in the discovery of new medical compounds should not be confused with the role of quality control testing in the production of drugs in the pharmaceutical industry. In the former case the tests are used to determine how the drug may be safely administered, whereas in the latter they are used to monitor the quality of the drug as it leaves the production line. Drug discovery may also require fundamental research into the biological and chemical nature of the diseased state. These and other aspects of drug design and discovery require input from specialists in other fields and the medicinal chemist to have an outline knowledge of these fields.

This chapter seeks to give a broad overview of medicinal chemistry. It attempts to provide a framework for the topics discussed in greater depth in the succeeding chapters. In addition, it includes some topics of general interest to medicinal chemists.

1.2 What are Drugs and Why Do We Need New Ones?

Drugs are strictly defined as chemical substances that are used to prevent or cure diseases in humans, animals and plants. The **activity** of a drug is its pharmacological effect on the subject, for example, its analgesic or β-blocker action, whereas its **potency** is a quantitative measure of the effect (see section 8.6.1, Equation 8.25). Unfortunately the term drug is also used by the media and the general public to describe substances taken for recreational rather than therapeutic purposes. However, this does not mean that these substances cannot be used clinically as drugs. Heroin (diamorphine), for example, is a very effective pain-killer and is used to alleviate pain in terminal cancer cases.

Heroin (Diamorphine)

CH_3COO O NCH_3 CH_3COO

Drugs act by interfering with biological processes so no drug is completely safe. All drugs, including those non-prescription drugs, such as aspirin and paracetamol, that are commonly available over the counter, act as poisons if taken in excess. For example, overdoses of paracetamol can cause coma and death. Furthermore, in addition to their beneficial effects most drugs have non-beneficial biological effects. Aspirin, which is commonly used to alleviate headaches, can also cause gastric irritation and bleeding in some people. The non-beneficial effects of some drugs, such as cocaine and heroin, are so undesirable that the use of these drugs has to be strictly controlled by legislation. These unwanted effects are commonly referred to as **side effects**. However, side effects are not always non-beneficial. The term also includes biological effects that are beneficial to the patient. For example, the antihistamine promethazine is licensed for the treatment of hayfever but also induces drowsiness, which may aid sleep.

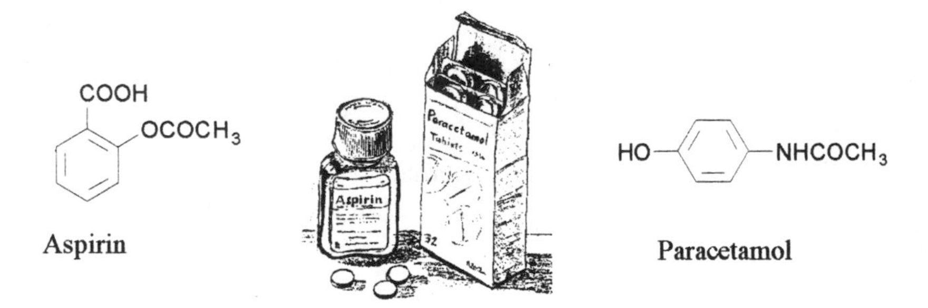

The overusage of the same drugs, such as antibiotics, can result in the development of resistance to that drug by the patients, microorganisms and the virus the drug is intended to control. Resistance occurs when a drug is no longer effective in controlling a medical condition. Drug resistance, or tolerance as it is known in humans, arises in people for a variety of reasons. For example, the effectiveness of barbiturates often decreases with repeated use because repeated dosing causes the body to increase its production of mixed-function oxidases in the liver that metabolise the drug thereby reducing the drug's effectiveness. An increase in the rate of production of an enzyme that metabolises the drug is a relatively common reason for drug resistance. Another general reason for drug resistance is the **downregulation** of receptors (see section 8.6.1). Downregulation occurs when repeated stimulation of a receptor results in the receptor being broken down. This results in the drug being less effective because there are fewer receptors available for it to act on. However, downregulation has been utilised therapeutically in a number of cases. The continuous use of gonadotrophin-releasing factor, for example, causes gonadotrophin receptors that control the menstrual cycle to be downregulated. This is why gonadotrophin-like drugs are used as contraceptives. Drug resistance may also be due to the appearance of a significantly high proportion of drug-resistant strains of microorganisms. These strains arise naturally and can rapidly multiply and become the currently predominant strain of that microorganism. For example, antimalarial drugs are proving less effective because of an increase in the proportion of drug-resistant strains of the malaria parasite.

New drugs are constantly required to combat drug resistance even though it can be minimised by the correct use of medicines by patients. They are also required for the improvement in the treatment of existing diseases, the treatment of newly identified diseases and the production of safer drugs by the reduction or removal of adverse side effects.

1.3 Drug Discovery and Design, a Historical Outline

Since ancient times the peoples of the world have had a wide range of natural products that they use for medicinal purposes. These products, obtained from animal, vegetable and mineral sources, were sometimes very effective. However, many of the products were very toxic and it is interesting to note that the Greeks used the same word *pharmakon* for both poisons and medicinal products. Information about these ancient remedies was not readily available to users until the invention of the printing press in the fifteenth century. This led to the widespread publication and circulation of herbals and pharmacopoeias, which resulted in a rapid increase in the use, and misuse, of herbal and other remedies. Misuse of tartar emetic (antimony potassium tartrate) was the reason for its use being banned by the Paris parliament in 1566, probably the first recorded ban of its type. The usage of such remedies reached its height in the seventeenth century. However, improved communications between practitioners in the eighteenth and nineteenth centuries resulted in the progressive removal of preparations that were either ineffective or too toxic from herbals and pharmacopoeias. It also led to a more rational development of new drugs.

The early nineteenth century saw the extraction of pure substances from plant material. These substances were of consistent quality but only a few of the compounds isolated proved to be satisfactory as therapeutic agents. The majority were found to be too toxic, although many, such as morphine and cocaine for example, were extensively prescribed by physicians. A change in attitude towards drug safety in the nineteenth century, coupled with the toxicity of compounds from natural sources, has drastically reduced their use in recent years.

The search to find less toxic medicines than those based on natural sources resulted in the introduction of synthetic substances as drugs in the late nineteenth century and their widespread use in the twentieth century. Initially, this development was centred around the natural products isolated from plant and animal material but as knowledge increased a wide range of synthetic compounds were evolved as drugs. The original pharmacologically active compound from which these synthetic analogues are developed is now known as the **lead** compound. The work of the medicinal chemist is centred around the discovery of new lead compounds with specific medical properties. It includes the development of more effective and safer analogues of these new and existing lead compounds. This usually involves synthesising and testing many hundreds of compounds before a suitable compound is produced. It is currently estimated that for every 10000 compounds synthesised, only one is suitable for medical use.

The first rational development of synthetic drugs was carried out by Paul Ehrlich and Sacachiro Hata, who produced arsphenamine in 1910 by combining synthesis with reliable biological screening and evaluation procedures. Ehrlich, at the begining of the twentieth century, had recognised that both the beneficial and toxic properties of a drug were important to its evaluation. He realised that the more effective drugs showed a greater selectivity for the target microorganism than its host. Consequently, to compare the effectiveness of different compounds, he expressed a drug's selectivity, and hence its effectiveness, in terms of its chemotherapeutic index, which he defined as:

$$\text{Chemotherapeutic index} = \frac{\text{Minimum curative dose}}{\text{Maximum tolerated dose}} \tag{1.1}$$

At the start of the nineteenth century Ehrlich was looking for a safer antiprotozoal agent with which to treat syphilis than the then currently used *Atoxyl*. He and Hata tested and catalogued in terms of his therapeutic index over 600 structurally related arsenic compounds. This led to their discovery in 1909 that arsphenamine (Salvarsan) could cure mice infected with syphilis. This drug was found to be effective in humans but had to be used with extreme care because it was very toxic. However, it was used up to the mid-1940s when it was replaced by penicillin.

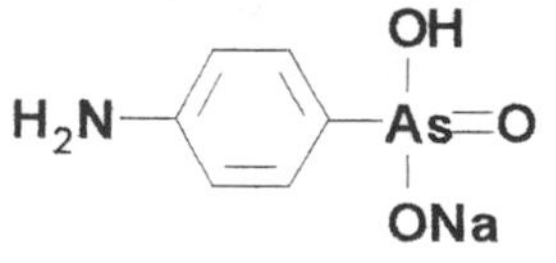

Atoxyl

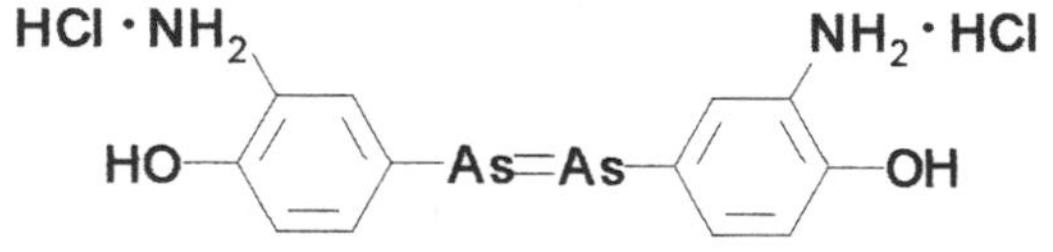

Arsphenamine (Salvarsan)

Ehrlich's method of approach is still one of the basic techniques used to design and evaluate new drugs in medicinal chemistry. However, his chemotherapeutic index has been updated to take into account the variability of individuals and is now defined as its reciprocal, the therapeutic index or ratio:

$$\text{Therapeutic index} = \frac{\text{Lethal dose required to kill 50\% of the test animals (LD}_{50})}{\text{The dose producing an effective therapeutic response in 50\% of the test sample (ED}_{50})} \tag{1.2}$$

In theory, the larger a drug's therapeutic index, the greater is its margin of safety. However, because of the nature of the data used in their derivation, therapeutic index values can only be used as a limited guide to the relative usefulness of different compounds.

The term **structure–activity relationship (SAR)** is now used to describe Ehrlich's approach to drug discovery, which consisted of synthesising and testing a series of structurally related compounds (see section 2.3). Although attempts to quantitatively relate chemical structure to biological action were first initiated in the nineteenth century, it was not until the 1960s that Hansch and Fujita devised a method that successfully incorporated quantitative measurements into structure–activity relationship determinations (see section 2.4). The technique is referred to as **QSAR (quantitative structure–activity relationships)** and QSAR methods have subsequently been expanded by a number of other workers. One of the most successful uses of QSAR has been in the development of the antiulcer agents cimetidine and ranitidine in the 1970s. Both SAR and QSAR are important parts of the foundations of medicinal chemistry.

$CH_2SCH_2CH_2NHC(=NCN)NHCH_3$ (on a 5-methylimidazol-4-yl ring)

Cimetidine

$(CH_3)_2NCH_2$–(furan)–$CH_2SCH_2CH_2NHC(=CHNO_2)NHCH_3$

Ranitidine

At the same time as Ehrlich was investigating the use of arsenical drugs to treat syphilis, John Langley formulated his theory of **receptive substances**. In 1905 Langley proposed that so-called receptive substances in the body could accept either a stimulating compound, which would cause a biological response, or a non-stimulating compound, which would prevent a biological response. These ideas have been developed by subsequent workers and the theory of receptors has become one of the fundamental concepts of medicinal chemistry. It is now accepted that the binding of a chemical agent – referred to as a **ligand** (see section 8.1) – to a receptor sets in motion a series of biochemical events that result in a biological or pharmacological effect. Furthermore, a drug is most effective when its structure or a significant part of its structure, both both regard to molecular shape and electron distribution (**stereoelectronic structure**), is complementary with the stereoelectronic structure of the receptor responsible for the desired biological action. The section of the structure of a ligand that binds to a receptor is known as its **pharmacophore**. Receptors (see Chapter 8) should not be confused with active sites (see Chapter 6), which are the regions of enzymes where metabolic chemical reactions occur.

The concept of receptors also gives a reason for side effects and a rational approach to ways of eliminating their worst effects. It is now believed that side effects can arise when the drug binds to either the receptor responsible for the desired biological response or to different receptors.

The mid- to late twentieth century has seen an explosion of our understanding of the chemistry of disease states, biological structures and processes. This increase in knowledge has given medicinal chemists a clearer picture of how drugs are distributed through the body, transported across membranes, their mode of operation and metabolism. This knowledge has enabled medicinal chemists to place groups that influence its absorption, stability in a biosystem, distribution, metabolism and excretion in the molecular structure of a drug. For example, the *in situ* stability of a drug and hence its potency could be increased by rationally modifying the molecular structure of the drug. Esters and N-substituted amides, for example, have structures with similar shapes and electron distributions (Figure 1.1a) but N-substituted amides hydrolyse more slowly than esters. Consequently, the replacement of an ester group by an N-substituted amide group **may** increase the stability of the drug without changing the nature of its activity. This **could possibly** lead to an increase in either the potency or time of duration of activity of a drug by improving its chances of reaching its site of action. However, changing a group or introducing a group may change the nature of the activity of the compound. For example, the change of the ester group in procaine to an amide (procainamide) changes the activity from a local anaesthetic to antirhythmic (Figure 1.1b).

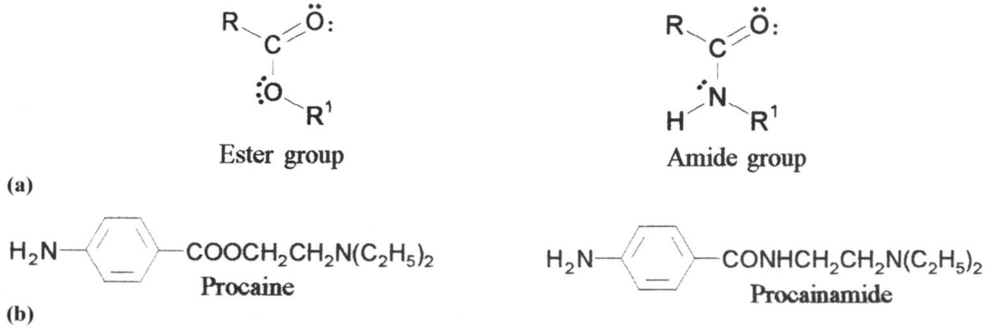

R, C, O, O, R^1 — **Ester group**

R, C, O, N, H, R^1 — **Amide group**

(a)

$H_2N-C_6H_4-COOCH_2CH_2N(C_2H_5)_2$ **Procaine**

$H_2N-C_6H_4-CONHCH_2CH_2N(C_2H_5)_2$ **Procainamide**

(b)

Figure 1.1. (a) The similar shapes and outline electronic structures (stereoelectric structures) of amide and ester groups (b) Procaine and procainamide.

Drugs normally have to cross non-polar lipid membrane barriers (see sections 4.2 and 4.3) in order to reach their site of action. As the polar nature of the drug increases, it usually becomes more difficult for the compound to cross these barriers. In many cases drugs whose structures contain charged groups will not readily pass through membranes. Consequently, charged structures can be used to restrict the distribution of a drug. For example, quaternary ammonium salts, which are permanently charged, can be used as an alternative to an amine in a structure in order to restrict the passage of a drug across a membrane. The structure of the anticholinesterase neostigmine, developed from physostigmine, contains a quaternary ammonium group that gives the molecule a permanent charge. This stops the molecule from crossing the blood–brain barrier, which prevents unwanted central nervous system (CNS) activity. However, its analogue miotine can form the free base. As a result, it is able to cross lipid membranes which can cause unwanted CNS side effects.

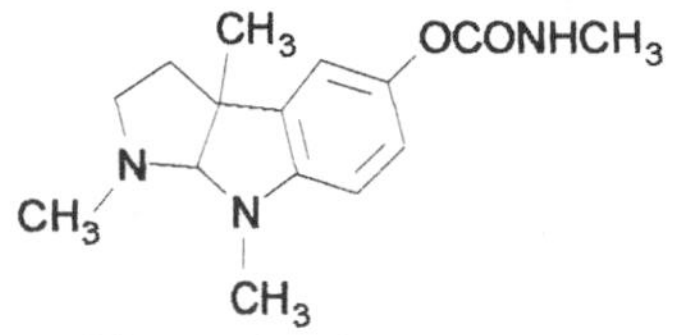

Physostigmine

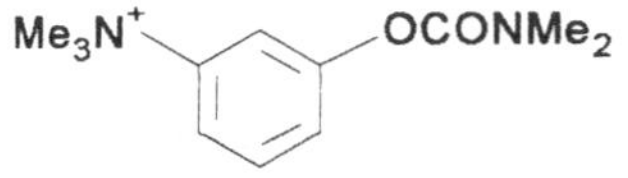

Neostigmine

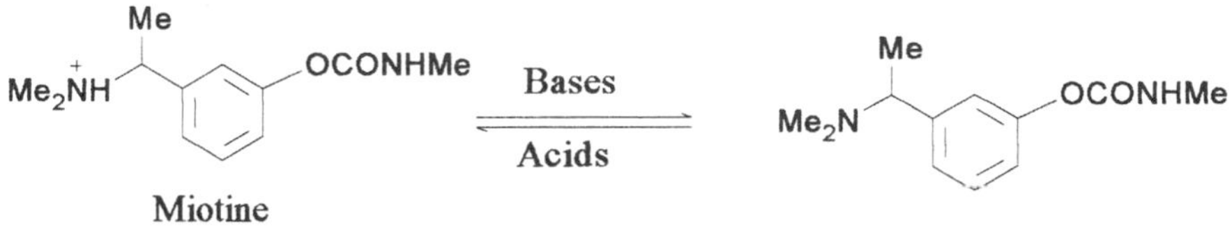

Bases

Acids

Miotine

Serendipity has always played a part in the development of drugs. For example, the development of penicillin by Florey and Chain was only possible because Alexander Fleming noted the inhibition of *Staphylococcus* by *Penicillium notatum*. In spite of our increased knowledge base, it is still necessary to pick the correct starting point if a successful outcome is to be achieved and luck still plays a part in selecting that point. This state of affairs will not change and undoubtedly luck will also lead to new discoveries in the future. However, modern techniques such as computer modelling (see section 2.5) and combinatorial chemistry (see section 2.6) introduced in the 1970s and 1990s, respectively, are likely to reduce the number of intuitive discoveries.

Computer modelling (see section 2.5) has reduced the need to synthesise every analogue of a lead compound. It is also often used retrospectively to confirm the information derived from other sources. Combinatorial chemistry (see section 2.6), originated in the field of peptide chemistry but has now been expanded to cover other areas. It is a group of related techniques for the simultaneous production of large numbers of compounds for biological testing. Consequently, it is used for structure–action studies and to discover new lead compounds. The procedures may be automated.

1.4 Methods and Routes of Administration, the Pharmaceutical Phase

The form in which a medicine is administered is known as its **dosage form**. Dosage forms can be subdivided according to their physical nature into liquid, semisolid and solid formulations. Liquid formulations include solutions, suspensions and emulsions. Creams, ointments and gels are normally regarded as semisolid formulations, whereas tablets, capsules and moulded products such as suppositories and pessaries are classified as solid formulations. These dosage forms normally consist of the active constituent and inert ingredients (**excipients**). Excipients can have a number of functions, such as fillers (bulk providing agent), lubricants, binders, preservatives and antioxidants. A change in the nature of the excipients can significantly affect the release of the active ingredient from the dosage form. Similarly, changes in the preparation of the active principle, such as the use of a different solvent for purification, can affect its bioavailability (see section 5.5) and consequently its effectiveness as a drug. This indicates the importance of a quality control procedure for all drugs, especially when they reach the manufacturing stage.

Phenytoin

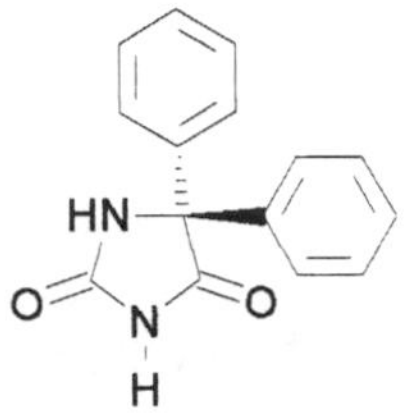

HN
O
N
H
O

Use of the correct dosage form is an important factor in the effectiveness of a drug. The use of an incorrect dosage form can render the medicine ineffective and potentially dangerous. The anticonvulsant phenytoin, for example, was found to be rapidly absorbed when lactose is used as a filler. This resulted in patients receiving toxic doses. In contrast, when calcium sulphate was used as a filler, the rate of absorption was so slow that the patient did not receive a therapeutic dose. The design of dosage forms lies in the field of the pharmaceutical technologist but it should also be considered by the medicinal chemist when developing a drug from a lead compound. It is no use having a wonder drug if it cannot be packaged in a form that makes it biologically available as well as acceptable to the patient.

Drugs are usually administered topically or systemically. The routes are classified as being either **parenteral** or **enteral** (Figure 1.2). Parenteral routes are those that avoid the gastrointestinal tract. The most usual method is intramuscular (i.m.) injection. However, some other parental routes are intravenous (i.v.) injection, subcutaneous (s.c.) injection and transdermal delivery systems. Nasal sprays and inhalers are also parenteral routes. The enteral route is where drugs are absorbed from the alimentary canal (p.o.), rectal and sublingual routes. In both parenteral and enteral routes a drug will only be effective if a sufficient concentration reaches the target area in the body for the period of time it takes to achieve the desired therapeutic effect. This means that the physiological stability of the drug, both when it is transported across a membrane (**absorption**) and in transit to the site of action (**distribution**), must be considered when designing or modifying the drug. Consequently, the route of administration can be of importance in reducing the problems encountered by the medicinal chemist when designing a drug. However, scientific factors are not the only considerations affecting the dosage form and administration route – the nature of the compliance, age and physical ability of the patient will all play a part. Elderly and mentally disabled people living in the community are amongst the group who are most likely to overdose or underdose. Schizophrenics and patients with conditions that require constant medication are particularly at risk in this respect. In these cases a slow-release intramuscular injection that need only be given once in every two

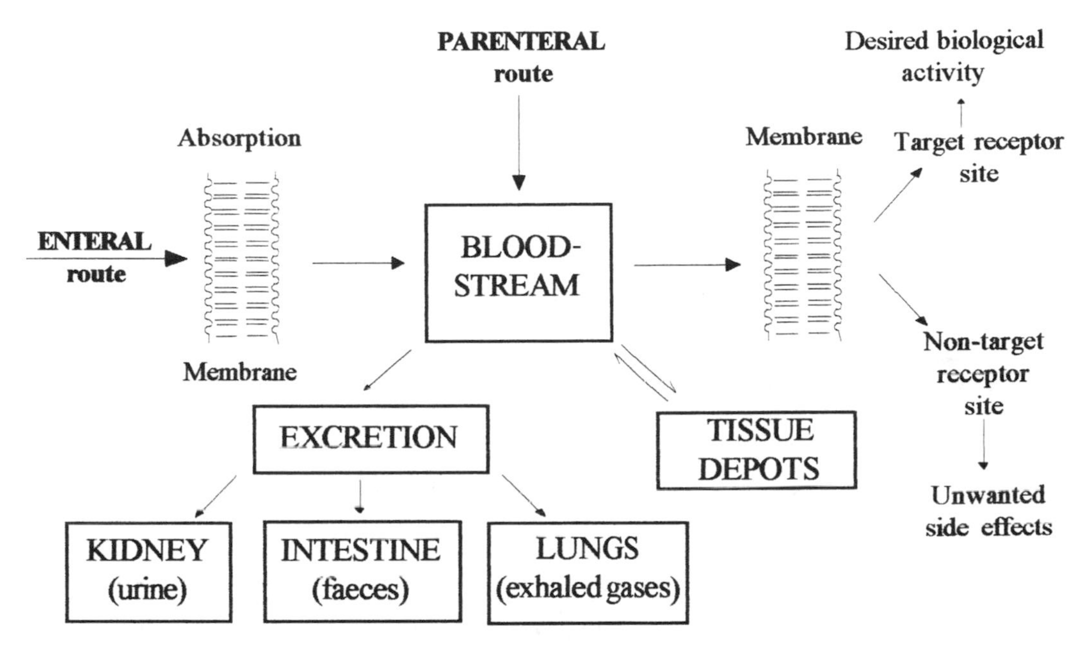

Figure 1.2. The main routes of drug administration and distribution in the body. The distribution of a drug is also modified by metabolism, which can occur at any point in the system.

to four weeks rather than a daily dose may be the most effective use of the medicine. Consequently, at an appropriately early stage in its development, the design of a drug should also take into account its target group. It is a waste of resources to find at a later date that a drug that is successful in the laboratory cannot be administered in a convenient manner to the patient.

Once the drug enters the bloodstream it is distributed around the body and so a proportion of the drug is lost by excretion, metabolism to other products or is bound to biological sites other than its target site. As a result, the dose administered is inevitably higher than that which would be needed if all the drug reached the appropriate site of biological action. The dose of a drug administered to a patient is the amount that is required to reach and maintain the concentration necessary to produce a favourable response at the site of biological action. Too high a dose usually causes unacceptable side effects, whereas too low a dose results in a failure of the therapy. The limits between which the drug is an effective therapeutic agent are known as its **therapeutic window** (Figure 1.3).

The dose of a drug and how it is administered is called the **dosage regimen**. Dosage regimens may vary from a single dose taken to relieve a headache, regular daily doses taken to counteract the effects of epilepsy and diabetes to continuous intravenous infusions for seriously ill patients. Regimens are designed to maintain the concentration of the drug within the therapeutic window at the site of action for the period of time that is required for therapeutic

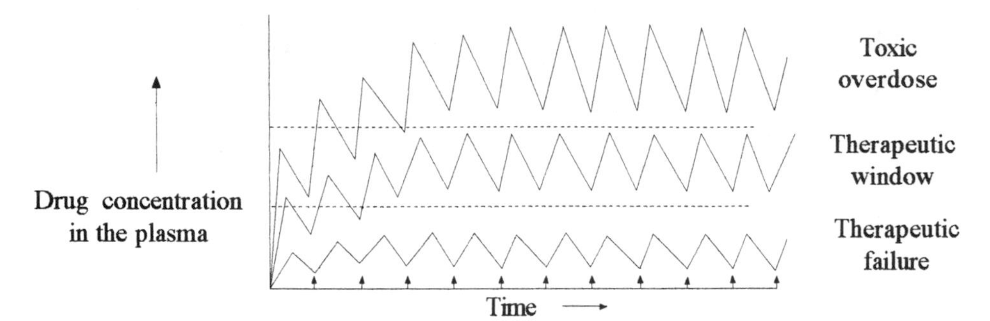

Figure 1.3. A simulation of a therapeutic window of a drug for a drug given in fixed doses at fixed time intervals (↑).

success. The form adopted depends on the nature of the medical condition and the medicant.

The design of an effective dosage regimen requires not just a knowledge of a drug's biological effects but also its **pharmacokinetic** properties, that is, its the rate of absorption, distribution, metabolism and elimination from the body. It is possible for a drug to be ineffective because of the use of an incorrect dosage regimen. When quinacrine was introduced as a substitute for quinine in the 1940s it was found to be ineffective at low dose levels or too toxic at the high dose levels needed to combat malaria. Quinacrine was only used successfully after its pharmacokinetic properties were studied. It was found to have a slow elimination rate and so in order to maintain a safe therapeutic dose it was necessary to use large initial doses but only small subsequent maintenance doses to keep the concentration within its therapeutic window. This dosage regimen reduced the toxicity to an acceptable level.

1.5 Introduction to Drug Action

Drug action is believed to be due to the interaction of the drug with enzymes, receptors and other molecules found in the biological system. The binding of a drug to the active or other sites (**allosteric sites**) of an enzyme usually has the effect of preventing the normal operation of that enzyme (see section 6.9). The drug's therapeutic effect will depend on the stability of the drug–enzyme complex as well as the fraction of active and allosteric sites occupied by the drug. The stronger the binding of the drug to the enzyme and the greater the number of sites occupied, the more effective the drug is likely to be in inhibiting the action of the enzyme.

Drugs act on receptors by binding to or near to a specific site on the receptor. This may either activate the receptor or prevent the binding of the receptor's normal substrate (**endogenous**

compounds) to that receptor. Ultimately, both of these actions can result in a physiological response that may have a therapeutic effect. The drug's effectiveness will, as in the case of drug–enzyme interaction, depend on the stability of the drug–receptor complex and the number of sites occupied by the drug (see section 8.6). Other targets for drug action include the nucleic acids (see sections 10. 12 and 10.13) and other naturally occurring molecules.

The degree of drug activity is directly related to the concentration of the drug in the aqueous medium in contact with the active or receptor site. The factors affecting this concentration in a biological system can be classified into the **pharmacokinetic phase** and the **pharmacodynamic phase** of drug action. The pharmacokinetic phase concerns the study of the parameters that control the journey of the drug from its point of administration to its point of action. The pharmacodynamic phase concerns the chemical nature of the relationship between the drug and its target, in other words, the effect of the drug on the body.

1.5.1 The Pharmacokinetic Phase

The pharmacokinetic phase of drug action includes the *A*bsorption, *D*istribution, *M*etabolism and *E*limination (ADME) of the drug. Many of the factors that influence drug action apply to all aspects of the pharmacokinetic phase. Solubility (see Chapter 3), for example, is an important factor in the absorption, distribution and elimination of a drug. Furthermore, the rate of drug dissolution controls its activity when the drug is administered by enteral routes (see section 1.4) as a solid or suspension.

1.5.1.1 Absorption

Absorption is the passage of the drug from its site of administration into the plasma after enteral administration. The use of the term does not apply to parenteral administration. Absorption involves the passage of the drug through the appropriate membranes. This passage can occur by a number of different mechanisms (see section 4.3). However, in general, neutral molecules are more readily absorbed through membranes than charged species. For example, the local anaesthetic benzocaine is absorbed into neurons as its free base even though its action is attributed to its charged salt form.

$$H_2N\text{—}C_6H_4\text{—}COOCH_2CH_3 \rightleftharpoons H_3\overset{+}{N}\text{—}C_6H_4\text{—}COOCH_2CH_3$$

Inactive form transported through membranes — Active form

Benzocaine

The degree of absorption can be related to such parameters as partition coefficient, solubility, pK_a, dosage form, excipients and particle size. For example, the ionisation of the analgesic aspirin is suppressed in the stomach by the acids produced from the parietal cells in the stomach lining. As a result, it is absorbed into the bloodstream in significant quantities in its unionised and hence uncharged form through the stomach membrane.

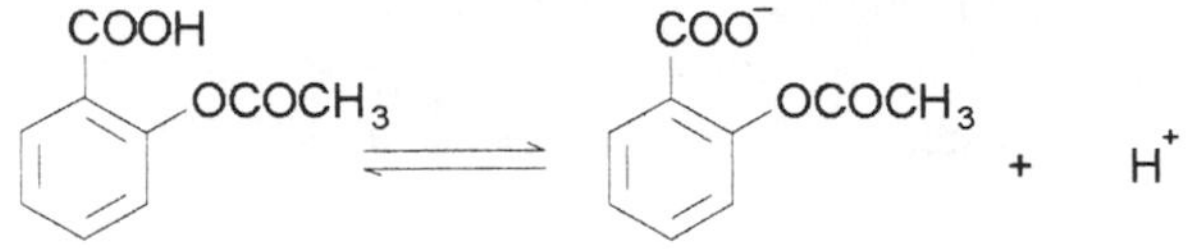

COOH OCOCH$_3$ ⇌ COO$^-$ OCOCH$_3$ + H$^+$

Aspirin

1.5.1.2 Distribution

Distribution is the transport of the drug from its initial point of administration or absorption to its site of action. The main route is through the circulation of the blood, however, some distribution does occur via the lymphatic system. Once the drug is absorbed it is rapidly distributed throughout all the areas of the body reached by the blood. This means that the chemical and physical properties of blood will have a considerable effect on the concentration of the drug reaching the receptor.

Drugs are transported in the bloodstream either in a 'free form' or bound to the plasma proteins. The binding of drugs to the serum proteins is usually reversible.

$$\text{Drug} \rightleftharpoons \text{Drug–Protein}$$

Drug molecules bound to plasma proteins have no pharmacological effect until they are released from those proteins. Consequently, this equilibrium can be an important factor in controlling a drug's pharmacological activity (see section 5.4.1.1). Furthermore, it is possible for one drug to displace another from a protein if it forms a more stable complex with that protein. This aspect of protein binding can be of considerable importance when designing drug regimens involving more than one drug. For instance, the displacement of antidiabetic agents by aspirin can trigger hypoglycaemic shock and so aspirin should not be used by patients taking these drugs. Moreover, low plasma protein concentrations can also affect the distribution of a drug in some diseases such as rheumatoid arthritis.

Major factors that influence distribution are the solubility and stability of drugs in the biological environment of the blood. Sparingly water-soluble compounds may be deposited in the blood vessels, leading to restriction in blood flow. This deposition may be influenced by the common ion effect (see sections 1.5.1.1 and 3.4.1.2). Drug stability is of particular importance in that serum proteins can act as enzymes that catalyse the breakdown of the drug. Decompositions such as these can result in a higher dose of the drug being needed in order to achieve the desired pharmacological effect. This increased dose increases the risk of toxic side effects in the patient. However, the active form of some drugs is produced by the decomposition of the administered drug. Drugs that function in this manner are known as **prodrugs** (see section 9.9). The first to be discovered, in 1935, was the bacteriacide prontosil. Prontosil itself is not active but is metabolised *in situ* to the antibacterial sulphanilamide. Its discovery paved the way to the devolvement of a wide range of antibacterial sulphonamide (sulfa) drugs. These were the only effective antibiotics available until the general introduction of penicillin in the late 1940s.

1.5.1.3 Metabolism

Drug metabolism is the biotransformation of the drug into other compounds (**metabolites**). Metabolites are usually more water soluble than their parent drug and are normally excreted in the urine. These biotransformations occur mainly in the liver but they can also occur in blood and other organs such as the brain, lungs and kidneys. Metabolism of a drug usually reduces the concentration of that drug in the systemic circulation, which normally leads either to a lowering or a complete suppression of pharmacological action and toxic effects. Exceptions are prodrugs such as prontosil, where metabolism produces the active form of the drug. A further complication is that metabolism can produce metabolites with either a different or similar activity to the parent drug, as well as toxic compounds (see section 9.2). Consequently, metabolism is an important factor that must be considered in the development of a potential drug.

1.5.1.4 Elimination

Elimination is the irreversible removal of the active form of the drug from the body by degradative metabolism and all forms of excretion. It reduces the effect of the drug by reducing its concentration at its site of action. A slow elimination process can result in a build-up of drug concentration in the body. This may benefit the patient in that the dose required to maintain the therapeutic effect can be reduced, which in turn lowers the chances of unwanted side effects. Conversely, the rapid elimination of a drug means that the patient has to receive either increased doses with a greater risk of toxic side effects or more frequent doses, which carries more risk of under- or overdosing. Consequently, the rate of elimination is an essential factor in assessing the performance of a drug and the subsequent development of more efficient analogues. It is an important consideration when designing an effective dosage regimen.

The main excretion route for drugs and their metabolites is through the kidney in solution in the urine. However, a significant number of drugs and their metabolic products are also excreted via the bowel in the faeces. Other forms of drug excretion, such as exhalation, sweating and breast-feeding, are not usually significant except in specific circumstances. Pregnant women and nursing mothers are recommended to avoid taking drugs because of the possibility of biological damage to the foetus and neonate. For example, the use of thalidomide by pregnant mothers in the 1960s resulted in the formation of drug-induced malformed foetus (**teratogenesis**). It has been estimated that the use of thalidomide led to the birth of 10000 severely malformed children.

In the kidneys drugs are eliminated by either **glomerular filtration** or **tubular secretion**. However, some of the species lost by these processes are reabsorbed by a recycling process known as **tubular reabsorption**. In the kidney, the glomeruli act as a filter allowing the passage of water, small molecules and ions but preventing the passage of large molecules and cells. Consequently, glomerular filtration eliminates small unbound drug molecules but not the larger drug–protein complexes. Tubular secretion on the other hand is an active transfer process (see section 4.3.5) and so both bound and unbound drug molecules can be excreted. However, both of these excretion systems have a limited capacity and not all the drug may be eliminated. In addition, renal disease can considerably increase or decrease the rate of drug elimination by the kidney.

Tubular reabsorption is a process normally employed in returning compounds such as water, amino acids, salts and glucose that are important to the well-being of the body from the urine to the circulatory system, but it will also return drug molecules. The mechanism of reabsorption is mainly passive diffusion (see section 4.3.3), but active transport (see section 4.3.5) is also involved, especially for glucose and lithium ions. The reabsorption of acidic and basic compounds is dependent on the pH (see section 3.4.1) of the urine. Consequently, it is possible to influence the activity of a drug by changing the pH of the urine. For example, making the urine alkaline in cases of poisoning by acidic drugs such as aspirin will cause these drugs to form ionic salts, which will result in a significantly lower tubular reabsorption because transport of the charged form of a drug across a lipid membrane is more difficult than the transport of the uncharged form of that drug. Similarly, in cases of poisoning by basic drugs such as amphetamines, acidification of the urine can, for a similar reason, reduce reabsorption.

Control of urinary pH is also required for drugs whose concentration reaches a level in the urine that results in crystallisation (crystalluria) in the urinary tract and kidney with subsequent tissue damage. For example, it is recommended that the urine is maintained at an alkaline pH and has a minimum flow of 190 ml h^{-1} when sulphonamides are administered.

Elimination occurs in the liver by **biliary clearance**. Very large molecules are metabolised to smaller compounds before being excreted. However, a fraction of some of the excreted drugs are reabsorbed through the **enterohepatic cycle**. This reabsorption can be reduced by the use of suitable substances in the dosage form: for example, the ion-exchange resin cholestyramine is used to reduce cholesterol levels by preventing its reabsorption.

1.5.2 Bioavailability of a Drug

The bioavailability of a drug is defined as the fraction of the dose of a drug that is found in general circulation (see section 5.5). It is influenced by such factors as absorption, distribution, metabolism and elimination. Bioavailability is not constant but varies with the body's physiological condition.

1.5.3 The Pharmacodynamic Phase

Pharmacodynamics is concerned with the result of the interaction of drug and body at the receptor site, that is, what the drug does to the body. It is now known that a drug is most effective when its shape and electron distribution, that is, its **stereoelectronic structure**, is complementary with the steroelectronic structure of the active site or receptor.

The role of the medicinal chemist is to design a drug structure that has the maximum beneficial effects with a minimum of toxic side effects. This design has to take into account the stereoelectronic characteristics of the target active or receptor site and also such factors as the drug's stability *in situ*, its polarity and its relative solubilities in aqueous media and lipids. The stereochemistry of the drug is particularly important because stereoisomers often have different biological effects that range from inactive to highly toxic (see Table 2.1).

1.6 Classification of Drugs

Drugs are classified in different ways depending on where and how the drugs are being used. The methods of interest to medicinal chemists are chemical structure and pharmacological action (which includes site of action and target system). However, it is emphasised that other classifications, such as the nature of the illness, are used both in medicinal chemistry and other fields, depending on what use is to be made of the information. In all cases, it is important to bear in mind that most drugs have more than one effect on the body and so a drug may be listed in several different categories within a classification scheme.

1.6.1 Chemical Structure

Drugs are grouped according to the structure of their carbon skeletons or chemical classifications, for example, steroids, penicillins and peptides. Unfortunately, in medicinal chemistry this classification has the disadvantage that members of the same group often exhibit different types of pharmaceutical activity. Steroids: for example, have widely differing activities: for example, testosterone is a sex hormone, spironolactone is a diuretic and fusidic acid is an antibacterial agent.

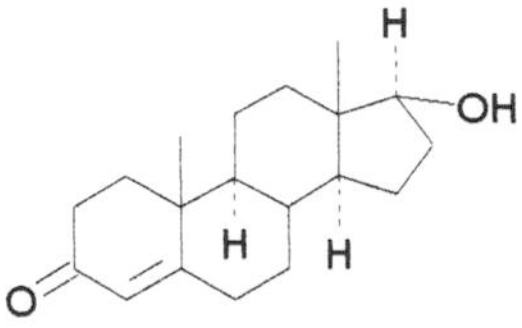

Testosterone 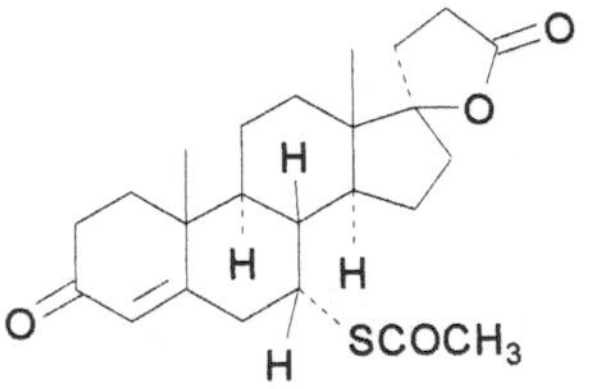 Spironolactone 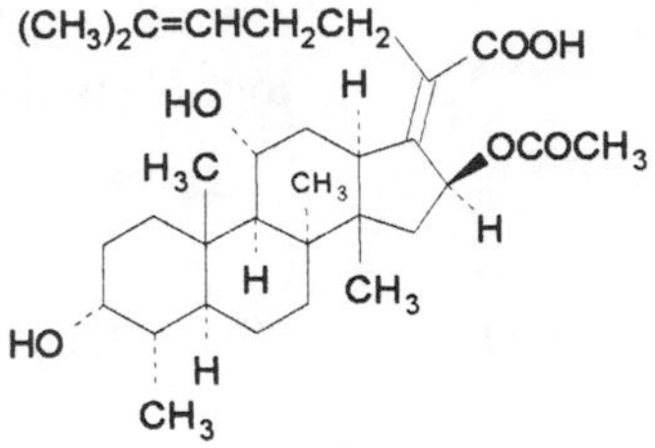 Fusidic acid

Classification by means of chemical structure is useful to medicinal chemists who are concerned with synthesis and structure–activity relationships.

1.6.2 Pharmacological Action

This classification lists drugs according to the nature of their pharmacodynamic behaviour, for example, diuretics, hypnotics, respiratory stimulants and vasodilators. This classification is particularly useful for doctors looking for an alternative drug treatment for a patient.

1.6.3 Physiological Classification

The World Health Organization (WHO) has developed a classification based on the body system on which the drug acts. This classification specifies 17 sites of drug action. However, a more practical method but less detailed system often used by medicinal chemists is based on four classifications, namely:

(i) *Agents acting on the CNS*. The central nervous system consists of the brain and spinal cord. Drugs acting on the CNS are the **psychotropic** drugs that effect mood and the **neurological** drugs required for physiological nervous disorders such as epilepsy and pain.

(ii) *Pharmacodynamic agents*. These are drugs that act on the body, interferring with the normal bodily functions. They include drugs such as vasodilators, respiratory stimulants and antiallergy agents.

(iii) *Chemotherapeutic agents*. Originally these were drugs such as antibiotics and fungicides that destroyed the microorganisms that were the cause of a disease in an unwitting host. However, the classification has also now become synonymous with the drugs used to control cancer.

(iv) *Miscellaneous agents*. This class contains drugs that do not fit into the other three categories, for example, hormones and drugs acting on endocrine functions.

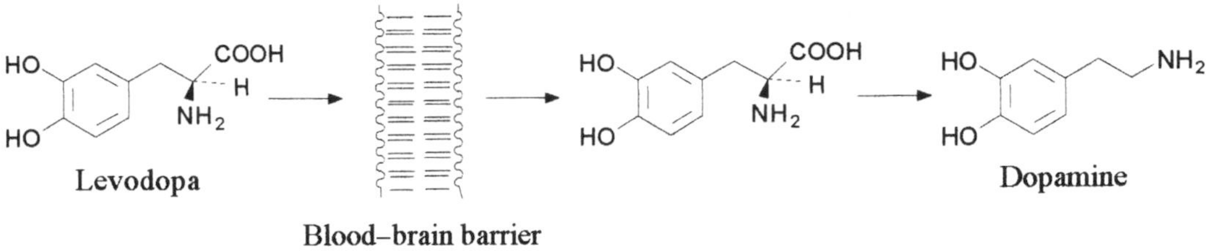

Figure 1.4. A schematic representation of the formation of dopamine from levodopa.

1.6.4 Prodrugs

Prodrugs are compounds that are pharmacologically inert but converted by enzyme or chemical action to the active form of the drug at or near their target site. For example, levodopa, used to treat Parkinson's syndrome, is the prodrug for the neurotransmitter dopamine. Dopamine is too polar to cross the blood–brain barrier but there is a transport system for amino acids such as levodopa. Once the prodrug enters the brain it is decarboxylated to the active drug dopamine (Figure 1.4).

1.7 Drug Stability

Drug stability can be broadly divided into two main areas; shelf-life and stability after administration. Shelf-life is the time taken for its pharmacological activity to decline to an unacceptable level. This level depends on the individual drug and so there is no universal specification. However, 10% decomposition is often taken as an acceptable limit provided that the decomposition products are not toxic.

Shelf-life deterioration occurs through microbial degradation and adverse chemical interactions and reactions. Microbial deterioration can be avoided by preparing the dosage form in the appropriate manner and storage under sterile conditions. It can also be reduced by the use of antimicrobial excipients. Adverse chemical interactions between the components of a dosage form can also be avoided by the use of suitable excipients. Decomposition by chemical reaction is usually brought about by heat, light, atmospheric oxidation, hydrolysis by atmospheric moisture and racemisation. These may be minimised by correct storage by the use of refrigerators, light-proof containers, air-tight lids and the use of the appropriate excipients.

Drug stability *in situ* is essential if a therapeutic dose is going to reach the receptor. The main method of increasing drug stability in the biological system is to prepare a more stable analogue with the same pharmacological activity. For example, pilocarpine, which is used to control glaucoma, rapidly loses it activity because the lactone ring readily opens under physiological conditions. Consequently, the lowering of intraocular pressure by pilocarpine lasts for about three hours, necessitating administration of 3–6 doses a day. However, the replacement of the C-2 of pilocarpine by a nitrogen yields an isosteric carbamate that has the same potency as pilocarpine but is more stable. Although this analogue was discovered in 1989 it has not been accepted for clinical use.

C-2

Hydrolysis

$HOOC$ HO

Pilocarpine **Carbamate analogue**

The *in situ* stability of a drug may also be improved by forming a complex with a suitable reagent. For example, complexing with hydroxypropyl-β-cyclodextrin is used to improve both the stability and solubility of thalidomide, which is still used to inhibit rejection of bone marrow transplants in the treatment of leukaemia. The half-life of a dilute solution of the drug is increased from 2.1 to 4.1 h whereas its aqueous solubility increases from 50 to 1700 $\mu g\,ml^{-1}$.

Cyclodextrins are bottomless flower-pot-shaped cylindrical oligosaccharides consisting of about 6–8 glucose units. The exterior of the 'flower-pot' is hydrophilic in character whereas the interior has a hydrophobic nature. Cyclodextrins are able to form inclusion complexes in which part of the guest molecule is held within the flower-pot structure (Figure 1.5). The hydrophobic nature of the interior of the cyclodextrin structure probably means that hydrophobic interaction plays a large part in the formation and stability of the complex. Furthermore, it has been found that the stability of a drug *in situ* is often improved when the active site of a drug lies within the cylinder and decreased when it lies outside the cylinder. In addition, it has been

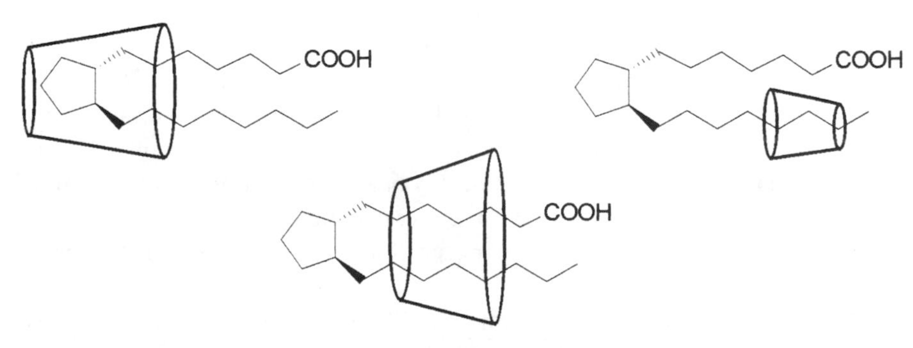

Figure 1.5. Schematic representations of the types of inclusion complexes formed by cyclodextrins and prostaglandins. The type of complex formed is dependent on the cavity size.

noted that the formation of these complexes may also improve the water solubility, bioavailability and pharmacological action and also reduce the side effects of some drugs. However, a high concentration of cyclodextrins in the bloodstream can cause nephrotoxicity

Prodrug formation can also be used to improve drug stability. For example, cyclophosphamide, which is used to treat a number of carcinomas and lymphomas, is metabolised in the liver to the corresponding phosphoramidate mustard, the active form of the drug.

Cyclophosphamide → **Phosphoramidate mustard**

The highly acidic gastric fluids can cause extensive hydrolysis of a drug in the gastrointestinal tract. This will result in poor bioavailability. However, drug stability in the gastrointestinal tract can be improved by the use of enteric coatings, which dissolve only when the drug reaches the small intestine.

1.8 Sources of Drugs

Originally drugs were derived from natural sources. Medical substances and potions were prepared from materials taken from the surrounding countryside. These natural sources are still important sources of lead compounds and new drugs. Today, the majority of lead compounds are discovered and prepared in the laboratory.

The discovery of a new drug is part luck and part structured investigation (see section 2.1). Many discoveries start with biological testing by pharmacologists of the potential source. These **bioassays**, or **screening** programmes as they are known, are used to determine the nature of the pharmacological activity of the material as well as its potency. Random screening programmes where all the substances and compounds available are tested regardless of their structures have been used extensively to discover lead compounds. The random screening of soil samples, for example, led to the discovery of the streptomycin and tetracycline antibiotics as well as many other lead compounds. Random screening is still employed but the use of more focused screening procedures where specific structural types are tested is more common.

Once a screening programme has identified materials of pharmacological activity of interest, the compound responsible for this activity is isolated and used as a lead compound for the production of related analogues. These compounds are subjected to further screening tests. Analogues are made of the most promising of these compounds and they, in turn, are subjected to the screening procedure. This sequence of selective screening and synthesis of analogues may be repeated many times before a potentially useful drug is found. Often the sequence has to be abandoned as being either unproductive or too expensive.

1.8.1 Plant Sources

The large diversity of plants in the world makes the technique of random screening a rather hit or miss process. In the past, pharmacologists have used taste and skin reaction to detect active materials but these methods are neither reliable nor safe. However, the screening of local folk remedies (**ethnopharmacology**) offers the basis of a more systematic approach. In the past, this has led to the discovery of many important therapeutic agents, for example, the antimalarial quinine from cinchona bark, the cardiac stimulant digitalis from foxgloves (Figure 1.6) and the antidepressant reserpine isolated from *Rauwolfia serpentina*.

A different approach to identifying useful sources is that used by Hostettmann and Marston, who deduced that because of the climate African plants must be resistant to constant fungal attack because they contain biologically active constituents. This line of reasoning led them to discover a variety of active compounds (Figure 1.7).

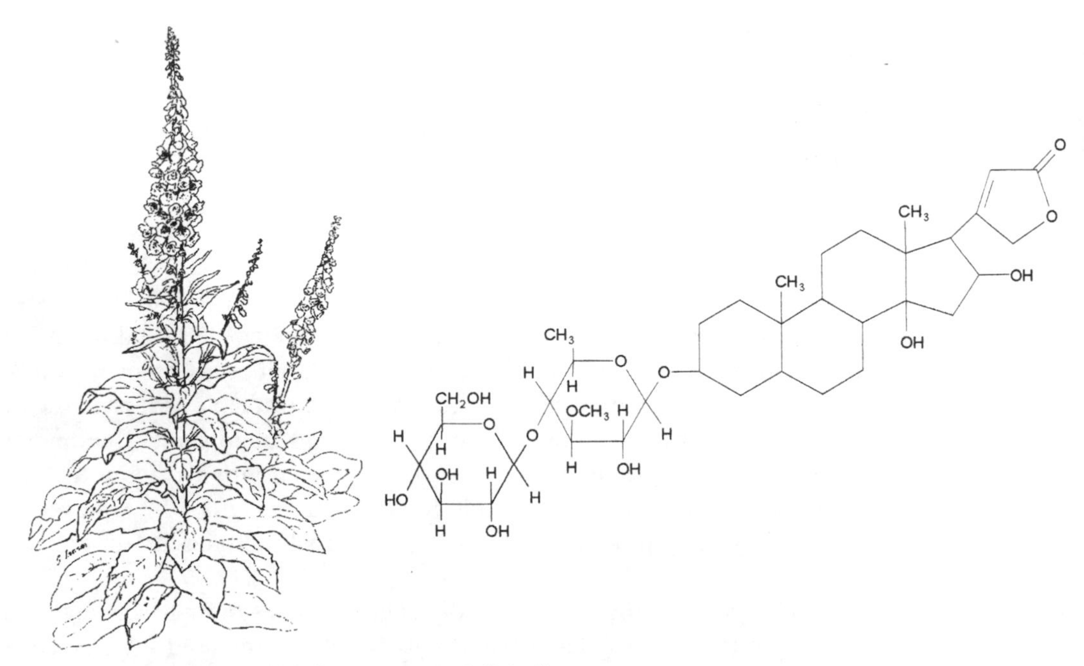

Figure 1.6. The common foxglove and the structure of digitalis.

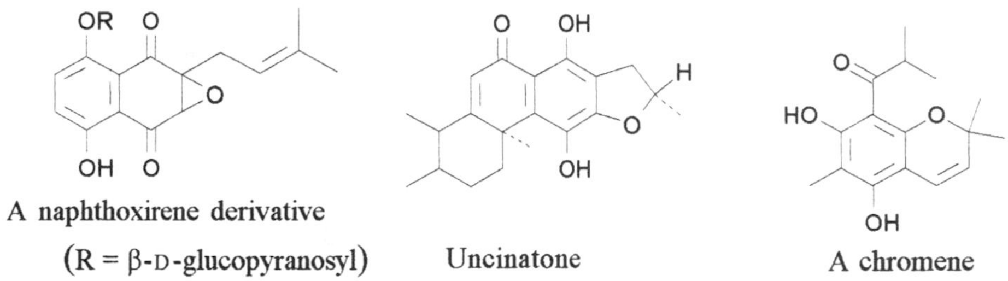

A naphthoxirene derivative
(R = β-D-glucopyranosyl)

Uncinatone

A chromene

Figure 1.7. Examples of the antifungal compounds discovered by Hostettmann and Marston.

Once screening identifies a material containing an active compound, the problem becomes one of extraction, purification and assessment of the pharmacological activity. However, the isolation of a pure compound of therapeutic value can cause ecological problems. The promising anticancer agent Taxol, for example, was isolated from the bark of the Pacific Yew tree. Its isolation from this source requires the destruction of this slow-growing tree. Consequently, the production of large quantities of Taxol from the Pacific Yew could result in the wholesale distruction of the tree, a state of affairs that is ecologically unacceptable. It is vitally important that plant, shrub and tree sources of the world are protected from further erosion because there is no doubt that they will yield further useful therapeutic agents in the future.

1.8.2 Marine Sources

Prior to the mid-twentieth century little use was made of marine products in either folk or ordinary medicine. In the last 40 years these sources have yielded a multitude of active compounds (Figure 1.8) with potential medical use. These compounds exhibit a range of biological activities and are an important source of new lead compounds and drugs. However, care must be taken so that exploitation of a drug does not endanger its marine sources such as shellfish, sponges, plants and sea snakes. Marine sources also yield the most toxic compounds known to man. Some of these toxins, such as tetrodotoxin and saxitoxin, are used as tools in neurochemical research work, investigating the molecular nature of action potentials and Na^+ channels. Although tetrodotoxin and saxitoxin are structurally different, they are both believed to block the external opening of these channels.

1.8.3 Microorganisms

The inhibitory effect of microorganisms was observed as long ago as 1877 by Louis Pasteur, who showed that microbes could inhibit the growth of anthrax bacilli in urine. Later in 1920 Fleming demonstrated that *Penicillin notatum* inhibited *Staphylococcus* cultures, which resulted in the isolation of penicillin by Chain and Florey in 1940. In 1941 Dubos isolated a pharmacologically active protein extract from *Bacillus brevis*, which was shown to contain the

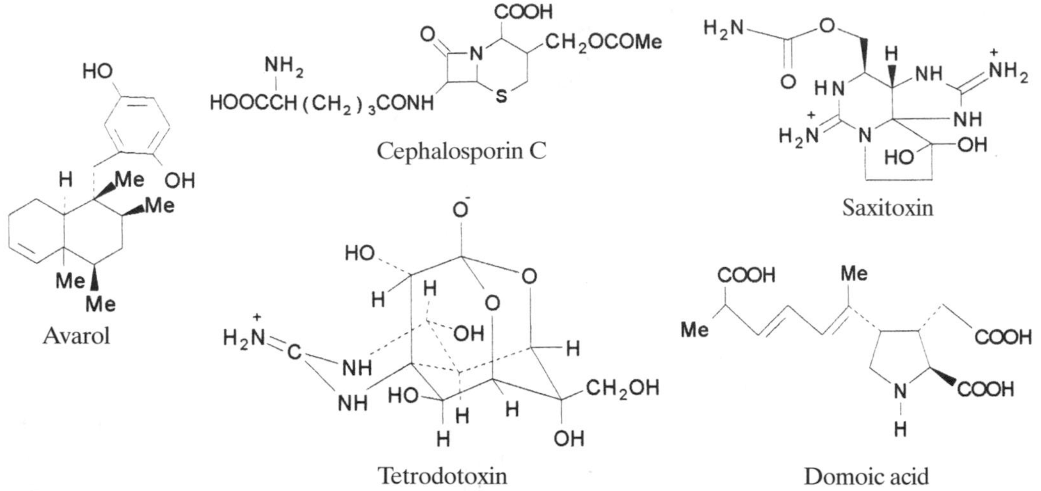

Avarol

Cephalosporin C

Saxitoxin

Tetrodotoxin

Domoic acid

Figure 1.8. Examples of active compounds isolated from marine sources (Me represents a methyl group). Avarol is reported to be an immunodeficiency virus inhibitor. It is extracted from the sponge *Disidea avara*. The antibiotic cephalosporin C was isolated from the fungus *Cephalosporin acremonium*. Domoic acid, which has anthelmintic properties, is obtained from *Chondria armata*. Tetrodotoxin and saxitoxin exhibit local anaesthetic activity but are highly toxic to humans. Tetrodotoxin is found in fish of the order of *Tetraodontiformis* and saxitoxin is isolated from some marine dinoflagellates.

antibiotic gramicidine. This was concurrent with Waksman who postulated that soil bacteria should produce antibiotics because the soil contains few pathogenic bacteria from animal excreta. His work on soil samples eventually led Schatz and co-workers, to the discovery and isolation in 1944 of the antibiotic streptomycin from the actinomycete *Streptomyces griseus*. This discovery triggered the current worldwide search for drugs produced by microorganisms. To date, several thousand active compounds have been discovered from this source, for example, the antibiotic chloramphenicol (*Streptomyces venezuelae*), the immunosuppressant cyclosporin A (*Tolypocladium inflatum* Gams) and the antifungal griseofulvin (*Penicillium griseofulvum*). An important advantage of using microorganisms as a source is that, unlike many of the marine and plant sources, they are easily collected, transported and grown.

1.8.4 Drug Synthesis

The need for drugs with lower toxicity than the naturally occurring compounds coupled with the devolvement of synthetic organic chemistry led to the start of the production of synthetic analogues of naturally occurring compounds in the late nineteenth and early twentieth centuries. In the first decade of the twentieth century, synthesis using basic SAR principles (see section 1.3) resulted in the development of a wide variety of compounds that were marketed for medical purposes. Notable amongst these compounds was the development of the local anaesthetics benzocaine and procaine from cocaine.

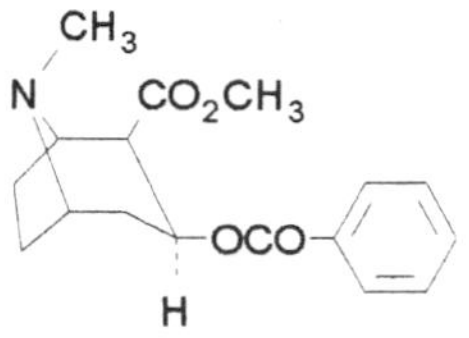

Cocaine

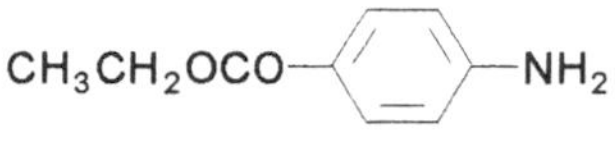

$CH_3CH_2OCO-C_6H_4-NH_2$

Benzocaine

$(C_2H_5)_2NCH_2CH_2OCO-C_6H_4-NH_2$

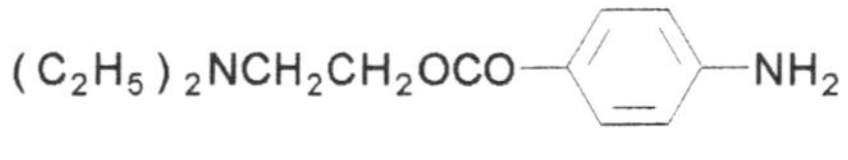

Procaine

The most popular approach to drug design is to start with the chemistry of the diseased state and determine the point where intervention is most likely to be effective (see section 2.1). This enables the medicinal chemist to suggest possible lead compounds, which are then synthesised so that their pharmacological action can be evaluated. Once a suitable lead is found, structural analogues of that lead are produced and screened. Hopefully this procedure will eventually lead to a compound that is suitable for clinical use. Obviously this approach is labour intensive and a successful outcome depends a great deal on luck. Various modifications to this approach have been introduced to reduce this element of luck (see Chapter 2).

1.8.5 Market Forces and 'Me-too Drugs'

The cost of introducing a new drug to the market is extremely high and continues to escalate. One has to be very sure that a new drug is going to be profitable before it is placed on the market. Consequently, the board of director's decision to market a drug or not depends largely on information supplied by the accountancy department rather than ethical and medical considerations. One way of cutting costs is for companies to produce drugs with similar activities and molecular structures to their competitors. These drugs are known as the 'me-too drugs'. They serve a useful purpose in that they give the practitioner a choice of medication with similar modes of action. This choice is useful in a number of situations, for example, when a patient suffers an adverse reaction to a prescribed drug or on the rare occasion that a drug is withdrawn from the market.

1.9 Drug Development and Production

Compounds that have been found to have the desired pharmacological activity with a suitable degree of potency are considered by pharmaceutical companies for further evaluation and development. This initially consists of examining the financial viability of the drug. In other words, will offsetting the cost of research and large-scale production against potential sales yield a reasonable profit in a suitable time? If production is not going to yield a reasonable profit, it is unlikely that further development will occur.

Development is the assessment of the purity, stability, potency and bioavailability of the new drug. It requires the collaboration of teams of workers from many different disciplines and its success is dependent on their skills and judgement. The process requires an investigation of the ADME of the drug (see section 1.5.1), its stability as well as its safety, production method, product formulation, packaging and marketing methods. After its release onto the market, the drug will need the ongoing monitoring of its impact, safety and quality control when it is being used by a large group of patients. All these investigations require the development of reliable analytical and microbiological protocols for the drug. The development of many of these protocols will lie in the field of the medicinal chemist.

The activities of all pharmaceutical companies are controlled to some extent from the research to the marketing stage by the country's regulatory authority. As a result, the pharmaceutical company producing a new product or the person ordering its production must obtain permission to produce the product from the statutory licensing body for the country in which the drug is to be sold. These bodies, which are essentially consumer protection agencies, issue a so-called 'product licence' when they are satisfied as to the method of production, efficacy, safety and quality of the product. This gives the pharmaceutical company or the person ordering its production the right to produce and sell the new product in the issuing country. Licences may be revoked if the producer does not strictly keep to the conditions laid down in the licence.

1.9.1 Trials and their Significance

The safety of a new drug is assessed in two stages, namely, **preclinical** and **clinical trials**. These trials assess the risks involved with the use of the new drug. They may also provide vital information concerning the ADME of the drug, which can be used in other areas of the development. The trials to be undertaken are detailed in the application for a product licence that is submitted to the licensing body. The licensing body will either approve of the trials programme or modify it. To obtain a product licence the producer must comply with all the terms of the licence.

Preclinical trials are essentially toxicity and other biological tests carried out by microbiologists on bacteria and by pharmacologists on tissue samples and animals to determine whether it is safe to test the drug on humans. The tests are carried out under both *in vivo* (in the living organism) and *in vitro* (in an artifical environment) conditions. They also provide other information concerning the drug's ADME. Relating animal tests to humans is difficult (see section 5.7) and the results are only acceptable if the dose–organ toxicity findings include a substantial safety margin. Once the drug has passed the preclinical trials it undergoes clinical trials in humans. These trials can raise legal and ethical problems and so must be approved by the appropriate legal and ethical committees before the trials are conducted. In most countries this approval requires the issuing of a certificate or licence by the appropriate medicines control agency.

In order to assess accurately the results of a clinical trial, the results must be compared with the normal situation and so in the trials conducted on healthy humans 50% of the subjects are normally given an inactive substance in a form that cannot be distinguished from the test substance. This inactive dosage form is known as a **placebo**. Furthermore, the results of a trial must be reliable and not subject to influence by either the person conducting the trial or the recipient of the drug. Consequently, it is now common practice to carry out a **double-blind** procedure where both the administrator of the drug and the recipient are unaware whether they are dealing with the drug itself or a placebo.

Trials conducted on healthy subjects do not always demonstrate the beneficial action of the new drug. It is necessary to carry out double-blind trials on unhealthy patients to assess its efficacy. However, the use of a placebo with patients who are ill raises moral and ethical considerations. Placebos may still be used if the withdrawal of therapy causes no lasting harm to patients. If this is not possible, the effect of the new drug is compared to that of an established drug used to treat the medical condition. The standard drug should be carefully selected. It should not be chosen so as to give the new drug an inflated degree of potency that could be used to give the manufacturer an unfair commercial advantage and the patient an inaccurate idea of the medicine's effectiveness.

The first clinical trials (**Phase I** trials) are usually conducted on small groups of healthy volunteers. These volunteers undergo an exhaustive medical examination before the tests and are strictly monitored at all times during the trial. The trials are conducted in either house-clinics or specialist outside facilities. The objective of these trials on healthy humans is to ascertain the behaviour of the new drug in the human body. They also yield information on the dosage form, absorption, distribution, bioavailability, elimination and side effects of the new drug. The relation of the side effects to specific metabolites allows medicinal chemists to eliminate the side effects by designing new analogues that do not give rise to that metabolite (see section 9.8). In addition, the trials give information concerning the level of the drug in the blood after intravenous and oral dosing, rate of excretion in the urine and via the bowel and the effect of gender on these parameters. At all times in the trials the functions of the kidney and other organs in the body are monitored for adverse reactions. The dose administered to the volunteers is initially a small fraction of that administered by the same route to animals.

Once the drug has been shown to be safe to use in human beings, the testing programme moves to **Phase II** trials. These are conducted on small numbers of patients having the condition that the drug has been designed to treat. These trials assess the drug's effectiveness in treating the condition and also help to establish a dose level and dosage regimen for the drug. The success of this phase leads to **Phase III** trials where the the new product is tried out on large numbers of patients. They are carried out using both placebos and comparison standards. Phase III trials are particularly useful for obtaining safety and efficacy data in order to satisfy the product licencing authorities. In both Phase II and III trials a few subjects will exhibit **adverse drug reactions** (ADRs). These are described as responses that are either unwanted or abnoxious that occur at the doses used for therapy. They exclude therapy failure. These ADRs are noted and added to the drug data sheet. Unless a high percentage of subjects exhibit the same ADR, they do not usually result in the drug being taken off the market.

When the new drug has been released onto the market the performance of the drug is monitored using very large numbers of patients, both in hospital and general practice. This monitoring is often referred to as the **Phase IV** trials. It provides more information about the drug's safety and efficiacy. In addition, trials are conducted on specific aspects of the drug's use with smaller specialist groups, for example, its kinetics in the elderly, infants, neonates and ethnic groups.

The interpretation of the results of all trials requires the close collaboration of clinicians and statisticians. Reliable results are only obtained if at least the minimum number of patients for statistical viability are involved in the preliminary trials. It is often difficult to measure precisely the parameter chosen for assessment. Consequently, results are usually quoted in terms of a probability coefficient: the lower the value of this coefficient, the more accurate the results. However, very reliable results will only be obtained from clinical trials if large groups of patients are tested. This is seldom feasible and manufacturers usually settle for the best statistical compromise. Because some adverse effects do not manifest themselves for years, it is necessary for constant monitoring of the drug (Phase IV trials) after it has been released for general use.

1.9.2 Scaling Up for Production

The medicinal chemist will have chosen the most convenient synthetic route for small-scale laboratory production. Scaling up to a viable production level for clinical trials may need a different synthetic approach as well as the construction of a suitable pilot plant. If a suitable production level at an economic cost cannot be achieved, the drug may not be developed.

1.9.3 Production and Quality Control

The manufacture of the new drug must be carried out under the conditions laid down in the product licence. Because it is not usually practical for manufacturers to dedicate a plant to the production of one particular drug, it is essential that the equipment used is cleaned and tested for adulterants before use. Many pharmaceutical manufacturers estimate that production line equipment is only used to produce the product for about 10% of its time. For most of the remaining time it is being stripped down, cleaned and reassembled.

The quality control of drugs and medicines during and after production is essential for their safe use. The importance of quality control was recognised by the Ancient Greeks, but the proper control of medicines was only achieved when accurate analytical methods were developed in the mid-nineteenth century. The development of reliable analytical methods led to the publication of national pharmacopoeias and other documents that specified the extent and the nature of the identification tests and quantitative assessments required to ensure that the product that reaches the public is fit for its purpose. These documents now cover the production, storage and application of pharmaceutical products. They are the subject of constant review but unfortunately this does not completely prevent the occurrence of medical problems. However, the continual updating of these documents does reduce the possibility of similar problems occuring in other products. It is gratifying to note that since the thalidomide disaster very few drugs have been removed from the market on safety grounds.

1.9.4 Patent Protection

The high cost of drug development and production makes it essential for a company to maximise its returns from a new drug. This can only be achieved by preventing rivals from unrestricted copying of a new product. Patents are used to prevent rival companies from manufacturing and marketing a product without the permission of the originator of the product. However, many companies do market other manufacturers' products under licence.

Patents have been used, in one form or another, as a means of industrial protection from the early fourteenth century to the present day. Originally, they were intended to encourage the development of new industries and products by granting the developer or producer the monopoly to either use specific industrial equipment or produce specific goods for a limited period. This monopoly, enforced by the appropriate government office, enabled innovators to obtain a just reward for their efforts. In most countries the awarding of a patent prevents third parties from manufacturing and selling the product without the consent of the innovator. However, patents do encourage and protect the development of new ideas by the publication of new knowledge.

Originally each state issued its own patent laws but by the mid-eighteenth century it was recognised that patent rights should extend beyond national boundaries. The first international agreement was the **Paris Convention** of 1883. This has been revised on numerous occasions, the current treaties in operation being the European Union's European Patent Convention (EPC) of 1978 and the Patent Cooperation Treaty (PTC) signed in Washington in 1970. The former is only open to European countries and administered by the European Patents Office (EPO). The latter is open to all countries of the world and is administered through the national patent offices of the country subscribing to the treaty.

The pharmaceutical industry is dependent on the existence of patents. The protection offered by a patent is of paramount importance to a company at the research, development and production stages of drug production. It is essential that a patent is filed as soon as new compounds have been made and shown to have interesting properties, otherwise a rival company working in the same field might pre-empt the patent, which leads to a large amount of expensive research work being unproductive in respect to company profits. Thus, it is particularly important that the patent covers the relevant field and does not give a rival manufacturer an exploitable loophole.

The time required for the development of a drug from discovery to production can take at least 7–15 years. In many countries, patents normally run for 20 years from the date of application. Consequently, the time available for a manufacturer to recoup the cost of development and show a profit is rather limited, which accounts for the high cost of some new drugs. Furthermore, it also means that some compounds are never developed because the patent-protected production time available to recoup the cost of development is too short.

1.10 Summary

Medicinal chemistry is the design, discovery and development of new drugs. The normal first step in the discovery of a new drug is to select a disease state and identify a target area where drug intervention is likely to be profitable in both senses of the word. The next step is to identify so-called **lead compounds** that may bring about the desired effect. These lead compounds are synthesised and tested. It is unlikely that they will be acceptable as drugs. However, lead compounds with suitable pharmacological activities act as a starting point for the development of the new drug. Development is the determination of the optimum structure for activity and safety. This normally includes synthesising structural analogues of the lead and assessing their pharmacological activity and safety. Various methods of approach are used to select the analogues most likely to give the required result. These include the use of SAR, QSAR, combinatorial chemistry and computer modelling.

Drugs usually act by binding to a target site, which is usually a receptor or active site. **Receptor sites** are areas in or on a cell where the binding of a drug molecule or other substrate initiates a physiological response, whereas **active sites** are the regions of enzymes where chemical reactions are catalysed. A drug is most effective when its stereochemical structure is complementary to the stereochemical structure of its target site.

The physical form in which the drug is administered to the patient is known as its **dosage form**. In most dosage forms the drug is usually mixed with excipients, which are pharmacologically inactive substances added to a dosage form in order to expedite its administration. **Administration** may be by the **parenteral route** or the **enteral route**. The dosage regimen is the prescribed manner in which a drug is taken by a patient.

Drug action can be classified into its pharmacokinetic and pharmacodynamic phases. The **pharmacokinetic phase** of drug action covers the effect of the body on the drug. It describes how the drug is influenced by such properties as absorption, distribution, metabolism and elimination, which influence the transport of the drug through the body from its site of action to its site of action. The **pharmacodynamic phase** of drug action is the pharmacological effect of the drug on the body, which is dependent on the structure of the drug molecule. For the maximum chance of activity the structure of the drug or its pharmacophore should be complementary to that of its target site. Each of the stereoisomers of a compound will exhibit its own pharmacological activity. This can range from similar through inert to toxic.

Screening is the programme of tests that are used to assess the pharmacological activity and safety of drugs. The screening programme for a new drug is divided into preclinical trials (Phase I trials) and clinical trials (Phases II, III and IV trials). **Preclinical trials** are the test programmes conducted on animals, tissue samples and bacteria to determine whether a new drug can be tested on humans. **Clinical trials** are test programmes that are carried out on humans to determine the safety and behaviour of the new drug *in situ*. All tests should be conducted under double-blind conditions if their results are to be valid. **Double-blind** procedures are screening test programmes in which both the administrator and the recipient are unaware of whether they are dealing with the drug or the placebo.

1.11 Questions

(1) Predict, giving a reason for the prediction, the **most likely** general effect of the stated structural change on either the *in situ* stability or the pharmacological action of the stated drug.
 (a) The introduction of *ortho*-ethyl groups in dimethylaminoethyl 4-aminobenzoate.
 (b) The replacement of the amino group in the CNS stimulant amphetamine ($PhCH_2CH(NH_2)CH_3$) by a trimethylammonium chloride group.
 (c) The replacement of the ester group in the local anaesthetic ethyl-4-aminobenzoate (benzocaine) by an amide group.

(2) State the general factors that need to be considered when designing a drug.

(3) Explain the meaning of the terms: (a) lead compound, (b) dosage form, (c) enteral administration of drugs, (d) drug regimen and (e) excipient.

(4) Define the meaning of the terms pharmacokinetic phase and pharmacodynamic phase in the context of drug action. List the main general factors that affect these phases.

(5) The drug amphetamine ($PhCH_2CH(NH_2)CH_3$) binds to the protein albumin in the bloodstream. Predict how a reduction in pH would be expected to influence this binding? Albumin is negatively charged at pH 7.4 and electrically neutral at pH 5.0.

(6) Discuss the general effects that stereoisomers could have on the activity of a drug.

(7) Suggest strategies for improving the stability of compound A in the gastrointestinal tract. What could be the general effect of these strategies on the pharmaceutical action of compound A (structure, page 30)?

(A) 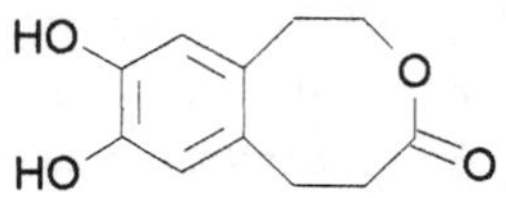

(8) Explain the differences between (a) preclinical and clinical trials and (b) Phase I and Phase II trials.
(9) What is a patent? Why is it necessary to patent drugs?

2 Drug Discovery by Design

2.1 Introduction

Drug discovery is part luck and part structured investigation. At the begining of the nineteenth century it was largely carried out by individuals but it now requires teamwork, the members of the team being specialists in various fields, such as medicine, biochemistry, chemistry, computerised molecular modelling, pharmaceutics, pharmacology, microbiology, toxicology, physiology and pathology.

The drugs used in medicine are developed from so-called **lead compounds** (Figure 2.1). These compounds were originally discovered by investigating local folk remedies, and natural products such as plants (land and marine), trees, microorganisms and animals. These are still useful sources of lead compounds but many are now produced as a result of investigations into the nature of disease states. Lead compounds are often unsuitable for clinical use because they may be either too toxic or have serious unwanted side effects. However, their structures serve as the starting points for the synthesis of so-called **analogues**, one or more of which may be suitable for clinical use (Figure 2.1).

The approach to drug design depends on the objectives of the design team. These objectives can range from changing the pharmacokinetics of an existing drug to designing a completely new compound. In all cases the team will start by devising an outline of the intended investigation. For example, if the objective is to modify the pharmacokinetics of an existing drug the design team has to decide what structural modifications need to be investigated in order to achieve that objective. On the other hand, if the objective is to find a new drug for a specific medical problem the starting point is a knowledge of the biochemistry of the condition and/or the microorganism responsible for the condition. This may require basic research into the disease-causing process before initiating the drug design investigation (Figure 2.1). The information obtained is used by the team to decide where intervention would be most likely to bring about the desired result. Once the point of intervention has been selected the team has to decide on the structure of a lead compound that could possibly bring about the required change. A number of candidates are usually considered but the expense of producing drugs dictates that the team has to choose only one or two of these compounds to either act as the lead or to be the inspiration for the lead compound. The final selection depends on the experience of the team.

One approach to lead compound selection is to carry out a comprehensive literature and database search to identify compounds found in the organism (**endogenous compounds**) and compounds that are not found in the organism (**exogenous compounds**) that have some biological

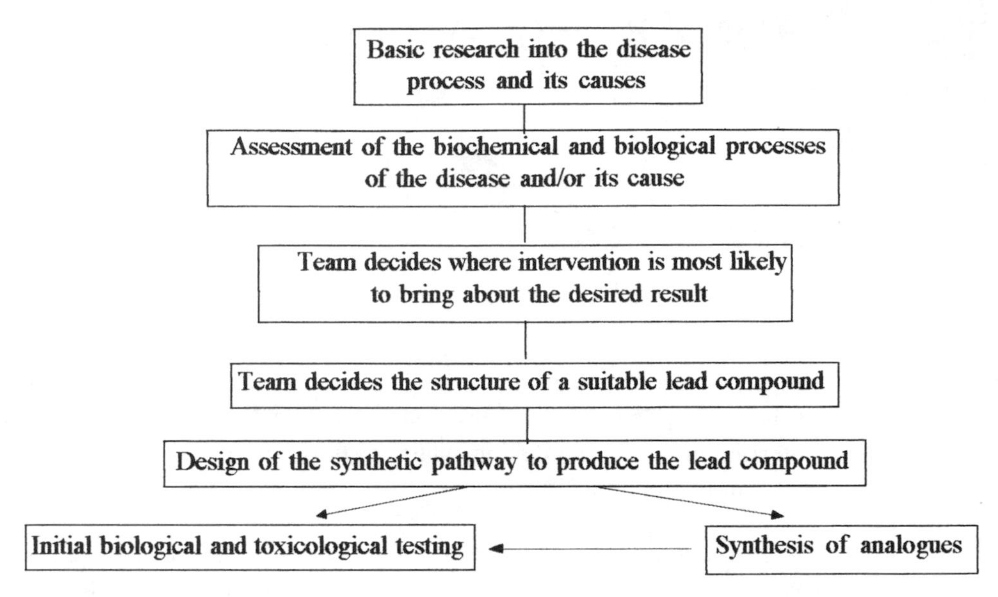

Figure 2.1. The general steps in the design of a new drug.

effect at the intervention site. These compounds are used as leads and modified in an appropriate manner.

Modern approaches to lead compound discovery include combinatorial chemistry (see section 2.6) and computer modelling techniques (see section 2.7). The former uses a simultaneous multiple synthesis technique to produce large numbers of potential leads. These potential leads are subjected to rapid high-throughput biological screening to identify the most active lead compounds. Once identified, these lead compounds are subject to further development. Computer modelling is normally used to check whether the three-dimensional structure of the proposed lead is complementary to that of its receptor domain. However, this does require a detailed knowledge of the three-dimensional structures of the ligand and target site.

A major consideration in the selection of a lead is its stereochemistry. It is now recognised that the biological activities of the individual enantiomers and their racemates may be very different. Consequently, it is necessary to evaluate pharmacologically the individual enantiomers as well as any racemates. However, it is often difficult to obtain specific enantiomers in a pure state (see section 12. 3). Both of these considerations make the production of optically active compounds expensive and so medicinal chemists often prefer to synthesise lead compounds that are not optically active. However, this is not always possible.

Once the structure of the proposed lead has been decided it becomes the responsibility of the medicinal chemist to devise a synthetic route and prepare a sample of this compound for testing. Once synthesised, the compound undergoes initial pharmacological and toxicological testing. The results of these tests enable the team to decide whether it is profitable to continue development by preparing analogues (Figure 2.1) because it is unlikely that the lead compound itself will be suitable for use as a drug. The usual scenario is to prepare a series of analogues and analyse the results of their biological testing to determine the structure with optimum activity. This analysis may make use of SAR (see section 2.3), QSAR (see section 2.4), computational chemistry (see section 2.5) and combinatorial chemistry (see section 2.6) to help discover the nature of this structure.

The selection of a lead compound and the development of a synthetic pathway for its preparation (see Chapter 12) is not the only consideration at the start of an investigation. It is no use preparing a series of compounds if there is no suitable testing procedure. Researchers must also devise suitable *in vivo* and *in vitro* tests to assess the activity and toxicity of the compounds produced. There is no point in carrying out an expensive synthetic procedure if at the end of the day it is impossible to test the product.

The processes of drug discovery outlined in this section are time consuming and expensive. It takes about 10 years for a drug to reach the general public and only one in about 10 000 of the compounds prepared is ever used.

2.2 Stereochemistry and Drug Design

It is now well established that the shape of a molecule is normally one of the most important factors that affect drug activity. Consequently, the overall shape of the structure of a molecule is an important consideration when designing an analogue. Some structural features impose a considerable degree of rigidity into a structure whereas others make the structure more flexible. Other structures give rise to stereoisomers that can exhibit different potencies, types of activity and unwanted side effects (see section 1.5.3). It is necessary to consider all these stereochemical features in the design of a potential analogue. However, the extent to which one can exploit these structural features will depend on our knowledge of the structure and biochemistry of the target biological system.

2.2.1 Structurally Rigid Groups

Groups that are structurally rigid are unsaturated groups of all types and saturated ring systems (Figure 2.2). The former include esters and amides as well as aliphatic conjugated systems and aromatic and heteroaromatic ring systems. The binding of these rigid structures to a target site can give information about the shape of that site as well as the nature of the interaction between the site and the ligand. Rigid structures may also be used to determine the conformation assumed by a ligand when it binds to its target site (see section 2.2.2). Furthermore, the fact that the structure is rigid means that it may be replaced by alternative rigid structures of a similar size and shape to form analogues that may have different binding characteristics and possibly, as a result, a different activity or potency.

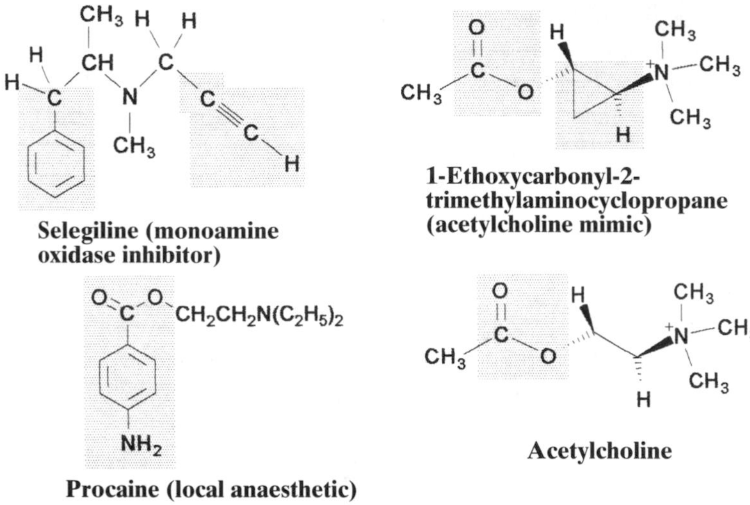

Selegiline (monoamine oxidase inhibitor)

1-Ethoxycarbonyl-2-trimethylaminocyclopropane (acetylcholine mimic)

Procaine (local anaesthetic)

Acetylcholine

Figure 2.2. Examples of structural groups that impose a rigid shape on sections of a molecule. The shaded areas represent the rigid sections of the molecule.

2.2.2 Conformation

Early work in the 1950s and early 1960s by Schueler and Archer suggested that the flexibility of the structures of both ligands and receptors accounted for the same ligand being able to bind to different subtypes of a receptor (see section 8.3). Archer also concluded that a ligand appeared to assume different conformations when it bound to the different subtypes of a receptor. For example, acetylcholine exhibits both muscarinic and nicotinic activity. Archer and collegues suggested that the muscarinic activity was due to the *anti* or staggered conformation whereas the nicotinic activity was due to the *syn* or eclipsed form (Figure 2.3). These workers based this suggestion on their observation that the *anti* conformation of 2-tropanyl ethanoate methiodides preferentially binds to muscarinic receptors whereas the *syn* conformation binds preferentially to nicotinic receptors. The structures of both of these compounds contain an acetylcholine residue locked in the appropriate conformation by the ring structure. This and subsequent investigations led to the conclusion that the development of analogues with restricted or rigid conformations could result in the selective binding of drugs to target sites that could result in very active drugs with reduced unwanted side effects.

The main methods of introducing conformational restrictions are by using bulky substituents, unsaturated structures or small ring systems. Small ring systems are usually the most popular choice (Figure 2.4). In all cases the structures used must be chosen with care because there will always be the possibility that steric hindrance will prevent the binding of the analogue to the target. However, if sufficient information is available, computer modelling can be of considerable assistance in the choice of structures.

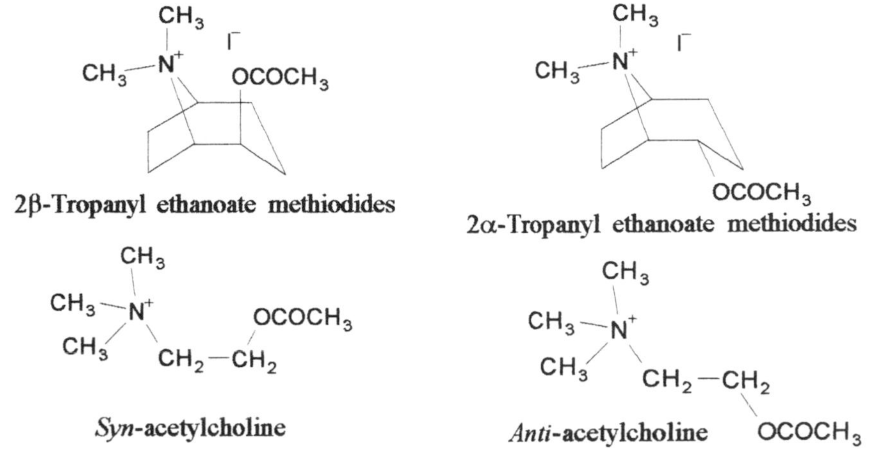

2β-Tropanyl ethanoate methiodides

2α-Tropanyl ethanoate methiodides

Syn-acetylcholine

Anti-acetylcholine

Figure 2.3. *Syn* and *anti* conformers of acetylcholine and 2-tropanyl ethanoate methiodides.

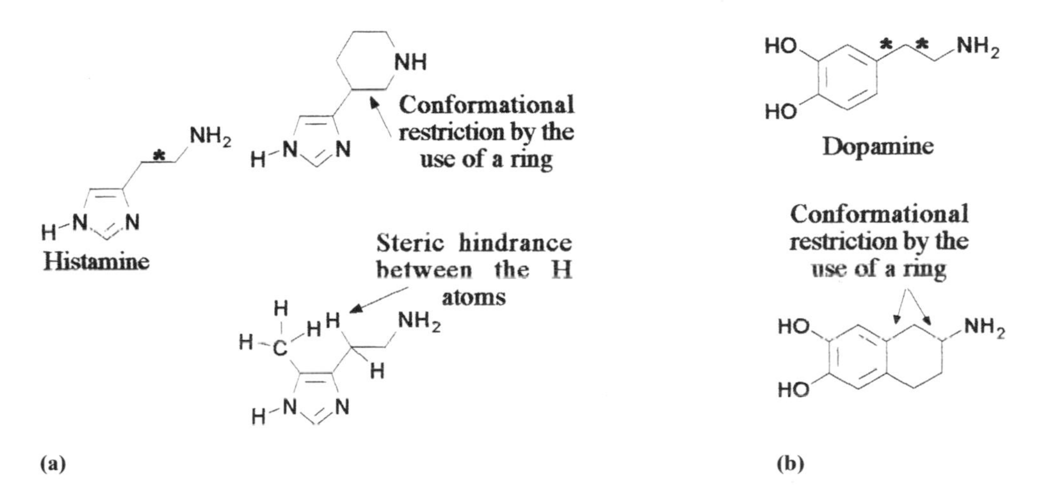

Figure 2.4. Examples of the use of conformational restrictions to produce analogues of (a) histamine and (b) dopamine. Bonds marked with an asterisk can exhibit free rotation and form numerous conformers.

A further limitation is knowing which bond to restrict. Even in simple molecules numerous eclipsed, staggered and gauche conformations are possible (Figure 2.5).

The biological data obtained using restricted conformation analogues can be of use in determining the most bioactive conformation of the ligand. If the analogue exhibits either the same or a greater degree of activity as the lead compound it may be concluded that the analogue has the correct conformation for binding to that site. However, if the analogue exhibits no activity the result could be due to either steric hindrance between the restricting group and the target or the analogue having the incorrect conformation. In this case computer modelling may be of some assistance.

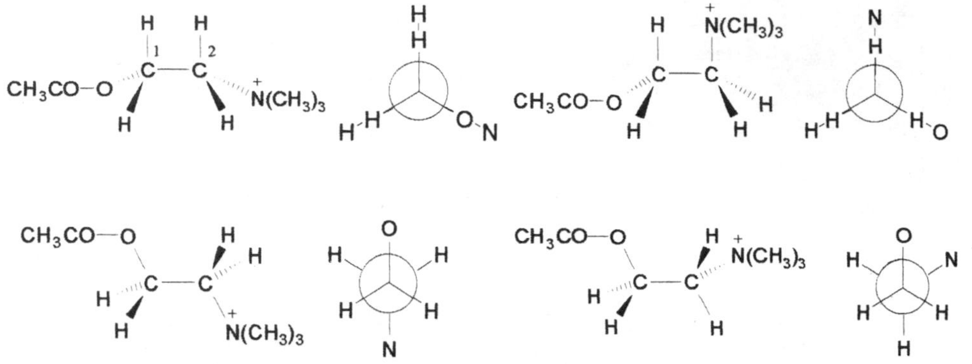

Figure 2.5. Examples of some of the conformations of the C1–C2 bond of acetylcholine. Other conformations occur about the C–N and C–O single bonds.

2.2.3 Configuration

Configurational centres impose a rigid shape on sections of the molecule in which they occur. However, their presence gives rise to geometric and optical isomerism. Because these stereoisomers have different shapes, biologically active stereoisomers will often exhibit differences in their potencies and/or activities (Table 2.1). These pharmacological variations are particularly likely when a chiral centre is located in a critical position in the structure of the molecule. The consequence of these differences is that it is now necessary to make and test separately all the individual stereoisomers of a drug.

As well as an effect on the activity, different stereoisomers will also change other physiochemical properties such as absorption, metabolism and elimination. For example, (−)-norgestrel is absorbed at twice the rate of (+)-norgestrel through buccal and vaginal membranes. The plasma half-life of *S*-indacrinone is 2–5 h whereas the value for the *R* isomer is 10–12 h.

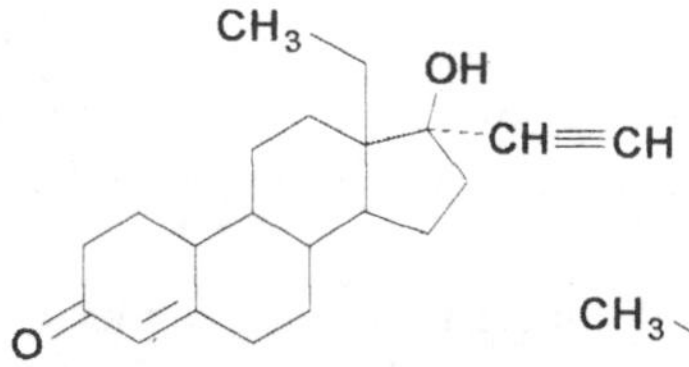

(−)-Norgestrel

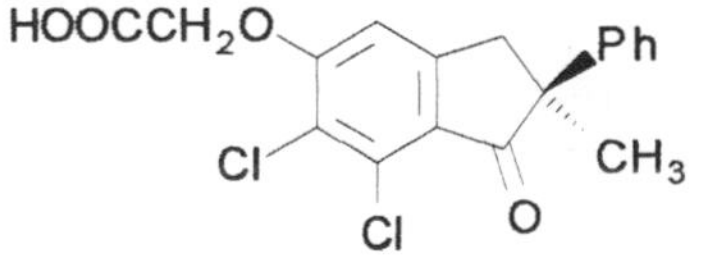

(+)-Norgestrel

***S*-Indacrinone**

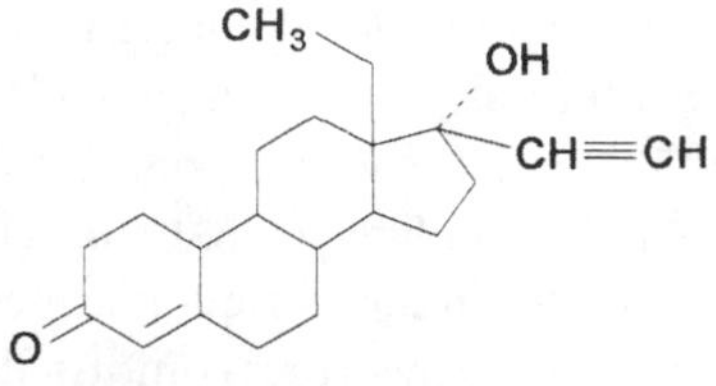

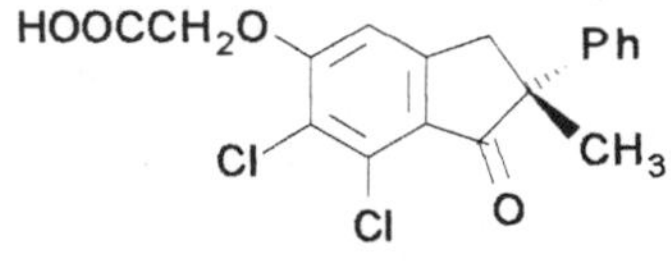

***R*-Indacrinone**

Table 2.1. Variations in the biological activities of stereoisomers.

First stereoisomer	Second stereoisomer	Example
Active	Activity of same type and potency	The *R*- and *S*-isomers of the antimalarial chloroquine have equal potencies 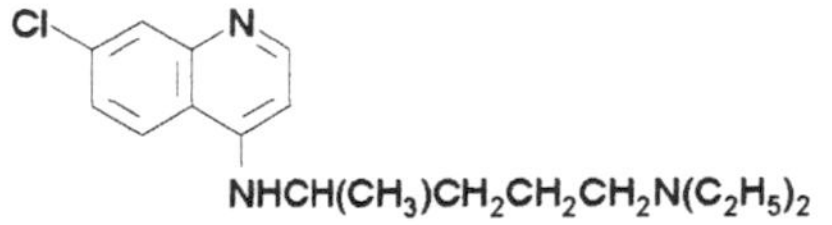 $NHCH(CH_3)CH_2CH_2CH_2N(C_2H_5)_2$
Active	Activity of same type but weaker	The *E*-isomer of diethylstilbestrol, an oestrogen, is only 7% as active as the *Z*-isomer 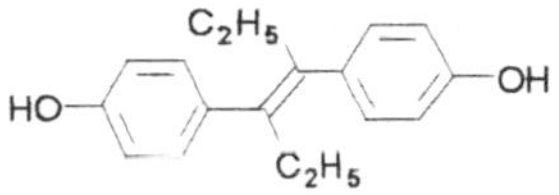
Active	Activity of a different type	*S*-Ketamine is an anaesthetic *R*-Ketamine has little anaesthetic action but is a psychotic Ketamine 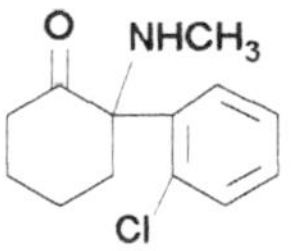
Active	No activity	*S*-α-Methyldopa is a hypertensive drug but the *R*-isomer is inactive *S*-α-Methyldopa 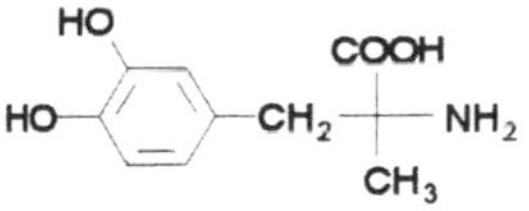
Active	Active but different side effects	Thalidomide, the *S*-isomer, is a sedative and has teratogenic side effects. The *R*-isomer is also a sedative but has no teratogenic activity *S*-Thalidomide 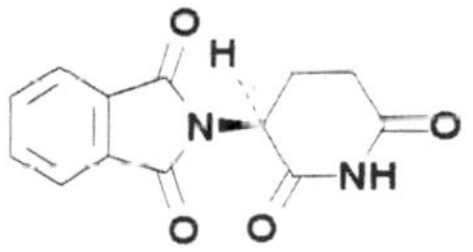

2.3 Structure–Activity Relationships (SAR)

Most drugs act at a specific site such as an enzyme or receptor. Compounds with similar structures often tend to have similar pharmacological activity. However, they usually exhibit differences in potency and unwanted side effects and in some cases different activities. These structurally related differences are commonly referred to as structure–activity relationships (SAR). A study of the structure–activity relationships of a lead compound and its analogues can be used to determine the parts of the structure of the lead that are responsible for its biological activity, that is, its **pharmacophore** and also its unwanted side effects. This information is subsequently used to develop a new drug that has increased activity (optimise its SAR), a different activity from an existing drug, fewer unwanted side effects and improved ease of administration to the patient.

Structure–activity relationships are usually determined by making minor changes to the structure of the lead and assessing the effect that this has on biological activity. Traditional SAR investigations are carried out by making large numbers of analogues of the lead and testing them for biological activity. Over the years numerous lead compounds have been investigated and from the mass of data it is possible to make some broad generalisations about the biological effects of specific structural changes. These changes may be conveniently classified as: (1) the size and shape of the carbon skeleton (see section 2.3.1), (2) the nature and degree of substitution (see sections 2.3.2 and 2.3.3), and (3) the stereochemistry of the lead (see section 2.2.3).

The selection of the changes required to produce analogues of a particular lead is made by considering the activities of compounds with similar structures and also the possible chemistry and biochemistry of the intended analogue. For example, replacing a hydroxy group with a methyl group could reduce the water solubility of the analogue and also its ability to hydrogen-bond. The former could reduce its ease of absorption whereas the latter could affect its ability to bind to its target site. It could improve the transport of the drug through membranes and also introduce changes in the metabolism of the drug. For example, oxidation of the methyl group to a carboxylic group could increase the rate of metabolism. All these effects could result in a loss of activity or a reduction in an unwanted side effect. A further consideration is the size of the analogue. Changing the structure of the lead could result in an analogue that is too big to fit its intended target site. Computerised molecular modelling can be used to check this provided that the structure of the target is known or can be simulated with some degree of accuracy.

Traditional SAR investigation procedures are useful tools in the search for new drugs. However, they are expensive in both personnel and materials. Consequently, a number of attempts have been made to improve on traditional structure–activity investigations with varying degrees of success (see sections 2.4, 2.5 and 2.6).

2.3.1 Changing Size and Shape

The shapes and sizes of molecules can be modified in a variety of ways, such as:

(i) changing the number of methylene groups in chains and rings;
(ii) increasing or decreasing the degree of unsaturation;
(iii) introducing or removing a ring system.

2.3.3.1 Changing the Number of Methylene Groups in a Chain

Increasing the number of methylene groups in a chain or ring increases the size and the lipid nature (lipophilicity) of the compound. The biological response curves associated with this increase in size can assume a variety of shapes (Figure 2.6a). It is believed that the increase in activity with increase in the number of methylene groups is probably due to an increase in the lipid solubility of the analogue, which gives a better membrane penetration. Conversely, a decrease in activity (Figure 2.6b) with an increase in the number of methylene groups is attributed to a reduction in the water solubility of the analogues. This reduction in water solubility can result in the poor distribution of the analogue in the aqueous media as well as the trapping of the analogue in biological membranes (see section 3.10.2). A further problem with large increases in the numbers of inserted methylene groups in chain structures is micelle formation (see section 3.11.1). Micelle formation produces large aggregates that, because of their shape, cannot bind to active sites and receptors.

Introducing chain branching, different sized rings and the substitution of chains for rings and *vice versa* may also have an effect on the potency and type of activity of analogues. For example, the replacement of the sulphur atom of the antipsychotic chlorpromazine by —CH_2—CH_2— produces the antidepressant clomipramine.

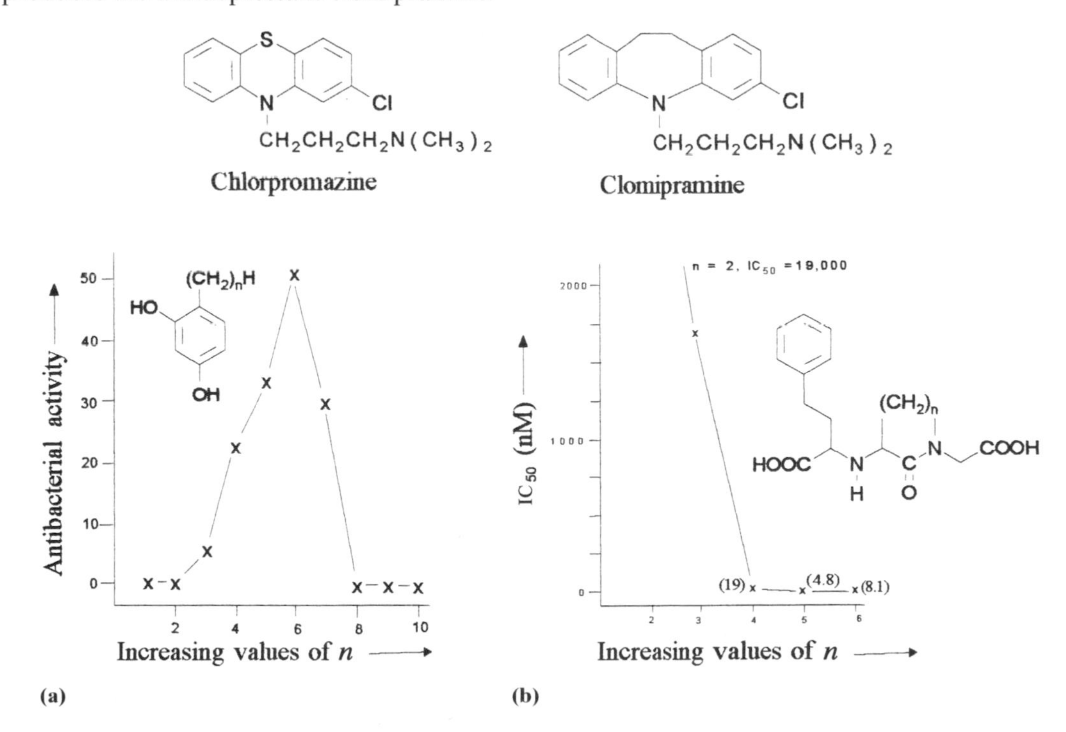

Figure 2.6. Examples of the variation of response curves with increasing numbers of inserted methylene groups. (a) A study by Dohme and collegues on the variation of antibacterial activity of 4-alkyl-substituted resorcinols. (b) Inhibition of angiotensin-converting enzyme by enalaprilat analogues (Thorsett). The values in parentheses are IC_{50} values for those analogues.

2.3.1.2 Changing the Degree of Unsaturation

The removal of double bonds increases the degree of flexibility of the molecule, which may make it easier for the analogue to fit into active and receptor sites by taking up a more suitable conformation. However, an increase in flexibility could also result in a change or loss of activity.

The introduction of a double bond increases the rigidity of the structure. It may also introduce the complication of *E*- and *Z*-isomers, which could have quite different activities (see Table 2.1). The analogues produced by the introduction of unsaturated structures into a lead compound may exhibit different degrees of potency or different types of activities. For example, the potency of prednisone is about 30 times greater than that of its parent compound cortisol, which does not have a 1–2 C=C bond. The replacement of the S atom of the antipsychotic phenothiazine drugs by a —CH=CH— group gives the antidepressant dibenzazepine drugs, such as protriptyline.

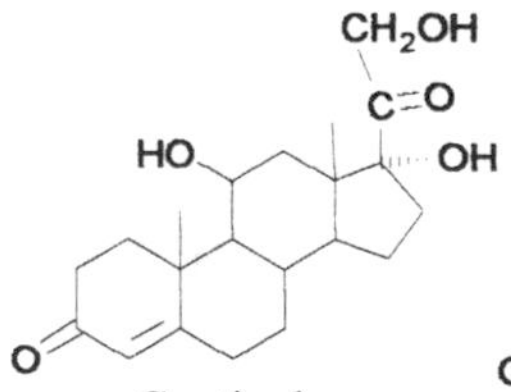

Cortisol 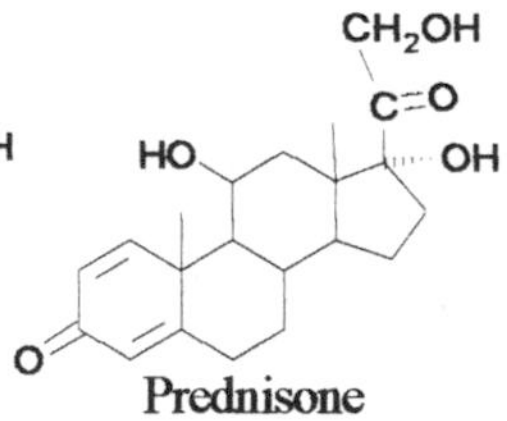 **Prednisone** 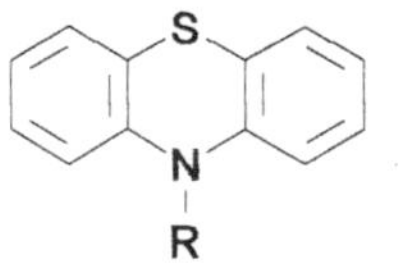 **Phenothiazine drugs** 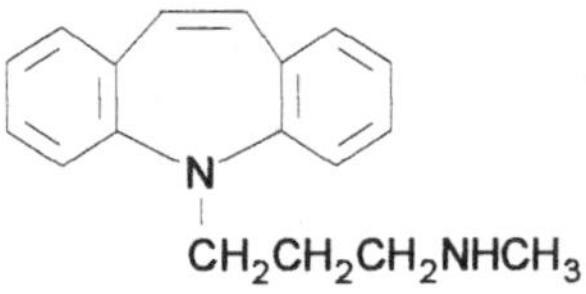 **Protriptyline (Vivactil)**

The introduction of a C=C group will often give analogues that are more sensitive to metabolic oxidation. This may or may not be a desirable feature for the new drug. Furthermore, the reactivity of C=C frequently causes the analogue to be more toxic than the lead.

2.3.1.3 Introduction or Removal of a Ring System

The introduction of a ring system changes the shape and increases the overall size of the analogue. The effect of these changes on the potency and activity of the analogue is not generally predictable. However, the increase in size can be useful in filling a hydrophobic pocket in a target site, which might strengthen the binding of the drug to the target. For example, it has been postulated that the increased inhibitory activity of the cyclopentyl analogue (rolipram) of 3-(3,4-dimethyloxyphenyl)-butyrolactam towards cAMP phosphodiesterase is due to the cyclopentyl group filling a hydrophobic pocket in the active site of this enzyme.

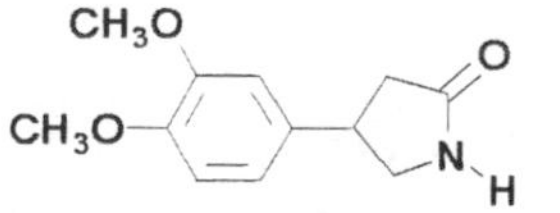

3-(3,4-Dimethoxyphenyl)-butyrolactam

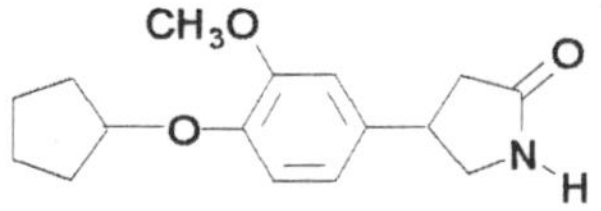

Rolipram, an antidepressant, is 10 times more active than 3-(3,4-dimethoxyphenyl)-butyrolactam

The incorporation of smaller, as against larger, alicyclic ring systems into a lead structure reduces the possibility of producing an analogue that is too big for its target site. It also reduces the possibility of complications caused by the existence of conformers. However, the selection of the system for a particular analogue may depend on the objective of the alteration. For example, the cyclopropane ring is usually more stable than the ethylenic C=C group and so could be used to replace this group if a more stable compound of a similar size is required. For example, the antidepressant tranylcypromine is more stable than its analogue 1-amino-2-phenylethene.

NH_2 NH_2

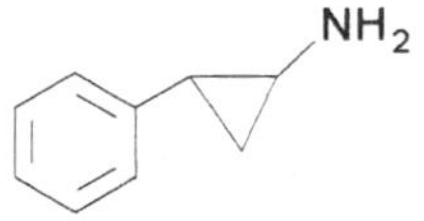

Tranylcypromine 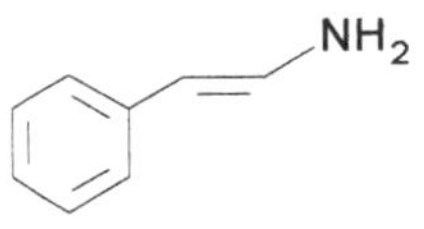 **1-Amino-2-phenylethene**

The insertion of aromatic systems into the structure of the lead will introduce rigidity into the structure as well as increase the size of the analogue. The latter means that small aromatic systems such as benzene and five-membered heterocyclic sytems are often preferred to larger systems. However, the π electrons of aromatic systems may or may not improve the binding of the analogue to its target site. Furthermore, heterocyclic aromatic systems will also introduce extra functional groups into the structure, which could also affect the potency and activity of the analogue. For example, the replacement of the *N*-dimcthyl group of chlorpromazine by an *N*-methylpiperazine group produces an analogue (prochlorperazine) with increased antiemetic potency but reduced neuroleptic activity. It has been suggested that this change in activity could be due to the presence of the extra tertiary-amine group.

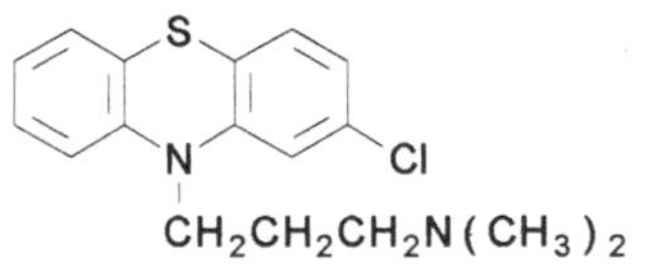

S N Cl $CH_2CH_2CH_2N(CH_3)_2$ S N Cl $CH_2CH_2CH_2-N$ $N-CH_3$

Clorpromazine 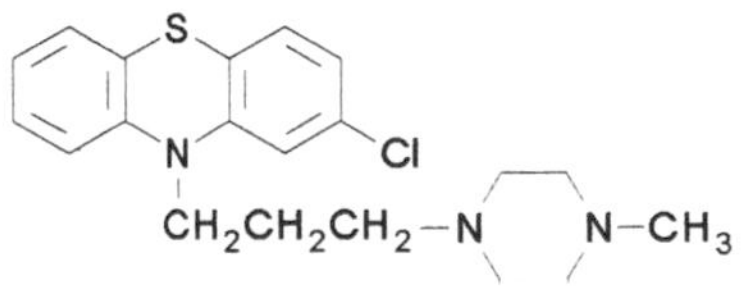 **Prochlorperazine**

The incorporation of ring systems, especially larger systems, into the structure of a lead can be used to produce analogues that are resistant to enzymic attack by sterically hindering the access of the enzyme to the relevant functional group. For example, the resistance of diphenicillin to β-lactamase is believed to be due to the diphenyl group preventing the enzyme from reaching the β-lactam. It is interesting to note that 2-phenylbenzylpenicillin is not resistant to β-lactamase attack. In this case, it appears that the diphenyl group is too far away from the β-lactam ring to hinder the attack of the β-lactamase.

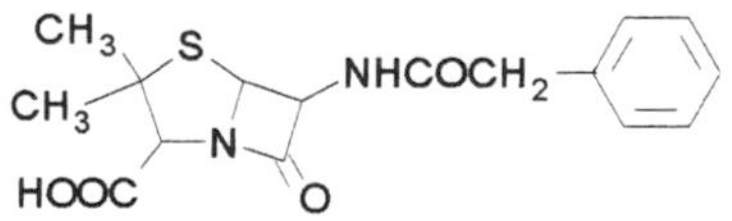

CH_3 S CH_3 $NHCOCH_2$ N HOOC O

Benzylpenicillin (not β-lactamase resistant)

CH_3 S CH_3 $NHCOCH_2$ N HOOC O

2-Phenylbenzylpenicillin (not β-lactamase resistant)

CH_3 S CH_3 NHCO N HOOC O

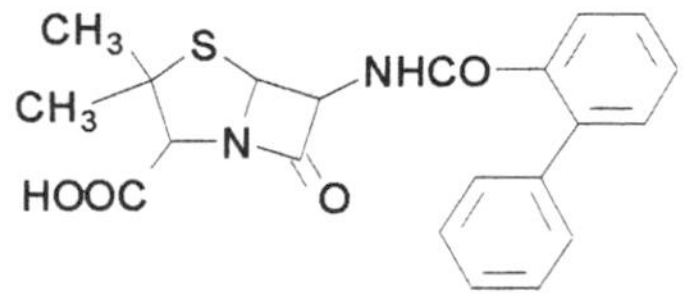

Diphenicillin (β-lactamase resistant)

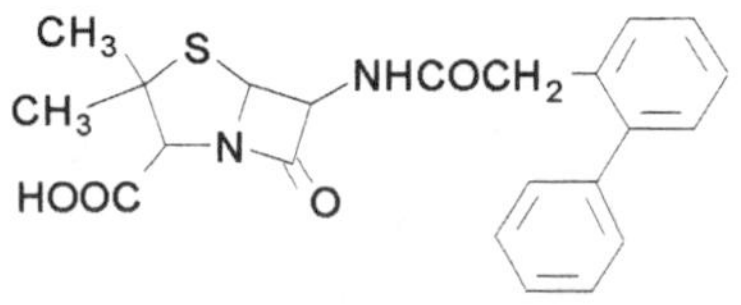

Many of the potent pharmacologically active naturally occurring compounds, such as the alkaloids morphine and curare, have such complex structures that it would not be economic to synthesise them on a large scale. Furthermore, they also tend to exhibit unwanted side effects. However, the structures of many of these compounds contain several ring systems. In these cases, one approach to designing analogues of these compounds centres around determining the pharmacophore and removing any surplus ring structures. It is hoped that this will also result in the loss of any unwanted side effects. The classic example illustrating this type of approach is the development of drugs from morphine (Figure 2.7).

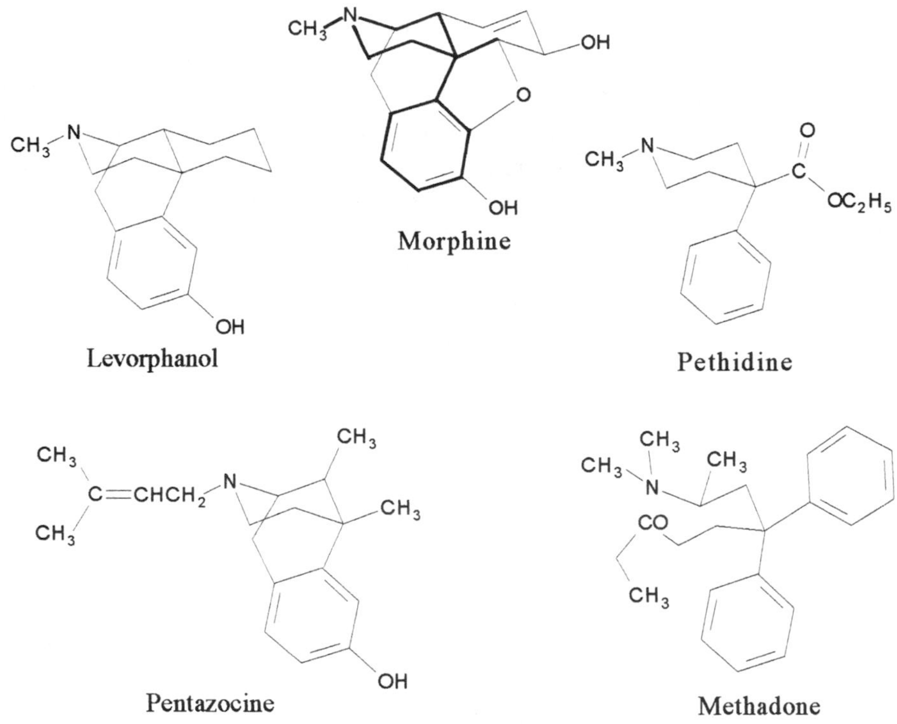

Morphine

Levorphanol

Pethidine

Pentazocine

Methadone

Figure 2.7. The pharmacophore of morphine was found to be the structure represented by the bonds in heavy type. Pruning and modification of the remaining structure of morphine resulted in the development of (a) the more potent but still highly addictive levorphanol, (b) the very much less potent pethidine, (c) the less potent and less addictive pentazocine and (d) the equally potent but much less addictive methadone, amongst other drugs.

2.3.2 Introduction of New Substituents

The formation of analogues by the introduction of new substituents into the structure of a lead may result in an analogue with significantly different chemical and hence pharmacokinetic properties. For example, the introduction of a new substituent may cause significant changes in lipophilicity that affect transport of the analogue through membranes and the various fluids found in the body. It would also change the shape, which could result in conformational restrictions that affect the binding to the target site. In addition, the presence of a new group may introduce a new metabolic pathway for the analogue. These changes will in turn affect the pharmacodynamic properties of the analogue. For example, they could result in an analogue with either increased or decreased potency, duration of action, metabolic stability and unwanted side effects. The choice of substituent will depend on the properties that the development team decide to enhance in an attempt to meet their objectives. Each substituent will impart its own characteristic properties to the analogue. However, it is possible to generalise about the effect of introducing a new substituent group into a structure but there will be numerous exceptions to the predictions.

2.3.2.1 Methyl Groups

The introduction of methyl groups usually increases the lipophilicity of the compound and reduces its water solubility (Table 2.2). It should improve the ease of absorption of the analogue into a biological membrane but will make its release from biological membranes into the aqueous media more difficult (see section 3.10).

Table 2.2. The change in the partition coefficients (P) of some common compounds when methyl groups are introduced into their structures. The greater the value of P the more lipid-soluble the compound. Benzene and toluene values were measured using an *n*-octanol/water system and the remaining values were measured using an olive oil/water system.

Compound	Structure	P	Analogue	Structure	P
Benzene		135	Toluene	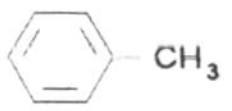 $-CH_3$	490
Acetamide	CH_3CONH_2	83	Proprionamide	$CH_3CH_2CONH_2$	360
Urea	NH_2CONH_2	15	*N*-Methylurea	$CH_3NHCONH_2$	44

The incorporation of a methyl group can impose steric restrictions on the structure of an analogue. For example, the *ortho*-methyl analogue of diphenhydramine exhibits no antihistamine activity. Harmes and collegues suggest that this could be due to the *ortho*-methyl group restricting rotation about the C—O bond of the side chain. This prevents the molecule from adopting the conformation necessary for antihistamine activity. It is interesting to note that the *para*-methyl analogue is 3.7 times more active than diphenhydramine.

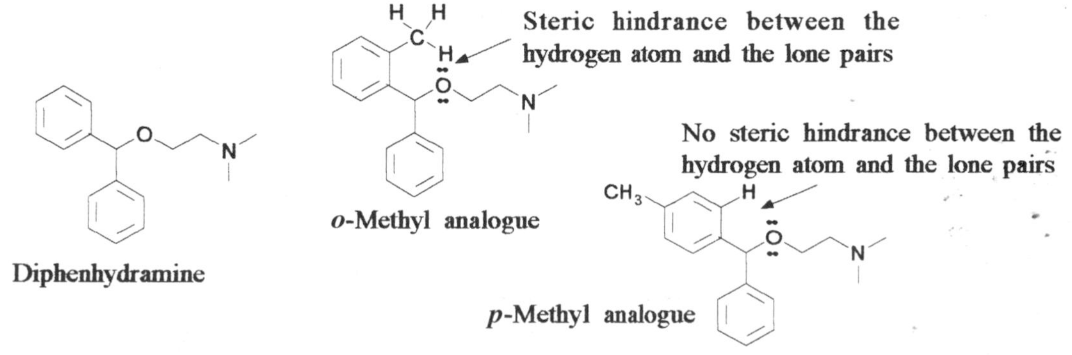

Steric hindrance between the hydrogen atom and the lone pairs

No steric hindrance between the hydrogen atom and the lone pairs

Diphenhydramine

***o*-Methyl analogue**

***p*-Methyl analogue**

The incorporation of a methyl group can have one of three general effects on the rate of metabolism of an analogue: (i) an increased rate of metabolism due to oxidation of the methyl group; (ii) an increase in the rate of metabolism due to demethylation by the transfer of the methyl group to another compound; or a reduction in the rate of metabolism of the analogue.

(i) A methyl group bound to an aromatic ring or a structure which increases its reactivity may be metabolised to a carboxylic acid, which can be eliminated more easily. For example, the antidiabetic tolbutamide is metabolised to its less toxic benzoic acid derivative. The introduction of a reactive C—CH_3 group offers a detoxification route for lead compounds that are too toxic to be of use.

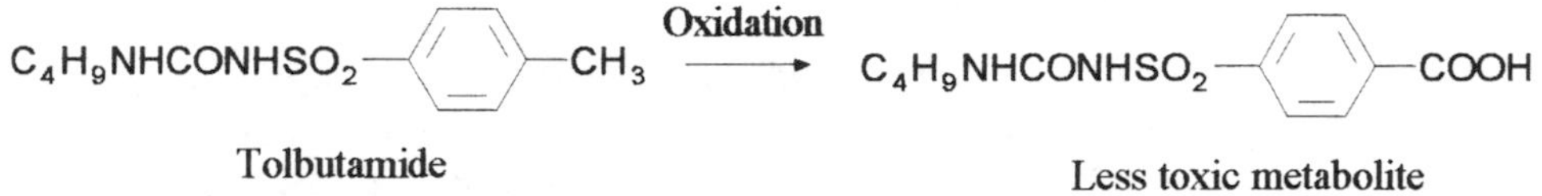

$C_4H_9NHCONHSO_2$—C_6H_4—CH_3 —Oxidation→ $C_4H_9NHCONHSO_2$—C_6H_4—COOH

Tolbutamide **Less toxic metabolite**

(ii) Demethylation is more likely to occur when the methyl group is attached to positively charged nitrogen and sulphur atoms, although it is possible for any methyl group attached to a nitrogen, oxygen or sulphur atom to act in this manner (see sections 9.5.9, 9.5.10, 9.5.14 and 9.5.19). A number of methyl transfers have been associated with carcinogenic action.

(iii) Methyl groups can reduce the rate of metabolism of a compound by masking a metabolically active group, thereby giving the analogue a slower rate of metabolism than the lead. For example, the action of the agricultural fungicide nabam is due to it being metabolised to the active diisothiocyanate. N-Methylation of nabam yields an analogue that is inactive because it cannot be metabolised to the active diisothiocyanate.

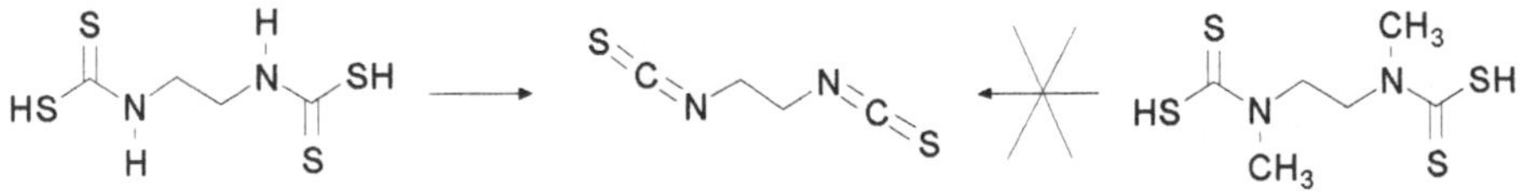

Nabam **1,2-Diisothiocyanoethane** **Dimethyl analogue of nabam**

Methylation can also reduce the unwanted side effects of a drug. For example, mono- and di-*ortho*-methylation with respect to the phenolic hydroxy group of paracetamol produce analogues with reduced hepatotoxicity. It is believed that this reduction is due to the methyl groups preventing metabolic hydroxylation of these *ortho* positions.

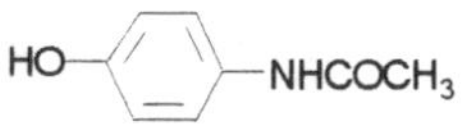

Paracetamol 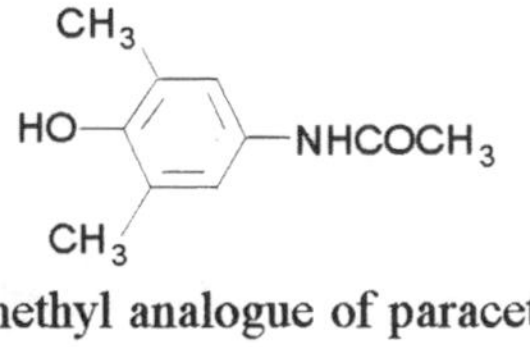 ***o,o'*-Dimethyl analogue of paracetamol**

Larger alkyl groups will have similar effects. However, as the size of the group increases, the lipophilicity will reach a point where it reduces the water solubility to an impractical level. Consequently, most substitutions are restricted to methyl and ethyl groups.

2.3.2.2 Halogen Groups

The incorporation of halogen atoms into a lead results in analogues that are more lipophilic and so less water soluble. Consequently, halogen atoms are used to improve the penetration of lipid membranes. However, there is an undesirable tendency for halogenated drugs to accumulate in lipid tissue.

The chemical reactivity of halogen atoms depends on both their point of attachment to the lead and the nature of the halogen. Aromatic halogen groups are far less reactive than aliphatic halogen groups, which can exhibit considerable chemical reactivity. For aliphatic carbon–halogen bonds the C—F bond is the strongest and usually less chemically reactive than aliphatic C—H bonds. The other aliphatic C–halogen bonds are weaker, their reactivity increasing down the periodic table. They are usually more chemically reactive than aliphatic C—H bonds. Consequently, the most popular halogen substitutions are the less reactive aromatic fluorine and chlorine groups. However, the presence of electron-withdrawing ring substituents may increase their reactivity to unacceptable levels. Trifluorocarbon groups ($—CF_3$) are sometimes used to replace chlorine because these groups are of a similar size. These substitutions avoid introducing a very reactive centre and hence a possible site for unwanted side reactions into the analogue. For example, the introduction of the more reactive bromo group can cause the drug to act as an alkylating agent.

The changes in potency caused by the introduction of a halogen or halogen-containing group will, as with substitution by other substituents, depend on the position of the substitution. For example, the antihypertensive clonidine with its *o,o′*-chloro substitution is more potent than its *p,m*-dichloro analogue. It is believed that the bulky *o*-chlorine groups impose a conformational restriction on the structure of clonidine, which probably accounts for its increased activity.

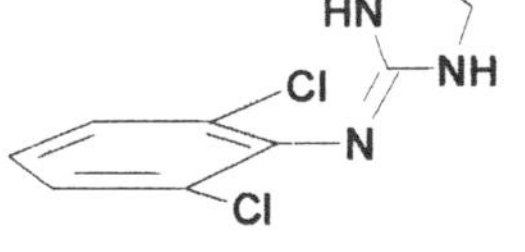

Clonidine ED_{20} **0.01 mg kg^{-1}**

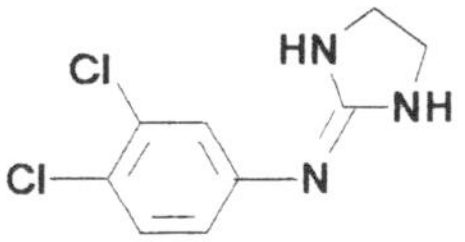

ED_{20} **3.00 mg kg^{-1}**

2.3.2.3 Hydroxy Groups

The introduction of hydroxy groups into the structure of a lead will normally produce analogues with an increased hydrophilic nature and a lower lipid solubility. It also provides a new centre for hydrogen bonding, which could influence the binding of the analogue to its target site. For example, the *ortho*-hydroxylated minaprine analogue binds more effectively to M_1-muscarinic receptors than many of its non-hydroxylated analogues.

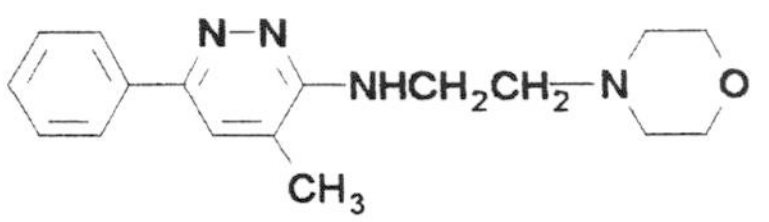

Minaprine

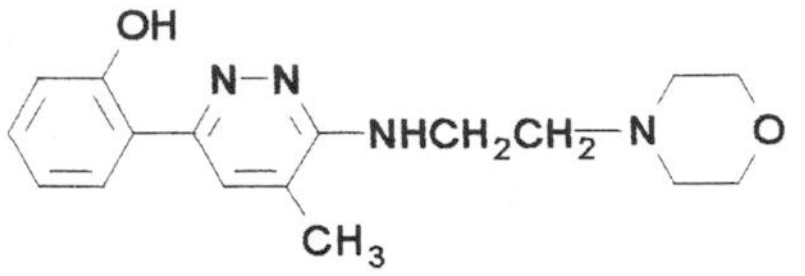

The *ortho*-hydroxylated analogue

The introduction of a hydroxy group also introduces a centre that, in the case of phenolic groups, could act as a bacterioside whereas alcohols have narcotic properties. However, the presence of hydroxy groups opens a new metabolic pathway (see sections 9.5.8, 9.6.2 and 9.6.4) that can either act as a detoxification route or prevent the drug from reaching its target.

2.3.2.4 Basic Groups

The basic groups usually found in drugs are amines, including some ring nitrogen atoms, amidines and guanidines. All these basic groups can form salts in biological media. Consequently, incorporation of these basic groups into the structure of a lead will produce analogues that have a lower lipophilicity but an increased water solubility (see section 3.6). This means that the more basic an analogue, the more likely it will form salts and the less likely it will be transported through a lipid membrane.

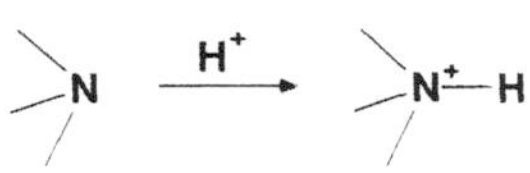

All types of amine 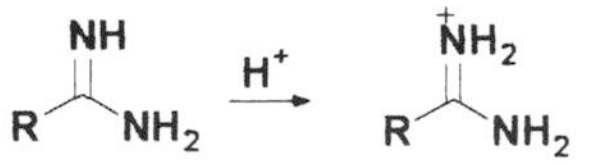 Amidines 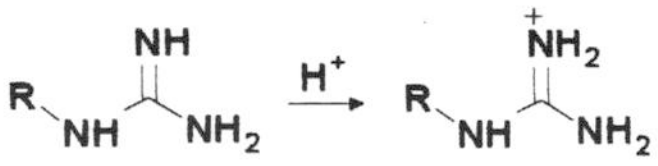 Guanidines

The introduction of basic groups may increase the binding of an analogue to its target by hydrogen bonding between that target and the basic group (Figure 2.8a). However, a number of drugs with basic groups owe their activity to salt formation and the enhanced binding that occurs due to the ionic bonding between the drug and the target (Figure 2.8b). For example, it is believed that many local anaesthetics are transported to their site of action in the form of their free bases but are converted to their salts which bind to the appropriate receptor sites.

The incorporation of aromatic amines into the structure of a lead is usually avoided because aromatic amines are often very toxic and are often carcinogenic.

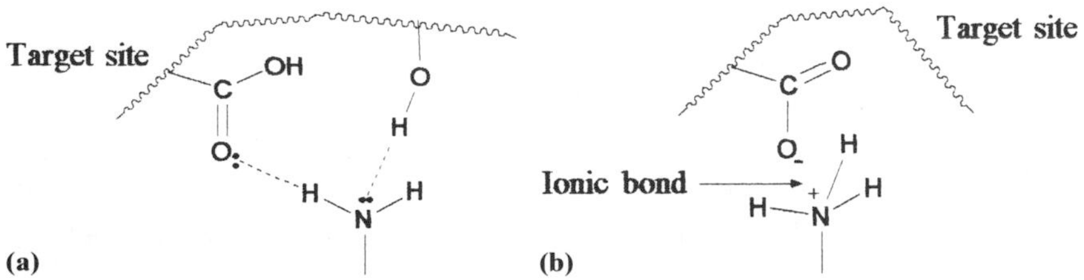

Target site (a) Target site Ionic bond (b)

Figure 2.8. Possible sites of (a) hydrogen bonding between a target site and amino groups and (b) ionic bonding between amine salts and a target site.

2.3.2.5 Carboxylic and Sulphonic Acid Groups

The introduction of acid groups into the structure of a lead usually results in analogues with an increased water but reduced lipid solubility. This increase in water solubility may be enhanced subsequently by *in vivo* salt formation. In general the introduction of carboxylic and sulphonic acid groups into a lead produces analogues that can be eliminated more readily (see section 9.6.3).

The introduction of carboxylic acid groups into small lead molecules may produce analogues that have a very different type of activity or are inactive. For example, the introduction of a carboxylic acid group into phenol results in the activity of the compound changing from being a toxic antiseptic to the less toxic anti-inflammatory salicylic acid. Similarly, the incorporation of a carboxylic acid group into the sympathomimetic phenylethylamine gives phenylalanine, which has no sympathomimetic activity. However, the introduction of carboxylic acid groups appears to have less effect on the activity of large molecules.

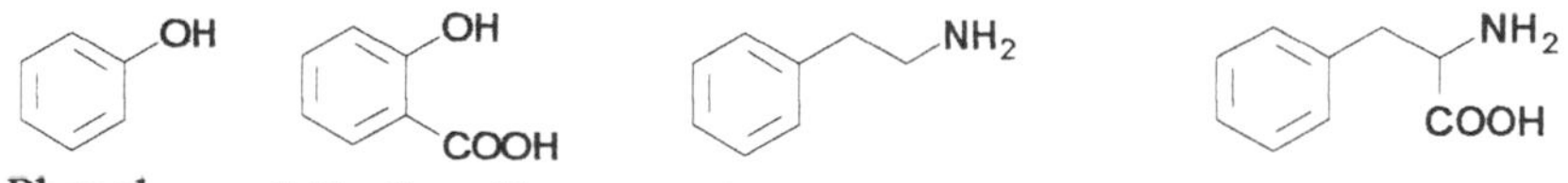

OH — Phenol; OH, COOH — Salicylic acid; NH_2 — Phenylethylamine; NH_2, COOH — Phenylalanine

Sulphonic acid groups do not usually have any effect on the biological activity but will increase the rate of elimination of an analogue.

2.3.2.6 Thiols, Sulphides and Other Sulphur Groups

Thiol and sulphide groups are not usually introduced into leads in SAR studies because they are readily metabolised by oxidation (see sections 9.5.20 and 9.5.19). However, thiols are sometimes introduced into a lead structure when improved metal chelation is the objective of the SAR study. For example, the antihypertensive captopril was developed from the weakly active carboxyacylprolines by replacement of their terminal carboxylic acid group (which is only a weak ligand for forming complexes with metals) by a thiol group (see section 6.12.2).

The introduction of thiourea and thioamide groups is usually avoided because these groups may produce goitre, which is a swelling on the neck due to enlargement of the thyroid gland.

2.3.3 Changing the Existing Substituents of a Lead

Analogues can also be formed by replacing an existing substituent in the structure of a lead by a new substituent group. The choice of group will depend on the objectives of the design team. It is often made using the concept of **isosteres**. Isosteres are groups that exhibit some similarities in their chemical and/or physical properties. As a result, they can exhibit similar

Table 2.3. Examples of bioisosteres. Each horizontal row represents a group of structures that are isosteric.

Classical isosteres	Bioisosteres
$—CH_3$, $—NH_2$, —OH, —F, —Cl. —Cl, —SH —PH_2	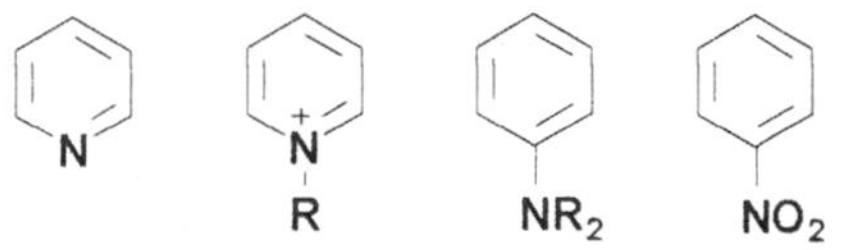
—Br, isopropyl $—CH(CH_3)_2$	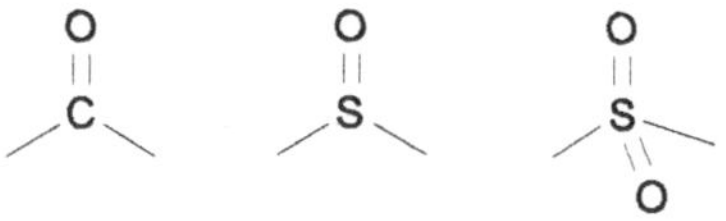
$—CH_2—$, —NH—, —O—, —S—	
$—COCH_2R$, —CONHR, —COOR, —COSR	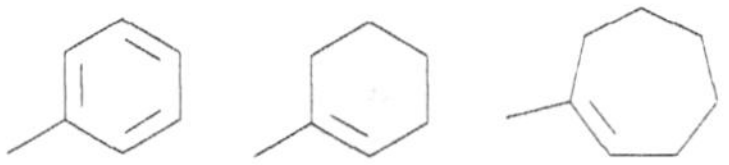
—HC=, —N= *In rings:* —CH=CH—, —S—	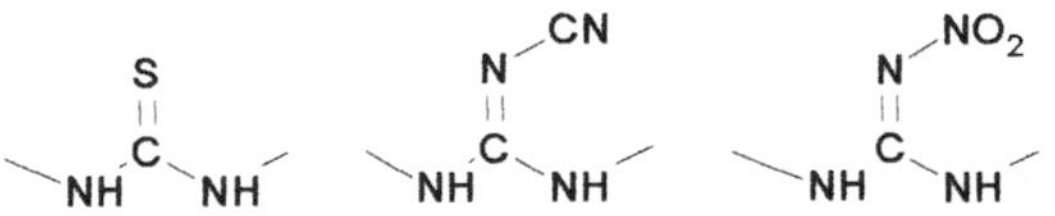
—O—, —S—, $—CH_2—$, —NH— —CH=, —N—	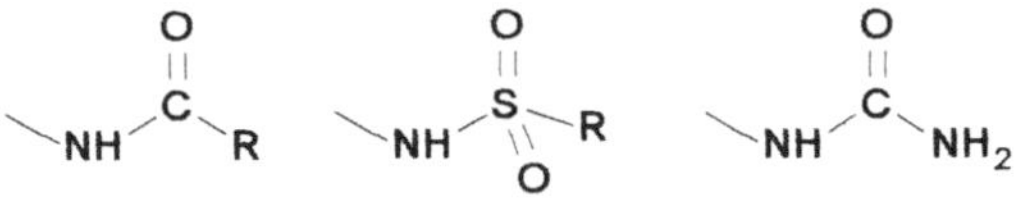

pharmacokinetic and pharmacodynamic properties. In other words, the replacement of a substituent by its isostere is more likely to result in the formation of an analogue with the same type of activity as the lead than the totally random selection of an alternative substituent. However, luck still plays a part and an isosteric analogue may have a totally different type of activity from its lead.

Classical isosteres were originally defined by Erlenmeyer as being atoms, ions and molecules that had identical outer shells of electrons. This definition has now been broadened to include groups that produce compounds that can sometimes have similar biological activities (Table 2.3). These groups are frequently referred to as **bioisosteres** in order to distinguish them from classical isosteres.

A large number of drugs have been discovered by isosteric and bioisosteric interchanges. For example, the replacement of the 6-hydroxy group of hypoxanthine by a thiol group gave the antitumour drug 6-mercaptopurine whereas the replacement of hydrogen in the 5-position of uracil by fluorine resulted in fluorouracil, which is also an antitumour agent. However, not all isosteric changes yield compounds with the same type of activity: the replacement of the —S— of the neuroleptic phenothiazine drugs by either —CH=CH— or $—CH_2CH_2—$ produces the dibenzazepines, which exhibit antidepressant activity (Figure 2.9).

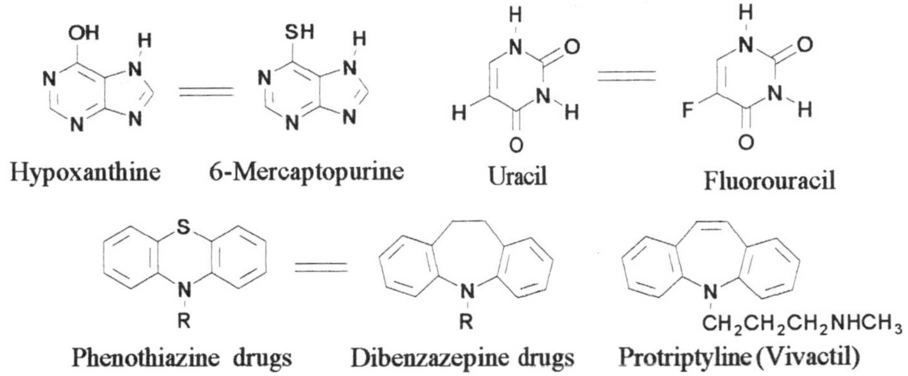

Hypoxanthine = 6-Mercaptopurine   Uracil = Fluorouracil

Phenothiazine drugs = Dibenzazepine drugs   Protriptyline (Vivactil) ($CH_2CH_2CH_2NHCH_3$)

Figure 2.9. Examples of drugs discovered by isosteric replacement.

2.4 Quantitative Structure–Activity Relationships (QSAR)

The success of the SAR approach to drug design depends not only on the knowledge and experience of the design team but also on a great deal of luck. QSAR is an attempt to remove the element of luck from drug design by establishing a mathematical relationship in the form of an equation between biological activity and measurable physicochemical parameters that **represent** properties such as lipophilicity, shape and electron distribution, which have a major influence on the drug's activity. These parameters are what are currently thought to be a satisfactory measure of the effect of these properties on the activity of a compound. They are normally defined so that they are in the form of numbers that are derived from practical data believed to be related to the property that the parameter represents. Consequently, it is possible to either measure or calculate these parameters for a group of compounds and relate their values to the biological activity of these compounds by means of mathematical equations using statistical methods such as regression analysis (see Appendix 1).

Large numbers of these parameters have now been measured and recorded in the literature. Consequently, it is often possible to calculate the theoretical value of a specific parameter for an, as yet, unsynthesised compound and, using the experimentally established equation relating activity to that parameter, predict the possible activity of this unknown compound. Alternatively the medicinal chemist could determine from the equation the value of the parameter, and hence the structures, that would give optimum activity. Predictions of these types allow the medicinal chemist to make a more informed choice as to what analogues to prepare. This could considerably cut down the cost of drug development.

The properties that have been found to influence a drug's activity are quite diverse, the major ones being lipophilicity, electronic effects and shape. The parameters commonly used to represent these properties are partition coefficients for lipohilicity (see section 2.4.2.1), Hammett σ constants for electronic effects (see section 2.4.2.2) and Taft M_s steric constants for steric effects (see section 2.4.3). Consequently, this text will be largely restricted to a discussion of the use of these constants. However, the other parameters mentioned in this and other texts are normally used in a similar fashion.

QSAR investigations are normally carried out on groups of related compounds that vary only in the nature of their substituents; very few have been carried out using structurally diverse compounds. In QSAR relationships the activity is normally expressed as 1/(a concentration term), usually C the minimum concentration required to cause a defined biological response. This means that an increase in activity will correspond to an increase in the value of 1/C. Furthermore, in most cases, to make the numbers easier to handle, activity is usually defined as log(1/C). As a result, QSAR relationships usually take the general form:

$$\text{Log}(1/C) = \text{Function}\{\text{parameter(s)}\} \tag{2.1}$$

2.4.1 The Partition Parameters

Two parameters are commonly used to relate distribution with biological activity, namely, the partition coefficient (P) and the lipophilicity substituent constant (π). The former parameter refers to the whole molecule whereas the latter is related to substituent groups.

2.4.1.1 Partition Coefficients (P)

A drug has to pass through a number of biological membranes in order to reach its site of action. Consequently, partition coefficients (see section 3.10) were the obvious parameter to use as a measure of the movement of the drug through these membranes. The nature of the relationship obtained depends on the range of P values for the compounds used. If this range is small the results may, by the use of regression analysis (see Appendix 1), be expressed as a straight line equation having the general form:

$$\log(1/C) = k_1 \log P + k_2 \tag{2.2}$$

where k_1 and k_2 are constants. This equation indicates a linear relationship between the activity of the drug and its partition coefficient. A number of examples of this type of correlation are known (Table 2.4).

Over larger ranges of P values the graph of log(1/C) against log P often has a parabolic form (Figure 2.10) with a maximum value ($\log P^o$). The existence of this maximum value implies that there is an optimum balance between aqueous and lipid solubility for maximum biological

Table 2.4. Examples of linear relationships between log(1/C) against log P. Equations (ii) and (iii) are adapted from C. Hansch, DRUG DESIGN I, ed. E. J. Ariens, copyright ©1971 by Academic Press, reproduced by permission of Academic Press Inc. and C. Hansch. r = the regression constant (see Appendix 1), n = the number of compounds tested and s = the standard deviation.

(i) Toxicity of alcohols to red spiders:
$\log(1/C) = 0.69 \log P + 0.16 \quad r = 0.979, n = 14, s = 0.087$

(ii) The binding of miscellaneous neutral molecules to bovine serum:
$\log(1/C) = 0.75 \log P + 2.30 \quad r = 0.96, n = 42, s = 0.159$

(iii) The binding of miscellaneous molecules to haemoglobin:
$\log(1/C) = 0.71 \log P + 1.51 \quad r = 0.95, n = 17, s = 0.16$

(iv) Inhibition of phenols on the conversion of P450 to P420 cytochromes:
$\log(1/C) = 0.57 \log P + 0.36 \quad r = 0.979, n = 13, s = 0.132$

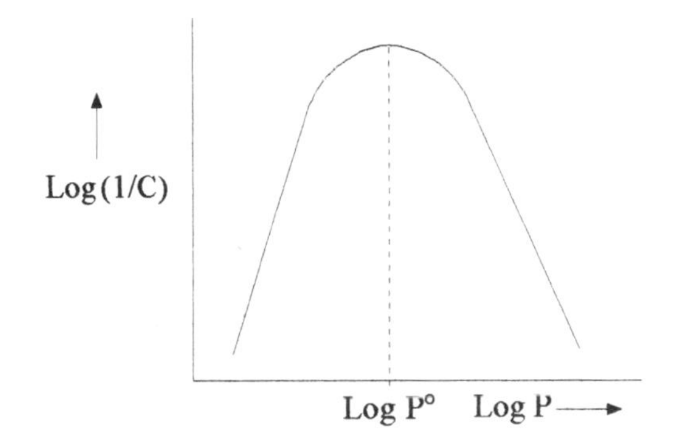

Figure 2.10. A parabolic plot for log(1/C) against log P.

activity. Below P° the drug will be reluctant to enter the membrane, whereas above P° the drug will be reluctant to leave the membrane. Log P° represents the optimum partition coefficient for biological activity. This means that analogues with partition coefficients near this optimum value are likely to be the most active and worth further investigation. Hansch and collegues showed that many of these parabolic relationships could be represented reasonably accurately by equations of the form:

$$\log(1/C) = -k_1(\log P)^2 + k_2 \log P + k_3 \tag{2.3}$$

The values of the constants in Equation (2.3) are normally determined by either regression analysis (see Appendix 1) or other statistical methods. For example, a study of the inducement of hypnosis in mice by a series of barbiturates showed that the correlation could be expressed by the equation:

$$\log(1/C) = -0.44(\log P)^2 + 1.58 \log P + 1.93 \quad (r = 0.969) \tag{2.4}$$

This equation has a maximum log P° at about 2.0. Hansch and collegues showed that a range of non-specific hypnotic drugs with widely different types of structure were found to have log

P values around 2. This implies that it is the solubility of these different drugs in the membrane rather than their structures that is the major factor in controlling their activity. On the basis of these and other partition studies, Hansch suggested in the mid-1960s that any organic compound with a P value of approximately 2 would, provided that it was not rapidly metabolised or eliminated, have some hypnotic properties and be rapidly transported into the CNS. Subsequent practical evidence gives some support to this assertion. The fact that the thiobarbiturates have log P values of about 3.1 suggests that these drugs probably have a different site of action from those of the barbiturates. The larger value also suggests that a more lipophilic receptor is involved.

The accuracy of the correlation of drug activity with partition coefficient will also depend on the solvent system used as a model to measure the partition coefficient values. The *n*-octanol/water system is frequently chosen because it has the most extensive database. However, more accurate results may be obtained if the organic phase is matched to the area of biological activity being studied. For example, *n*-octanol usually gives the most consistent results for drugs absorbed in the gastrointestinal (GI) tract, whereas less polar solvents such as olive oil frequently give more consistent correlations for drugs crossing the blood–brain barrier. More polar solvents such as chloroform give more consistent values for buccal absorption (soft tissues in the mouth). However, when correlating P values with potency of activity it should be borne in mind that the partition coefficient of a drug is usually only one of a number of parameters influencing its activity. Consequently, in cases where there is a poor correlation between the partition coefficient and the drug's activity, other parameters must be playing a more important part in the action of the drug.

2.4.1.2 Lipophilic Substituent Constants (π)

Lipophilic substituent constants are also known as hydrophobic substituent constants. They represent the contribution that a group makes to the partition coefficient and were defined by Hansch and co-workers by the equation:

$$\pi = \log P_X - \log P_H \tag{2.5}$$

where P_H and P_X are the partition coefficients of the standard compound and its monosubstituted derivative, respectively. For example, the value of π for the chloro group of chlorobenzene could be calculated from the partition coefficients for benzene and chlorobenzene:

$$\pi_{Cl} = \log P_{(C_6H_5Cl)} - \log P_{(C_6H_6)} \tag{2.6}$$

substituting the appropriate P values:

$$\pi_{Cl} = 2.84 - 2.13 = 0.71$$

The values of π will vary depending on the solvent system used to determine the partition coefficients. However, most π values are determined using the *n*-octanol/water system. A pos-

Table 2.5. Examples of the variations of π values with chemical structure.

Substituent X	Aliphatic systems R-X	C_6H_5-X	O_2N-C_6H_4-X	HO-C_6H_4-X
—H	0.00	0.00	0.00	0.00
—CH_3	0.50	0.56	0.52	0.49
—F	−0.17	0.14		0.31
—Cl	0.39	0.71	0.54	0.93
—OH	−1.16	−0.67	0.11	−0.87
—NH_2		−1.23	−0.46	−1.63
—NO_2		−0.28	−0.39	0.50
—OCH_3	0.47	−0.02	0.18	−0.12

itive π value indicates that a substituent has a higher lipophilicity than hydrogen and so will probably increase the concentration of the compound in the *n*-octanol layer and by inference its concentration in the lipid material of biological systems. Conversely, a negative π value shows that the substituent has a lower lipophilicity than hydrogen and so probably increases the concentration of the compound in the aqueous media of biological systems.

The π value for a specific substituent will vary with its structural environment (Table 2.5). Consequently, average values or the values relevant to the type of structures being investigated may be used in determining activity relationships. Where several substituents are present the value of π for the compound is the sum of the π values of each of the separate substituents. For example, the π value for the two methyl groups in 1,3-dimethylbenzene is:

$$\pi = 2 \times 0.56 = 1.12$$

Lipophilic substituent constants can be used as an alternative to the partition coefficient when dealing with a series of analogues in which only the substituents are different. This usage is based on the assumption that the lipophilic effect of the unchanged part of the structure is similar for each of the analogues. Consequently, the π values of substituents indicate the significance of the contribution of that substituent to the lipophilicity of the molecule. Furthermore, biological activity – π relationships that have high regression constants and low standard deviations demonstrate that the substituents are important in determining the lipophilic character of the drug.

Lipophilic substituent constants can also be used to calculate theoretical partition coefficients for whole molecules. These calculated values are often in good agreement with the experimentally determined values provided that the substituents are not sterically crowded. For example, the calculated value of the *n*-octanol/water partition coefficient for 1,3-dimethylbenzene is 3.25, which is in good agreement with the experimentally determined value of 3.20. However, poorer agreements are usually found when substituents are located close to each other in the molecule. In addition, strong electron interactions between substituent groups can result in inaccurate theoretical P values.

2.4.2 Electronic Parameters

The distribution of the electrons in a drug molecule will have a considerable influence on the distribution and activity of a drug. In order to reach its target a drug normally has to pass through a number of biological membranes (see section 4.3). As a general rule, non-polar and polar drugs in their unionised form are usually more readily transported through membranes than polar drugs and drugs in their ionised forms (see section 4.3.3). Furthermore, once the drug reaches its target site the distribution of electrons in its structure will control the type of bonds it forms with that target (see section 8.2), which in turn affects its biological activity.

The first attempt to quantify the electronic effects of groups on the physicochemical properties of compounds was made by Hammett in 1940. He studied the effect of substituents on the position of equilibria of benzoic acid derivatives. His proposals are now used extensively as a measure of the electronic effects of the structure of a drug on its activity.

2.4.2.1 The Hammett Constant (σ)

The distribution of electrons within a molecule depends on the nature of the electron-withdrawing and election-donating groups found in that structure. For example, benzoic acid is weakly ionised in water:

$$C_6H_5COOH \rightleftharpoons C_6H_5COO^- + H^+ \qquad K = \frac{[PhCOO^-][H^+]}{[PhCOOH]}$$

Substitution of a ring hydrogen by an electron-withdrawing substituent (X), such as a nitro group, will weaken the O—H bond of the carboxyl group and stabilise the carboxylate anion (Figure 2.11). This will move the equilibrium to the right, which means that the substituted compound is stronger than benzoic acid ($K_X > K$). Conversely, the introduction of an electron donor substituent (X), such as a methyl group, into the ring strengthens the acidic O—H group and reduces the stability of the carboxylate anion. This moves the equilibrium to the left which means that the compound is a weaker acid than benzoic acid ($K > K_X$).

Hammett used these concepts to calculate **substituent constants** (σ_X) for a variety of ring substituents (X) of benzoic acid using this acid as the reference standard (Table 2.6). He used these constants, which are now known as Hammett substituent constants or simply Hammett constants, to describe the relationship between linear free energy and structure.

Hammett constants were defined as:

$$\sigma_X = \log \frac{K_X}{K} \qquad (2.7)$$

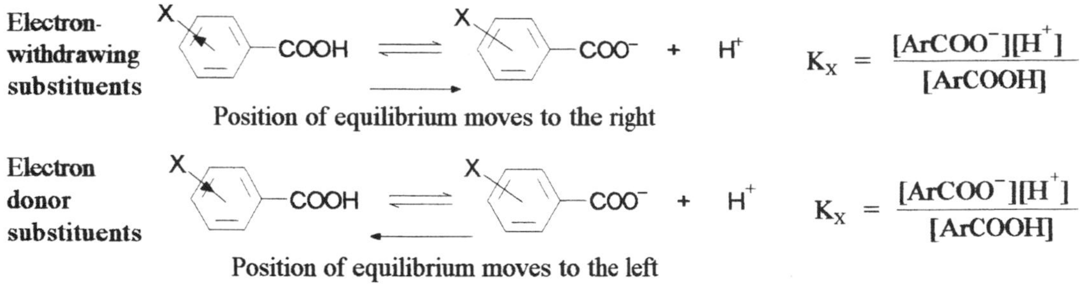

Electron-withdrawing substituents

$X\text{-}C_6H_4\text{-}COOH \rightleftharpoons X\text{-}C_6H_4\text{-}COO^- + H^+ \qquad K_X = \frac{[ArCOO^-][H^+]}{[ArCOOH]}$

Position of equilibrium moves to the right

Electron donor substituents

$X\text{-}C_6H_4\text{-}COOH \rightleftharpoons X\text{-}C_6H_4\text{-}COO^- + H^+ \qquad K_X = \frac{[ArCOO^-][H^+]}{[ArCOOH]}$

Position of equilibrium moves to the left

Figure 2.11. The effect of electron-withdrawing and donor groups on the position of equilibrium of substituted benzoic acids.

Table 2.6. Examples of the different electronic substitution constants used in QSAR studies. Inductive substituent constants (σ_I) are the contribution that the inductive effect makes to Hammett constants and can be used for aliphatic compounds. Taft substitution constants (σ^*) refer to aliphatic substituents. The Swain–Lupton constants represent the contributions due to the inductive (F) and mesomeric or resonance (R) components of Hammett constants. Adapted from *An Introduction to the Principles of Drug Design and Action by Smith and Williams*, 3rd Edn, ed. H. J. Smith, 1998. Reproduced by permission of Harwood Academic Publishers.

Substituent	Hammett constants		Inductive constants σ_I	Taft constants σ^*	Swain–Lupton constants	
	σ_m	σ_p			F	R
—H	0.00	0.00	0.00	0.49	0.00	0.00
$—CH_3$	−0.07	−0.17	−0.05	0.00	−0.04	−0.13
$—C_2H_5$	−0.07	−0.15	−0.05	−0.10	−0.05	−0.10
—Ph	0.06	−0.01	0.10	0.60	0.08	−0.08
—OH	0.12	−0.37	0.25	–	0.29	−0.64
—Cl	0.37	0.23	0.47	–	0.41	−0.15
$—NO_2$	0.71	0.78	–	–	0.67	0.16

that is:

$$\sigma_X = \log K_X - \log K \tag{2.8}$$

and so, as $pK_a = -\log K_a$:

$$\sigma_X = pK - pK_X \tag{2.9}$$

A negative value for σ_X indicates that the substituent is acting as an electron donor group because $K_X > K$. Conversely, a positive value for σ_X shows that the substituent is acting as an electron-withdrawing group because $K > K_X$. Its value varies with the position of the substituent in the molecule. Consequently, this position is usually indicated by the use of the subscripts o, m and p. Where a substituent has opposite signs depending on its position on the ring it means that in one case it is acting as an electron donor and in the other as an electron-withdrawing group. This is possible because the Hammett constant includes both the inductive and mesomeric (resonance) contributions to the electron distribution. For example, the σ_m Hammett constant for the methoxy group of *m*-methoxybenzoic acid is 0.12, whereas for *p*-methoxybenzoic acid it is –0.27. In the former case the electronic distribution is dominated by the inductive (I or F) contribution whereas in the latter case it is controlled by the mesomeric (M) or resonance (R) effect.

m-Methoxybenzoic acid I > M

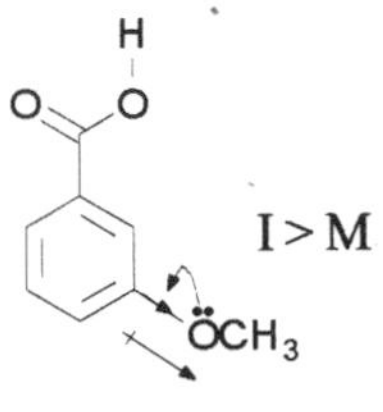

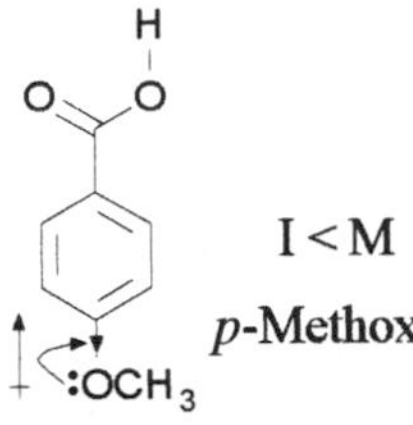

p-Methoxybenzoic acid I < M

Hammett postulated that the σ values calculated for the ring substituents of a series of benzoic acids could also be valid for those ring substituents in a different series of similar aromatic compounds. This relationship has been found to be in good agreement for the *meta* and *para* substituents of a wide variety of aromatic compounds but not for their *ortho* substituents. The latter is believed to be due to steric hindrance and other effects, such as intramolecular hydrogen bonding, playing a significant part in the ionisations of compounds with *ortho* substituents. Hammett substitution constants also suffer from the disadvantage that they only apply to substituents directly attached to a benzene ring. Consequently, a number of other electronic constants (Table 2.6) have been introduced and used in QSAR studies in a similar manner to the Hammett constants. However, Hammett substitution constants are probably still the most widely used electronic constants for QSAR studies.

Attempts to relate biological activity to the values of Hammett substitution and similar constants have been largely unsuccessful because electron distribution is not the only factor involved (see section 2.3.3). One of the successful attempts to relate biological activity to structure using Hammett constants was the investigation by Fukata and Metcalf into the effectiveness of diethyl aryl phosphates for killing fruit flies. This investigation showed that the activity of these compounds is dependent only on electron distribution factors. Their results may be expressed by the relationship:

$$\log(1/C) = 2.282\sigma - 0.348 \qquad (2.10)$$

Diethyl aryl phosphate insecticides

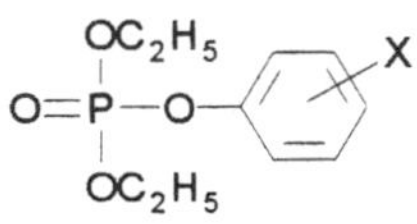

This equation shows that the greater the positive value for σ, the greater the biological activity of the analogue. This type of knowledge enables one to predict the activities of analogues and synthesise the most promising rather than spend a considerable amount of time synthesising and testing all the possible analogues.

2.4.3 Steric Parameters

In order for a drug to bind effectively to its target site the dimensions of the pharmacophore of the drug must be complementary to those of the target site (see sections 6.3 and 8.2). The Taft steric parameter (E_s) was the first attempt to show the relationship between a measurable parameter related to the shape and size (bulk) of a drug and the dimensions of the target site and a drug's activity. This has been followed by Charton's steric parameter (ν), Verloop's steric parameters and the molar refractivity (MR) amongst others. The most used of these additional parameters is probably the molar refractivity. However, in all cases the required parameter is calculated for a set of related analogues and correlated with their activity using a suitable statistical method such as regression analysis (see Appendix 1). The results of individual investigations have shown varying degrees of success in relating the biological activity to the parameter. This is probably because little is known about the finer details of the three-dimensional structures of the target sites.

2.4.3.1 The Taft Steric Parameter (E_s)

Taft in 1956 used the relative rate constants of the acid-catalysed hydrolysis of α-substituted methyl ethanoates to define his steric parameter because it had been shown that the rates of these hydrolyses were almost entirely dependent on steric factors. He used methyl ethanoate as his standard and defined E_s as:

$$E_s = \frac{\log k_{(XCH_2COOR)}}{\log k_{(CH_3COOR)}} = \log k_{(XCH_2COOR)} - \log k_{(CH_3COOR)} \tag{2.11}$$

where k is the rate constant of the appropriate hydrolysis and $E_s = 0$ when X = H. It is assumed that the values for E_s (Table 2.7) obtained for a group using the hydrolysis data are applicable to other structures containing that group. The methyl-based E_s values can be converted to H-based values by adding −1.24 to the corresponding methyl-based values.

Taft steric parameters have been found to be useful in a number of investigations. For example, regression analysis Appendix 1 has shown that the antihistamine effects of a number of related analogues of diphenhydramine (Figure 2.12) are related to their biological response (BR) by

Table 2.7. Examples of the Taft steric parameter E_s.

Group	E_s	Group	E_s	Group	E_s
H—	1.24	F—	0.78	CH_3O—	0.69
CH_3—	0.00	Cl—	0.27	CH_3S—	0.19
C_2H_5—	−0.07	F_3C—	−1.16	$PhCH_2$—	−0.38
$(CH_3)_2CH$—	−0.47	Cl_3C—	−2.06	PhOCH—	−0.33

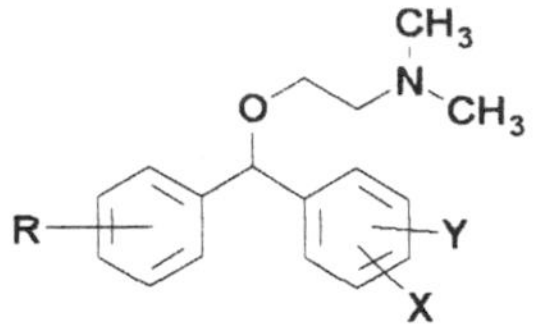

CH3 N CH3 O R Y X

Figure 2.12. The general formula of diphenhydramine analogues.

Equation (2.12), where E_s is the sum of the *ortho* and *meta* E_s values in the most highly substituted ring.

$$\log BR = 0.440E_s - 2.204 \quad (n = 30, s = 0.307, r = 0.886) \tag{2.12}$$

Regression analysis also showed that the biological response was related to the Hammett constant by the relationship:

$$\log BR = 2.814\sigma - 0.223 \quad (n = 30, s = 0.519, r = 0.629) \tag{2.13}$$

A comparison of the standard deviations (s) for Equations (2.12) and (2.13) shows that the calculated values for the Hammett constants σ for each of the analogues are more scattered than the calculated values for the corresponding Taft E_s values. Furthermore, although both the r and s values for Equation (2.12) are reasonable, those for Equation (2.13) are unacceptable. This indicates that the antihistamine activity of these analogues appears to depend more on steric than electronic effects. This deduction is supported by the fact that using regression analysis to obtain a relationship involving both the Hammett and Taft constants does not lead to a significant increase in the r and s values (Equation 2.14).

$$\log BR = 0.492E_s - 0.585\sigma - 2.445 \quad (n = 30, s = 0.301, r = 0.889) \tag{2.14}$$

Taft constants suffer from the disadvantage that they are determined by experiment. Consequently, the difficulties in obtaining the necessary experimental data have significantly limited the number of values recorded in the literature.

2.4.3.2 Molar Refractivity (MR)

The molar refractivity is a measure of both the volume of a compound and how easily it is polarised. It is defined as:

$$MR = \frac{(n^2 - 1)M}{(n^2 + 2)\rho} \tag{2.15}$$

where n is the refractive index, M is the relative molecular mass and ρ is the density of the compound. The M/ρ term is a measure of the molar volume whereas the refractive index term

Table 2.8. Examples of calculated MR values. Reproduced by permission of John Wiley and Sons Ltd. from C. Hansch and A.J. Leo, *Substituent Constants for Correlation Analysis in Chemistry and Biology*, 1979.

Group	MR	Group	MR	Group	MR
H—	1.03	F—	0.92	CH_3O—	7.87
CH_3—	5.65	Cl—	6.03	HO—	2.85
C_2H_5—	10.30	F_3C—	5.02	CH_3CONH—	14.93
$(CH_3)_2CH$—	14.96	O_2N—	7.63	CH_3CO—	11.18

is a measure of the polarisability of the compound. Although MR is calculated for the whole molecule, it is an additive parameter and so the MR values for a molecule can be calculated by adding together the MR values for its component parts (Table 2.8).

2.4.3.3 The Other Parameters

These can be broadly divided into those that apply to sections of the molecule and those that involve the whole molecule. The former include parameters such as van der Waals' radii, Charton's steric constants and the Verloop steric parameters. The latter range from relative molecular mass (RMM) and molar volumes to surface area. They have all been used to correlate biological activity to structure with varying degrees of success.

2.4.4 Hansch Analysis

Hansch analysis is based on the attempts by earlier workers, notably, Richardson (1867), Richet (1893), Meyer (1899), Overton (1901), Ferguson (1939) and Collander (1954), to relate drug activity to measurable chemical properties. Hansch and co-workers in the early 1960s proposed a multiparameter approach to the problem based on the lipophilicity of the drug and the electronic and steric influences of groups found in its structure. They realised that the biological activity of a compound is a function of its ability to reach and bind to its target site. Hansch proposed that drug action could be divided into two stages:

(i) the transport of the drug to its site of action;
(ii) the binding of the drug to the target site.

He stated that the transport of the drug is like a *random walk* from the point of administration to its site of action. During this *walk* the drug has to pass through numerous membranes and so the ability of the drug to reach its target is dependent on its lipophilicity. Consequently, this ability could be expressed mathematically as a function of either the drug's partition coefficient or the π value(s) of appropriate substituents (see sections 2.3.1 and 3.10). However, on reaching its target the binding of the drug to the target site depends on the shape, electron distribution and polarisability of the groups involved in the binding. A variety of parameters are now used to describe each of these aspects of drug activity, the most common ones being the Hammett electronic σ (see section 2.4.2.1) and Taft E_s constants (see section 2.4.3.1).

Hansch postulated that the biological activity of a drug could be related to all or some of these factors by simple mathematical relationships based on the general format:

$$\log\frac{1}{C} = k_1(\text{partition parameter}) + k_2(\text{electronic parameter}) + k_3(\text{steric parameter}) + k_4 \quad (2.16)$$

where C is the minimum concentration required to cause a specific biological response, and k_1, k_2, k_3 and k_4 are numerical constants obtained by feeding the data into a suitable computer statistical package. For example, the general equation relating activity to all these parameters often takes the general form:

$$\log\frac{1}{C} = k_1P - k_2P^2 + k_3\sigma + k_4E_S + k_5 \quad (2.17)$$

where other parameters could be substituted for P, σ and E_S. For example, π may be used instead of P and MR for E_S. However, it is emphasised that these equations, which are collectively known as Hansch equations, do not always contain the main three types of parameter (Table 2.9). The numerical values of the constants in these equations are obtained by feeding the values of the parameters into a suitable computer package. These values are obtained either from the literature (eg. π, σ, E_s) or determined by experiment (eg. C, P, etc.).

Table 2.9. Examples of simple Hansch equations.

Compound	Activity	Hansch equation
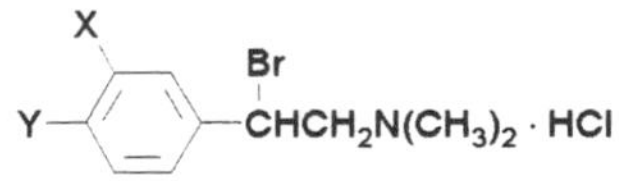 Y-, X-substituted phenyl–CH(Br)CH$_2$N(CH$_3$)$_2$ · HCl	Antiadrenergic	$\log 1/C = 1.22\pi - 1.59\sigma + 7.89$ ($n = 22; s = 0.238; r = 0.918$)
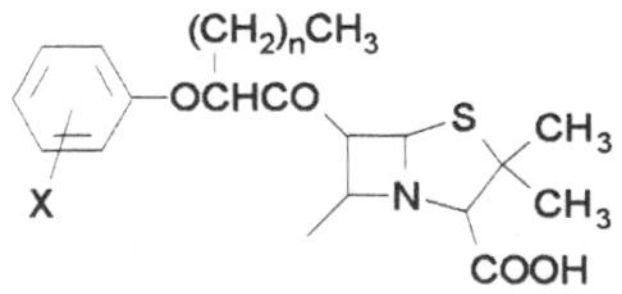 X-substituted phenyl–CH((CH$_2$)$_n$CH$_3$)OCO– penicillin nucleus (S, CH$_3$, CH$_3$, N, COOH)	Antibiotic (*in vivo*)	$\log 1/C = -0.445\pi + 5.673$ ($n = 20; r = 0.909$)
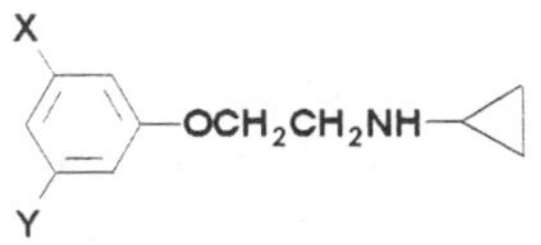 X-, Y-substituted phenyl–OCH$_2$CH$_2$NH–cyclopropyl	Monoamine oxidase inhibitor (humans)	$\log 1/C = 0.398\pi + 1.089\sigma + 1.03E_s + 4.541$ ($n = 9; r = 0.955$)
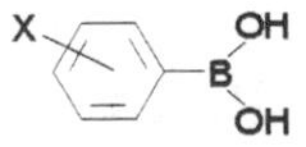 X-substituted phenyl–B(OH)$_2$	Concentration (C_b) in the brain after 15 min	$\log C_b = 0.765\pi - 0.540\pi^2 + 1.505$

Many QSAR investigations involve varying more than one ring substituent. In these cases the values of the same parameter for each substituent are expressed in the Hansch equation as either the sum of the individual parameters or as independent individual parameters. For example, in the hypothetical case of a benzene ring with two substituents X and Y the Hammett constants could be expressed in the Hansch equation as either $k_1\Sigma(\sigma_X + \sigma_Y)$ or $k_1\sigma_X + k_2\sigma_Y$. A comprehensive list of many of the parameters used in Hansch analysis may be found in a review by Tute (*Advances in Drug Research*, **6**, 1 (1971)).

Hansch equations can be used to give information about the nature of the mechanism by which drugs act, as well as predict the activity of as yet unsynthesised analogues. In the former case the value of the constant for a parameter gives an indication of the importance of the influence of that parameter in the mechanism by which the drug acts. Consider, for example, a series of analogues whose activity is related to the parameters π and σ by the hypothetical Hansch equation:

$$\log\frac{1}{C} = 1.78\pi - 0.12\sigma + 1.674 \qquad (2.18)$$

The small value of the coefficient for σ relative to that of π in Equation (2.18) shows that the electronic factor is not an important feature of the action of these drugs. Furthermore, the value of the regression coefficient (r) for the equation will indicate whether suitable parameters were used for its derivation. Values of r that are significantly lower than 0.9 indicate that either the parameter(s) used to derive the equation was unsuitable or that there is no relationship between the compounds used and their activity. This suggests that the mechanisms by which these compounds act may be very different and therefore unrelated.

Predictions of the activities of as yet unsynthesised analogues are useful in that they allow the medicinal chemist to make a more informed choice as to which analogues to synthesise. However, these predictions should only be made within the limits used to establish that relationship. For example, if a range of partition coefficients with values from 3 to 8 were used to obtain an activity relationship – partition coefficient equation then this equation should not be used to predict activities of compounds with partition coefficients of less than 3 and greater than 8.

Hansch equation activity predictions that are widely different from the observed values suggest that the activity of a compound is affected by factors that have not been included in the derivation of the Hansch equation. The discovery of this type of anomaly may give some insight into the mechanism by which the compound acts. For example, a study by Hansch on the activity of penicillins against a strain of *Staphylococcus aureus* in mice gave the *in vivo* relationship:

$$\log 1/C = -0.445\pi + 5.673 \quad (n = 20, s = 0.191, r = 0.909) \qquad (2.19)$$

This relationship (2.19) predicts that a penicillin with branched side chain $PhOCH(CH_3)CO—$ should be less active than a penicillin with the similar sized unbranched side chain

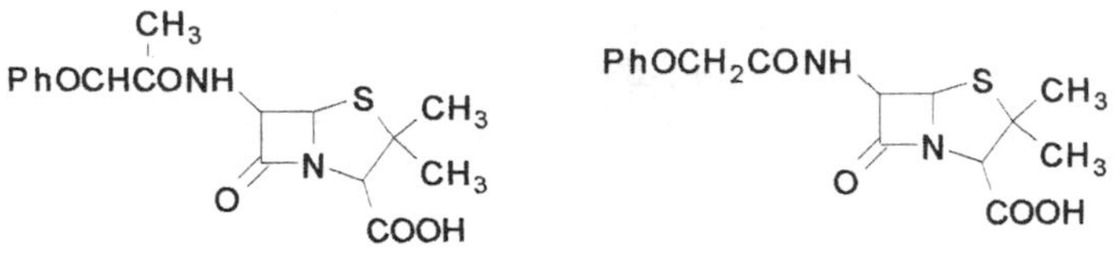

Figure 2.13. Penicillin analogues.

$PhOCH_2CO$— (Figure 2.13) because the branched side chain has a higher π value. However, Hansch found that the penicillin with the branched side chain was the more active. He suggested that this anomaly could be explained by the fact that branched side chains were more resistant to metabolism than straight side chains.

The accuracy of Hansch equations and hence the success of QSAR investigations depend on using enough analogues, the accuracy of the data and the choice of parameters:

(i) The greater the number of analogues used in a study, the greater the probability of deriving an accurate Hansch equation. A rough rule of thumb is that the minimum number of compounds used in the study should be not less than 5x, where x is the number of parameters used to obtain the relationship. Where substituents are being varied it is also necessary to use as wide a variety of substituents in as large a range of different positions as possible.

(ii) The accuracy of the biological data used to establish a Hansch equation will affect the accuracy of the derived relationship because its value depends to some extent on the subject used for the measurement. Consequently, it is necessary to take a statistically viable number of measurements of any biological data such as activity and use an average value in the derivation of the Hansch equation. Furthermore, extreme parameter values should not be used because they are likely to dominate the regression analysis and so give less accurate Hansch equations. Their presence suggests that the parameter is either not suitable or there is no possible correlation.

(iii) The choice of parameter is important because one parameter may give an equation with an acceptable regression constant whereas another would give an equation with an unacceptable regression constant. For example, it may be possible to correlate successfully the dipole moments of a series of analogues with biological activity but fail to obtain a correlation for the same series of analogues using Hammett constants as a parameter. Furthermore, some parameters like π and σ are interrelated and so their use in the same equation can lead to confusion in interpretating the equation.

2.4.4.1 Craig Plots

Craig plots are two-dimensional plots of one parameter against another (Figure 2.14). The plot is divided into four sections corresponding to the positive and negative values of the parameters. They are used, in conjunction with an already established Hansch equation for a series of related aromatic compounds, to select the best aromatic substituents that are likely to produce

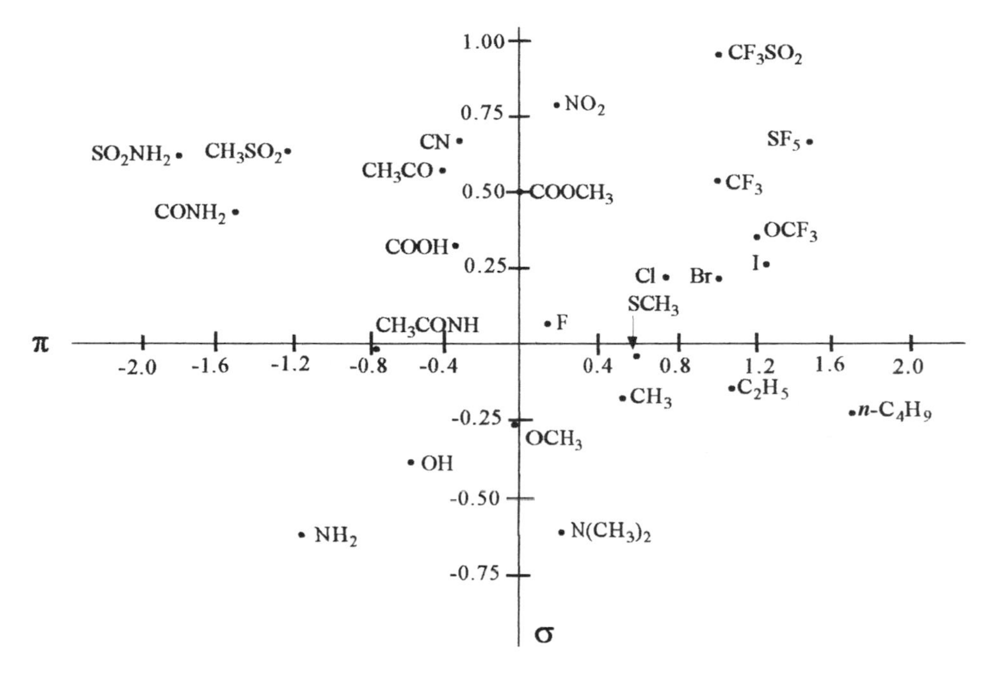

Figure 2.14. An example of a Craig plot of *para* Hammett constants against *para* π values. Reprinted by permission of John Wiley and Sons, Inc. from P. N. Craig, in *Burger's Medicinal Chemistry*, ed, M. E. Wolff, 4th Edn. Part 1, p. 343, 1980. Copyright © John Wiley & Sons Inc.

new highly active analogues. For example, suppose that a Hansch analysis carried out on a series of aromatic compounds yields the Hansch equation:

$$\log 1/C = 2.67\pi - 2.56\sigma + 3.92 \tag{2.20}$$

To obtain a high value for the activity (1/C) and, as a result, a low value for C, it is necessary to pick substituents with a positive π value and a negative σ value. In other words, if high-activity analogues are required, the substituents should be chosen from the lower right-hand quadrant of the plot. However, it is emphasised that the use of a Craig plot does not guarantee that the resultant analogues will be more active than the lead because the parameters used may not be relevant to the mechanism by which the analogue acts.

2.4.5 The Topliss Decision Tree

The Topliss decision tree is essentially a flow diagram that in a series of steps directs the medicinal chemist to produce a series of analogues, some of which should have a greater activity than the lead used to start the tree. It is emphasised that only some of the compounds will be more active than the lead compound. The method is most useful when it is not possible to make the large number of compounds necessary to produce an accurate Hansch equation. It effectively reduces the number of compounds that a medicinal chemist needs to synthesise in order to discover potent analogues.

Two decision trees are in use, one for substituents directly attached to an aromatic ring and the other for aliphatic side-chain substituents of an aromatic ring system (Figure 2.15). They

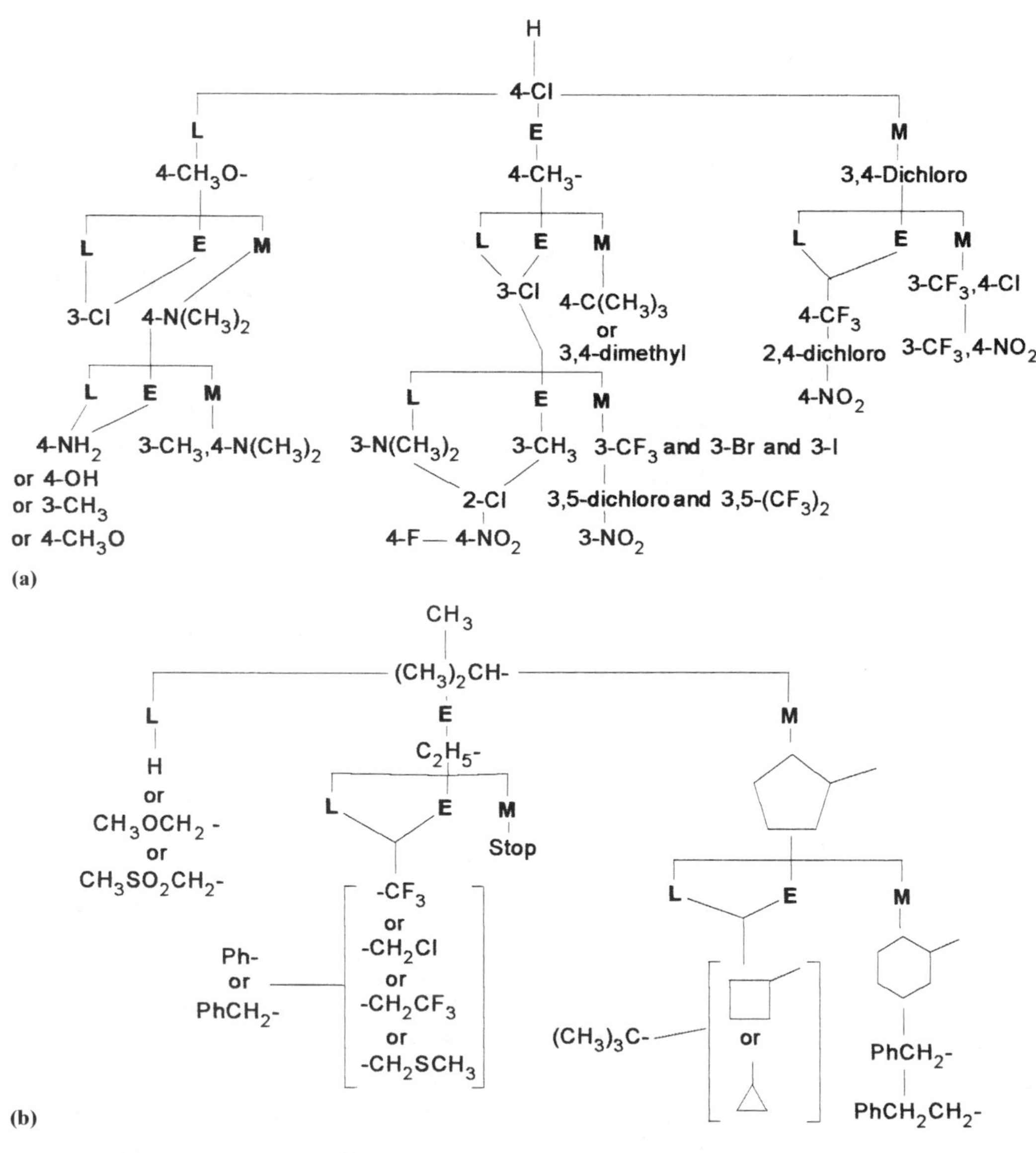

Figure 2.15. The Topliss decision trees for (a) an unfused aromatic ring and (b) an aliphatic side chain (L = significantly lower activity, E = about the same activity and M = significantly greater activity). Reprinted with permission from J. G. Topliss, Utilization of operational schemes for analog synthesis in drug design. *Journal of Medicinal Chemistry*, **15** (10), 1006 (1972). Copyright 1972 American Chemical Society.

were constructed using parameters similar to those described earlier in this chapter and are both used in a similar manner. However, the use of the aromatic substituent decision tree requires:

(i) the lead compound to have an unfused aromatic ring;
(ii) the biological activity of the lead compound and its analogues to be readily determinable.

To use the aromatic decision tree the 4-chloro analogue of the lead is synthesised and its activity compared with that of its lead. This activity may be significantly less (L), approximately the same (E) or significantly greater (M) than that of the original lead. If the activity is greater than that of the lead the next analogue to be prepared is the next one on the M route, namely the 3,4-dichloro analogue. Alternatively, if the activity of the analogue is less than that of the original lead the next step is to produce the 4-methoxy analogue as indicated by the L route on the tree. Similarly, if the activity is about the same as that of the original lead the E route is followed and the 4-methyl analogue synthesised. This procedure is repeated, the activity of each new analogue being compared with that of the preceding compound in order to determine which branch of the tree gives the next analogue. Suppose, for example, that a compound A (Figure 2.16) is active against *S. aureus* and activity of this compound and its analogues can be readily assessed by a biological method. The first step in the Topliss approach is to synthesise the 4-chloro derivative (B) of A. Suppose that the activity of B is greater than that of A, then following the M branch of the Topliss tree (Figure 2.16) indicates that the next analogue to produce is the 3,4-dichloro derivative (C) of A. Once again suppose that the biological assay of C was less than that of B. In this case, the Topliss tree shows that the next most promising analogue is the 4-trifluromethyl derivative of (D) of A. At this point one would also synthe-

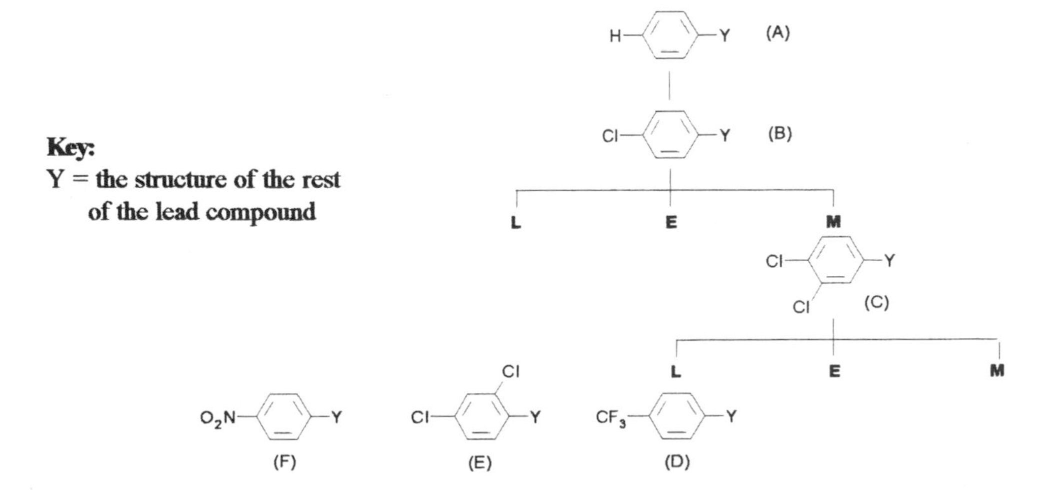

Figure 2.16. A hypothetical example of the use of the Topliss decision tree. The compounds are synthesised in the order A, B, C . . . etc. It should be realised that only some of the compounds synthesised will be more potent than the original lead A.

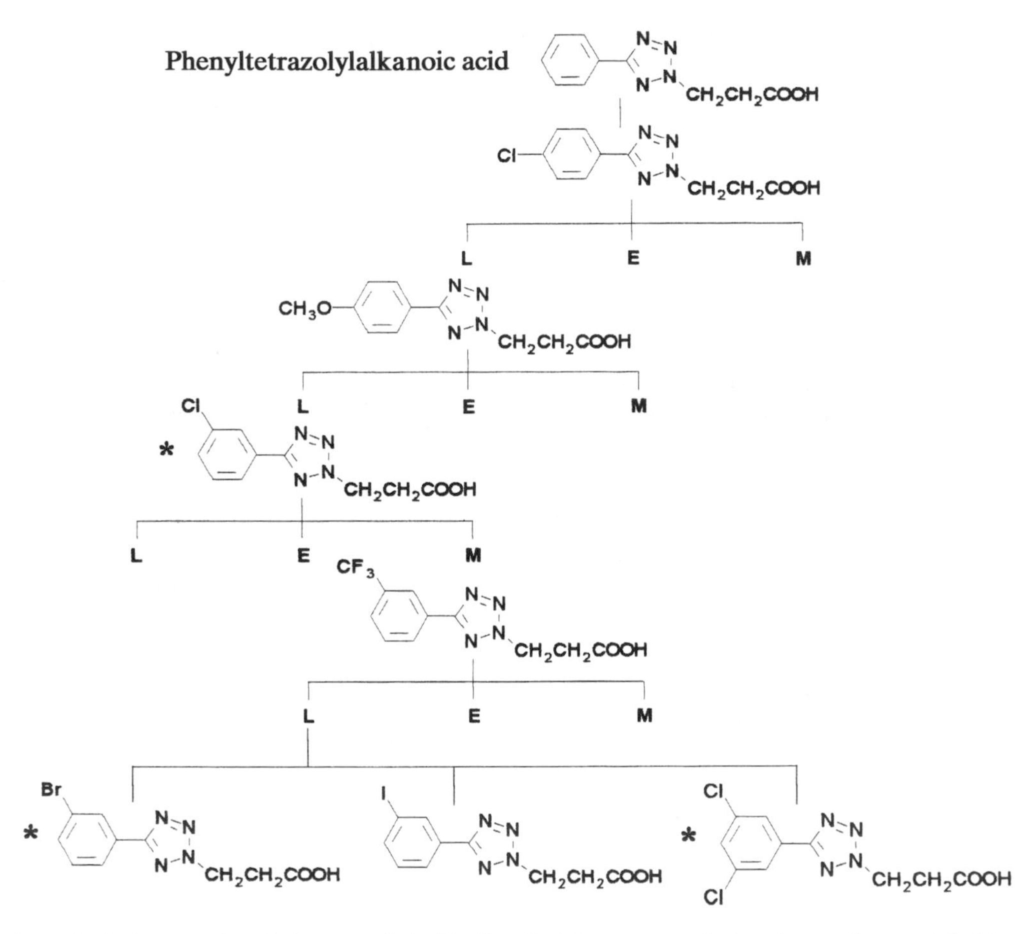

Figure 2.17. A hypothetical example of the use of the Topliss decision tree to find active aryltetrazolylalkanoic acids. The analogues found by a traditional approach to have a higher activity and potency than the lead are indicated by an asterisk.

sise and biologically test the 2,4-dichloro (E) and the 4-nitro analogues (F) of A. It is emphasised that the decision tree is not a synthetic pathway for the production of each of the analogues. It simply suggests which of the substituents would be likely to yield a more potent analogue. The synthetic route for producing each of the suggested analogues would vary for each analogue and would use the most appropriate starting materials.

The Topliss decision tree does not give all the possible analogues but it is likely that a number of the most active analogues will be found by this method. Consider, for example, the anti-inflammatory aryltetrazolylalkanoic acids (Figure 2.17). The traditional approach required the synthesis of 28 analogues to discover the four most active analogues. However, the use of the Topliss decision tree would have yielded three of the most active components by the synthesis of only eight analogues, a considerable saving on time and money.

A limitation of the Topliss method is the requirement that the lead must contain an unfused aromatic ring system. However, because many of the drugs in current usage fulfil this requirement, the Topliss decision tree method can be of some considerable use in the discovery of more effective new analogues.

2.5 Computer-aided Drug Design

The development of desk-top computers has given the medicinal chemist a powerful tool to use in the development of drugs. A wide variety of computer programs and methods have been developed to visualise the three-dimensional shapes of both the ligands and their target sites. In addition, sophisticated graphics packages also allow the medicinal chemist to evaluate the interactions between a compound and its target site before synthesising that compound. This means that the medicinal chemist need only synthesise and test the most promising of the compounds, which considerably increases the chances of discovering a potent drug. It also significantly reduces the cost of development.

2.5.1 Modelling Drug–Receptor Interactions

The three-dimensional shapes of both ligand and target site may be determined by X-ray crystallography or computational methods based on either quantum or molecular mechanics. Quantum mechanical calculations require considerably more computing power and so molecular mechanics is a more useful source of three-dimensional structures. If sufficient crystal data are available a three-dimensional computer-generated structure may be built up using three-dimensional structures obtained from databases, such as the Cambridge and Brookhaven databases. The three-dimensional structures obtained by these methods may be displayed as either stick or space fill pictures on computer screens (Figure 2.18). However, in all cases, the accuracy of the structures obtained will depend on the accuracy of the data used in their determination.

The three-dimensional structures produced on a computer screen may be manipulated on the screen to show different views of the structures. Furthermore, with more complex programs it is possible to superimpose one structure on top of another. In other words it is possible to superimpose the three-dimensional structure of a potential drug on its possible target site. This process is known as **docking** (Figure 2.19). It enables the medicinal chemist to evaluate the fit of potential drugs to their target site. Furthermore, the use of a colour code to indicate the nature of the atoms and functional groups present in the three-dimensional structures also enables the medicinal chemist to investigate the binding potential of the ligand to the target site. However, it must be remembered that in many cases this binding should be weak because the drug has to be able to leave the receptor site after it has activated that site (see section

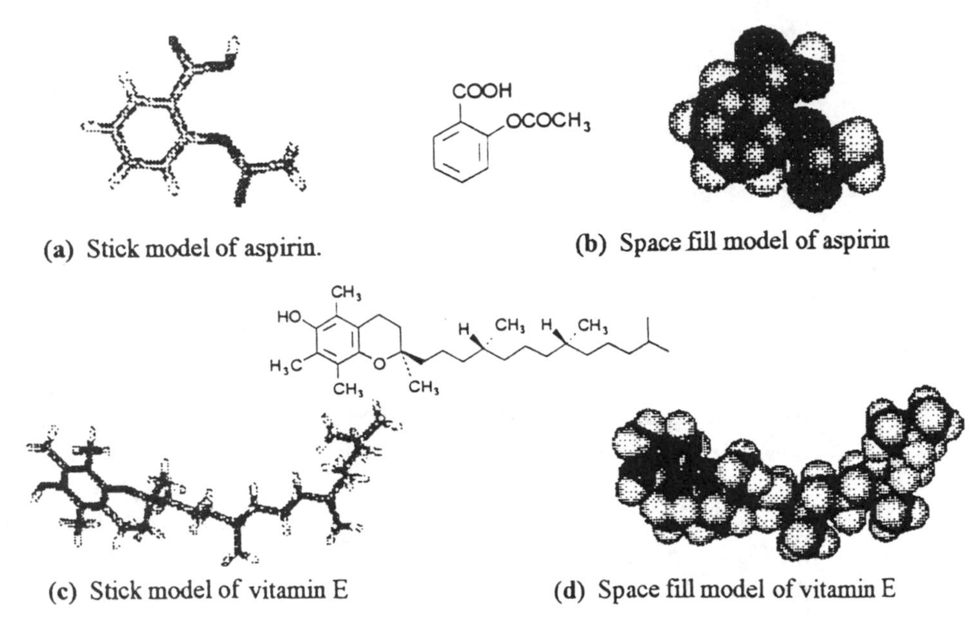

Figure 2.18. Examples of the stick and space fill models displayed by graphics packages.

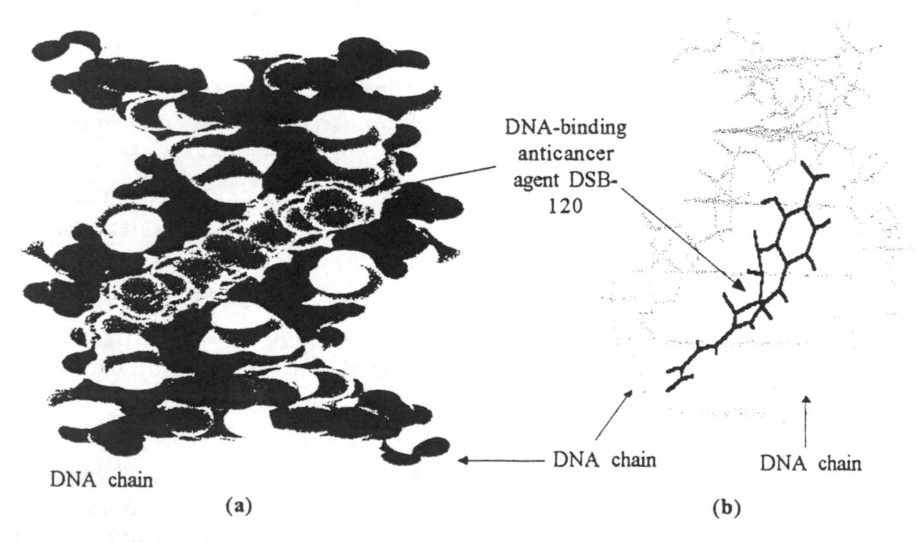

Figure 2.19. The docking of DBS-120 to a fragment of DNA: (a) CPK model; (b) Dreiding model. (Courtsey of Professor D. Thurston, University of Nottingham.)

8.6.1). If a ligand is a good fit and its functional groups are positioned so that they can interact with the structure of the proposed target site, it is likely to be biologically active.

A major problem with this approach is that many molecules do not have the rigid structures implied by their traditional molecular formulae. Saturated ring systems and conjugated systems give the structure of a molecule a degree of rigidity but the presence of simple single bonds results in a flexible structure whose shape (conformations) will depend on the energy of the molecular environment at the time. This means that the conformations assumed by the ligand and target site are not necessarily the same as those determined by X-ray crystallography or computer methods. However, computational chemistry has been applied to conformational analysis and a number of techniques, such as the Metropolis Monte Carlo method and Comparative Molecular Field Analysis (CoMFA), have been developed to determine the effect of conformational changes on the effectiveness of a docking procedure.

2.6 Combinatorial Chemistry

2.6.1 Introduction

The rapid increase in molecular biology technology has resulted in the development of rapid, efficient drug-testing systems. The techniques used by these systems are collectively known as **high-throughput screening**. High-throughput screening methods give accurate results even when extremely small amounts of the test substance are available. However, if it is to be used in an economic fashion as well as efficiently, this technology requires the rapid production of a large number of substances for testing, which cannot be met by traditional organic synthesis methods.

Traditional organic synthesis is basically the joining together of structural building blocks in a set sequence to form one product. (Figure 2.20). For example, structure A is joined to structure B to form the product A-B. This product is then reacted with a new reagent C to form the product A-B-C and so on to ultimately produce the final product. Using this slow, labour-intensive traditional approach a medicinal chemist is able to produce about 25 test compounds a year, which means that drug development is expensive. To be effective, high-throughput screening tests require the production of large numbers of test compounds. Combinatorial chemistry was developed to discover new lead compounds by satisfying this demand. In contrast to traditional synthetic methods, it allows the simultaneous synthesis of all the possible

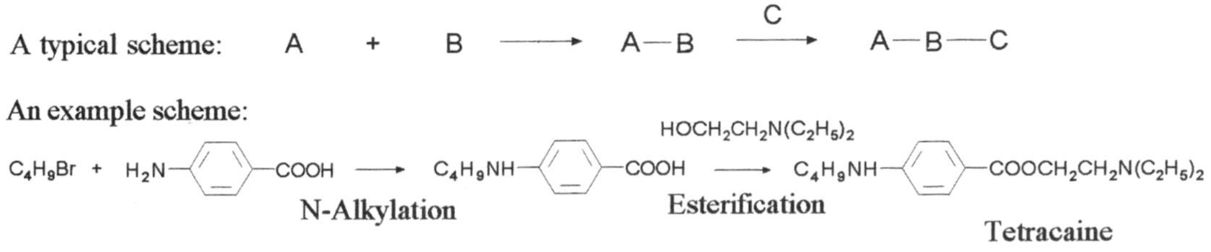

A typical scheme: A + B ⟶ A—B $\xrightarrow{C}$ A—B—C

An example scheme:

C_4H_9Br + $H_2N-C_6H_4-COOH$ ⟶ (N-Alkylation) $C_4H_9NH-C_6H_4-COOH$ $\xrightarrow{HOCH_2CH_2N(C_2H_5)_2}$ (Esterification) $C_4H_9NH-C_6H_4-COOCH_2CH_2N(C_2H_5)_2$

Tetracaine

Figure 2.20. A traditional stepwise organic synthesis scheme illustrated by the synthesis of the local anaesthetic tetracaine.

compounds that could be formed from a number of building blocks. The products of such a process are known as a **combinatorial library**. Libraries may be a collection of individual compounds or mixtures of compounds. Screening the components of a library for activity using high-throughput screening techniques enables the development team to select suitable leads for further investigation.

2.6.2 The Basic Concept of Combinatorial Chemistry

Combinatorial chemistry differs from traditional synthesis in that it involves the simultaneous reaction of a set of compounds with a second set of compounds to produce a set of products known as a **combinatorial library**. Consider, for example, the reaction of a set of three compounds (A_{1-3}) with a set of three building blocks (B_{1-3}). In combinatorial synthesis, A_1 would simultaneously undergo separate reactions with compounds B_1, B_2 and B_3, respectively (Figure 2.21). At the same time compounds A_2 and A_3 would also be undergoing reactions with compounds B_1, B_2 and B_3. These simultaneous reactions would produce a library of nine products. If this process is repeated by reacting these nine products with three new building blocks (C_{1-3}), a combinatorial library of 27 new products would be obtained.

The reactions used in such a synthesis usually involve the same functional groups, that is, the same reaction occurs in each case. Very few libraries have been constructed where different types of reaction are involved in the same stage. In theory this approach results in the formation of all the possible products that could be formed. For example, a two-stage synthesis with 10 starting compounds and 10 different building blocks at each stage would in theory yield a library of 1000 compounds. However, in practice some reactions may not occur. The techniques used in combinatorial chemistry to obtain this number of products are dealt with in section 2.6.4.

2.6.3 The Design of Combinatorial Syntheses

One of two general strategies may be followed when designing a combinatorial synthesis (Figure 2.22a). In the first case the building blocks are successively added to the preceding structure so that it grows in only one direction. It usually relies on the medicinal chemist finding suitable protecting groups so that the reactions are selective. This design approach is useful if the product is a polymer (**oligomer**) formed from a small number of monomeric units. Alternatively, the synthesis can proceed in different directions from an initial building block known as a **template**, provided that the template has either the necessary functional groups or they can be generated during the course of the synthesis (Figure 2.22b). This approach may also require the use of suitable protecting groups (see section 12.2.4).

The reactions used when designing a combinatorial sequence should *ideally* satisfy the following criteria:

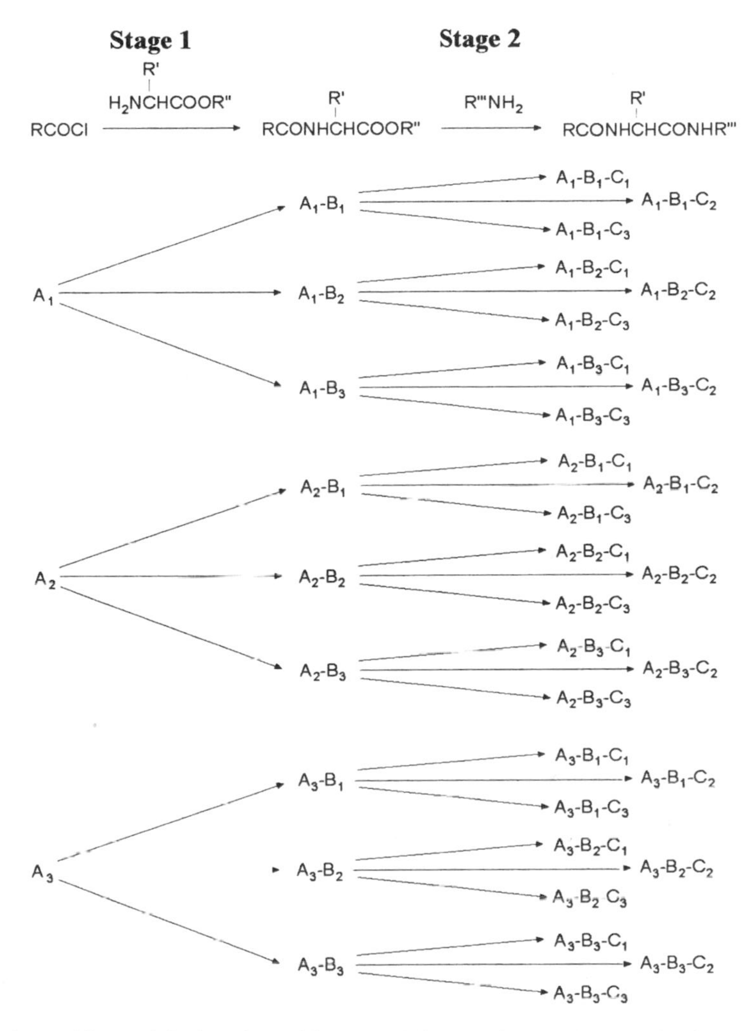

Figure 2.21. The principle of combinatorial chemistry illustrated by a scheme for synthesis of a hypothetical polyamide using three building blocks at each stage.

(i) the reactions should form a bond between the building blocks;
(ii) the reactions should be specific and give a high yield;
(iii) the reactions used in the sequence should allow for the formation of as wide a range of structures for the final products as possible, including all the possible stereoisomers;
(iv) the reactions should be suitable for use in automated equipment;
(v) the building blocks should be readily available;
(vi) the building blocks should be as diverse as possible so that the range of final products includes structures that utilise all the types of bonding (see section 8.2) to bind to or react with the target;
(vii) it must be possible to determine accurately the structures of the final products.

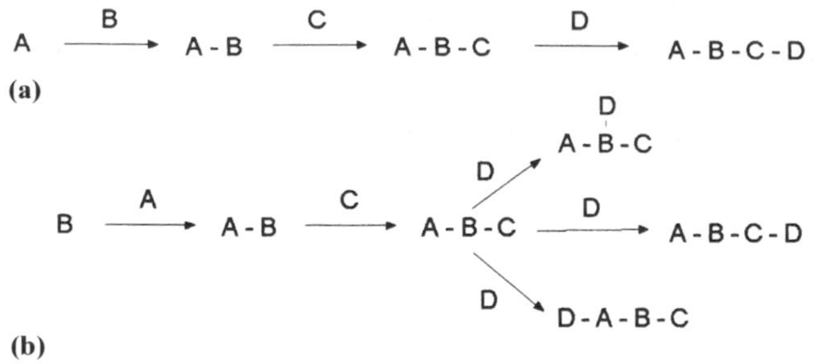

Figure 2.22. Strategies for designing a combinatorial synthesis. (a) The sequential attachment of building blocks. (b) The non-sequential attachment of building blocks using B as a template. Alternative pathways are given to illustrate the versatility of this approach (see section 2.1).

In practice it is not always possible to select reactions that meet all these criteria. However, the last criterion must be satisfied otherwise there is little point in carrying out the synthesis.

The degree of information available about the intended target will also influence the selection of the building blocks. If little is known, a random selection of building blocks is used in order to identify a lead. However, if a there is a known lead, the building blocks are selected so that they produce analogues that are related to the structure of the lead. This allows the investigator to study the SAR and/or determine the optimum structure for potency.

2.6.4 The General Techniques Used in Combinatorial Synthesis

Combinatorial synthesis may be carried out on a solid support (see section 2.6.5) or in solution (see section 2.6.6). In both cases synthesis usually proceeds using one of the strategies outlined in Figure 2.22. Both solid support and solution synthetic methods may be used to produce libraries that consist of either individual compounds or mixtures of compounds. Each type of synthetic method has its own distinct advantages and disadvantages (Table 2.10).

2.6.5 The Solid Support Method

This approach originated with the Merrifield solid support peptide synthesis. Merrifield used a solid support of polystyrene–divinylbenzene resin beads, each bead having a large number of monochlorinated methyl side chains. The C-terminal of the first amino acid in the peptide chain is attached by an S_N2 displacement of these chloro groups, which means that **one bead acts as the solid support for the formation of a large number of peptide molecules of the same type**. Additional amino acids are added to the growing peptide chain using the reaction sequence shown in Figure 2.23. This sequence, in common with other peptide syntheses, uses protecting groups such as *t*-butyloxycarbonyl (Boc) to control the position of amino acid coupling. To form the amide peptide link, the N-protected amino acids are converted to a more

Table 2.10. A comparison of the advantages and disadvantages of the solid support and solution techniques of combinatorial chemistry.

On a solid support	In solution
Reagents can be used in excess in order to drive the reaction to completion	Reagents cannot be used in excess, unless additional purification is carried out
Purification is easy, simply wash the support	Purification can be difficult
Automation is easy	Automation is difficult
Fewer suitable reactions	In theory any organic reaction can be used
Scale-up relatively expensive	Scale-up is easy and relatively inexpensive
Not well documented and time will be required to find a suitable support and linker for a specific synthesis	Only requires time for the development of the chemistry

active acylating derivative of dicyclohexylcarbodiimide (DCC), which reacts with an unprotected amino group to join the new amino acid residue to the growing peptide.

A Boc group

$(CH_3)_3C{-}O{-}CONHCH(R^2)COOH$

Boc-protected amino acid residue

+

Dicyclohexylcarbodiimide (DCC)

→

$(CH_3)_3C{-}O{-}CONHCH(R^2)COOC(=N{-}C_6H_{11})NH{-}C_6H_{11}$

Active acyl derivative of DCC

In 1985 Houghton introduced his **tea bag** method for the rapid solid-phase multiple peptide synthesis. In this technique the beads are contained within a porous polypropylene bag. All the reactions, including deprotections, are carried out by placing the bags in solutions containing the appropriate reagents. The use of the bag makes it easy to purify the resin beads by washing with the appropriate solutions. Furthermore, the method has considerable flexibility and has been partly automated.

Merrifield's original resin bead has been largely superseded by other beads such as the TentaGel resin bead. This bead is more versatile because it can be obtained with different functional groups at the end of the side chain (Figure 2.24). These functional groups are separated from the resin by a polyethyleneglycol (PEG) insert. As a result, the reacting groups of the side chains are further from the surface of the bead. It locates the reaction centre in the solvent, which facilitates reactant access and makes reaction easier. Like the Merrifield bead, each TentaGel bead contains a large number of side-chain functional groups. For example, the number of amine groups per bead is about 6×10^{13}. This means that in theory each bead could act as the support for the synthesis of up to 6×10^{13} molecules of the **same** compound. In peptide synthesis the amount of peptide found on one bead is usually sufficient for its structure to be determined using the Edman thiohydantoin microsequencing technique.

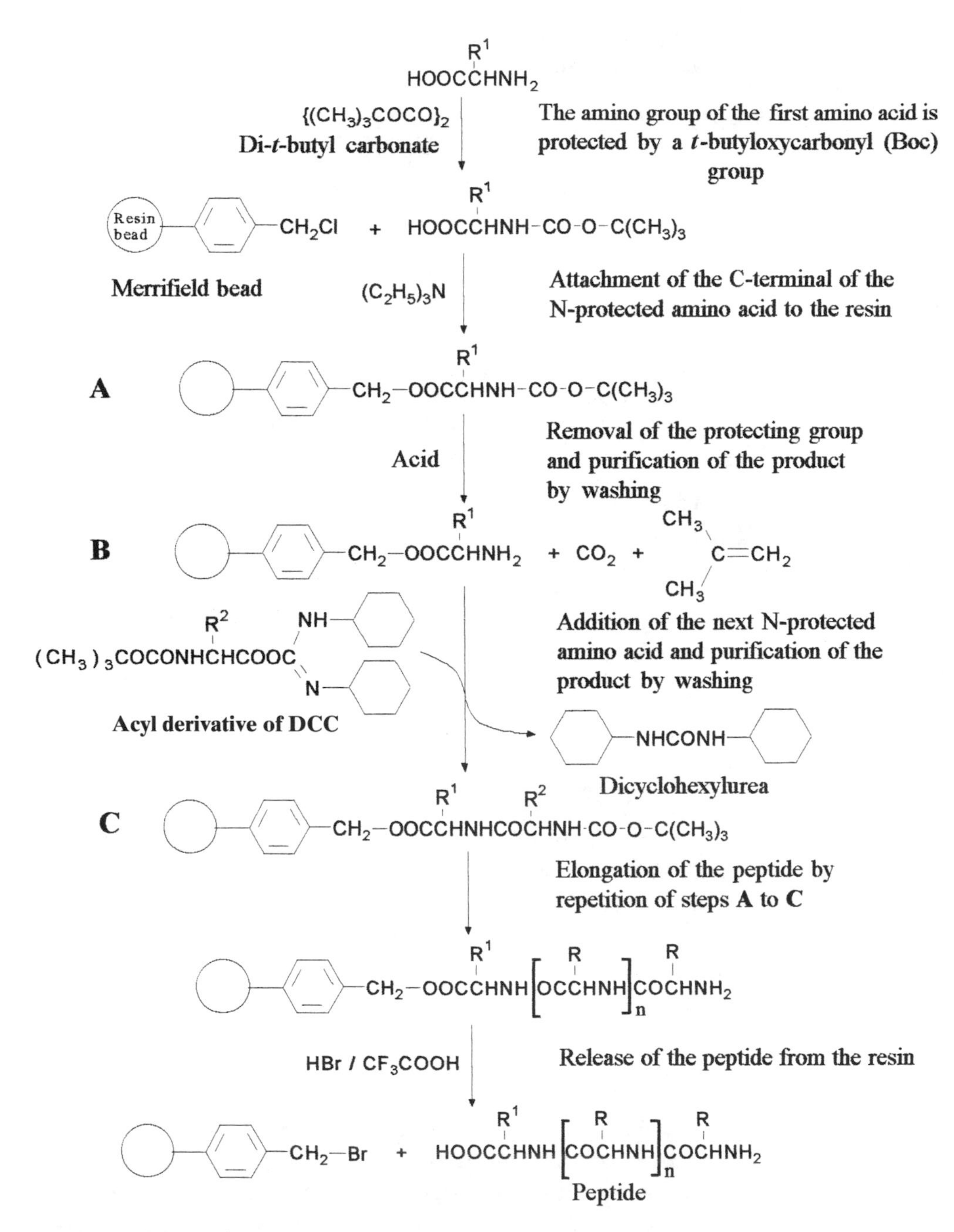

Figure 2.23. An outline of the Merrifield peptide synthesis, where R is any amino acid side chain.

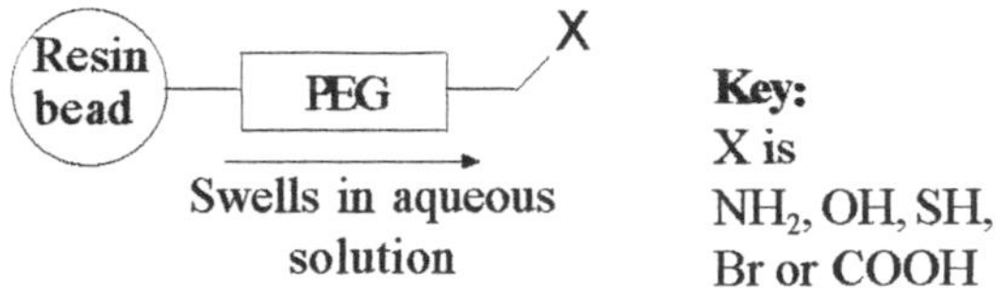

Resin bead — PEG — X

Swells in aqueous solution

Key:
X is
NH_2, OH, SH,
Br or COOH

Figure 2.24. TentaGel resin beads.

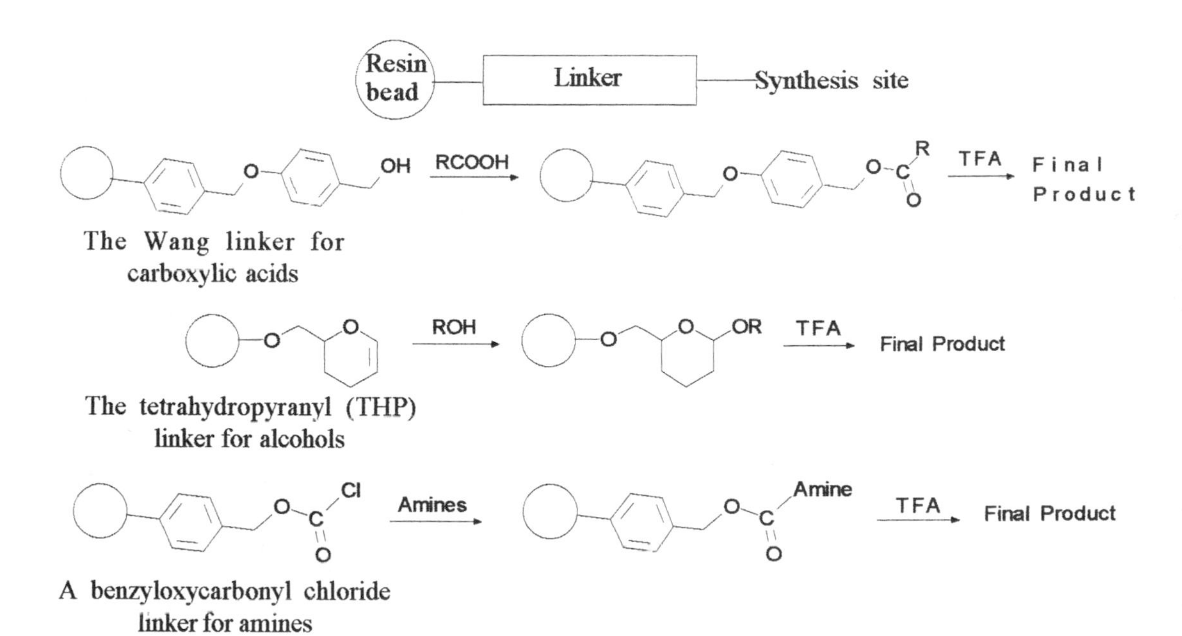

Figure 2.25. Examples of linkers and the reagents used to detach the final product. TFA = trifluoroacetic acid.

The group that anchors the compound being synthesised to the bead is known either as a **handle** or a **linker** (Figure 2.25). As well as modifying the properties of the bead, they move the point of substrate attachment further from the bead, making reaction easier. The choice of linker will depend on the nature of the reactions used in the proposed synthetic pathway. For example, an acid-labile linker, such as HMP (hydroxymethylphenoxy-resin), would not be suitable if the reaction pathway contained reactions that were conducted under strongly acidic conditions. Consideration must also be given to the ease of detaching the product from the linker at the end of the synthesis. The method employed must not damage the required product but must also lend itself to automation.

Combinatorial synthesis on solid supports is usually carried out either by using the parallel synthesis or the Furka split and mix procedures. The precise method and approach adopted when using these methods will depend on the nature of the combinatorial library being produced and also the objectives of the investigating team. However, in all cases it is necessary to determine the structures of the components of the library by either keeping a detailed record of the steps involved in the synthesis or giving beads a label that can be decoded to give the structure of the compound attached to that bead (see sections 2.6.5.3, 2.6.5.3 and 2.6.5.4). The method adopted to identify the components of the library will depend on the nature of the synthesis.

2.6.5.1 Parallel Synthesis

This technique is normally used to prepare combinatorial libraries that consist of separate compounds. It is not suitable for the production of libraries containing thousands to millions of compounds. In parallel synthesis the compounds are prepared in separate reaction vessels but at the same time, that is, in parallel. The array of individual reaction vessels often takes the form of either a grid of wells in a plastic plate or a grid of plastic rods called pins attached to a plastic base plate (Figure 2.26) that fits into a corresponding set of wells. In the former case the synthesis is carried out on beads placed in the wells, whereas in the latter case it takes place on so-called plastic 'crowns' pushed on to the tops of the pins, the building blocks being attached to these crown by linkers similar to those found on the resin beads. The position of each synthetic pathway in the array and hence the structure of the product of that pathway are usually identified by a grid code.

Both the well and pin arrays are used in the same general manner. However, the manipulation of the arrays is different. Consider the reaction between a set of eight building blocks (X1, X2, . . . X8) and a second set of twelve building blocks (Y1, Y2, . . . Y12) using a well array. The X compounds are placed so that each row of wells contains only one type of X compound. For example, row A will only contain compound X1, row B will only contain compound X2 and so

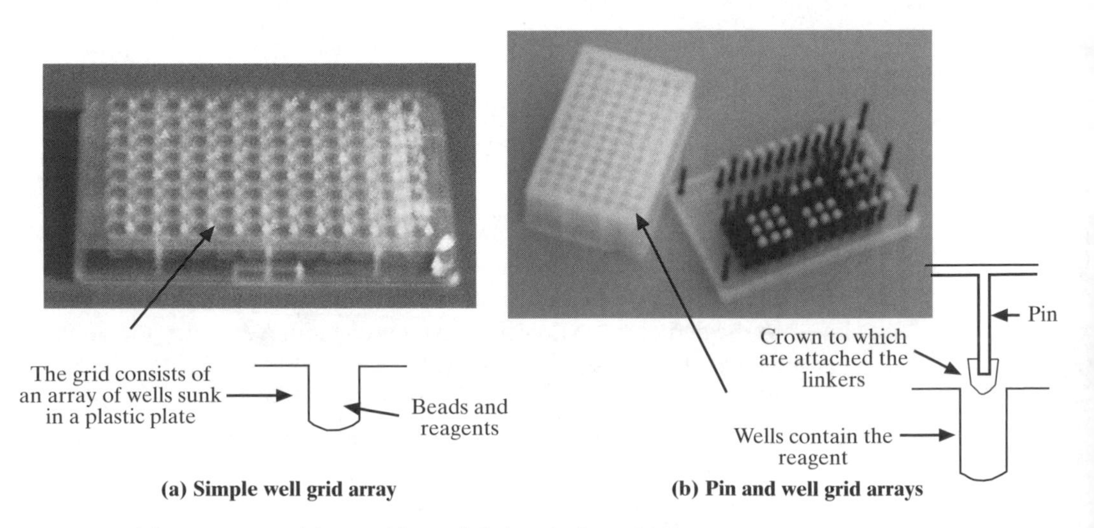

Figure 2.26. Examples of the arrays used in combinatorial chemical synthesis.

	A	B	C	D	E	F	G	H
1	X1	X2	X3	X4	X5	X6	X7	X8
2	X1	X2	X3	X4	X5	X6	X7	X8
3	X1	X2	X3	X4	X5	X6	X7	X8
4	X1	X2	X3	X4	X5	X6	X7	X8
5	X1	X2	X3	X4	X5	X6	X7	X8
6	X1	X2	X3	X4	X5	X6	X7	X8
7	X1	X2	X3	X4	X5	X6	X7	X8
8	X1	X2	X3	X4	X5	X6	X7	X8
9	X1	X2	X3	X4	X5	X6	X7	X8
10	X1	X2	X3	X4	X5	X6	X7	X8
11	X1	X2	X3	X4	X5	X6	X7	X8
12	X1	X2	X3	X4	X5	X6	X7	X8

(a) The placement of the first building blocks.

	A	B	C	D	E	F	G	H
1	X1-Y1	X2-Y1	X3-Y1	X4-Y1	X5-Y1	X6-Y1	X7-Y1	X8-Y1
2	X1-Y2	X2-Y2	X3-Y2	X4-Y2	X5-Y2	X6-Y2	X7-Y2	X8-Y2
3	X1-Y3	X2-Y3	X3-Y3	X4-Y3	X5-Y3	X6-Y3	X7-Y3	X8-Y3
4	X1-Y4	X2-Y4	X3-Y4	X4-Y4	X5-Y4	X6-Y4	X7-Y4	X8-Y4
5	X1-Y5	X2-Y5	X3-Y5	X4-Y5	X5-Y5	X6-Y5	X7-Y5	X8-Y5
6	X1-Y6	X2-Y6	X3-Y6	X4-Y6	X5-Y6	X6-Y6	X7-Y6	X8-Y6
7	X1-Y7	X2-Y7	X3-Y7	X4-Y7	X5-Y7	X6-Y7	X7-Y7	X8-Y7
8	X1-Y8	X2-Y8	X3-Y8	X4-Y8	X5-Y8	X6-Y8	X7-Y8	X8-Y8
9	X1-Y9	X2-Y9	X3-Y9	X4-Y9	X5-Y9	X6-Y9	X7-Y9	X8-Y9
10	X1-Y10	X2-Y10	X3-Y10	X4-Y10	X5-Y10	X6-Y10	X7-Y10	X8-Y10
11	X1-Y11	X2-Y11	X3-Y11	X4-Y11	X5-Y11	X6-Y11	X7-Y11	X8-Y11
12	X1-Y12	X2-Y12	X3-Y12	X4-Y12	X5-Y12	X6-Y12	X7-Y12	X8-Y12

(b) The placement of the second building blocks

Figure 2.27. The pattern of well loading for a simple 96-compound combinatorial library.

on (Figure 2.27a). Beads are added to each well and the array placed in a reaction environment that will join the X compound to the linker of the bead. The Y compounds are now added to the wells so that each numbered row at right angles to the lettered rows contains only one type of Y compound, that is, compound Y1 is only added to row 1, compound Y2 is only added to row 2 and so on (Figure 2.27b). The array is placed in a suitable reaction environment and the X and Y compounds are allowed to react to form 96 different products. The product is purified by washing and the wells loaded with the next reagent using the one compound per row concept, but in this and subsequent steps only either the numbered or lettered rows are used, not both unless a library of mixtures is required. Finally, the products are liberated from the resin by the appropriate linker cleavage reaction (see Figure 2.25), isolated and used in the appropriate screening processes. The structures of all the compounds are determined by instrumental methods (mainly NMR, GC, HPLC and MS) and by following the history of the synthesis using the grid references of the wells.

The pin array is used in a similar manner to the well array except that the array of crowns is inverted so that the crowns are suspended in the reagents placed in a corresponding array of wells (Figure 2.27). Reaction is brought about by placing the combined pin and well unit in a suitable reaction environment. The loading of the wells follows the pattern described in the previous paragraph. These types of techniques have been used to prepare libraries of up to several hundred compounds. For example, hydantoins have been shown to have some use in controlling epilepsy. Cody and co-workers at the Parke-Davis Pharmaceutical Research Division (Ann Arbor, Michigan) have produced a 39-compound hydantoin library by reacting a

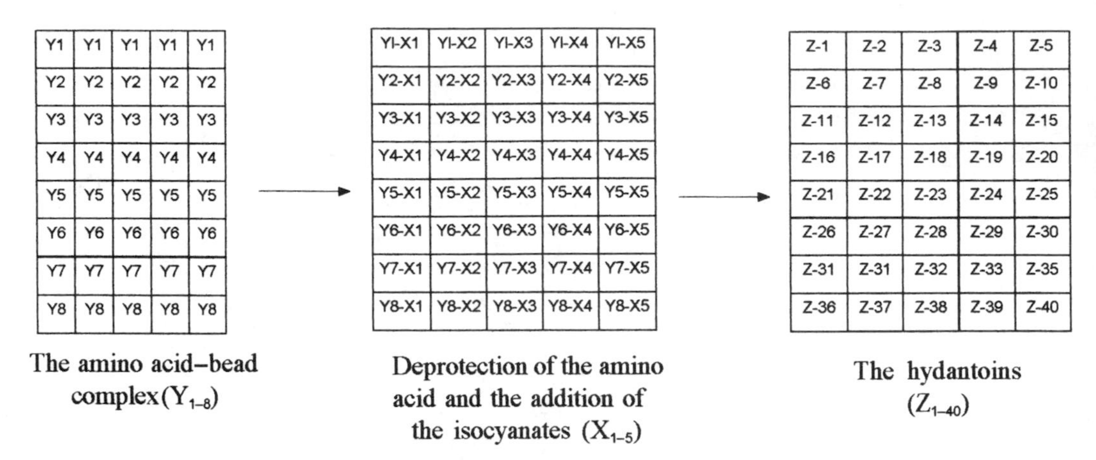

Figure 2.28. The pattern of loading the wells of a Parke-Davis Diversomer for the production of a hydantoin combinatorial library.

resin-bound amino acid with an isocyanate to form a substituted urea. Acid-catalysed cyclisation of this substituted urea produces the hydantoins in yields that range from about 5 to 80%.

Resin bead —$OCOC(R^1)(R^2)NHBOC$ —(TFA or Piperidine)→ Resin —$OCOC(R^1)(R^2)NH_2$ —(R^3NCO)→ Resin —$OCOC(R^1)(R^2)NHCONHR^3$ —(HCl, Heat)→ Hydantoins

A substituted urea   Hydantoins

The reactions were carried out using a Parke-Davis Diversomer apparatus with a reaction vessel grid of 5 × 8 rows. An N-protected amino acid was attached to a suitable resin. This was repeated with a further seven different N-protected amino acids to produce eight separate resin–amino acid adducts. Each of these samples was placed individually in one of the Y row wells of the grid (Figure 2.28). The amino acids were deprotected and reacted with five different isocyanates, a different isocyanate in each of the X rows of the grid. The product of this stage was treated with hot 6M hydrochloric acid to simultaneously form the final product of the synthesis and release it from the resin. Although a total of 40 compounds is possible, a library of only 39 compounds was produced because one reaction gave a zero yield.

Fodor's method for parallel synthesis. In theory almost any solid material can be used as the solid support for parallel combinatorial synthesis. Fodor and co-workers (1991) produced peptide libraries using a form of parallel synthesis that could be performed on a glass plate. The plate is treated so that its surface is coated with hydrocarbon chains containing a terminal amino group. These amino groups are protected by the UV-labile 6-nitroveratryloxycarbonyl (NVOC) group.

NVOC

Glass plate

UV

A

$NVOC-NHCH(R^1)COOR$

UV

B

Repeat steps A and B until the peptide synthesis is complete

A photolithography mask (M1) is placed over the plate so that only a specific area of the plate can be irradiated with UV light (Figure 2.29). This results in the removal of the NVOC-protecting group from the amino groups in the irradiated area. The entire plate is exposed to the first activated NVOC-protected amino acid. However, it will only bond to the amino groups

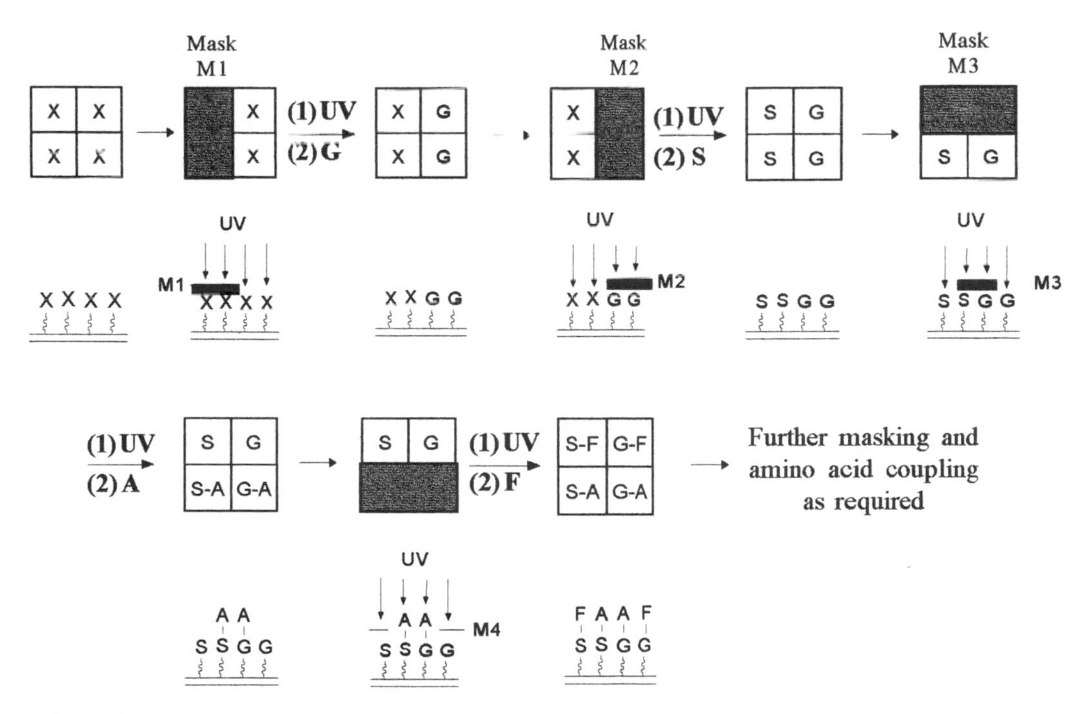

Figure 2.29. A schematic representation of the Fodor approach to parallel synthesis. X represents an NVOC-protected amino group attached to the glass plate. The other letters correspond to the normal code used for amino acids. Each of these amino acids is in its NVOC-protected form.

exposed in the irradiated area. The process is repeated using a new mask (M2) and a second activated NVOC-protected amino acid attached to the exposed amino groups. This process is repeated using different masks (M3. . . . etc.) until the desired library is obtained, the structure of the peptide occupying a point on the plate depending on the masks used and the activated NVOC-protected amino acid used at each stage in the synthesis. The technique is so precise that it has been reported that each compound occupies an area of about 50 μm × 50 μm. The way in which the masks are used will control the order in which the amino acids are added to the growing peptides on the plate.

2.6.5.2 Furka's Mix and Split Technique

This technique, which was developed by Furka and co-workers from 1988 to 1991, is used to make both large (thousands of compounds) and small (hundreds of compounds) libraries. Large libraries are possible because the technique produces one compound on each bead, that is, all the molecules formed on one bead are the same but different from those formed on all the other beads. The technique has the advantage that it reduces the number of reactions required to produce a large library. For example, if the synthetic pathway of a compound required three steps, it would require 30 000 separate reaction vessels to produce a library of 10 000 compounds if the reactions were carried out in separate reaction vessels using orthodox chemical methods. The Furka mix and split method reduces this to about 22 reactions. The Furka method produces the library of compounds on resin beads. The beads are divided into a number of equally sized portions corresponding to the number of initial building blocks. Each of the starting compounds is attached to its own group of beads using the appropriate chemical reaction (Figure 2.30). All the portions of beads are now mixed and separated into the number of equal portions corresponding to the number of reactants for the first step in the synthesis. One reactant building block is added to each portion and the reaction carried out by putting the mixture of resin beads and reactants in a suitable reaction vessel. After reaction all the beads are mixed before separating them into the number of equal portions corresponding to the number of reactants being used in the second step of the synthesis. Again one of the second-step building blocks is added to each of these new portions and the mixture allowed to react to produce the products of the next step in the synthesis. This process of mix and split is continued until the required library is synthesised.

Furka's mix and split method has been used extensively to prepare peptide libraries using schemes similar to that developed by Merrifield (Figure 2.23). In peptide and similar polymer library formation where the same building blocks are used at each step, the maximum possible number of compounds that can be synthesised for a given number of different building blocks (b) is given by:

$$\text{Number of compounds} = b^{x} \tag{2.21}$$

where x is the number of steps in the synthesis.

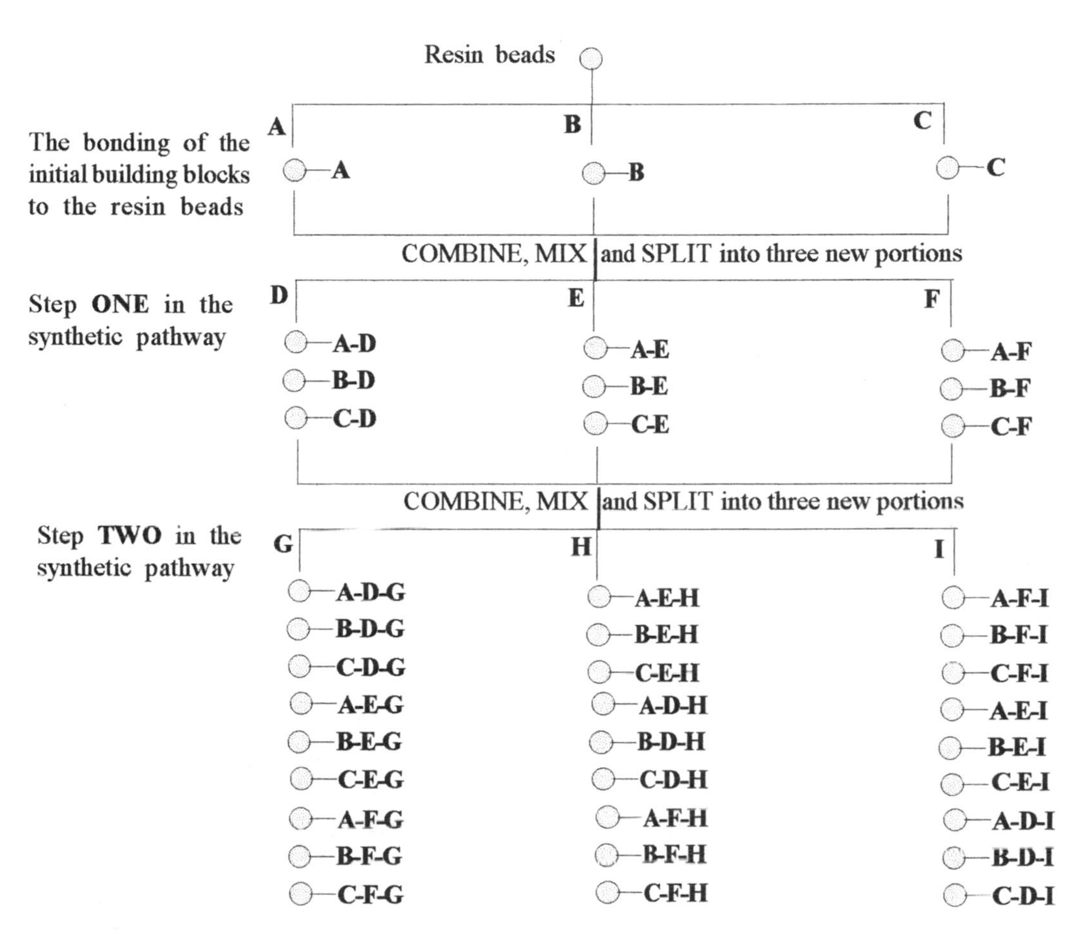

Figure 2.30. An example of the Furka approach to combinatorial libraries using a two-step synthesis involving three building blocks at each stage.

The identity of the compounds produced in a mix and split combinatorial synthesis is usually deduced using a combination of analytical methods such as NMR, MS, HPLC and GC and the reaction history of the compound. Unlike in parallel synthesis, the history of the bead cannot be traced from a grid reference; it has to be traced using a suitable encoding method (see sections 2.6.5.3, 2.6.5.4 and 2.6.5) or deconvolution (see section 2.6.5.7). Encoding methods use a code to indicate what has happened at each step in the synthesis, ranging from putting an identifiable tag compound on to the bead at each step in the synthesis to using computer-readable silicon chips as the solid support.

2.6.5.3 Sequential Chemical Tagging Methods

Sequential chemical tagging methods use specific compounds (tags) as a code for the individual steps in the synthesis. These tag compounds are sequentially attached in the form of a

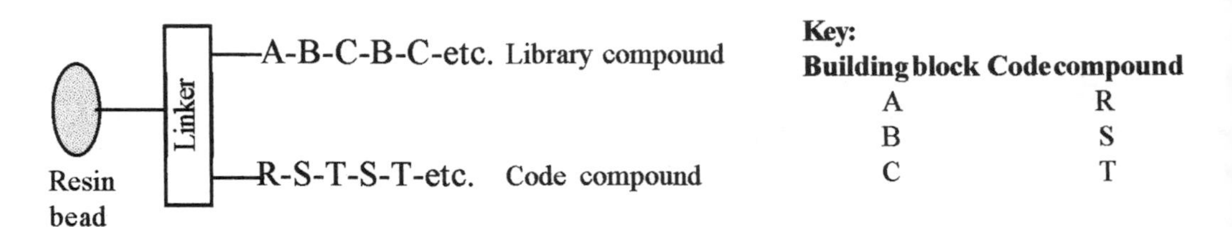

Figure 2.31. Chemical encoding of resin beads.

polymer-like molecule to the same linker or bead as the library compound at each step in the synthesis (Figure 2.31). At the end of the synthesis both the library compound and the tag compound are liberated from the bead. The tag compound is decoded to give the history and hence the possible structure of the library compound.

Compounds used for tagging must satisfy a number of criteria:

(i) the concentration of the tag should be just sufficient for its analysis, that is, the majority of the linkers should be occupied by the combinatorial synthesis;
(ii) the tagging reaction must take place under conditions that are compatible with those used for the synthesis of the library compound;
(iii) it must be possible to separate the tag from the library compound;
(iv) analysis of the tag should be rapid and accurate using methods that could be automated.

Many peptide libraries have been encoded using single-stranded DNA oligonucleotides as tags because these meet the conditions previously outlined. Each oligonucleotide acts as the code for one amino acid (Table 2.11). At each step in the peptide synthesis a second parallel reaction is carried out on the same bead to attach the corresponding oligonucleotide tag. In other words, two alternating parallel syntheses are carried out on the same bead (Figure 2.32). At the end of the synthesis the oligonucleotide tag can be isolated from its bead and its base sequence can be determined and decoded to give the history of the formation of the peptide attached to the bead.

Table 2.11. The use of oligonucleotides to encode amino acids in peptide synthesis.

Amino acid	Structure	Oligonucleotide code
Glycine (Gly)	NH_2–CH_2COOH	CACATG
Methionine (Met)	$CH_3SCH_2CH_2CH(NH_2)COOH$	ACGGTA

Normally a branched linker is used: one branch for the synthesis and the other for the tag. A polymerase chain reaction (PCR) primer is attached to the tag site so that at the end of the

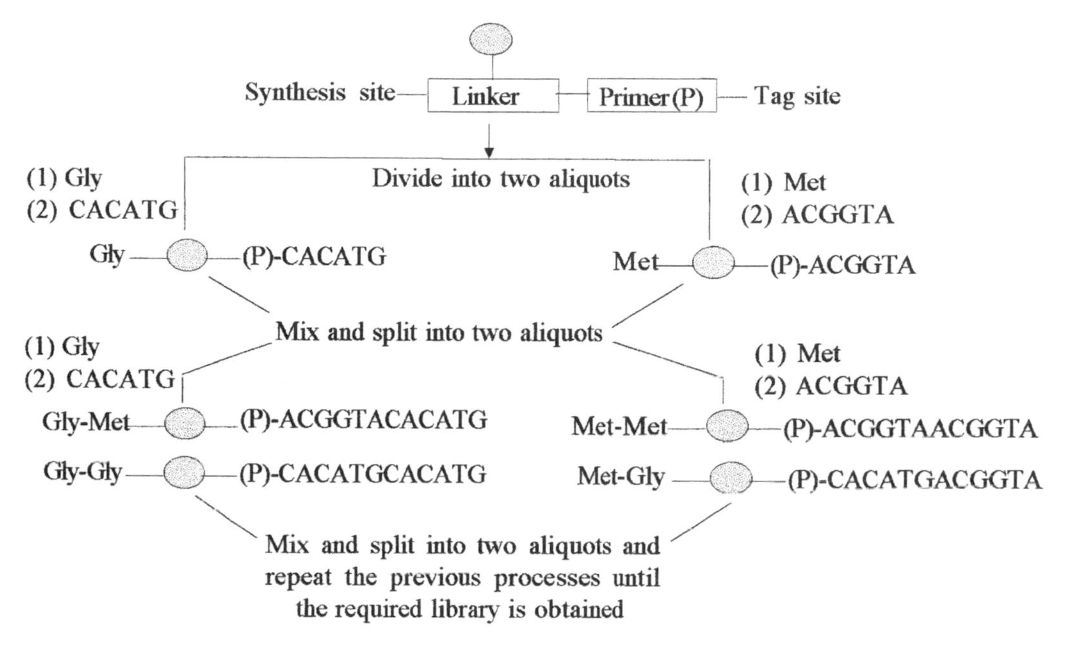

Figure 2.32. The use of oligonucleotides to encode a peptide combinatorial synthesis for a library based on two building blocks.

combinatorial synthesis the concentration of the completed tag DNA-oligonucleotide molecule may be increased using the *Taq* polymerase procedure. This amplification of the yield of the tag makes it easier to identify the sequence of bases, which leads to a more accurate decoding.

Peptides and individual amino acids have also been used to code for the building blocks in a synthesis because they can be sequentially joined. Syntheses are usually carried out using a branched linker so that the synthesis of the encoding molecule can be carried out in parallel to that of the combinatorial library molecule it encodes. For example, the Zuckermann approach uses a diamine linker protected at one end by an Fmoc group and at the other end by a Moz group (Figure 2.33). The Fmoc group was cleaved under basic conditions and the building block joined to the linker. The Moz group was removed under acid conditions and a suitably protected peptide attached. The process was repeated for the coupling of each building block to each portion of beads as the mix and split procedure progressed. At the end of the synthesis each bead is separated from its fellows, and the product and its encoding peptide liberated from that bead. It should be remembered that each bead will yield up to 6×10^{13} product molecules (see section 2.6.5), which is sufficent to carry out high-throughput screening procedures. In addition each bead will also produce sufficient of the tagging compound to deduce the structure of the product molecule. The detached product and tag are separated and the sequence of amino acids in the encoding peptide determined using the Edman sequencing method. This sequence is used to determine the history of the formation and hence the structure of the product found on that bead.

9-Fluorenylmethoxy chloroformate

4-Methoxybenzyloxy chloroformate

$-CH_2OCONH-$ ⊥ $-NHCOOCH_2-C_6H_4-OCH_3$

Base-labile Fmoc protecting group

Acid-labile Moz protecting group

(1) Basic Fmoc cleavage
(2) Building block (B^1) coupling

B^1NH- ⊥ $-NHCOOCH_2-C_6H_4-OCH_3$

(1) Acidic Moz cleavage
(2) Peptide tag (P^1) coupling

B^1NH- ⊥ $-NHP^1$

Repeat this sequence of deprotecting and coupling as appropriate in the mix and split procedure

Figure 2.33. An outline of the Zuckermann approach using peptides for encoding.

2.6.5.4 Still's Binary Code Tag System

A unique approach by Still was to give each building block its own chemical equivalent of a binary code for each stage of the synthesis using inert aryl halides (Figure 2.34a). One or more of these tags are attached to the resin using a photolabile linker at the appropriate points in the synthesis. They indicate the nature of the building block and the step at which it was incorporated into the solid support (Table 2.12). The amount of tag used at each step must be strictly controlled so that only about 1% of the available linker functional groups are occupied by a tag. Aryl halide tags are used because they can be detected in very small amounts by GC. They are selected on the basis that their retention times were roughly equally spaced (Figure 2.34b).

Table 2.12. A hypothetical tagging scheme for the preparation of tripeptides using binary combinations of six tags.

Step	Tag		
	Glycine (Gly)	Alanine (Ala)	Serine (Ser)
1	T1	T2	T1 + T2
2	T3	T4	T3 + T4
3	T5	T6	T5 + T6

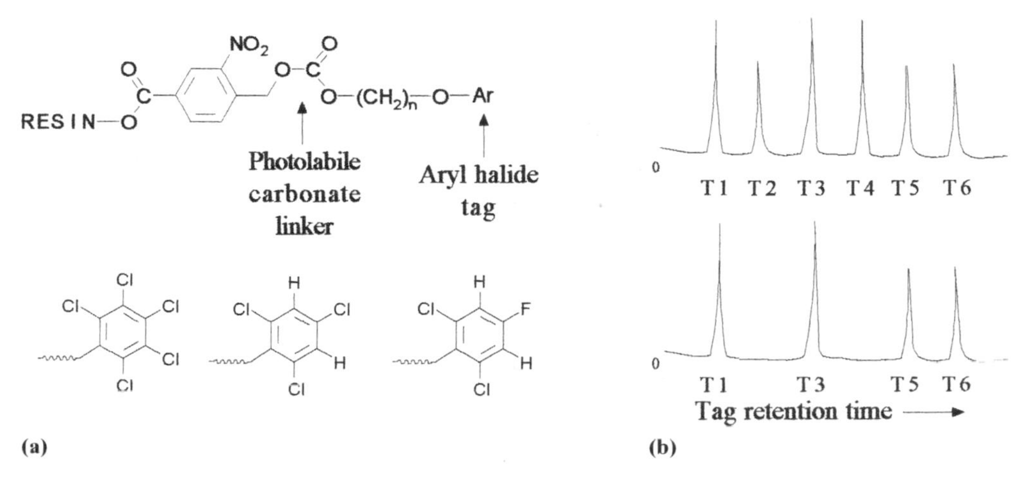

Figure 2.34. (a) Molecular tags used by Still. ~~~ Indicates the point at which the tag is attached to the linker. (b) A hypothetical representation of the GC plots obtained for some aryl halide tags.

At the end of the synthesis the tags are detached from the linker and are detected by GC. The gas chromatogram is read like a bar code to account for the history of the bead. Suppose, for example, that the formation of a tripeptide using the tagging scheme outlined in Table 2.12 gave the tag chromatogram shown in Figure 2.34b. The presence of T1 shows that the C-terminal of the peptide is glycine because this residue is attached to the amino group of the linker; the presence of T3 shows that the second residue is also glycine whereas the presence of T5 and T6 indicates that the third amino acid in the peptide is serine.

2.6.5.5 Computerised Tagging

Nicolaou has devised a method of using silicon chips to record the history of a synthesis. Silicon chips can be coded to receive and store radio signals in the form of a binary code. This code can be used as a code for the building blocks of a synthesis. The silicon chip and beads are placed in a container known as a **can** that is porous to the reagents used in the synthesis. This is effectively the tea bag approach to solid support synthesis. Each can is treated as though it were one bead in a mix and split synthesis. The cans are divided into the required number of aliquots corresponding to the number of building blocks used in the initial step of the synthesis. Each batch of cans is reacted with its own building block and the chip is irradiated with the appropriate radio signal for that building block. The mix and split procedure is followed and at each step the chips in the batch are irradiated with the appropriate radio signal. At the end of the synthesis the prepared library compound is cleaved from the chip, which is interrogated to determine the history of the compound synthesised on the chip. The method has the advantage of producing larger amounts of the required compounds than the normal mix and split approach because the same compound is produced on all the beads in a can.

2.6.6 Combinatorial Synthesis in Solution

The main problem with preparing libraries using solution chemistry is the difficulty of removing unwanted impurities at each step in the synthesis. Consequently, many of the strategies used for the preparation of libraries using solution chemistry are directed to the purification of the products of each step of the synthesis. This and other practical problems have usually restricted the use of solution combinatorial chemistry to synthetic pathways consisting of two or three steps.

Combinatorial synthesis in solution can be used to produce libraries that consist of single compounds or mixtures using traditional organic chemistry. Single compound libraries are prepared using the parallel synthesis technique (see section 2.6.5.1). Libraries of mixtures are formed by separately reacting each of the members of a set of similar compounds with the same mixture of all the members of the second set of compounds. Consider, for example, a combinatorial library of amides formed by reacting a set of five acid chlorides (A^1–A^5) with ten amines (B^1–B^{10}). Each of the five acid chlorides is reacted separately with an equimolar mixture of all ten amines and each of the amines is reacted with an equimolar mixture of all the acid chlorides (Figure 2.35). This produces a library consisting of a set of five mixtures based on individual acid halides and ten mixtures based on individual amines. This means that each compound in the library is prepared twice, once from the acid chloride set and once from the amine set. Consequently, determining the most biologically active of the mixtures from the acid halide set will define the acyl part of the most active amide and similarly identifying the most biologically active of the amine-based set of mixtures will identify the amine residue of that amide. Libraries used in this manner are often referred to as indexed libraries.

This method of identifying the structure of the most active component of combinatorial libraries of mixtures is known as **deconvolution** (see section 2.6.7). It depends on both of the mixtures containing the active compound giving a positive result for the assay procedure. It is not possible to identify the active structure if one of the sets of mixtures gives a negative result. Furthermore, complications arise if more than one mixture is found to be active. In this case all the possible structures have to be synthesised and tested separately. However, it is generally found that the activities of the library mixtures are usually higher than that exhibited by the individual compounds responsible for activity after they have been isolated from the mixture.

2.6.7 Screening and Deconvolution

The success of a library depends not only on it containing the right compounds but also on the efficiency of the screening procedure. A key problem with very large combinatorial libraries of mixtures is the large amount of work required to screen these libraries. Deconvolution is a method, based on the process of elimination, of reducing the number of screening tests required to locate the most active member of a library consisting of a mixture of all the components. It

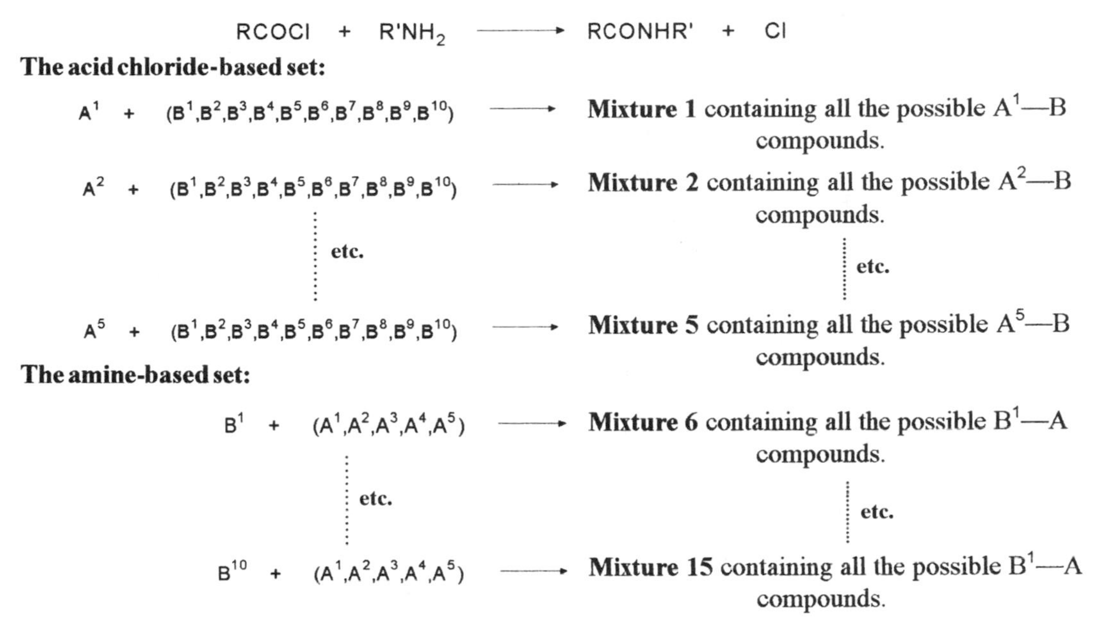
RCOCl + R'NH2 → RCONHR' + Cl
The acid chloride-based set:
A1 + (B1,B2,B3,B4,B5,B6,B7,B8,B9,B10) → Mixture 1 containing all the possible A1—B compounds.
A2 + (B1,B2,B3,B4,B5,B6,B7,B8,B9,B10) → Mixture 2 containing all the possible A2—B compounds.
etc.
etc.
A5 + (B1,B2,B3,B4,B5,B6,B7,B8,B9,B10) → Mixture 5 containing all the possible A5—B compounds.
The amine-based set:
B1 + (A1,A2,A3,A4,A5) → Mixture 6 containing all the possible B1—A compounds.
etc.
etc.
B10 + (A1,A2,A3,A4,A5) → Mixture 15 containing all the possible B1—A compounds.

Figure 2.35. A schematic representation of the formation of a fifteen mixture indexed library using a one step process.

is based on producing and biologically assaying similar secondary libraries that contain one less building block than the original library. It is emphasised that the biological assay is carried out on a mixture of all the members of the secondary library. If the secondary library is still as active as the original library the missing building block is not part of the active structure. Repetition of this process will eventually result in a library that is inactive, which indicates that the missing building block in this library is part of the active structure. This procedure is carried out for each of the building blocks at each step in the synthesis. Suppose, for example, that one has a tripeptide library consisting of a mixture of 1000 compounds. This library was produced from ten different amino acids (A^1–A^{10}) using two synthetic steps, each of which involved ten building blocks (Figure 2.36). The formation of a secondary library by omitting amino acid A^1 from the initial set of amino acids but reacting these nine with all ten amino acids in the first and second steps would produce 900 compounds. These compounds will not contain amino acid residue A^1 in the first position of the tripeptide. If the resulting library is biologically inactive the active compound must contain this residue at position one in the tripeptide. However, if the mixture is active the process must be repeated using A^1 but omitting a different amino acid residue from the synthesis. In the worst scenario it would mean that the 900-member library would have to be prepared ten times in order to determine the first residue of the most active tripeptide. Repeating this process of omission, combinatorial synthesis and biological testing but using groups of nine reactants for the first step will give the amino acid that occupies the second place in the peptide chain. Further repetition but using groups of nine amino acid reactants in the second step will identify the third amino acid in the chain.

Preparation of the original library	10	Ten amino acid reactants →	100	Ten amino acid reactants →	1000
The preparation of the first group of secondary libraries to find the first residue in the peptide	9	Nine amino acid reactants →	90	Ten amino acid reactants →	900
The preparation of the second group of secondary libraries to find the second residue in the peptide	10	Nine amino acid reactants →	90	Ten amino acid reactants →	900
The preparation of the third group of secondary libraries to find the third residue in the peptide	10	Ten amino acid reactants →	100	Nine amino acid reactants →	900

Figure 2.36. A schematic representation of deconvolution. The numbers represent the number of components in the mixture.

In order to be effective, deconvolution procedures require that both the synthesis and assay of the library be rapid. The procedure is complicated when there is more than one active component in the library. In this case it is necessary to prepare and test all the possible compounds indicated by deconvolution in order to identify the most active compound in the library.

2.7 Summary

Drug discovery is part luck and part structured investigation. It starts with the design team deciding on the objective of the investigation. This may be either to modify the pharmacokinetics of an existing drug or to find an entirely new drug. In the latter case a suitable structure for the lead compound is decided upon by the team after considering the biological evidence. In both cases the procedure is to synthesise analogues of either the existing drug or the lead compound for screening. It is unlikely that a lead compound will be suitable for clinical use. A number of strategies exist to help the medicinal chemist choose suitable analogues. Such strategies include SAR, QSAR, Hansch analysis, the Topliss decision tree, computer modelling and combinatorial chemistry.

The **shape of a molecule** often plays an important part in the nature of its activity. For example, stereoisomers may or may not have a similar activity and potency (Table 2.1). Consequently, the design of analogues should take into account structural features, such as the presence of structurally rigid groups in the lead compound as well as the conformation and configuration of its pharmacophore (see section 2.2).

Structure–activity relationships (SAR) are broad generalisations relating the effect of minor structural changes in a lead compound to the activity and potency of the resultant analogues (see section 2.3). These generalisations are used in drug development to predict the effect of a specific structural change in the lead on the biological activity of the resultant analogue. The results of these predictions are used, with varying degrees of success, to select the most profitable analogues for synthesis. SAR studies can also be used to identify the pharmacophore of a drug.

Quantitative structure–activity relationships (QSAR) is an attempt to relate the biological activities of compounds with similar structures to measurable parameters (see section 2.4). These parameters are a measure of the properties of a drug that have an influence on the activity of that drug. The three major influences and, in parentheses, the parameters commonly used to represent them are:

(i) **lipophilicity** (partition coefficient P and lipophilic constant π);
(ii) **steric effects** (the Taft steric constant E_s and the molar refractivity MR);
(iii) **electronic effects** (Hammett constants σ).

The mathematical relationship between one or more of these parameters and the activity of a compound is known as a **Hansch equation**. These equations are obtained from the relevant data using **regression analysis**. The value of the **regression constant (r)** for this analysis indicates whether suitable parameters have been used to derive the Hansch equation. Values of r greater than 0.9 are taken to indicate that suitable parameters have been used. Hansch equations may be used to predict the activities of compounds with similar structures to those used to determine the equation. However, predictions that differ considerably from experimentally determined values suggest that the activity of the compound is affected by factors that have not been included in the derivation of the equation. Hansch equations may also be used to indicate the relative importance of the various parameters on the activity of a drug. The smaller the value of the coefficient of a parameter in a Hansch equation, the smaller the influence that parameter has on the activities of the compounds used to obtain that equation.

The **Topliss decision tree** is a mechanical method of selecting the structures of analogues (see section 2.4.5). Its use requires the lead compound to possess an unfused aromatic ring system with or without an aliphatic side chain. In addition the biological activity of the lead and its analogues must be easily determined. The tree indicates which analogues to synthesise. It is used by comparing the activity of a specific analogue with the activity of the compound that preceded it in the tree. The result of this comparison indicates the next analogue to synthesise. This approach is particularly useful when it is not possible to make all the compounds required for a Hansch analysis.

Computer modelling (See section 2.5) is used to determine whether a structure will fit a specific target site. It can also show whether the functional groups of the drug can undergo a suitable interaction with the structure of the target site. Both stick and space fill models are used to represent the drug molecules and target sites. The structures of target sites may, if sufficient information is available, be built up from standard three-dimensional structures obtained from suitable databases. Furthermore, computer techniques have been developed to study the effect of conformational changes on the binding of the drug to its target site.

Combinatorial chemistry is used to find **lead compounds**, to carry out SAR studies and to determine which members of a series of analogues are biologically active (see section 2.6). Combinatorial chemistry is the simultaneous synthesis of large numbers of compounds known as a **combinatorial library**. A library may consist of collections of single compounds or groups of mixtures of compounds. These compounds may be synthesised on a solid support or in solution. The solid support methods are usually used to produce libraries of single compounds. The experimental procedures for the production of these libraries are based on the **Merrifield method of peptide synthesis** on resin beads. They use either the technique of **parallel synthesis** or **Furka's mix and split method**. In the former case, the coordinates of the reaction vessel gives the history of the synthesis of the product and hence its structure. However, in the latter case, sequential tagging methods are used to identify the structure of the product. These methods use either specific compounds (tags) as a code for the individual steps in the synthesis or record the history of the steps on a silicon chip.

Combinatorial synthesis in solution uses standard organic synthesis methods. It can be used to produce libraries of single compounds or groups of mixtures of compounds. Single compound libraries are produced using parallel synthesis techniques. Libraries of mixtures are formed by reacting a single compound with a mixture of building blocks using the parallel synthesis method. In the latter case the identity of the active products in a mixture may be deduced by a technique known as **deconvolution**.

2.8 Questions

(1) (a) Explain the meaning of the terms lead compound and analogue in the context of drug development.
    (b) Outline the general steps in the development of a new drug.

(2) Outline by means of suitable examples the significance that (a) structurally rigid groups, (b) conformations and (c) configuration have on the design of new drugs.

(3) State the full wording of the abbreviation 'SAR'. Describe the general way in which SAR is used to develop a drug. Illustrate the answer by reference to the changes in the activities of 4-alkylresorcinols caused by changes in the length of the 4-alkyl group.

(4) (a) Explain why chlorine and fluorine are normally the preferred halogen substituents for an SAR investigation.
    (b) What alternative halogen-containing group could be used in place of chlorine? Give one reason for the use of this group.

(5) Explain the meaning of the terms, (a) bioisostere and (b) pharmacophore.

(6) Suggest how the introduction of each of the following groups into the structure of a lead could be expected to affect the bioavailability of the resultant analogue: (a) a sulphonic acid group, (b) a methyl group and (c) a thiol group. Assume that these groups are introduced into the section of the leads structure that does not contain its pharmacophore.

(7) Outline the fundamental principle underlying the QSAR approach to drug design.

(8) Lipophilicity, shape and electron distribution all have a major influence on drug activity. State the parameters that are commonly used as a measure of these properties in the QSAR approach to drug design.

(9) (a) Describe the approach to drug design known as Hansch analysis.

(b) Phenols are antiseptics. Hansch analysis carried out on a series of phenols with the general structure A yielded the Hansch equation:

$$\text{Log}\,1/C = 1.5\pi - 0.2\sigma + 2.3$$
$$(n = 23, s = 0.13, r = 0.87)$$

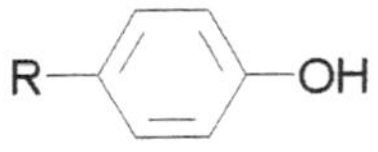

R—C_6H_4—OH

(A)

What is (a) the significance of the terms n. s and r, (b) the relative significance of the lipophilicity and electronic distribution of a phenol of type A on its activity and (c) the effect of replacing the R group of A by a more polar group.

(10) (a) Describe how the Topliss decision tree is used in drug design.

(b) What is a major limitation to its use?

(11) Outline the basic principle underlying combinatorial chemistry. What criteria should be satisfied by the building blocks used in a combinatorial synthesis?

(12) List the general considerations that should be taken into account when designing a combinatorial synthesis.

(13) Outline Merrifield's method of peptide synthesis.

(14) A linker requires the use of oxidation to release its peptides from the solid support. Comment on the suitability of this linker when preparing peptides that contain (a) alanine, (b) methionine and (c) serine.

(15) Describe Furka's mix and split technique for carrying out a combinatorial synthesis. How does this method differ from Fodor's parallel synthesis method?

(16)

$$H_2N\overset{R^1}{\overset{|}{C}}HCONH\overset{R^2}{\overset{|}{C}}HCOOH$$

(B)

Design a combinatorial synthesis for the formation of a combinatorial library of compounds with the general formula B. Outline any essential practical details.

(17) Outline the range of encoding methods used to deduce the structures of compounds produced in a Furka's mix and split combinatorial synthesis.
(18) Describe, in general terms, how the technique of deconvolution can be used to identify the most active component in a combinatorial library consisting of groups of mixtures of compounds.

3 Drug Solubility

3.1 Introduction

The solubility of drugs in the aqueous medium and lipid membranes plays a major part in their absorption and transport to their sites of action. A drug's solubility and behaviour in water are particularly important because water is the major constituent of all living matter as well as being essential for life. In living matter water acts as an inert solvent, a dispersing medium for colloidal solutions and as a nucleophilic reagent in numerous biological reactions. Normal metabolic activity can only occur when at least 65% of a cell is water. Both the solubility and ionisation of drugs in water, together with the partition of drugs between water and lipids, influence the ease of absorption into a cell membrane as well as the drug's transport from its site of administration to its site of action. In addition, the relation of the values of measurable physical constants such as solubility and partition coefficients to biological activity can be used to guide the design of new drug analogues (see section 2.4). Furthermore, hydrogen bonding and hydrophobic interactions in water influence the conformations of biological macromolecules, which in turn affect their biological behaviour.

3.1.1 The Importance of Water Solubility

Water is intimately involved in the transport of a drug from its site of administration to its site of action. The degree of water solubility of a drug is particularly important when a drug is administered orally as a solid or in suspension. In these cases the drug usually has to dissolve in the aqueous gastric fluid before it can be absorbed (see section 4.2.1) and transported via the systemic circulation to its site of action. However, poor absorption due to poor water solubility may be used in some instances to deliver the drug to the required site of action. For example, pyrantel embonate, which is used to treat pinworm and hookworm infestations of the gastrointestinal tract, is insoluble in water. This poor water solubility coupled with the polar nature of the salt (see section 4.3.3) means that the drug is poorly absorbed from the gut and so the greater part of the dose is retained in the gastrointestinal tract, the drug's site of action.

COOH, OH, N, N, S, CH_2, OH, CH_3, COOH

Pyrantel embonate **Embonate**

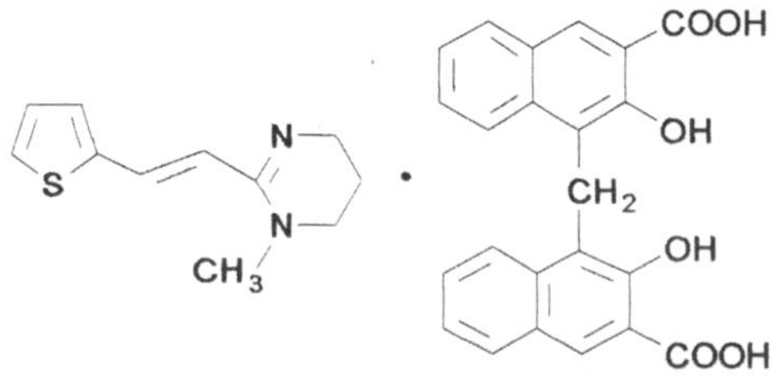

The low water solubility of a drug can also be used to produce a drug depot, produce chewable dosage forms and mask bitter tasting drugs because taste depends on the substance forming an aqueous solution. Consequently, one of the medicinal chemist's development targets for a new drug is to develop an analogue that has the required degree of water solubility for its use. The reactivity of water will also affect the stability of the drug in transit. Hydrolysis by water is one of the main routes for the metabolism of drugs containing ester, amide and other hydrolysable groups. For example, one of the metabolic pathways of the local anaesthetic lignocaine is hydrolysis to the amine.

Lignocaine $\xrightarrow[\text{Hydrolysis}]{H_2O}$ 2,6-Dimethylaniline

CH_3, CH_3, $-NHCOCH_2N$, C_2H_5, C_2H_5 ; CH_3, CH_3, $-NH_2$

3.2 The Structure of Bulk Liquids

It is believed that the molecules of a pure liquid are distributed throughout the liquid in a random fashion. Unlike a gas, the molecules of a liquid are close together and so can only travel very short distances before colliding with another molecule. Even so, there are sufficient empty spaces in the pure liquid for the molecules to move into, which gives the structure a fluid nature in keeping with the physical nature of liquids (Figure 3.1). As the molecules move through the bulk sample of the liquid they are temporarily bonded to other passing molecules by weak intermolecular forces of attraction, such as hydrogen bonds and dipole–dipole forces of attraction. It is these forces of attraction that are responsible for maintaining the molecules in the liquid state. However, these forces are not sufficient to prevent the loss of some molecules from the liquid by evaporation. In closed vessels some of the molecules that escape will also return to the liquid. In closed systems the rate of loss decreases as the concentration of the molecules in the area above the surface of the liquid increases. As a result, the system very quickly reaches an equilibrium where the rate of loss is the same as the rate at which the molecules return to the liquid. At equilibrium the pressure exerted by the escaped molecules is known as the vapour pressure of the liquid and is a constant at constant temperature for a particular liquid. Its value is of some importance in the design of chemical plant for the manufacture of drugs and other chemicals.

3.2.1 The Structure of Water

The structure of bulk samples of water has not been fully elucidated. Various models have been proposed to explain the unusual properties of water but so far there is not one model that satisfactorily explains all the physical properties of bulk samples of water. A popular simple

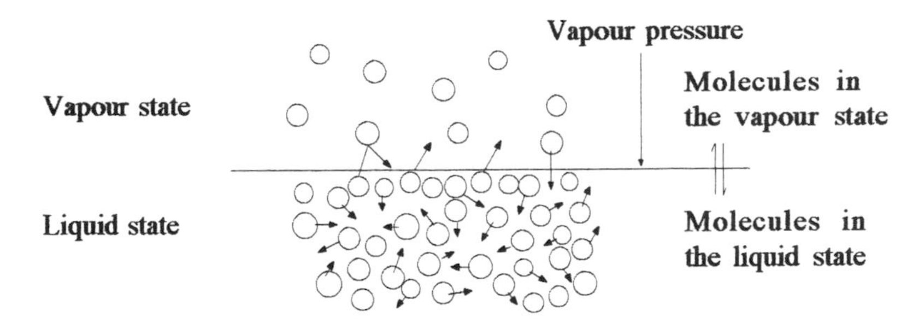

Figure 3.1. A representation of the structure of bulk liquids. The arrows indicate the random nature of the movement of the molecules of the liquid.

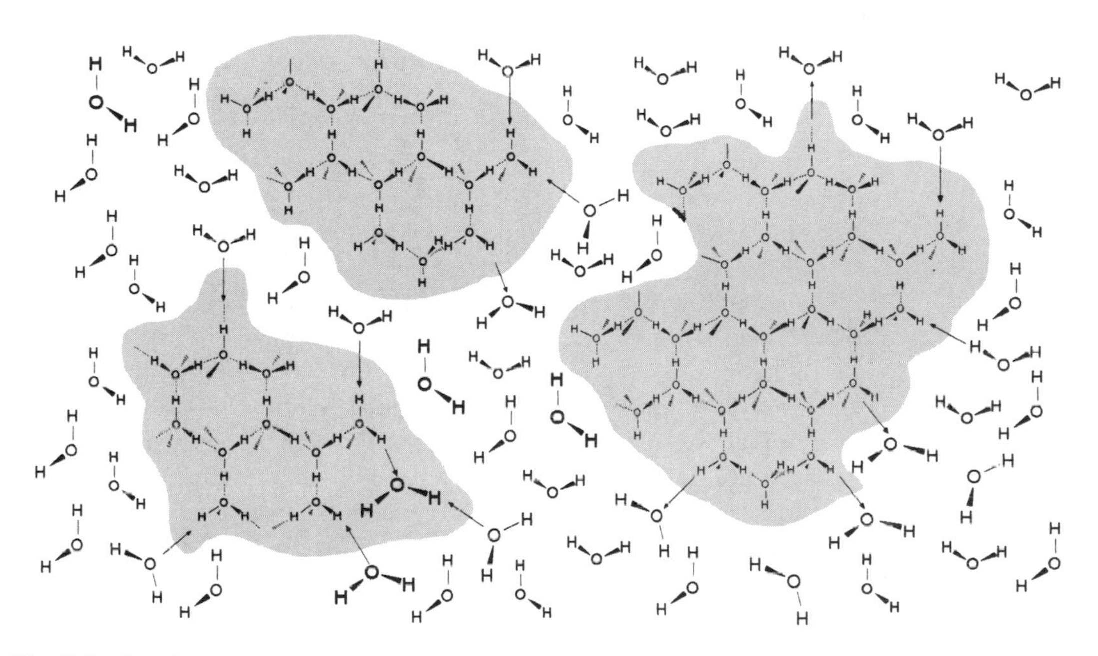

Figure 3.2. The flickering cluster model for the structure of water. Clusters are the shaded areas. The arrows show water molecules entering or leaving a cluster. Water can form a maximum of four hydrogen bonds but it is unlikely that this occurs at room temperature. It should be noted that the flickering cluster model is not accepted by all scientists.

picture is that of the **flickering cluster** model. In this model, at any one point in time, bulk samples of water consist of a haphazard array of individual non-hydrogen-bonded molecules. Randomly scattered through these unassociated water molecules are irregular clusters of hydrogen-bonded water molecules with a more ordered ice-like structure (Figure 3.2). These clusters have a short life span and almost as soon as they are formed they disintegrate and reform in new, different sized and shaped clusters by hydrogen bonding with the surrounding water molecules. The average life span of the hydrogen bonds within a cluster is estimated to be about 9.5 ps. This dynamic picture gives the structure of bulk samples of water a very fluid nature.

3.3 Solutions

A solution consists of particles, molecules or ions, usually of the order of 0.1–1 nm in size dispersed in a solvent. The small size of the solute means that it cannot be detected by the naked eye and so solutions have a uniform appearance. Solutes are often solvated in solution. **Solvation** is essentially the formation of a region around the solute particles where the solvent molecules move with the solute (see section 3.2.2.1). These solvent molecules are bound to the solute by a variety of weak attractive forces such as hydrogen bonding, van der Waals' forces and dipole–dipole interactions. Solvation is one of the factors that stabilise a solution.

The formation of a solution is normally accompanied by an increase in disorder and hence an increase in entropy. Many endothermic solvation processes are spontaneous because of the large increase in entropy that occurs when the solute dissolves.

Solutes are generally classified as either polar or non-polar. Polar solutes have permanent dipoles and so there are strong electrostatic attractive forces between their molecules and the polar water molecules (Figure 3.3), which accounts for polar compounds being considerably more water soluble than non-polar solutes. Non-polar solutes either have no dipole or one that is considerably smaller than those found in polar solutes. The attractive forces between non-polar molecules and water are likely to be weak and so non-polar molecules are more likely to be soluble in lipids than polar compounds. In practice, a rough guide to the solubility of a compound in a solvent may be predicted using the rule of thumb that ***'like dissolves like'***.

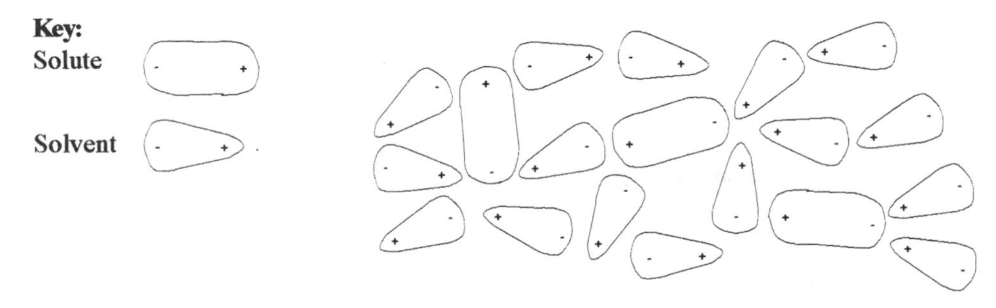

Figure 3.3. A representation of the dipole–dipole electrostatic attractions between molecules of the same and different types. Dipole–dipole attractions occur where the positive and negative ends of the dipoles are in close proximity.

A number of the physical properties of solutions vary with changes in physical conditions such as temperature and pressure. Solutions whose properties are **linearly** proportional to the concentration of the solute are referred to as **ideal solutions**. Conversely, solutions whose properties are not linearly proportional to the concentration of the solute are known as **non-ideal** solutions. This non-linear behaviour is due to the influence of intermolecular interactions in the solution. In general, the more concentrated a solution the more likely it will exhibit non-ideal behaviour. However, aparent non-ideal behaviour will also be exhibited by solutes that dissociate or associate in solution if the degrees of association and dissociation are not taken into account.

3.3.1 The Nature of Lipid Solutions

Little is known about the structure of the solutions formed when a solute dissolves in a liquid lipid. However, such solutions are believed to be formed because of hydrophobic interactions (see section 3.3.2.2), hydrogen bonding and other dipole–dipole attractive forces between the solute molecules and the lipid molecules.

3.3.2 The Nature of Aqueous Solutions

The general physical properties of aqueous solutions depend on the type of solute.

3.3.2.1 Polar Solutes

The solubility of all polar species in water is believed to be mainly due to the extra-large resultant dipole of water molecules (Figure 3.4). This large dipole allows water molecules to form dipole–dipole, ion–dipole and hydrogen bonds with solute particles. These attractive forces bind one or more layers of water molecules to the polar species to form effectively a more ordered and more stable but still temporary structure than the flickering clusters of pure water. The process of solvation is known as **hydration** and the structures formed are **hydrates**.

Ionic compounds dissolve in water because water is able to disrupt their structures and form stable hydrated ions in solution. The stability of these hydrated ions is believed to be due to the relatively high dielectric constant of water (78.54 at 25°C). Because Coulomb's law states that the force of attraction between two charges is inversely proportional to the dielectric constant of the solvent, the force of attraction between oppositely charged ions in water is small, which reduces the possibility of the ions recombining to form the unionised compound. Furthermore, the ions are stabilised by hydration. These hydrated ions are formed because of the relatively strong ion–dipole electrostatic attractive forces that exist between an ion and the surrounding water molecules. It has been suggested that there is a region (the primary region) immediately surrounding the ion where the water molecules are immobilised by the ion's electrostatic field. In this region the dipoles of the water molecules are orientated towards the ion

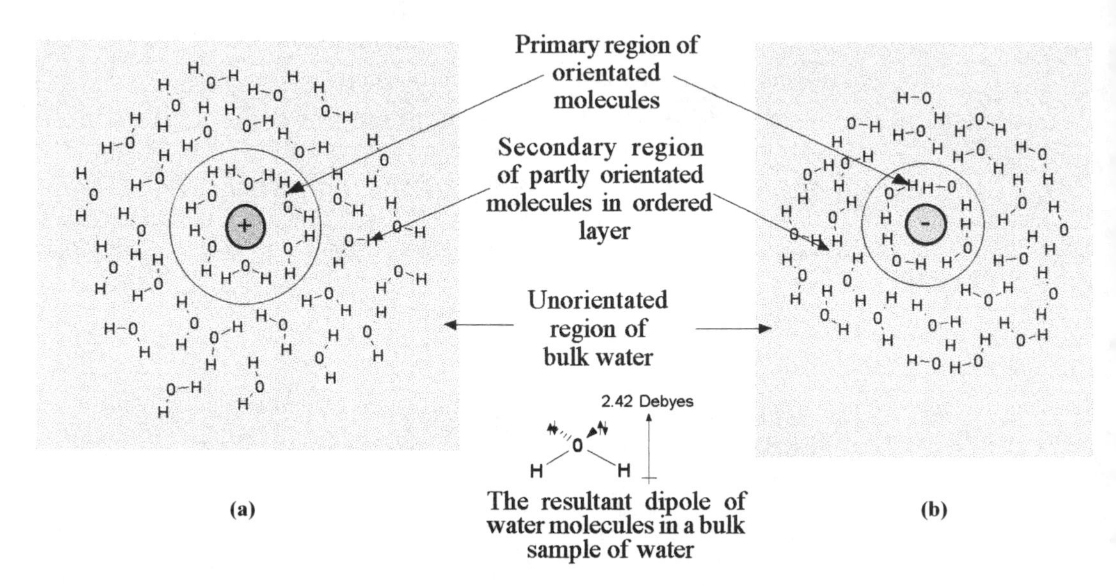

Figure 3.4. The hydration of (a) cations and (b) anions in water.

(Figure 3.4). These molecules move with the ion as it randomly moves through the water. The primary region is surrounded by the secondary region where the water molecules are attracted to the ion but do not necessarily move with the ion. Beyond this secondary region lies the bulk water region with its clusters and unorientated water molecules. The primary and secondary regions of water molecules effectively shield an ion from the other ions in solution, preventing coagulation into larger visible particles.

Hydrogen and hydroxide ions form hydrogen-bonded hydrates in dilute aqueous solution. Experimental evidence indicates that the number of hydrated water molecules associated with a hydrogen ion is at least four. However, for practical purposes H_3^+O is used for the hydrogen ion in aqueous solution. Similarly, OH^- is used for the hydroxide ion.

The solubility of non-ionic polar solutes in water is due mainly to the presence in the molecule of functional groups that can form hydrogen bonds (Figure 3.5) and dipole–dipole attractions with water molecules. Hydrogen bonding plays the major part in forming the temporary solute hydrates of these species: the greater the possibility of hydrogen bonding, the greater the degree of solubility of the solute.

3.3.2.2 Non-polar Solutes

Non-polar solutes are not very soluble in water. Thermodynamic calculations show that there is a drop in the entropy of the system when some non-polar solutes dissolve in water. Because it is more usual for entropy to increase when a solute dissolves, it has been suggested that this

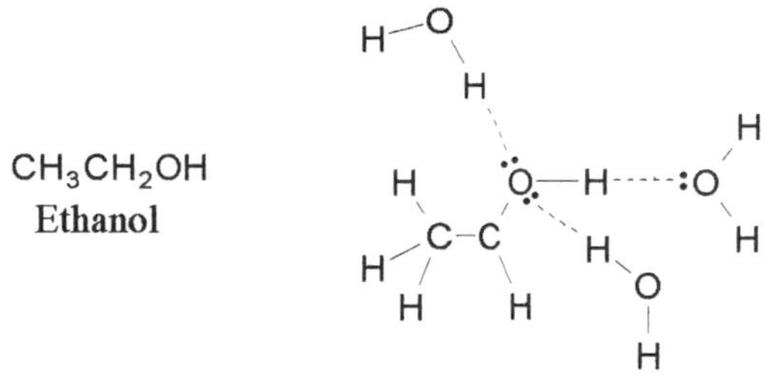

CH_3CH_2OH
Ethanol

Figure 3.5. Hydrogen bonding between water and ethanol.

drop in entropy occurs because the solute molecules and the water molecules in their immediate vicinity assume a more ordered structure. The molecules forming the solute particle are held together by van der Waals' forces (about 45% of the bonding) and attractive forces known as hydrophobic bonding (Figure 3.6).

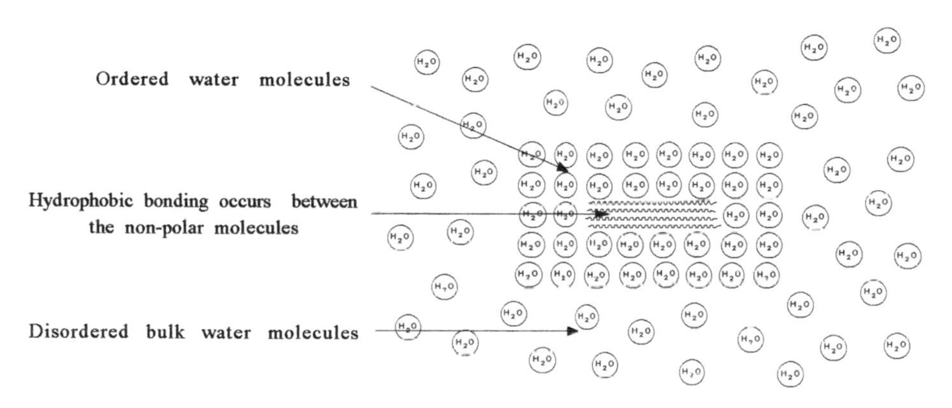

Figure 3.6. A representation of hydrophobic bonding.

3.4 Solubility

The solubility of a drug both in water and lipids is an important factor in its effectiveness as a therapeutic agent and in the design of its dosage form. It depends on both the structure of the compound (see section 3.5) and the nature of the solvent (see section 3.6). Several methods of predicting the solubility of a compound in a solvent have been proposed but none of these methods is accurate and comprehensive enough for general use. Consequently, the solubility of a drug is always determined by experiment. Because most drugs are administered at room temperature (25°C) and the body's temperature is 37°C, the solubility of drugs is usually measured and recorded at these temperatures. However, the correlation between the activities of a series of drugs with similar structures and their solubilities in one phase, normally water, is normally poor. This indicates that there are other factors playing important roles in controlling drug activity.

The solubility of a drug in a solvent is normally recorded as the amount of solute in either a specific volume or mass of solution or solvent. The units commonly used are:

(i) *Solubility*: The number of grams of solute dissolved in 100 g of solvent (% w/w). The mass (% w/v) and volume (% v/v) of solute in 100 cm^3 of solution are also used to record solubility.
(ii) *Molarity (M)*: The number of moles dissolved in one litre of solution (units: mol dm^{-3}). It should be noted that **concentration** is often measured in mol dm^{-3}.
(iii) *Molality*: The number of moles dissolved in 1000 g of solvent (units: mol kg^{-1})

Solubility measurements based on the volume of solvent or solution are temperature dependent because volume increases with increasing temperature. Consequently, these measurements are normally recorded at 25°C unless stated otherwise. However, solubilities based on unit mass are independent of temperature because mass does not change with changes in temperature.

3.4.1 Solids

The solubility of solids in all solvents is temperature dependent, usually increasing with increase in temperature. However, there are some exceptions (Figure 3.7). In medicinal chemistry the solubilities at room temperature (for drugs administered in solution) and body temperature are of primary importance. A plot of solubility against temperature for a compound is known as a solubility curve. Above the line the system consists of a saturated

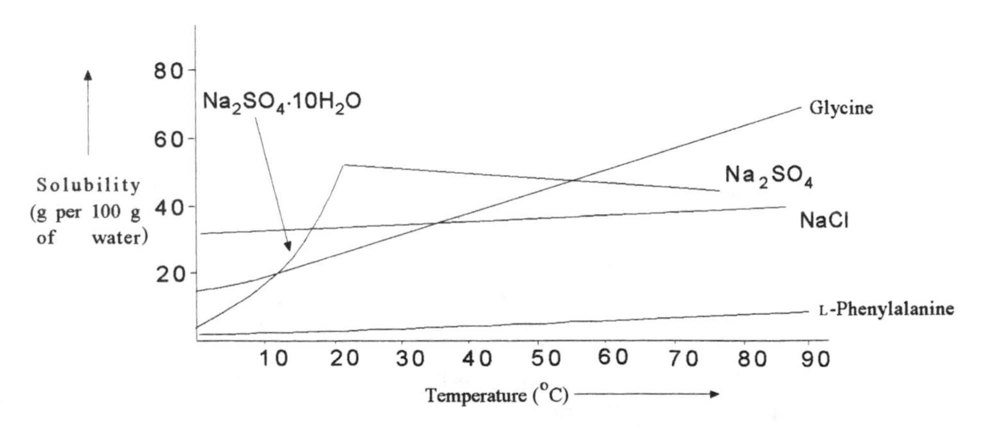

Figure 3.7. Examples of the variation of the solubility of some compounds in water with temperature.

solution in contact with undissolved solid, whereas below the line the system is a homogeneous mixture of the solute and solvent. Solubility curves are usually smooth curves and any abrupt changes in the slope of the curve indicates a change in the nature of the solute. For example, the solubility curve of sodium sulphate in water exhibits an abrupt change in slope at 32.5°C. Below 32.5°C sodium sulphate exists as the decahydrate and the dissolution is endothermic, but above 32.5°C sodium sulphate exists as the anhydrous salt and the dissolution is exothermic.

3.4.1.2 Sparingly Soluble Substances that Ionise in Water

The solubility of a sparingly soluble substance that ionises in water can be recorded in the usual solubility units (see section 3.4) but it may also be recorded in terms of its so-called **solubility product**. Consider an aqueous system where a **saturated** solution of a sparingly soluble, ionisable compound CA is in contact with solid salt.

$$CA_{(s)} \rightleftharpoons C^{+}_{(aq)} + A^{-}_{(aq)}$$

This mixture constitutes a heterogeneous equilibrium system and so, applying the law of mass action for ***dilute*** solutions at equilibrium at constant temperature:

$$K_{sp} = [C^{+}][A^{-}] \tag{3.1}$$

where K_{sp} is the equilibrium constant and is known as the solubility product of the compound. At constant temperature, the value of K_{sp} is a constant and is a measure of the limit of solubility of the compound. The term $[C^{+}][A^{-}]$ is known as the ionic product of the compound. It may be used to predict the physical nature of sparingly soluble salt–water systems. For example, if the calculated value of the ionic product for a theoretical concentration of an ionic compound exceeds that of the recorded value of K_{sp} for that compound, at equilibrium the system would consist of a saturated solution in contact with undissolved solid. If the calculated value of the ionic product is the same as the recorded value of K_{sp} the system will be a homogeneous saturated solution at equilibrium, but if the value of the ionic product is lower than the recorded value of K_{sp} it will be a homogeneous unsaturated solution of the compound.

Example 3.1. *Calculate the solubility of barium sulphate in* $g\,dm^{-3}$ *in a saturated solution at 25°C if the solubility product of barium sulphate at this temperature is* $1.08 \times 10^{-10}\,mol^{2}\,dm^{-6}$.

The equilibrium for the dissolution is:

$$BaSO_{4(solid)} \rightleftharpoons BaSO_{4(aq)} \rightleftharpoons Ba^{2+}_{(aq)} + SO^{2-}_{4(aq)}$$

Therefore, for this heterogeneous equilibrium:

$$K_{sp} = [Ba^{2+}_{(aq)}][SO^{2-}_{4(aq)}]$$

However, as barium sulphate is a strong electrolyte, all the dissolved compound will be in the form of ions. Therefore, if the solubility of a saturated solution of barium sulphate at 25°C is **Z** *mol dm*$^{-3}$*, the equation for the ionisation of the barium sulphate in solution shows that there will be* **Z** *mol dm*$^{-3}$ *of barium and* **Z** *mol dm*$^{-3}$ *of sulphate ions in solution, and so:*

$$K_{sp} = 1.08 \times 10^{-10} = \mathbf{Z}^2$$

$$\mathbf{Z} = \sqrt{1.08 \times 10^{-10}} = 1.04 \times 10^{-5}\,mol\,dm^{-3}$$

$$\mathbf{Z} = 1.04 \times 10^{-5} \times 233.4 = 2.4 \times 10^{-3}\,g\,dm^{-3}$$

Thus the solubility of barium sulphate in water is $2.4 \times 10^{-3}\,g\,dm^{-3}$.

Example 3.2. *Predict the physical nature of a mixture of* 1.18×10^{-5} *mol of barium sulphate and 1 dm*^{3} *of water. The solubility product of barium sulphate at 25°C is* 1.08×10^{-10}.

The previous example shows that the solubility of barium sulphate in water is $1.04 \times 10^{-5}\,mol\,dm^{-3}$*. The concentration of barium sulphate in the proposed mixture would be* $1.18 \times 10^{-5}\,mol\,dm^{-3}$ *if it all dissolved. This is in excess of that required for a saturated solution and so the mixture will consist of a saturated solution of barium sulphate in contact with undissolved solid.*

The concentration terms in the expression do not distinguish between the sources of an ion. Consequently, solubility product may be used to calculate the solubility of a sparingly soluble ionisable compound in a solution that contains similar ions from another source. Consider, for example, the solubility of barium sulphate in a 0.1 M sodium sulphate solution. The solubility product expression for barium sulphate dissolved in 0.1 M sodium sulphate is now:

$$K_{sp} = [Ba^{2+}][SO_4^{2-}]_{(Total)} \tag{3.2}$$

but, in this case, the total sulphate ion concentration $[SO_4^{2-}]_{(Total)}$ is given by:

$$[SO_4^{2-}]_{(Total)} = [SO_4^{2-}] \text{ from the sodium sulphate} + [SO_4^{2-}] \text{ from the barium sulphate}$$

Because sodium sulphate is a strong electrolyte it will be fully ionised in solution and so the concentration of sulphate ions in solution from a 0.1 M solution of sodium sulphate will be 0.1 mol. Therefore, if the solubility of a saturated solution of barium sulphate in a 0.1 M solution of sodium sulphate is y mol dm^{-3}, the concentrations of barium and sulphate ions from the barium sulphate in solution will both be y mol dm^{-3}. Substituting these values in Equation (3.2):

$$K_{sp} = 1.08 \times 10^{-10} = [y][y + 0.1]$$

and;

$$y^2 + 0.1y = 1.08 \times 10^{-10}$$

but y is very small compared to 0.1, therefore y^2 will be very small compared to 0.1y and so, to a first approximation:

$$0.1y = 1.08 \times 10^{-10}$$

and;

$$y = 1.08 \times 10^{-9}\,\text{mol}\,\text{dm}^{-3}$$

This calculation shows that the solubility of barium sulphate in 0.1 M aqueous sodium sulphate is less than its solubility in pure water because of the presence of sulphate ions from the sodium sulphate. The depression of the solubility of a compound by the presence of an identical ion from a different source is known as the **common ion effect**. It follows that the solubilities of all sparingly-water-soluble compounds that ionise in water will be reduced by the presence of a common ion. For example, the hydrogen ions produced in the stomach will reduce the ionisation of all sparingly-water-soluble acidic drugs in the stomach, which can improve their absorption into the bloodstream through the stomach wall by increasing the concentration of unionised drug molecules in solution. Uncharged molecules are normally transported more easily through biological membranes than charged molecules (see section 4.3.3). However, it must be realised that the degree of ionisation is not the only factor that could affect the absorption of a drug.

3.4.2 Liquids

The solubility of liquids in solvents usually increases with temperature. However, the situation is complicated by the fact that some solute–solvent systems can exist as two or more immiscible phases at certain combinations of temperature and composition. For example, the temperature–composition diagram (phase diagram) of the phenol/water system (Figure 3.8) is liquid above the line VWYZ and solid below this line. Above 66°C the mixture exists as one homogeneous liquid phase regardless of the relative concentrations of the components of the system. However, below 66°C the mixture can exist as one of two homogeneous phases depending upon its composition. The temperature of 66°C is known as the **upper critical solution temperature** (upper CST) of the phenol/water system. The CST is the minimum temperature above which a system can exist as a single homogeneous liquid phase. At other temperatures and compositions the appearance of the mixture will depend on the phase diagram of that system. In the phenol/water system, a mixture with the composition and temperature corresponding to point A is a homogeneous solution of phenol dissolved in water. The line WX is in fact the solubility curve for a solution of phenol in water. At point B the system consists of two phases: a saturated solution of water in phenol and a saturated solution of phenol in water. There will

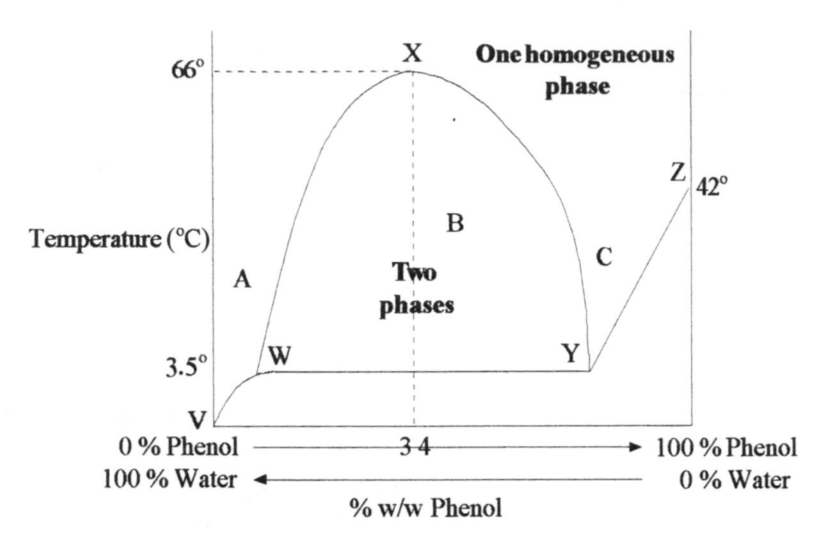

Figure 3.8. The phase diagram for the phenol/water system at 1 atmosphere.

be two liquid layers present in the containing vessel. However, at point C the mixture is now a single homogeneous phase consisting of a solution of water in phenol. The line XY represents the solubility curve of water in phenol.

In some liquid-solvent systems the mixture exists as a homogeneous liquid phase below a certain minimum temperature (Figure 3.9a). This minimum temperature is known as the **lower critical solution temperature** (lower CST). Above the lower CST the system consists of one or two phases depending on its temperature and composition. In the triethylamine/water system, for example, at the points marked A the system is a single phase consisting of a solution of triethylamine in water. However, at point B the system consists of two phases: a solution of triethylamine in water and a solution of water in triethylamine.

A number of systems are known to exhibit both a lower and an upper CST (Figure 3.9b). However, some systems that should exhibit both an upper and a lower CST do not exhibit the latter because a change of state occurs before the relevant CST is reached. For example, the ether/water system does not show a lower CST because water freezes before this temperature is reached.

The CST is sensitive to the presence of added substances, including impurities, that dissolve in both liquid phases. The CST may be raised or lowered depending on the relative solubilities of the added substance in the liquids comprising the original system. For example, the addition of 1% of a soap reduces the CST of the phenol/water system to 25°C, whereas the addition of 1% of potassium chloride raises the CST to 78°C. This type of behaviour can have a considerable effect on the design of liquid dosage forms.

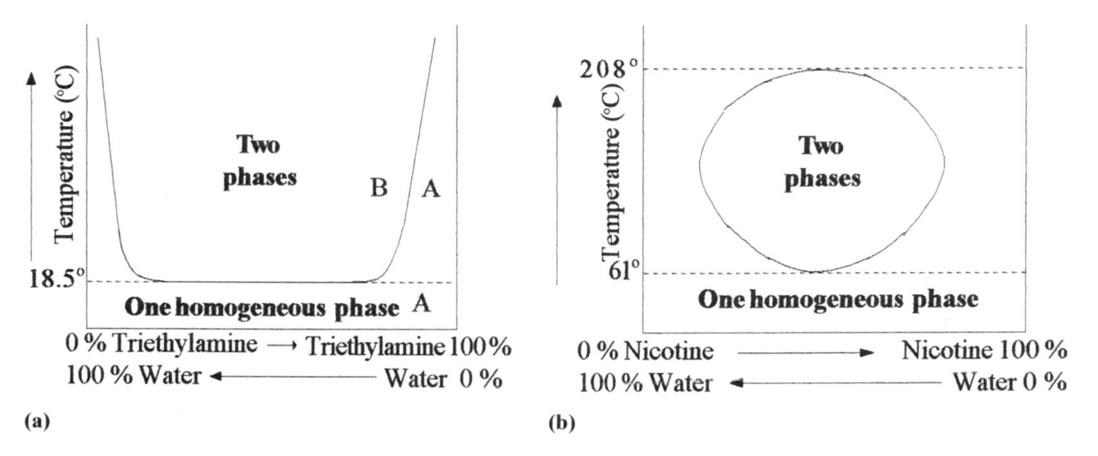

Figure 3.9. The phase diagrams for (a) the triethylamine/water system and (b) the nicotine/water system at 1 atmosphere.

3.4.3 The Solubility of Gases

The solubility of a gas in a liquid depends on the temperature and pressure of the gas, its structure (see section 3.5) and the nature of the solvent (see section 3.9). However, the rate at which a gas dissolves is also dependent on the surface area of the liquid in contact with the gas; the greater the surface area, the greater the rate of solution.

The solution of a gas in water is almost always an exothermic process:

$$\underset{\text{(undissolved)}}{\text{Gas}} + \text{Liquid} \rightleftharpoons \text{Solution} \quad \Delta H \text{ is negative}$$

Applying Le Chatelier's principle to this process, an increase in temperature moves the equilibrium to the left, which means that as the temperature rises the solubility of almost all gases decreases. This decrease in solubility explains why air and carbon-dioxide-free water can be prepared by boiling the water to remove these gases.

At constant temperature the solubility (C_g) of a gas that does not react with the liquid is directly proportional to its partial pressure (P_g). This relationship is expressed mathematically by Henry's Law:

$$C_g = K_g P_g \tag{3.3}$$

where K_g is a constant at constant temperature. The value of K_g is a characteristic property of the gas.

Henry's Law applies separately to each of the components of a mixture of gases and not the mixture as a whole. It is obeyed by many gases over a wide range of pressures. However, gases that react with the liquid usually show wide deviations from Henry's Law. For example, at constant temperature sparingly soluble gases such as hydrogen, nitrogen and oxygen dissolve in water according to Henry's Law but very soluble gases such as ammonia and hydrogen chloride, which react with water, do not obey Henry's Law. Sparingly soluble gases that slightly react with water show small deviations from Henry's Law.

A consequence of Henry's Law is that the concentration of a gas dissolved in a biological fluid is often expressed in terms of its partial pressure as against the more conventional mass-based units (see section 3.4). For example, in healthy people the partial pressure of oxygen (pO_2) in arterial blood is about 100 mmHg, whereas in venous blood it is about 40 mmHg.

The increase in the solubility of a gas with increase in pressure is the basis of hyperbaric medicine. In some situations the cells of the body cannot obtain sufficient oxygen even when the patient breathes in pure oxygen. For example, in carbon monoxide poisoning the carbon monoxide binds to the haemoglobin, which prevents it taking up oxygen in the lungs. As a result, the tissues are starved of oxygen and the patient could die. To alleviate this condition pure oxygen is administered under pressure to the patient. The increase in pressure results in oxygen directly dissolving in the plasma, which keeps the tissues sufficiently supplied for recovery to occur. However, pure oxygen is also highly toxic and must only be administered under strictly controlled conditions.

3.5 Solubility and the Structure of the Solute

The solubility of a compound depends on its degree of solvation in the solvent. Structural features in a solute molecule that improve the degree of solvation will result in a more soluble solute. Both water and lipid solubility are important in drug action. However, a reasonable degree of water solubility is normally regarded as an essential requirement for a potential drug. Not only is water solubility usually essential for drug action but it also makes drug toxicity testing, bioavailability evaluation and clinical application easier. The production of compounds with the required degree of water solubility early in the development of a new drug can considerably reduce the overall cost of development because it does not cause any delays in the later stages of development. If it was found to be necessary to produce a water-soluble analogue of a lead compound at a later stage in the development the new analogue would have to be put through the same comprehensive testing procedure as the lead compound or the lead compound would have to be formulated in such a way that the problem of its water solubility is overcome. In the former case this would necessitate the repeating of expensive toxicity and bioavailability trials, whereas the latter could result in an expensive delay in the trials programme and possibly production.

Polar groups such as hydroxy, amine and aldehyde groups, which form hydrogen bonds with water molecules, and functional groups that ionise in water will improve the water solubility of a compound because these functional groups are more able to form hydrates with water molecules (see section 3.3.2.1). However, water solubility also depends on the size and nature of the carbon–hydrogen skeleton of the compound. The higher the ratio of carbon atoms to polar groups in the molecule, the lower the water solubility of the compound. Furthermore, aromatic compounds tend to be less soluble in water than the corresponding non-aromatic compounds. Compounds whose structures do not contain polar groups have a low water solubility. Using these general rules it is possible to predict the relative solubilities of compounds with similar carbon–hydrogen skeletons. However, the more complex the structure, the less accurate are these predictions.

The water solubility of a lead compound can be improved by three general methods: salt formation (see section 3.6); by incorporating water-solubilising groups into its structure (see section 3.7); and the use of special dosage forms (see section 3.8). In salt formation, the activity of the drug is normally unchanged although its potency may be different. However, when new structural groups are incorporated into the structure of a drug the activity of the drug could be changed. Consequently, it will be necessary to carry out a full trials programme on the new analogue. Both of these modifications can be a costly process if they have to be carried out at a late stage in drug development. The use of specialised dosage forms does not usually need extensive additions to the trials programme but these formulation methods are only suitable for use with some drugs. However, in some circumstances it should be noted that poor water solubility is a desirable property for a drug (see section 3.6.1).

3.6 Salt Formation

Salt formation usually improves the water solubility of acidic and basic drugs. This improvement occurs because the salts of these drugs dissociate in water to produce hydrated ions (see section 3.3.2.1). Acidic drugs are usually converted to their metallic or amino salts, whereas the salts of organic acids are normally used for basic drugs (Table 3.1).

The degree of water solubility depends on the structure of the acid or base used to form the salt. For example, the salts of the acids lactic, citric and tartaric and the base diethanolamine (Figure 3.10) contain functional groups that can form hydrogen bonds with water, which enhances their water solubility. Consequently, acids and bases that can form hydrogen bonds with water would form salts that are normally more soluble in water than the salts of acids and bases that cannot form hydrogen bonds. However, if a drug is too water soluble it will not dissolve in lipids, which normally results in either its activity being reduced or the time for its onset of action being increased. It should be noted that the presence of a high concentration

Table 3.1. Examples of the acids and bases used to form the salts of drugs.

Anions and anion sources	Cations and cation sources
Ethanoic acid ~ ethanoate (CH_3COO^-)	Sodium ~ sodium ion (Na^+)
Citric acid ~ citrate (see Fig. 3.10)	Calcium ~ calcium ion (Ca^{2+})
Lactic acid ~ lactate (see 3.10)	Zinc ~ zinc ion (Zn^{2+})
Tartaric acid ~ tartrate (see Fig. 3.10)	Diethanolamine ~ $R_2NH_2^+$ (see Fig. 3.10)
Hydrochloric acid ~ chloride (Cl^-)	*N*-Methylglucamine ~ $R\overset{+}{N}H_2CH_3$ (see Fig. 3.10)
Sulphuric acid ~ sulphate (SO_4^{2-})	Aminoethanol ($HN_2CH_2CH_2OH$) ~ RNH_3^+
Sulphuric acid ~ hydrogen sulphate (HSO_4^-)	

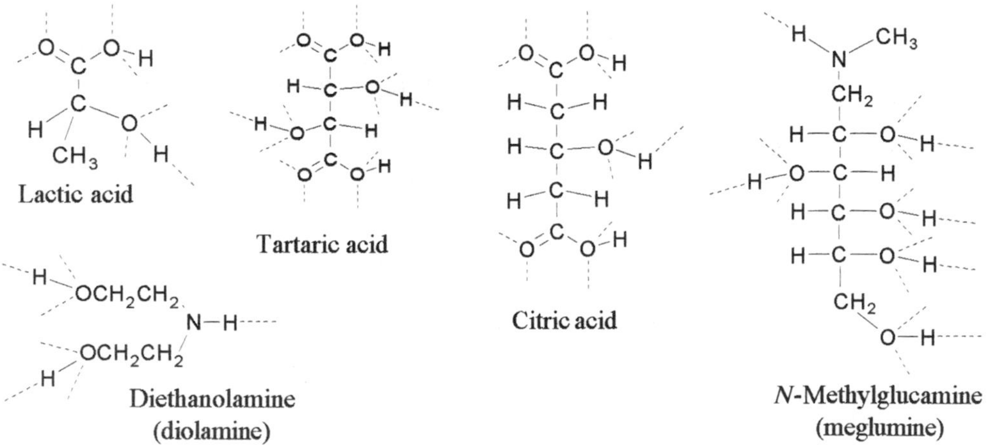

Lactic acid

Tartaric acid

Citric acid

N-Methylglucamine (meglumine)

Diethanolamine (diolamine)

Figure 3.10. Hydrogen bonding of some acids and bases with water. The possible positions of hydrogen bonds are shown by the dashed lines; lone pairs are omitted for clarity. At room temperature it is highly unlikely that all the possible hydrogen bonds will be formed. ***Note:*** hydrogen bonds are not shown for the acidic protons of the acids as these protons are donated to the base on salt formation. Similarly no hydrogen bonds are shown for the lone pairs of the amino groups because these lone pairs accept a proton in salt formation.

of chloride ions in the stomach will reduce the solubility of sparingly soluble chloride salts because of the common ion effect (see section 3.4.1.2).

Water-insoluble salts are often less active than the water-soluble salts because it is more difficult for them to reach their site of action (see section 3.2.1). However, in some cases this insolubility can be utilised in delivering the drug to its site of action (see section 3.1.1).

Some water-insoluble salts dissociate in the small intestine to liberate a component acid or base. This property has been utilised in drug delivery, for example, erythromycin stearate dissociates in the small intestine to liberate the antibiotic erythromycin, which is absorbed as the free base. Furthermore, salts with a low water solubility can be used as a drug depot. For example, penicillin G procaine has a solubility of about 0.5 g in 100 g of water. When this salt is administered as a suspension by intramuscular injection it acts as a depot by slowly releasing penicillin. Salt formation is also used to change the taste of drugs to make them more palatable to the patient. For example, the antipsychotic chlorpromazine hydrochloride is water soluble but has a very bitter taste, which is unacceptable to some patients. However, the water-insoluble embonate salt is almost tasteless and so is a useful alternative because it can be administered orally in the form of a suspension.

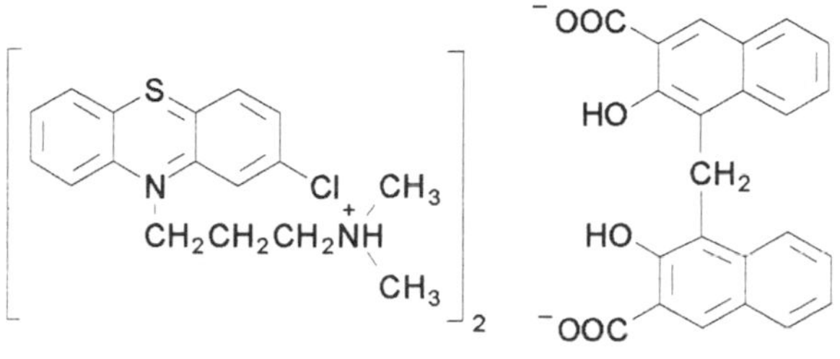

Chlorpromazine embonate

Salts dissolve by ionising in water:

$$\text{Salt} \rightleftharpoons \text{Cation} + \text{Anion}$$

Hydrogen ions will disturb this equilibrium if they combine with the anion to form a less soluble acid. Similarly, the presence of hydroxide ions can also disturb this equilibrium if they associate with the cation to form a less soluble base. Consequently, in these cases the solubility of the salt will depend on the pH of the solvent system. However, it is not possible to predict the extent of the solubility of a particular salt but, in general, increasing the hydrophilic nature of the salt should increase its water solubility. There are numerous exceptions to this generalisation and each salt should be treated on its merits.

3.7 The Incorporation of Water-solubilising Groups in a Structure

The discussion of the introduction of water-solubilising groups into the structure of a lead compound can be conveniently broken down into four general areas:

(i) the type of group introduced;
(ii) whether the introduction is reversible or irreversible;
(iii) the position of incorporation;
(iv) the chemical route of introduction.

3.7.1 The Type of Group

The incorporation of polar groups into the structure of a compound will normally result in the formation of an analogue with a better water solubility than its parent lead compound. Groups that either ionise or are capable of relatively strong intermolecular forces of attraction with water will usually result in analogues with an increased water solubility. For example, the incorporation of strongly polar alcohol, amine, amide, carboxylic acid, sulphonic acid and phosphorus oxyacid groups, which form relatively stable hydrates with water, would be expected to result in analogues that are more water soluble than those formed by the introduction of the less polar ether, aldehyde and ketonic functional groups. The introduction of acidic and basic groups is particularly useful because these groups can be used to form salts (see section 3.6), which would give a wider range of dosage forms for the final product. However, the formation of zwitterions by the introduction of either an acid group to a structure containing a base or a base group into a structure containing an acid group can reduce water solubility. Introduction of weakly polar groups such as carboxylic acid esters, aryl halides and alkyl halides will not significantly improve water solubility and can result in enhanced lipid solubility. As well as individual functional groups, multifunctional group structures such as glucose residues can also be introduced. In all cases the degree of solubility obtained by the incorporation cannot be predicted accurately because it also depends on other factors. Consequently, the type of group introduced is generally selected on the basis of previous experience.

3.7.2 Reversible and Irreversible Groups

The type of group selected also depends on the degree of permanency required. Groups that are bound directly to the carbon skeleton of the lead by less reactive C–C, C–O and C–N bonds are likely to be irreversibly attached to the lead structure. Groups that are linked to the lead by ester, amide, phosphate, sulphate and glycosidic links are more likely to be metabolised from the resulting analogue to reform the parent lead because the analogue is transferred from its point of administration to its site of action. Compounds with this type of solubilising group are acting as prodrugs (see section 9.9) and so their activity is more likely to be the same as the parent lead compound. However, the rate of loss of the solubilising group will depend on the nature of the transfer route and this could affect the activity of the drug.

3.7.3 The Position of the Water-solubilising Group

The position of the new group will depend on the reactivities of the relevant areas of the sections of the structure of the lead compound and the part of the molecule that is the pharmacophore. The former requires a knowledge of the general chemistry of the functional groups found in the lead compound. For example, if the lead structure contains aromatic ring systems it may undergo electrophilic substitution in these ring systems, whereas aldehyde groups are susceptable to oxidation, reduction, nucleophilic addition and condensation. This general reactivity is matched with the general methods used to introduce the desired water-solubilising group. The actual method selected will be governed by the location of the pharmacophore in the lead compound. In order to preserve the type of activity exhibited by the lead, the introduced water-solubilising group should be attached to a part of the structure that is not involved in the drug–receptor interaction. Consequently, the method used should not involve this region.

The incorporation of acidic residues into a lead structure is less likely to change the type of activity but can result in the analogue exhibiting haemolytic properties. Furthermore, the introduction of an aromatic acid group usually results in anti-inflammatory activity, whereas carboxylic acids with an alpha functional group may act as chelating agents. It also means that the formulation of the analogue as its salt is restricted to mainly metallic cations. This can result in a surfeit of these ions in the patient, which could be detrimental. In addition, the introduction of an acid group into drugs whose structures contain basic groups would result in zwitterion formation in solution with a possible reduction in solubility.

Basic water-solubilising groups have a tendency to change the mode of action because bases often interfere with neurotransmitters and biological processes involving amines. However, their incorporation does mean that the analogue can be formulated as a wide variety of acid salts. The introduction of a basic group into drugs whose structures contain acid groups could result in zwitterion formation in solution with a possible reduction in solubility. Non-ionisable groups do not have the disadvantages of acidic and basic groups.

3.7.4 Methods of Introduction

Water-solubilising groups may be introduced at any stage in the synthesis of a drug. The reaction used to introduce the group depends on both the type of group being introduced and the chemical nature of the target structure. Many methods involve the use of protecting groups for either the water-solubilising entity or groups already present in the substrate structure. N-Alkylations, N-acylations, O-alkylations and O-acylations of alcohols, phenols and amines are frequently used to introduce all types of water-solubilising groups into a structure (Figure 3.11). However, N-alkylation can yield mixtures of secondary and tertiary amines, their salts and the corresponding quaternary salt, depending on the type of amine alkylated.

3.7.4.1 Carboxylic Acid Groups by Alkylation

Carboxylic acid groups can be introduced by alkylation of alcohols, phenols and amines with suitably substituted acid derivatives. O-Alkylation may be achieved by a Williamson's synthesis using both hydroxy- and halo-substituted acid derivatives, whereas only halo-substituted derivatives are used for N-alkylation (Figure 3.12). In all cases, the presence of more than one reactive group in the lead structure will necessitate the use of protecting groups in order to prevent unwanted reactions.

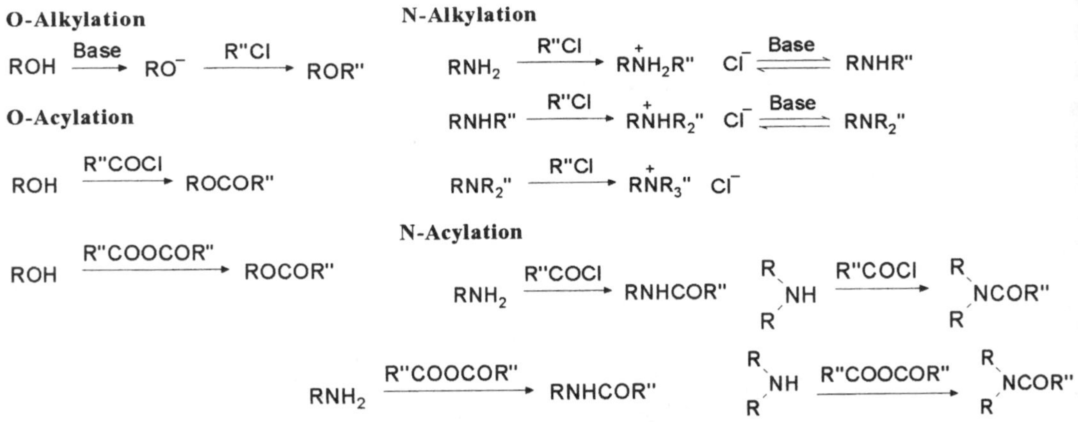

O-Alkylation

$$ROH \xrightarrow{Base} RO^- \xrightarrow{R''Cl} ROR''$$

O-Acylation

$$ROH \xrightarrow{R''COCl} ROCOR''$$

$$ROH \xrightarrow{R''COOCOR''} ROCOR''$$

N-Alkylation

$$RNH_2 \xrightarrow{R''Cl} R\overset{+}{N}H_2R''\ Cl^- \underset{}{\overset{Base}{\rightleftharpoons}} RNHR''$$

$$RNHR'' \xrightarrow{R''Cl} R\overset{+}{N}HR_2''\ Cl^- \underset{}{\overset{Base}{\rightleftharpoons}} RNR_2''$$

$$RNR_2'' \xrightarrow{R''Cl} R\overset{+}{N}R_3''\ Cl^-$$

N-Acylation

$$RNH_2 \xrightarrow{R''COCl} RNHCOR'' \qquad R_2NH \xrightarrow{R''COCl} R_2NCOR''$$

$$RNH_2 \xrightarrow{R''COOCOR''} RNHCOR'' \qquad R_2NH \xrightarrow{R''COOCOR''} R_2NCOR''$$

Figure 3.11. N- and O-Alkylations and acylations. Tertiary amines are not normally acylated. R can be either an aromatic or aliphatic system.

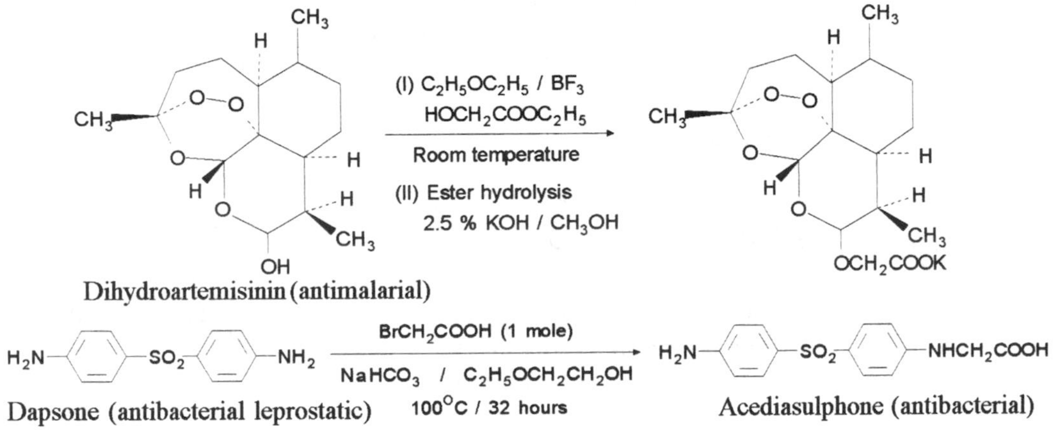

Dihydroartemisinin (antimalarial) $\xrightarrow[\text{(II) Ester hydrolysis, 2.5 \% KOH / } CH_3OH]{\text{(I) } C_2H_5OC_2H_5 / BF_3,\ HOCH_2COOC_2H_5,\ \text{Room temperature}}$ (OCH_2COOK derivative)

Dapsone (antibacterial leprostatic) $H_2N{-}C_6H_4{-}SO_2{-}C_6H_4{-}NH_2 \xrightarrow[NaHCO_3\ /\ C_2H_5OCH_2CH_2OH,\ 100^{\circ}C\ /\ 32\ hours]{BrCH_2COOH\ (1\ mole)} H_2N{-}C_6H_4{-}SO_2{-}C_6H_4{-}NHCH_2COOH$ Acediasulphone (antibacterial)

Figure 3.12. Examples of the introduction of residues containing carboxylic acid groups by alkylation of alcohols and amines.

3.7.4.2 Carboxylic Acid Groups by Acylation

Acylation of alcohols, phenols and amines with the anhydride of the appropriate dicarboxylic acid is used to introduce a side chain containing a carboxylic acid group into the lead structure. For example, succinic anhydride is used to produce chloramphenicol sodium succinate and succinylsulphathizole. However, if the lead structure contains more than one labile group it will be necessary to use protecting groups in order to prevent unwanted reactions.

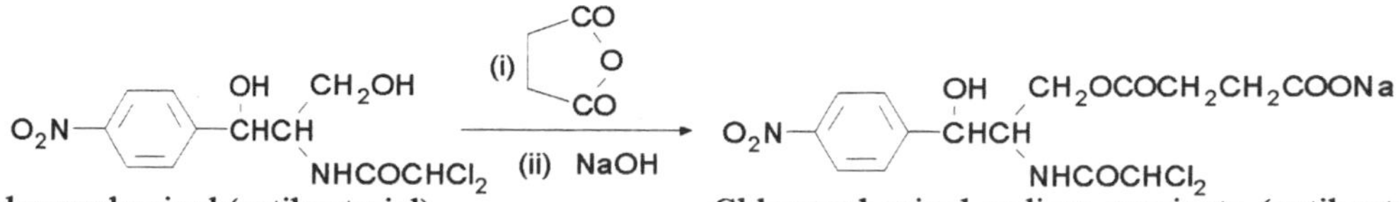

$O_2N-C_6H_4-CH(OH)CH(CH_2OH)NHCOCHCl_2$ $\xrightarrow{\text{(i) succinic anhydride, (ii) NaOH}}$ $O_2N-C_6H_4-CH(OH)CH(CH_2OCOCH_2CH_2COONa)NHCOCHCl_2$

Chloramphenicol (antibacterial) → Chloramphenicol sodium succinate (antibacterial)

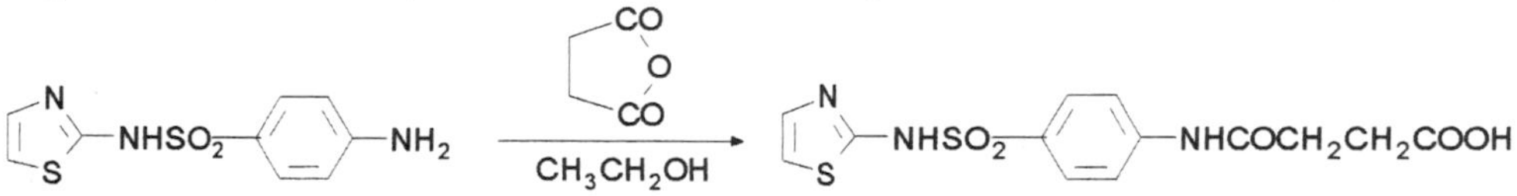

Thiazolyl$-NHSO_2-C_6H_4-NH_2$ $\xrightarrow[CH_3CH_2OH]{\text{succinic anhydride}}$ Thiazolyl$-NHSO_2-C_6H_4-NHCOCH_2CH_2COOH$

Sulphathioazole (antibacterial) → Succinylsulphathioazole (antibacterial)

The resulting acid esters and amides can be formulated as their metallic or amine salts. However, because the esters are liable to hydrolysis in aqueous solution, the stability of the resulting analogue in aqueous solution must be assessed. For example, chloramphenicol sodium succinate is so unstable in aqueous solution that it is supplied as a lyophilised powder that is only dissolved in water when it is required for use. The solution must be used within forty-eight hours.

3.7.4.3 Phosphate Groups

Phosphoric acid halide derivatives have been used successfully to introduce phosphate groups into drug structures. The hydroxy groups of the acid halide must normally be protected by a suitable protecting group (Figure 3.13). These protecting groups are removed in the final stage of the synthesis to reveal the water-solubilising phosphate ester. The resulting phosphate esters tend to be more stable in aqueous solution than carboxylic acid esters.

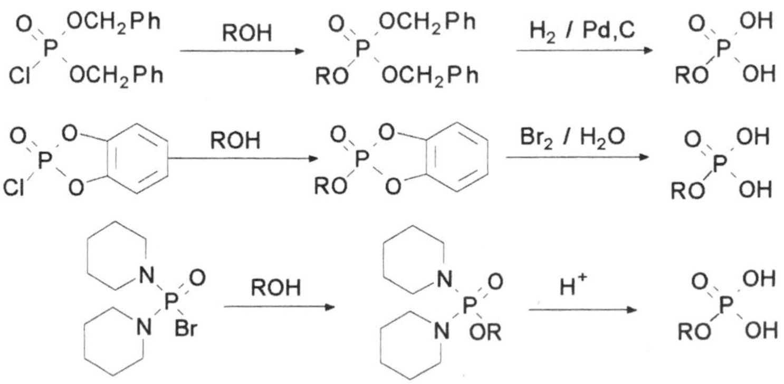

$Cl-P(=O)(OCH_2Ph)_2 \xrightarrow{ROH} RO-P(=O)(OCH_2Ph)_2 \xrightarrow{H_2\ /\ Pd,C} RO-P(=O)(OH)_2$

$Cl-P(=O)(\text{catechol}) \xrightarrow{ROH} RO-P(=O)(\text{catechol}) \xrightarrow{Br_2\ /\ H_2O} RO-P(=O)(OH)_2$

$Br-P(=O)(N(CH_2)_5)_2 \xrightarrow{ROH} RO-P(=O)(N(CH_2)_5)_2 \xrightarrow{H^+} RO-P(=O)(OH)_2$

Figure 3.13. Some general methods of phosphorylating alcohols.

3.7.4.4 Sulphonic Acid Groups

Sulphonic acid groups can be incorporated into the structures of lead compounds by direct sulphonation with concentrated sulphuric acid.

8-Hydroxyquinoline $\xrightarrow{H_2SO_4}$ (5-SO_3H derivative) $\dashrightarrow$ Various steps $\rightarrow$ 8-Hydroxy-7-iodo-5-quinolinesulphonic acid (topical antiseptic)

Alternative routes are the addition of sodium bisulphite to conjugated C═C bonds and the reaction of primary and secondary amines with sodium bisulphate and methanal.

N^4-Cinnamylidenesulphanilamide $\xrightarrow{NaHSO_3}$ Noprylsulphamide (antibacterial)

Noraminopyrine (analgesic, antipyretic) $\xrightarrow[HCHO]{NaHSO_3}$ Dipyrone (analgesic, antipyretic)

3.7.4.5 Incorporation of Basic Groups

Water-solubilising groups containing basic groups can be incorporated into a lead structure by the alkylation and acylation of alcohols, phenols and amines. Alkylation is achieved by the use of an alkyl halide whose structure contains a basic group, whereas acylation usually involves the use of acid halides and anhydrides. In both alkylation and acylation the basic groups in the precursor structure must be either unreactive or protected by suitable groups (Figure 3.14).

Amide derivatives are usually more stable than esters. Esters are often rapidly hydrolysed in serum, the reaction being catalysed by serum esterases. In these cases the analogue effectively acts as a prodrug (see section 8.9). The introduction of water-solubilising amino acid residues using standard peptide chemistry preparative methods has also been used successfully to introduce basic residues (Figure 3.15).

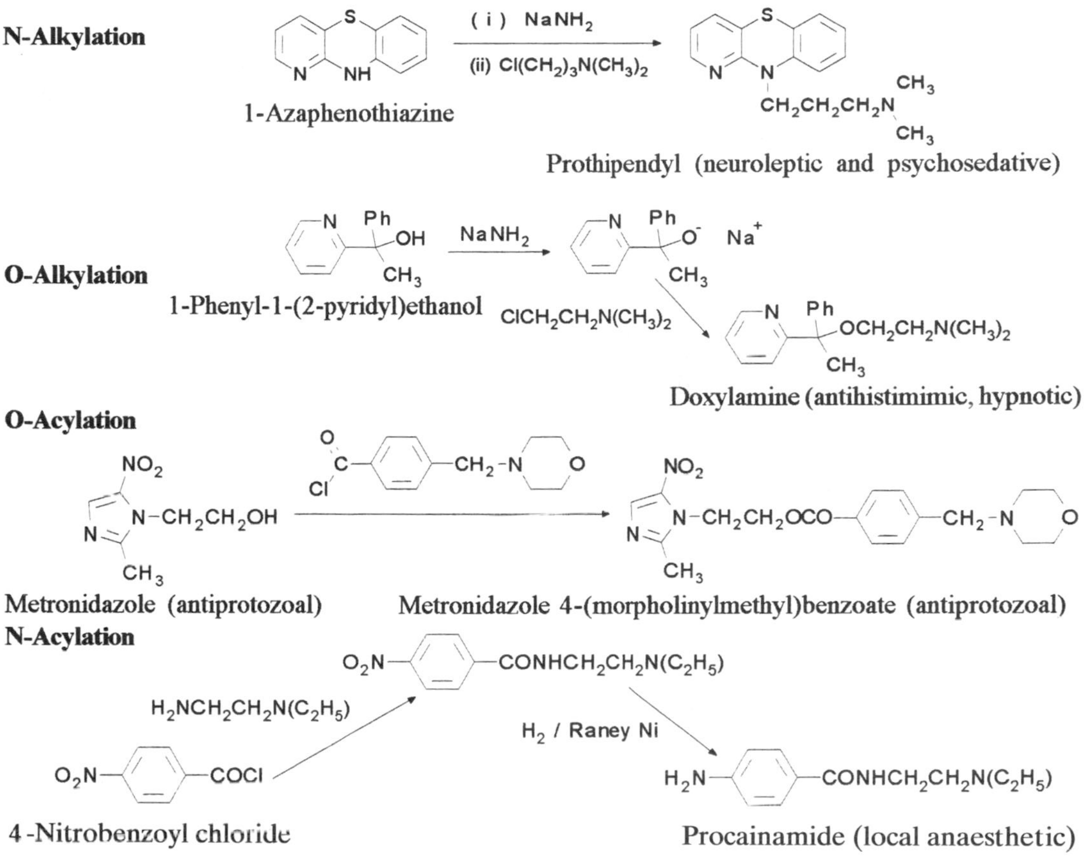

N-Alkylation

1-Azaphenothiazine — (i) $NaNH_2$; (ii) $Cl(CH_2)_3N(CH_3)_2$ → Prothipendyl (neuroleptic and psychosedative)

O-Alkylation

1-Phenyl-1-(2-pyridyl)ethanol — $NaNH_2$; $ClCH_2CH_2N(CH_3)_2$ → Doxylamine (antihistimimic, hypnotic)

O-Acylation

Metronidazole (antiprotozoal) → Metronidazole 4-(morpholinylmethyl)benzoate (antiprotozoal)

N-Acylation

4-Nitrobenzoyl chloride — $H_2NCH_2CH_2N(C_2H_5)$; H_2 / Raney Ni → Procainamide (local anaesthetic)

Figure 3.14. Examples of the introduction of basic groups by alkylation and acylation. These basic groups are used to form the more water-soluble salts of the drug, in which form it is usually administered.

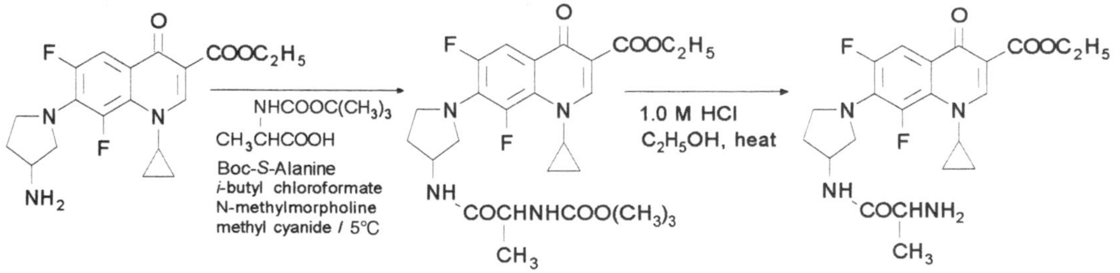

Boc-S-Alanine, *i*-butyl chloroformate, N-methylmorpholine, methyl cyanide / 5°C; 1.0 M HCl, C_2H_5OH, heat

Figure 3.15. The introduction of amino acid residues using *t*-butyloxycarbonyl (BOC) as a protecting group. The second step is the removal of the protecting group.

The Mannich reaction is also frequently used to introduce basic groups into both aromatic and non-aromatic structures.

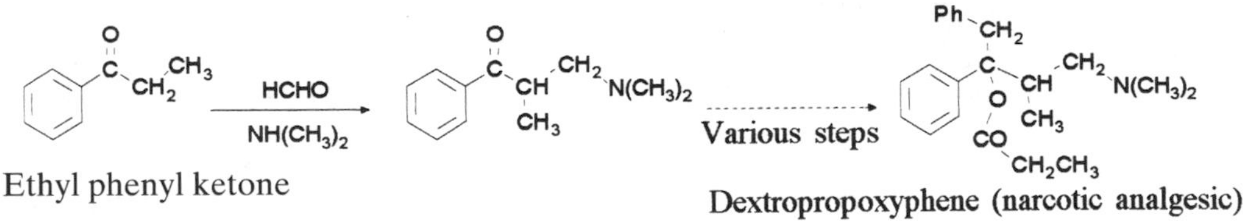

Ethyl phenyl ketone $\xrightarrow[NH(CH_3)_2]{HCHO}$ $PhCOCH(CH_3)CH_2N(CH_3)_2$ $\xrightarrow{\text{Various steps}}$ Dextropropoxyphene (narcotic analgesic)

3.7.4.6 Polyhydroxy and Ether Residues

The introduction of polyhydroxy and ether chains has been used in a number of cases to improve water solubility. 2-Hydroxyethoxy and 2,3-dihydroxypropoxy residues have been introduced by reaction of the corresponding monochlorinatedhydrin and the use of suitable epoxides, amongst other methods (Figure 3.16).

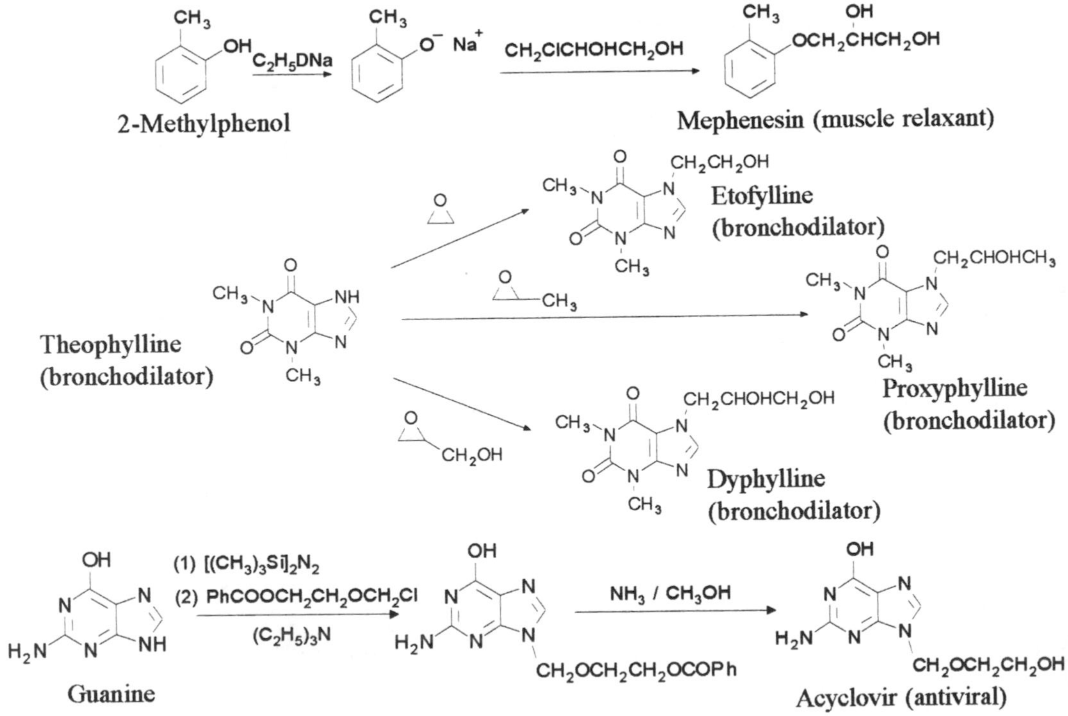

2-Methylphenol $\xrightarrow{C_2H_5ONa}$ sodium salt $\xrightarrow{CH_2ClCHOHCH_2OH}$ Mephenesin (muscle relaxant)

Theophylline (bronchodilator) + ethylene oxide → Etofylline (bronchodilator)

Theophylline + propylene oxide → Proxyphylline (bronchodilator)

Theophylline + glycidol (CH_2OH epoxide) → Dyphylline (bronchodilator)

Guanine $\xrightarrow[(C_2H_5)_3N]{(1)\ [(CH_3)_3Si]_2N_2\ \ (2)\ PhCOOCH_2CH_2OCH_2Cl}$ $CH_2OCH_2CH_2OCOPh$ derivative $\xrightarrow{NH_3\ /\ CH_3OH}$ Acyclovir (antiviral)

Figure 3.16. Examples of the methods of introducing polyhydroxy and ether groups into a lead structure.

Sugar residues as water-solubilising groups are rarely incorporated by O-glycosidic links but they have been attached to a number of drugs through N-glycosidic links involving the

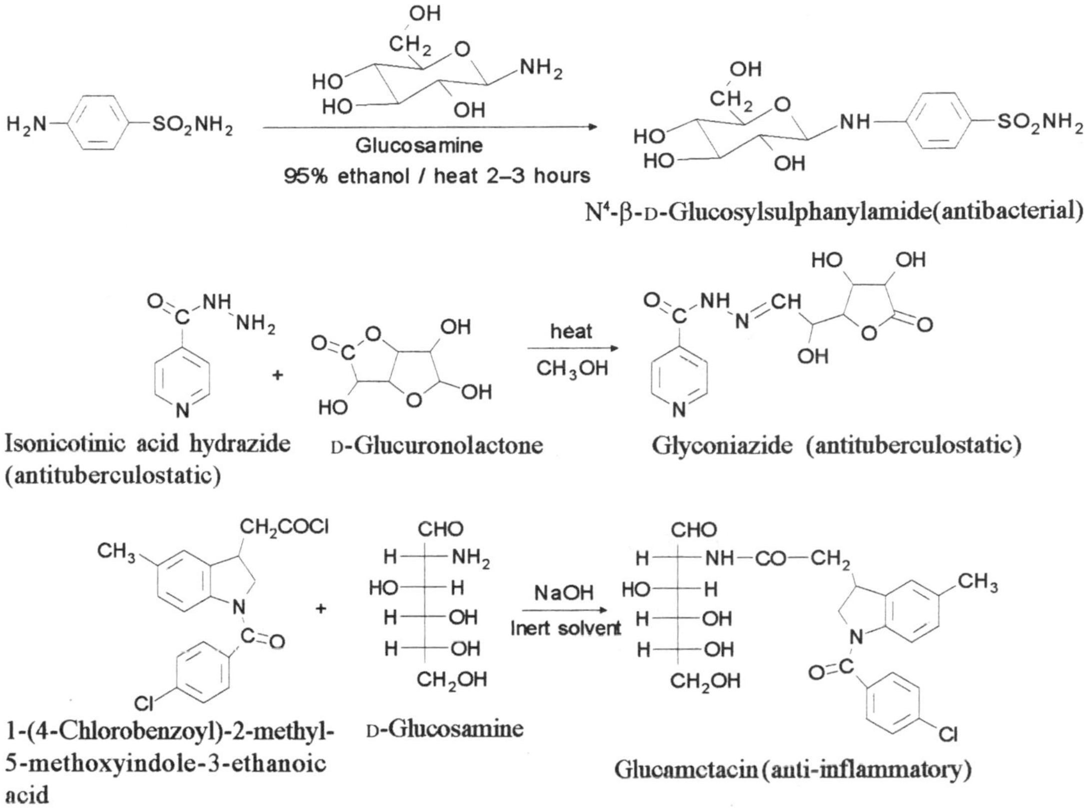

Glucosamine
95% ethanol / heat 2–3 hours

N[4]-β-D-Glucosylsulphanylamide(antibacterial)

heat
CH_3OH

Isonicotinic acid hydrazide (antituberculostatic) D-Glucuronolactone Glyconiazide (antituberculostatic)

NaOH
Inert solvent

1-(4-Chlorobenzoyl)-2-methyl-5-methoxyindole-3-ethanoic acid D-Glucosamine Glucametacin (anti-inflammatory)

Figure 3.17. Examples of the use of sugar residues to increase water solubilities of drugs.

nitrogen atoms of amines and hydrazide groups (Figure 3.17) as well as N-acylation of amino sugars by suitable acid halides.

3.8 Formulation Methods of Improving Water Solubility

The delivery of water-insoluble or sparingly soluble drugs to their site of action can be improved by the formulation of the dosage form. Techniques include the use of cosolvents, colloidal particles, micelles (see sections 3.11.1.1 and 3.11.1.2), liposomes (see section 3.11.1.3) and complexing with water-soluble compounds such as the cyclodextrans (see section 1.7).

3.8.1 Cosolvents

The addition of a water-soluble solvent (cosolvent) can improve the water solubility of a sparingly soluble compound. Cosolvents for pharmaceutical use should have minimal toxic effects and not affect the stability of the drug. Consequently, the concentration of the cosolvent used must be within the acceptable degree of toxicity associated with that cosolvent. In pharmaceutical preparations, the choice of solvents is usually limited to alcohols such as ethanol, propan-2-ol, 1,2-dihydroxypropane, glycerol and sorbitol and some low-molecular-mass polyethylene glycols; however, other solvents are sometimes used. Mixtures of cosolvents are also used to achieve the required degree of solubility. For example, paracetamol is formulated as an elixir in an aqueous sucrose solution by the use of a mixture of ethanol and 1,2-dihydroxypropane.

3.8.2 Colloidal Solutions

Sparingly-water-soluble drugs and potential drugs can be dispersed in an aqueous medium as colloidal-sized particles with diameters of 1–1000 nm. Colloidal solutions are known generally as **sols**. Water-based colloids are often referred to as **hydrosols**. Hydrosols offer a potential method for the formulation of drugs that are sparingly soluble in water. For example, hydrosols of the poorly-water-soluble drugs cyclosporin and isradipine have been prepared and found to be stable for five or more days. Hydrosols stored as spray-dried powders have been successfully reconstituted as liquid sols after several years of storage under cool dry conditions.

Hydrosols used for parenteral delivery systems normally contain colloidal particles with a diameter of less than 200 nm because colloidal particles below this size will not block small capillaries. They may be prepared by dissolving a high concentration of the drug in an organic solvent that is miscible with water. This concentrated solution is rapidly mixed with an aqueous solution containing a suitable stabiliser. These stabilisers may be either an electrolytic salt or a polymer. Both of these types of stabiliser act by being adsorbed onto the surface of the colloidal particles. In the case of electrolytic stabilisers the colloidal particles gain an electrostatic charge that repels other colloidal particles, thus preventing them from coagulating into particles large enough to precipitate. Polymer stabilisers prevent coagulation by steric hindrance. The degree of stabilisation will depend on the nature of both the hydrosol and the stabiliser. However, all hydrosols will slowly crystallise and because of this are normally stored as either freeze-dried or spray-dried powders.

Colloidal solutions can also be prepared by mechanically grinding larger particles into colloidal-sized particles. This process is not very efficient: in most cases only about 5% of the material is converted into a sol. It also produces particles with very different sizes.

Sparingly-water-soluble drugs may also be delivered using so-called **nanoparticles**. These are colloidal particles that are either solid (nanospheres) or hollow (nanocapsules). The drug is either adsorbed on the exterior or trapped in the interior of both types of nanoparticle. To date they have been used only for experimental and clinical testing.

Nanoparticles are formed by a variety of methods using a wide range of materials, for example, polysaccharides, proteins, polyacrylates and polyamides. The wide range of precursors means that nanoparticles may be produced with properties that can be used to test specific aspects of drug activity.

3.8.3 Emulsions

Emulsions are systems in which one liquid is dispersed as fine droplets with diameters of 0.1–100 μm. in a second liquid. Dosage forms usually consist of either oil-in-water (o/w), where the dispersed medium is the oil, or water-in-oil (w/o), where water is the dispersed phase. More complex systems where water droplets are encased in oil drops dispersed in water (w/o/w) or where oil droplets are encased in water droplets dispersed in an oil medium (o/w/o) are also known. These emulsion systems are intrinsically unstable and so the formation of a stable emulsion normally requires additional components known as emulsifying agents. These are surfactants (see section 3.11) such as glyceryl monostearate and polyoxyethylene sorbitan monooleate (Tween 80), which dissolve in both water and oils. Emulsions are frequently made using mixtures of emulsifying agents because this has been found to give more stable mixtures than those prepared using a single emulsifying agent.

All types of emulsion can be used as a delivery vehicle for liquid drugs that are not very soluble in water. However, this form of drug delivery is not widely used, although there is some evidence that administering some drugs as emulsions improves their bioavailability. For example, griseofulvin orally administered to rats is more readily absorbed from a corn oil-in-water emulsion than other types of orally administered dosage forms (Figure 3.18).

3.9 The Effect of the pH on the Solubility of Acidic and Basic Drugs

Aqueous biological fluids are complex systems that contain a variety of different solutes. These species will have an effect on each other's relative solubilities and the solubilities of drugs or xenobiotics introduced into the fluid. They also control the pH of the fluid, which is of considerable importance to the bioavailability of acidic and basic drugs. An acidic or

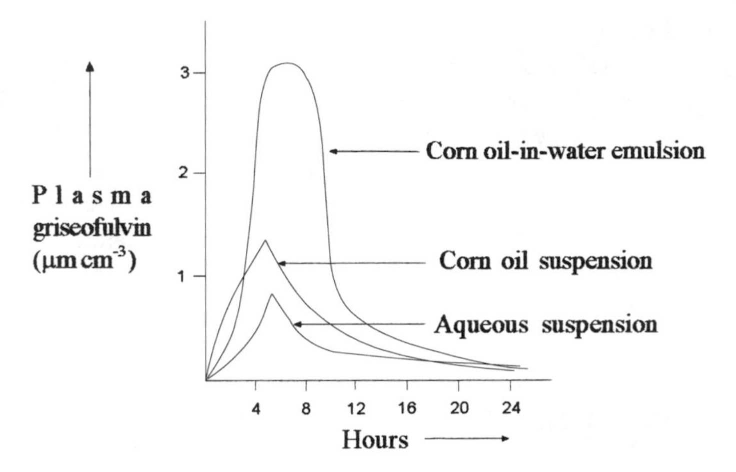

Figure 3.18. The change in plasma concentrations of griseofulvin after administration using different dosage forms. Reprinted by permission of Liley-Liss, Inc., a subsidiary of John Wiley & Sons, Ltd., from P. J. Carrigan and T. R. Bates, Biopharmaceutics of drugs administered in lipid-containing dosage forms. I: GI absorption of griseofulvin from an oil-in-water emulsion in the rat. *Jarmal of Pharmaceutical Sciences*, **62,** 1476 (1973).

basic pH will either enhance or reduce the ionisation of these drugs, with subsequent changes in their solubility and absorption through membranes. The degree of ionisation of weak monobasic acidic drugs at different pH values may be calculated using the Henderson–Hasselbalch equation:

$$pK_a = pH + \log\frac{[\text{Non-ionised form}]}{[\text{Ionised form}]} \tag{3.4}$$

Example 3.3. *The pK_a of aspirin, a weak acid, is 3.5. Calculate the degree of ionisation of aspirin in the (i) stomach and (ii) intestine if the pH of the contents of the stomach is 1 and the pH of the contents of the intestine is 6.*

Substitute the values of pH and pK_a into Equation (3.4) for each of the two problems:

(i) $3.5 = 1 + \log\frac{[\textit{Unionised form}]}{[\textit{Ionised form}]}$

therefore:

$$\log\frac{[\textit{Unionised form}]}{[\textit{Ionised form}]} = 3.5 - 1 = 2.5$$

(ii) $3.5 = 6 + \log\frac{[\textit{Unionised form}]}{[\textit{Ionised form}]}$

therefore:

$$\log\frac{[\textit{Unionised form}]}{[\textit{Ionised form}]} = 3.5 - 6 = -2.5$$

and

$$\frac{[Unionised\ form]}{[Ionised\ form]} = antilog\ 2.5$$

$$= 316.23$$

and

$$\frac{[Unionised\ form]}{[Ionised\ form]} = antilog\ -2.5$$

$$= \frac{1}{antilog\ 2.5}$$

$$= \frac{1}{316.23}$$

These figures show that aspirin is only slightly ionised in the stomach (one ionised molecule for every 316 unionised molecules) but is almost completely ionised (316 ionised molecules for every one unionised molecule) in the intestine. As a result, under the pH conditions specified in Example 3.3 aspirin will be absorbed more readily in the stomach than in the intestine because drugs are more easily transferred through a membrane in their unionised form. The value from (i) also accounts for the relative ease of aspirin absorption from the stomach even though aspirin is almost insoluble in water at 37°C. However, the degree of ionisation is not the only factor influencing a drug's absorption and although it may offer a good explanation for the behaviour of one drug it does not explain the rate and ease of absorption of all drugs. Other factors may be more significant. For example, the weakly acidic drugs thiopentone, secobarbitone and barbitone (barbital) have pK_a values of 7.6, 7.9 and 7.8, respectively. Consequently, their degrees of ionisation in water are almost the same but their rates of absorption from the stomach are very different. This indicates that other factors besides ionisation influence the transport of these drugs through the stomach membrane. This observation is supported by the fact that the partition coefficients (see section 3.7) for the chloroform/water system of these drugs are considerably different (thiopentone > 100, secobarbitone 23 and barbitone 0.7).

The degree of ionisation of weak monoacidic bases at different pH values may also be calculated in a similar manner using the Henderson–Hasselbalch equation for the monoacidic bases:

$$pK_a = pH + \log\frac{[\text{Ionised form}]}{[\text{Non-ionised form}]} \tag{3.5}$$

The presence of soluble acids and bases in biological fluids can result in the formation of the corresponding salts of acidic and basic drugs, which usually have different solubilities to that of the parent drug. Salt formation is frequently used to enhance the water solubility of a sparingly soluble drug (see section 3.5.1). However, salt formation does not always result in an increase in solubility.

The solubilities of compounds such as peptides and proteins that contain both acidic and basic groups is complicated by internal salt formation. Peptides and proteins have their lowest

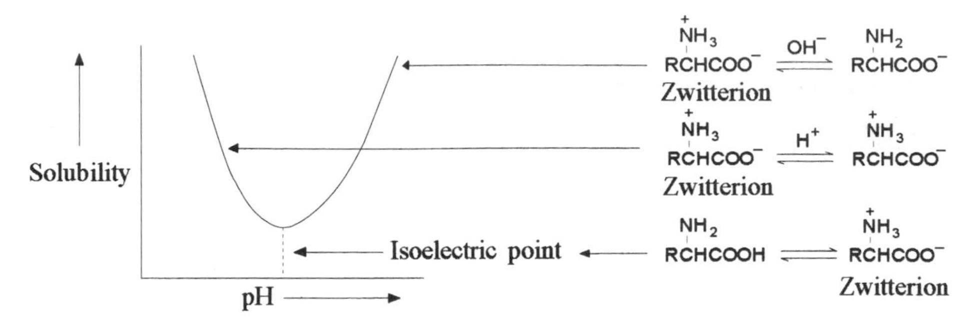

Figure 3.19. A typical solubility curve for a peptide. R represents the rest of the structure of the peptide or protein. It should be realised that the acidic and basic groups do not have to be adjacent to each other in a peptide or protein and that these compounds usually have a number of free acid and base groups.

solubilities near their isoelectric points (Figure 3.19) where the solution contains the internal salt (zwitterion). On the lower pH value side of the isoelectric point a peptide or protein will form the corresponding cation, whereas on the higher pH value side they will form the corresponding anion. Both the cationic and anionic forms of the peptide will exhibit a higher solubility than its zwitterion. These variations of solubility will effect the therapeutic effectiveness of the peptide and so will influence the design of their dosage forms.

Biological fluids, such as plasma, contain dissolved electrolytes and non-electrolytes. The presence of an electrolyte in an aqueous solution will usually increase the solubilities of any other electrolytes in solution provided that the electrolytes do not have any ions in common (see section 3.4.1.2). However, the presence of electrolytes reduces the solubility of non-electrolytes in aqueous solution. The cations and anions from the electrolyte form stronger bonds with the water molecules and form hydrates more readily with water than the non-electrolyte molecules. As a result, the ions displace the non-electrolyte molecules from their weaker hydrates with a subsequent reduction in the solubility of the non-electrolyte.

$$\text{Non-electrolyte hydrate} + \text{Ion} \rightarrow \text{Ion hydrate} + \text{Non-electrolyte}$$

If sufficient electrolyte is added to an aqueous solution the non-electrolyte is precipitated out of solution. The process is referred to as 'salting out'. Peptides and proteins are especially sensitive to salting out and the technique is often used to isolate these compounds from solution. Consequently, the solubility of peptide drugs in electrolytes is a factor that needs to be considered in the design of the dosage forms used in trials and with patients. The presence of dissolved non-electrolytes reduces the solubility of electrolytes. This is because the presence of the non-electrolyte reduces the dielectric constant of the system (see section 3.3.2.1). This reduces the degree of ionisation of the electrolyte, which results in a corresponding decrease in the solubility of the electrolyte.

3.10 Partition

The transport of a drug to its site of action normally involves the drug having to pass through a large number of lipid membranes. Consequently, the relative solubilities of a drug in the aqueous medium and lipids are of considerable importance in the transport of that drug to its site of action, especially at the aqueous medium/lipid interface. Partition (distribution) coefficients are a measure of the way a compound distributes itself between two immiscible solvents and so attempts have been made to correlate the activities of drugs with their lipid/water partition coefficients. These correlations have been used with some degree of success to predict the activities of compounds that could be potential drugs. However, the results are only valid in situations where solubility and transport by diffusion through a membrane are the main factors controlling drug action.

It is not easy to measure partition coefficients *in situ* and so instead the less accurate organic solvent/aqueous solution model systems are used. The partition coefficients of these model systems are usually calculated assuming ideal solutions, using Equation (3.6).

$$\text{The partition coefficient (P)} = \frac{[\text{Drug in the organic phase}]}{[\text{Drug in an aqueous phase}]} \qquad (3.6)$$

The partition coefficient P is a constant for the specified system provided that the temperature is kept constant and ideal dilute solutions are used. However, a 5°C change in temperature does not normally cause a significant change in the value of the partition coefficient. Because the charged form of a drug is not easily transferred through a membrane, a more useful form of Equation (3.6) for biological investigations is:

$$\text{P} = \frac{[\text{Non-ionised drug in the organic phase}]}{[\text{Non-ionised drug in an aqueous phase}]} \qquad (3.7)$$

The values of partition coefficients are normally measured at either 25°C or 37°C. *n*-Octanol is the most commonly used organic phase in pharmacological investigations but other organic solvents such as butanol, chloroform and olive oil are also used. The aqueous phase is either water or a phosphate buffer at pH 7.4, the pH of blood. A high P value for the partition coefficient indicates that the compound will readily diffuse into lipid membranes and fatty tissue. The compound is said to be hydrophobic (water-hating) and have a high hydrophobicity. A low P value indicates that the compound is reluctant to enter lipid material and prefers to stay in the more polar aqueous medium. Compounds of this type are said to be hydrophilic (water-loving) and have a low hydrophobicity. Hydrophobicity can have a significant effect on biological activity. For example, compounds with relatively high hydrophobicity (relatively high P value) will easily enter a membrane but will be reluctant to leave and so will not be readily transported through the membrane if diffusion is the only transport mechanism for the drug. This means that the drug could fail to reach its site of action in sufficient quantity to be effective (see section 5.1) unless hydrophobicity is the dominant factor in its action. Conversely, analogues with relatively low hydrophobicities will not easily diffuse into a membrane and so could also fail to reach their site of action in effective quantities.

If hydrophobicity is the most important factor in drug action, an increase in hydrophobicity usually results in an increase in action. For example, general anaesthetics are believed to act by dissolving in cell membranes. The octanol/water partition coefficients of the anaesthetics diethyl ether, chloroform and halothane are 0.98, 1.97 and 2.3, respectively, which indicates that halothane would be the most soluble of these in lipid membranes. This corresponds to their relative activities, with halothane being the most potent. Consequently, the hydrophobicity of a compound in terms of its partition coefficient is one of the factors considered in the QSAR approach to drug design (see section 2.4.1.1).

3.10.1 Theoretical Determination of Partition Coefficients

The practical measurement of P values is not always as easy as Equation (3.6) would suggest. Consequently, a number of theoretical methods have been developed to calculate P values. The most popular of these methods is based on measuring the P values of a large number of compounds and using statistical analysis methods to relate differences in their structures to differences in their P values. This approach, initially developed by Rekker, has been extended by Hansch and co-workers by the use of computer programs. The CLOGP program developed by Hansch and co-workers divides the structure of a molecule into suitable fragments, extracts the corresponding numerical values from its database and calculates the P value of the compound.

The partition coefficients of a compound in an organic/water phase system may be calculated from the P value of the same compound in a different organic/water phase system. Experimental work has shown that for ideal solutions the two values are related by the expression:

$$\text{Log}\,\text{P}' = a\log\text{P} + b \tag{3.8}$$

where P′ is the partition coefficient in the new organic solvent/water system, P is the partition coefficient in the octanol/water system and a and b are constants characteristic of the new organic solvent/water system. The values of a and b are properties of the system and in ideal solutions are independent of the nature of the solute (Table 3.2). Consequently, the values of a and b for a particular solvent system may be calculated by measuring the values of P and P′ for a compound and assigning standard values for a and b in one of the solvent systems. For example, when $a = 1$ and $b = 0$ for the water/octanol system, values of $a = 1.13$ and $b = -0.17$ have been obtained for the diethyl ether/water system.

Table 3.2. The values of a and b for various organic/water phase systems relative to the octanol/water system where $a = 1$ and $b = 0$.

Solvent/water system relative to the octanol/water system	a	b
Butanol	0.70	0.38
Chloroform	1.13	−1.34
Cyclohexane	0.75	0.87
Diethyl ether	1.13	−0.17
Heptane	1.06	−2.85
Oleyl alcohol	0.99	−0.58

Table 3.3. Examples of surfactants. Reproduced from G. Thomas, *Chemistry for Pharmacy and the Life Sciences including Pharmacology and Biomedical Science*, 1996, permission of Prentice Hall, a Pearson Education Company.

Compound	Structural formula Hydrophobic end—Hydrophilic end
Cationic surfactants	
Sodium stearate	$CH_3(CH_2)_{16}$—$COO^- Na^+$
Anionic surfactant	
Dodecylpyridinium chloride	$C_{12}H_{25}$—$C_5H_5N^+ Cl^-$
Dodecylamine hydrochloride	$CH_3(CH_2)_{11}$—$NH_3^+ Cl^-$
Ampholytic surfactants	
Dodecyl betaine	$C_{12}H_{25}N^+(CH_3)_2CH_2$—$COO^-$
Non-ionic surfactants	
Heptaoxyethylene monohexyldecyl ether	$CH_3(CH_2)_{15}$—$(OCH_2CH_2)_7OH$
Polyoxyethylene sorbitan monolaurate	The polyoxethylene ethers of the lauric acid esters of sorbitan

3.11 Surfactants, Structure and Action

Surfactants are compounds that lower the surface tension of water. Their structures contain both strong hydrophilic and strong hydrophobic groups (Table 3.3) and so they dissolve in both polar and non-polar solvents. They are classified as cationic, anionic, ampholytic and non-ionic surfactants, depending on the nature of their hydrophilic groups. Cationic and anionic surfactants ionise to form the appropriate ions in water. Ampholytic surfactants have electrically neutral structures that contain both positive and negative charges. They are zwitterions. Non-ionic surfactants do not form ions in solution.

Surfactants are frequently used to prepare aqueous solutions of compounds that are insoluble or sparingly soluble in water. The hydrophobic part of the structure of the surfactant binds to the compound and the strong water affinity of the hydrophilic part of the surfactant effectively

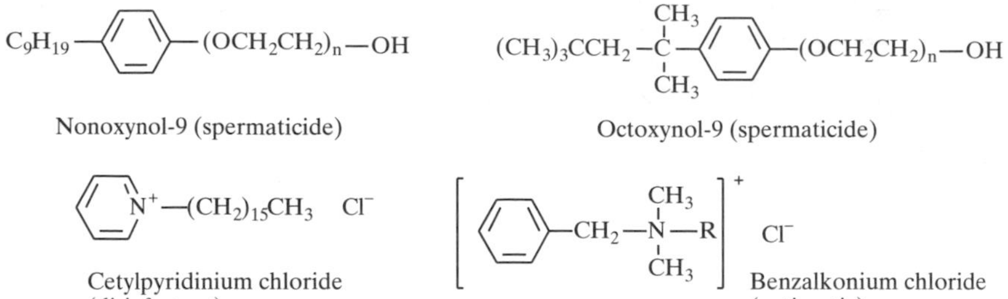

C_9H_{19}—⬡—$(OCH_2CH_2)_n$—OH

Nonoxynol-9 (spermaticide)

$(CH_3)_3CCH_2$—$C(CH_3)_2$—⬡—$(OCH_2CH_2)_n$—OH

Octoxynol-9 (spermaticide)

N^+—$(CH_2)_{15}CH_3$ Cl^-

Cetylpyridinium chloride (disinfectant)

$[$⬡—CH_2—$N(CH_3)_2$—R$]^+$ Cl^-

Benzalkonium chloride (antiseptic)

Figure 3.20. The structures of some biologically active surfactants. A number of nonoynols and octoxynols are known, their structures containing different numbers of ethylene oxide units. In each case the number after the name indicates the average number of ethylene oxide units (n) per molecule in the prepared sample. Benzalkonium chloride is a mixture of compounds where R ranges from C_8 to C_{18}.

pulls the compound into solution in the water. This behaviour is also the basis of detergent action. The surfactant molecules (detergent) bind to the dirt particles and the strong water affinity of their polar groups pulls the dirt particles into suspension in water. This solubilising effect of surfactants is also of considerable importance in the design of dosage forms.

In biological systems surfactants dissolve in both the aqueous medium and lipid membranes and so tend to accumulate at the interfaces between these phases. This property is the reason for the antiseptic and disinfectant action of some non-ionic and quaternary ammonium surfactants. Surfactants such as cetylpyridinium chloride and octoxynol-9 (Figure 3.20) partially dissolve in the lipid membrane of the target cell. This lowers the surface tension of the cell membrane, which results in lysis and the death of the cell (see section 4.4.2.3).

Octoxynol-9 and nonoxynol-9 are different from other surfactant antiseptics in that they do not dissolve in the cell membranes of pathogens. They dissolve in the cell membranes of spermatozoa. This immobilises the sperm, which allows these compounds to be used as spermicides in birth control.

Naturally occurring surfactants are involved in a number of bodily functions. For example, bile salts, which are produced in the liver, play an essential part in the digestion of lipids in the intestine. Surfactants produced in the membranes of the aveoli prevent the accumulation of water and mucus in the lungs. Furthermore, a number of drugs with structures that contain a suitable balance between hydrophobic and hydrophilic groups have also been reported to exhibit surfactant properties (Figure 3.21).

As the concentration of a surfactant dissolved in water increases, the system changes from a true solution to a colloidal solution. The surfactant molecules find it more energetically favourable to form colloidal aggregates known as micelles in which the hydrophobic sections of the molecules effectively form a separate organic phase with the hydrophilic part of the molecule in the aqueous medium (Figure 3.22). The concentration at which micelles begin to form

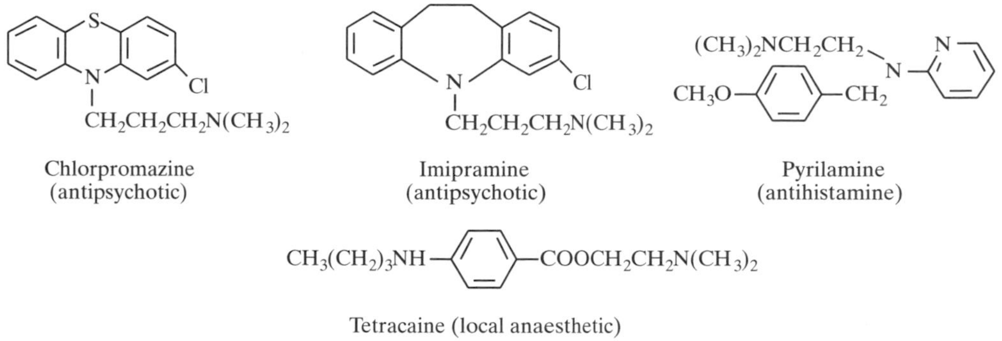

Chlorpromazine (antipsychotic)

Imipramine (antipsychotic)

Pyrilamine (antihistamine)

Tetracaine (local anaesthetic)

Figure 3.21. The structures of some drugs that have been reported to exhibit surfactant properties as well their normal pharmacological activity.

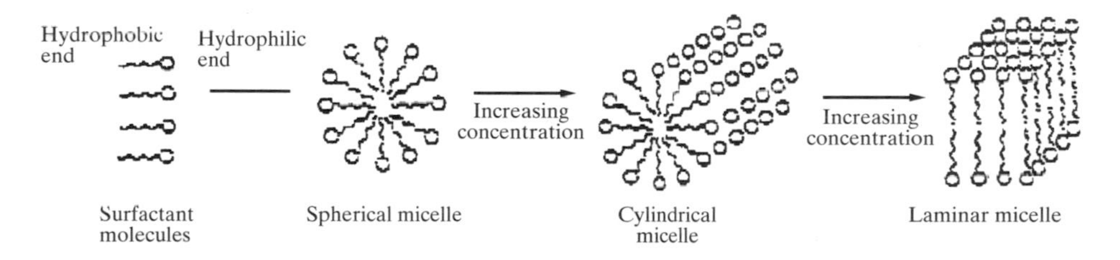

Figure 3.22. Micelle structure and its variation with concentration. Reproduced from G. Thomas *Chemistry for Pharmacy and the Life Sciences including Pharmacology and Biomedical Science*, 1996, by permission of Prentice Hall, a Pearson Education Company.

is known as the **critical micelle concentration** (CMC). For example, the CMC for sodium dodecyl sulphate is about 0.08 mol dm^{-3}. Many of the physical properties of surfactant solutions, such as surface tension, electrical conductivity and osmotic pressure, exhibit an abrupt change at the CMC point. Consequently, changes in these physical measurements can be used to indicate the onset of micelle formation.

3.11.1 Micelles

The shape of a micelle depends on the concentration of the surfactant in solution. At concentrations just above the critical micelle concentration the micelles tend to be spherical in shape. As the concentration increases the micelle changes from a sphere to cylindrical, laminar and other forms (Figure 3.22).

3.11.1.1 Drug Solubilisation

Incorporation into suitable micelles can be used to solubilise water-insoluble and sparingly-water-soluble drugs. The way in which a drug is incorporated into the micelle depends on the

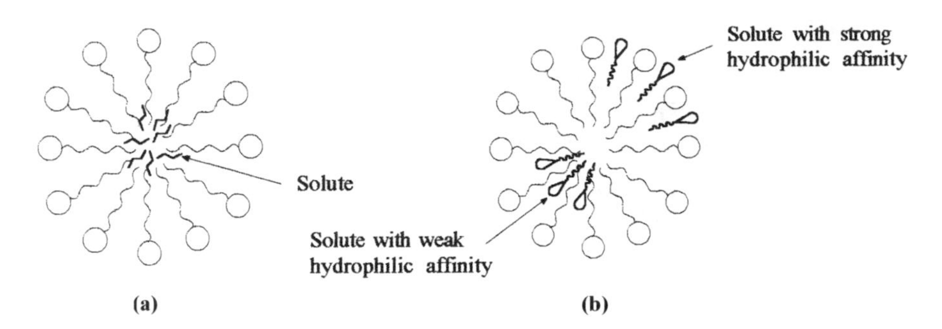

Figure 3.23. Solubilisation of solutes in a micelle: (a) non-polar solutes; (b) polar solutes.

structure of the drug (Figure 3.23a). Non-polar compounds tend to accumulate in the hydrophobic core of the micelle whereas water-insoluble polar compounds are orientated with their polar groups towards the surface of the micelle (Figure 3.23b). The position of the polar compound in the micelle will depend on the relative affinities of the polar group of the solute molecule for the aqueous medium and the non-polar sections of the molecule for the hydrophobic core of the micelle. A relatively strong affinity for the aqueous medium will result in the polar group of the solute being near or on the surface of the micelle, whereas a weak affinity for the aqueous medium will result in the polar group being located further into the interior of the micelle. In all cases, the solute molecules are held in the micelle by intermolecular forces of attraction such as hydrogen bonds and hydrophobic bonding.

Incorporation into micelles is used to deliver water-insoluble and sparingly-water-soluble compounds to their site of action both for clinical use and testing purposes. This technique has a number of disadvantages. The drug's absorption and hence its activity is dependent on it being released from the micelle. Surfactants also suffer from the disadvantage that they frequently irritate mucous membranes and many are haemolytically active. Ionic surfactants can react with anionic and cationic drug substances and so non-ionic surfactants are more widely used in the preparation of dosage forms. Cationic surfactants are, however, used as preservatives. Incorporation of a drug into a micelle can also result in a more rapid decomposition of some compounds because their molecules are in close proximity to each other in the micelle. However, this is utilised in synthetic chemistry to promote reactions.

Incorporation into micelles has been shown to reduce the rates of hydrolysis and oxidation of susceptable drugs. All types of surfactant have been shown to improve the stability of these drugs, the degree of protection usually increasing the further the drug penetrates into the core of the micelle. For example, benzocaine has been shown to be more stable to alkaline hydrolysis than homatropine in the presence of non-ionic surfactants. This is probably because benzocaine is less polar than homatropine and its molecules are located deeper in the core of the micelle.

H_2N– –$COOC_2H_5$ $\xrightarrow[\text{Hydrolysis}]{NaOH/H_2O}$ H_2N– –$COONa$ + C_2H_5OH

Benzocaine

CH_3 CH_3

OH OH

OCOCH $\xrightarrow[\text{Hydrolysis}]{NaOH/H_2O}$ + CHCOONa

OH

Homatropine

3.11.1.2 Mixed Micelles as Drug Delivery Systems

Mixed micelles are formed by mixtures of surfactants. A suitable selection of the surfactants results in a mixed micelle that has low haemolytic and membrane irritant actions. For example, diazepam has been solubilised and stabilised by mixed micelles produced from lecithin and sodium cholate. Sodium cholate is very haemolytically active but this action is considerably reduced by the presence of the lecithin, which has no haemolytic activity. It appears that the mixing of haemolytically active and haemolytically non-active surfactants either reduces or stops this form of biological activity.

3.11.1.3 Liposomes

Phospholipids can act as surfactants. They will, when dispersed in water, spontaneously organise themselves into aggregates called liposomes provided that the temperature is below the so-called chain melting temperature of the lipid. This is the temperature at which the lipid changes from a solid to a liquid crystal. In their simplest form, liposomes consist of a roughly spherical bilayer of phospholipid molecules surrounding an interior core of water (Figure 3.24). The polar heads of the exterior and interior lipid molecules are orientated towards the exterior and interior water molecules. More complex liposomes have structures that consist of a number of concentric bilayer shells.

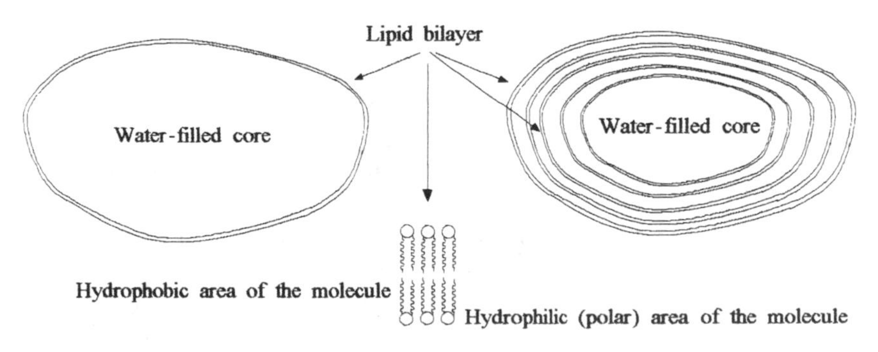

Figure 3.24. The structures of liposomes.

Liposomes are used as drug delivery systems for a wide variety of agents. The drug is incorporated into the interior of the liposome: hydrophilic drugs usually occupy the aqueous core and hydrophobic drugs are normally found in the double layers. The liposomes used are stable at normal body temperatures but break down at higher temperatures. Consequently, in theory, the drug is protected from degradation and safely transported through the biological system until the liposome reaches an area of infection where the temperature is higher. Here the liposome decomposes, releasing the drug hopefully at the desired site of action. In spite of this specific action only a few drugs, such as the antimycotic amphotericin B and the anticancer drugs doxorubicin and daunorubicin, have been successfully delivered to their sites of action in this manner.

Liposomes have a double-layer structure similar to biological membranes. Consequently, they have been used as model membranes to study the diffusion of substances into and out of the liposome with a view to correlating these observations to the *in situ* behaviour of drugs.

3.12 Summary

The **structures of bulk samples** of water and liquid lipids are not understood. However, the relative solubility of a compound in water and lipid media is an important factor in the bioavailability and action of a drug. **Drugs** with a **low water solubility** are usually less easily absorbed and transported to their site of action. Drugs with a **high lipid solubility** are more easily absorbed but do not reach their site of action in sufficient concentration because they become trapped in lipid membranes or fatty depots. For **optimum activity** it is necessary to have a balance between water and lipid solubility.

A **solution** consists of solute molecules dispersed in a solvent. The solute molecules are held in this dispersed state in a liquid by forming weak transient intermolecular forces of attraction, such as hydrogen bonding and other types of dipole–dipole attractions with the solvent molecules. **Polar compounds** are more water soluble than **non-polar compounds** because they form stable hydrates with water molecules.

The **water solubility** of a lead compound may be improved by introducing polar groups such as amine salts and hydroxy, carboxylic and sulphonic acid groups that increase its degree of solvation. These polar groups should not be incorporated in the region of the pharmacophore if the type of activity of the lead is to be retained. They usually increase solvation by hydrogen bonding to water and/or introducing ionisable groups, which results in the formation of ion–dipole bonds between water and the solute. The introduction of non-polar groups reduces water solubility and increases lipid solubility. A wide variety of chemical methods are available for the introduction of polar and non-polar groups into a lead structure.

The **solubility** of a compound in biological fluids also varies with temperature and the nature of the chemical constituents in the fluid. It is usually expressed in terms of either moles or grams per unit volume or mass. **Solubility product** is used as a measure of the solubility of sparingly soluble solutes. In general, the solubility of liquids and solids increases whereas that of gases decreases with increasing temperature. The **pH of the medium** will have a significant effect on the ionisation of acidic and basic solutes, which in turn affects their solubility. The presence of *common ions* will reduce a solute's solubility.

The **water solubility** of a drug can be improved by the use of either a cosolvent or administration in the form of hydrosols and emulsions. **Surfactants** can also be used to prepare aqueous solutions of poorly soluble compounds. The hydrophobic parts of a number of surfactant molecules bind to a solute particle. This gives the solute particle a hydrophilic coating, which effectively pulls it into solution. **Micelles**, formed from surfactant molecules, have also been used as drug delivery vessels for poorly-water-soluble drugs.

Drugs usually have to pass through a number of membranes in order to reach their site of action and so the relative solubilities of the drug in the aqueous medium and lipids are important factors in the mode of action of that drug. **The partition coefficient (P)** of a compound is used as a measure of how a drug distributes itself between two immiscible liquids. It is difficult to measure in biological systems, consequently organic liquid/water systems such as the octanol/water system are used as models.

3.13 Questions

(1) Why is water solubility an important factor in the action of drugs?
(2) Describe the flickering cluster model for the structure of bulk samples of water.
(3) (a) Define the meaning of the terms (i) ideal solution and (ii) polar solute.
(b) Describe, with the aid of suitable diagrams, how ions dissolve in water.
(4) The solubility of calcium oxalate is $7\,mg\,dm^{-3}$ at room temperature. Calculate the solubility product of calcium oxalate. (The relative molecular mass of calcium oxalate is 128.)
(5) The solubility of the monohydrochloride salt of an antimalarial drug A in water at 37°C is $6.2 \times 10^{-4}\,mol\,dm^{-3}$. Calculate the approximate solubility of the drug in 1 M hydrochloric acid. How would this data be used to explain the poor absorption of A from the stomach?
(6) The solubility (% w/w) of pure dry oxygen at a pressure of 1 atmosphere and a temperature of 20°C in water is 4.25 mg. Calculate the solubility of oxygen in water when the water is in contact with air saturated with water vapour at a pressure of 1 atmosphere and a temperature of 20°C if the partial pressure of oxygen in water saturated air is 160 mmHg.

(7) List the structural features that would indicate whether a compound is likely to be reasonably water soluble. Illustrate the answer by reference to suitable examples.

(8) Suggest general methods by which the water solubility of a compound could be improved without affecting its type of biological action.

(9) Suggest, by means of chemical equations, one route for the introduction of each of the following residues into the structure of 4-hydroxybenzenesulphonamide:
   (a) an acid residue,
   (b) a basic residue,
   (c) a neutral polyhydroxy residue.

(10) Calculate the degree of ionisation of codeine, pK_a 8.2, in a solution with a pH of 2. Predict how the degree of ionisation could affect the ease of absorption of codeine in (a) the stomach and (b) the intestine when the pH of the stomach fluids is 2.0 and the pH of the intestinal fluid is 6.

(11) The partition coefficients (P) together with the minimum concentrations to just bring about local anaesthesia in humans of a hypothetical series of compounds with similar structures are given in the following table:

Compound	**A**	**B**	**C**	**D**	**E**	**F**	**G**	**H**
Partition coefficient	3.162	39.81	316.2	1995	50 120	158 500	316 200	2 239 000
Concentration that causes the same biological response ($mmol\,dm^{-3}$)	0.3162	0.0562	0.0177	0.0100	0.0063	0.0149	0.0237	0.1000

   (a) Determine, by graphical or other means, the optimum P value for maximum biological activity.
   (b) The minimum concentration of a compound Z with a similar structure to those used in the table was found to be $0.1683\,mmol\,dm^{-3}$. Its P value was 525. Explain the significance of this observation.

(12) What general structural features are characteristic of surfactants.

(13) Suggest the most appropriate organic/aqueous medium for use in determining P values in the following cases:
   (a) antral nervous system activity,
   (b) gastrointestinal tract absorption,
   (c) buccal absorption.

(15) Explain how micelles and liposomes could be used as drug delivery vehicles.

4 Biological Membranes

4.1 Introduction

All cells have a membrane, known as the **cytoplasmic** or **plasma membrane**, that separates the internal medium of a cell (**intracellular fluid**) from its surrounding medium (**extracellular fluid**). In the cells (**eukaryotes**) of higher organisms, membranes also form the boundaries of the internal regions that retain the intracellular fluids in separate compartments (Figure 4.1a, b). Those compartments that can be recognised as separate entities, such as the nucleus, mitochondria and lysosomes, are known collectively as **organelles**. Organelles carry out specialised tasks within the cell. However, in the **prokaryotic** cells (Figure 4.1c) of simpler organisms where there are no organelles, the plasma membrane is also involved in many of the functions of the organelles. The more fragile plasma membranes of plant cells and bacteria are also protected by a more rigid external covering known as the cell wall.

The primary function of the plasma membrane is to maintain the integrity of the cell in its environment. It is now also known that the membranes of all types of cell regulate the transfer of substances in and out of the cell and between its internal compartments. This movement controls the health as well as the flow of information between and within cells. The plasma membrane of a cell is also involved in both the generation and receipt of chemical and electrical signals, cell adhesion (which is responsible for tissue formation), cell locomotion, biochemical reactions and cell reproduction. The internal cell membranes have similar functions and, in addition, are often actively involved in the function of organelles. Most drugs either interact with the receptors and enzymes attached to the membrane or have to pass through a membrane in order to reach their site of action.

The role of membranes and cell walls in maintaining cell integrity and their involvement in cellular function makes these areas of cells potential targets for drug action. However, in order to design new and better drugs it is necessary to have a detailed picture of the structures of cell membranes and walls as well as a comprehensive knowledge of the chemistry of the biochemical processes that occur in these regions. This chapter attempts to give a broad picture of the current relevance of plasma membrane and cell wall structure to drug action and design.

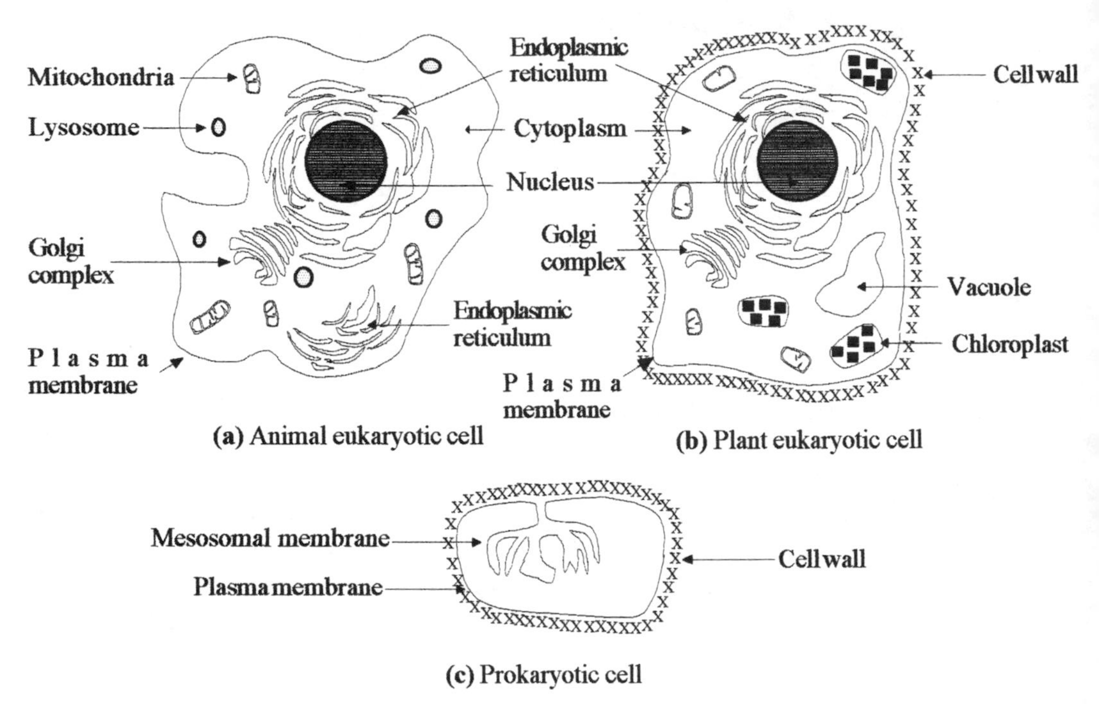

Figure 4.1. A diagrammatic representation of the structures of (a) animal and (b) plant eukaryote cells showing the principal cell organelles. These organelles can also have internal plasma membranes. (c) A diagrammatic representation of the structure of prokaryotic cells such as bacteria (see also Figure 4.11).

4.2 The Plasma Membrane

The currently accepted structure of membranes (Figure 4.2), based on that originally proposed by Singer and Nicolson in 1972, is a fluid-like bilayer arrangement of phospholipids with proteins and other substances such as steroids and glycolipids either associated with its surface or embedded in it to varying degrees. This structure is an intermediate state between the true liquid and solid states, with the lipid and protein molecules having a limited degree of rotational and lateral movement.

X-ray diffraction studies have shown that many naturally occurring membranes are about 5 nm thick. Experimental work has also shown that a potential difference exists across most membranes due to the movement of ions through ion channels (see section 4.2.2) and pumps in the membrane, the intracellular face of the membrane being the negative side of the membrane. For most membranes at rest, that is, not undergoing cellular stimulation, the potential difference (**resting potential**) between the two faces of the membrane varies from –20 to –200 mV.

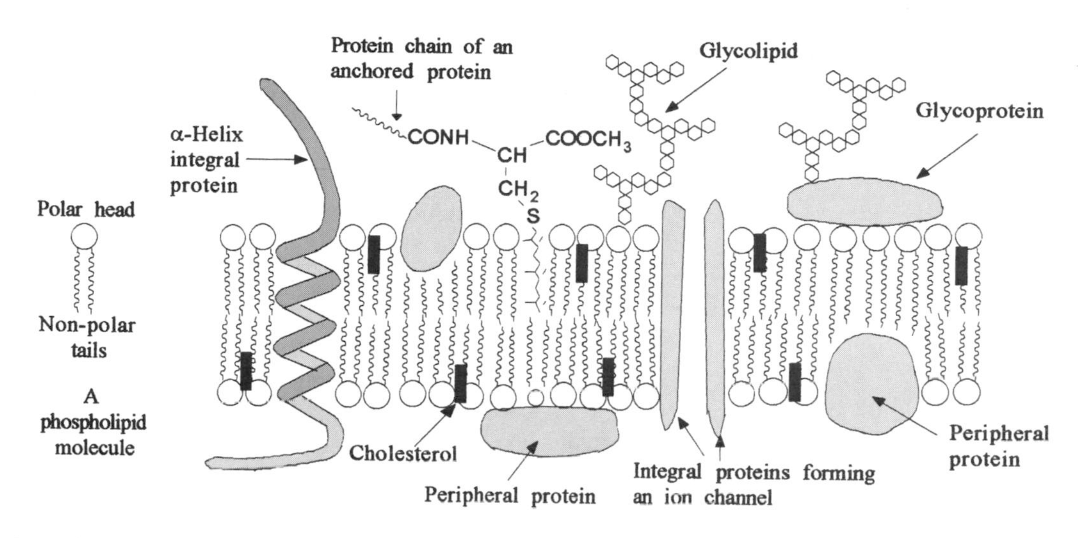

Figure 4.2. The fluid mosaic model of membranes.

4.2.1 Lipid Components

The lipid component of the plasma membrane of mammals is mainly composed of glycerophospholipids, sphingolipids and cholesterol. Glycerophospholipids (phosphatides and plasmalogens) are glycerol-based lipids whereas, the sphingolipids (sphingomyelins) are based on ceramide, a derivative of the amino alcohol sphingosine.

CH_2OH–$CHOH$–CH_2OH

Glycerol

$CH_3(CH_2)_{12}CH{=}CH{-}CH{-}OH$, $H_3\overset{+}{N}{-}CH$, CH_2OH

Sphingosine

$CH_3(CH_2)_{12}CH{=}CH{-}CH{-}OH$, $RCONH{-}CH$, CH_2OH

Ceramide (R = fatty acid residue)

Each of the phospholipid molecules found in plasma membranes has a polar region (**hydrophilic head**) and a long hydrocarbon chain non-polar region (**hydrophobic tail**) (Figure 4.3). These lipid molecules are held together by weak hydrophobic bonding and van der Waals' forces, which give the structure liquid-like properties. The lipid molecules are aligned in the membrane so that their polar heads form the surfaces of the membrane that are in contact with either the extracellular or intracellular aqueous fluid. This means that the interior of a membrane is non-polar (hydrophobic) in nature. Consequently, non-polar compounds will diffuse into the membrane more readily than polar compounds. However, in order to be absorbed into the membrane, a compound must have some water solubility otherwise it will be repelled by the polar nature of the membranes surface.

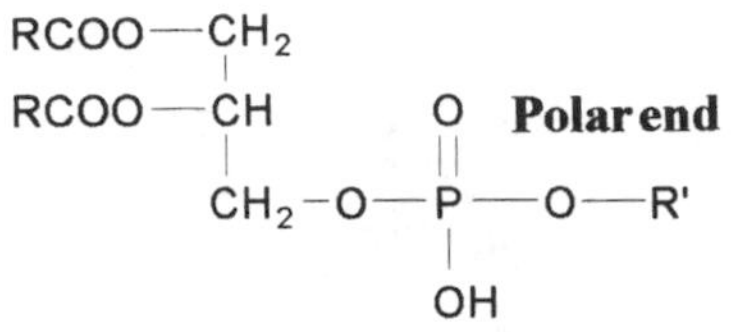

$RCOO—CH_2$
$RCOO—CH$ O **Polar end**
$CH_2—O—P(=O)(OH)—O—R'$

The phosphatidyl phospholipids

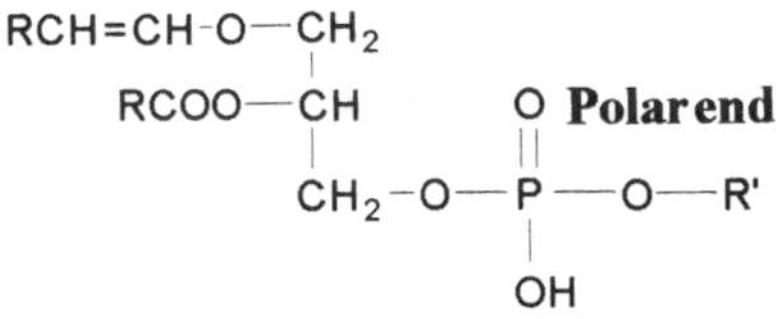

$RCH{=}CH—O—CH_2$
$RCOO—CH$ O **Polar end**
$CH_2—O—P(=O)(OH)—O—R'$

Plasmalogens

Key: R′ =

A choline residue, α-lecithins (phosphatidyl cholines (PC)), the main phospholipids found in mammals. A glycerol residue (phosphatidyl glycerol (PG)), the main plant phosphatidyl lipid. Phosphatidyl phosphate residues are attached to the 1,3-positions of the R′ glycerol group (diphosphatidyl glycerol).

An ethanolamine residue, α-cephalins (phosphatidyl ethanolamine (PE)), a major mammal phospholipid. A serine residue (phosphatidyl serine (PS)), small amounts present in many tissues.

An inositol residue (phosphatidyl inositols (PI)), small amounts present in all membranes.

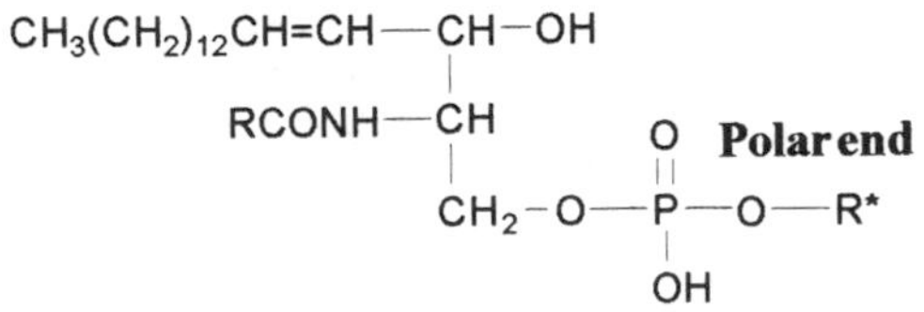

$CH_3(CH_2)_{12}CH{=}CH—CH—OH$
$RCONH—CH$ O **Polar end**
$CH_2—O—P(=O)(OH)—O—R^*$

Sphingomyelins (nerve and muscle membranes of animals)

Figure 4.3. Some of the classes of phospholipids found in membranes. R groups are long-chain fatty acid residues that may or may not be the same. R′ can have one of the structures listed and R* is either a choline or an ethanolamine residue.

The acyl and ether hydrocarbon chains of relevant lipid molecules usually contain between 14 and 24 carbon atoms, the most common being 16 and 18. These chains may be saturated or unsaturated and are usually unbranched. The saturated sections of the chains appear to be largely in the eclipsed (or anti) conformation, giving sections of the molecule a relatively straight structure that is approximately aligned across the membrane. The unsaturated bonds may have either a *cis* or a *trans* configuration. The more common *cis* configuration will introduce a bend in the orientation of the lipid within the membrane (Figure 4.4). Living cells normally contain a higher proportion of unsaturated to saturated fatty acid residues.

The glycerol residues of phospholipids are aligned so that the three carbon atoms are roughly in line with the terminal fatty acid chain. This means that the fatty acid attached to carbon 2 has to make an approximately 90° bend in its chain in order to line up with the other fatty acid residue.

A generally small but important percentage of the constituents of plasma membranes are the sphingolipids. These compounds are carbohydrate derivatives of ceramide in which either a monosaccharide or oligosaccharide residue is attached by a β-glycosidic link to the hydroxy group at position 1. Their main subgroups are the **cerebrosides**, **sulphatides** and **gangliosides** (Figure 4.5). The **cerebrosides** have one glucose or galactose residue, the **sulphatides** have a sulphate ester at position 3 of a galactose residue and the **gangliosides** have a polysaccharide

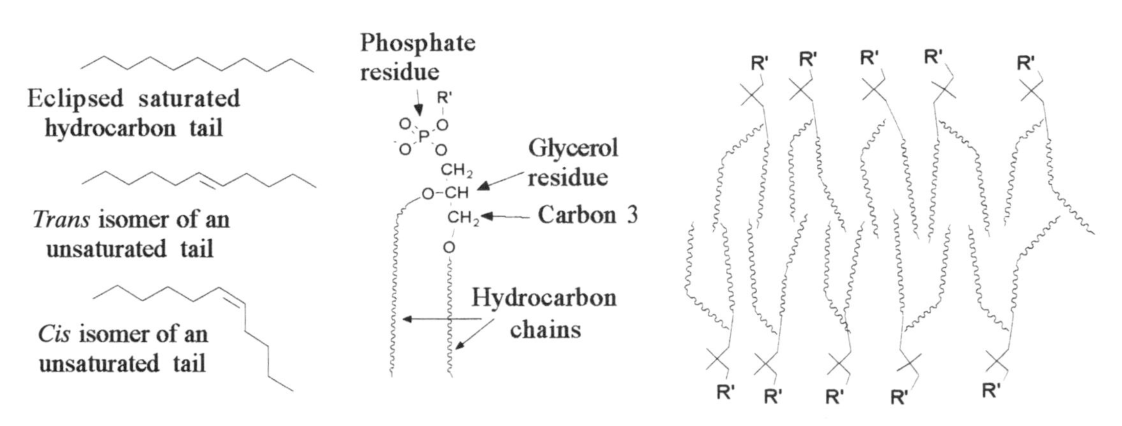

Figure 4.4. A schematic arrangement of the hydrocarbon tails in the lipid bilayer.

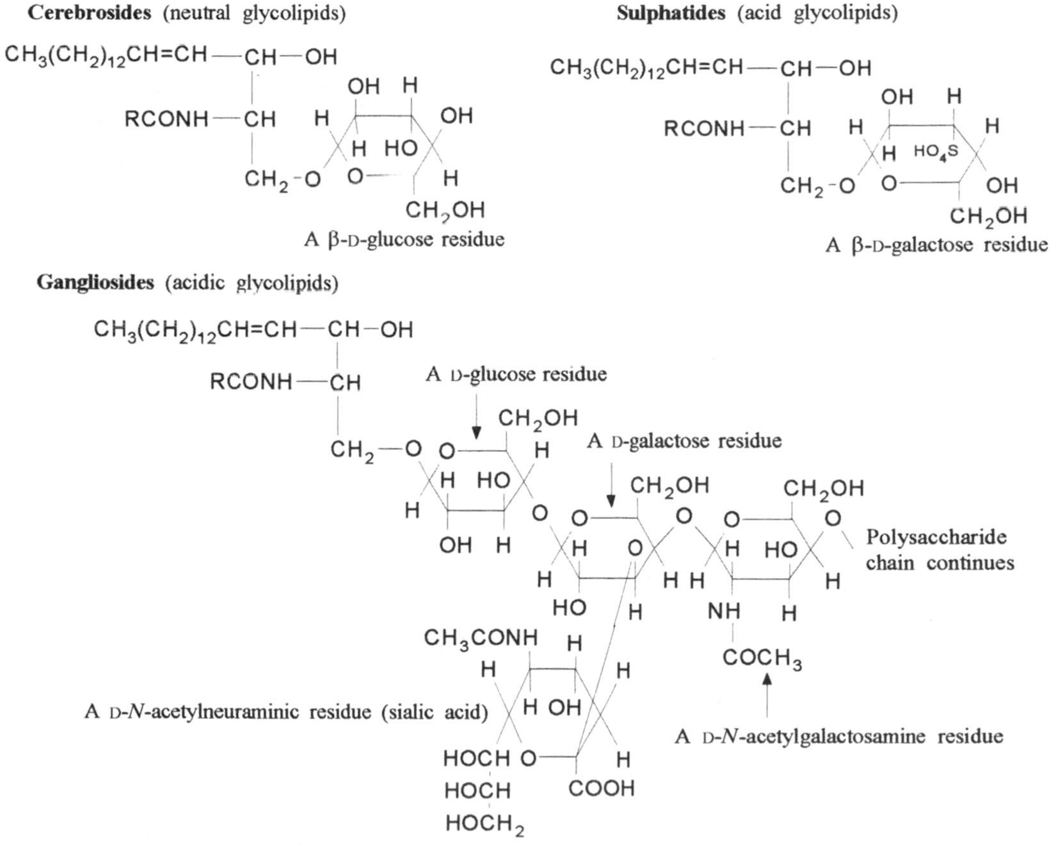

Cerebrosides (neutral glycolipids)

$CH_3(CH_2)_{12}CH{=}CH$—CH—OH

RCONH—CH

CH_2-O

A β-D-glucose residue

Sulphatides (acid glycolipids)

$CH_3(CH_2)_{12}CH{=}CH$—CH—OH

RCONH—CH

CH_2-O

HO_4S

A β-D-galactose residue

Gangliosides (acidic glycolipids)

$CH_3(CH_2)_{12}CH{=}CH$—CH—OH

RCONH—CH

CH_2—O

A D-glucose residue

A D-galactose residue

Polysaccharide chain continues

CH_3CONH

A D-*N*-acetylneuraminic residue (sialic acid)

HOCH

HOCH

$HOCH_2$

COOH

$COCH_3$

A D-*N*-acetylgalactosamine residue

Figure 4.5. The main subgroups of sphingolipids.

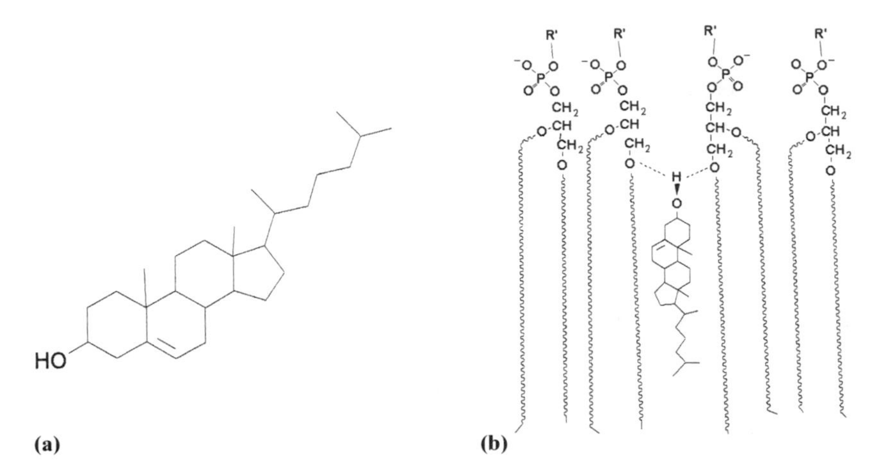

Figure 4.6. (a) Cholesterol and (b) a schematic representation of the hydrogen bonding of cholesterol in membranes.

chain. The ganglioside polysaccharide chain contains amino sugar derivatives such as **sialic acid** as well as glucose and galactose residues.

The glycosphingolipids are only present in small amounts in plasma membranes, but they are associated with a number of important cell functions. Specific glycosphingolipids appear to be involved in cell to cell recognition, tissue immunity blood grouping and also in the transmission of nerve impulses between neurons because large concentrations are found at nerve endings. The accumulation of certain gangliosides in tissue, due to a deficiency of the enzymes required for their degradation, is related to a number of genetically transmitted conditions such as Tay-Sachs disease, Gaucher's disease, Krabbe's leukostrophy, Fabry's disease and Niemann–Pick disease.

Cholesterol (Figure 4.6) molecules are found embedded in both surfaces of animal plasma membranes and to a lesser extent in the membranes of their organelles. The molecule occurs in the membrane with its hydrocarbon side chain lined up alongside the hydrocarbon chains of the phospholipids. Unlike the phospholipids, the structure of cholesterol is relatively rigid and so it stiffens the membrane by hydrogen bonding to the adjacent oxygen atoms of the lipid esters. This helps to prevent the membrane from acting as a true liquid. Cholesterol also affects other physical properties of membranes, for example, its relatively non-polar nature means that membranes that are rich in cholesterol and relatively saturated glycosphingolipids are resistant to the passive passage of water and ions.

4.2.2 Protein Components

Cell membranes, with the exception of those of the Schwann cells of the myelin sheath of neurons (nerve cells), usually contain more protein than lipid in terms of the total dry mass of

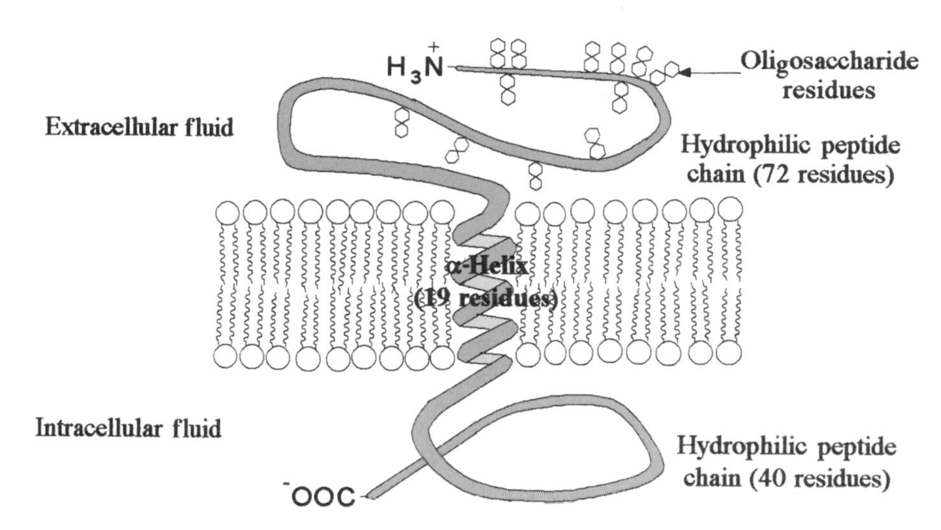

Figure 4.7. A representation of the orientation of glycophorin in animal membranes.

a membrane. These proteins are responsible for carrying out many of the active functions of membranes, such as acting as receptors and transportation routes for various substances across the membrane. They are normally classified as **integral** (**intrinsic**), **peripheral** (**extrinsic**) or **lipid-anchored** proteins. Those proteins whose chains either span the bilayer or are partly embedded in the bilayer usually have their N-terminal in the extracellular fluid and their C-terminal in the intracellular fluid.

Integral proteins are strongly associated with the membrane and are either deeply embedded in or pass right through the membrane. They can only be displaced from the membrane by disrupting its structure using solvents, disruptive enzymes and detergents. Integral proteins can be roughly divided into two types: those where most of the protein is embedded in the bilayer and those where part of the protein is embedded in the lipid layer but the greater part extends into either the extracellular or intracellular fluid or both. The former are often involved in the transport of species across the membrane (see sections 4.3.4 and 4.3.5). The latter usually has oligosaccharides attached to the section protruding into the extracellular fluid. These oligosaccharides have a variety of functions. In glycophorin (Figure 4.7), for example, the oligosaccharides act as the receptors for the influenza virus and also constitute the ABO and MN blood groups.

Both types of integral protein are **amphiphilic**: the surfaces of the protein segments in the extra- and intracellular fluids are hydrophilic in nature and the surfaces of the segment within the membrane are hydrophobic. The sections of proteins that cross a membrane are usually in the form of an α-helix with the polar groups orientated towards the centre of the helix and the non-polar groups on the outer surface of the helix.

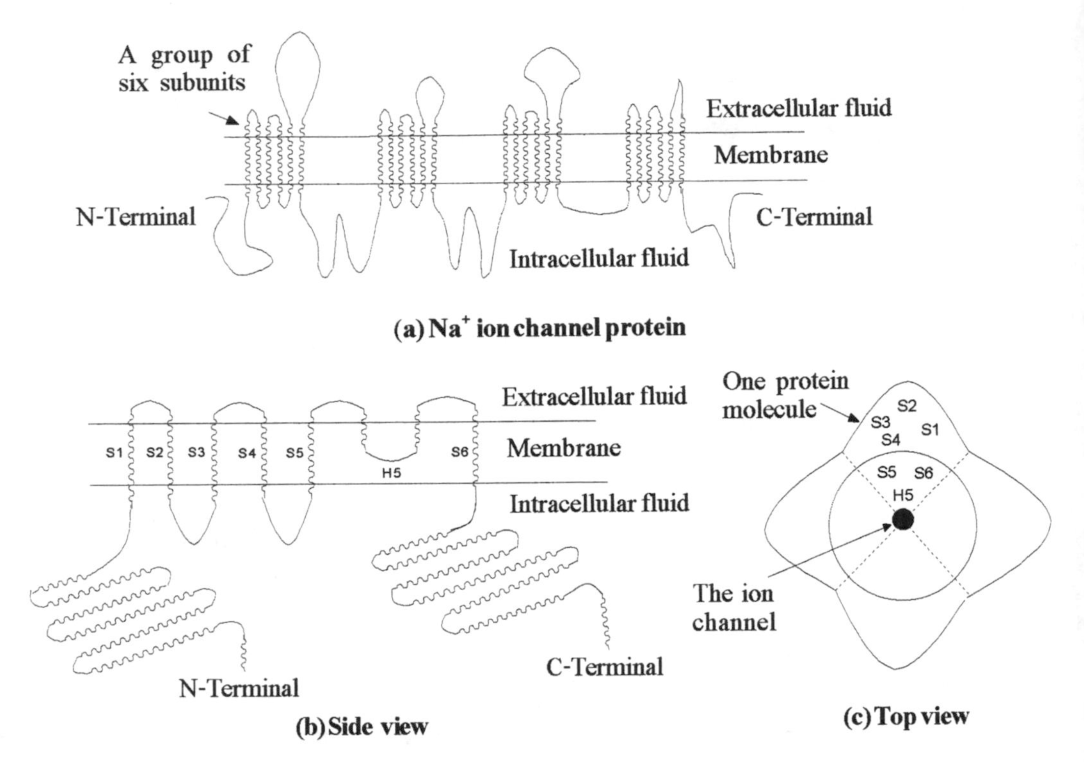

Figure 4.8. **(a)** A schematic representation of the transmembrane protein responsible for forming Na^+ ion channels in rat brain. The H5 domain in the membrane is not shown. The four groups of six transmembrane spans and one H5 unit form the Na^+ ion channel. **(b, c)** A schematic representation of the protein forming K^+ ion channels. Each subunit consists of six α-helical transmembrane spans plus an H5 domain. Four of these subunits are grouped together to form a K^+ ion channel.

Transmembrane integral proteins may cross a membrane several times. The sections of the protein that cross the membrane are referred to as **transmembrane domains** or **spans**. A single protein molecule may have several groups of transmembrane spans (Figure 4.8a), which may be grouped together in a membrane in such a way that they form a water-filled pore through the membrane. However, it is more common for several transmembrane integral protein molecules to be grouped together to form a pore through the membrane. Each separate molecule in such a group is known as a **subunit**. For example, the protein molecules that form a pore large enough to allow the passage of K^+ ions have six transmembrane spans (Figure 4.8b). Four of these subunits are grouped together to form the pore (Figure 4.8c). Each subunit also has a smaller H5 domain that is embedded in the extracellular surface of the membrane so that it narrows the extracellular exit of the pore. The position of this domain explains why these pores are wider on the intracellular side of the membrane. This has a considerable bearing on the action of some drugs (see section 4.4.3). Pores that allow the passage of ions through a membrane are known as **ion channels**.

A wide variety of ion channels have been identified. Ion channels are usually fairly selective, allowing the passage of a specific ion but opposing the passage of other ions. Consequently, ion channels are referred to as Na^+, K^+, Cl^- . . . ion channels according to the nature of the ion allowed through the channel. However, each type of ion channel has a number of subtypes attributed to differences in the proteins forming the channel. For example, four different Ca^{2+} ion channels are known, namely: the L type, found in skeletal, cardiac and smooth muscle tissue; the T type, found in pacemaker cells; the N type, found in neurons; and the P type, found in Purkinje cells. Furthermore, the proteins forming some channels undergo conformational changes that effectively open and close the channel like the action of a gate. These so-called gates are controlled by either changes in the membrane potential (**voltage-gated channels**) or the binding of a ligand to a receptor (**ligand-gated channel**). The movement of ions in and out of a cell through ion channels is an important process in cell function. The prevention of this movement is one of the ways in which drugs act on cell membranes (see section 4.4.3).

Peripheral proteins are probably attached to the surface of the membrane by electrostatic, hydrophobic and hydrogen bonding. These bonds are readily broken by metal chelating agents, changes in pH and ionic strength, which explains why many peripheral proteins are able to migrate over the surface of the membrane. Peripheral proteins have a variety of biological functions, including enzyme and antibody activity, whereas the intracellular peripheral proteins, actin and spectrin, form part of the cell's cytoskeleton. This is the network of protein filaments that is thought to determine the shape of the cell and control its ability to move. It also controls the movement of organelles within the cell and is involved in cell division.

Some proteins are attached to the membrane by so called lipid anchors that form part of the lipid bilayer (Figure 4.9). Four different anchor systems have been identified, namely: the amide-linked myristoyl, the thioester-linked fatty acid, the thioether-linked prenyl and the glycosyl phosphatidylinositol (GPI) anchors. All these types of anchor have been found in eukaryotic cells but GPI anchors have not been found in prokaryotic cells. A wide variety of proteins are attached to membranes as lipid-anchored proteins. They include the gag proteins of certain retroviruses (see section 10.14.2.2), the transferrin receptor protein, yeast mating factors, the α-subunit of G proteins (see section 8.4.2), surface antigens and cell surface hydrolases.

4.2.3 The Carbohydrate Component

Significant amounts of carbohydrate are associated with both the plasma and internal cell membranes. It takes the form of both small and large heterosaccharide chains consisting mainly of sialic acid, fucose, galactose, mannose, *N*-acetylglucosamine and *N*-acetylgalactosamine. The chains can also contain small amounts of glucose. Experimental evidence indicates that sialic acid is often the terminal sugar at the unattached end of the chain. These heterosaccharide chains, which are attached to lipids and surface proteins, form an integral part of the structure of the membrane and are not easily removed from the cell surface. They have a variety of biological functions. For example, they act as receptors for a variety of viruses, they are involved

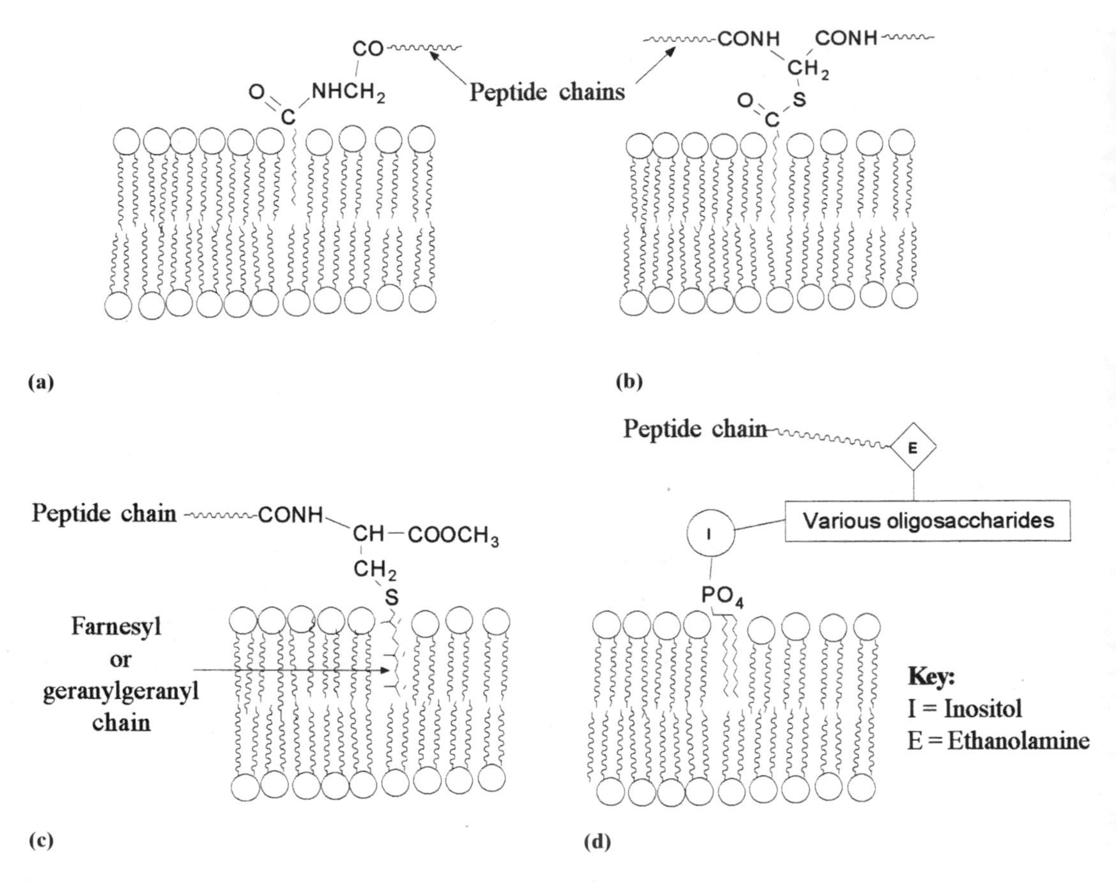

Figure 4.9. Anchor systems for membrane lipid-anchored proteins. (a) Amide-linked myristoyl anchor. The protein is anchored by forming an amide link with the fatty acid myristol. (b) Thioester-linked fatty acid anchor. An *S*-palmitoyl ester is formed between the palmitic anchor and a cysteine residue. (c) Thioether-linked prenyl anchor. The anchor is either a farnesyl or geranylgeranyl group linked to a protein with the C-terminal sequence CAAX, where C is the methyl ester of cysteine, A is usually an aliphatic amino acid residue and X is Ala, Met or Ser for a farnesyl group and Leu for a geranylgeranyl group. (d) *C*-Terminal sequence glycosyl phosphatidylinositol (GPI) anchor groups.

in cell–cell interaction and tissue immunity receptor control and they determine a person's blood group. The carbohydrate chains are also potential targets for drug action.

Sialic acid (*N*-Acetylneuraminic acid) α-L-Fucose β-D-Galactose β-D-*N*-Acetylgalactose β-D-Mannose β-D-*N*-Acetylglucosamine β-D-Glucose

4.2.4 Similarities and Differences between Plasma Membranes in Different Cells

The chemical nature of the lipid components of the plasma membranes of different cells varies both in composition and concentration according to the cell function and the nature of the organism (Table 4.1). Differences in composition and concentration also occur between plasma membranes and the organelle membranes in the same cell. For example, some organelle membranes tend to contain a higher percentage of unsaturated fatty acyl and ether residues. However, the membranes of the organelles of plants and animals that have the same function have similar lipid compositions except that plant membranes contain stigmasterol or cytosterol and animal membranes contain cholesterol (Figure 4.10).

Membranes normally contain the same proteins. However, the amount of each protein shows considerable variation. For example, the concentrations of actin and myosin are much higher in muscle cells than in other types of eukaryotic cell.

4.2.5 Cell Walls

Plants, fungi and most bacteria have a well-defined cell wall that covers the outer surface of the plasma membrane of the cell. This is a rigid structure that protects the fragile interior of the cell from damage by the surrounding environment as well as cementing cells together to form larger organisms such as plants. The cell wall consists mainly of a complex polypep-

Table 4.1. The most abundant components of membranes. The organism or organelle membrane is given in parentheses. (For abbreviations, see Figure 4.3.)

Compound	Animals	Bacteria	Fungi	Plants
Steroids	Cholesterol	Rare	Ergosterol (yeast and other fungi)	Stigmasterol Sitosterol Cytosterol
Phospholipids	PC, PE, DPG, SM	PC, PE, PI (cyanobacteria)		PC, PE, PI, DPG (chloroplasts)

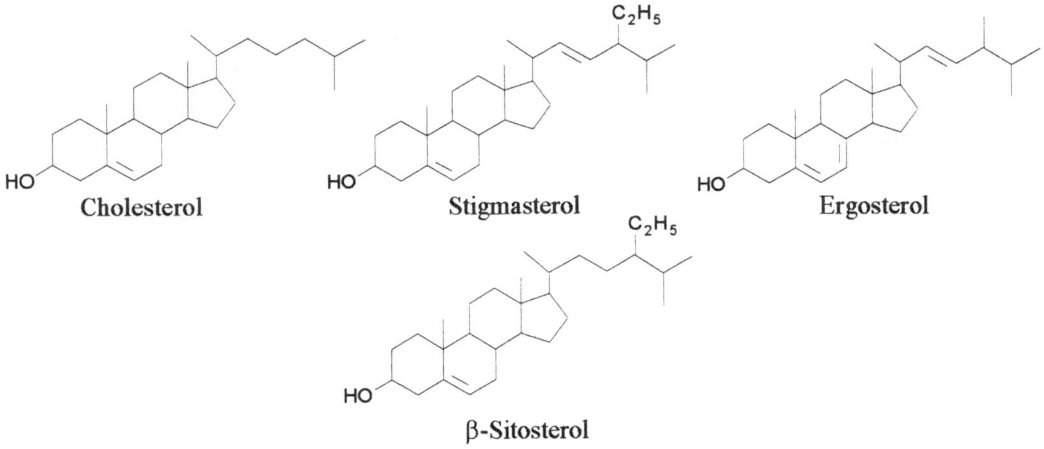

HO Cholesterol

HO Stigmasterol C_2H_5

HO Ergosterol

HO β-Sitosterol C_2H_5

Figure 4.10. The principal steroids found in membranes.

tide–polysaccharide matrix generally referred to as a **peptidoglycan**. Its components depend on the nature of the organism, for example, in plants its polysaccharide components are mainly cellulose whereas in bacteria a wider variety of sugar residues are found. Cell walls are continually being replaced and so this process offers a potential target for drugs. For example, a number of antibiotics act by preventing cell wall synthesis (see section 4.4).

4.2.5.1 Bacterial Cell Walls

Bacteria have a high internal osmotic pressure. Their strong rigid cell walls maintain their shape and integrity by preventing either the swelling and bursting (lysis) or the shrinking of the bacteria when the osmotic pressure of the surrounding solution changes. It enables the bacteria to survive in hypotonic and hypertonic environments. The cell walls of bacteria are broken down by enzymes in their surrounding medium and so they are continuously being rebuilt. A number of antibiotics such as penicillin act by preventing this renewal of the cell wall, which results in the lysis of the bacteria because of its high internal osmotic pressure (see section 4.4.4).

Bacteria are commonly classified as being either Gram-positive or Gram-negative, depending on their response to the Gram stain test. The cell walls of Gram-positive bacteria are about 25 nm thick and consist of up to 20 layers of the peptidoglycan (Figure 4.11). In contrast, the cell walls of Gram-negative bacteria are only 2–3 nm thick and consist of an outer lipid bilayer attached through hydrophobic proteins and amide links to the peptidoglycan. This lipid–peptidoglycan structure is separated from the plasma membrane by an aqueous compartment known as the **periplasmic space**. This space contains transport sugars, enzymes and other substances. The complete structure separating the cytoplasm from its surroundings is known as the **cell envelope**.

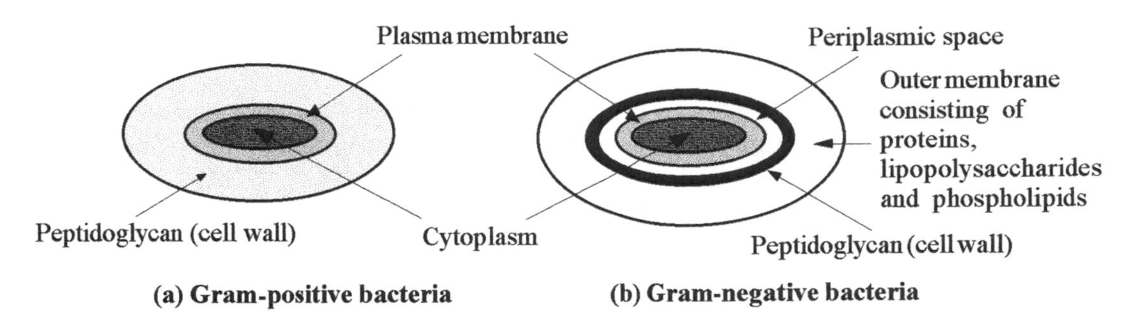

Figure 4.11. Schematic cross-sections of the cell envelopes of (a) Gram-positive and (b) Gram-negative bacteria.

The peptidoglycans found in Gram-positive and Gram-negative bacteria are commonly known as mureins. They are polymers composed of polysaccharide and peptide chains that form a single, net-like molecule that completely surrounds the cell. The polysaccharide chains consist of alternating 1–4-linked β-*N*-acetylmuramic acid (NAM) and β-*N*-acetylglucosamine (NAG) units (Figure 4.12a). Tetrapeptide chains are attached through the lactic acid residues of the NAM units of these polysaccharide chains.

In Gram-positive bacteria, the tetrapeptide chains of one polysaccharide chain are cross-linked from the γ-amino group of lysine by pentaglycine peptide bridges to the terminal alanine of the tetrapeptide chains of a second polysaccharide chain to form a net-like polymer (Figure 4.12b). The pentapeptide occasionally contains other residues. In Gram-negative bacteria the tetrapeptide chains are directly linked by peptide bond (amide group) bridges (Figure 4.12c). The structure is denser than that found in the Gram-positive cell wall because the peptidoglycan chains are closer together.

4.2.6 Bacterial Cell Exterior Surfaces

The exterior surface of Gram-positive bacteria is covered by **teichoic acids**. These are ribitol-phosphate or glycerol-phosphate polymer chains that are frequently substituted by alanine and glycosidically linked monosaccharides (Figure 4.13). They are attached to the peptidoglycan by a phosphate diester link. Teichoic acids can act as receptors to bacteriophages and some appear to have antigenic properties.

The exterior surface of Gram-negative bacteria is more complex than that of the Gram-positive bacteria. It is coated with **lipopolysaccharides**, which largely consist of long chains of repeating oligosaccharide units that are attached to the outer membrane by a core oligosaccharide (Figure 4.14). These lipopolysaccharides often contain monosaccharide units such as abequose and 2-keto-3-deoxyoctanoate (KDO) that are rarely found in other organisms. The repeating units, which are known as **O-antigens**, are unique to a particular type of bacteria. Experimental evidence suggests that they play a part in the bacteria's recognition of host cells.

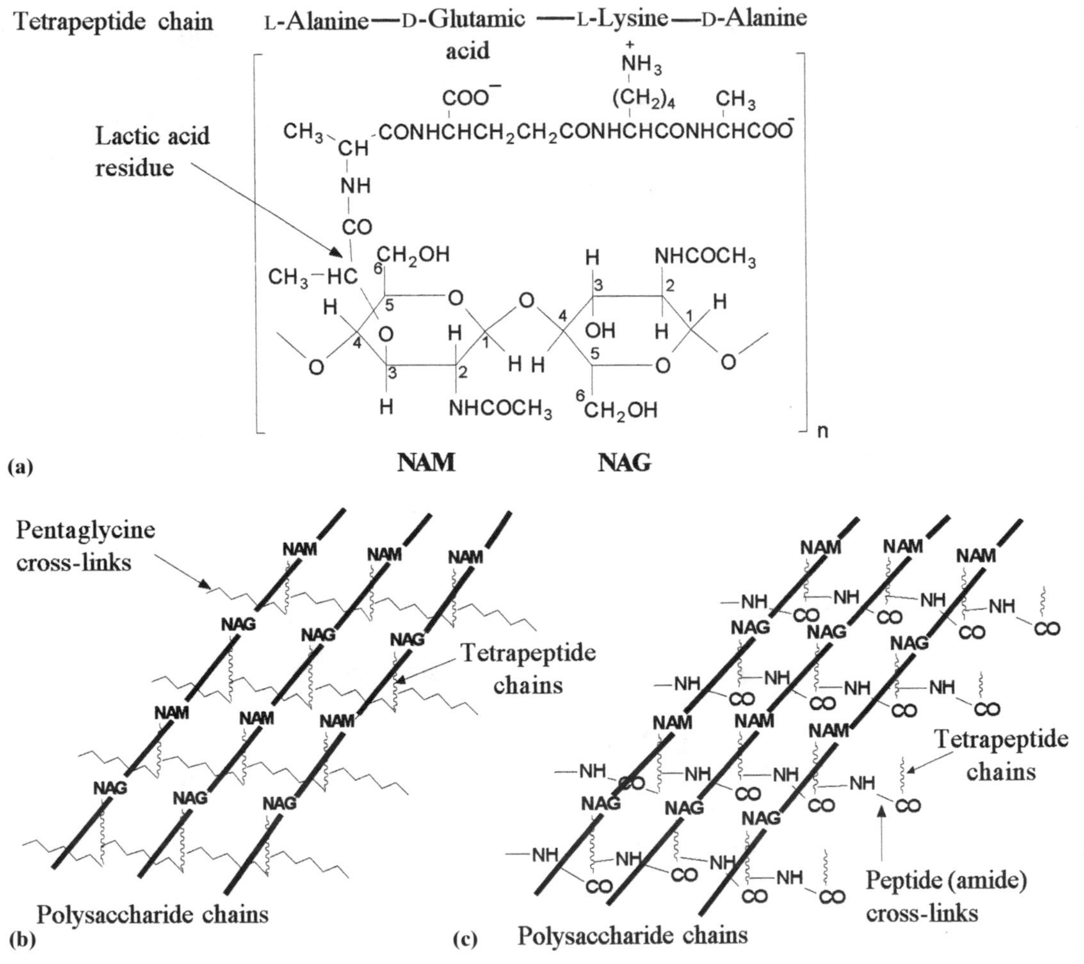

Figure 4.12. (a) The structure of the monomer unit of the polyglycan of *Staphylococcus aureus*. The cross-links between chains are from the γ-amino group of the lysine to the carboxylate of the C-terminal D-alanine. The tetrapeptide can contain other amino acid residues. (b, c) Schematic representations of the structure of the peptidoglycan of Gram-positive (b) and Gram-negative (c) bacteria.

It is also believed that they enable the host's immunological system to identify the invading bacteria and produce antibodies that destroy the bacteria. However, a particular bacterial cell can have a number of different O-antigens and it is this diversity that allows some bacteria to evade a host's immune system.

4.2.7 Animal Cell Exterior Surfaces

The surfaces of animal cells play an essential part in the function of the cells. Numerous experiments have shown that cells are able to communicate with each other. For example, normal

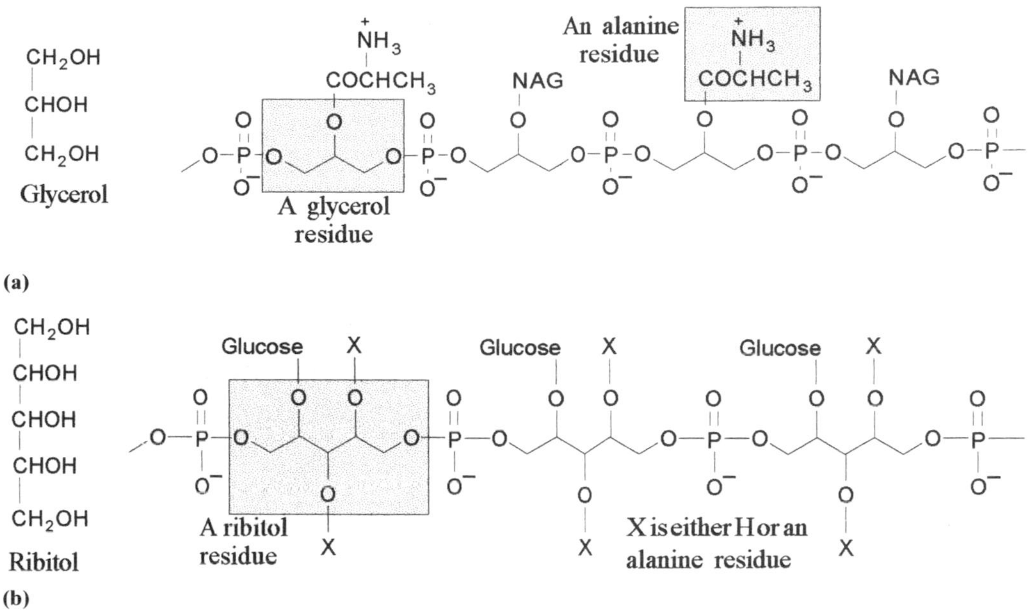

Figure 4.13. (a) A glycerol-based teichoic acid. (b) A ribitol-based teichoic acid. In many teichoic acids the monosaccharide residues are glucose and *N*-acetylglucosamine.

healthy cells stop growing when their surfaces come into contact. This is known as **contact inhibition**. One of the characteristics of cancerous cells is that they lose this contact inhibition and do not stop growing when they make contact with other cells. It is now thought that information is also passed by the interaction of glycoproteins on the cell surface with the proteins, especially the protoglycans found in the extracellular fluid between the cells. Protoglycans are a family of glycoproteins in which the main carbohydrates are glycoaminoglycans. Surface carbohydrate moieties have also been shown to be involved in many cell processes ranging from the aggregation of cells to form organs to the infection of organisms by bacteria.

4.2.8 Viruses

A discussion of the structure of viruses is given in section 10.14.1.

4.2.9 Tissue

Cells are the basic building blocks for all known life forms. These cells occur in a huge variety of sizes and shapes and have a tremendously varied range of functions. Tissue is the biological structure formed by groups of cells adhering together. Its physical and biochemical properties will depend on the types of cell forming the tissue. However, all tissues have certain

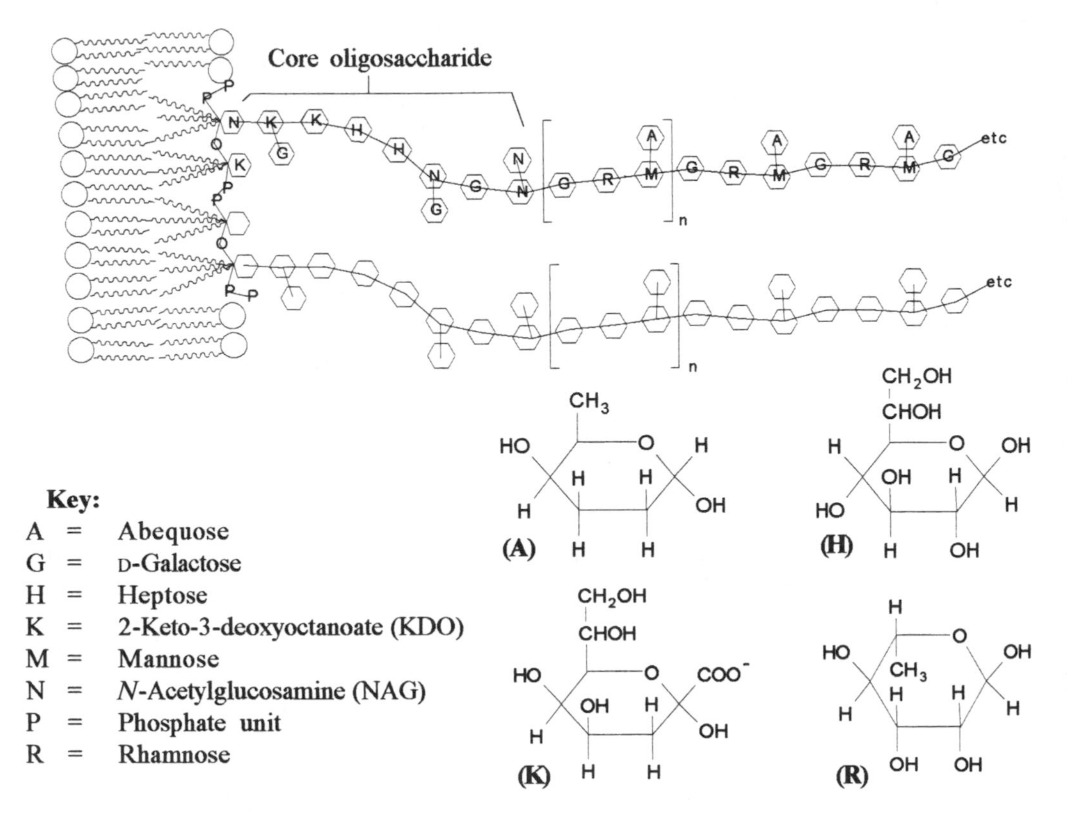

Figure 4.14. The structure of the surface lipopolysaccharides of Gram-negative bacteria.

features in common, such as a supporting framework, blood vessels to supply nutrients and remove waste products and a nerve system to transmit relevant information to the cells forming the tissue. Ancillary cells, such as macrophages, melanocytes and lymphocytes, enter the tissue from other sources, either during its formation or continuously during its lifetime. All the cells in the tissue have a limited life span and are continually dying and being replaced by new cells.

The spaces between cells vary depending on the type of tissue. For example, the gaps between the endothelial cells forming the stomach lining are very small in order to prevent the leakage of hydrochloric acid into the underlying tissues. These gaps are known as **tight junctions**. Tight junctions are also found between the endothelial cells lining the interior surfaces of *many* of the blood vessels of the circulatory system. This means that the gaps between these endothelial cells are too small to allow the passage of xenobiotics to their target sites in the underlying tissue. This forces drug molecules to pass through a large number of cell membranes in order to reach their sites of action. Consequently, it is important to ensure that potential drugs are able to cross plasma membranes and also, where necessary, penetrate cell walls. However, in some conditions such as inflammation, the cells lining the blood vessels move apart and allow the leakage of unwanted fluid into the underlying tissues, causing them to swell (oedema). These conditions could also allow xenobiotics to penetrate to the underlying tissues.

The gaps between the endothelial cells that surround the capillaries of the brain are extremely small. They form a structure that is known as the **blood–brain barrier**. The extremely small gaps in this barrier mean that almost all the substances entering the brain have to pass through the endothelial cell membranes. This makes it more difficult for polar substances to enter the brain unless they are actively transported (see section 4.3.5). Consequently, this factor must be taken into account when designing drugs to target the brain. Another factor is that the blood–brain barrier also contains enzymes that protect the brain.

The gaps between endothelial cells can also be quite large. For example, the gaps (up to 1 μm) between the endothelial cells lining the capillaries of the liver and spleen are large enough to allow the passage of protein molecules.

4.2.10 Skin

This is a complex mixture of cell types. The epidermis consists of dead cells that have migrated to the surface undergoing **apoptosis** *en route*. Apoptosis is the process by which cells undergo 'programmed cell death' in order to maintain a healthy body. The outermost layers of the epidermis, the **stratum corneum**, consist of these dead cells laid down in a keratin matrix. This results in a strong structure with no gaps between the dead cells. It protects the body by preventing excess water loss and the entry of infectious organisms.

The structure of skin affects the design of topically applied drugs in that the more lipid soluble a drug is, the more likely it is to penetrate the stratum corneum and be absorbed by the underlying blood vessels. Water-soluble drugs are only able to penetrate an intact stratum corneum through the hair follicles and sweat glands and so only very small amounts of these drugs can reach the underlying tissues. However, large quantities of water-soluble drugs will be rapidly absorbed through lesions of the skin because these drugs will now be in direct contact with the underlying tissue.

4.3 The Transfer of Species Through Cell Membranes

4.3.1 Osmosis

Water is able to diffuse through membranes when there is a solute particle concentration gradient across the membrane. All the fluid-containing compartments of the body are either apparently or almost iso-osmotic. This iso-osmotic equilibrium is time dependent and so if there is a sudden change in composition of a fluid in a compartment a concentration gradient may be formed across any relevant membranes. This will result in a net movement of water from the area of lower solute particle concentration to the area of higher solute particle concentration, which can lead to cells either contracting (crenation) or swelling with perhaps subsequent lysis. Consequently, in sensitive areas of the body such as the eye, the production of a solute particle concentration gradient by the introduction of a xenobiotic can result in unwanted tissue damage.

Changes in solute particle concentration may be brought about by the introduction of substances into a body compartment. However, metabolism of the substance may reduce the effect. For example, a 5% solution of glucose is initially isotonic with plasma. Therefore, initially, introduction of this solution into a body compartment will not change the osmotic pressure of the system. However, as the glucose is metabolised the solution becomes hypotonic, water will flow into the compartment and the subsequent change in osmotic pressure could cause cell damage. Nevertheless, to reduce cell damage due to unwanted osmosis, it is important that liquid formulations of drugs are designed to be isotonic with the relevant body fluids.

4.3.2 Filtration

The channels formed by some integral proteins (see section 4.2.2) act like the pores in a filter paper and allow the passage of small molecules in and out of the cell. Filtration occurs when solute particles are forced through these channels in the membrane by the external pressure and no other agents. Most pores have a diameter of 0.6–0.7 nm, which allows the passage of species with relative molecular masses up to about 100. Because drugs usually have relative molecular masses above this value, it means that most drugs cannot filter through a membrane. However, they may be carried through the gaps between cells by the pressure gradients produced by the heart. Even so, these gaps are narrow and so will not usually allow the passage of large protein molecules and thus the passage of drugs bound to these proteins will also be restricted.

4.3.3 Passive Diffusion

Passive diffusion is the process whereby a solute diffuses through a membrane from a high to a low concentration without the membrane actively participating in that process. It is a major route for the transfer of **uncharged** and **non-polar** solutes that readily dissolve in lipids through membranes. The process initially requires the partition of the solute between the donating (solute concentration c_1) aqueous fluid and the lipids of the membrane, diffusion of the solute through the membrane from one side to the other and finally partition of the solute between the lipid membrane and the receiving (solute concentration c_2) aqueous fluid (Figure 4.15). The whole process is driven solely by the concentration gradient ($c_1 - c_2$) between the aqueous fluids on either side of the membrane and only ceases when the concentrations of the solute are the same on both sides of the membrane. However, in many cases, diffusion does not stop

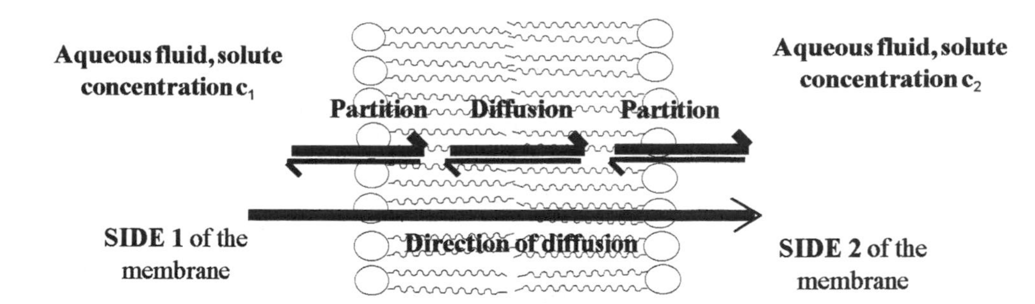

Figure 4.15. A diagrammatic representation of the mechanism of passive diffusion through a membrane where $c_1 > c_2$.

because the solute is carried away or metabolised or both when it has passed into the aqueous fluid on the receiving side of the membrane. This ensures that there is always a concentration gradient across the membrane.

The passage of more polar uncharged compounds by passive diffusion is restricted or prevented by most membranes. Notable exceptions are the outer membranes of Gram-negative bacteria and mitochondria, both of which show a high non-discriminatory permeability to uncharged polar molecules due to the presence of large pores formed by the transmembrane protein, porin. Charged species are too polar to dissolve in the lipid membrane and so do not normally pass through a membrane by passive diffusion, but small ions can diffuse through the water-filled protein channels. Small anions will diffuse to positively and small cations to negatively charged areas, that is, down their respective electrical gradients. In this case, diffusion ceases when the electrical gradient is neutralised.

The ease of diffusion of polar compounds whose structures contain ionisable acidic and basic groups is dependent on both their degree of ionisation at physiological pH and the lipid solubility of the unionised form. Because a membrane will not usually allow the passage of charged organic compounds, the lower the degree of ionisation the more likely the compound will be transferred through a membrane. In addition, the compound must have an optimum solubility in the lipid membrane because if the solute is too soluble it will readily enter the membrane but will be reluctant to leave. For example, acids with pK_a values greater than 7.5 are almost unionised at physiological pH. Consequently, they are likely to be rapidly transferred across a membrane provided that their unionised form is reasonably but not too lipid soluble. Acids with a pK_a of 2.5 or less will be almost fully ionised at physiological pH. Consequently, it is likely that their lipid solubility will be low and their diffusion rate slow. The transfer of bases by passive diffusion can be analysed in a similar manner.

The degree of ionisation may be calculated using the Henderson–Hasselbalch equation (see section 3.9). It is not practical to measure the solubility of a species in a membrane but an estimate of the degree of solubility can be obtained from the partition coefficients of model systems such as the octanol/water system (see section 3.10). The accuracy of these predictions depends on the accuracy of the relationship between the model system and the membrane. As a general rule of thumb, solutes with high partition coefficient values will be more likely to be lipid soluble whereas those with low partition coefficient values will be less likely to be soluble in the membrane. However, it should be noted that a high degree of lipid solubility (high partition coefficient value) can be counter-productive (see section 3.10.2).

At constant temperature the rate of diffusion of an uncharged species through a membrane is dependent on the partition coefficients for the entry (K_1) and exit of the drug (K_2), the thickness of the membrane (x), the concentration gradient across the membrane ($c_1 - c_2$), and the surface area (S) of the membrane involved in the diffusion. The relationship between the rate of diffusion and these parameters is summarised in Fick's first law of diffusion:

$$J = \frac{DS(K_1c_1 - K_2c_2)}{x} \tag{4.1}$$

where J is the rate of appearance of the drug in the fluid on side 2 (Figure 4.15) of the membrane, D is a constant known as the diffusion coefficient and c_1 and c_2 are the solute concentrations defined in Figure 4.15. The diffusion coefficient is a characteristic property of the diffusing substance–membrane system. It makes allowance for the chemical and physical states of the diffusing species and the membrane. The term D/x is known as the permeability coefficient (P) for the membrane and the diffusing substance. It is a measure of the solute's ability to move from the aqueous medium into the membrane. This deduction is supported by the fact that experimental evidence shows that for a number of compounds the value of P increases with an increase in the value of their non-polar solvent/water partition coefficients.

4.3.4 Facilitated Diffusion

Facilitated diffusion involves a carrier protein that transports the substance through the membrane (Figure 4.16). It only occurs in the direction of the concentration gradient, that is, from high to low concentration and so does not require energy supplied by the cell. The rate of transport will be rapid for low concentrations of solute but will slow down when the solute concentration reaches the point where it saturates all the available carriers. Facilitated diffusion appears to play a minor role in the transport of xenobiotics through a membrane but is the mechanism by which glucose is transported through most cell membranes.

4.3.5 Active Transport

Active transport is the transport of a solute through a membrane by a so-called **carrier protein**. The solute combines with a specific protein causing this protein to change its conformation. This change results in the transport of the solute from one side of the membrane to the other, where it is released into the aqueous medium (Figure 4.16). The process usually operates against the concentration gradient, with the solute travelling from a low concentra-

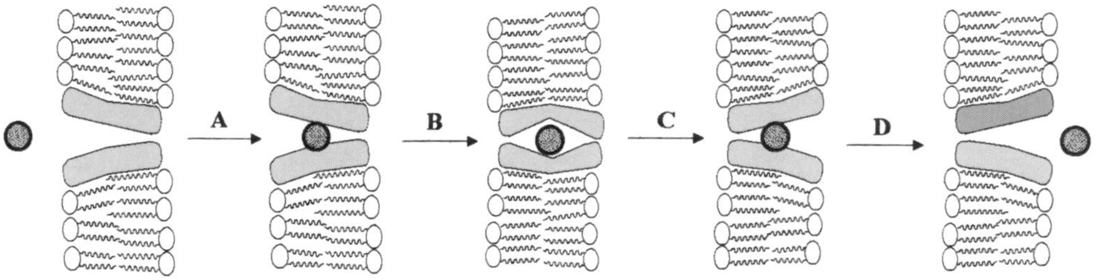

Figure 4.16. A schematic representation of carrier protein transport: (A) the substrate binds to the carrier protein; (B, C) the conformation of the carrier protein changes; (D) the substrate is released on the other side of the membrane.

tion to a high concentration. This process requires the cell to expend considerable quantities of energy.

The carrier proteins are highly selective, transporting solutes with specific chemical structures. They are involved in the transport of many naturally occurring compounds and so will also transport drugs with structures related to these natural products. For example, the drug levodopa, which is used to treat Parkinson's syndrome, is an amino acid. It is transported by the same active transport system as the naturally occurring amino acids tyrosine and phenylalanine. This type of structural relationship can be use in the approach to the design of new drugs.

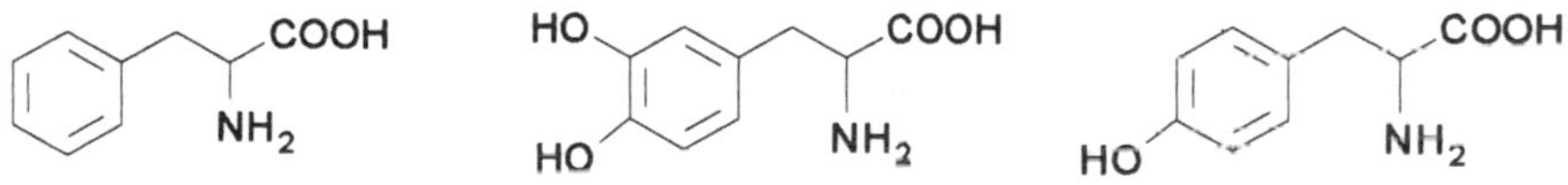

Phenylalanine Levodopa Tyrosine

The rate of active transport is dependent on the concentration of the solute at the absorption sites. Transport follows first-order kinetics at concentrations less than those required to saturate the available carriers but changes to zero order at concentrations above the saturation point. Consequently, increasing a drug's concentration in the extracellular fluid above this saturation concentration does not increase the rate at which the drug is delivered to its site of action in the cell provided that the drug is transferred through the cell membrane by an active transport mechanism. One of the most important active transport systems is the sodium pump. This is a protein that moves Na^+ into a cell when the cell is deficient in those ions. A number of drugs, such as digitalis (see Figure 1.6) act by interfering with active transport.

4.3.6 Endocytosis

Large molecules, fragments of dead tissue, whole bacteria and other particles that are visible under a microscope can pass through a membrane into a cell by a process known as **endocytosis** (Figure 4.17). In endocytosis, contact of the substance being transported with the membrane causes the membrane to form a pocket (invaginate). The substance enters the pocket,

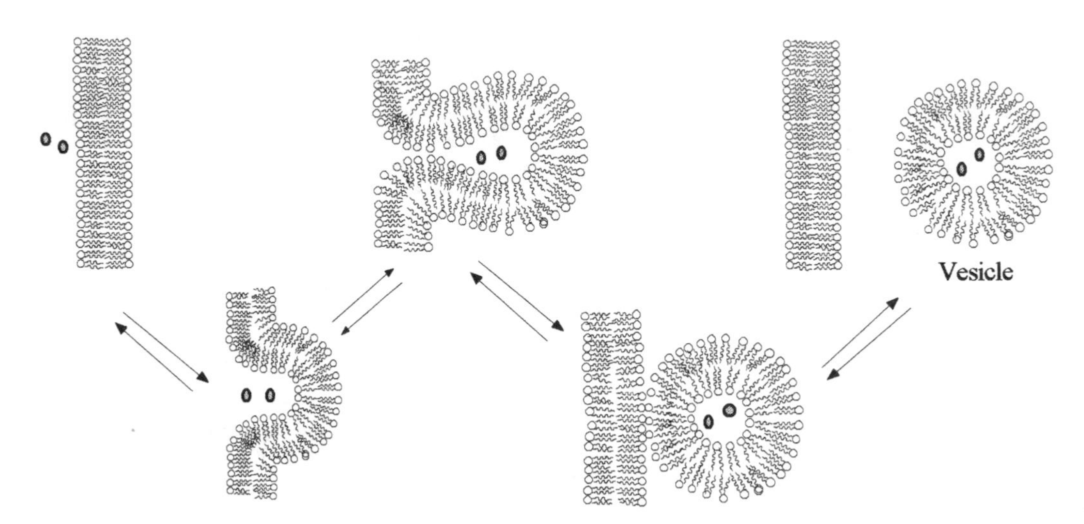

Figure 4.17. Schematic representations of endocytosis (left to right) and exocytosis (right to left). The extracellular fluid is on the left of each membrane and intracellular fluid is on the right of each membrane.

which closes around the substance to form a bag-like structure. This structure separates from the membrane and passes into the intracellular fluid. Once in the intracellular fluid the membrane of the vesicle has to disperse in order to release its contents into the intracellular fluid. In this independent form the structure is known as a **vesicle**. Endocytosis has the potential to decrease the surface area of the membrane but this is balanced by coupling with exocytosis (see section 4.3.7).

Endocytosis can be classified into two general types: **phagocytosis** and **pinocytosis**. Phagocytosis (cell eating) is concerned with the transport of particles that are visible under a microscope through the membrane. Pinocytosis (cell drinking) is the same process, except that the substances being transported are in solution. In both cases two different mechanisms have been discovered: **constitutive endocytosis** and **receptor-mediated endocytosis**. The former is a non-induced continuous process whereas the latter occurs mainly through receptors located either in the membrane or in either clathrin- (a polypeptide) or caveolin- (a protein) coated indentations in the membrane. In these cases the cytoplasmic surface of the vesicle is coated with the appropriate protein. Receptor-mediated endocytosis is more specific and occurs more rapidly than constitutive endocytosis. It is responsible for the transport of low-density lipoproteins, vitamins, insulin, iron, toxins, viruses and growth factors into a cell.

4.3.7 Exocytosis

Exocytosis is essentially the reverse of endocytosis. The substance that is to be transferred out of the cell is packaged in the form of a vesicle, which fuses with the membrane. This fusion results in the formation of a pocket open to the extracellular fluid. The substance diffuses out of the pocket but on the extracellular side of the membrane (Figure 4.17). The process requires the expenditure of energy by the cell and the presence of Ca^{2+} ions but details of the mechanism are not fully known. However, it has been shown to occur by either a **non-constitutive** or a **constitutive** pathway. The former pathway uses vesicles that are lined with the protein clathrin by the appropriate organelles whereas the latter utilises unlined vesicles. Exocytosis has the potential to increase the total quantity of the membrane of a cell. This potential increase is balanced in healthy cells by the coupling of exocytosis with the removal of excess membrane by endocytosis.

Exocytosis is used to secrete digestive enzymes into the gastrointestinal tract and release many hormones and neurotransmitters into the extracellular fluid.

4.4 Drug Action that Affects the Structure of Cell Membranes and Walls

Most drugs act on the enzymes and receptors found in cell membranes and walls. However, a number of drugs act by either, disrupting the structure of the cell membranes and walls, or inhibiting the formation of cell membranes and walls or blocking ion channels. For example, the venoms of both the eastern diamondback rattlesnake and the Indian cobra contain the enzyme phospholipase A_2. This enzyme catalyses the hydrolysis of the C_2 fatty acid residue from phosphatidyl lipids. The phospholipid product of this hydrolysis acts as a detergent, breaking down the membranes of red blood cells and causing them to burst, usually with fatal results to the infected mammal.

$$\begin{array}{l} CH_2\text{-O-P(=O)(O}^-\text{)-O-X} \\ | \\ RCO\text{-O-}CH \\ | \\ R'CO\text{-O-}CH_2 \end{array} \xrightarrow[\text{Phospholipase } A_2]{H_2O} \begin{array}{l} CH_2\text{-O-P(=O)(O}^-\text{)-O-X} \\ | \\ HO\text{-}CH \\ | \\ R'CO\text{-O-}CH_2 \end{array} + RCOO^-$$

Acts as a detergent

In general, drugs that either disrupt the structure of cell membranes and walls or their synthesis appear to act by:

(i) inhibiting the action of enzymes and other substances in the cell membrane involved in the production of compounds necessary for maintaining the integrity of the cell;
(ii) inhibiting processes involved in the formation of the cell wall, resulting in an incomplete cell wall that leads to loss of vital cellular material and subsequent death of the cell;

(iii) forming channels through the cell wall or membrane, making it more porous and thus resulting in the loss of vital cellular material and the death of the cell;
(iv) making the cell more porous by breaking down sections of the membrane.

All microorganisms have plasma membranes that have characteristics in common. Consequently, drugs can act by the same mechanism on quite different classes of microorganism. For example, griseofulvin is both an antifungal and an antibacterial agent (see section 4.4.2.1). However, the membranes of prokaryotic cells exhibit a number of significantly different characteristics to those of eukaroytic cells (see section 4.1). It is these differences that must be exploited by medicinal chemists if they are to find new drugs to treat microbiological infestations. They also account for the selectivity of current drug substances when used on humans, animals and plants.

4.4.1 Antifungal Agents

Fungal infections usually involve the skin and mucous membranes of the body. The fungal microorganisms are believed to damage the cell membrane, leading to a loss of essential cellular components. Antifungal agents counter this attack by both **fungistatic** and **fungicidal** action. Fungistatic action occurs when a drug prevents the fungi from reproducing, with the result that it dies out naturally, whereas fungicidal action kills the fungi. The suffixes *-static* and *-cide* are widely used to indicate these general types of action.

4.4.1.1 Azoles

The azoles are a group of substituted imidazoles that exhibit fungistatic activity at nanomolar concentration and fungicidal activity at higher micromolar concentrations (Figures 4.18 and

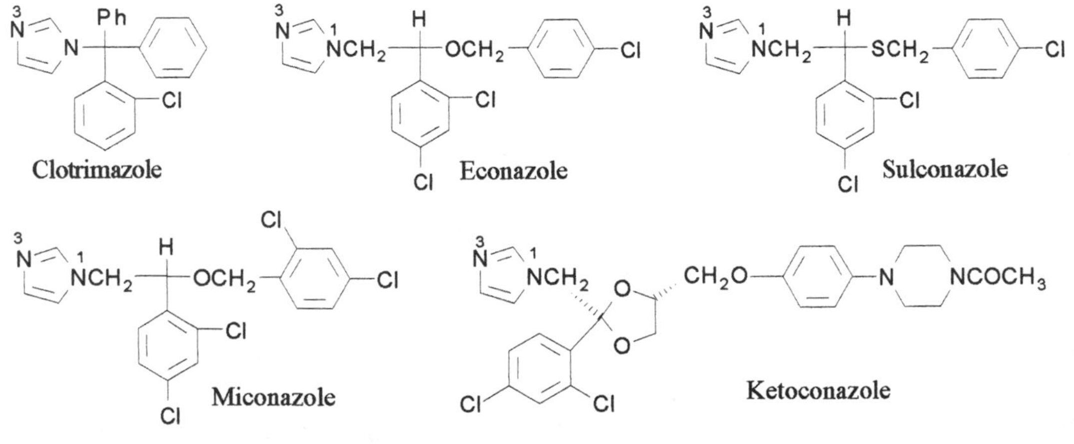

Clotrimazole

Econazole

Sulconazole

Miconazole

Ketoconazole

Figure 4.18. Examples of the structures of some active 1,3-diazoles. Note the common structural features.

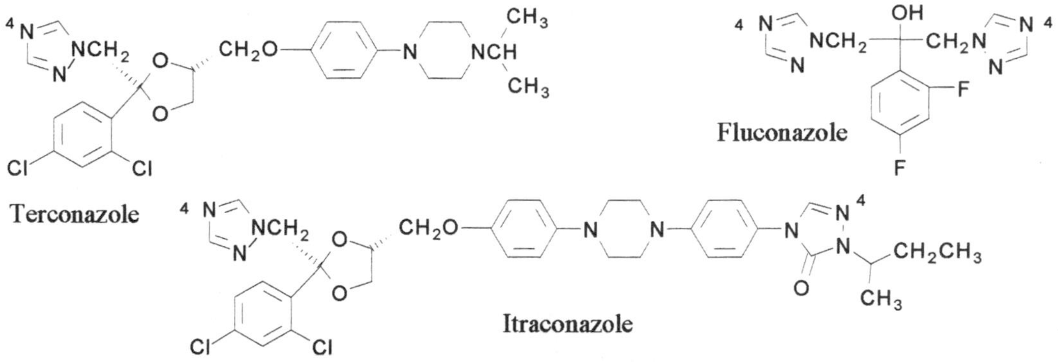

Figure 4.19. Examples of azoles based on 1,2,4-triazole ring systems.

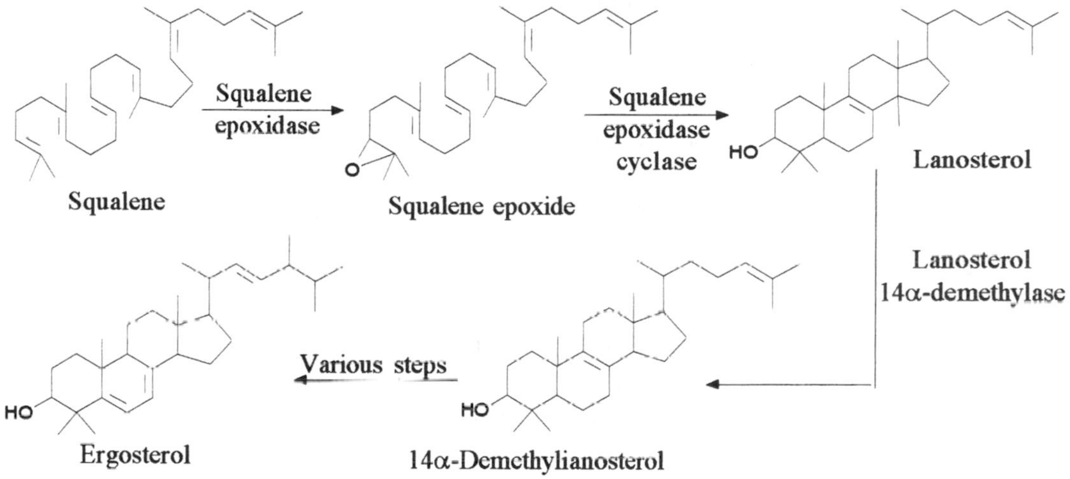

Figure 4.20. An outline of the biosynthesis of ergosterol.

4.19). They are active against most fungi that infect the skin and mucous membrane. Azoles are also active against some systemic fungal infections, bacteria, protozoa and helminthic species.

In common with most drugs, the azoles are believed to act at a number of different sites, all of which contribute to their fungicidal action. However, their main point of action is believed to be the inhibition of some of the cytochrome P-450 oxidases found in the membranes of the microorganisms. In particular, azoles have been linked to the inhibition of the enzyme 14α-sterol demethylase ($P\text{-}450_{DM}$), which is essential for the biosynthesis of ergosterol, the main sterol found in the fungal cell membranes (Figure 4.20). It is believed that nitrogen at position 3 of the imidazole rings (Figure 4.18) and nitrogen at position 4 of the triazole rings (Figure

4.19) bind to the iron of the haem units found in the enzyme, thereby blocking the action of the enzyme. This appears to lead to an accumulation of 14α-methylated sterols such as lanosterol in the membrane, which is thought to increase the membrane's permeability, allowing essential cellular contents to leak and cause irreversible cell damage and death. However, the precise details of the mode of action of azoles have yet to be fully elucidated. Azoles also inhibit the P-450 oxidases found in mammalian steroid biosynthesis, but in mammals much higher concentrations than those necessary for the inhibition of the fungal sterol 14α-demethylases are usually required.

Structure–action studies have shown that a weakly basic imidazole or 1,2,4-triazole rings substituted only at the N-1 position are essential for activity. The substituent must be lipophilic in character and usually contains one or more five- or six-membered ring systems, some of which may be attached by an ether, secondary amine or thioether group to the carbon chain. The more potent compounds have two or three aromatic substituents, which in the more potent compounds are singly or multichlorinated or fluorinated at positions 2, 4 and 6. These non-polar structures give the compounds a high degree of lipophilicity and hence membrane solubility.

4.4.1.2 Allylamines and Related Compounds

Allylamines are synthetic derivatives of 3-aminopropene (Figure 4.21). They are weak bases, their hydrochlorides being only slightly soluble in water. The allylamine group appears to be essential for activity.

Allylamines are believed to act by inhibiting squalene epoxidase, the enzyme for the squalene epoxidation stage in the biosynthesis of ergosterol in the fungal membrane (Figure 4.20). This leads to an increase in squalene concentration in the membrane with subsequent loss of membrane integrity, which allows the loss of cell contents to occur. Tolnaftate, although it is not an allylamine, appears to act in a similar fashion. However, allylamines do not appear to inhibit significantly the mammalian cholesterol biosynthesis.

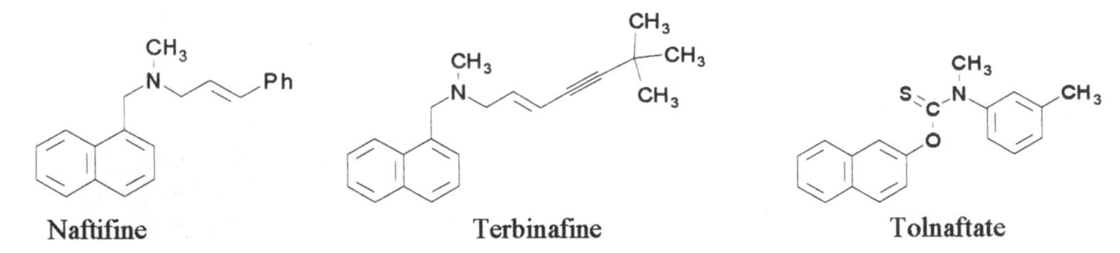

Figure 4.21. Examples of the structures of allylamines. Naftidine and terbinafine are used as fungicides to treat dermatophytes and filamentous fungi but only have a fungistatic action against pathogenic yeasts.

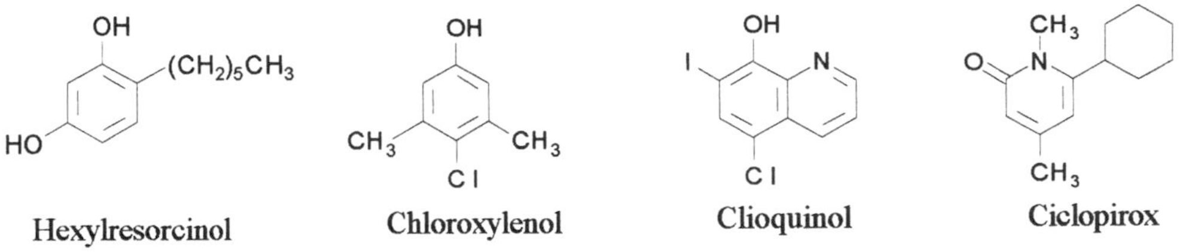

Hexylresorcinol Chloroxylenol Clioquinol Ciclopirox

Figure 4.22. Examples of phenolic compounds used as antifungal agents.

4.4.1.3 Phenols

There are numerous phenolic antifungal agents (Figure 4.22). They are believed to destroy sections of the cell membrane, which results in the loss of the cellular components and the death of the cell. The mechanism by which this destruction occurs is not known. Ciclopirox is not a phenol but appears to have a similar action. However, at low concentrations it has been shown to block the movement of amino acids into susceptible fungal cells.

4.4.2 Antibacterial Agents

Antibacterial antibiotics act at a variety of sites (see Appendix 2). However, in many cases, they act by either making the plasma membrane of bacteria more permeable to essential ions and other small molecules by ionophoric action (see section 4.4.2.1) or by inhibiting cell wall synthesis (see section 4.4.2.2). Those compounds that act on the plasma membrane also have the ability to penetrate the cell wall structure. In both cases, the net result is a loss in the integrity of the bacterial cell envelope, which leads to irreversible cell damage and death.

4.4.2.1 Ionophoric Antibiotic Action

Ionophores are substances that can penetrate a membrane and increase its permeability to ions. They may be naturally occurring compounds such as the antibiotic gramicidin A produced by *Bacillus brevis* and valinomycin obtained from *Streptomyces fulvissimus* or synthetic compounds like the crown and cryptate compounds (Figure 4.23). However, ionophores transport ions in both directions across a membrane. Consequently, they will only reduce the concentration of a specific ion until its concentration is the same on both sides of a membrane. However, a number of drugs are believed to owe their action to the ionophoric transfer of essential substances out of the cell.

Ionophores operate in two ways:

(i) they form channels across the membrane through which ions can diffuse down a concentration gradient (see Figure 4.24a);

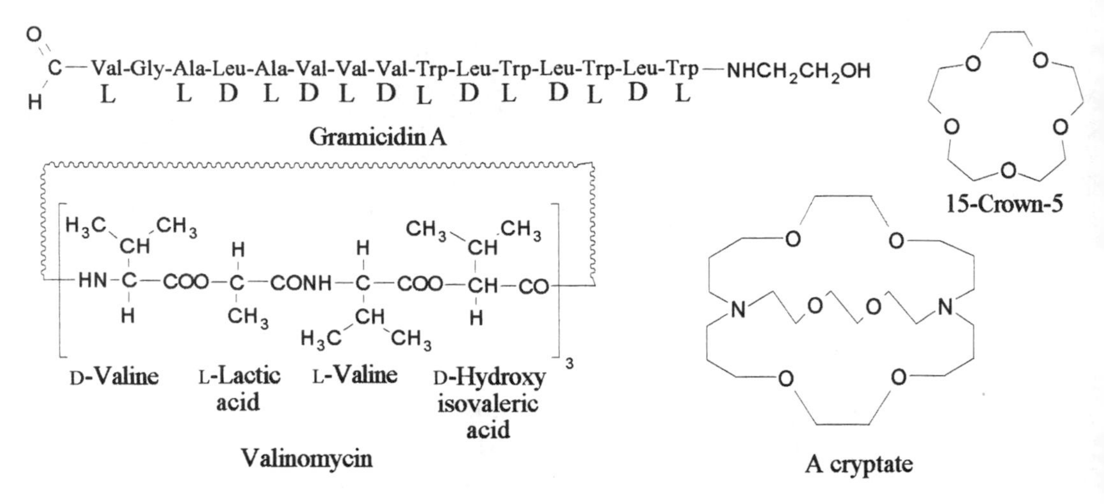

Figure 4.23. Examples of naturally occurring and synthetic ionophores.

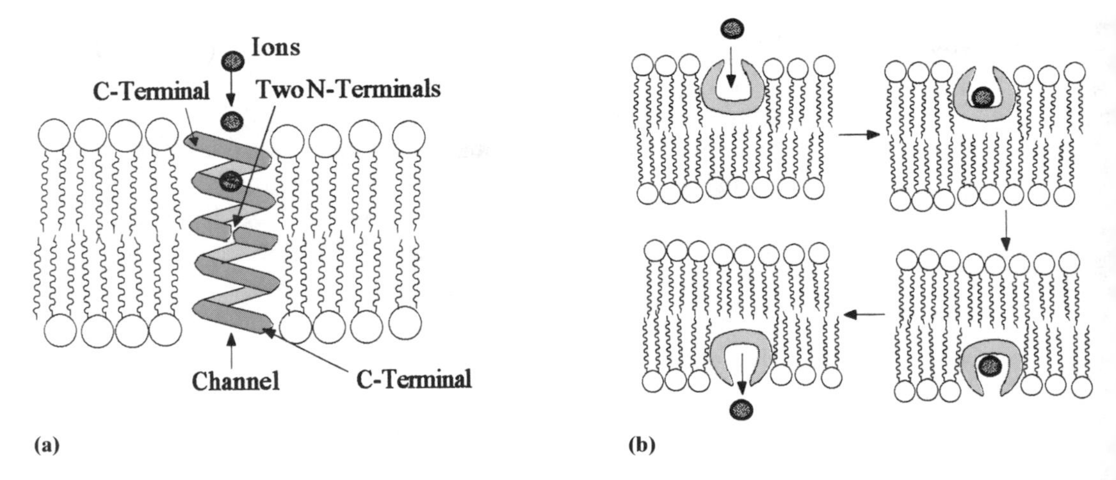

Figure 4.24. The general mode of action of ionophores in ion transport. (a) A channel formed by two gramicidin A molecules, N-terminal to N-terminal. (b) The sequence of events in the operation of a carrier ionophore such as valinomycin.

(ii) they act as carriers that pick up the ion on one side of the membrane, transport it across and release it into the fluid on the other side of the membrane (Figure 4.24b).

Channel ionophores. The structure of each channel is characteristic of the channel-former. For example, gramicidin forms a channel composed of two molecules whose N-terminals meet in the middle of the membrane. Each gramicidine molecule is in the form of a left-handed helix, which results in the polar groups lining the interior of the channel. This facilitates the transfer of polar ions through the channel. A single gramicidin channel can allow the transport of up to 10^7 K^+ ions per second.

Carrier ionophores. Carrier ionophores are specific for particular ions. For example, valinomycin will transport K^+ ions but not Na^+ or Li^+ ions. It is believed to form an octahedral complex with six carbonyl-group oxygen atoms acting as ligands. The resulting chelation complex has a hydrophobic exterior, which allows the complex to diffuse through the membrane. However, the rigid nature of the molecule coupled with its size makes the binding site of valinomycin too large to form octahedral complexes with Na^+ and Li^+ ions. Consequently, it is more energetically favourable for Na^+ and Li^+ ions to remain in solution as their hydrated ions.

Ionophores are mainly active against Gram-positive bacteria. However, until now, most of the compounds examined do not differentiate significantly between bacterial and mammalian membranes and so are of little clinical use. However, they are of considerable use as research tools in the investigation of drug action.

4.4.2.2 Cell Wall Synthesis Inhibition

The cell walls of all bacteria are being replaced continuously because they are continuously being broken down by enzymes in the extracellular fluid. Antibacterial agents can inhibit this replacement biosynthesis of the cell wall at any stage in its formation. Investigations using *Staphylococcus aureus* have yielded a great deal of detail about the biosynthesis of its cell wall but there are still areas of the biosynthesis that have not yet been fully elucidated. Experimental investigations of the cell wall synthesis of other bacterial species suggest that similar routes are followed. A detailed knowledge of the route followed and the enzymes involved is an extremely useful prerequisite in the quest for new drug substances.

It is convenient to introduce the subject of antibacterial action due to inhibition of cell wall synthesis by dividing the synthesis into three stages:

(i) the formation of precursor starting materials;
(ii) the formation of the peptidoglycan chains;
(iii) the cross-linking of the peptidoglycan chains.

However, it should be realised that not only can an antibiotic inhibit cell wall formation but it may also have other areas of action such as the plasma membrane of a bacteria. Furthermore, it is emphasised that the biochemical pathways discussed are a simplification based on experimental evidence. However, it is likely that the drugs act in the same manner on other susceptible bacteria.

Drugs that inhibit the formation of the starting compounds. A convenient starting point for cell wall synthesis is *N*-acetylglucosamine-1-phosphate (NAG-1-P), which is found in all life forms. This compound is believed to react with uridine triphosphate (UTP) to form uridine

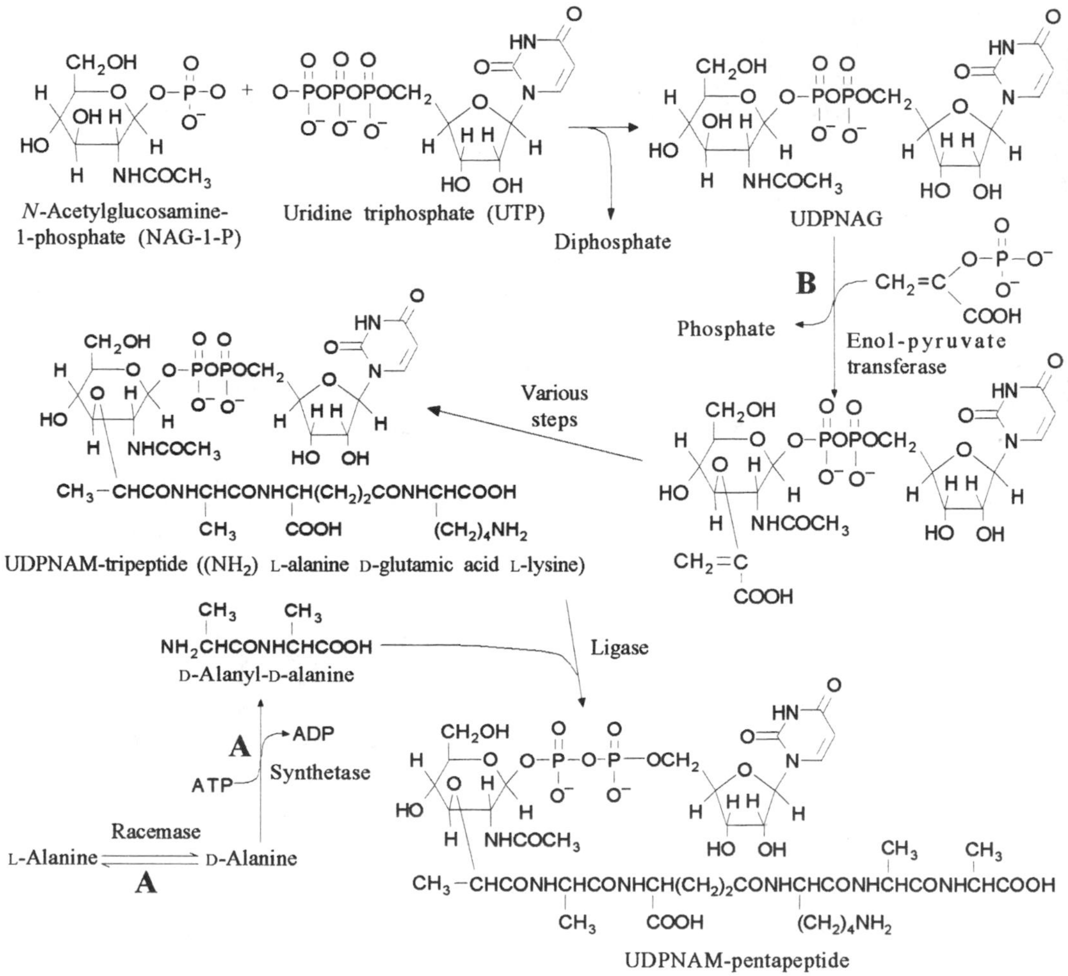

Figure 4.25. An outline of the biosynthesis of the precursors of the peptidoglycan chain of the cell wall of *Staphylococcus aureus*.

diphospho-*N*-acetylglucoamine (UDPNAG), one of the precursors of the peptidoglycan chain (Figure 4.25). Some of the UDPNAG is further converted by a series of steps into the uridine diphospho-*N*-acetylmuramic acid pentapeptide derivative (UDPNAM-pentapeptide), the second precursor of the peptidoglycan polymer chain. Drug action can inhibit any of the steps in the formation of both UDPNAG and UDPNAM-pentapeptide. However, inhibition of the synthesis of the latter is likely to be potentially more rewarding because its formation requires a larger number of steps, which gives a wider scope for intervention. Drugs in clinical use that act by inhibiting different processes in this stage of cell wall synthesis are cycloserine and fosfomycin (Figure 4.26).

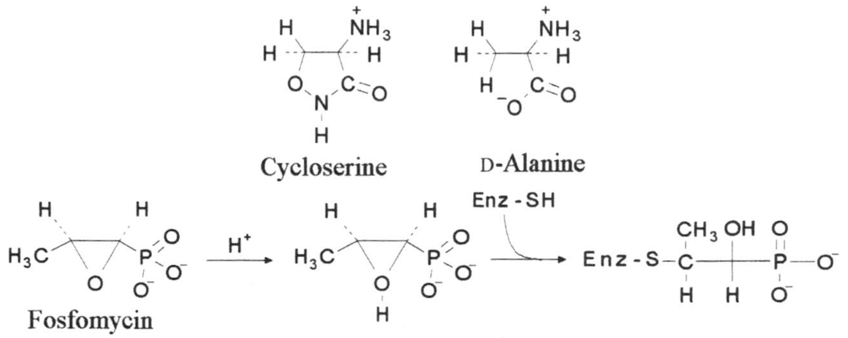

Cycloserine

D-Alanine

Fosfomycin

Enz-SH

Figure 4.26. Cycloserine and fosfomycin.

Cycloserine is a broad-spectrum antibiotic produced by *Streptomyces orchidaceus*. The drug is used mainly as a second-line antitubercular agent. It enters the bacteria by active transport systems, which results in a high concentration in the bacterial cell, a primary requirement for activity. D-Cycloserine inhibits both alanine racemase and D-alanyl-D-alanyl synthetase, which blocks the conversion of the tripeptide to the pentapeptide at two places (A in Figure 4.25). The affinity of the enzymes in *Staphylococcus aureus* for the drug has been found to be 100 times higher than the affinity for its natural substrate D-alanine. This affinity is believed to depend on the isoxazole ring, whose shape corresponds to one of the conformations of D-alanine. It is believed that the rigid structure of the isoxazole ring gives the drug a better chance of binding to the active sites of the enzymes than the more flexible structure of D-alanine.

Fosomycin, produced by a number of *Streptomyces* species, is active against both Gram-positive and Gram-negative bacteria. However, it is used mainly to treat Gram-positive infections. The drug acts by inhibiting the enol-pyruvate transferase (B in Figure 4.25) that catalyses the incorporation of phosphoenolpyruvic acid (PEP) into the UDPNAG molecule. However, the drug does not inhibit other enol-pyruvate transferases used to incorporate PEP in a number of other biosynthetic reactions. Consequently, it appears that the activity of the drug is due to it forming an inactive product with the enzyme. It has been suggested that this product is formed by the acid-catalysed nucleophilic substitution of the oxiran ring by the sulphydryl groups of the cysteine residues in the active site of the enzyme.

Drugs that inhibit the synthesis of the peptidoglycan chain. The sequence of reactions, starting from UDPNAM-pentapeptide and UDPNAG, to form the peptidoglycan chain (Figure 4.27) are not completely known although the main stages have been identified. However, it is known that the reactions are catalysed by membrane-bound enzymes. A number of antibiotics, such as bacitracin, are believed to inhibit some of the stages of the biosynthesis of the peptidoglycan chains.

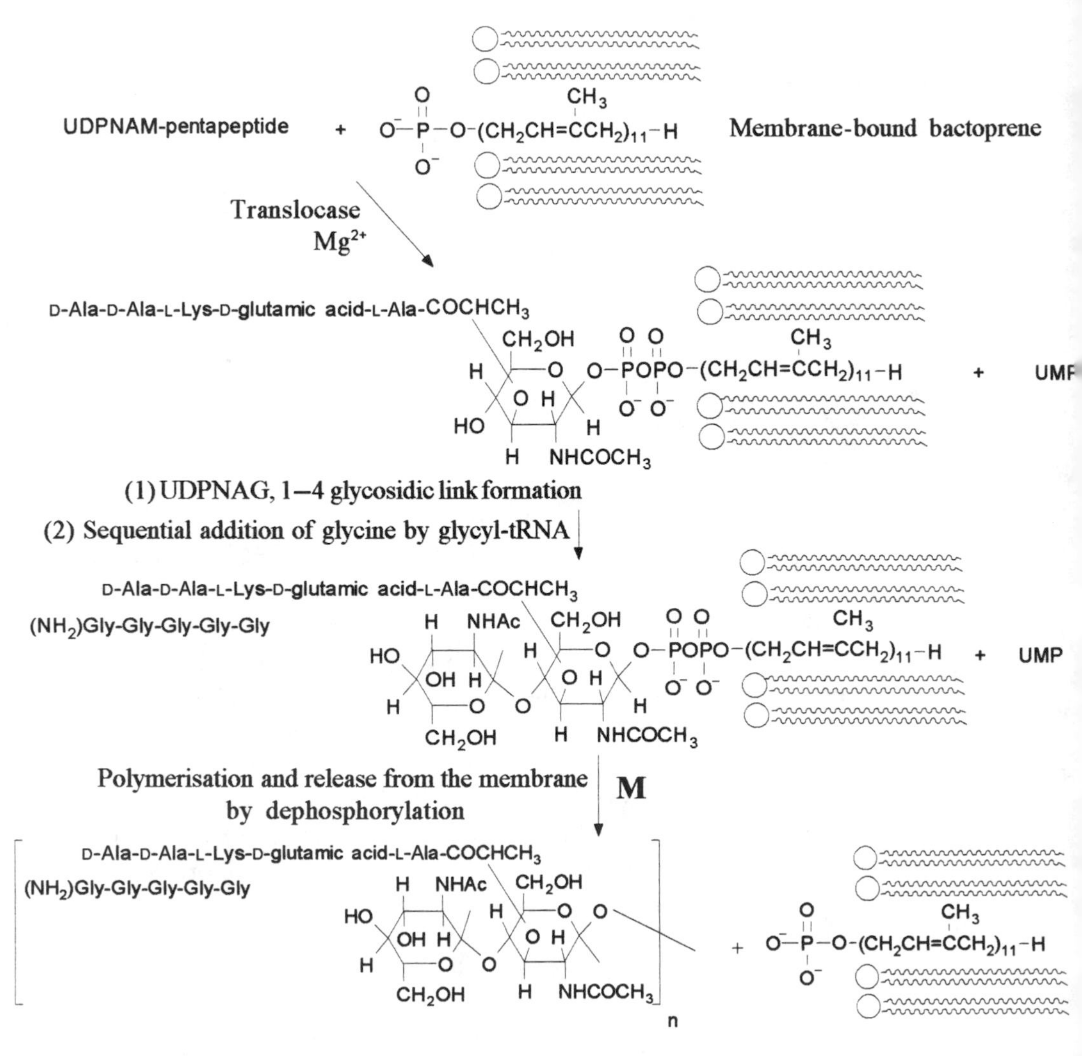

Figure 4.27. An outline of the formation of the peptoglycan chains of *Staphylococcus aureus* from UDPNAM-pentapeptide and UDPNAG. UMP = uridine monophosphate.

Key
→ =CONH

D-Asn → L-Asp
L-His → L-Asp → L-Lys ← L-Ile ← D-Glu ← L-Leu—OC (thiazoline ring, S, N) — CH(NH$_2$)—CH(CH$_3$)CH$_2$CH$_3$
D-Phe → L-His
L-Ile → D-Phe
D-Orn → L-Ile
L-Lys → D-Orn

Bacitracin A

Bacitracin is a mixture of similar peptides produced from *Bacillus subtilis*. The main component of this mixture is bacitracin A, which is active against Gram-positive bacteria. However, its high degree of neuro- and nephrotoxicity means that the drug is seldom used and even then

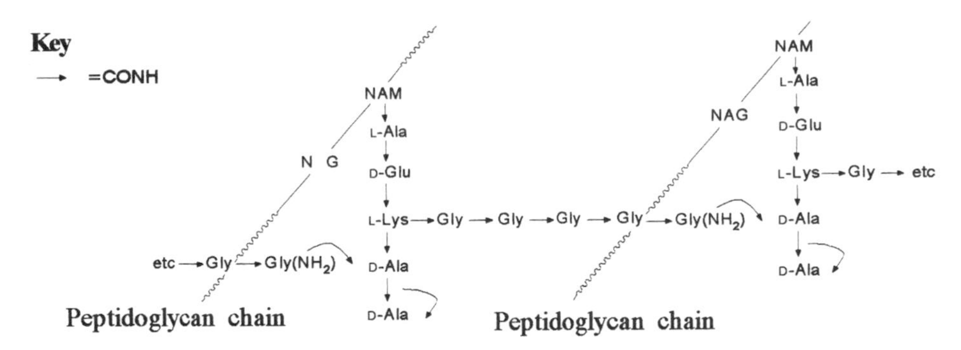

Figure 4.28. A schematic outline of the formation of the peptide cross-links in the formation of the cell wall of *Staphylcoccus aureus*.

somewhat cautiously. Its main site of action appears to be the inhibition of the dephosphorylation of the membrane-bound phospholipid carrier bactoprene (step M in Figure 4.27). Its action is enhanced by the presence of zinc ions.

Drugs that inhibit the cross-linking of the peptidoglycan chains. The final step in the formation of the cell wall is the completion of the cross-links. This converts the water-soluble and therefore mobile peptidoglycans into the insoluble stationary cell wall. Investigations using *Staphylcoccus aureus* indicated that the cross-linking is brought about by a multistep displacement of the terminal alanine of the peptide attached to one peptidoglycan chain and its replacement by the terminal glycine of the peptide attached to a second peptidoglycan chain (Figure 4.28). This reaction is catalysed by transpeptidases.

The β-lactam group of antibiotics inhibit cell wall synthesis by inhibiting the transpeptidases responsible for the cross-linking between the peptidoglycan chains. This group of antibiotics, named after the β-lactam ring that they all have in common, includes the widely used penicillins and the cephalosporins (Figure 4.29). Both of these groups of β-lactam antibiotics are more effective against Gram-positive bacteria than Gram-negative bacteria. However, some cephalosporins, such as ceftazidime, which is administered intravenously, are very effective against Gram-negative bacteria.

The β-lactam antibiotics have to reach the plasma membrane of the bacteria before they can act. Because the outer surfaces of Gram-positive bacteria are covered by a thin layer of teichoic acids (see section 4.2.5.1) they offer less resistance to drug penetration than Gram-negative bacteria where the drug has to penetrate both the outer membrane and the periplasmic space (see Figure 4.11) before it can interfere with cell wall synthesis. In Gram-negative bacteria the drug diffuses through the outer membrane through **porin channels** formed by integral trimeric proteins. A large number of these channels, with diameters of about 1.2 nm, are found in each bacterial cell wall. However, not all porin channels are able to transport β-lactam drugs. Some bacterial genera such as *Pseudomonas* are resistant to β-lactam antibiotics because their porin channels will not allow the transport of these drugs. Once through the outer mem-

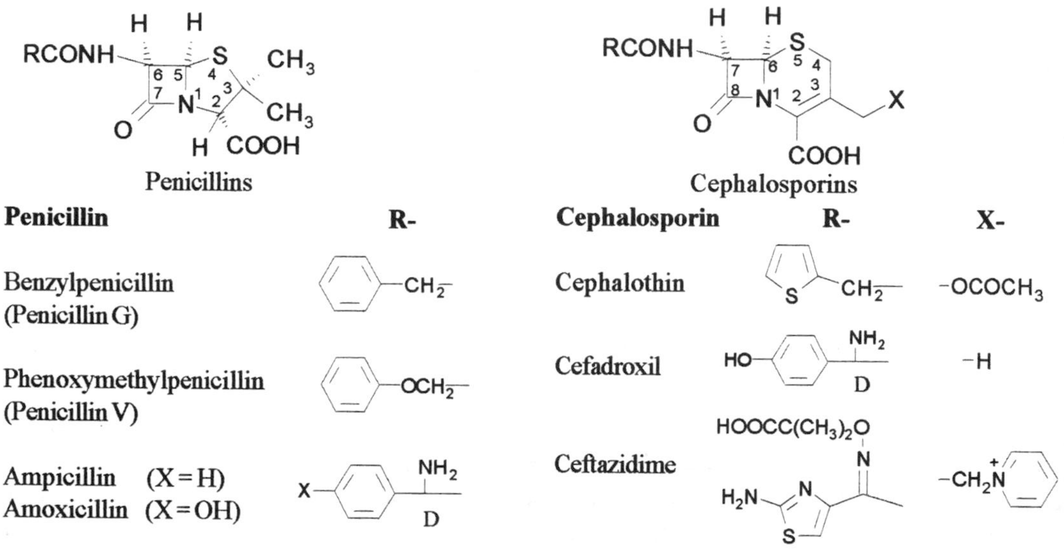

Penicillins

Cephalosporins

Penicillin	R-
Benzylpenicillin (Penicillin G)	$C_6H_5CH_2-$
Phenoxymethylpenicillin (Penicillin V)	$C_6H_5OCH_2-$
Ampicillin (X=H) Amoxicillin (X=OH)	$X-C_6H_4-CH(NH_2)-$ (D)

Cephalosporin	R-	X-
Cephalothin	2-thienyl-CH_2-	$-OCOCH_3$
Cefadroxil	$HO-C_6H_4-CH(NH_2)-$ (D)	$-H$
Ceftazidime	$HOOCC(CH_3)_2O-N=C(-)$ 2-amino-thiazol-4-yl	$-CH_2N^+$ (pyridinium)

Figure 4.29. Examples of penicillins and cephalosporins. The R residues of ampicillin, amoxicillin and ceftazidime have D-configurations.

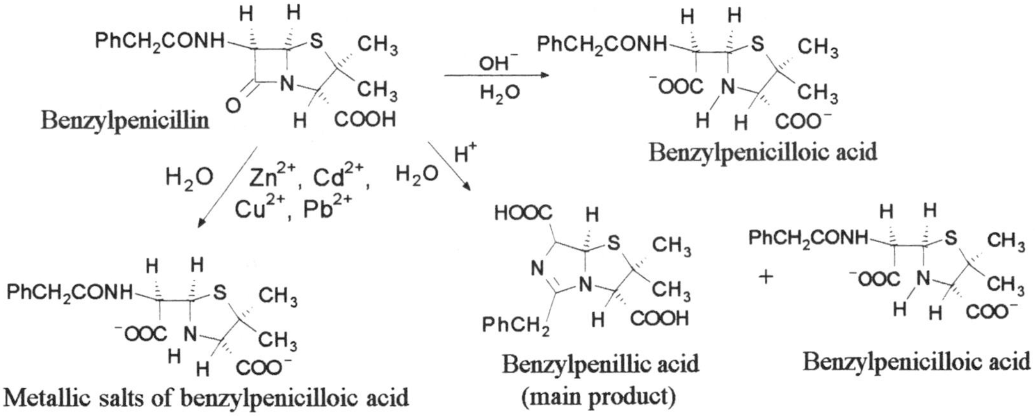

Benzylpenicillin → (OH^-, H_2O) → Benzylpenicilloic acid

Benzylpenicillin → (H_2O, Zn^{2+}, Cd^{2+}, Cu^{2+}, Pb^{2+}) → Metallic salts of benzylpenicilloic acid

Benzylpenicillin → (H_2O, H^+) → Benzylpenillic acid (main product) + Benzylpenicilloic acid

Figure 4.30. Some of the decomposition routes of the β-lactam ring of benzylpenicillin.

brane the drug diffuses across the periplasmic space, which contains β-lactamases that can inactivate the drug. Gram-positive bacteria also produce these enzymes, which they release into the extracellular fluid. Finally, the drug penetrates to the outer surface of the plasma membrane where it binds to, and blocks the action of, the transpeptidases and other proteins involved in cell wall synthesis. The precise nature of the blocking mechanism has not yet been fully elucidated but appears to involve the β-lactam ring system. This ring system is very reactive and is easily decomposed by acid- and base-catalysed hydrolysis (Figure 4.30), the rate of which depends on the structure of the penicillin.

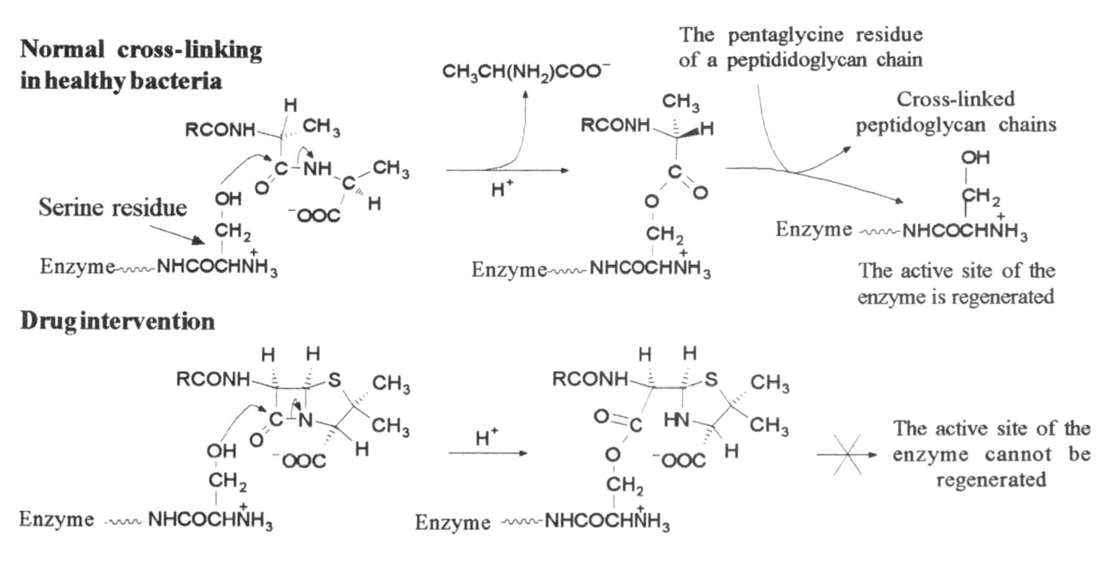

Figure 4.31. A schematic outline of the chemistry proposed for the action of penicillins.

The final stage in the formation of the cross-links between the peptidoglycan chains in bacteria is catalysed by a glycopeptide transpeptidase. It is believed that the hydroxyl group of a serine residue in this enzyme displaces the last alanine residue from the tetrapeptide chain. The displaced alanine diffuses away from the reaction site, allowing attack by the amino group of the terminal glycine of the pentaglycine chain on the alanine bound to the enzyme to complete the peptide linkage and regenerate the enzyme (Figure 4.31). However, it is thought that because the geometry of the penicillins resembles that of the alanyl-alanyl unit the bacterium mistakes the drug for its normal substrate. The β-lactam ring reacts with the enzyme to form a covalently bound acyl derivative. The 1,3-thiazolidine ring of this derivative is believed to prevent a pentaglycine unit from attacking the enzyme–acyl linkage and regenerating the active site of the enzyme.

The increasing numbers of bacteria resistant to β-lactam antibiotics are slowly becoming a major problem. Resistance to penicillins and cephalosporins by some bacteria is mainly due to inactivation of the drug by the β-lactamases produced by that bacterium. Both Gram-positive and Gram-negative bacteria produce β-lactamases. In the former case the enzyme is liberated into the medium surrounding the bacteria. This results in the hydrolysis of the β-lactam ring and inactivation of the penicillin and other β-lactam drugs before the drug reaches the bacteria. However, with Gram-negative bacteria, the hydrolysis takes place within the periplasmic space. In addition some Gram-negative bacteria produce *acylases*, which can cleave the side chains of penicillins. Bacteria that have developed a resistance to β-lactam antibiotics are treated using either a dosage form incorporating a β-lactamase inhibitor such as clavulanic acid or sulbactam or a lactamase-resistant drug.

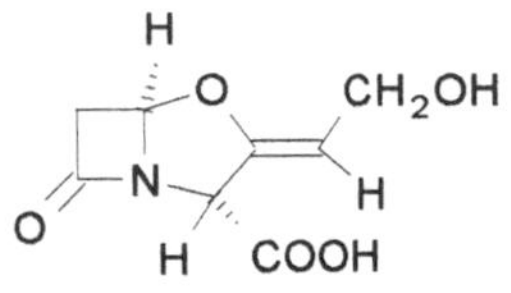

Clavulanic acid

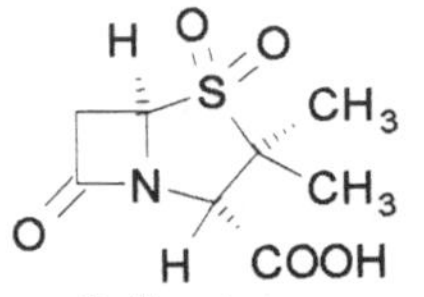

Sulbactam

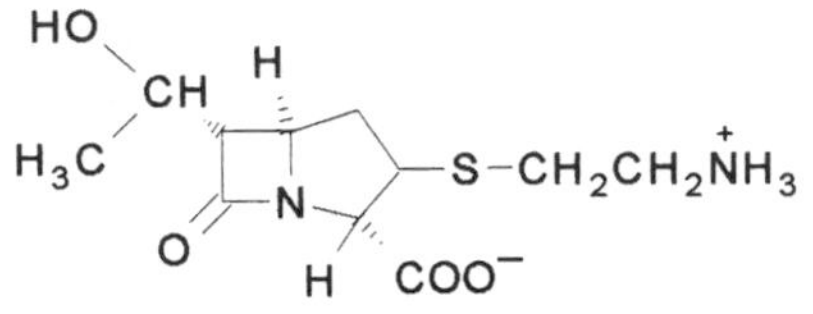

Thienamycin

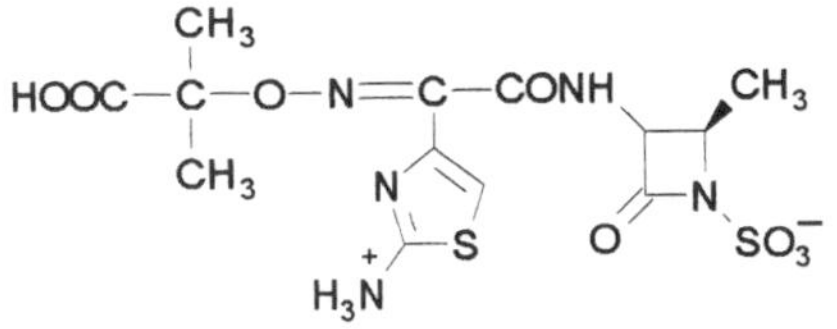

Azthreonam

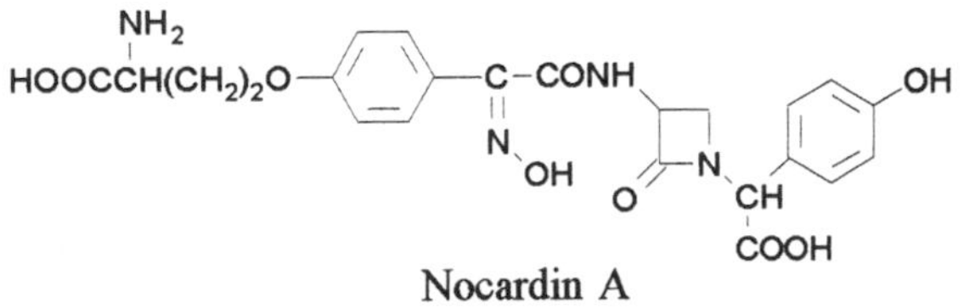

Nocardin A

The relationship between the structures of β-lactams and their activity has been the subject of much discussion. Originally it was believed that the amide-linked side chain, the carboxylic acid at position 2 and the fused thio-ring systems were all essential for the pharmacological activity of β-lactam antibiotics. However, the discovery of β-lactams such as thienamycin, azthreonam (for synthesis, see Figure 12.13) and nocardin A whose structures do not contain all these functional groups suggests that the β-lactam ring is the only essential requirement for activity.

Vancomycin (Figure 4.32), a glycopeptide antibiotic isolated from *Streptomyces orientalis* in 1955, also inhibits the formation of the peptide links between the peptidoglycan chains. In spite of the extensive use of the drug, very little bacterial resistance to vancomycin has developed. It is mainly used for Gram-positive infections but is irritating on intravenous injection. Oral administration does not give useful blood levels. However, the drug is used orally to treat pseudomembranous enterocolitis caused by high concentrations of *Clostridium difficile* in the intestine.

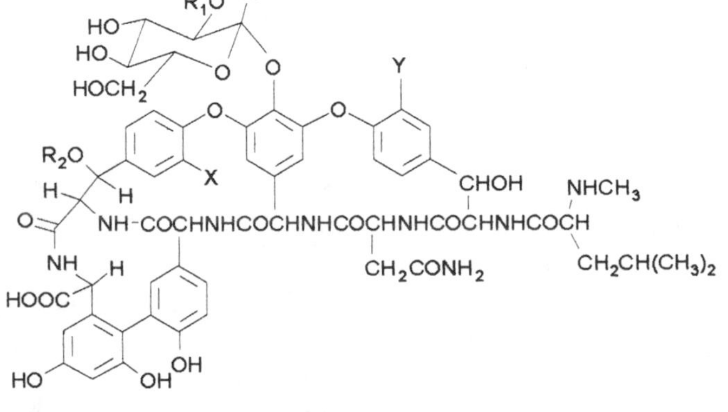

Figure 4.32. Vancomycin and some related compounds. Vancomycin: X = Cl, Y = Cl, R_1 = α-vancosaminyl, R_2 = H. Decaplanin: X = H, Y = Cl, R_1 = α-rhamnosyl, R_2 = α-epivancosaminyl. Orienticin A: X = Cl, Y = H, R_1 and R_2 = α-epivancosaminyl.

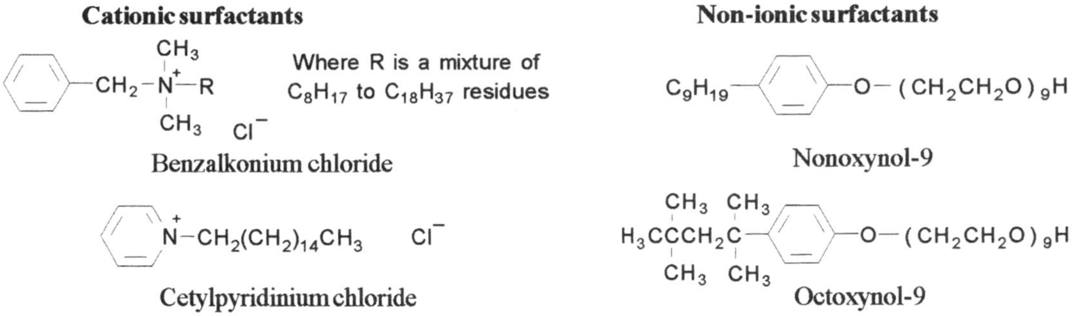

Cationic surfactants

Where R is a mixture of C_8H_{17} to $C_{18}H_{37}$ residues

Benzalkonium chloride

Cetylpyridinium chloride

Non-ionic surfactants

Nonoxynol-9

Octoxynol-9

Figure 4.33. Surfactants used as antibacterial agents.

The structure of vancomycin (Figure 4.32) is based on a tricyclic ring system of aliphatic and aromatic amino acids with a disaccharide side chain. This structure is rigid with a peptide-lined pocket that has a strong affinity for D-Ala-D-Ala residues. Vancomycin inhibits cell wall synthesis by binding to the D-Ala-D-Ala end group of the pentapeptide chain of the peptidoglycan cell wall precursor. NMR spectroscopy and molecular modelling suggest that the D-Ala-D-Ala residue is multiple hydrogen bonded to the vancomycin in this pocket. This inhibits the formation of the peptide cross-links between the polyglycan chains (Figure 4.28), which results in a loss of bacterial cell wall integrity.

Those bacteria that exhibit resistance to the drug appear to have replaced the D-Ala-D-Ala unit of the pentapeptide residue by a D-Ala-D-lactate unit. This structural change is believed to reduce the number of hydrogen bonds between the drug and the D-Ala-D-lactate of the peptidoglycan precursor by one. However, this small change is sufficient to prevent the drug from operating efficiently.

4.4.2.3 Surface-active Agents

Surface-active agents (see section 3.11) disrupt cell membranes because they dissolve in both the aqueous extracellular fluid and the lipid membrane. This lowers the surface tension of the membrane, which allows water to flow into the cell and ultimately results in lysis and the bactericidal action. In all cases a balance between the hydrophilic and lipophilc sections of the molecule is essential for action. Both cationic and non-ionic surfactants are used (Figure 4.33). In addition, detergent surfactants, such as sodium dodecyl sulphate, are also used to remove proteins from cell membranes.

4.4.3 Local Anaesthetics

Local anaesthetic agents block the nerves that transmit the sensation of pain. They mainly act on the cell membrane of the nerve cells (**neurons**; Figure 4.34). This anaesthetic action is reversible. However, although local anaesthetic agents are administered locally, they can also

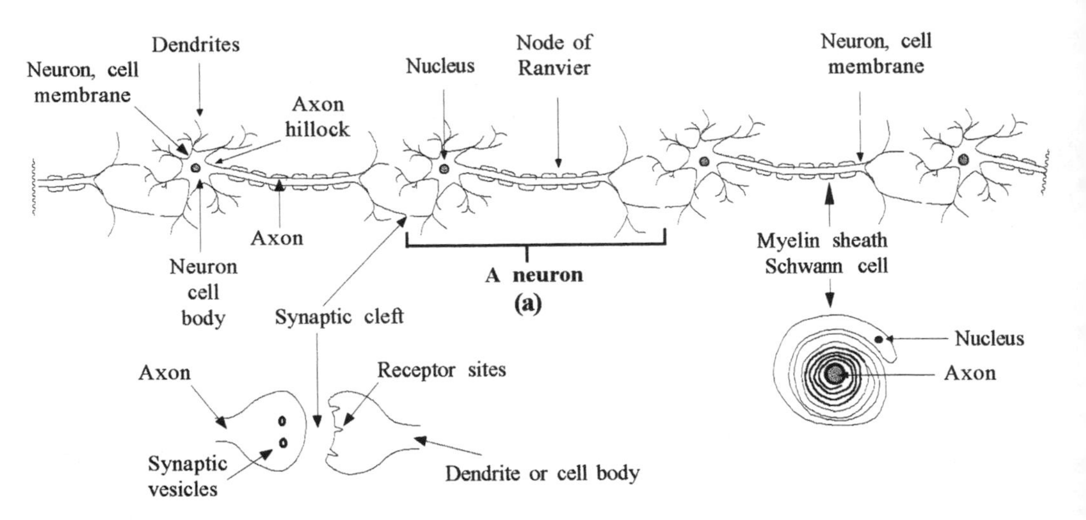

Figure 4.34. A schematic representation of a section of a nerve fibre. Nerve fibres consist of chains of cells known as neurons (a) where the axon of one neuron is separated from either the end of a dendrite or the cell body of a second neuron by the synaptic cleft. The synaptic vesicles contain the chemical messengers (neurotransmitters) that transmit the nerve impulse by diffusing across the synaptic cleft to the receptors. Nerves consist of bundles of nerve fibres enclosed in various protective tissues.

exert systemic effects because they can be transported from the point of application to other organs of the body by the bloodstream.

Neurons are activated by chemical, electrical and mechanical stimuli. They transmit the information provided by these sources as electrical signals known as the **action potential** or **nerve impulse**, which travels along the length of the cell at speeds of up to $120\,m\,s^{-1}$ in myelinated axons but only $10\,m\,s^{-1}$ in unmyelinated axons. At the synaptic knob the electrical signal releases chemical messengers, which cross the synaptic cleft and trigger the appropriate electrical signal in the next neuron.

The transmission of the signal along the neuron is due to successive changes in the potential difference across the membrane (**membrane potential**). For cells at rest, that is, cells that are undergoing no stimulation, the interior surface of the plasmic membrane is the negative side of the potential difference. The so-called resting potential of such cells can vary from –20 to about –90 mV. This potential is mainly due to the passive transport of small inorganic ions such as Na^+, K^+, Ca^{2+} and Cl^- ions through the membrane via ion channels (see section 4.2.2). The number per unit area of these ion channels varies: the highest density is at the nodes of Ranvier whereas there are considerably fewer in the myelinated sections of the neuron. Ion channels are selective in that they will only allow the passage of certain ions. Consequently, their classification is based on the type of ions they transport. Some ion channels are not permanently open but open briefly when changes occur in the conformations of the proteins forming the channel. These channels are referred to as **gated** channels. Opening of a channel may be initiated by either the presence of specific ligands (**ligand-gated channels**) or changes in the

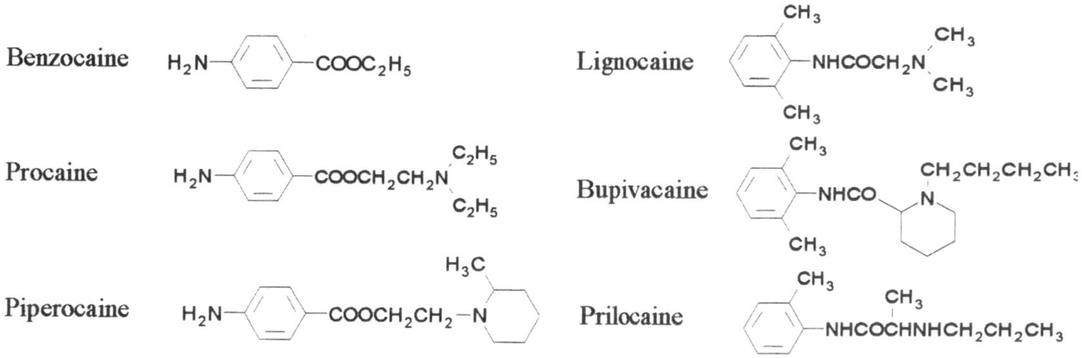

Figure 4.35. Examples of ester- and amide-based local anaesthetic agents in clinical use.

membrane potential (**voltage-gated channels**). The automatic closing of a gated channel after a brief time interval is an essential part of the mechanism of action potential transmission.

Stimulation of a nerve is believed to result in the opening of the Na^+ gated channels in the plasma membrane. This allows a large number of Na^+ ions to flood into the neuron before the gated channels automatically close. These Na^+ ions neutralise the negative potential of the inner surface of the membrane, which allows K^+ ions to flow out of the neuron through ungated K^+ leak channels. This results in the membrane being depolarised to the extent that its potential reaches values of about +40 mV relative to the exterior of the neuron. However, the automatic closure of the Na^+ channels and the action of the sodium pump in removing the excess Na^+ ions from the neuron repolarises the membrane by deneutralising the negative charges on the inner surface of the plasmic membrane. These negative charges attract K^+ ions back into the axon so that the membrane returns to its resting state, the complete process of depolarisation and repolarisation usually being accomplished in about 4 ms.

The initial action potential of a neuron either originates in the cell body or is generated by messages from the dendrites. It stimulates the sequence of changes described in the preceding paragraph. The action potential generated by these changes will trigger a similar series of changes in an adjacent section of the membrane, and so on. However, the automatic closing of the gated Na^+ channels means that the action potential usually moves along the axon away from the cell body. It also means that the nerve is ready to transmit a new action potential from the cell body within milliseconds of a previous transmission.

The main local anaesthetic agents in clinical use are basic compounds that can be broadly classified as either aromatic amide or ester derivatives (Figure 4.35). Cocaine and other alkaloid local anaesthetic drugs are only used under strictly controlled conditions because of their addictive properties. However, both the alkaloid and basic aromatic amide and ester local anaesthetic agents are believed to act by blocking the Na^+ ion channels, which prevents the initial influx of Na^+ ions into the neuron. This blocking action is thought to occur in three ways:

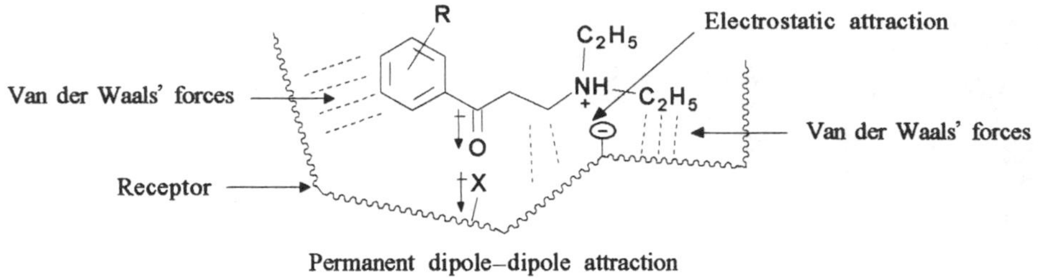

Figure 4.36. A schematic representation of the binding of ester-type local anaesthetic agents to a receptor site.

(i) the agent blocks the external entry to the Na^+ ion channel;
(ii) the drug enters the channel and acts as a stopper, closing the channel about midway along its length;
(iii) the drug binds to the proteins forming the channel and as a result distorts their structures to the extent that the channel will not allow the passage of Na^+ ions.

The structures of most of the local anaesthetic agents in clinical use contain basic groups such as tertiary amine groups, which are ionised at physiological pH:

$$RN(R')_2 + H^+ \rightleftharpoons RNH(R')_2$$

Because neutral molecules diffuse into the non-polar lipid membrane more easily than charged species, the local anaesthetic must enter the lipid membrane in its uncharged form. However, once inside the neuron, experimental evidence suggests that it is the charged form of the drug that binds to the receptor site(s). This is believed to occur by van der Waals' forces, dipole–dipole attractions and electrostatic forces (Figure 4.36).

Alkaloid local anaesthetic agents are believed to act by blocking the external entries of the Na^+ ion channels, whereas it is thought that the main action of basic amide- and ester-based agents is to internally block these channels. In the latter case, experimental evidence shows that the drug follows either a hydrophobic or a hydrophilic route to its site of action. In the hydrophobic route, the drug passes into the membrane where it diffuses within the membrane to its site of action (Figure 4.37). This route is particularly important because it offers a route for the blocking of closed Na^+ ion channels by local anaesthetic agents. In the hydrophilic route, the drug passes through the membrane into the intracellular fluid where it forms its conjugate acid, which is the active form of the drug. This species diffuses into the Na^+ ion channel to its site of action. Experimental evidence has shown that drugs following the hydrophilic route cannot enter Na^+ ion channels from the extracellular fluid because the channel entrances are too narrow. They have to enter via the wider interior entrances. Furthermore, experimental studies of the action of benzocaine on the giant squid axon have indicated that there are at least two sites of action of this drug, which implies that other local anaesthetic agents have more than one site of action inside the Na^+ channels.

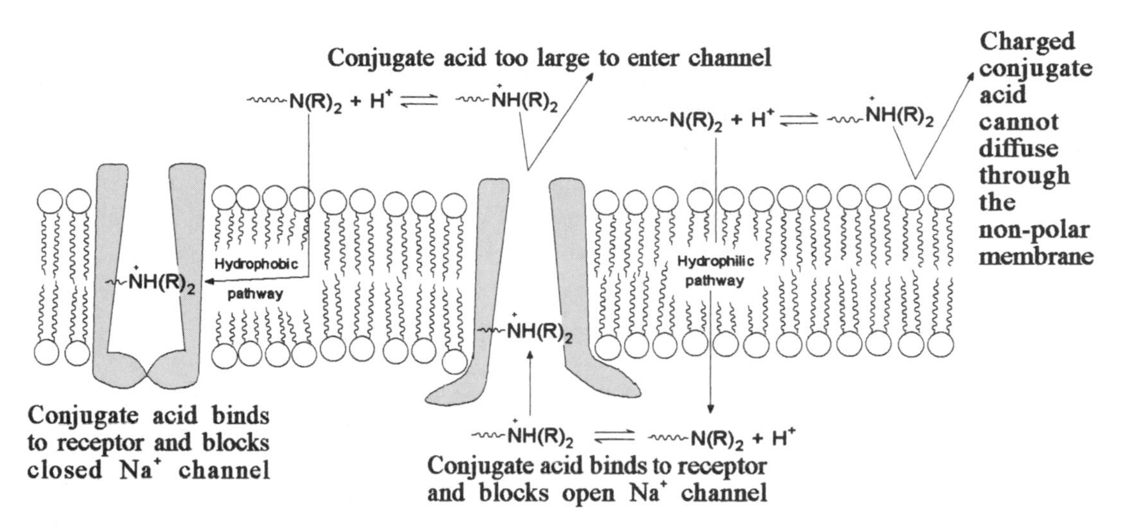

Figure 4.37. A representation of the hydrophobic and hydrophilic routes for blocking both open and closed ion Na^+ channels.

The structure–activity relationships of local anaesthetic agents are well documented. Active amide- and ester-based drugs have hydrophilic and lipophilic centres separated by a structure containing an ester or amide group (Figure 4.38). The hydrophilic centre normally contains a basic group, usually a secondary or tertiary amine, whereas the lipophilic centre usually consists of a substituted carbocyclic or heterocyclic ring system. As well as an ester or amide group, the structure linking the hydrophilic and lipophilic centres can contain a short hydrocarbon chain, nitrogen, oxygen and sulphur atoms. The most common links are based on hydrocarbon chains.

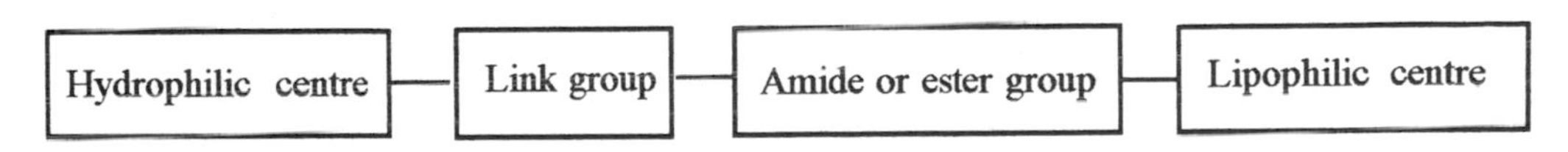

Figure 4.38. The common features of the structures of most local anaesthetics in clinical use.

The hydrophilic centre gives the local anaesthetic agent its water solubility whereas the lipophilic centre confers lipid solubility on the molecule. Water solubility is essential for transporting the drug to the membrane and, once in the neuron, to its site of action. Lipid solubility is also essential so that the local anaesthetic agent can penetrate the non-polar lipid membrane and reach its site of action. It follows that the best local anaesthetic action will occur with compounds that have the correct balance between their hydrophilic and lipophilic centres. A measure of this balance is the degree of ionisation of the compound in water. Because local anaesthetic agents are bases, it is possible to relate the degree of ionisation to their pK_a values; the higher the pK_a value, the higher the compound's degree of ionisation. Compounds with a high degree of ionisation will be more soluble in water but will be less able to reach their site of action because transport through the cell membrane will be more difficult due to a reduced lipid solubility. Most of the clinically useful local anaesthetic agents have pK_a values in the range 7.5–9.5. The pH of the extracellular fluid can significantly affect the action of local anaesthetics whose mode of action depends on the molecule penetrating the membrane. Conditions, such as inflammation, that give the extracellular fluid an acidic pH will increase the ionisation of the basic local anaesthetic molecules, forming an equilibrium mixture with a higher proportion of charged molecules. These charged molecules do not penetrate the neuron, which reduces the proportion of neutral local anaesthetic molecules that can penetrate the membrane and act on the ion channels.

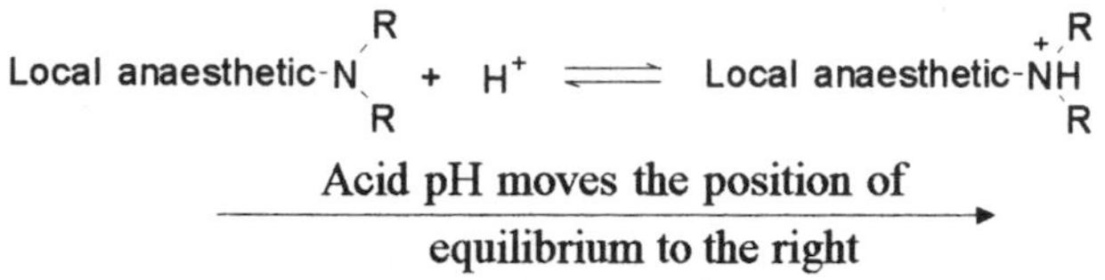

$$\text{Local anaesthetic-}NR_2 + H^+ \rightleftharpoons \text{Local anaesthetic-}\overset{+}{N}HR_2$$

Acid pH moves the position of equilibrium to the right →

Substitution of the aromatic ring of the lipophilic centres of some ester-based local anaesthetics by electron donor groups resulted in increased local anaesthetic activity, whereas substitution by electron acceptor groups gave reduced activity. It has been suggested that electron donor groups increase the strength of the dipole of the carbonyl group by increasing the polarisation of the carbonyl group (Figure 4.39). This subsequently increases the strength of the carbonyl group–receptor binding. In contrast, electron acceptor substituents lower the strength of the carbonyl's dipole by decreasing the polarisation of the group. This reduces the strength of the carbonyl group's dipole, which in turn reduces the strength of the dipole–dipole binding to the receptor.

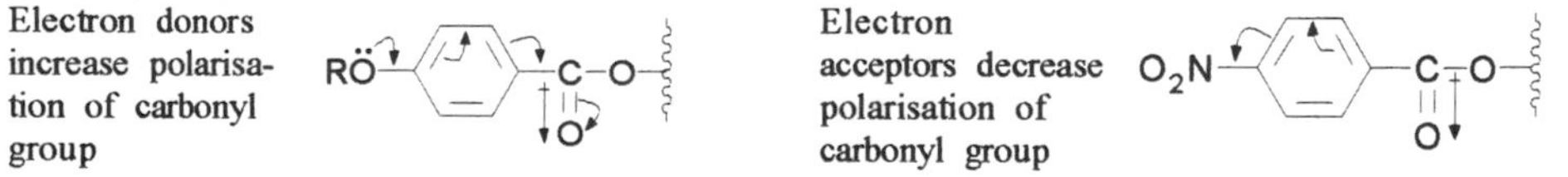

Electron donors increase polarisation of carbonyl group — RÖ–C₆H₄–C(=O)–O–

Electron acceptors decrease polarisation of carbonyl group — O_2N–C₆H₄–C(=O)–O–

Figure 4.39. The effect of electron donor and acceptor substituents on the binding of some local anaesthetics to their receptor sites.

4.5 Summary

The **plasma** or **cyctoplasmic membrane** separates the intracellular fluid from the extracellular fluid in all types of cell. It is a fluid-like **bilayer** arrangement of **phospholipids** in which proteins and other substances are either embedded in the membrane or attached to its surface. Significant amounts of carbohydrate residues are attached to membrane proteins and lipids. These carbohydrates are not easily removed from the membrane. They are involved in a variety of cell functions. Some cells also have a rigid cell wall that protects the more fragile plasma membrane.

The **lipid components** of the plasma membranes of mammals are mainly **phospholipids**, **sphingolipids** and **steroids**. Their polar heads form the surfaces of the lipid bilayer. A small proportion of **glycosphingolipids** are also found in the lipid bilayer. These lipids are involved in cell recognition, tissue immunity and other important cell functions. The **sterols** stiffen a membrane by hydrogen bonding to adjacent lipid molecules. **Cholesterol** is the sterol found in the lipid bilayer of mammals whereas **ergosterol** occurs in the lipid membranes of fungi.

Membrane proteins have a wide variety of biological functions, such as acting as receptors, enzymes and transportation routes for various species across membranes. They are classified as integral, peripheral and anchored proteins. **Integral proteins** are either deeply embedded in or pass right through a membrane. They are not easily detached from the membrane. Integral proteins that pass through a membrane usually have their C-terminal in the intracellular fluid and their N-terminal in the extracellular fluid. The section of an integral protein that crosses a membrane is known as a transmembrane domain or span. **Transmembrane spans** normally take the form of an **α-helix**. They may be grouped together to form water-filled pores through the membrane known as **ion channels**. These ion channels allow specific ions to pass from one side of the membrane to the other.

Proteins that are attached to the surface of a membrane are known as **peripheral proteins**. They are attached to the surface of the membrane by weak bonds and so are relatively easy to detach. Consequently, many peripheral proteins are able to migrate over the surface of the membrane.

Anchored proteins are attached to lipid molecules that form part of the lipid bilayer. Four different anchor systems have been identified, namely myristoyl, thioester-linked fatty acids, thioether-linked prenyl and glycosylphosphatidylinositol.

There are significant chemical differences between the cell envelopes of mammals, plants, bacteria and fungi. These differences can be exploited as targets in the design of drugs.

Tissue consists of groups of cells held together by a variety of forces. The sizes of the spaces between the cells vary depending on the type of tissue. **Tight gaps** such as those found in the endothelial cells of the gastrointestinal tract and central nervous system (**the blood–brain barrier**) force drugs to pass through membranes in order to reach their site of action. However, the gaps between some endothelial cells, such as those of the kidney and liver, are quite large.

Substances are transported through membranes by osmosis, filtration, passive diffusion, facilitated diffusion, active transport, endocytosis and exocytosis. **Passive diffusion** is the major route for the transport of substances across membranes. It occurs down a concentration gradient. In passive diffusion the drug dissolves in the lipid membrane, diffuses across the membrane and dissolves out of the membrane into the aqueous medium. Because the interior surfaces of membranes are non-polar, passive diffusion is most effective for uncharged non-polar species. Charged and highly polar species are not easily transported through a membrane by passive diffusion. However, a potential drug must have sufficient polar character to penetrate the hydrophilic surface of the membrane. Consequently, the structures of potential drugs must contain a balanced ratio of lipid-solubilising regions to water-solubilising regions if a sufficient concentration of the compound is to cross the membrane by passive diffusion.

A number of drugs act by affecting the structure of membranes and cell walls. The main ways by which they act are by:

(i) inhibiting the action of enzymes and other substances that are involved in the production of the compounds that are required for producing and maintaining the integrities of plasma membranes and cell walls;
(ii) inhibiting the processes that are involved in either the formation or regeneration of the cell wall;
(iii) forming channels through the membrane that allow the loss of vital cellular materials;
(iv) making the membrane more porous by breaking down sections of the membrane;
(v) blocking ion channels.

For example, **azoles** inhibit cytochrome P-450 oxidases in fungi cell walls, which leads to a build-up of 14α-methylated sterols in the membrane and increases the membranes porosity. **Allylamines** inhibit squalene epoxidase in fungi cell walls, which leads to a build-up of squalene in the membrane and increases the membrane's porosity. **Phenols** are believed to destroy sections of the fungi cell membranes. **Some antibiotics** inhibit cell wall synthesis by inhibiting:

(i) the formation of the precursors (e.g. **cycloserine** and **fosfomycin**);
(ii) the formation of the peptidoglycan chains (e.g. **bactracin**);
(iii) the cross-linking of the peptidoglycan chains (e.g. **penicillins**, **cephalosporins** and **vancomycin**).

Ionophores are substances that can penetrate a membrane and make it more porous by either forming channels through the membrane (e.g. **gramicidin A**) or acting as carriers (**valinomycin**).

Surfactants, used as disinfectants, lower the surface tension of lipid membranes, which increases the osmosis of water into the cell. This increases the osmotic pressure in the cell, which results in lysis.

Local anaesthetics in clinical use normally act by blocking ion channels about midway along the length of the channel. Most local anaesthetics have structures that consist of hydrophilic and lipophilic centres separated by either an ester or an amide group and a short hydrocarbon chain.

4.6 Questions

(1) Briefly define the meaning of each of the following terms: (a) extracellular fluid, (b) cytoplasmic membrane, (c) organelle and (d) prokaryotic cell.

(2) Outline the fluid mosaic model of the plasma membrane structure. Explain why the interior surfaces of most membranes are negatively charged with respect to the exterior surface.

(3) Draw the general structural formulae of each of the following compounds: (a) SM, (b) PE, (c) PI (d) PC and (e) PS.

(4) Distinguish carefully between the terms cell wall, plasma membrane and outer membrane in the context of bacteria. Ilustrate the answer by a description of the general chemical structures of each feature.

(5) Describe the structure of the cell envelope of (a) Gram-positive bacteria and (b) Gram-negative bacteria.

(6) Describe the main differences between each of the following:
   (a) fungicidal and fungistatic drugs;
   (b) exocytosis and endocytosis;
   (c) eukaryotic and prokaryotic cells;
   (d) active transport and facilitated diffusion.

(7) Calculate the degree of ionisation of each of the following local anaesthetics at pH 7.4. Use the data to suggest a potential order of activity of these compounds. On what principle is this order of activity based?
   Benzocaine pK_a 2.5; Cocaine pK_a 5.59 and Procaine pK_a 8.15.

(8) Briefly explain how active transport could influence the design of a drug. Illustrate the answer by reference to one suitable example.

(9) What are ionophores? How do they act and when does their action cease?

(10) Suggest two reasons for Gram-negative bacteria being more resistant to treatment with penicillin than Gram-positive bacteria.

(11)

(A) CH_3 C_3H_7 —NHCO N CH_3

Describe the essential structural features that should be incorporated into the design of a potential local anaesthetic agent. Compound A is the local anaesthetic Ropivacaine. Suggest, by drawing its molecular structure, a logical analogue of A that might also be expected to exhibit local anaesthetic activity.

(12) Suggest a **feasible** chemical explanation of how a cephalosporin might act on bacteria cell walls.

(13) Drug B is readily absorbed from the gastrointestinal tract by passive diffusion. Predict the most likely effect of each of the structural changes on the absorption of the resulting analogues of B, assuming that each of the analogues is still absorbed by passive diffusion:
   (a) incorporating a carboxylic acid group into the structure of B;
   (b) incorporating an amino group into the structure of B;
   (c) esterifying an existing carboxylic acid group in the structure of B.

5 Pharmacokinetics

5.1 Introduction

The action of a drug is initially dependent on it reaching its site of action in sufficient concentration for a long enough period of time for a significant pharmacological response to occur. This build-up in drug concentration and the maintenance of this concentration over a period of time depends on the route of drug administration, the efficiency of drug absorption, the rate at which the drug is transported to its site of action, the rate of drug metabolism on route and, at the site of action, the rate of drug excretion as well as the age, gender and physiological state of the patient. Pharmacokinetics is the study of the relationships between drug response in the patient and these factors.

Pharmacokinetics is based on the hypothesis that the magnitude of the responses to a drug, both therapeutic and toxic, is a function of its concentration at its site of action. The relationship of this concentration to the dose administered is not simple. Once a drug is absorbed into the body it must find its way to its site of action. In the course of this transportation some of the drug will be metabolised (see section 9.2) and some will be irreversibly excreted by the liver and/or kidneys and/or lungs. The **irreversible** processes by which a drug is prevented from reaching its site of action are collectively referred to as **elimination**. Uptake into the tissues is not regarded as an elimination process because it is usually reversible, the drug returning to the general circulation system (systemic circulation) in the course of time.

The process of elimination means that the concentration of the drug reaching the desired site of action may not be high enough to provide the required therapeutic effect. Consequently, it is important to have a method of monitoring the concentration of a drug in contact with its site of action. However, because the precise site of action is often unknown it is not usually possible to determine the concentration of a drug at its site of action and so the pharmacokinetic behaviour of a drug is usually monitored by following the concentration of the drug in the plasma and other suitable body fluids (Figure 5.1). These measurements are statistically correlated with the effects of the drug on patients. However, it is often difficult to obtain data using humans so many investigations are carried out using animals. The results of these experiments have been extrapolated to humans (see section 5.7) with varying degrees of success. Consequently, it is necessary to carry out trials on humans before the drug is released for clinical use.

The relationship between drug concentration and patient response shows that there is a range of plasma concentrations over which a drug is therapeutically successful. This range is referred

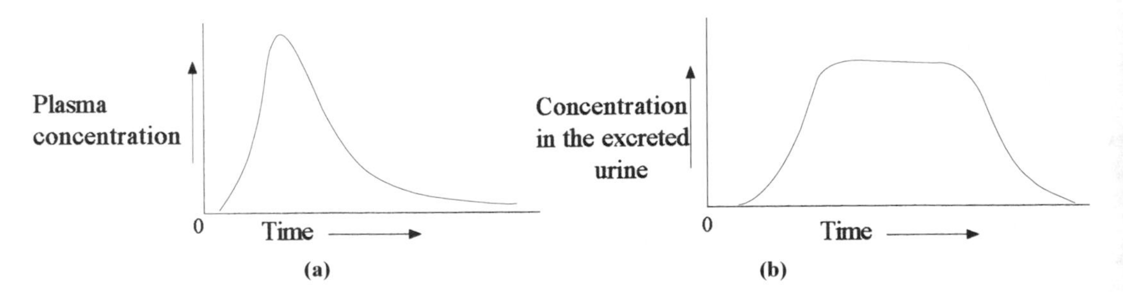

Figure 5.1. Typical variations in the concentration of a drug with time in samples of (a) plasma and (b) urine after the administration of a single oral dose of the drug at time zero. In both cases the precise shape of the graph will depend on the drug being studied.

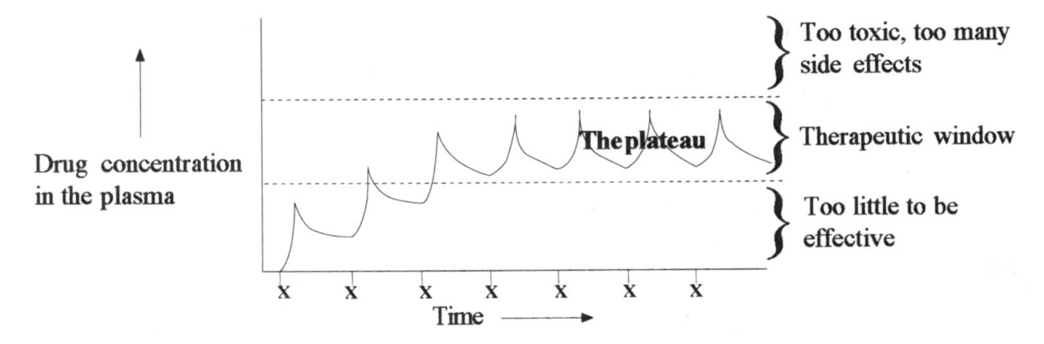

Figure 5.2. A schematic representation of a therapeutic window. Successive orally administered doses are given at the points X. Too high a dose results in the plateau being above the therapeutic window, whereas too low a dose gives a plateau below the therapeutic window.

to as the **therapeutic window** of the drug (Figure 5.2). At concentrations higher than the upper limit of the window the toxic effects outweigh the therapeutic value of the drug, whereas at concentrations below the lower limit of the window the drug concentration is too low to be totally effective. However, it should be realised that the therapeutic window of a drug can vary considerably from patient to patient. Consequently, the clinically acceptable values used for a dosage form can still cause unacceptable toxic effects in some individuals.

When a drug is given orally to a patient, its concentration in the plasma gradually increases to a maximum as it is absorbed from the gastrointestinal (GI) tract. Once in the plasma the processes of elimination start to reduce the concentration of the drug. However, in most cases absorption is faster than elimination. Consequently, a succession of doses at regular time intervals will usually result in a build-up of the drug concentration to the desired level in the plasma and by inference the correct concentration at the site of action for therapeutic success. Because, in the normal course of events, the concentration of the drug never reaches zero before the next dose is administered, one would expect a steady rise in the drug's concentration in the plasma. However, there is a limit to the amount of drug that the plasma can contain and

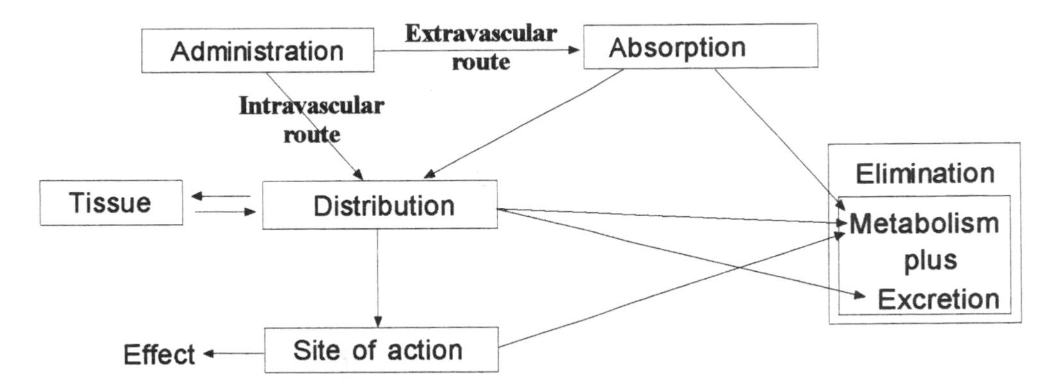

Figure 5.3. The general stages and their relationships in the life cycle of a drug after administration.

this, coupled with the elimination processes, results in the drug concentration reaching a plateau. If the dose is correct, this plateau will lie within the therapeutic window of the drug (Figure 5.2).

5.1.1 General Classification of Pharmacokinetic Properties

The pharmacokinetic behaviour of a drug after administration is broadly classified into the general regions of absorption, distribution, metabolism and excretion (Figure 5.3). The parameters associated with each of these general regions are used to specify the pharmacokinetic properties of drugs in the body. The methods used to calculate these parameters are independent of the method of administration although the values obtained will depend on the administrative route. For example, intravascular routes will not normally give values for absorption parameters. However, intravascular routes do give higher concentrations of the drug in the general circulatory system (**systemic circulatory system**).

Each of the general regions of pharmacokinetics and their associated parameters will be discussed later in this chapter under the most appropriate method of administration. However, it is emphasised that the parameters apply across the board and are not confined to the method of administration under which they are introduced.

5.1.2 Drug Regimens

A drug regimen is the way in which a drug is administered to a patient. It normally includes the method of administration, the dose, its frequency of administration and the duration of the treatment. The response of individual patients to the same doses of a drug can be very varied (Figure 5.4). It will depend on the age and weight of the subject as well as the severity of the disease. Consequently, to use any drug effectively, the drug regimen should be tailored to an individual's requirements. This would require extensive and expensive investigations and is not

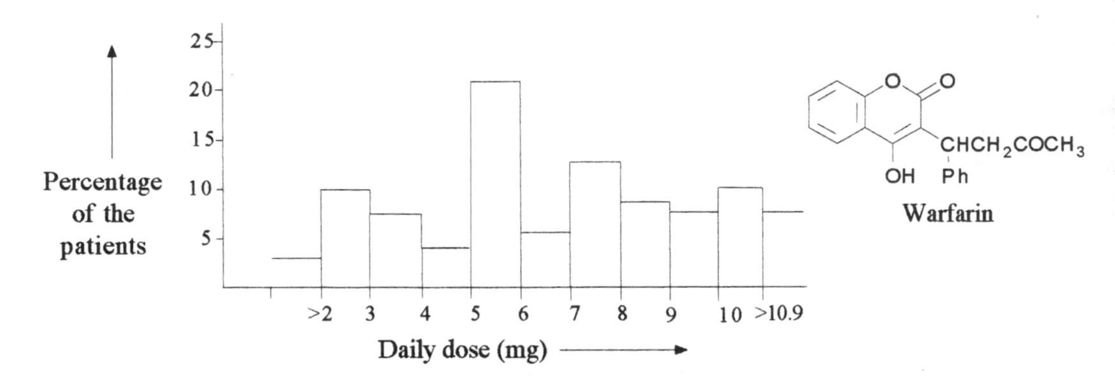

Figure 5.4. The variation of the daily dose of warfarin needed to produce similar prothrombin times in adult patients. The prothrombin time is the time taken for citrated plasma treated with calcium and standardised reference thromboplastin to clot. (Data from J. Koch-Wesser, The serum level approach to individualisation of drug dosage. *European Journal of Clinical Pharmacology*, **9**, 1–8 (1975). Copyright 1975 Springer-Verlag.

normally necessary for most patients suffering from medical complaints such as the common cold, thrush, diarrhoea and bronchitis. However, it can lead to more effective and less expensive treatments in more critical medical conditions. In these cases, pharmacokinetic investigations aim to determine the most successful course of treatment by balancing the desirable therapeutic effects against undesirable side effects. The optimum drug regimen for any drug is the one that maintains the concentration of the drug within the therapeutic window for the patient.

5.1.3 The Importance of Pharmacokinetics in Developing a Drug

Pharmacokinetics is important in a number of aspects of drug design. It can prevent the discarding of what could be an important drug. For example, quinacrine, developed during the Second World War as an alternative to quinine, was found to be too toxic when administered in high enough doses to be effective. However, a study of its pharmacokinetic properties showed that it had a slow elimination rate and so rapidly accumulated to toxic levels in the plasma. Once this elimination pattern was discovered it was a simple matter to use high initial doses and to follow this with smaller doses just sufficient to maintain the drug's concentration within its therapeutic window.

CH_3
$NHCH(CH_2)_3N(C_2H_5)_2$
OCH_3
Cl N
Quinacrine

$CH_2=CH$
N
CHOH
OCH_3
N
Quinine

Cys-Tyr-Ile-Gln-Asn-Cys-Pro-Leu-Gly(NH_2)
Oxytocin

A study of the pharmacokinetic properties of a compound indicates which properties need to be modified in order to produce a more effective analogue. Consider, for example, a drug that is not suitable for development because it has too short a duration of action. The logical way forward is to determine which analogues have a slower elimination rate by testing on suitable animal models. The analogues selected will depend on what is believed to be the structural cause of this rapid elimination. If, for example, the drug is rapidly metabolised by esterases in the plasma, compounds that are more stable to these enzymes would be tested. Alternatively, the drug may be very water soluble and as a result poorly absorbed, in which case the approach is to produce and test less-water-soluble analogues.

A study of pharmacokinetics forms the basis of the design of the dosage forms. Drugs with a narrow therapeutic window require smaller and more frequent doses or a change in the method of administration. If a potential drug is administered incorrectly it will be ineffective in trials and consequently the drug will be discarded even though it could have been of considerable therapeutic value if it had been administered correctly. For example, oxytocin could have been overlooked if it had not been administered correctly in trials. Investigations of its pharmacokinetic properties showed that it has a narrow therapeutic window and is eliminated within minutes of entering the systemic circulation. In addition, it is also metabolised by enzymes in the GI tract enzymes. These characteristics make it impossible to administer the drug orally. Consequently, the only way of maintaining an accurate oxytocin concentration in the plasma is to administer it by carefully monitored continuous intravenous infusion.

5.2 Drug Concentration Analysis and its Therapeutic Significance

The determination of drug concentration in the body requires taking samples of biological fluids from patients and test animals. Samples are taken by either **invasive** or **non-invasive** methods. Invasive methods include the removal of blood, spinal fluid and tissue (biopsy) samples whereas non-invasive methods include collecting urine, faeces, expired air and saliva samples. The information obtained from these samples will depend on their source. For example, changes in the drug concentration in plasma are usually good indications of the changes occurring in the tissue whereas the concentration of a drug in the faeces may either indicate the degree of biliary excretion of the drug or, when compared with the orally administered dose, show the degree of absorption of the drug. However, most analytical methods are designed for plasma analysis and so plasma concentrations are the most commonly reported measurement.

The cells in tissue are in contact with the extracellular fluid, which has a similar composition to that of blood plasma. Consequently, the concentration of a drug in the blood plasma is found, in many cases, to be a good measure of that drug's concentration at its receptor sites, which are usually found in tissue cells. This means that monitoring the level of a drug in a patient's plasma is often a good method of checking that the dose level is correct for effective therapeutic action. However, for some drugs, such as those used in cancer chemotherapy, the concentration of the drug in the plasma is not a good indicator of pharmacological response. Consequently, clinical decisions should not be based solely on blood plasma levels.

The assessment of the pharmacokinetic behaviour of a drug in a patient requires an accurate method for assaying that drug. Many drug substances are mixtures of isomers or compounds with similar structures. The components of these mixtures can have similar or completely different pharmacological actions (see section 1.5.3). In mixtures where the components have significantly different pharmacological activities, it is necessary to either develop assay methods that distinguish between these components or separate the components and use the relevant pure substance in the investigation. However, separation of components is not always possible and can be very expensive. Consequently, medicinal chemists try to produce drugs that are not optically active or mixtures of compounds.

To obtain meaningful pharmacokinetic data it is necessary to analyse for the metabolites of the drug as well as for the drug itself. This is particularly important if the metabolites are either active or toxic. For example, the activity of a drug may be due to a large increase in concentration of an active metabolite rather than the drug itself (see section 9.2.2). Similarly, the unrecognised increase in concentration of a toxic metabolite could lead to the unnecessary abandonment of what could have been a useful drug. This type of problem has been reduced in a number of cases by the use of a suitable dosage form. For example, 2-mercaptoethanesulphonate (MESNA) reduces the toxic effects of the antineoplastic agent cyclophosphamide by increasing the rate of elimination of the toxic metabolites of this drug.

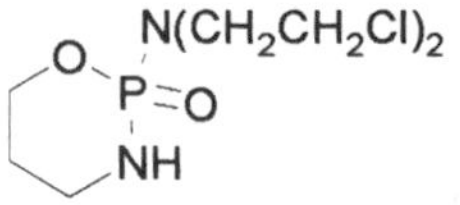

$N(CH_2CH_2Cl)_2$ / O / P=O / NH

Cyclophosphamide

$HSCH_2CH_2SO_3^-$

2-Mercaptoethanesulphonate

Identification of the metabolites is also of importance when assessing the plasma concentration of a drug using methods based on the use of radioactive isotopes such as ^{14}C and ^{3}H. The use of these isotopes makes it possible to rapidly assay compounds in many areas of the body and so follow the route taken by the radioactive drug. However, metabolism of the drug can result in all the labelled atoms being transferred to a metabolite. Consequently, it is important that the chemical identity of the substance containing the tracer is known if the data are to be interpreted accurately. This is particularly important if the potential drug is acting as a prodrug.

The rate of change of plasma concentration with time is normally recorded as a plot of concentration against time (Figure 5.4). In practice, the shape of the graph usually indicates that the process exhibits either zero- or first-order kinetics. This means that the data can be

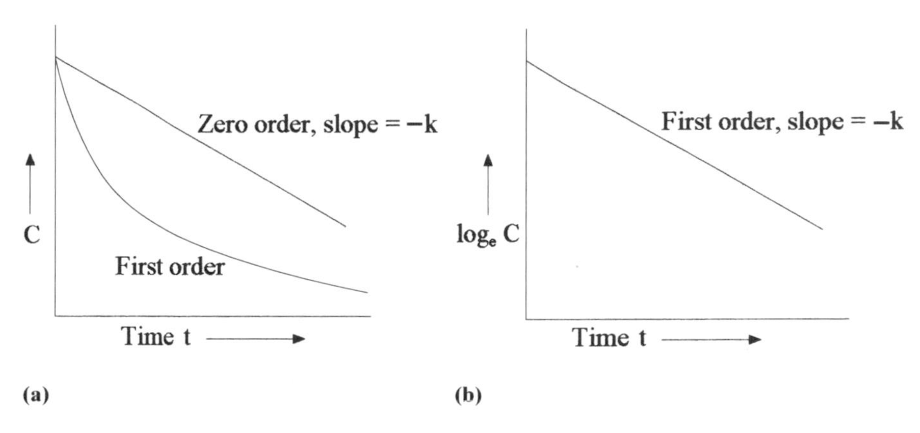

Figure 5.5. (a) Zero- and first-order concentration–time plots and (b) first-order log$_e$ concentration–time plots.

expressed mathematically using Equations (5.1) and (5.2), where k is the rate constant and C is the concentration of the drug normally associated with these kinetic processes. For first-order processes, a plot of log$_e$ C against t will be a straight line with a slope of k.

$$\text{Zero order} \qquad \text{Rate} = k \qquad \text{and} \qquad C = C_o - kt \tag{5.1}$$

$$\text{First order} \qquad \text{Rate} = kC \qquad \text{and} \qquad C = C_o e^{-kt} \tag{5.2}$$

5.3 Pharmacokinetic Models

Accurate assessment of the results of a pharmacokinetic investigation requires the use of mathematical methods. In order to apply these methods to the behaviour of a drug in what is a complex biological system, it is necessary to use so-called **model** systems. These models simulate the rate relationships between drug absorption, distribution, response and elimination in the various sections of the biological system. The accuracy of all pharmacokinetic models in describing the drug concentration changes and relating these changes to pharmacological and toxic responses depends on the accurate assay of drug concentrations in the plasma and tissues. Because it is often impossible to obtain the required samples from human subjects, pharmacokinetic models are often developed from data obtained from animals.

Pharmacokinetic models enable the medicinal chemist to use mathematical equations to describe the relationships between the concentrations of a drug in different tissues and, as a result, predict the concentrations of a drug in a tissue for any drug regimen. This information has a wide variety of uses, such as correlating drug doses with pharmacological and toxic responses and determining an optimum dose level for an individual. However, because these

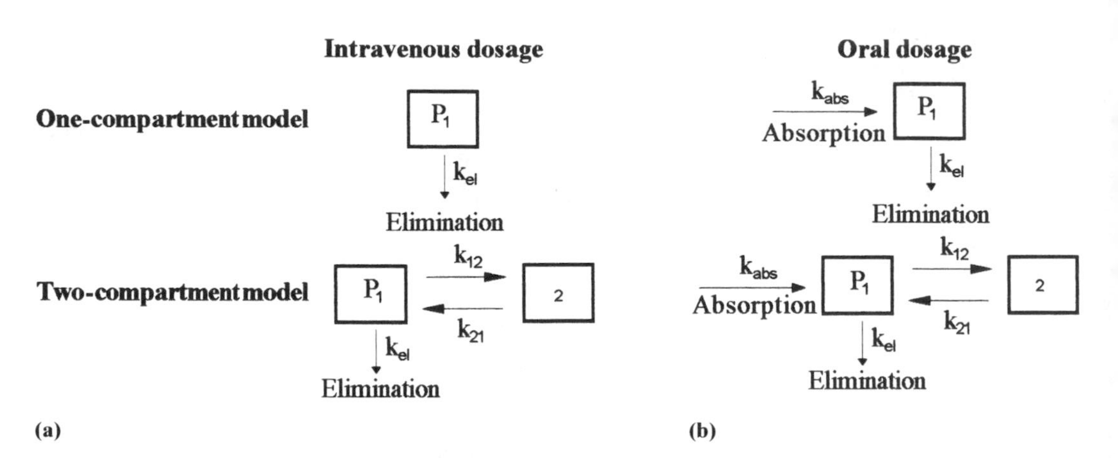

Figure 5.6. Compartmental models in which a drug is administered by (a) intravenous injection and (b) orally. The rate constant for the appropriate movement of the drug is k with the appropriate subscript. P_1 is the plasma and highly perfused tissues compartment, which is the central compartment in all compartmental models. It is the first destination of a drug.

models are based on a simplification of what is a very complex system, it is necessary to treat the results of these analyses with some degree of caution until the model has been rigorously tested.

A commonly used pharmacokinetic model is the **compartmental model**. These models visualise the biological system as a series of interconnected compartments (Figure 5.6), which allows the biological relationships between these compartments to be expressed in the form of mathematical equations. In all compartmental models, a compartment is defined as a group of tissues that have a similar blood flow and drug affinity. It is not necessarily a defined anatomical region in the body but all compartmental model systems assume that:

(i) the compartments communicate with each other by reversible processes;
(ii) rate constants are used as a measure of the rate of entry and exit of a drug from a compartment;
(iii) the distribution of a drug within a compartment is rapid and homogeneous;
(iv) each drug molecule has an equal chance of leaving a compartment.

Compartmental models are normally based on a central compartment that represents the plasma and the highly perfused tissues (Figure 5.6). Elimination of a drug is assumed to occur only from this compartment because the processes associated with elimination occur mainly in the plasma and the highly perfused tissues of the liver and kidney. Other compartments are connected to the central compartment as required by the nature of the investigation. The simplest compartmental model is the one-compartment model, in which the compartment represents the circulatory system and all the tissues perfused by the drug (Figure 5.6). This is the model used in this text.

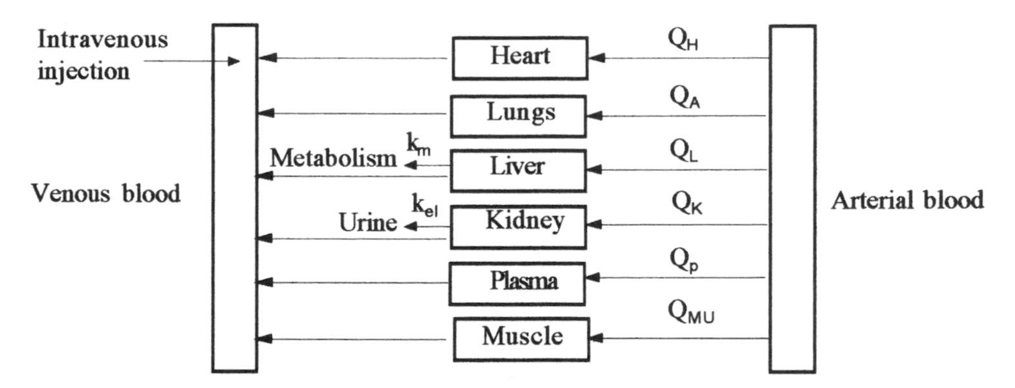

Figure 5.7. A simple drug perfusion model for an intravenous injection of a drug. Each box represents either an organ or a tissue type. Tissues and organs that are impervious to the drug are not included. Tissues with similar degrees of blood perfusion are grouped together. k_m and k_{el} the rate constants for removal of the drug and Q represents the rate of blood perfusion of the tissue.

An alternative model that is also widely used is the **flow or perfusion model**. This model uses the compartment concept but the compartments represent the anatomical regions of the body (Figure 5.7). The perfusion model uses blood flow to assess the distribution of the drug to the various organs in the body and the degree of tissue binding as a measure of the uptake of the drug into an organ. This means that the drug concentration in an organ is determined using the blood flow, the size of the organ and the partition of the drug between the blood and the organ. However, these factors may be affected by the physiological state of the subject and this must be taken into account when drawing general conclusions from investigations using perfusion models. An advantage of the perfusion model is that the deductions from animal studies can in some cases be extrapolated accurately to humans. For further details of the perfusion model and its uses, the reader is referred to more specialised texts because this introductory text will only deal with the one-compartment model.

5.4 Intravascular Administration

The main methods of intravascular administration are intravenous (i.v.) injection and infusion. When a single dose of a drug is administered to a patient by intravenous injection, the dose is usually referred to as an **i.v. bolus**. Intravascular administration places the drug directly in the patient's circulatory system, which bypasses the body's natural barriers to drug absorption. Once it enters the circulatory system the drug is rapidly distributed to most tissues because a dynamic equilibrium is speedily reached between the drug in the blood and the tissue. This means that a fast i.v. bolus injection will almost immediately give a high initial concentration of the drug in the circulatory system but this will immediately start to fall because of elimination processes (Figure 5.8a). However, the plasma concentration of a drug administered by

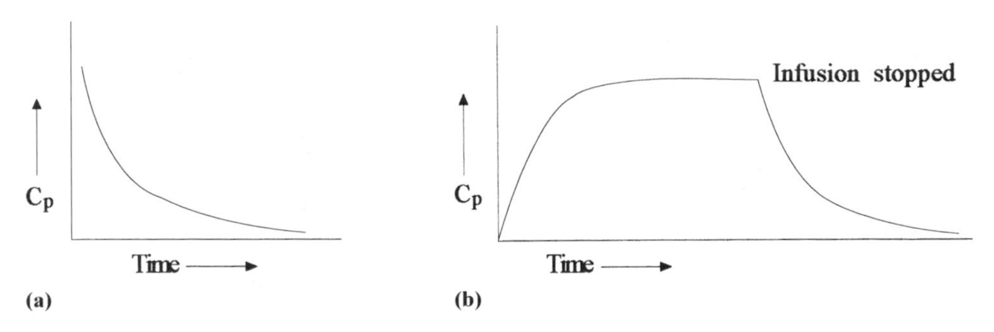

Figure 5.8. The variation of the concentration of a drug in the plasma (C_p) with time when administered by (a) a rapid single intravenous injection and (b) intravenous infusion. With rapid intravenous injections the graph does not show the time taken to carry out the injection; it is normally taken as being spontaneous. In these cases the curve starts at the point where the first plasma concentration measurements were taken.

intravenous infusion will increase with time until the rate of infusion is equal to the rate of elimination (Figure 5.8b). At this point the drug plasma concentration remains constant until infusion is stopped, whereupon it falls. Administration by intravenous infusion can be used to maintain accurately the required concentration of a drug in the blood and tissues.

5.4.1 Distribution

Once a drug is absorbed it is distributed to all the accessible tissues of the body. Consequently, the use of the term **distribution** in pharmacokinetics refers to the transfer of the drug from its site of absorption to its site of action. The main drug distribution route is the circulatory system. The rate and extent of drug distribution will depend on the chemical structure of the drug, the rate of blood flow (Q), the ease of transport of the drug through membranes (see section 4.3), the binding of the drug to the many proteins found in the blood (see section 5.4.1.1) and the elimination processes (see section 3.1) that occur on its route. Only a small proportion of the administered dose will reach the site of action; the remainder will either undergo elimination or be absorbed into the tissues it meets on route. The former processes are irreversible but the latter are reversible and so will gradually release the drug back to the general circulatory system and its site of action.

5.4.1.1 The Binding of Drugs to Plasma Proteins

A proportion of the drug molecules that enter the general circulatory system bind to proteins by electrostatic interactions, hydrogen bonding, dipole–dipole interactions and other types of van der Waals' forces. The binding is reversible, the bound and free drug forming a dynamic equilibrium mixture in the blood:

$$\textbf{Free drug} \rightleftharpoons \textbf{Bound drug}$$

The degree of binding is defined as:

$$\text{Percentage of protein binding} = \frac{\text{Concentration of bound drug} \times 100}{\text{Total concentration of bound and free drug}} \quad (5.3)$$

The most important protein with regard to binding is albumin, but drugs also bind to the α_1-acid glycoproteins and lipoproteins. Weakly acidic drugs tend to bind to albumin whereas weakly basic drugs prefer to bind to the α_1-acid glycoproteins. The reversible binding of drugs to plasma proteins has a significant effect on a number of pharmacokinetic parameters because these parameters are dependent on the concentration of the free drug in the plasma. The bound drug has no effect. For example, drugs with a low percentage of protein binding will have a higher plasma concentration of the free drug and so will be more readily available for elimination and to exert either a therapeutic or toxic effect than drugs with a high percentage of plasma protein binding. However, in the latter case the high percentage of protein binding could result in a longer duration of action as the drug is released from the protein. Pathological states such as renal failure and inflammation, or a change in physiological status due to pregnancy, fasting and malnutrition, can have a considerable effect on plasma protein binding. Competition from other drugs can also alter binding.

5.4.1.2 Intravenous Injection (i.v. Bolus)

The administration of a drug by a rapid intravenous injection places the drug in the blood where its concentration immediately begins to decline (Figure 5.9a). This decline, which is due to metabolism and excretion via the kidney and liver, is usually followed by plotting a graph of the plasma concentration of the drug against time. However, it should be realised that numerical information calculated from plasma concentrations may not be the same as that obtained using whole-blood concentrations.

The one-compartment model of drug distribution assumes that the administration and distribution of the drug in the plasma and associated tissues is instantaneous. This does not happen in practice and is one of the possible sources of error when using this model to analyse experimental pharmacokinetic data. The fall in the plasma concentration of a drug is usually close to an exponential curve (Figure 5.8a), that is, the process approximately follows first-order kinetics:

$$\text{Rate of elimination} = k_{el}C \quad (5.4)$$

and the curve can be represented to an acceptable degree of accuracy by equations of the form:

$$C = C_o e^{-k_{el}t} \quad (5.5)$$

where: C_o is the plasma concentration of the drug in the body at a time $t = 0$, that is, the concentration immediately following bolus injection; C is the plasma concentration of the drug in

Table 5.1. The values of the pharmacokinetic parameters of some common drugs recorded in the literature (various sources).

Drug	V_d ($l kg^{-1}$)	$t_{1/2}$ (h)	% Plasma binding
Aspirin	0.1–0.2	0.28	90% below 100 μg, 50% above 4 μg
Ampicillin	0.3	1–2	About 20%
Cimetidine	2.1	1–3	13–26%
Chlorpromazine	20	Mean 15–30	95–98%
Diazepam	1.0	Mean 48	98–99%
Ibuprofen	0.14	2	99%
Propranolol	3.9	2–6	90%
Morphine	3.0–5.0	2–3	20–35%
Warfarin	0.15	Mean 42	97–99%

the body at a time t; t is the lapsed time between the administration and the measurement; and k_{el} is the rate constant for the irreversible elimination of the drug.

The plasma concentration, C, of a drug in the system is related to the total amount of the drug in the system, D, by the relationship:

$$C = D/V_d \tag{5.6}$$

where V_d is the apparent volume of distribution. The value of V_d is a measure of the total volume of the body perfused by the drug. It is an apparent volume because it is the volume of plasma that is equivalent to the total volume of the tissue and circulatory system perfused by the drug. This means that the values of V_d for a drug can be considerably higher than the volume of the blood in the circulatory system, which is usually about 5 litres for a 70-kg person. Values of V_d are usually recorded in terms of litres per kilogram (Table 5.1), which gives a value of 0.071 ($l kg^{-1}$) for a 70-kg person. A value of less than 0.071 for V_d indicates that the drug is probably distributed mainly within the circulatory system whereas a value greater than 0.071 indicates that the drug is distributed in both the circulatory system and specific tissues.

The value of V_d may be calculated by substituting the values for the dose D_o administered and the plasma concentration at the time t = 0 in Equation (5.6). The time taken to administer a drug and for that drug to achieve a homogeneous distribution throughout the system means that it is not possible to measure the total amount of the drug present in the plasma at the time t = 0. However, a theoretical value for C_o at the time t = 0 can be obtained by extrapolating a plot of $\log_e C$ against t to t = 0 (Figure 5.9b).

5.4.1.3 Elimination and Elimination Half-life

Elimination is the term used to represent the **irreversible** processes that remove a drug from the body during its journey to its site of action and after its action. The processes

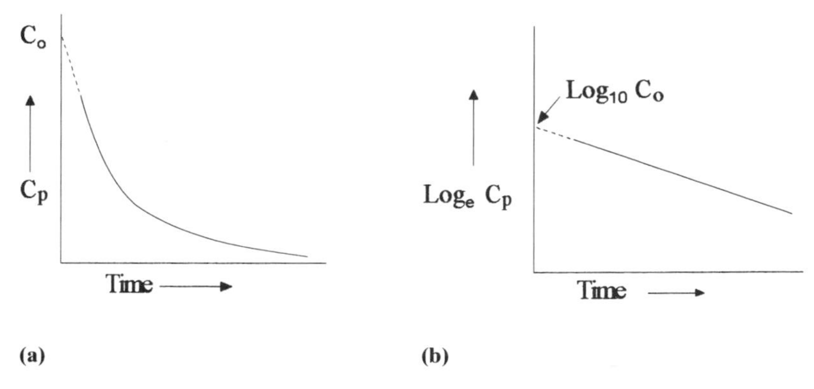

Figure 5.9. Extrapolation of the plots of (a) concentration against time and (b) log C_e against time for the changes in plasma concentration of a drug with time.

involved are metabolism and excretion. The main centre for metabolism is the liver. Excretion occurs mainly through the kidney and liver but some also occurs via the lungs and sweat. Excretion through the kidney occurs by glomerular filtration and tubular secretion. Tubular secretion occurs by active transport (see section 4.3.5) and will remove protein-bound drugs as well as free drugs. However, in tubular secretion protein-bound drugs have to dissociate from the protein before being transported and ultimately excreted in the urine. Tubular reabsorption often returns a considerable proportion of a drug and other substances to the systemic circulation. The degree of reabsorption will depend on the drug's ability to cross membranes. Consequently, drugs that are polar or highly ionised at the pH of the urine are less likely to be reabsorbed. For example, the polar β-lactams are not readily reabsorbed. Furthermore, acidic urine will enhance the ionisation of basic drugs and so reduce their reabsorption. Similarly, basic urine will have the same effect on acidic drugs. The reabsorption of drugs can be enhanced by changing the pH of a patient's urine using oral doses of sodium hydrogen carbonate or potassium citrate to make the urine more basic and ammonium chloride to make the urine more acidic. Biliary excretion involving the liver is not fully understood. The process is specific, only occurring for compounds with relative molecular masses greater than about 500.

The rate of elimination is an important characteristic of a drug. Too rapid an elimination necessitates frequent repeated administration of the drug if its concentration is to reach its therapeutic window. Conversely, too slow an elimination could result in the accumulation of the drug in the patient which might give an increased risk of toxic effects. Most drug eliminations follow first-order kinetics (Equation 5.2) but there are some notable exceptions, such as ethanol, which exhibits zero-order kinetics (Equation 5.1). This allows the calculation of blood alcohol levels at any time after drinking the alcohol even though the blood sample was taken some time after the alcohol was drunk. The fact that most drug elimination follows first-order kinetics means that it is not usually possible to determine the time at which a dose of a drug would be completely cleared from the system because the plasma

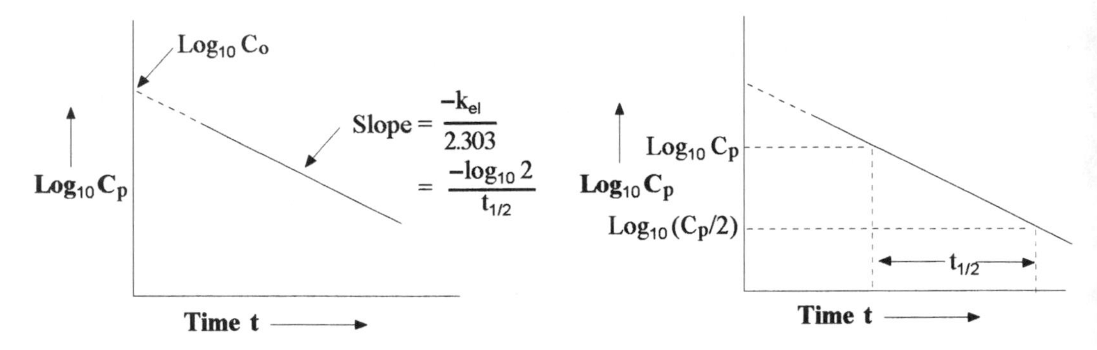

Figure 5.10. Determination of the values of $t_{1/2}$ and k_{el} from logarithmic plots of plasma concentration against time. The logarithm plot for log to base ten is shown but a natural logarithmic plot would be similar except that the slope would be equal to k_{el}.

concentration–time curves are exponential (Figure 5.8a). Consequently, both the biological half-life and the elimination rate constants (k_{el}) are used as indicators of the rate of elimination of a drug from the system. Half-lives are normally quoted in the literature and k_{el} values are calculated as required.

Half-life ($t_{1/2}$). This is the time taken for the concentration of a drug to fall to half its original value. Consequently, for $t_{1/2}$:

$$0.5 = \frac{C_{p2}}{C_{p1}} \tag{5.7}$$

where C_{p1} is the initial plasma concentration and C_{p2} is the plasma concentration after a lapsed time equal to the half-life $t_{1/2}$. Therefore, for **first-order** elimination processes, substituting Equation (3.7) in Equation (3.2) gives:

$$C_{p2} = C_{p1}e^{-k_{el}t_{1/2}}$$

and

$$\frac{C_{p2}}{C_{p1}} = 0.5 = e^{-k_{el}t_{1/2}}$$

and, taking log to base e gives:

$$\ln 0.5 = -k_{el}t_{1/2}$$

$$-0.693 = -k_{el}t_{1/2}$$

$$t_{1/2} = \frac{0.693}{k_{el}} \tag{5.8}$$

Elimination rate constant. For **first-order** elimination processes the value of the elimination rate constant may be calculated by interpreting the experimental data using the logarithmic forms of Equation (5.2).

for natural logarithms:
$$\ln C_p = \ln C_o - k_{el}t \tag{5.9}$$

for logarithms to base 10:
$$\log_{10} C_p = \log_{10} C_o - \frac{k_{el}t}{2.303} \tag{5.10}$$

Both of these equations give straight-line plots (Figure 5.10) and so it is possible to obtain a value of k_{el} by measuring the slope of the graph provided that the experimental data give a reasonable straight line. Half-life values may also be calculated from these graphs. It is advisable to take an average of several measurements of $t_{1/2}$ made from different initial values of C_p in order to obtain an accurate value for the half-life. Alternatively $t_{1/2}$ may be calculated by substituting the value of k_{el} in Equation (5.8).

The measurement of the half-life of a potential drug may give the medicinal chemist a reference point when developing drugs from a lead compound. It might enable the pharmacological effect of a lead to be compared with its analogues on a numerical basis and, as a result, provide an indication of the best course of action to take for the successful development of a useful drug. For example, if a lead has a short duration of action, analogues with larger $t_{1/2}$ and smaller k_{el} values than those of the lead are more likely to give the required pharmacological effect. Similarly, if the lead is too toxic, analogues with smaller $t_{1/2}$ and larger k_{el} values need to be developed. It is emphasised that $t_{1/2}$ and k_{el} data are not infallible and should not be considered in isolation. The more and wider the range of information one has, the more likely it is that a successful course of action will be pursued.

5.4.1.4 Clearance and its Significance

Clearance (Cl) is the volume of blood in a defined region of the body that is cleared of a drug in unit time. For example, total clearance (Cl_T) is the volume of blood in the whole body cleared of the drug in unit time, and hepatic clearance (Cl_H) is the volume of blood passing through the liver that is cleared in unit time. Clearance is an artificial concept in that it is not possible for a drug to be removed from only one part of the total volume of the blood in the body or organ. It is the parameter that relates the rate of elimination of a drug from a defined region of the body to the plasma concentration of that drug. For example, the rate of elimination of a drug from the whole of the body is related to the drug plasma concentration by the total clearance (Equation 5.11):

$$\text{Rate of elimination of a drug from the whole body} = Cl_T C_P \tag{5.11}$$

The rate of elimination of a drug is usually first order, therefore making the assumption that:

$$\text{Rate of elimination} = k_{el}D \tag{5.12}$$

where D is the amount of the drug in the body at time t. Substituting for D from Equation (5.6) gives:

$$\text{Rate of elimination} = k_{el}V_dC_P \tag{5.13}$$

therefore, substituting Equation (5.13) in Equation (5.11) gives:

$$Cl_TC_P = k_{el}V_dC_P$$

and

$$Cl_T = k_{el}V_d \tag{5.14}$$

and substituting Equation (5.8) in Equation (5.14) gives:

$$Cl_T = V_d\frac{0.693}{t_{1/2}} \tag{5.15}$$

For first-order elimination, clearance is a constant because both $t_{1/2}$ and k_{el} are constant. However, should the order of the elimination change due to a change in the biological situation, such as the drug concentration increasing to the point where it saturates the metabolic elimination pathways, then clearance may not be constant.

The clearance (Cl) of a drug from a region of the body is the sum of all the clearances of all the contributing processes in that region. For example, hepatic clearance is the sum of the clearances due to metabolism (Cl_M) and excretion (Cl_{Bile}) in the liver, that is:

$$Cl_H = Cl_M + Cl_{Bile} \tag{5.16}$$

For an i.v. bolus, which places the drug directly in the circulatory system, total clearance (Cl_T) of the drug from the body can also be determined from blood plasma measurements. The area under the curve (AUC) of a plot of C_p against t represents the total amount of the drug that reaches the circulatory system in time t. It can be used to calculate total clearance because it is related to the dose administered by the relationship:

$$\text{Dose} = Cl_T \cdot \text{AUC} \tag{5.17}$$

which may be derived from Equation (5.11) as follows:

$$\text{Rate of elimination} = Cl_T \cdot C_P$$

$$dD/dt = Cl_T \cdot C_P$$

and so:

$$dD = Cl_T \cdot C_P dt \quad (5.18)$$

Integrating Equation (5.18) between the limits D and 0 and t and 0 gives:

$$D = Cl_T \cdot C_P t \quad (5.19)$$

For intravenous doses the amount of the drug in the body at t = 0 is the dose administered. Therefore, because C_Pt is the area under the plasma concentration–time curve for the drug, this gives Equation (5.17).

This relationship can also be used to calculate the clearance that occurs in a specific time by simply measuring the AUC for that time. Furthermore, analysis of the total amount of a drug in the urine can be used to estimate renal clearance (Cl_R) because the total amount of unchanged drug found in the urine (U) can be related to its plasma concentration by a similar mathematical expression to Equation (5.17):

$$U = Cl_U \cdot AUC \quad (5.20)$$

Equation (5.17) holds true regardless of the way in which a single dose of the drug is administered. However, for enteral routes the dose is the amount absorbed (see section 5.5.1) and not the dose administered.

Clearance will vary with body weight and so for comparison purposes values are normally quoted per kilogram of body weight (Table 5.2). It also varies with the degree of protein binding. A large proportion of a drug with a high degree of protein binding will not be available to take part in the metabolic and excretion processes. In other words, it will not be so readily available for elimination as a drug with a low degree of protein binding.

Table 5.2. Clearance values of some drugs.

Drug	Clearance ($cm^3 min^{-1} kg^{-1}$)	Drug	Clearance ($cm^3 min^{-1} kg^{-1}$)
Atropine	8	Bumetamide	3
Bupivacaine	8	Caffeine	1–2
Disopyramide	0.5–2 (dose dependent)	Ethambutol	9
Mepivacaine	5	Pentobarbitone	0.3–0.5
Rantidine	About 10	Vancomycin	About 1

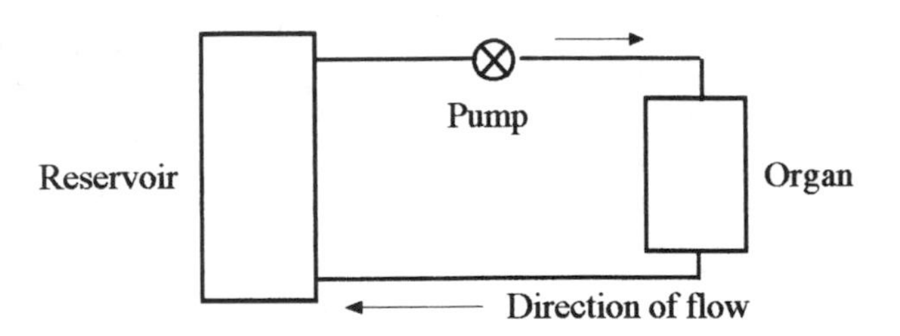

Figure 5.11. A simple extraction system.

Clearance is a more useful concept in pharmacokinetics than either $t_{1/2}$ or k_{el}. It enables blood flow rate (Q), which controls the rate at which a drug is delivered to a specific region of the body, to be taken into account when assessing the pharmacokinetic behaviour of the drug in that region of the body (see section 5.5). Clearance values enable the medicinal chemist to compare the effect of structural changes on drug behaviour and, as a result, decide which analogues might yield drugs with the desired pharmacokinetic properties.

All drugs administered by i.v. bolus injection are carried to their site of action by the blood. In order to reach its destination, a drug may have to pass through several organs where some of the drug may be lost by elimination. This loss is known as **extraction**. The effect of extraction on the distribution of the drug may be rationalised by the use of the concept of clearance. Consider, for example, a closed system in which a pump is pumping a fluid from a reservoir through an organ before being returned to the reservoir (Figure 5.11). This system may be regarded as being analogous to the heart pumping the blood through an organ. If the organ eliminates some of the drug from the blood by excretion and metabolism, the proportion of the drug removed by a single transit of the total dose of the drug through the organ is defined as:

$$E = \frac{C_{in} - C_{out}}{C_{in}} \quad (5.21)$$

where E is known as the extraction ratio. Because the liver is a major site of metabolism and excretion, the hepatic extraction values (E_H) values of many drugs have been determined (Table 5.3).

The extraction ratio has no units. Its values range from 0 to 1. A value of 0.4 means that 40% of the drug is irreversibly removed as it passes through the organ. Consequently, as a drug is delivered to an organ via the circulatory system, the clearance of the organ is related to the blood flow rate by the relationship:

$$Cl = Q \cdot E \quad (5.22)$$

where Q (volume per unit time) is the rate of blood flow.

Table 5.3. The hepatic extraction values of some drugs.

Drug E_H value <0.3 (low)	Drug E_H value 0.3–0.7	Drug E_H value >0.7 (high)
Antripyrine	Aspirin	Cocaine
Diazepam	Codeine	Lignocaine
Nitrazepam	Nifedipine	Nicotine
Warfarin	Nortriptyline	Propranolol

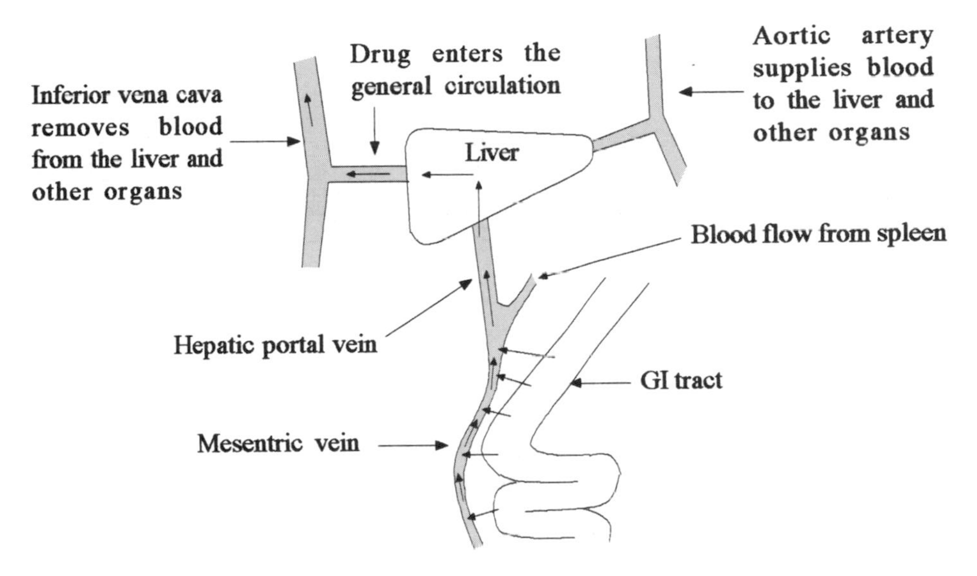

Figure 5.12. A schematic outline of the route of drugs (←) absorbed from the GI tract.

Hepatic extraction ratios are important in the design of dosage forms because the liver plays a major part in the extraction of drugs from the circulatory system. For example, propranolol has an E value for hepatic extraction of about 0.7. If this drug is administered by i.v. bolus injection, most of the drug would be distributed throughout the general circulatory system before reaching the liver. However, if the drug is administered orally, 70% of the dose would never reach the general circulatory system. This is because orally administered drugs that are absorbed from the GI tract pass through the liver before they enter the general circulatory system (Figure 5.12). Consequently, the hepatic extraction ratio of a drug is useful in determining its dose level and how it is to be administered. For example, if a drug has a high hepatic extraction ratio then a much higher dose of the drug must be used if it is to be given orally than if it were given by intravenous injection. This would increase the risk of toxic side effects.

It is difficult to obtain drug concentration data by sampling the human hepatic portal vein and so investigations are normally carried out on animals. Unfortunately, animals can behave

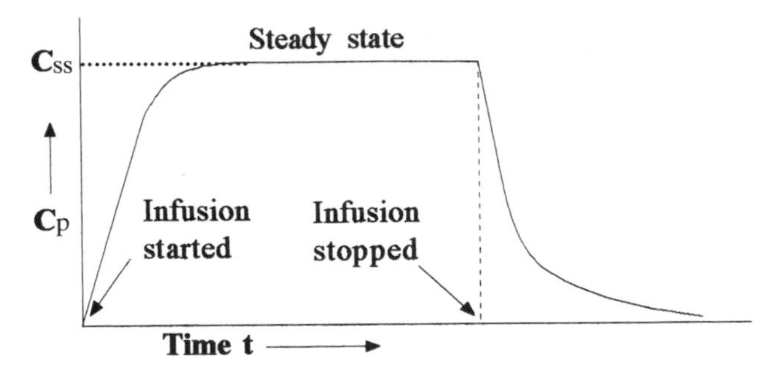

Figure 5.13. Plasma concentration changes with time in intravenous infusion.

in quite a different fashion to humans and so the conclusions drawn from this source can only be used as a rough guide to drug behaviour. However, extraction ratios are another piece of measurable information that can be used to link the required desirable characteristics of a potential new drug with the chemical structures of the analogues of a lead. Furthermore, it can lead to the avoidance of the loss of a potential drug by indicating the most effective dosage form.

5.4.2 Intravenous Infusion

In intravenous infusion, the drug is infused into the vein at a steady rate. Initially the plasma concentration of the drug increases because the amount of the infused drug exceeds the amount of the drug being eliminated (Figure 5.13). However, as the concentration of the drug in the plasma increases, the rate of elimination also increases until the rate of infusion equals the rate of elimination, at which point the concentration of the drug in the plasma remains constant. As long as the infusion rate is kept constant, the drug plasma concentration will remain at this steady-state level (C_{ss}). When infusion is stopped, the drug plasma concentration will fall, usually in an exponential curve because the biological situation is now the same as if a dose of the drug had been given at that time by i.v. bolus injection (see section 5.4).

The concentration of a drug is normally maintained at a constant value during intravenous infusion. This means that the infusion process is zero order (see Equation 5.1). Because elimination can be assumed to be first order (see section 5.4.1), the rate of change of the plasma concentration of a drug can be described by the mathematical relationship:

Rate of change of plasma concentration = Rate of infusion − Rate of elimination

that is:

$$\frac{dC_p}{dt} = k_o - k_{el}C_p \qquad (5.23)$$

where k_o is the rate constant for the infusion. The value of the rate of elimination ($k_{el}C_p$) increases, as the plasma concentration increases, until it equals the rate of infusion and there is no change in the drug plasma concentration, that is: $dC_p/dt = 0$. At this point the plasma concentration has reached the steady state and so:

$$\text{Rate of elimination} = \text{Rate of infusion}$$

that is:

$$k_{el}C_p = k_o \tag{5.24}$$

but at the steady state:

$$C_{ss} = C_p \tag{5.25}$$

and so:

$$C_{ss} = \frac{k_o}{k_{el}} \tag{5.26}$$

Because k_{el} is related to the total clearance (see section 5.14) by:

$$Cl_T = k_{el}V_d \tag{5.27}$$

substituting for k_{el} in Equation (5.26) gives:

$$C_{ss} = \frac{k_oV_d}{Cl_T} = \frac{k_o^*}{Cl_T} \tag{5.28}$$

where k_o^* is the amount of drug infused per unit time. Equations (5.26) and (5.28) can be used to calculate the rate of infusion required to achieve a specific steady-state plasma concentration. Furthermore, because these equations are independent of time, an increase in the rate of infusion and a subsequent increase in the value of k_o will not result in a reduction of the time taken to reach a specific value of C_{ss}. It will simply increase the value of C_{ss} because the rate of elimination and hence k_{el} will remain constant. Consequently, too high a rate of infusion could increase the steady-state plasma concentration of the drug to a value above the top limit of its therapeutic window for the drug, which in turn would increase the chances of a toxic response from the patient.

A time-dependent relationship can be determined for the initial part of the infusion before the plasma concentration reaches the steady state. Integration of Equation (5.23) gives the relationship:

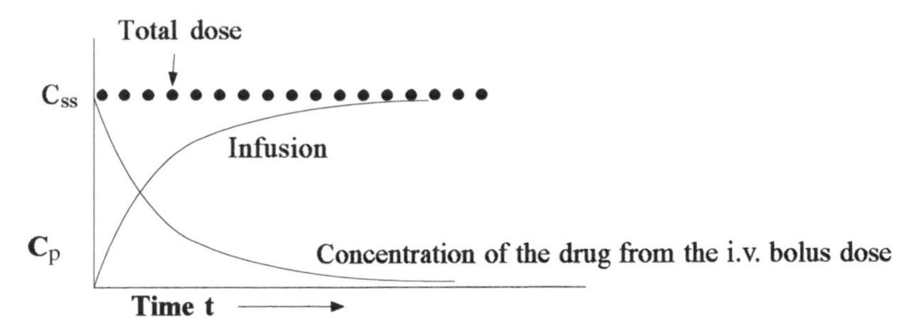

Figure 5.14. The effect of a single i.v. bolus on the plasma concentration of a drug administered by intravenous infusion.

$$t = \frac{-1}{k_{el}} \ln(1 - C_p/C_{ss}) \tag{5.29}$$

where t is the time taken to reach the drug plasma concentration C_p. The initial stage of the intravenous infusion normally follows first-order kinetics. Consequently, substituting Equation (5.8) in Equation (5.29) leads to the relationship:

$$t = \frac{-t_{1/2}}{0.693} \ln(1 - C_p/C_{ss}) \tag{5.30}$$

that is:

$$t = -t_{1/2} 1.44 \ln(1 - C_p/C_{ss}) \tag{5.31}$$

Equation (5.31) allows the calculation of the time taken to reach the effective therapeutic plasma concentration, which is normally taken as being 90% of the C_{ss} value. This time is dependent on the half-life value: the shorter the half-life, the sooner the C_{ss} plateau is reached. For example, when infused at a suitable rate, the antibiotic penicillin G has a half-life of about 30 min and reaches its effective therapeutic value in 100 min, whereas procainamide with a half-life of 2.8 h requires 9 h to reach its effective therapeutic value. Calculations, based on Equation (5.31), can be used to devise the best dosage form for potential and existing drugs.

To reduce the time required to obtain an effective therapeutic plasma concentration, a single i.v. bolus injection may be given in conjunction with an intravenous infusion. As a result, the C_{ss} drug plasma concentration is reached almost immediately. However, once the drug enters the blood-stream it undergoes elimination regardless of its original source: i.v. bolus or infusion. This elimination is compensated for by a build-up in the drug concentration from the intravenous infusion (Figure 5.14). At any time t the total concentration of the drug in the system is the sum of the drug from the i.v. bolus and the infusion. Its value is equal to the steady-state concentration C_{ss} of the drug. The net result is that the patient almost immediately receives an almost constant effective therapeutic dose of the drug. The pharmacokinetics of biological situations, such as the inhalation of a general anaesthetic gas and the slow release of a drug from an implant, in which a drug is introduced into the blood at a constant rate may also be described using a similar approach to that described for intravenous diffusion.

5.5 Extravascular Administration

The most common form of enteral dosage form is oral administration. Consequently, this section will mainly discuss the pharmacokinetics of orally administered drugs and will largely ignore other enteral routes. Most orally administered drugs are absorbed through the membranes of the GI tract. The rate of absorption will depend mainly on the rate of dissolution of the dosage form, that is, the rate at which the drug solid dosage form passes into solution. This will depend on the chemical structure of the drug (see section 4.3.3), the pH of the medium containing the drug (see section 3.9), the lipid/aqueous medium partition coefficient of the drug and the surface area of the absorbing region of the GI tract. The small surface area and the short time that a drug normally takes to pass through the stomach means that drug absorption is often less in this region of the GI tract than in the small intestine, which has a far larger surface area. However, absorption in the stomach is better if the drug is taken after a meal because the presence of food in the stomach slows down the passage of the drug through the stomach.

Drugs absorbed from the GI tract must pass through the liver in order to reach the general circulation system (Figure 5.12). As a result, a fraction of the drug is lost by metabolism and excretion as it passes through the GI tract membrane, liver and other organs before it reaches the systemic circulation. These losses are referred to as the **first-pass effect or first-pass metabolism**. First-pass metabolism is effectively the elimination of the drug before it enters the general circulatory system. In this circuit the main areas of excretion and metabolism are the enzyme-rich liver and lungs and so the term **first-pass metabolism** is usually taken to refer to the elimination of a drug by these two organs. However, because the liver is the first organ that the drug passes through after absorption from the GI tract and because it is also the principal area of metabolism, the effect of the lungs is often ignored and the term first-pass metabolism is frequently used as though it involves only the liver.

The physiology of drug absorption from the GI tract has a direct effect on the **bioavailability** of a drug. Bioavailability is defined as the fraction of the dose of a drug (F) that enters the general circulatory system, that is:

$$F = \frac{\text{Amount of drug that enters the general circulatory system}}{\text{Dose administered}} \tag{5.32}$$

Since the area under the plasma concentration–time curve (AUC) for a drug is a measure of the concentration of a drug reaching the general circulatory system, the bioavailability of a drug may also be defined in terms of the AUC as:

$$F = AUC/Dose \tag{5.33}$$

Bioavailability is related to the extraction of the drug by the relationship:

$$F = 1 - E \tag{5.34}$$

For orally administered drugs, Equation (5.34) approximates to:

$$F = 1 - E_H \tag{5.35}$$

where E_H is the hepatic extraction ratio. Therefore, drugs with high hepatic extraction values ($E_H \sim 1$) will seldom reach the general circulatory system in sufficient quantity to be therapeutically effective if administered orally. Furthermore, for these drugs, because $E \sim 1$, Equation (5.22) becomes:

$$Q_H = Cl_H \tag{5.36}$$

that is, the hepatic clearance of substances that undergo high first-pass effects approaches the rate of blood flow to the liver. Consequently, the determination of the hepatic clearance of a potential drug after intravascular administration using animal experiments is used to determine whether a potential drug will have a high first-pass metabolism effect.

Bioavailability studies are used to compare the efficiency of the delivery of the dosage forms of a drug to the general circulatory system as well as the efficiency of the route of administration for both licensed drugs and new drugs under development. Two useful measurements are **relative** and **absolute bioavailability**.

Relative bioavailability. Relative bioavailability may be used to compare the relative absorptions of the different dosage forms of the same drug and also the relative availabilities of two different drugs with the same action when delivered using the same type of dosage form. It is defined for equal doses as:

$$\text{Relative bioavailability} = \frac{\text{AUC for drug A (or dosage form A)}}{\text{AUC for drug B (or dosage form B)}} \tag{5.37}$$

Percentage relative bioavailability figures may be obtained by multiplying Equation (5.37) by 100. A correction must be made if different drug doses are used, in which case Equation (5.37) becomes:

$$\text{Relative bioavailability} = \frac{(\text{AUC for drug A or dosage form A})/\text{Dose A}}{(\text{AUC for drug B or dosage form B})/\text{Dose B}} \tag{5.38}$$

Example 5.1. *The variation of the plasma concentration with time for a single dose of a drug administered orally using either one 100-mg tablet or one 5-cm^3 dose of a linctus containing 100 mg of the drug was determined for a number of healthy volunteers. The results of the study showed that the tablet curve had an AUC of 43.7 $\mu g\,h\,cm^{-3}$ whereas the linctus curve had an AUC of 42.5 $\mu g\,h\,cm^{-3}$. Because the doses of the drugs are the same, substituting these experimental values in Equation (5.38) gives:*

$$\textit{Relative bioavailability} = \frac{AUC\ tablet}{AUC\ linctus} = \frac{43.7}{42.5} = 1.028$$

This means that there was almost no difference between the two dosage forms as far as absorption (bioavailability) is concerned but drug absorption was slightly better from the tablet than the linctus. A value significantly higher than unity would have suggested that the bioavailability of the drug from the tablet is much better than from the linctus, whereas a value significantly less than unity would have indicated that the reverse was true.

This type of calculation is useful in drug design because it ensures that the dosage forms used in trials are effective in delivering the drug to the general circulation. It is also used by licensing authorities as a check on the efficacy of products when manufacturers change the dosage form of a drug in clinical use.

Absolute bioavailability. Absolute bioavailability is used as a measure of the efficiency of the absorption of the drug. It is defined in terms of the total dose of the drug that the body would receive if the drug was placed directly in the general circulation by an i.v. bolus injection, that is:

$$\text{Absolute bioavailability}(F) = \frac{\text{AUC for oral dosage form /oral dose}}{\text{AUC for i.v. dosage form /i.v. dose}} \tag{5.39}$$

Example 5.2. *The change in plasma concentration with time for a series of analogues of a lead compound was followed using a number of healthy volunteers and the data obtained were recorded in Table 5.4. Predict from this data which analogues would be the most profitable to develop.*

For analogue A, substituting in Equation (5.38):

Table 5.4. Experimental results of plasma concentration–time studies for three analogues.

Analogue	Oral dose (mg)	AUC ($\mu g\,h\,cm^{-3}$)	i.v. Dose (mg)	AUC ($\mu g\,h\,cm^{-3}$)
A	100	67.3	50	47.8
B	100	49.2	50	46.5
C	100	91.2	50	46.7

$$\text{Absolute bioavailability } (F) \text{ of } A = \frac{67.3/100}{47.8/50} = 0.71$$

Similarly the absolute bioavailabilities of analogues B and C are 0.53 and 0.97, respectively. This means that analogue C would have the best absorption and on this basis would give the best chance of a successful outcome. However, it should be realised that the final decision as to which analogue to develop would not be based solely on its bioavailabity. It would be based on a consideration of all the pharmacokinetic and pharmacodynamic data obtained for the three analogues.

5.5.1 Single Oral Dose

When a single dose of a drug is administered orally its plasma concentration increases to a maximum value (C_{max}) at t_{max} before falling with time (Figure 5.15). The increase in plasma concentration occurs as the drug is absorbed. It is accompanied by elimination, which starts from the instant when the drug is absorbed. The rate of elimination increases as the concentration of the drug in the plasma increases to the maximum absorbed dose. At this point, the rate of absorption equals the rate of elimination. Once absorption ceases, elimination becomes the dominant pharmacokinetic factor and plasma concentration falls.

The change in the plasma concentration–time curve for a single oral dose is useful in a number of ways. It shows the time taken for the drug to reach its therapeutic window concentration (Figure 5.15) and the period of time the plasma concentration lies within the therapeutic window. The latter is an important consideration when selecting the time interval (t_{di}) for administering repeat doses in order to maintain the plasma concentration within the therapeutic window. It also shows whether the drug reaches toxic values above the upper limit of the therapeutic window. Moreover, because the time (t_{max}) taken to reach the maximum plasma concentration (C_{max}) is an approximate measure of the rate of absorption of the drug, the value of t_{max} can be used to compare the absorption rates of that drug from different dosage forms. However, in view of the difficulty in taking serum samples at exactly the right time, both t_{max} and C_{max} are normally determined by calculation (see section 5.5.2). Both of these factors are important in the selection of drugs for further development and the design of their dosage forms.

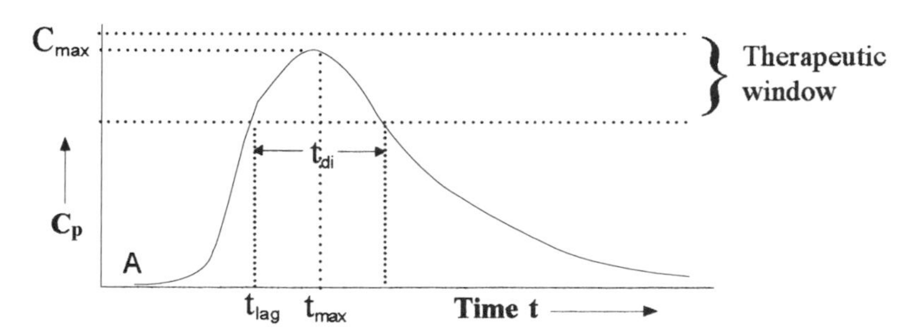

Figure 5.15. The change in plasma concentration of a drug with time due to a single oral dose of the drug. A small time lapse A occurs before the drug reaches the blood. This is mainly the time taken for the drug to reach its site of absorption from the mouth.

The absorption of a drug into the general circulatory system is a complex process. Drugs given orally dissolve in the GI tract fluids before being absorbed through the GI tract membrane. The rate of absorption depends on both the drug's chemical nature, which is discussed in Chapter 4, and the physical conditions at the site of absorption. However, most drugs exhibit approximately first-order absorption kinetics except when there is a high local concentration that saturates the absorptive mechanism, in which case zero-order characteristics are often found. Elimination also normally follows first-order kinetics but can change if the elimination processes are saturated by a high drug concentration due to the use of a high dose.

The rate of change of the amount (A) of an orally administered drug in the body will depend on the relative rates of absorption and elimination, that is:

$$\frac{dA}{dt} = \text{Rate of absorption} - \text{Rate of elimination} \qquad (5.40)$$

This equation may be used to calculate the changes in a drug's plasma concentration with time. The nature of the calculation, which is beyond the scope of this text, will depend on the order of the absorption and elimination processes and the type of compartmental model used. For example, for a one-compartment model in which the drug exhibits first-order absorption and elimination (Figure 5.16), by using Equation (5.40) it is possible to show that:

$$C_p = \frac{FD_o}{V_d} \cdot \frac{k_{ab}}{(k_{ab} - k_{el})}(e^{-k_{el}t} - e^{-k_{ab}t}) \qquad (5.41)$$

where k_{ab} and k_{el} are the absorption and elimination rate constants, respectively and D_o is the dose administered.

The elimination rate constant, like other pharmacokinetic parameters, is in theory independent of the method of administration. Therefore, provided that the dose is below the

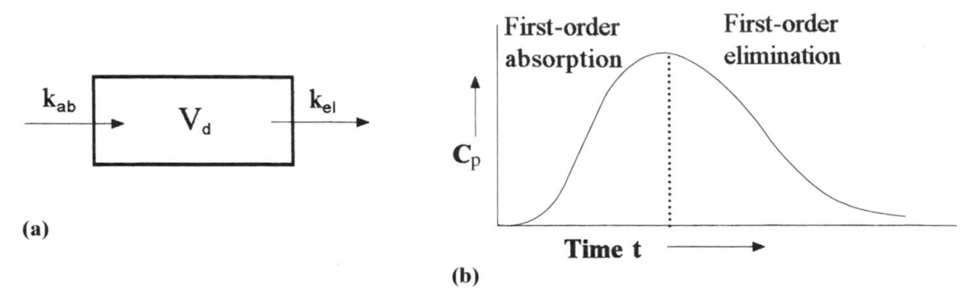

Figure 5.16. (a) A one-compartment model for a single orally administered dose. (b) The plasma concentration–time curve for a drug that exhibits first-order kinetics for both its absorption and elimination.

saturation concentration of the elimination processes, the value of k_{el} may be determined by an independent experiment using a single i.v. bolus of the drug (see section 5.4.1.5). Substituting this value of k_{el} together with the values of F (see Equation 5.32), V_d (see section 5.6) D_o, C_p and t in Equation (5.41) will give a value for k_{ab}. It is necessary to make several calculations using different values of C_p in order to obtain an accurate figure. The absorption rate constant k_{ab} may be used to compare the relative rates of absorption of drugs with the same action as well as the relative rates of absorption of the same drug from different dosage forms.

The elimination rate constant may also be calculated from the elimination section of the drug plasma concentration–time curve for the oral dose of the drug (Figure 5.16). In this section of the curve the rate of absorption is zero and so $k_{ab} = 0$. Therefore Equation (5.41) becomes:

$$C_p = \frac{FD_o}{V_d} \cdot \frac{k_{ab}}{(k_{ab} - k_{el})}(e^{-k_{el}t}) \qquad (5.42)$$

taking logarithm to base e:

$$\ln C_p = \ln\left(\frac{FD_o}{V_d} \cdot \frac{k_{ab}}{(k_{ab} - k_{el})}\right) - k_{el}t \qquad (5.43)$$

Expressing Equation (5.43) in terms of logarithm to base 10:

$$\log C_p = \log\left(\frac{FD_o}{V_d} \cdot \frac{k_{ab}}{(k_{ab} - k_{el})}\right) - \frac{k_{el}t}{2.303} \qquad (5.44)$$

Hence a plot of $\log C$ against t will be a straight line (Figure 5.17) with a slope of $-k_{el}/2.303$ and an intercept on the y-axis equal to:

$$\log\left(\frac{FD_o}{V_d} \cdot \frac{k_{ab}}{(k_{ab} - k_{el})}\right) \qquad (5.45)$$

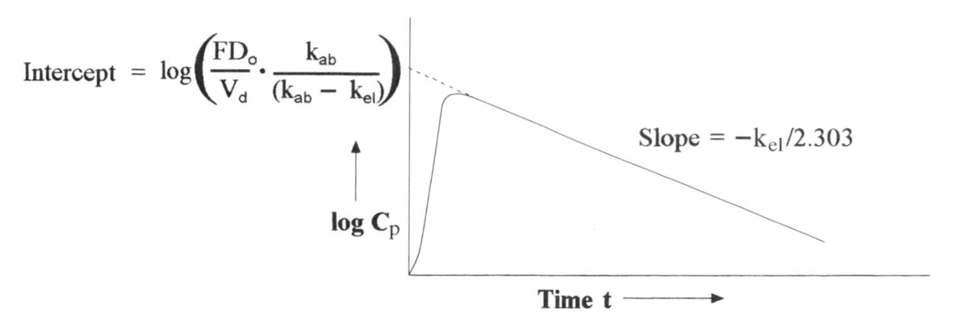

Figure 5.17. The $\log C_p$ against t plot for the elimination stage of a first-order absorption and elimination of a drug.

The rate constant k_{el} for the elimination is calculated from the slope of the graph. Its value is used to calculate the rate constant k_{ab} for the absorption of the drug by substituting it in either Equation (5.42) or the expression for the intercept, Equation (5.45).

The mathematical approach outlined in this section may be generally applied to drugs that do not exhibit first-order absorption and elimination processes. Consider, for example, a drug that exhibits zero-order absorption and first-order elimination kinetics. The rate of change of the drug in the body will be given by substituting the relevant rate expressions in Equation (5.40), which gives:

$$\frac{dA}{dt} = k_o - k_{el}A \tag{5.46}$$

where k_o is the zero rate constant for the absorption process. However, it should be realised that some drug absorption and elimination processes do not exhibit zero- or first-order kinetics and so these processes cannot always be so easily quantified.

5.5.2 The Calculation of t_{max} and C_{max}

At t_{max} the rate of change of the plasma concentration of the drug is zero, that is:

$$\frac{dC_p}{dt} = 0 \tag{5.47}$$

The mathematical nature of the calculation of the values of t_{max} and C_{max} will depend on the order of the kinetics exhibited by the drug. For example, for drugs that exhibit first-order absorption and elimination kinetics, the plasma concentration at any time t is given by Equation (5.41). Differentiation of this equation with respect to time, in order to obtain the rate of change of the plasma concentration of the drug with time, gives:

$$\frac{dC_p}{dt} = \frac{FD_o}{V_d} \cdot \frac{k_{ab}}{(k_{ab} - k_{el})}(-k_{el}e^{-k_{el}t} + k_{ab}e^{-k_{ab}t}) \quad (5.48)$$

therefore, at t_{max}:

$$\frac{FD_o}{V_d} \cdot \frac{k_{ab}}{(k_{ab} - k_{el})}(-k_{el}e^{-k_{el}t_{max}} + k_{ab}e^{-k_{ab}t_{max}}) = 0 \quad (5.49)$$

Simplifying:

$$-k_{el}e^{-k_{el}t_{max}} + k_{ab}e^{-k_{ab}t_{max}} = 0 \quad (5.50)$$

$$k_{ab}e^{-k_{ab}t_{max}} = k_{el}e^{-k_{el}t_{max}} \quad (5.51)$$

and taking logs to base e:

$$\ln k_{ab} - k_{ab}t_{max} = \ln k_{el} - k_{el}t_{max} \quad (5.52)$$

gives:

$$t_{max} = \frac{\ln k_{el} - \ln k_{ab}}{k_{el} - k_{ab}} \quad (5.53)$$

Once t_{max} has been calculated C_{max} can be found by substitution of the values of F (see Equation 5.32), V_d (see section 5.4.1.2), D_o, k_{ab} and k_{el} for the drug in Equation (5.41). Because F and V_d are both constants, C_{max} is proportional to the dose administered: the larger the dose, the greater C_{max}.

5.5.3 Repeated Oral Doses

In order for a drug to be therapeutically effective, its plasma concentration must be maintained within its therapeutic window for a long enough period of time to obtain the desired therapeutic effect. This can only be achieved by the use of repeat doses at regular time intervals. Initially, for each dose, the rate of absorption will exceed the rate of elimination and so the plasma concentration of the drug will steadily increase as the number of doses increases (Figure 5.18). However, as the plasma concentration increases so does the rate of elimination and eventually the plasma concentration will reach a plateau. The value of the plateau plasma concentration will vary between a maximum and minimum; the time interval between these values depends on the time interval between the doses. The values of k_{ab} and k_{el} may be used to calculate the plateau maximum and minimum drug plasma concentration values. This is useful in designing multiple dosage form regimens.

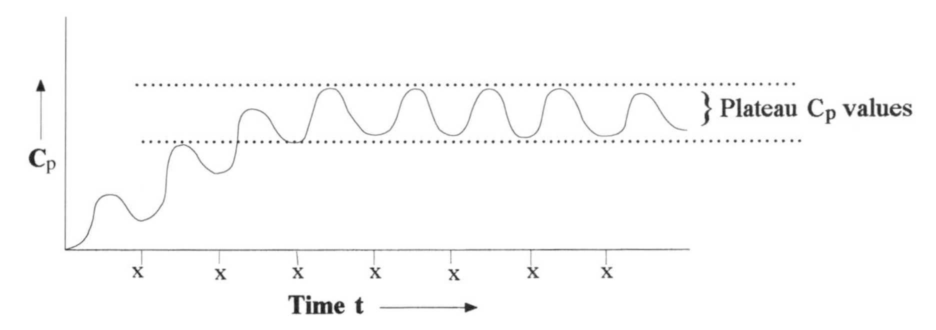

Figure 5.18. The general changes in plasma concentration with time for repeated oral doses. Repeat doses were administered at regular time intervals x.

The time taken to achieve a plateau concentration may be reduced by using a larger than usual initial dose. This **loading dose,** as it is known, gives a relatively high initial plasma concentration, which acts as an elevated starting point for the succeeding normal doses. This reduces the time taken to reach the plateau concentration, that is, the therapeutic window, and so can be particularly useful in cases of serious illness.

5.6 The Use of Pharmacokinetics in Drug Design

Pharmacokinetic data is used to differentiate between active substances with good and poor pharmacokinetic characteristics. For example, substances with poor absorption, high first-pass metabolism and an unsuitable half-life (too long or too short) will normally be discarded in favour of substances with more appropriate pharmacokinetic properties.

Pharmacokinetics is used in all the development stages of a drug from preclinical to phase IV trials. Legislation normally demands that the absorption, elimination, distribution, clearance, bioavailability, $t_{1/2}$ and V_d of all existing and new drugs must be defined using preclinical trials. In theory all these parameters can be scaled up from animal experiments to predict the behaviour of the substance in humans but the correlation is usually only approximate (see section 5.7). The same parameters are required for Phase I trials. Phase I results for bioavailability and $t_{1/2}$ are used by the pharmaceutical industry as part of the evidence for deciding whether further investigation of a potential drug would be justified. For example, if the first-pass elimination of the potential drug is high (low bioavailability), the drug is not likely to be therapeutically effective using an oral dosage form. This poses questions such as: will the community allow i.v. bolus and infusion methods? Is the action of the drug unique enough to justify the expense of these dosage forms? A positive answer to these and other questions means that further development could occur. However, a negative answer means that the development is stopped and a new analogue would be selected for investigation.

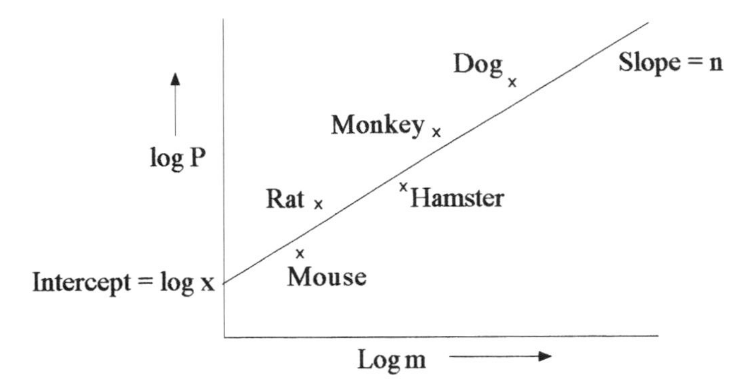

Figure 5.19. A simulation of the logarithmic change of a pharmacokinetic parameter with body mass for various animals.

Phase I trials are also used to predict appropriate dose levels for the expected patient community.

Once the drug has been found to be safe and effective in the Phase I trials, the evaluation of pharmacokinetic parameters is required for the diseased state, age (young and old) and gender in the Phase II trials. It is especially important to determine the effect of reduced liver and kidney function on the elimination of the new drug in order to avoid toxic effects due to the use of too high a dose level in patients with these conditions. All this information is used to develop effective safe dosage forms for new drugs and to check new dosage forms for existing drugs.

5.7 Extrapolation of Animal Experiments to Humans

The extrapolation of the results of animal experiments to human can be made using the relationship:

$$P = xm^n \tag{5.54}$$

where P is the parameter being extrapolated, such as Cl, V_d or $t_{1/2}$, m is the body mass of the animal used to determine P and x and n are constants characteristic of the parameter and compound being investigated. This relationship may also be expressed in the form:

$$\log P = \log x + n \log m \tag{5.55}$$

Because Equation (5.55) corresponds to that for a straight line ($y = mx + c$), the values of n and x may be determined by plotting a graph of $\log P$ against $\log m$ for a number of animals with different body weights. The value of n is the slope of the graph while $\log x$ is the intercept on the y-axis. For example, animal experiments (Figure 5.19) have shown that the plasma clearance ($P = Cl_P$) for cyclophosphonamide is related to body weight by the equation:

$$Cl_P = 16.7\,m^{0.754} \tag{5.56}$$

In general, experimental work has shown that the values of n tend to be of the order of 0.75 for clearance, 0.25 for half-life and 1.0 for volume of distribution.

5.8 Summary

Pharmacokinetics is the study of how the body handles a drug. It attempts to quantify drug absorption, distribution, metabolism and elimination. **Pharmacological response** is presumed to be a function of the drug concentration at its site of action. The drug concentration at the site of action is assumed, with some exceptions, to be related to the concentration of the drug in the plasma. It is related to the total amount of the drug in the body at a time t by the *apparent* **volume of distribution (V_d)** (see Equation 5.6). The value of V_d for a drug includes the readily perfused tissues as well as the general circulatory system. The range of plasma concentration over which a drug is therapeutically effective is known as the drug's **therapeutic window**.

The rate of change of the plasma concentration of a drug with time often exhibits zero- or first-order kinetics and so can be described by the mathematics associated with these orders (see Equations 5.1 and 5.2). This requires the use of model systems. Two general types of model most commonly used are the compartmental model and the perfusion model. Only the former is used in this text. All compartmental models are based on a central compartment that represents the systemic circulatory system and readily perfused tissues. Absorption and elimination are normally associated with the central plasma and perfused tissues compartment. Elimination includes drug removal by both metabolism and excretion.

Distribution is the transport of a drug from its point of absorption to all the accessible body compartments. Its rate and extent will depend on the properties of the drug, the method of administration and the rates of blood flow and elimination of the drug.

Elimination is the irreversible removal of the drug from the body by excretion and metabolism, k_{el} being the rate constant for the process. The rate of elimination of a drug often exhibits first-order kinetics (see Equation 5.4). The **elimination half-life** of a drug ($t_{1/2}$) is the time taken for the concentration of the drug to fall to half its original value. It is a constant for drugs that exhibit first-order elimination kinetics (see Equation 5.8). The elimination rate constant for a drug that exhibits first-order kinetics may be determined from a graph of the logarithm of the plasma concentration against time (see Equation 5.10). First pass metabolism is effectively the elimination that occurs before the drug does its first circuit around the body in the general circulatory system.

Clearance is the parameter that relates the rate of elimination to the drug plasma concentration (see Equation 5.11). It is defined as the volume of biological fluid that is completely cleared of a drug in unit time. For drugs that exhibit first-order elimination, clearance is a constant independent of plasma concentration. It is an additive property, the total clearance being the sum of all the clearances contributing to the elimination process (see Equation 5.16), except in the case of the lungs. Clearance can be determined by measuring the area under the plasma concentration–time curve (AUC) for a drug (see Equation 5.20).

Extraction is the loss of a drug as it passes through an organ. The extraction ratio (E) is the ratio of the drug removed by the organ to the drug entering the organ (see Equation 5.21). The clearance of an organ is given by: $Cl = Q \cdot E$. The main methods of **intravascular administration** of a drug are intravascular injection (i.v. bolus) and intravenous infusion. For a **single i.v. bolus injection** the rate of change of the plasma concentration falls with time. With many drugs this fall follows first-order kinetics (see Equation 5.4). In **intravenous infusion** the rate of infusion follows zero-order kinetics. The plasma concentration of a drug reaches a steady-state value (C_{SS}) after an initial period in which it increases. At the steady state in intravenous infusion, the rate of infusion equals the rate of elimination and the value of C_{SS} can be calculated using Equation (5.26). The value of C_{SS} may be calculated from the plasma clearance using Equation (5.28). The **effective therapeutic dose** of a drug is usually taken as being 90% of its C_{SS} value.

The bioavailability of a drug is the fraction of the administered dose that reaches the general circulatory system (see Equation 5.32). The **relative bioavailability** of a drug is the bioavailability of the drug in an orally administered dosage form compared to its bioavailability from a different type of orally administered dosage form (see Equation 5.38). Relative bioavailability is used to compare dosage forms of the same drug. **Absolute bioavailability** is the bioavailability of the drug in an orally administered dosage form compared to its bioavailability from an i.v. bolus (see Equation 5.39). It is used as a measure of the efficiency of the absorption of a drug.

The main **extravascular administration** route is oral administration. For a **single oral dose** of a drug the plasma concentration of the drug will increase with time to a maximum (C_{max}) and then decrease. Initially absorption is the dominant process and plasma concentration increases. However, as the concentration of the drug in the general circulatory system increases, elimination processes becomes more important and the plasma concentration decreases. The time taken for the concentration of a single orally administered drug to reach its maximum plasma concentration (C_{max}) is t_{max}. The values of t_{max} and C_{max} are recorded as pharmacokinetic characteristics of a drug. They can only be determined by calculation, the nature of the calculation depending on the order of the absorption and elimination processes (see section 3.5.2). **Repeated oral doses** are used to maintain the plasma concentration of a drug within its therapeutic window. In repeated oral dosing the plasma concentration of a drug reaches a plateau, which consists of a series of maximum and minimum values whose degree of separation depends on the dose interval. At the plateau the rate of absorption equals the rate of elimination.

Pharmacokinetic parameters determined using **animals** can be extrapolated to give values applicable to humans (see Equation 5.54). This extrapolation is usually not very accurate.

5.9 Questions

(1) Explain the meaning of each of the following terms: (a) therapeutic window; (b) drug regimen; (c) i.v. bolus; (d) clearance; (e) first-pass effect.

(2)

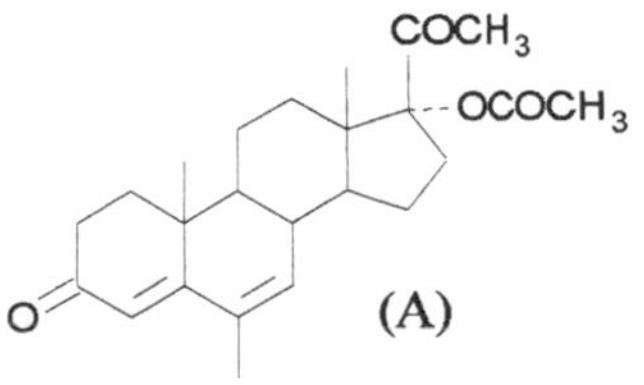

$COCH_3$ $OCOCH_3$ O

(A)

Megestrol acetate (A) is an oral contraceptive. What pharmacokinetic parameters should be determined for other potential oral contraceptives in order to compare their actions with compound A? Give reasons for choosing your selected parameters.

(3) Discuss the general importance of quantitative analysis in pharmacokinetic studies. Include in the discussion a brief description of the problems that can give rise to inaccurate interpretation of kinetic data.

(4) Digoxin and cyclosporin have narrow therapeutic windows. The rate of elimination of cyclosporin is much faster than that of digoxin. If the doses required for therapeutic success are similar, how does this information affect their drug regimens?

(5) A patient was given a single dose of 30 mg of a drug by an i.v. bolus injection. The drug plasma concentration was determined at set time intervals, the data obtained being recorded in Table 5.5. Calculate:

(a) the value of the elimination constant of the drug;
(b) the apparent volume of distribution of the drug;
(c) the clearance of the drug;

What fundamental assumption has to be made in order to calculate these values?

Table 5.5

Time (h)	Plasma concentration ($\mu g\,cm^{-3}$)
1	5.9
2	4.7
3	3.7
4	3.0
5	2.4
6	1.9

(6) The clearance of a drug from a body compartment is $5\,cm^3\,min^{-1}$. If the compartment originally contained 50 mg of this drug, calculate the amount of drug remaining in the system after 1, 2, 3, 4, 5, 6, 7, 8, 9 and 10 min if the volume of the compartment was $50\,cm^3$. Plot a graph of concentration against time and determine the values of $t_{1/2}$ and k_{el} for the compartment and the drug.

(7) A dose of 50 mg of the drug used in question 5 was administered orally to the same patient who received the i.v. bolus in question 5. Plasma samples were taken at regular time intervals and a graph of plasma concentration against time was plotted. If the area under this curve was 5.01, calculate the absolute bioavailability of the drug in the patient, assuming first-order absorption and elimination. Comment on the value of the figure obtained. You may use the i.v. data recorded in question 5.

Table 5.6.

Analogue	Elimination rate constant (m^{-1})	AUC i.v. bolus (30-mg dose) ($\mu g\,m\,cm^{-3}$)	AUC single oral dose (30-mg dose) ($\mu g\,m\,cm^{-3}$)
A	0.1386	30.4	31.5
B	0.0277	31.2	47.6
C	0.0462	100.3	81.4
D	0.0173	69.7	81.9

(8) The data in Table 5.6 are based on plasma concentration–time curves for a number of analogues of a lead compound. Calculate the relevant pharmacokinetic parameter(s) and indicate the best analogue for further investigation if a drug with a reasonable duration of action is required. Assume that the absorption and elimination of the drug follow first-order kinetics.

(9) A patient is being treated with morphine by intravenous infusion. The steady-state plasma concentration of the drug is to be maintained at $0.04\,\mu g\,cm^{-3}$. Calculate the rate of infusion necessary-assuming a first-order elimination process. (for morphine, V_d is $4.0\,dm^3$ and $t_{1/2}$ is 2.5 h).

6 Enzymes

6.1 Introduction

Enzymes act as catalysts for almost all of the chemical reactions that occur in all living organisms. Their important general characteristics are:

(i) mild conditions are required for the enzyme action;
(ii) they have a big capacity, with a minute amount of enzyme rapidly producing a large amount of a product;
(iii) they usually exhibit a high degree of specificity;
(iv) their activities can be controlled by substances other than their substrates.

Enzymes are usually large protein molecules that are sometimes referred to as **apoenzymes**. However, some RNA molecules, known as **ribozymes** (see section 6.14), can also act as enzymes. The structures of a number of enzymes contain groups of metal ions, known as **clusters** (see section 7.2.4), coordinated to the peptide chain. These enzymes are often referred to as **metalloenzymes**.

Some enzymes require the presence of organic compounds known as **coenzymes** (Figure 6.1) and/or metal ions and inorganic compounds referred to as **cofactors** before the enzyme will function. Coenzymes and cofactors are separate chemical species that are bound to the apoenzyme by electrostatic bonds, hydrogen bonds and van der Waals' forces. These composite active enzyme systems are known as **holoenzymes**.

$$\underset{(inactive)}{\text{Apoenzyme}} + \underset{(inactive)}{\text{Coenzyme and/or Cofactor}} = \underset{(active)}{\text{Holoenzyme}}$$

However, the term enzyme is commonly used to refer to both holoenzyme systems and those enzymes that do not require a coenzyme and/or cofactor.

Enzymes are widely distributed in the human body. It has been reported that there are over 3000 in a single cell. They are found embedded in cell walls and membranes as well as in the various biological fluids in living organisms. All enzymes are produced by cells and mainly function within that cell. It is often difficult to isolate and purify enzymes, but several thousand enzymes have been purified and characterised to some extent.

A number of enzymes are produced by the body from inactive protein precursors. These precursors are known as **proenzymes** or **zymogens**. The formation of the enzyme often requires

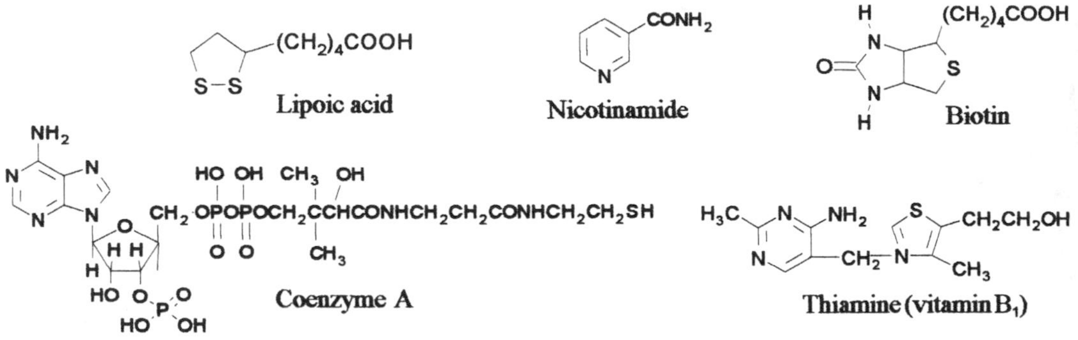

Lipoic acid

Nicotinamide

Biotin

Coenzyme A

Thiamine (vitamin B_1)

Figure 6.1. Examples of the varied nature of coenzymes.

the removal of part of the peptide chain of the precursor protein. For example, the active enzyme trypsin is produced from the proenzyme trypsinogen by the loss of a six amino acid residue chain from the N-terminal of trypsinogen. This loss is accompanied by a change in the conformation of the resultant protein to form the active form of the enzyme. Proenzymes allow the body to produce and control enzymes that could be harmful in the wrong place. For example, trypsin catalyses the breakdown of proteins. Consequently, the formation of active trypsin in the body could pose a problem and so the body produces the inactive form that is only converted to the active form when it reaches the intestine, where it catalyses the breakdown of proteins in our food.

Enzymes with different structural forms can catalyse the same reactions. These enzymes are usually isolated from different tissues but can be found in the same tissue. Where enzymes have the same catalytic activities they are referred to as **isoenzymes** or **isozymes**. Isoenzyme structures often contain the same protein subunits but either in a different order or ratio. For example, the structure of lactate dehydrogenase, which catalyses the conversion of lactate to pyruvate and vice versa, is a tetramer consisting of four so-called H and M subunits. The ratio of H to M subunits in the enzyme will depend on its source. Lactate dehydrogenase isolated from the heart contains mainly H subunits whereas that from the liver and skeletal muscle contains mainly M subunits. Although isoenzymes catalyse the same reaction they can exhibit significantly different properties such as thermal stability, the values of pharmacokinetic parameters (see Chapter 5), electrophoretic mobilities and the effects of inhibitors.

Enzymes that catalyse the same reactions but have significantly different structures are known as **isofunctional** enzymes. Isofunctional enzymes are produced by different tissues and different organisms. Their different origins and structures mean that they can act as selective targets for drugs. For example, the enzyme dihydrofolate reductase catalyses the interconversion of dihydrofolate to tetrahydrofolate in both humans and bacteria. The antibiotic trimethoprim is more effective in inhibiting this conversion in bacteria than in humans, which makes it useful for the treatment of bacterial infections in humans. Nevertheless, resistance to this drug is found in some species of bacteria (see section 6.8).

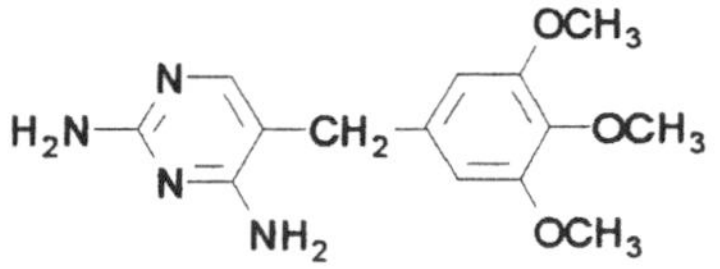

Trimethoprim

6.2 Classification and Nomenclature

Enzymes are broadly classified into six major types (Table 6.1).

The International Union of Biochemistry has recommended that enzymes have three names, namely: a systematic name that shows the reaction being catalysed and the type of reaction based on the classification in Table 6.1; a recommended trivial name; and a four-figure Enzyme Commission code (EC code) also based on the classification in Table 6.1. Nearly all systematic and trivial enzyme names have the suffix **-ase.**

Systematic names show, often in semichemical equation form, the conversion that the enzyme promotes and the class of the enzyme. For example, L-aspartate: 2-oxoglutarate aminotransferase catalyses the transfer of an amino group from L-aspartic acid to 2-oxoglutaric acid, whereas D-glucose ketoisomerase catalyses the conversion of D-glucose to D-fructose.

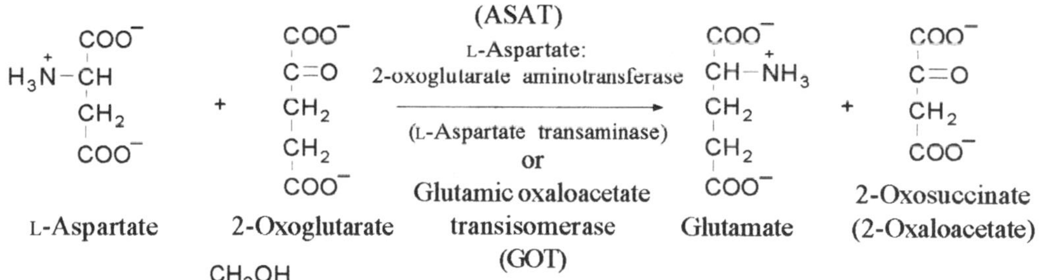

L-Aspartate + 2-Oxoglutarate → Glutamate + 2-Oxosuccinate (2-Oxaloacetate)

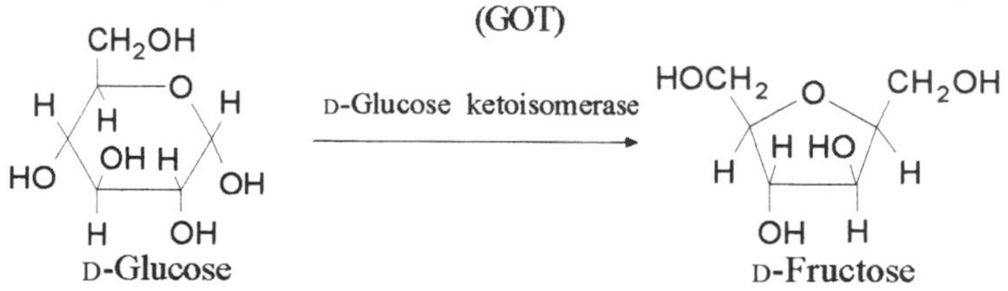

(ASAT) L-Aspartate: 2-oxoglutarate aminotransferase (L-Aspartate transaminase) or Glutamic oxaloacetate transisomerase (GOT)

D-Glucose → D-Fructose (D-Glucose ketoisomerase)

Trivial names are usually based on the function of the enzyme but may also include or be based on the name of the substrate. For example, alcohol dehydrogenase is the trivial name for alcohol NAD^+ oxidoreductase, the enzyme that catalyses the oxidation of ethanol to ethanal, whereas glucose-6-phosphatase catalyses the hydrolysis of glucose-6-phosphate to glucose and phosphate ions.

Table 6.1. The classification of enzymes.

Code	Classification	Type of reaction catalysed
1	Oxidoreductases	Oxidations and reductions
2	Transferases	The transfer of a group from one molecule to another
3	Hydrolases	Hydrolysis of various functional groups
4	Lyases	Cleavage of a bond by non-oxidative and non-hydrolytic mechanisms
5	Isomerases	The interconversion of all types of isomer
6	Ligases (synthases)	The formation of a bond between molecules

$$NAD^+ + \underset{\text{Ethanol}}{CH_3CH_2OH} \xrightarrow{\text{Alcohol dehydrogenase}} \underset{\text{Ethanal}}{CH_3CHO} + NADH + H^+$$

$$\text{Glucose-6-phosphate} \xrightarrow{\text{Glucose-6-phosphatase}} \text{Glucose} + PO_4^{3-}$$

However, some trivial names in current use are historical and bear no relationship to the action of the enzyme or its substrate. For example, pepsin and trypsin are the names commonly used for two enzymes that catalyse the breakdown of proteins during digestion.

The Enzyme Commission's code is unique for each enzyme. It is based on the classification in Table 6.1 but further subdivides each class of enzyme according to how it functions. For example, the Enzyme Commission code for the lactate dehydrogenase that catalyses the oxidation of L-lactate to pyruvate is EC 1.1.1.27. The letters EC show that the numbers are an Enzyme Commission code for an enzyme. The first number indicates that the enzyme is an oxidoreductase, the second that it is acting on the CH—OH bond of electron donors and the third that NAD^+ is the electron acceptor. The fourth number identifies the enzyme as a specific oxidoreductase, namely lactate dehydrogenase. The full code is given in the International Union of Biochemistry publication entitled *Enzyme Nomenclature*.

$$NAD^+ + \underset{\text{L-Lactate}}{HO{-}CH(COO^-)(CH_3)} \xrightarrow{\text{Lactate dehydrogenase}} \underset{\text{Pyruvate}}{CH_3{-}C({=}O){-}COO^-} + NADH + H^+$$

This text uses trivial names because they are usually easier to pronounce. In addition, some letter abbreviations will also be used.

6.3 Active Sites and Catalytic Action

The reactant(s) whose reaction is catalysed by an enzyme is known as the **substrate(s)**. All naturally occurring enzyme-controlled processes can be regarded as equilibrium processes. However, under body conditions many processes appear only to proceed in one direction because as soon as they are formed the products of the process are either immediately used as the substrates of a subsequent enzyme-catalysed reaction or removed.

In its simplest form, the catalytic action of enzymes is believed to depend on the substrate or substrates binding to the surface of the enzyme. This binding usually occurs on a specific part of the enzyme known as its **active site**. Once the substrate (S) or substrates are bound to the surface of the enzyme (E), reaction takes place and the products (P) are formed and released whilst the enzyme is recycled into the system.

$$E + S \rightleftharpoons \text{E-S Complex} \rightleftharpoons \text{E-P Complex} \rightleftharpoons P + E$$

Recycled (E from products returned to the start)

Active sites are usually visualised as pockets, clefts or indentations in the surface of the enzyme. These physical features are the result of conformations of the protein's peptide chain. Consequently, the amino acid residues forming the site can be located some distance apart in the peptide chain but are brought together by the folding of the peptide chain. For example, the amino acid residues that form the active site of lactate dehydrogenase occupy positions 101, 171 and 195 in the peptide chain. The amino acid residues often involved in forming active sites are serine, histidine, arginine, cysteine, lysine, aspartic acid and glutamic acid.

It is believed that the enzyme reduces the activation energy required for each of the stages in an enzyme-controlled reaction (Figure 6.2).

The specific nature of most enzyme action has been explained by a number of models, the earliest being the lock and key. In this model the enzyme acts as the lock and the reactants act as the key. Just as the key has to be of the correct shape to turn in the lock, the reactant molecules must have a shape that fits the configuration of the active site in order to react (Figure 6.3). This model assumes that the enzyme structure is rigid and that the active site will always have the correct shape for that reaction. However, it is now known that the structures of enzymes are not rigid and can change to accommodate the substrate. In spite of this, the lock and key model is still a useful concept even though it is an over-simplification that does not satisfactorily explain all enzyme behaviour.

The discovery that the binding of a substrate to an active site changes the conformation of the site led Daniel Koshland to propose the **induced fit hypothesis** of enzyme action. He proposed that the enzyme and substrate interact with each other, changing their conformations so that the substrate fits the active site more accurately (Figure 6.4). This allows the substrate to bind

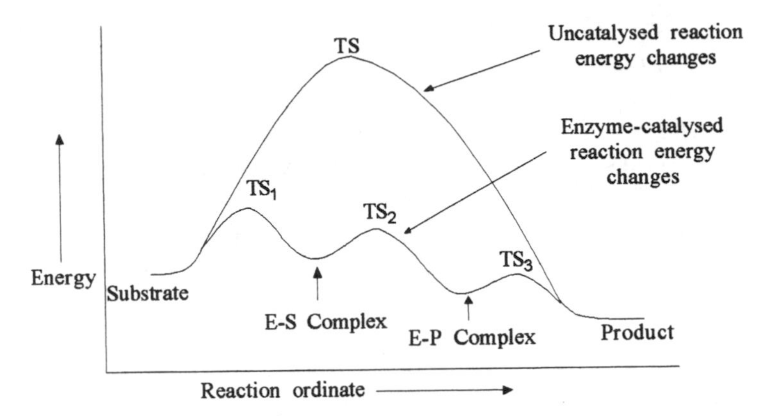

Figure 6.2. The effect of an enzyme on the minimum energy pathway of a simple enzyme-catalysed reaction involving a single substrate (TS = transition state). The heights of the energy barriers TS_1, TS_2 and TS_3 will vary depending on which step in the enzyme-controlled route is the rate-controlling step

in the exact position necessary for reaction, which explains why enzymes are more effective than chemical catalysts in increasing the rates of reactions.

Coenzymes are essential components of the active sites of many enzymes. They appear to act by binding to the apoenzyme to form the active site of the enzyme (Figure 6.5). This binding is often very strong and can involve covalent bonds as well as the full range of electrostatic forces.

The active site is not the only place where compounds may bind to an enzyme. These alternative sites are known as **allosteric sites**. The binding of a compound to an allosteric site of an

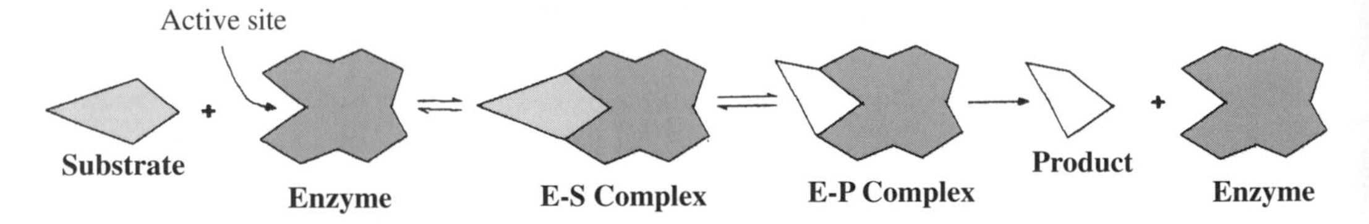

Figure 6.3. The lock and key model of enzyme action.

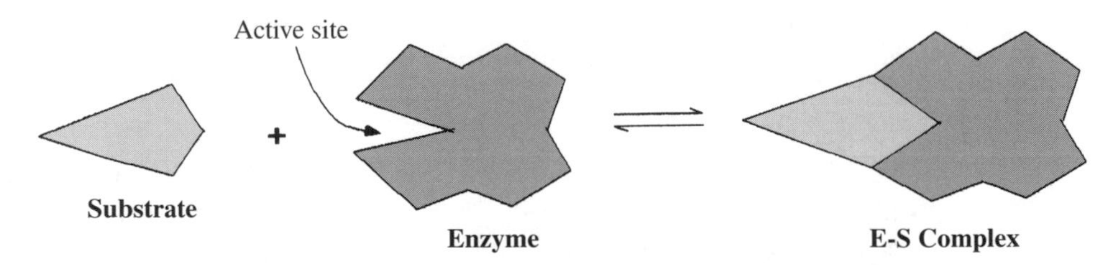

Figure 6.4. A schematic representation of the induced fit hypothesis of enzyme action.

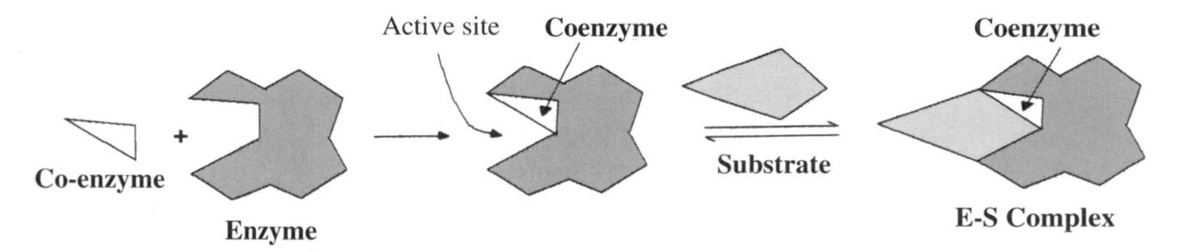

Figure 6.5. A schematic representation of the action of the coenzymes using the induced fit model.

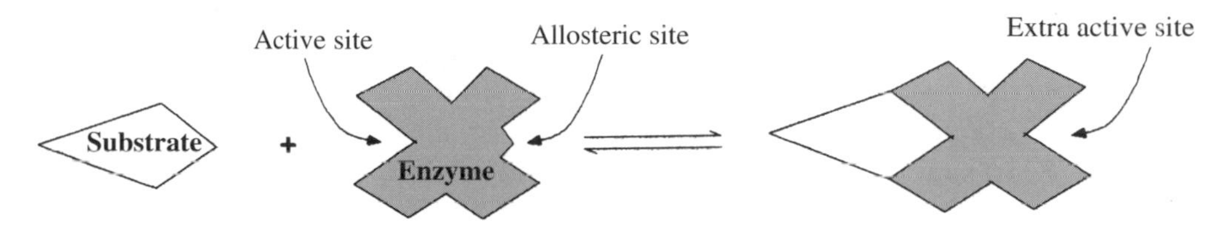

Figure 6.6. A schematic representation of allosteric activation.

enzyme may cause a change in its conformation (Figure 6.6). These changes may result in either the activation (see section 6.2.1) or the inhibition of an enzyme (see section 6.4), or have no effect at all.

6.3.1 Allosteric Activation

The change in conformation caused by a substrate binding to the active site of an enzyme can result in the formation of another active site on the enzyme (Figure 6.6). This behaviour is known as allosteric activation. It occurs with increasing concentration of the substrate and increases the capacity of the enzyme to process the substrate.

6.4 Regulation of Enzyme Activity

The chemical activity of a cell must be controlled in order for the cell to function correctly. Enzymes offer a means of controlling that activity in that they can be switched from active to inactive states by changes induced in their conformations. Compounds that modulate the action of enzymes and other molecules by these conformational changes are known as **regulators** or **effectors**. These compounds may switch on (activators) or switch off (inhibitors) an enzyme. They may be generally classified for convenience as: (i) compounds that covalently modify the enzyme and (ii) allosteric regulators.

6.4.1 Covalent Modification

This form of regulation involves the attachment of a chemical group to the enzyme by covalent bonding. Attachment may either inactivate or activate an enzyme. The process is catalysed by so-called **modifying** or **converter** enzymes. It is reversible, the reverse process being catalysed by different enzymes from those required for the attachment. For example, glycogen phosphorylase, the enzyme that catalyses the formation of glucose-1-phosphate from glycogen, is inactive until the hydroxy group of a serine residue is phosphorylated. This conversion is itself catalysed by kinases which are themselves usually activated by Ca^{2+} ions.

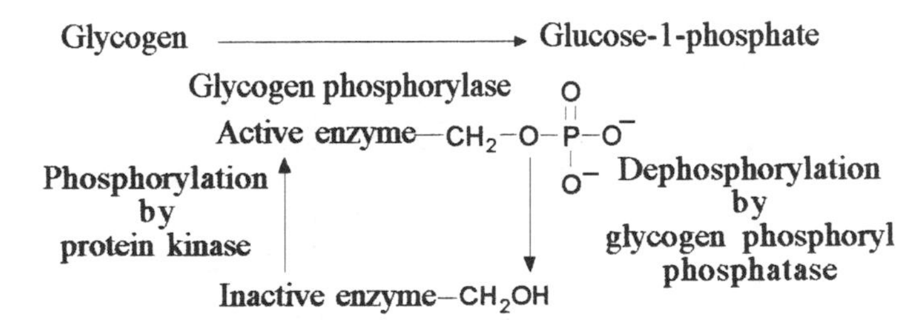

6.4.2 Allosteric Control

This type of control involves the reversible binding of a compound to an allosteric site on the enzyme. These compounds may be either one of the compounds involved in the metabolic pathway (feedback regulators) or a compound that is not a product of the metabolic pathway. In both cases the binding usually results in conformational changes that either activate or deactivate the enzyme (Figure 6.7). Compounds responsible for these changes are known as **effectors**, **modulators** or **regulators** and the allosteric sites are referred to as **regulatory sites**.

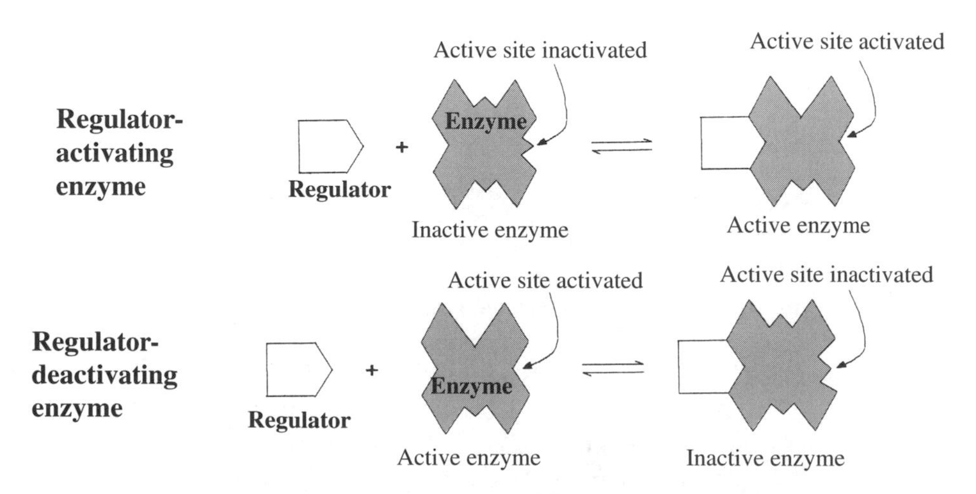

Figure 6.7. A schematic representation of allosteric control.

Feedback control is an automatic form of allosteric regulation. It often, but not always, utilises the end product of the metabolic pathway to regulate its production. Consider, for example, the metabolic pathway:

Can inhibit E_1

$$\text{Earlier steps in the metabolic pathway} \xrightarrow{E} A \xrightarrow{E_1} B \xrightarrow{E_2} C \xrightarrow[\text{A number of steps}]{E_{various}} D$$

where E, $E_1 \ldots E_n$ are the enzymes used in the metabolic pathway. When the concentration of D is low the enzymes operate to increase the production of D. As the concentration of D increases, it increasingly inhibits the action of E_1. Eventually the enzyme stops acting and no more B and therefore no more D is produced. As D is utilised by the body, its concentration decreases and as a consequence its inhibiting effect on E_1 also decreases.

Some regulators that activate an enzyme need to be activated themselves before they can act. For example, calmodulin (CaM) is a small dumb-bell-shaped protein with four binding sites that has a high affinity for Ca^{2+} ions. It activates a number of enzymes such as ATPase and nitric oxide synthase (see section 11.4). CaM is inactive until the concentration of Ca^{2+} ions in the surrounding medium increases. As the concentration of Ca^{2+} ions in the medium increases, the number of sites occupied by ions increases until all four are occupied. Binding of Ca^{2+} ions to all four sites changes CaM's conformation, which allows it to bind to and activate the enzyme. Conversely, a decrease in the calcium ion concentration of the surrounding medium results in the release of Ca^{2+} ions from CaM, which results in it becoming inactive with a subsequent deactivation of the enzyme system.

$$\text{CaM} + \text{Ca}^{2+} \rightleftharpoons \text{CaM-Ca}^{2+} \rightleftharpoons \underset{\textbf{(Active enzyme)}}{\text{Enzyme-CaM-Ca}^{2+}}$$

Where a regulator inhibits an enzyme by binding to a regulatory site it may be removed by a second modulator. For example, the regulator of protein kinase, an enzyme that catalyses the transfer of phosphate groups, is a protein molecule that binds to the regulatory site of the kinase to form an inactive complex. This complex is activated by cyclic adenosine monophosphate (cAMP), which removes the regulator (Figure 6.8). cAMP is known as a **positive modulator** because it activates the enzyme.

Proenzymes are also a form of enzyme control. The active form of the enzyme is produced from the proenzyme where and when it is required. This often requires the removal of a section of the protein chain by a reaction controlled by another enzyme. This second enzyme is known as an **activator** because, unlike a modulator, it does not bind to the enzyme and cannot switch off an enzyme. For example, the active form of the digestive enzyme chymotrypsin is produced from its inactive proenzyme chymotrypsinogen by cleavage of an arginine–isoleucine peptide

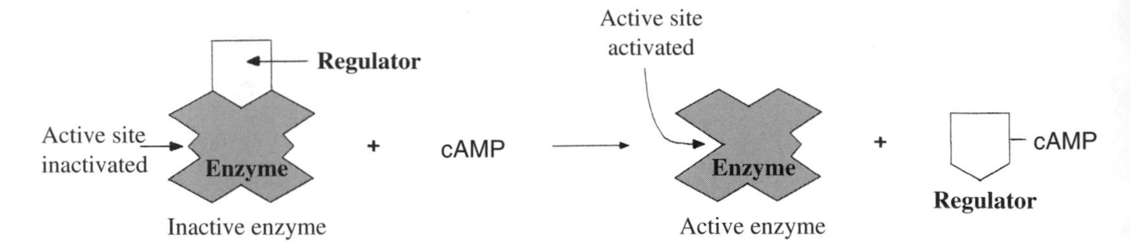

Figure 6.8. A schematic representation of the action of cAMP as a positive modulator.

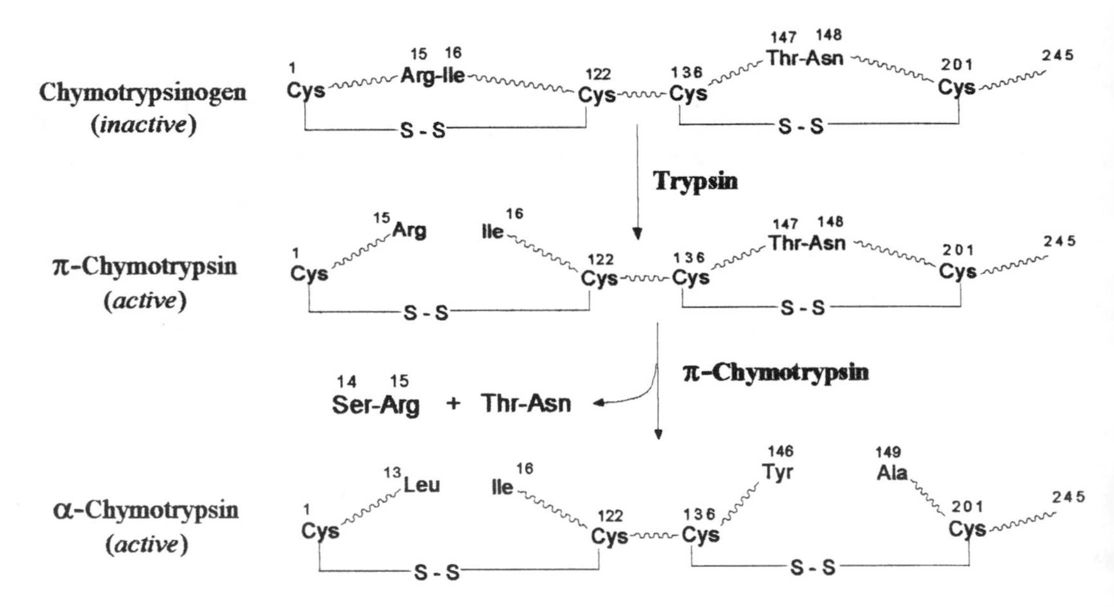

Figure 6.9. The production of active chymotrypsin from its proenzyme chymotrypsinogen. The enzymes for each stage of the process are in bold type.

bond by trypsin to form the active π-chymotrypsin (Figure 6.9). π-Chymotrypsin autocatalytically removes two dipeptides from itself to form α-chymotrypsin, which is the active form normally referred to as chymotrypsin. The bond fission and the removal of the two dipeptides allow the protein molecule to assume its active conformation.

6.5 The Specific Nature of Enzyme Action

The specific action of enzymes can be broadly described in terms of the nature and functional groups of compounds and their stereochemistry. Most enzymes will only act on a specific functional group in groups of structurally similar compounds of roughly similar size. For example,

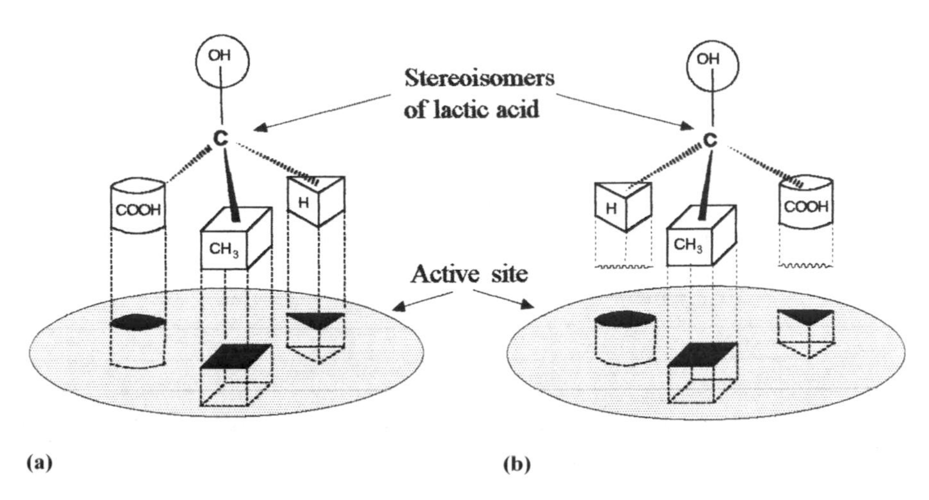

Figure 6.10. A schematic illustration of the stereospecific nature of enzymes. The groups of the two stereoisomers of lactic acid are represented by the three-dimensional shapes shown and the sections of the active site that bind to these groups are represented as the correspondingly shaped sockets. In (a) the relevant groups fit into these sockets, that is, the structures are complementary and the groups of L-lactic acid are correctly aligned for binding. However, in (b) two of the groups will be misaligned. The wedge and cylinder are unable to fit their sockets and so the groups of D-lactic acid will not be correctly aligned for binding to the active site.

alcohol dehydrogenase will catalyse the oxidation of a number of other small primary and secondary alcohols such as methanol, ethanol and propan-2-ol to the corresponding aldehydes and ketones. The rates and yields of these reactions vary considerably. However, some enzymes are selective in that they mainly catalyse reactions involving one particular compound whereas others will catalyse the reactions of large numbers of different substrates with the same functional group. For example, the digestive enzyme carboxypeptidase will remove all types of C-terminal amino acid residues from proteins except arginine, lysine and proline. Enzymes are also often stereospecific because they can form asymmetric active sites. This means that an active site will have a strong affinity for the complementary, shaped compound (Figure 6.10) but not its stereoisomers. For example, trypsin will catalyse the hydrolysis of L-Arg–L-Ile peptide links in proteins but will not catalyse the hydrolysis of D-Arg–D-Ile peptide links. Most enzymes exhibit this type of stereospecific behaviour.

6.6 The Mechanisms of Enzyme Action

The mechanisms of the reactions at the active sites of enzymes have been widely investigated. There is now no doubt that the increased speed of enzyme-catalysed reactions is due in part to the close proximity and the correct alignment of the reactants at the active site. A detailed knowledge, obtained from X-ray crystallography, of the three-dimensional structures of the

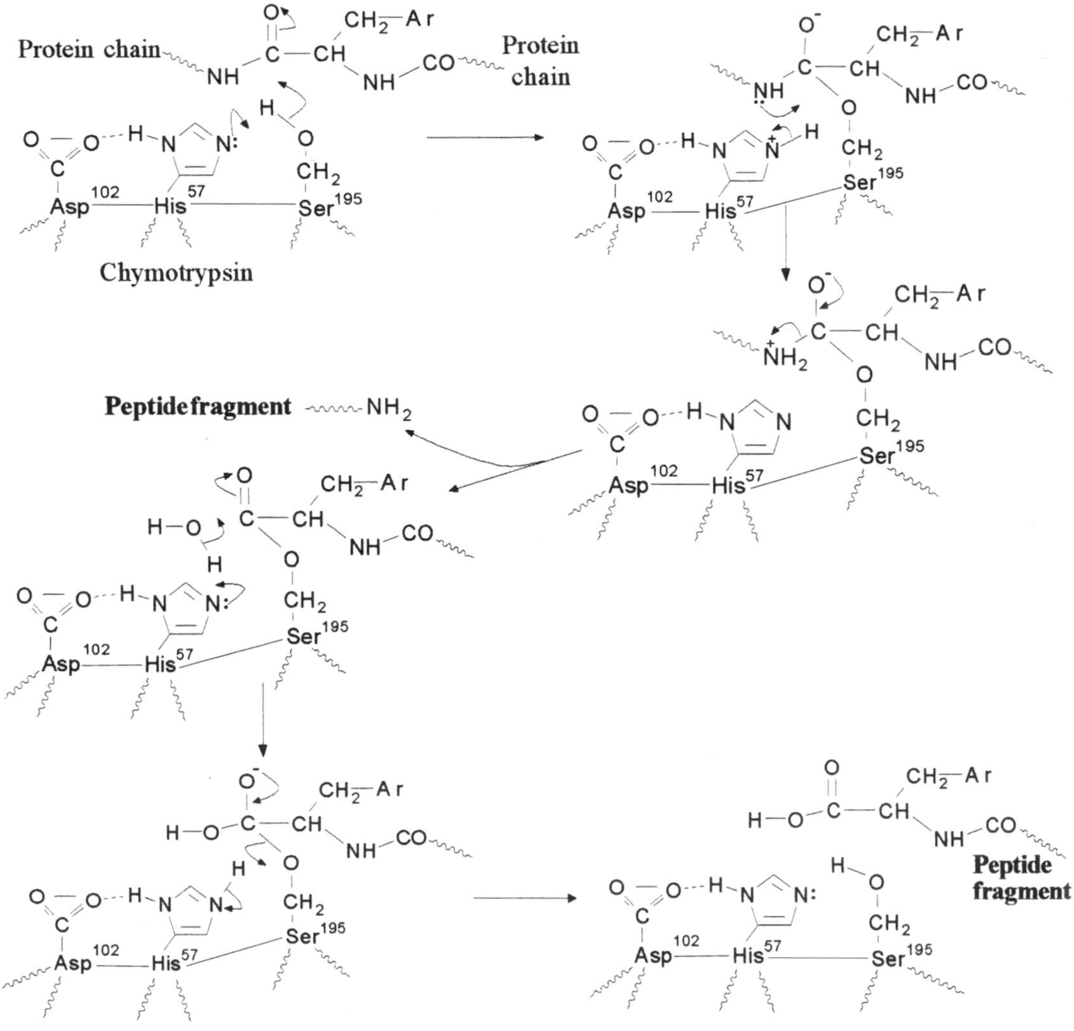

Figure 6.11. A simplification of the proposed mechanism for the cleavage of peptide links by chymotrypsin. The active site of this enzyme consists of reactive serine (195), histidine (57) and aspartic acid (102) residues. The numbers in parentheses indicate the relative positions of these residues in the peptide chain of the protein.

enzyme and bound substrate plus other experimental evidence means that in many cases the details of enzyme action have been established. This has resulted in the proposal of reaction mechanisms for a number of processes. However, it should be realised that each mechanism is specific for the enzyme–substrate system under consideration (Figure 6.11). Consider, for example, the serine proteases, a family of enzymes whose active sites contain a reactive serine residue. This family, which includes chymotrypsin, trypsin, thrombin and elastin, catalyses the hydrolysis of peptide links (amide bonds) in proteins. However, the bond cleaved will depend on the enzyme. For example, both trypsin and chymotrypsin cleave the peptide link on the carbonyl group side of the amino acid residues in the peptide chain, but trypsin cleaves the peptide

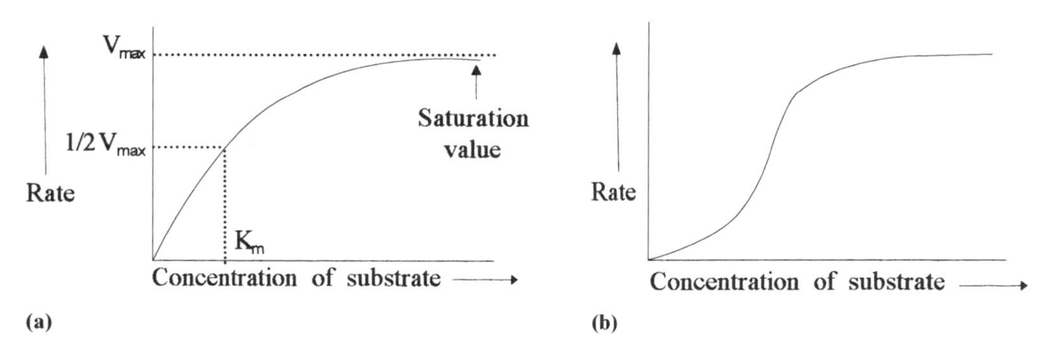

Figure 6.12. The effect of concentration on the rate of allosteric (a) modulated and (b) feedback-regulated enzyme-catalysed reactions.

links of basic amino acid residues and chymotrypsin cleaves those of aromatic amino acid residues.

6.7 The General Physical Factors Affecting Enzyme Action

The main physical factors affecting enzyme action are the relative concentrations of the enzyme and substrate, the pH and the temperature. If the concentration of the substrate is kept constant, the rate of the reaction increases with increase in enzyme concentration. However, if the concentration of the enzyme is kept constant, as is likely in biological processes, and the concentration of the substrate is increased, the rate of reaction increases to a maximum value known as the **saturation value** (Figure 6.12a). This maximum rate corresponds to the saturation of all the available active sites of the enzyme. In the case of enzymes that are under feedback control (see section 6.4.2) the curve is sigmoidal (Figure 6.12b), as opposed to hyperbolic for enzymes not under feedback control but still regulated by a modulator, that is, under allosteric control.

Enzymes are only usually effective over specific ranges of pH characteristic of the enzyme (Figure 6.13). These ranges usually correspond to the pH of the environment in which the enzyme occurs. Outside this range the enzyme may undergo irreversible denaturation with subsequent loss of activity. Initially an increase in temperature will usually result in an increase in the rate of an enzyme-controlled reaction. However, enzymes are sensitive to temperature and once it increases beyond a certain point the protein irreversibly denatures and the enzyme becomes inactive. An enzyme's operating temperature range will normally correspond to about that of its normal environment.

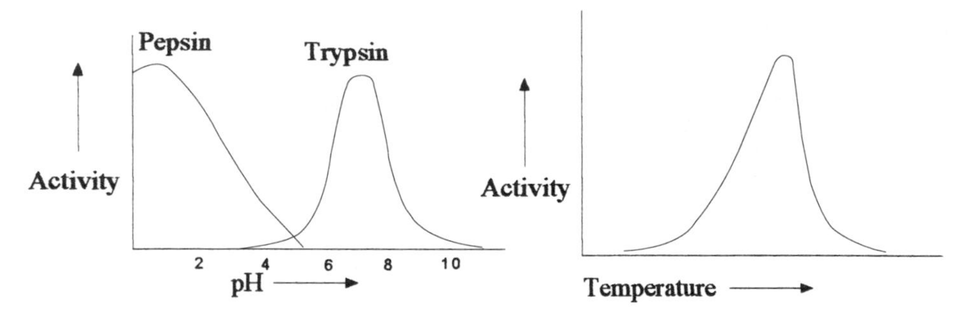

Figure 6.13. The effect of pH and temperature on the rates of enzyme-catalysed reactions.

6.8 Enzyme Kinetics

6.8.1 Single Substrate Reactions

In these reactions a single substrate is converted to the products:

$$E + S \rightleftharpoons E\text{-}S \rightleftharpoons E\text{-}P \rightarrow E + P$$

The mathematical relationship between the rate of the enzyme-catalysed reaction and the concentration of the single substrate depends on the nature of the enzyme process. Many processes exhibit first-order kinetics at low concentrations of the substrate, which changes to zero order as the concentration increases, that is, a hyperbolic relationship exists between rate and substrate concentration (Figure 6.12a). Processes under allosteric control have a sigmoid relationship between reaction rate and the substrate concentration (Figure 6.12b). This suggests second and higher orders of reaction, that is, the rate is proportional to $[S]^n$ where $n > 1$. Only the kinetics of enzyme processes that exhibit hyperbolic rate–substrate concentration relationships will be discussed in this text.

The kinetics of enzyme processes with a single substrate that exhibit a hyperbolic rate–substrate concentration relationship may often be described by the Michaelis–Menten equation:

$$V = \frac{V_{max}[S]}{K_m + [S]} \tag{6.1}$$

where K_m is a constant known as the **Michaelis constant**, [S] is the concentration of the substrate, V is the rate of the reaction and V_{max} is the rate of the reaction when the concentration of the substrate approaches infinity. When $V = V_{max}/2$, it follows that $K_m = [S]$, that is, the value

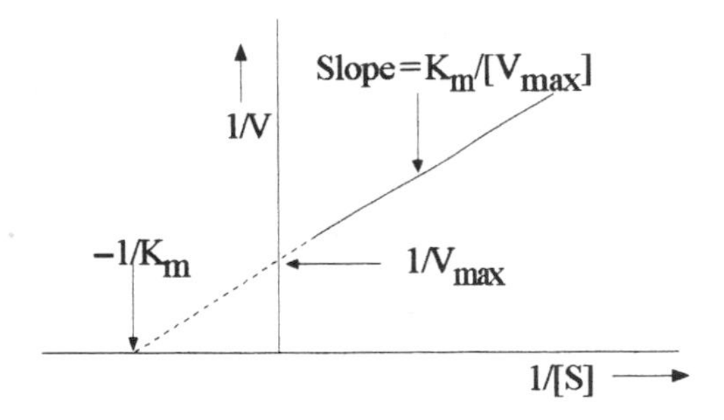

Figure 6.14. The Lineweaver–Burk double reciprocal plot of enzyme processes with a hyperbolic rate curve.

Table 6.2. Examples of K_m values for some enzymes and their substrates.

Enzyme	Substrate	K_m (mM)
Chymotrypsin	Acetyl-L-trypohamide	5
	N-Acetyltyrosinamide	32
	Glycyltyrosinamide	122
Glutamate dehydrogenase	Glutamate	0.12
	NADH	0.018
Penicillinase	Benzylpenicillin	0.05

of the Michaelis constant is the concentration of the substrate at $V_{max}/2$ (Figure 6.12a). Rearrangement of the Michaelis–Menten equation gives the more useful Lineweaver–Burk equation:

$$\frac{1}{V} = \frac{K_m}{V_{max}[S]} + \frac{1}{V_{max}} \tag{6.2}$$

This equation is of the form y = mx + c. Consequently, a graph of 1/V against 1/[S] will be a straight line with a slope of K_m/V_{max} and an intercept on the y-axis of $1/V_{max}$ (Figure 6.14). Furthermore, when V = 0 the intercept on the x-axis will be $-1/K_m$.

The values of K_m and V_{max} are characteristic properties of the enzyme and substrate system under the specified conditions (Table 6.2). The value of K is a measure of the affinity of the substrate for the enzyme.

The use of reciprocal values leads to an increased error for low values when plotting any data. Consequently, several other rearrangements of the Michaelis–Menten equation have been developed to convert this equation into a straight-line format (Figure 6.15). The Eadie–Hofstee plot in particular is good for detecting deviations from a straight line, that is, deviations from Michaelis–Menten kinetics. These deviations may indicate that the enzyme is under feedback control. However, the Lineweaver–Burk plot will be used in this text.

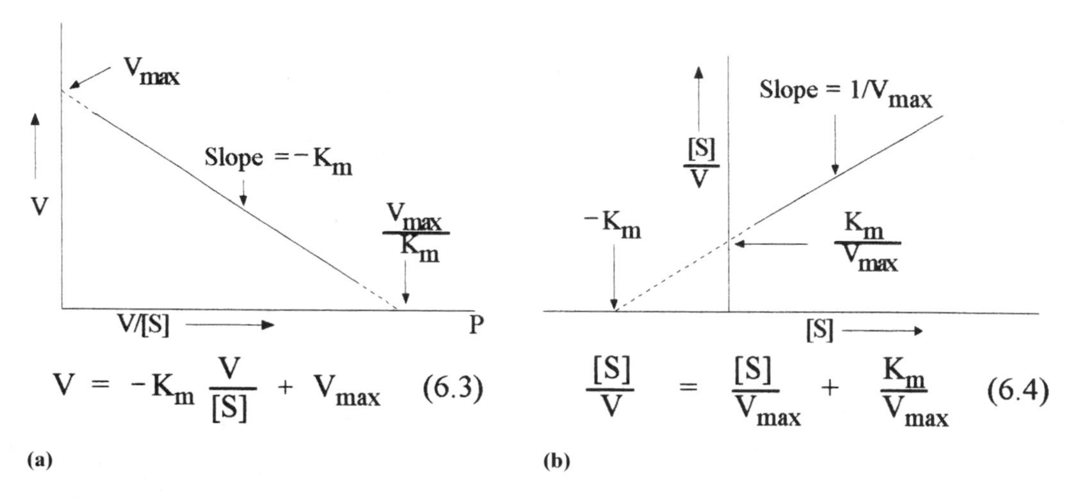

Figure 6.15. The (a) Eadie–Hofstee and (b) Hanes–Wolf plots for enzyme processes with a hyperbolic rate curve.

6.8.2 Multiple Substrate Reactions

Enzymes often catalyse reactions that involve more than one substrate. The kinetics of these reactions will depend on the number of substrates and the general mechanism of enzyme action. However, the kinetics of many of these reactions will still follow the mathematics discussed in section 6.8.1 provided the concentrations of all the substrates except the one being studied are kept constant. For example, reactions involving two substrates, A and B (bisubstrate reactions) often proceed by one of two general routes; the sequential and the double displacement routes.

6.8.2.1 The Sequential or Single-displacement Reactions

These reactions can follow two general routes. In the first case the order in which the substrates A and B bind to the enzyme has no effect (random order) but in the second case one substrate known as the leading substrate, must bind to the enzyme before the other substrate can bind to the enzyme (ordered process). Once both substrates are bound to the enzyme reaction to the products occurs.

Random order process $\quad E + A + B \rightleftharpoons A\text{-}E\text{-}B \rightleftharpoons P\text{-}E\text{-}N \rightarrow E + P + N$

Ordered process $\quad E + A \rightleftharpoons A\text{-}E + B \rightleftharpoons A\text{-}E\text{-}B \rightleftharpoons P\text{-}E\text{-}N \rightarrow E + P + N$

Both these types of sequential reaction routes exhibit the characteristic type of Lineweaver-Burk plot shown in Figure 6.16a, when the concentrations of one of the substrates and the enzyme are kept constant and the concentration of the other substrate varied.

6.8.2.2 Double-displacement or Ping-pong Reactions

In reactions following this route a substrate binds to the enzyme E and forms a complex that forms a product P and a modified form of the enzyme E′. At this point the second substrate binds to the modified enzyme and reacts to form a second product N and regenerate the original form of the enzyme.

$$\mathrm{E} + \mathrm{A} \rightleftharpoons \mathrm{E\text{-}A} \rightleftharpoons \mathrm{E'\text{-}P} \xrightarrow{\ \ \mathrm{P}\ \uparrow\ \ } \mathrm{E'} \xrightarrow[\mathrm{B}]{} \mathrm{E'\text{-}B} \rightleftharpoons \mathrm{E'\text{-}N} \longrightarrow \mathrm{E} + \mathrm{N}$$

Reactions proceeding by this type of route exhibit characteristic Lineweaver–Burk plots of the type shown in Figure 6.16b, when the concentration of one substrate is varied and the concentrations of the enzyme and other substrate are kept constant.

6.9 Enzyme Inhibitors

The use of compounds to inhibit enzyme action is an important possibility for therapeutic intervention. This approach has been used to disrupt essential steps in a key metabolic pathway responsible for a pathological condition (see section 10.12) as well as the cell wall and plasma membrane synthesis of microorganisms (see section 4.4). Compounds used in this way have a wide range of structures. Those whose structures closely resemble that of the normal substrate for an enzyme are called **antimetabolites** (see section 10.12).

Inhibitors (I) may have either a reversible or irreversible action. Reversible inhibitors tend to bind to an enzyme (E) by electrostatic bonds, hydrogen bonds and van der Waals' forces and

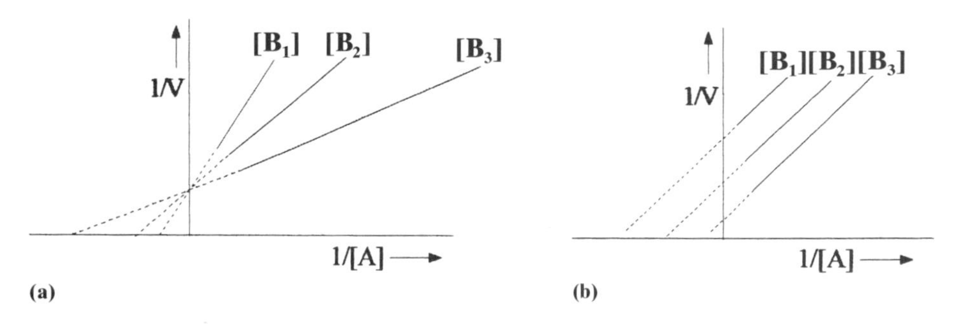

Figure 6.16. The Lineweaver–Burk plots of reactions for (a) the sequential and (b) double-displacement routes for enzyme-catalysed reactions. The lines correspond to different constant concentrations of substrate B, where $[B_3] > [B_2] > [B_1]$.

so tend to form an equilibrium system with the enzyme. A few reversible inhibitors bind by weak covalent bonds but this is the exception rather than the rule. **Irreversible inhibitors** usually bind to an enzyme by strong covalent bonds. However, it should be realised that in both reversible and irreversible inhibition the inhibitor does not need to bind to all of the active site in order to prevent enzyme action. It is only necessary for the inhibitor to partially block the active site in order for it to inhibit the reaction.

Reversible: $E + I \rightleftharpoons$ E-I complex

Irreversible: $E + I \rightarrow$ E-I complex

6.9.1 Reversible Inhibitors

These are inhibitors that form a dynamic equilibrium system with the enzyme. Reversible inhibitors are normally time dependent because the removal of unbound inhibitor from the vicinity of its site of action will disturb this equilibrium. This allows more enzyme to become available, which decreases the inhibition of the process. Consequently, the inhibitory effect of reversible inhibitors is time dependent. Furthermore, drugs acting as reversible enzyme inhibitors will only be effective for a specific period of time.

Most reversible inhibitors may be further classified as being competitive, non-competitive or uncompetitive.

6.9.1.1 Competitive Inhibition

In competitive inhibition the inhibitor *usually* binds by a reversible process to the same active site of the enzyme as the substrate. This competition between the inhibitor and substrate for the same site results in some enzyme molecules being inhibited whereas others function as normal. The net result is an overall reduction in the rate of conversion of the substrate to the products (Figure 6.17). Because the binding of the inhibitor is reversible, its effect will be reduced as the concentration of the substrate increases until, at high substrate concentrations, the inhibitory effect is completely prevented. This allows V_{max} for the process to be reached. However, it also follows that as the concentration of the inhibitor increases, its inhibiting effect increases. Consequently, for a competitive inhibitor to be effective a relatively high concentration has to be maintained in the region of the enzyme.

For single substrate enzyme processes exhibiting simple Michaelis–Menten kinetics the Lineweaver–Burk plots for competitive inhibition show characteristic changes when different concentrations of the inhibitor I are used. These changes do not affect the value of V_{max} but can be used to diagnose the possible presence of a competitive inhibitor (Figure 6.18).

Because the substrate and inhibitor compete for the same active site, it follows that they will probably be structurally similar. For example, succinate dehydrogenase, which catalyses the

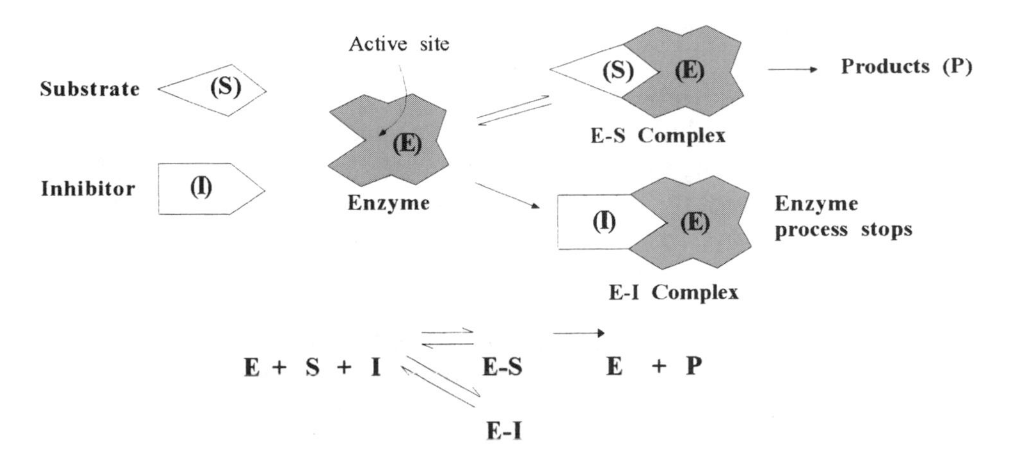

Figure 6.17. A schematic representation of competitive inhibition.

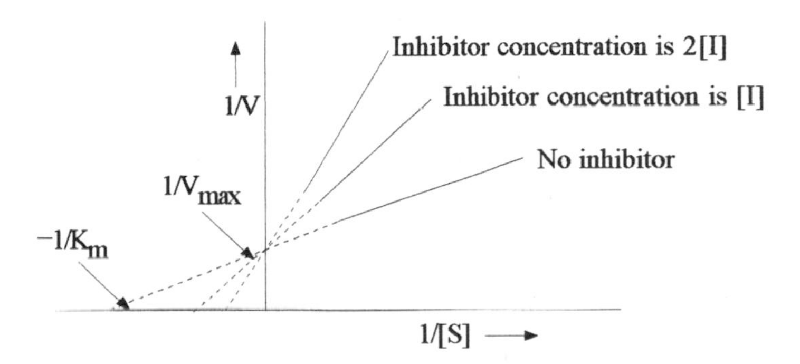

Figure 6.18. The Lineweaver–Burk plot for competitive inhibition.

conversion of succinate to fumarate, is inhibited by malonate, which has a similar structure to succinate (Figure 6.19).

The structural relationship between substrates and competitive inhibitors offers a rational approach to drug design in this area (see section 6.6).

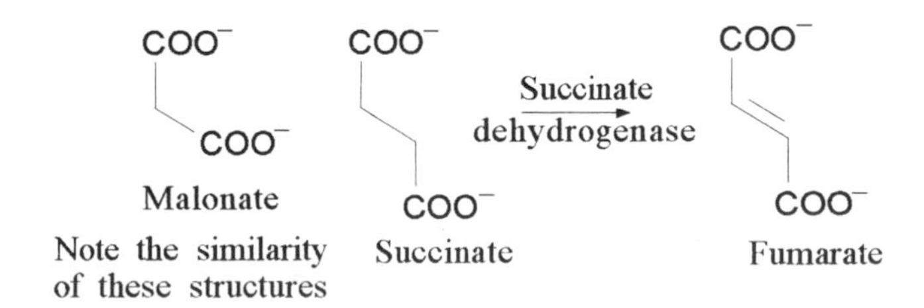

Figure 6.19. The similarity of the structures of malonate and succinate explains why malonate inhibits succinate dehydrogenase.

6.9.1.2 Non-competitive Inhibition

Non-competitive inhibitors bind reversibly to an allosteric site on the enzyme (Figure 6.20). In **pure** non-competitive inhibition, the binding of the inhibitor to the enzyme does not influence the binding of the substrate to the enzyme. However, this situation is uncommon and the binding of the inhibitor usually causes conformational changes in the structure of the enzyme, which in turn affect the binding of the substrate to the enzyme. This is known as **mixed** non-competitive inhibition.

In both types of non-competitive inhibition the binding of the inhibitor to the enzyme ultimately results in the formation of an enzyme–substrate–inhibitor (I-E-S) complex that will not yield the normal products for the enzyme-catalysed process. However, some normal products will be formed because not all the E-S complex formed will be converted to the I-E-S complex (Figure 6.20). Furthermore, because non-competitive inhibition is reversible the E-S complex can be reformed by the loss of the inhibitor. Consequently, effective inhibition will depend on maintaining a relatively high concentration of the inhibitor in contact with the enzyme in order to force the equilibria to favour the formation of I-E and I-E-S.

Where single substrate enzyme processes exhibit Michaelis–Menten kinetics, both types of non-competitive inhibition exhibit characteristic changes to their Lineweaver–Burk plots (Figure 6.21). Both types of non-competitive inhibitors decrease the value of V_{max}. Further-

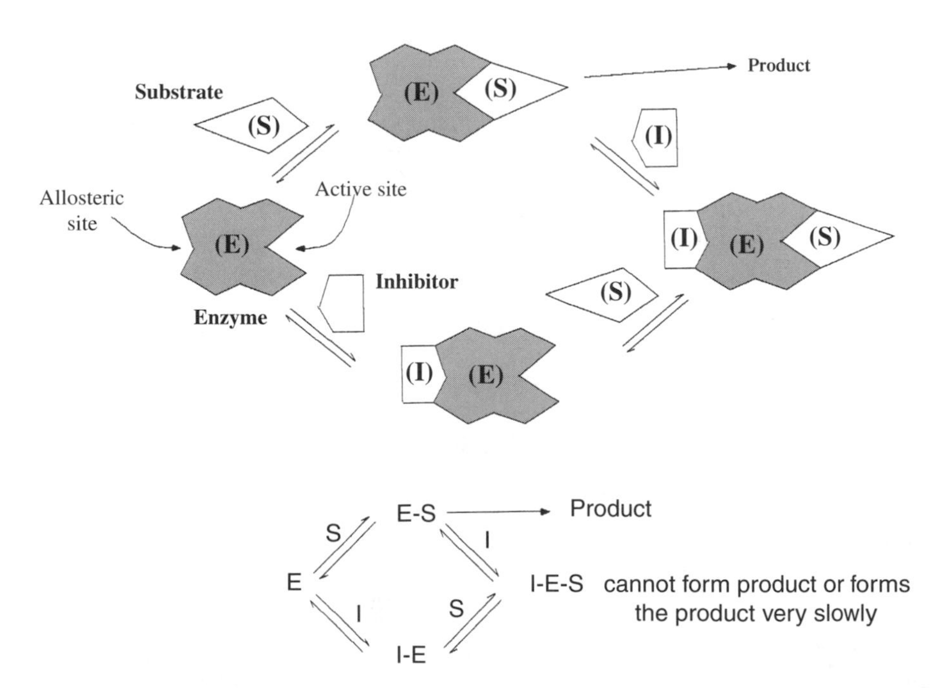

Figure 6.20. Non-competitive inhibition.

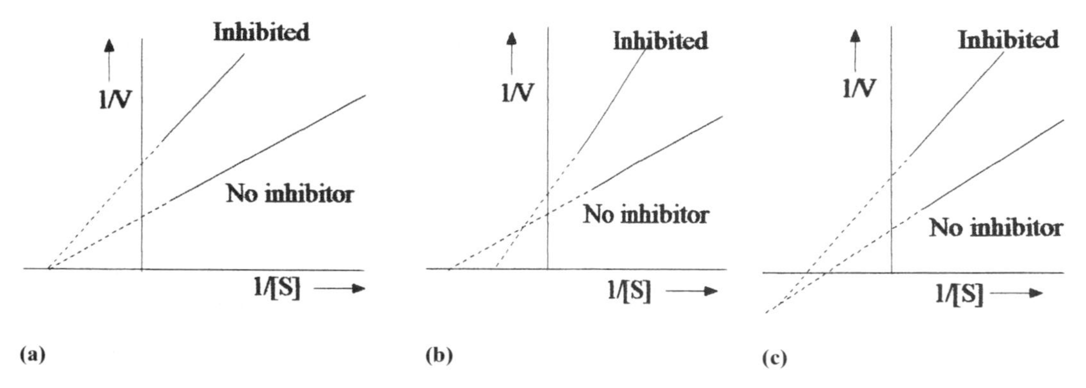

Figure 6.21. The Lineweaver–Burk plots for (a) pure and (b) the two general cases of mixed non-competitive inhibition.

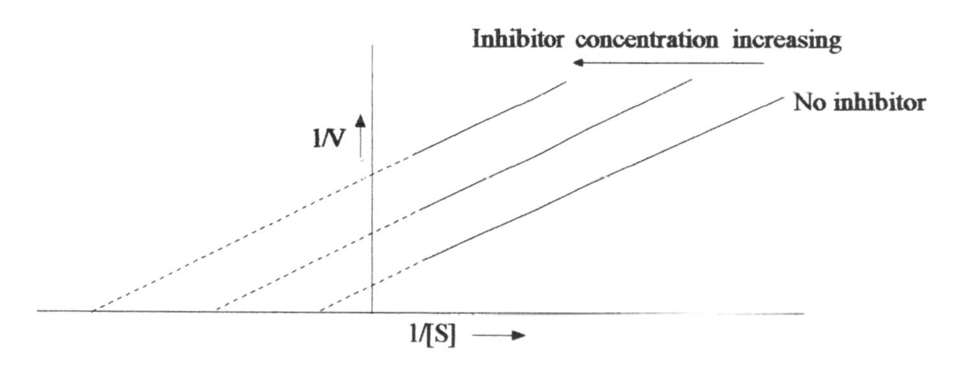

Figure 6.22. The changes in the Lineweaver–Burk plot for increasing amounts of an uncompetitive inhibitor.

more, K_m remains the same for pure non-competitive inhibition but changes for mixed non-competitive inhibition.

The fact that in non-competitive inhibition the inhibitor does not bind to the active site of the enzyme means that the structure of the substrate cannot be used as the basis of designing new drugs that act in this manner to inhibit enzyme action.

6.9.1.3 Uncompetitive Inhibition

Uncompetitive inhibitors are believed to form a complex with the enzyme–substrate complex.

$$E + S \rightleftharpoons E\text{-}S + I \rightleftharpoons I\text{-}E\text{-}S$$

The substrate residue in this I-E-S complex is unable to react and form its normal product. An uncompetitive inhibitor does not change the slope of its Lineweaver–Burk plot of 1/V against [S] but moves it towards the left-hand side of the plot as the concentration of the inhibitor increases (Figure 6.22).

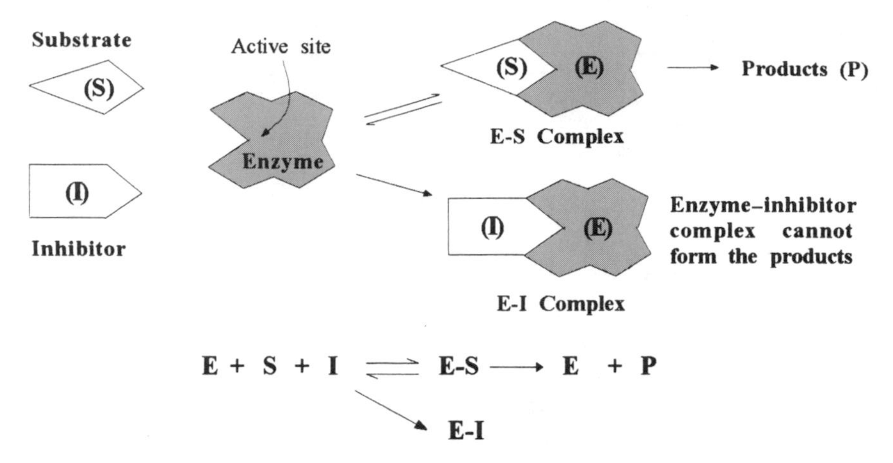

Figure 6.23. Irreversible inhibition.

6.9.2 Irreversible Inhibition

Irreversible inhibitors bind to the enzyme by either strong non-covalent or strong covalent bonds. Compounds bound by strong non-covalent bonds will slowly dissociate, releasing the enzyme to carry out its normal function. However, whatever the type of binding the enzyme will resume its normal function once the organism has synthesised a sufficient number of additional enzyme molecules to overcome the effect of the inhibitor.

Irreversible inhibitors may be classified for convenience as active site-directed inhibitors (ASDI) and suicide or irreversible mechanism-based inhibitors (IMBI).

6.9.2.1 Active Site-directed Inhibitors

Active site-directed inhibitors are compounds that bind at or near to the active site of the enzyme. These inhibitors usually form strong covalent bonds with either the functional groups that are found at the active site or close to that site (Figure 6.23). The inhibitor effectively reduces the concentration of the active enzyme, which in turn reduces the rate of formation of the normal products of the enzyme process.

The action of many irreversible inhibitors depends on the inhibitor possessing a functional group that can react with a functional group at the active site of the enzyme. Consequently, as the functional groups found at the active sites of enzymes are usually nucleophiles, the incorporation of strongly electrophilic groups in the structure of a substrate can be used to develop new inhibitors (Table 6.3). Furthermore, this approach can also be used to enhance the action of a known inhibitor. In all cases the product of the reaction of the inhibitor with the enzyme must be relatively stable if inhibition is to be effective.

Table 6.3. Examples of the electrophilic groups used to produce active site-directed inhibitors.

Nucleophilic group of enzyme (E)	Electrophilic group		Product	
E—NH_2	Anhydrides	RCOOCOR	Amides	RCONH—E
	Ketones	>C=O	Imines	>C=N—E
	Arenesulphonyl halides	RSO_2X	Arenesulphonamides	RSO_2NH—E
E—COOH	Epoxides	R-epoxide (O)	Hydroxyesters	$RCH(OH)CH_2CO_2$—E
	α-Haloacetates	X—$CH_2CO_2^-$	Half esters	$^-O_2CCHCO_2$—E
E—OH	Phosphoryl halides	$(RO)_2PO(X)$	Phosphates	$(RO)_2OPO$—E
	Carbamates	RNHCOOR	Carbamates	RNHCOO—E
E—SH	α-Haloesters	X—CH_2CO_2R	Sulphides	$ROCOCH_2$—S—E
	α-Haloacetates	X—$CH_2CO_2^-$	Sulphides	$ROCOCH_2$—S—E
	Epoxides	R-epoxide (O)	Hydroxy sulphides	$RCH(OH)CH_2$—S—E

Many of the inhibitors developed in this way are too reactive and as a result too toxic to be used clinically. However, they have been used to identify the amino acid residues forming the active site of an enzyme. The inhibitor, when it irreversibly binds to a functional group at the active site, effectively acts as a chemical label for the amino acid residue containing that group. Identification of the labelled residue indicates which residues are involved in forming the active site of the enzyme.

Most of the active site-directed irreversible inhibitors in clinical use were not developed from a substrate. They were obtained or developed by other routes and only later was their mode of action discovered. For example, aspirin, first used clinically at the end of the nineteenth century, was a direct development from the use of salicylic acid as an antipyretic by Carl Buss in 1875.

Aspirin is believed to irreversibly inhibit prostaglandin synthase (cyclooxygenase), the enzyme that catalyses the conversion of arachidonic acid to PGG_2, which acts as a source for a number of other prostaglandins (Figure 6.24). Prostaglandins are local hormones that experimental work has indicated to be associated with the pathology of inflammation and fever.

Experimental evidence suggests that aspirin acts by acetylating serine hydroxy groups at the enzyme's active site, probably by a transesterification mechanism.

Prostaglandin synthase

NHCOCHNHCO / CH_2 / :OH + COOH, O–C(=O)–CH_3 (aspirin) ⟶ NHCOCHNHCO / CH_2 / $^+$O–H / C–CH_3 (O–, O$^-$), COOH ⟶ COOH, O$^-$ (salicylate) + NHCOCHNHCO / CH_2 / O / C(=O)CH_3 + H^+

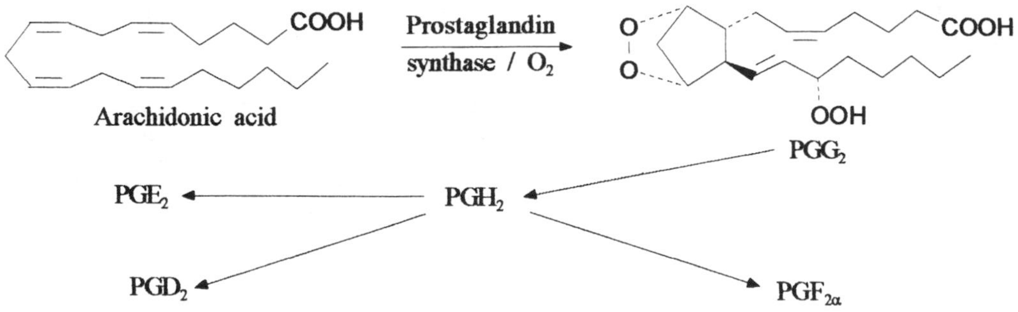

Figure 6.24. An outline of the biosynthesis of prostaglandins.

6.9.2.2 Suicide Inhibitors

Suicide inhibitors, alternatively known as K_{cat} or irreversible mechanism-based inhibitors (IMBI), are irreversible inhibitors that are often analogues of the normal substrate of the enzyme. The inhibitor binds to the active site where it is modified by the enzyme to produce a reactive group that reacts irreversibly to form a stable inhibitor–enzyme complex (Figure 6.25). This subsequent reaction may or may not involve functional groups at the active site. These mechanisms mean that suicide inhibitors are likely to be reasonably specific in their action because they can only be activated by a particular enzyme. Consequently, this specificity means that drugs designed as suicide inhibitors could exhibit a lower degree of toxicity.

The action of suicide inhibitors usually involves either the prosthetic group of the enzyme or its coenzyme. For example, 7-aminomethylisatoic anhydride is a suicide inhibitor of serine protease thrombin that catalyses the hydrolysis of peptides and proteins. This catalysis involves the rapid hydrolysis of acyl–enzyme intermediates. The inhibitor reacts with the hydroxy group of a serine residue of thrombin to form an acyl derivative, which rapidly decarboxylates to form a hydrolysis-resistant complex stable enough to inhibit the enzyme. It is possible that the resistance to hydrolysis may be due to hydrogen bonding between the *ortho*-amino group and the oxygen atom bound to the enzyme (E).

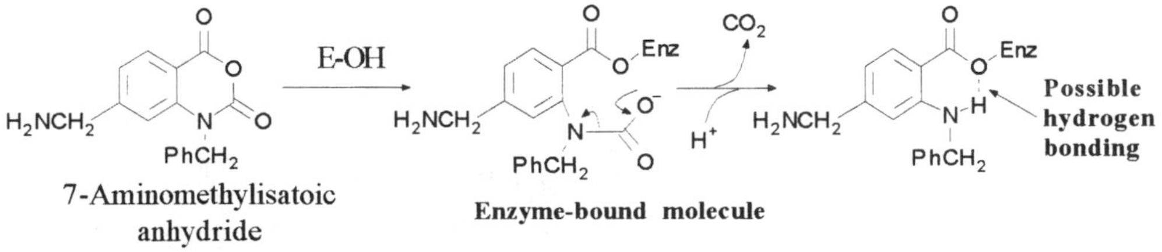

Since enzymes normally possess nucleophilic groups, suicide inhibitors usually react with enzymes to form electrophilic centres which can react with these nucleophilic groups. These centres often take the form of α,β-unsaturated carbonyl compounds and imines. They react by

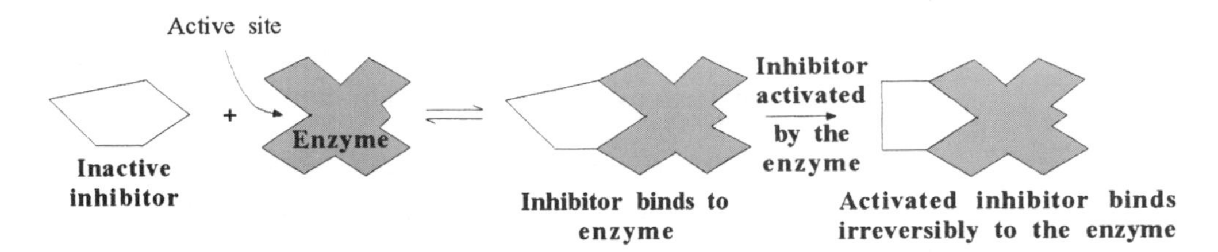

Figure 6.25. A schematic representation of suicide inhibition in which the enzyme activates the inhibitor to form a strong covalent bond with the enzyme.

a type of Michael addition with nucleophilic groups (Nu), such as OH of serine residues, the SH of cysteine residue and the ω-NH_2 of lysine residues frequently found at the active sites of enzymes.

Enz—Nu → Enz—Nu X⁻ → Enz—Nu X → Enz—Nu X

X = O, S or NR

The formation of the α,β-unsaturated carbonyl compound or imine often requires the reverse of a Michael addition at the active site of the enzyme.

X = O, S or NR
L = a stable leaving group

→ L + (X) + Enz-$\overset{+}{B}$H

A wide variety of structures have been found to act as sources of the electrophilic groups of suicide inhibitors (Figure 6.26). However, these structures will only give rise to an electrophilic group if the compound containing the structure can act as a substrate for the enzyme.

The effect of suicide inhibitors increases as the concentration of the inhibitor increases relative to the concentration of the enzyme. Consequently, for processes involving single substrates that follow simple Michaelis–Menten kinetics, the Lineweaver–Burk plot for different concentrations of the inhibitor is a series of straight lines (Figure 6.20).

6.10 Transition State Inhibitors

The substrate in an enzyme-catalysed reaction is converted to the product through a series of transition state structures (Figure 6.2). Although these transition state structures are transient, they bind to the active site of the enzyme and therefore must have structures that are compatible with the structure of the active site. Consequently, it has been proposed that *stable com-*

pounds with structures similar to those of these transition state structures could bind to the active site of an enzyme and act as inhibitors for that enzyme. Compounds that fulfil this requirement are known as **transition state inhibitors**.

A number of transition state inhibitors have been identified. For example, the antibiotic coformycin, which inhibits adenosine deaminase, has a structure similar to that of the transition state of adenosine in the enzyme process for conversion of adenosine to inosine.

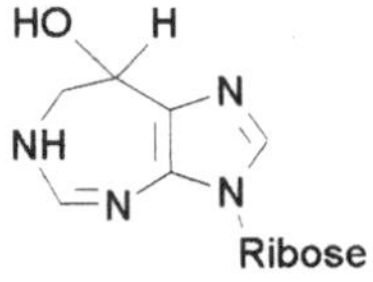

Coformycin

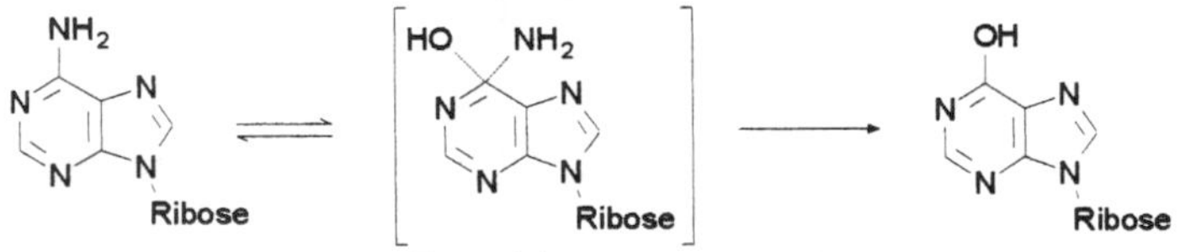

Adenosine | Transition state on adenosine deaminase | Inosine

Transition state inhibitors can act in a reversible or irreversible manner. Their possible structures can be deduced using classical chemistry and mechanistic theory. In addition, structures can be visualised using computer graphics. Consequently, a consideration of the transition state structure offers a possible method of approach when designing new drugs. This method of approach was used to design the experimental anticancer drug sodium *N*-phosphonoacetyl-L-aspartate (PALA, Figure 6.27). In the early 1950s it was observed that some rat liver tumours appeared to utilise more uracil in DNA formation than healthy liver. Consequently, one approach to treating cancers was to develop drugs that inhibited the formation of uracil.

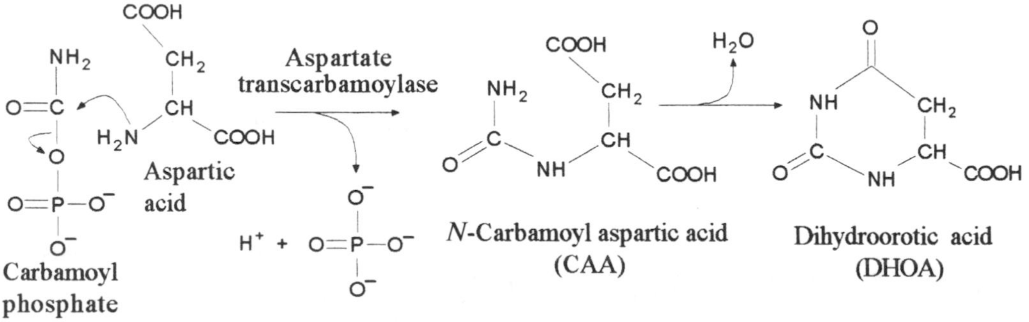

Carbamoyl phosphate + Aspartic acid → (Aspartate transcarbamoylase) → *N*-Carbamoyl aspartic acid (CAA) + H^+ + phosphate → (H_2O) → Dihydroorotic acid (DHOA)

The first step in the biosynthesis of pyrimidines is the condensation of aspartic acid with carbamoyl phosphate to form *N*-carbamoyl aspartic acid, the reaction being catalysed by aspartate transcarbamoylase. It has been proposed that the transition state for this conversion (Figure 6.27a) involves the simultaneous loss of phosphate with the attack of the nucleophilic amino group of the aspartic acid on the carbonyl group of the carbamoyl phosphate. Conse-

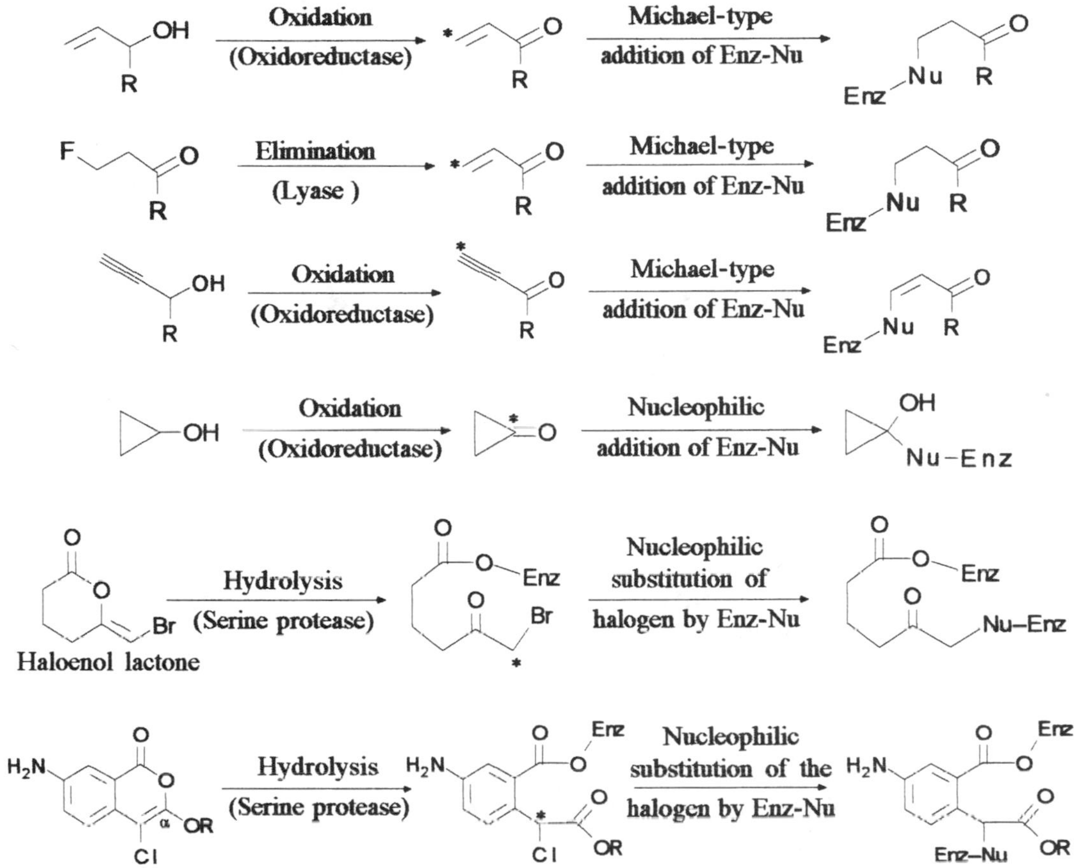

Oxidation (Oxidoreductase)
Michael-type addition of Enz-Nu
Elimination (Lyase)
Michael-type addition of Enz-Nu
Oxidation (Oxidoreductase)
Michael-type addition of Enz-Nu
Oxidation (Oxidoreductase)
Nucleophilic addition of Enz-Nu
Haloenol lactone
Hydrolysis (Serine protease)
Nucleophilic substitution of halogen by Enz-Nu
Hydrolysis (Serine protease)
Nucleophilic substitution of the halogen by Enz-Nu

Figure 6.26. Examples of the reactions to enhance or form the electrophilic centres* of suicide inhibitors and their subsequent reaction with a nucleophile at the active site of the enzyme. The general structures used for the enzyme–inhibitor complexes are to illustrate the reactions. The enzyme ester groups may or may not be present in the final complex.

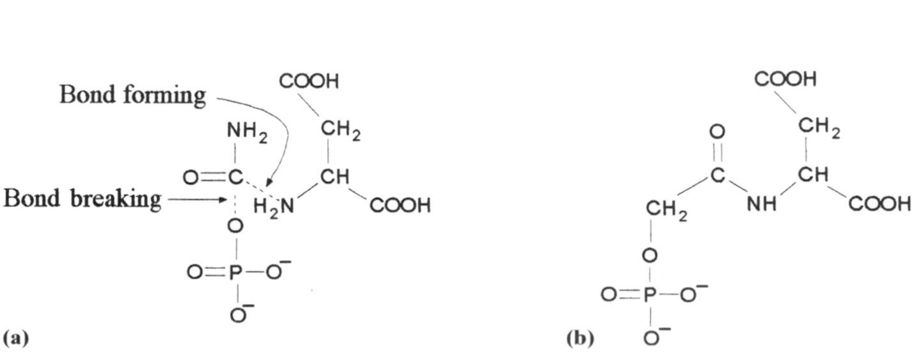

Bond forming
Bond breaking
(a)
(b)

Figure 6.27. The structures of (a) the proposed transition state for the carbamoyl phosphate/aspartic acid stage in pyrimidine synthesis and (b) sodium *N*-phosphonoacetyl-L-aspartate (PALA).

quently, PALA (Figure 6.28b) was designed to have a similar structure to this transition state but without the amino group necessary for the next stage in the synthesis, which is the conversion of *N*-carbamoyl aspartic acid to dihydroorotic acid.

It was found that PALA bound 10^3 times more tightly to the enzyme than the normal substrate. Furthermore, the drug was found to be effective against some cancers in rats.

6.11 Enzymes and Drug Design: Some General Considerations

Enzymes are obvious targets in drug design. Inhibition offers a method of either preventing or regulating cell growth. However, the design team must first decide whether this strategy will achieve the desired therapeutic result. This decision requires a detailed knowledge of both the chemistry and pharmacology of the condition for which the drug is intended as well as commercial considerations. In many cases, the detailed chemical and pharmacological knowledge required is not available and so the discovery of new lead compounds relies on the random screening of collections of compounds, combinatorial chemistry (see section 2.6) or intuitive guess-work.

One advantage of targeting enzymes is their diversity. Consequently, an enzyme process that occurs in a pathogen may not occur in humans. This means that an inhibitor active in a pathogen should not inhibit the same process in humans (see section 6.7.1). However, the use of this inhibitor in humans does not preclude the occurrence of unwanted side effects. In spite of this potential disadvantage it does offer a possible route to more effective drugs.

Non-competitive inhibition offers the least rational approach to drug design because it does not affect the active site. Consequently, the structure of the substrate and its chemical behaviour cannot be used as the basis for the design of a new drug. Conversely, with inhibition involving the active site of the enzyme, the structure of the substrate and its chemical reactivity can be used as the basis of rational approaches to the design of drug substances. A problem with reversible inhibitors is that the inhibition can be reversed by increasing the concentration of the substrate, which automatically starts to occur naturally as soon as the enzyme is inhibited. The reversible nature of the inhibition means that as the substrate's concentration increases it is better able to compete with the inhibitor for the active site. In addition, it is more likely to displace the inhibitor from the active site. These problems do not occur with irreversible inhibitors. However, high concentrations of irreversible inhibitors are usually required if they are to be effective, which also increases the risk of unwanted side effects. A further complication in designing inhibitors is that isoenzymes often respond differently to inhibitors. In extreme cases an inhibitor will inhibit one enzyme but not all its isoenzymes. Consequently, an inhibitor may not have the desired therapeutic effect.

Table 6.4. Examples of drugs used as enzyme inhibitors.

Enzyme	Inhibitor	Condition or action
Dihydropteroate synthetase	Sulphamethoxazole	Bacteriostatic
Dihydrofolate reductase	Methotrexate	Cancer
Thymidylate synthase	Fluorouracil	Cancer
Angiotensin-converting enzyme	Captopril	Hypertension
β-Lactamase	Penicillins	Bacteriocide
HIV reverse transcriptase	Zidovudine	AIDS
Cyclooxygenase	Aspirin	Analgesic
Xanthine oxidase	Allopurinol	Gout

Reduction

Prontosil

p-Aminobenzenesulphonamide

Figure 6.28. The conversion of prontosil into the active form of the drug: sulphanilamide.

6.12 Examples of Drugs Used As Enzyme Inhibitors

Enzyme inhibitors are used for a wide range of medical conditions (Table 6.4). This section briefly describes the discovery and development of two classes of inhibitors used in this capacity as being representative of the drugs in current clinical use. Other clinically used inhibitors are described as they arise naturally in the text.

6.12.1 Sulphonamides

In 1935 the German bacteriologist Gerhard Domagk, who had been evaluating the effect of a variety of substances on streptococci strains, successfully treated his daughter for a virulent streptococcal infection using the azo dye prontosil (Figure 6.28). Subsequent investigation showed that this drug was reduced *in vivo* to the active agent *p*-aminobenzenesulphonamide (sulphanilamide). Prontosil was in fact acting as a prodrug (see section 9.9). This discovery of this lead compound led in the next decade to the synthesis and testing of large numbers of sulphonamides for antibacterial action.

Sulphonamides are competitive inhibitors of folic acid synthesis. They are bacteriostatics preventing the replication of bacteria. This control of the spread of the bacteria enables the body's natural defences to gain strength and destroy the microorganism. Sulphonamides act by inhibiting dihydropteroate synthetase, which catalyses the reaction of 2-amino-4-hydroxy-6-

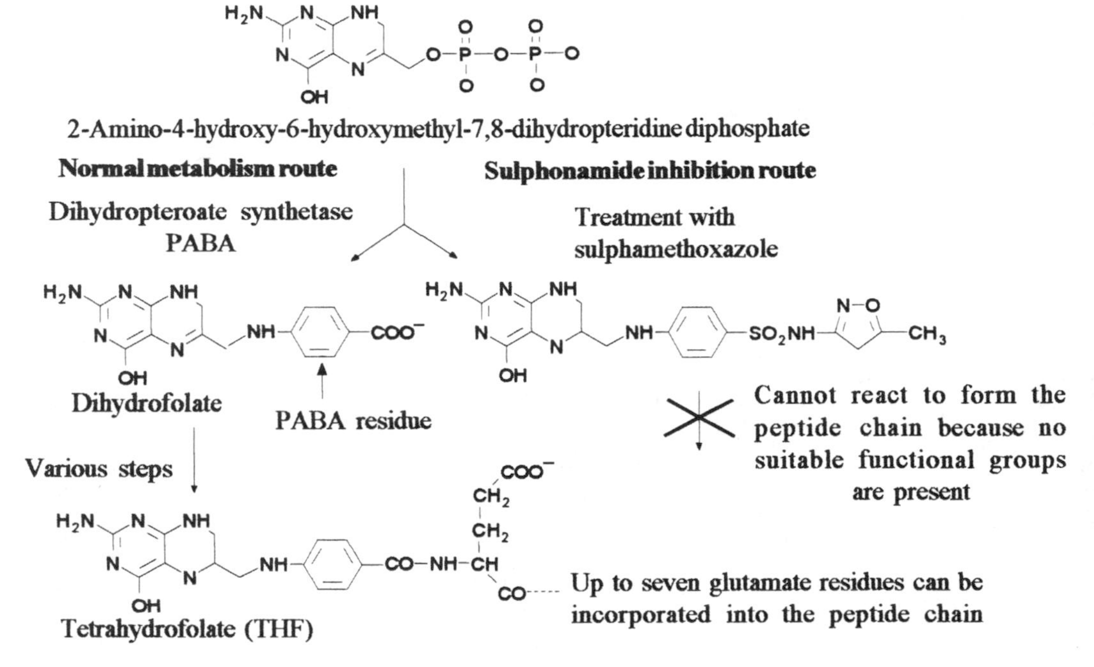

Figure 6.29. An outline of the chemistry of the inhibition of bacterial folic acid synthesis by sulphamethoxazole. PABA = *p*-aminobenzoic acid.

hydroxymethyl-7,8-dihydropteridine diphosphate with *p*-aminobenzoic acid (PABA) to form dihydrofolate (Figure 6.29). The sulphonamide binds to the active site and reacts with the 2-amino-4-hydroxy-6-hydroxymethyl-7,8-dihydropteridine diphosphate to form a product that cannot be converted into folic acid. As a result, the bacteria cannot produce enough tetrahydrofolate (THF), which is used as a coenzyme in the production of purines required for DNA synthesis and subsequent cell reproduction.

COOH / NH_2	SO_2NH_2 / NH_2	SO_2-NH-(N–O isoxazole)-CH_3 / NH_2
p-Aminobenzoic acid (PABA)	*p*-Aminobenzenesulphonamide	Sulphamethoxazole

Sulphonamides have a similar structure to PABA. Consequently they are believed to act by competing for the same active site as PABA.

It is interesting to note that humans are not able to synthesise folic acid and have to obtain it from their diet. Consequently, dihydropteroate inhibitors should have no effect on humans. However, because of the diversity of the chemical processes occurring in the human body these inhibitors may still give rise to unwanted side effects because they are able to bind to active sites and receptor sites controlling other processes.

Human angiotensinogen —Renin→ **Asp-Arg-Val-Try-Ile-His-Pro-Phe-His-Leu**
Angiotensin I
↓ Angiotensin-converting enzyme (ACE)
Asp-Arg-Val-Try-Ile-His-Pro-Phe + His-Leu
Angiotensin II

Figure 6.30. The conversion of angiotensinogen to angiotensin II.

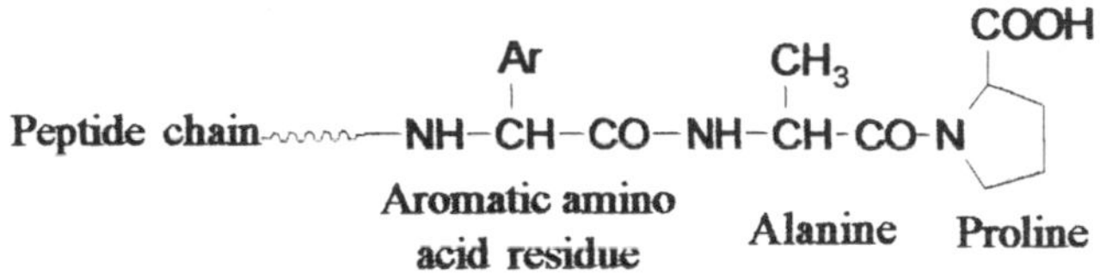

Peptide chain~~~NH–CH(Ar)–CO–NH–CH(CH_3)–CO–N(proline ring, COOH)

Aromatic amino acid residue **Alanine** **Proline**

Figure 6.31. The general structure of the C-terminal of the most effective inhibitors of peptide inhibitors of ACE.

6.12.2 Captopril and Its Derivatives

Angiotensin II, an octapeptide, is a potent vasoconstrictor responsible for increasing blood pressure. It also acts as a hormone, controlling electrolyte balance and other biological processes. Angiotensin II is produced from angiotensinogen, an α-globulin produced by the liver, by a series of hydrolyses catalysed by the enzymes renin and angiotensin-converting enzyme (Figure 6.30). Overproduction can result in hypertension (high blood pressure). Angiotensin-converting enzyme (ACE) is also involved in a number of other processes that can result in hypertension.

Initial experimental work showed that ACE could be inhibited by small peptides, the most effective of these peptides having a proline residue at the C-terminal, alanine at the penultimate position and an aromatic amino acid residue at the antepenultimate position (Figure 6.31).

These peptides were not suitable for oral administration as drugs because they underwent extensive hydrolysis in the stomach and gastrointestinal tract. Further investigation of the structure of ACE showed that it contained one zinc ion per molecule of the enzyme, which was thought to coordinate with the carbonyl of the peptide bond (Figure 6.32). This coordination is believed to increase the polarisation of the carbonyl group, facilitating the nucleophilic attack of a water molecule bound to the enzyme on a peptide bond of the Angiotensin I peptide chain.

Experimental investigations also indicated that the structure of ACE was similar to carboxypeptidase A, a zinc-containing pancreatic digestive enzyme whose structure had been determined by X-ray crystallography. This enzyme was found to be strongly inhibited by 2*R*-

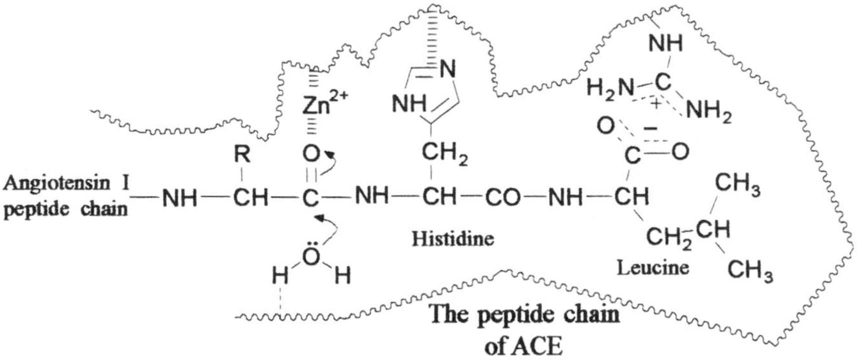

Figure 6.32. A schematic representation of the binding of the C-terminal of angiotensin I to the active site of ACE.

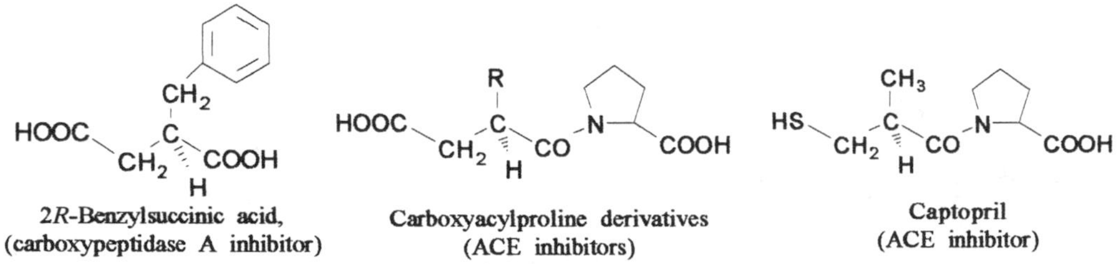

Figure 6.33. Compounds used to inhibit carboxypeptidase A and ACE.

benzylsuccinic acid (Figure 6.33). Consequently, using all this information and computer simulation of the active site of carboxypeptidase A, a series of carboxyacylproline derivatives (Figure 6.33) have been synthesised and tested. All of the compounds synthesised were found to be weak inhibitors of ACE but subsequent replacement of the terminal carboxylate group by a thiol group resulted in compounds that were strong ACE inhibitors. Thiols are better ligands than carboxylate ions for coordinating with zinc(II) ions, which probably accounts for the increase in potency. The most potent of these thiol derivatives was captopril (Figure 6.33), which is now an important drug in the treatment of hypertension.

Captopril is a competitive ACE inhibitor. It has a number of unwanted side effects including skin rashes, dry coughs, palpitations and renal impairment. This, coupled with the fact that thiols are readily oxidised *in vivo* to disulphides, has led to the development of a number of other ACE inhibitors with structures similar to captopril (Figure 6.34). The approach adopted was to increase the potency of the carboxyacylproline derivatives by increasing their degree of binding to the enzyme by making them more like the products of ACE hydrolysis (Figure 6.30). The methylene group of the chain was replaced by a secondary amine (—NH—) to make the structure look more like a peptide. Because the third residue of angiotensin I was phenylalanine (Phe), an aromatic residue was incorporated to increase the binding to the active site. It is interesting to note that in the most effective peptide inhibitors of ACE the third amino acid

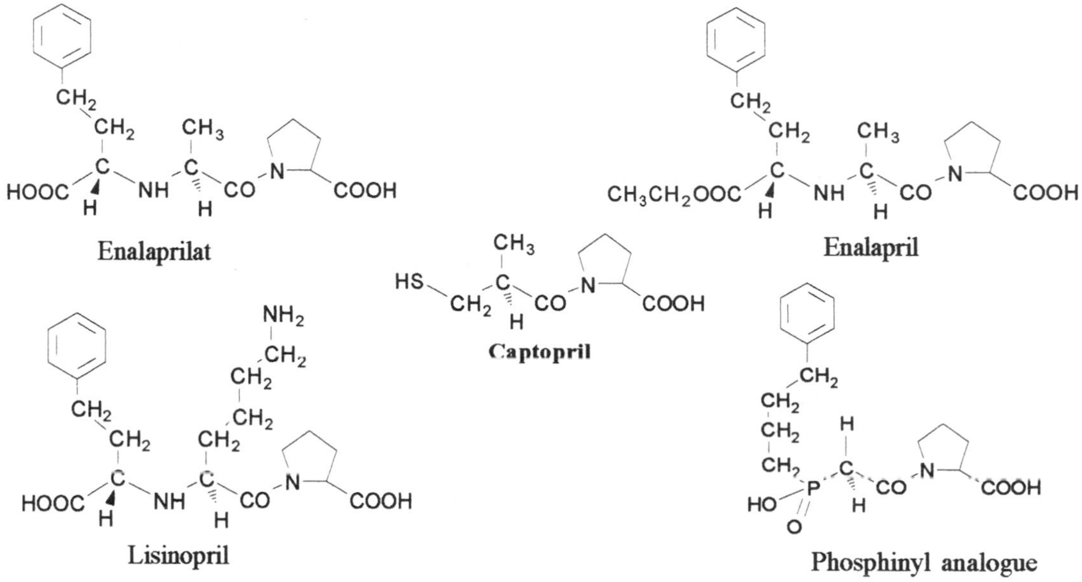

Figure 6.34. ACE inhibitors.

was an aromatic residue (Figure 6.31). The result was the eventual production of enalaprilat and the clinically used compounds enalapril and lisinopril, all of which have structures related to that of captopril (Figure 6.34). However, they were all stronger inhibitors of ACE than captopril. Although enalaprilat is more effective than captopril it is poorly absorbed when administered orally because of its strong polar nature. As a result, it can only be given by intravenous injection. Reduction in its polar nature by converting the carboxyl group to its ethyl ester produced enalapril, which is absorbed and is also more potent than enalaprilat. Unfortunately both enalapril and lisinopril still suffer from the same side effects as captopril but they do have the advantage that lower doses of the drug are required, which reduces the possibility of side effects. Strong ACE inhibitors in which a phosphinyl group have been used to replace the secondary amino group have also been synthesised.

6.13 Enzymes and Drug Resistance

Drug resistance occurs when a drug no longer has the desired clinical effect. This may be due to either a natural inbuilt resistance in some individuals and organisms or may arise naturally in the course of treatment. The former is probably due to differences in the genetic code of individuals within a species whereas the latter arises because of natural selection. In natural selection the drug kills the susceptible strains of an organism but does not affect other strains of the same organism. Consequently, these immune strains multiply and become the common strain of the organism, which subsequently results in ineffective drug treatment.

Resistance occurs on an individual basis and so is not usually detected until a wide sample of the population has been treated with or indirectly exposed to the drug. Its detection necessitates the discovery of new drugs to treat the condition. Its emergence is probably due to the widespread and poorly controlled use of a drug. For example, the generous use of antibiotics in farming is strongly suspected to be the reason for an increase in antibiotic-resistant strains of bacteria in humans. The response of medicinal chemists to resistance is either to devise new or to modify existing drugs. This approach suffers from the high probability of being unsuccessful as well as being time consuming and expensive. In the light of human experience it would be better if in future we reduced the possibility of resistance by using the effective existing drugs more intelligently.

Drug resistance can be linked to a change in either the permeability of the membranes of the organism or an enzyme system(s) of the organism. Enzymes may be involved in drug resistance in a number of ways (see *β-lactam antibiotics*, section 4.4.2.2) but in many cases resistance may be due to several different processes occurring at approximately the same time.

6.13.1 Changes in Enzyme Concentration

A significant increase or decrease from the normal concentration of an enzyme can result in resistance to a drug. The overproduction of an enzyme can have two effects:

(i) The target process catalysed by the enzyme will not be inhibited because excess enzyme is produced For example, the resistance of malarial parasites is believed to be due to overproduction of dihydrofolate reductase as a result of the drug stimulating the parasites' RNA.

(ii) Increased production of enzymes that inactivate the drug: for example, β-lactamases inactivate most penicillins and cephalosporins by hydrolysing their β-lactam rings (see section 4.4.2.2(iii)). A number of enzymes deactivate inhibitors by incorporating (conjugation) phosphate by phosphorylation of hydroxyl groups, adenine by adenylation of hydroxyl groups or acetyl by acetylation of amino groups in the inhibitor's structure (Figure 6.35). ATP is believed to be the usual provider of phosphate and adenylic acid whereas acetyl coenzyme A is thought to be the normal source of acetyl groups. For example, resistance to the antibiotic kanamycin A can occur by all three routes (Figure 6.35) although kanamycin-A-resistant bacteria do not normally use all three routes.

Many aminoglycoside antibiotics are susceptible to this second type of enzyme inhibition. However, amikacin (AK), where the C_1-NH_2 of kanamycin A has been acylated by *S*-α-hydroxy-γ-aminobutanoic acid, is not susceptible to 3′-phosphorylation and 2″-adenylation.

The underproduction of an enzyme could result in insufficient enzyme being present to produce the active form of a drug from a prodrug. For example, the resistance to the anti-leukaemia drug 6-mercaptopurine is caused by a reduced production in hypoxanthine–guanine

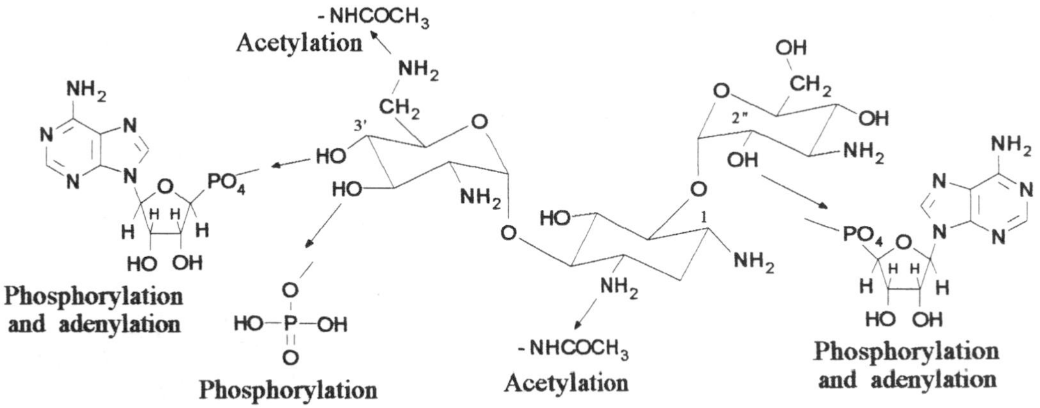

Figure 6.35. The inhibition of kanamycin A by enzymatic inactivation.

phosphoribosyltransferase, the enzyme required to convert the prodrug to its active ribosyl 5′-monophosphate derivative.

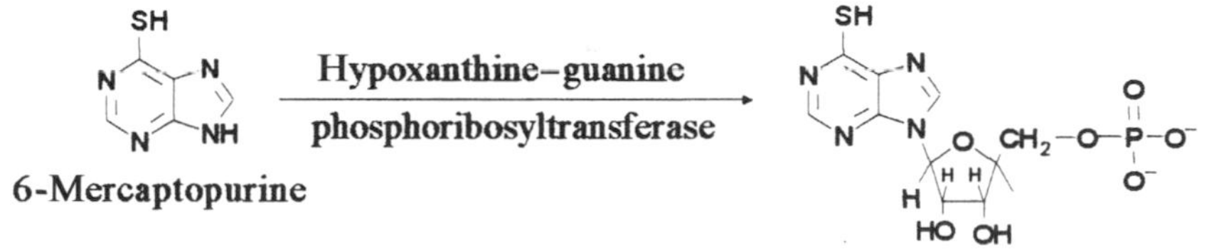

6-Mercaptopurine —Hypoxanthine–guanine phosphoribosyltransferase→

These changes in the production of the enzyme are believed to be due to genetic changes in the organism.

6.13.2 An Increase in the Production of the Substrate

Increased production of the substrate can prevent competitive reversible inhibitors from binding to the active site in sufficient quantities to be effective (see section 6.9.1). The high concentration of the substrate moves the position of equilibrium to favour the formation of the E-S complex:

$$\text{I-E} \rightleftharpoons \text{I} + \text{E} + \text{S} \rightleftharpoons \text{E-S}$$

For example, the inhibition of dihydropteroate synthetase by sulphonamides (see section 6.12.1) results in a build-up of PABA. This increase in substrate concentration prevents sulphonamides from inhibiting dihydropteroate synthetase, which is a key enzyme in the production of the RNA necessary for bacterial reproduction. Similarly, a build-up of angiotensin I will overcome the effect of ACE inhibitors. This is thought to be the reason for the concentration of plasma angiotensin II returning to normal in some cases where there has been a chronic administration of ACE inhibitors.

6.13.3 Changes in the Structure of the Enzyme

Changes in the structure of the target enzyme result in a structure that is not significantly inhibited by the drug. However, the modified enzyme is still able to produce the normal product of the reaction, which allows the unwanted metabolic pathway to continue to function. For example, resistance to the antibiotic trimethoprim is believed to be due to a plasmid (see section 10.14.1), directed change in the structure of dihydrofolate reductase in the bacteria.

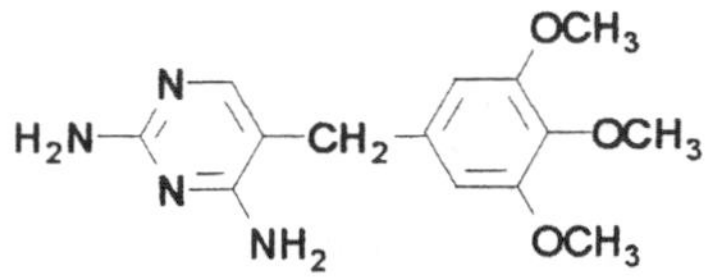

Trimethoprim

Similarly, the resistance of the *Escherichia coli* strains to sulphonamides has been shown to be due to their containing a sulphonamide-resistant dihydropteroate synthase.

6.13.4 The Use of an Alternative Metabolic Pathway

The blocking of a metabolic pathway by a drug can result in the opening of a new pathway controlled by a different enzyme that is not inhibited by the same drug.

6.14 Ribozymes

A number of biological reactions in which certain RNA molecules act as catalysts have been discovered. These catalytic RNAs exhibit many of the same general properties as protein-based enzymes. For example, they are substrate specific, they increase reaction rate and they reappear unchanged at the end of the reaction. However, in a number of cases their action appears to be enhanced significantly by the presence of protein subunits. These subunits do not act as catalysts for the reaction in the absence of the ribozyme.

6.15 Summary

Enzymes are compounds that catalyse almost all the reactions that occur in all living organisms. They are found embedded in cell walls and membranes as well as in the various biological fluids found in living organisms. Many enzymes consist of a protein molecule known as the **apoenzyme** and a smaller organic molecule known as **coenzymes** and/or metal ions known as **cofactors**. However, many enzymes do not need a coenzyme to be active.

Enzymes are classified into six major types – oxidoreductases, transferases, hydrolases, lyases, isomerases and ligases (synthases) – according to the type of reaction they catalyse. Those that contain clusters of metal atoms in their structures are known as **metalloenzymes** whereas **proenzymes** are proteins that give rise to enzymes by losing part of their amino acid chain. **Isoenzymes** and **isofunctional enzymes** are groups of enzymes with similar functions. However, isoenzymes have very similar structures but isofunctional enzymes have very different structures.

Three enzyme **nomenclature** systems are in common use, namely: a systematic name indicating the process that the enzyme catalyses and also the class to which the enzyme belongs; a trivial name; and a unique four-figure code allocated by the Enzyme Commission. This code is prefixed by the letters EC. All enzyme names except the EC code usually end with the suffix **-ase.**

The region of the enzyme where the substrate reaction occurs is known as an **active site**. These reactions occur relatively easily because it is believed that enzymes reduce the activation energies of the reactions they catalyse. The original theory to explain the action of enzymes was the **lock and key theory**. This has been superseded by **Koshland's induced fit hypothesis**, which states that the conformations of the substrate and active site undergo slight changes so that the substrate fits the active site. The amino acid residues that form the active site are often well separated in the peptide chain but are brought together by the folding of the chain. Coenzymes and cofactors usually form part of the active site of an enzyme.

Allosteric sites are sites where a molecule other than the substrate can bind to the enzyme. An enzyme's capacity to deal with a substrate may be enhanced by the conversion of an allosteric site to an active site (**allosteric activation**). This may occur when a substrate binds to its active site. It results in a conformation change that opens a new additional active site.

Enzyme activity is regulated by either **covalent modification** or **allosteric control**. Covalent modification occurs when a chemical group is reversibly attached to the enzyme by covalent bonding. Allosteric control occurs when either a compound, usually the final product, formed in the pathway (**feedback control**) or another compound reversibly inhibits an enzyme in the pathway. Allosteric control occurs when a compound reversibly inhibits the enzyme by binding to an allosteric site. Compounds responsible for allosteric control are known as **modulators**, **regulators** and **effectors**.

The **action** of most enzymes is specific with regard to the **stereochemistry** and the group of the substrate affected. Enzymes are only effective over limited ranges of pH and temperature. For a constant concentration of the enzyme the **rate of enzyme-controlled reactions (V)** increases up to a maximum value (V_{max}) with increase in substrate concentration (see Figure 6.12a). For processes under allosteric control, the rate of increase of reaction rate with increase in the concentration of a substrate is sigmoid (see Figure 6.12b).

The **kinetics of enzyme processes** with a single substrate may be described by the **Michaelis–Menten equation** (Equation 6.1). Rearrangement of the Michaelis–Menton equation gives the more useful **Lineweaver–Burk equation** (Equation 6.2). For single substrate processes that exhibit Michaelis–Menten kinetics, the Lineweaver–Burk plot of 1/[V] against 1/[S] is a straight line with a slope of $K_m/[V_{max}]$ and intercepts on the x- and y-axis of $-1/K_m$ and $1/V_{max}$x, respectively. The values of K_m and V_{max} are characteristic properties of an enzyme process; K_m is the substrate concentration at $V_{max}/2$ and is a measure of the affinity of the substrate for the enzyme.

The **kinetics** of each of the substrates of many multiple substrate processes fit the Michaelis–Menten equation provided that the concentrations of the other substrates are kept constant. **Multisubstrate enzyme processes** usually proceed by one of two general routes known as either sequential or double-displacement reactions. In **sequential enzyme processes** either the order in which the substrates bind to the enzyme is not significant or one substrate binds to the enzyme before the other substrates can bind to the enzyme. However, in sequential reactions, the products can only be formed when all the substrates are bound to the enzyme. In **double-displacement reactions** a substrate binds to the enzyme and reacts to form a product and modify the structure of the enzyme before a second substrate can bind to the modified enzyme and form the second product so regenerating the enzyme back to its original form. Sequential and double-displacement reactions exhibit characteristic Lineweaver–Burk plots (see Figure 6.16).

Enzyme inhibitors may be classified as being either reversible or irreversible. **Reversible inhibitors** are subclassified as competitive, non-competitive and uncompetitive inhibitors. They form an equilibrium system with the enzyme binding to the enzyme by means of weak electrostatic interactions, weak covalent bonding, hydrogen bonding and van der Waals' forces. **Competitive inhibitors** compete with the substrate for the active site. Consequently, the structure of the substrate can be used as the basis of the design of reversible inhibitors. **Non-competitive** and **uncompetitive inhibitors** block enzyme action by binding to an allosteric site. This causes conformation changes at the active site that may prevent the substrate binding to the enzyme.

Irreversible inhibitors bind by non-reversible processes to the enzyme. They are classified as **active site-directed inhibitors** (ASDI) and **suicide** or **Irreversible mechanism-based inhibitors** (IMBI). Both types of irreversible inhibitor bind to the enzyme by strong covalent and non-covalent bonds bonds. **Active site-directed inhibitors** bind at or near the active site, which prevents the substrate from binding to the active site. The incorporation of electrophilic groups (Table 6.3) into the structure of the substrate can be used as the basis of the design of new inhibitors. **Suicide inhibitors** are compounds that reversibly bind to the active site. This binding process unmasks or enhances a reactive group that reacts to make the binding irreversible.

Many of these reactive groups are electrophiles, which react by a type of Michael addition with nucleophilic groups at the active site.

Transition state inhibitors are stable compounds whose structure resembles that of the transient structure of all the species present at the transition state of the process.

Drug resistance is said to occur when a drug no longer has the desired clinical effect. It occurs by either natural selection or changes in the nature of the enzyme systems controlling vital metabolic pathways due to genetic changes in the organism. Drug resistance linked to changes in the nature of the enzyme system may be due to:

(i) changes in enzyme concentration;
(ii) an increase in substrate concentration;
(iii) changes in the structure of the enzyme;
(iv) the use of an alternative metabolic pathway.

Ribozymes are RNA molecules that act as enzymes.

6.16 Questions

(1) Distinguish between each of the following terms: (a) isoenzyme and isofunctional enzyme; (b) enzyme and proenzyme; (c) active and allosteric site; (d) modulator and activator; and (e) single- and double-displacement reactions in the context of enzyme reactions.
(2) What information can be obtained from the systematic name of an enzyme?
(3) A Ca^{2+} ion channel blocker is administered to a patient. What would be the expected effect of this treatment on NO synthase.
(4) Outline, in general terms, the part played by the active site of an enzyme when it catalyses a reaction.
(5) Explain how the Lineweaver–Burk plots can be used to indicate the general type of mechanism employed by an enzyme inhibitor. What major assumptions is your explanation based on?
(6) Describe the main features of competitive, non-competitive and irreversible inhibition of enzymes.
(7) Compounds A and B are inhibitors of a monosubstrate (S) enzyme process that exhibits simple Michaelis–Menten kinetics. Use the data in Table 6.5 to determine the general type of enzyme inhibition processes exhibited by A and B. Decide, on the basis of this information, whether A or B could be considered a suitable candidate for further investigation into their suitability as lead compounds for the development of a new drug. Give an explanation for your decision.

Table 6.5. Experimental results (V in $\mu mol\, s^{-1}$ and [S] in $mmol\, dm^{-3}$).

No inhibitor		**[A] ($mmol\, dm^{-3}$)**				**[B] ($mmol\, dm^{-3}$)**			
		0.102		**0.217**		**0.137**		**0.315**	
[S]	**V**	**[S]**	**V**	**[S]**	**V**	**[S]**	**V**	**[S]**	**V**
0.083	0.122	0.125	0.063	0.333	0.062	0.091	0.091	0.087	0.060
0.125	0.155	0.153	0.072	0.500	0.075	0.118	0.101	0.143	0.079
0.200	0.194	0.200	0.095	1.000	0.094	0.222	0.122	0.250	0.089
0.500	0.259	0.500	0.139	20.00	0.137	0.666	0.144	0.500	0.095

(8) (a) Explain the meaning of the term 'suicide inhibitor'.

(A)

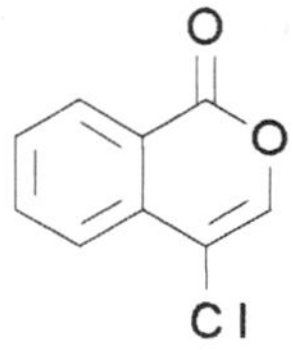

(b) Compound A acts as a suicide inhibitor of a serine protease. Outline a possible mechanism for its action and suggest modifications to its structure that might improve its action.

(9) (a) What is the natural substrate of lactate dehydrogenase?

(b) How could this substrate be modified to produce a possible active site-directed inhibitor for lactate dehydrogenase? Outline the reasoning behind your modification.

(10) Explain how the transition state of a reaction can be used to design an enzyme inhibitor. The transition state of an enzyme-catalysed reaction involves pyrrole binding to the active site of the enzyme. Suggest modifications to the structure of pyrrole that would (a) increase the binding and (b) decrease the binding of the modified molecule to the active site. Explain the electronic basis of your choices.

(11) What is drug resistance and how does it arise? Outline the ways by which enzymes are linked to drug resistance. How can drug resistance be minimised?

7 Complexes and Chelating Agents

7.1 Introduction

Metallic elements are essential components of many of the processes that are necessary for healthy human and animal life. A number of metallic elements, namely, sodium, potassium, calcium and magnesium, are required in bulk and are sometimes referred to as **minerals** although non-metallic elements such as phosphorus, chlorine (as chloride) and sulphur are also covered by this term. Other essential metallic elements, referred to as **trace elements**, are present in amounts that are less than 0.01% of the average human body mass. Examples of trace elements are lithium, chromium, manganese, molybdenum, cobalt, nickel, selenium, iodine, tin, and zinc. However, besides trace and mineral elements, the healthy human body normally contains small amounts of other elements, such as silicon, arsenic and boron, which do not appear to be essential for maintaining health although current research may ultimately determine their purpose.

Metal ions and atoms occur in biological systems as either free ions or covalently bound in the structures of complex organic compounds. Free ions fulfil a variety of functions, for example, the passage of sodium and potassium ions through the membranes of neurons is responsible for the transmission of an electrical impulse through the neuron (see section 4.4.3) whereas the movement of calcium into a cell can activate the enzyme system that initiates the formation of the messenger molecule nitric oxide (see section 11.4). Covalently bonded metal ions are involved in a variety of biological functions. For example, zinc is the reactive centre of a number of enzymes that initiate the cleavage of bonds (see Figure 6.33) whereas iron and copper are the reactive centres of many electron-transfer enzyme systems. Covalently bound metal ions can also act in a structural role holding the enzyme structure in the correct conformation to activate a distant active site. For example, zinc is believed to act in this capacity in bovine erythrocyte superoxide dismutase (Cu-Zn BESOD), which has a copper–zinc bimetallic site (Figure 7.1). The zinc ion holds the structure in the correct conformation for the copper to act as the active centre for the conversion of superoxide to hydrogen peroxide and oxygen. Other proteins store or transport metals by forming complexes with the metal.

The concentration of essential metals in human and animal bodies is critical for healthy life. Too low a concentration causes deficiency diseases whereas too high a concentration is toxic. For example, too low a concentration of selenium would result in liver necrosis and white muscle disease but too high a concentration can cause the development of cancers and a disease in cattle known as the blind staggers, characterised by impaired vision and muscle weakness. Consequently, the control of metal ion concentration to prevent or alleviate related pathological conditions is an important aspect of medicine. Too high a concentration can sometimes be

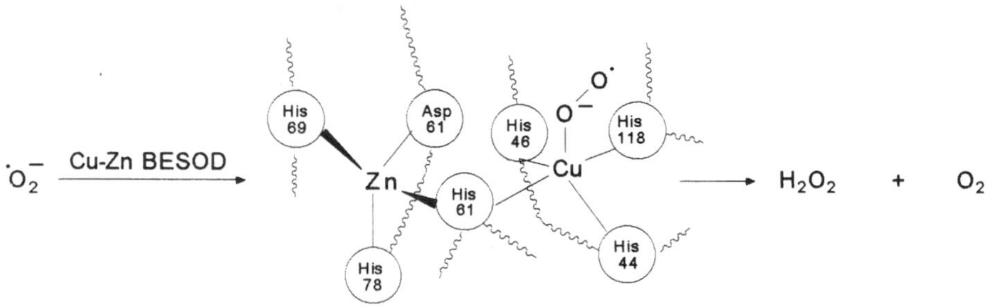

$\dot{O}_2^-$ Cu-Zn BESOD → H_2O_2 + O_2

Figure 7.1. A representation of the structure of BESOD and its involvement in the conversion of superoxide to oxygen and hydrogen peroxide.

reduced by either increasing the rate of excretion by the formation of a suitable complex or removing the source of the problem. Similarly, too low a concentration can be increased by treatment with metal ion supplements. These often taken the form of complexes because metallic ions are more easily absorbed in this form.

Living organisms can also absorb non-essential metals in concentrations that are not beneficial to the welfare of that organism. These metals are absorbed because of pollution of the atmosphere and food chain. For example, lead, mercury and other heavy metals may be found in humans eating plants grown on contaminated land. Treatment of so-called heavy metal poisoning usually makes use of drugs that act as chelating agents (see section 7.5.1).

Metal complexes are also used as therapeutic agents. For example, cisplatin and auranofin are major drugs in the treatment of genitourinary, head and neck tumours and rheumatoid arthritis, respectively. A number of other therapeutic agents are believed to involve complex formation in their action. However, few tailor-made complexes have yet been developed. Complexes such as ^{99m}Tc-mercaptoacetylglycine, with radioactive metallic isotopes in their structures, are widely used as diagnostic aids.

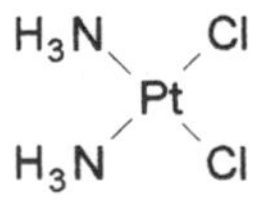

***cis*-Diamminedichloroplatinium(II)**
(cisplatin or *cis*-DDP)

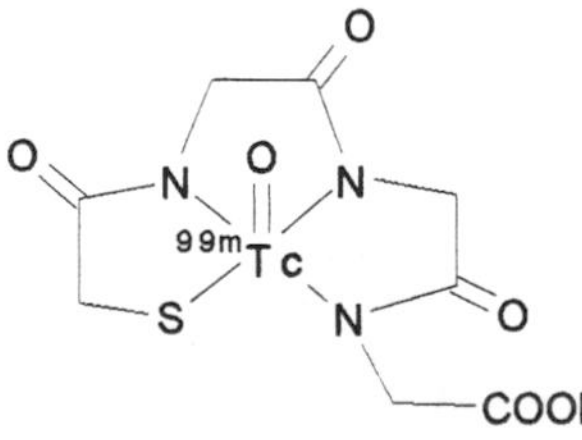

^{99m}Tc-Mercaptoacetylglycine
(^{99m}Tc-MAG3)

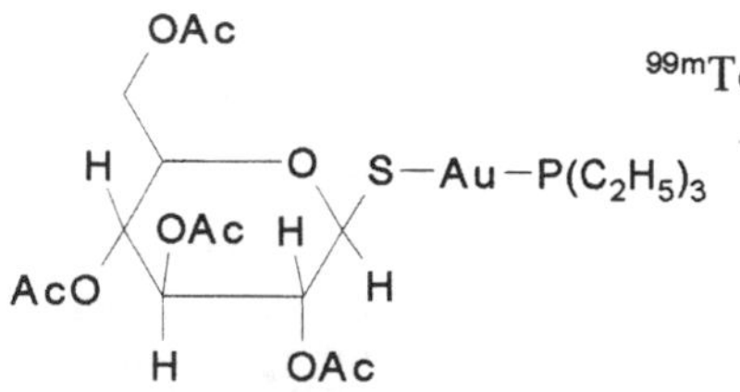

2,3,4,6-Tetra-*O*-acetyl-1-1-β-D-thioglucose(triethylphosphine)gold(I)
(auranofin)

Table 7.1. Examples of ligands. Positively charged ligands are rare. The atoms forming the bonds with the metal atom are indicated by an asterisk.

Classification	Example	Name (nomenclature)
Monodentate (unidentate)	*CN⁻*	Cyanide ion (cyano); complexes via C or N
	Cl^-*	Chloride ion (chloro)
	H_2O*	Water (aqua)
	*NH_3	Ammonia (ammine)
Bidentate	$H_2\overset{*}{N}CH_2CH_2\overset{*}{N}H_2$	1,2-Diaminoethane (ethylenediamine, en)
	$^-\overset{*}{O}OC\text{-}CO\overset{*}{O}^-$	Oxalate (ethanedioate)
Terdentate (tridentate)	$H_2\overset{*}{N}CH_2CH_2\overset{*}{N}HCH_2CH_2\overset{*}{N}H_2$	Diethylenetriamine (dien)
	[terpyridine structure: three pyridine rings, N* on each]	Terpyridine (terpy)
Quadridentate (tetradentate)	H_2N^*–CH_2–CH_2–HN^*–CH_2–CH_2–$^*NH$–CH_2–CH_2–$^*NH_2$	Triethylenetetramine (trien)

7.2 The Shapes and Structures of Complexes

The term complex is normally used in chemistry to denote a compound whose structure contains one or more metal atoms or ions to which are bonded electrically neutral or charged species called **ligands** (Table 7.1). **Organometallic** compounds are complexes in which the metal is directly bonded to a carbon atom. The shapes of complexes about the metal are normally determined by X-ray crystallography. It is difficult to determine the shapes of biological molecules *in situ* but the advent of specialist nuclear magnetic resonance (NMR) techniques is providing some information in this respect.

Stable complexes are formed when the electronic configuration of the bonded metal corresponds to that of the nearest noble gas in the periodic table. This means that complexes formed by main group metals are stable when the sum of the electrons in their outermost shells and those supplied by the ligands equals eight (the **octet rule**). Similarly, stable transition group metal complexes are formed when the number of electrons in the outermost shell of the metal plus the electrons supplied by the ligand equals eighteen (the **eighteen-electron rule**). These rules can be used to predict the number of ligands that could bond to a metal to form a stable complex. However, there are numerous exceptions to these rules.

The structures of complexes are usually explained in terms of molecular orbital (MO) theory. These explanations are usually based on a traditional σ covalent bond being formed between the metal and the ligand plus, in many cases, an extra dative π type of bond formed between the metal and the ligand (see section 7.2.1.3). The latter means that, in many cases, the single lines used to represent covalent bonds between the metal and the ligand in the structural formulae of complexes do not represent a standard two-electron covalent bond. However, the MO pictures of many complexes are not available because the appropriate calculations are so large that they have not yet been made.

7.2.1 Ligands

Ligands can be classified in a number of ways:

(i) the number of atoms donating electrons to a metal atom;
(ii) the number of electrons they contribute to a metal atom;
(iii) the type of bond they form with a metal atom.

7.2.1.1 The Number of Atoms Forming Bonds with the Metal Atom

The number of donor atoms associated with a metal atom is known as the **coordination number** of the metal atom. The most common coordination numbers are four and six.

Ligands are classified according to their coordination number of donor atoms as uni-, di, tri-, tetradentate, etc. or using the Greek prefixes mono-, bi-, ter-, quadri dentate, etc. (Table 7.1). Both systems are used in the literature.

The geometric arrangements of the coordinated atoms of the ligands bonded to the metal in naturally occurring complexes usually approximate to those shown in Table 7.2. For example, the atoms bonding to a four-coordinate zinc ion are usually arranged in a distorted rather than a regular tetrahedral shape about the zinc ion. However, it is possible for an element to exhibit more than one geometry in both different complexes and in the same complex (see section 7.2.4).

Multidentate ligands that form ring structures in which the ligand is bonded by more than one atom to a single metal atom are known as **chelating agents**. Chelating agents that bind strongly to metal cations to form stable water-soluble complexes are known as **sequestering agents**. The complexes produced by chelating agents are known as **metal chelates** or **chelation compounds.** The formation of ring systems may impose restrictions on the stereochemistry of the complex. For example, the flexible diethylenetriamine forms rings in which the three bonding atoms and the metal atom do not have to be in the same plane. However, rings will only be formed by the rigid, fully conjugated terpyridine if the bonding atoms and the metal atom are all in the same plane.

Table 7.2. The common geometrical arrangements of coordinated atoms (L) about a central metal ion.

Coordination number	Common geometric arrangements	Examples of metals that can exhibit this arrangement
2	L—M—L Linear	Au(I), Ag(I), Hg(II), Cu(I)
3	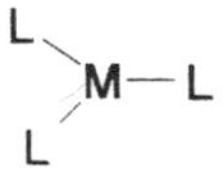 Trigonal planar	Cr(III), Fe(III)
4	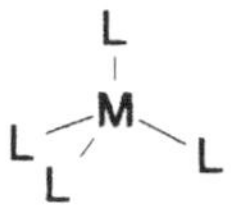 Tetrahedral; 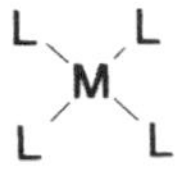 Square planar	**Tetrahedral:** Co(II), Cu(II), Zn(II), Fe(III), Co(IV), Ti(IV), Ni(II). **Square planar:** Cu(II), Pt(II), Ni(II), Cr(II), Mn(III)
5	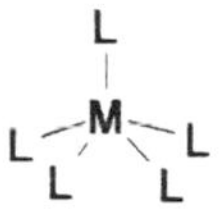 Square pyramidal; 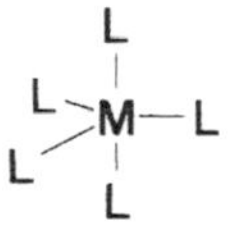 Trigonal bipyramidal	**Square pyramidal:** Mo(IV), Cu(II), Fe(II) **Trigonal bipyramidal:** V(IV), Nb(IV), Ta(V)
6	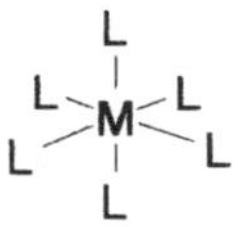 Octahedral	Co(IV), Fe(II), Mg(II), Cr(III)

7.2.1.2 The Number of Electrons Donated to the Metal Atom

Ligands are classified as one, two, three . . . eight electron donors to the metal atom. For example, ligands such as methyl, fluorine and hydroxide ions that form a normal covalent bond with the metal are classified as one-electron donors whereas ligands that form dative bonds by the donation of two electrons from the ligand are said to be two-electron donors, and so on (Table 7.3). However, with some ligands, the number of electrons donated will depend on the nature of the complex in which they occur. For example, bromine normally acts as a one-electron donor but in the form of a ligand bridge it acts as a three-electron donor (see section 7.2.2). However, the symbol η^n before the name of a ligand implies that **n atoms** of the ligand are involved in the bonding but not necessarily n electrons..

7.2.1.3 The Type of Bond Formed with the Metal Atom

This system classifies ligands as simple or π-acceptor ligands.

Table 7.3. Examples of the classification of ligands according to the number of electrons donated by the ligand. Electrons are counted as one for unpaired and two for lone pairs and π bonds.

Classification	Ligand example	Structure
One-electron donor	Bromine	•Br
Two-electron donor	Ammonia	$:NH_3$
Three-electron donor	η^3-Allyl	$H_2C{=}CH{-}\dot{C}H_2$
Four-electron donor	η^4-Conjugated dienes	$CH_2{=}CH{-}CH{=}CH_2$
Five-electron donor	η^5-Cyclopentadienyl (Cp)	
Six-electron donor	η^6-Arene	R

Simple ligands. Simple ligands, such as methyl, fluorine and ammonia, form classical covalent bonds with the metal atom by sharing an electron with the metal atom or dative bonds by donating a pair of electrons to the metal atom.

π-Acceptor ligands. These ligands form a σ bond to the metal by donating electrons to the metal atom. The metal atom also donates electrons to an energetically favourable empty π or π* molecular orbital in the ligand. This reduces the electron density of the metal.

Two important biologically active compounds that form complexes whose structures can be explained in this manner are carbon monoxide and nitric oxide (see Chapter 11). In the case of the structure of the metal–carbon monoxide bond in metal carbonyls the molecular orbital picture of carbon monoxide (Figure 7.2) shows that carbon has a lone pair that occupies a σ* molecular orbital. This lone pair forms a dative σ bond with a vacant σ molecular orbital in the metal. The structure of carbon monoxide also has two energetically favourable vacant π* molecular orbitals that can interact with filled metal d orbitals to give a significant degree of metal to carbon monoxide π–π* interaction. This type of π bonding is often referred to as **back-bonding**. In carbon monoxide, the total degree of back-bonding from the metal to the ligand almost balances the electron donation from the ligand to the metal. Consequently, the polarisation of the M–CO bond is low, which accounts for the low dipole moment of M–CO bonds (about 0.5 D). The strength of the π–π* interaction also accounts for the M–CO bond's chemical stability.

Nitric oxide has one more electron than carbon dioxide (Figure 11.1). This electron occupies a 2π* molecular orbital and enables nitric oxide to act as a three-electron donor, unlike carbon monoxide, which is a two-electron donor. However, the structure of the M–NO bond is very

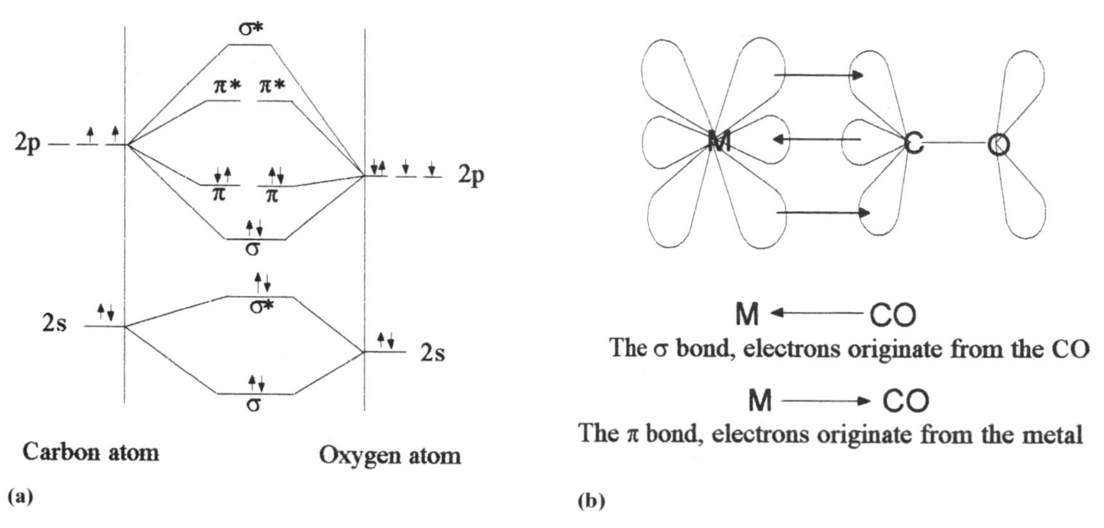

Figure 7.2. (a) The molecular orbital picture of carbon monoxide.The 1s orbitals are not shown because they do not contribute to the bonding. (b) The structure of the metal–carbonyl bond.

similar to that of the M–CO bond. Like carbon monoxide, nitric oxide donates its lone pair of σ* electrons to the metal to form a σ bond but, unlike the M–CO bond, the metal atom only contributes three electrons to the M–NO π–π* interaction, the fourth electron being supplied by the nitric oxide. The net result is that the metal π–π* nitric oxide interaction has a full set of four electrons, three originating from the metal and one from the nitric oxide.

The high degree of interaction between the metal and the ligand explains the strength of metal–carbon monoxide and metal–nitric oxide bonds. However, the metal–nitric oxide bond appears to be stronger because carbon monoxide is displaced in preference to nitric oxide in complexes containing both NO and CO ligands. The strengths of these metal–ligand bonds account for the ease and strength with which both nitric oxide and carbon monoxide bind to iron and other metals in biological molecules.

Many ligand–metal bonds are explained in terms of synergic back-bonding. However, each case must be considered on its own merits. For example, the full π molecular orbitals of alkenes and conjugated alkenes form bonds with the metal atom or ion by interacting with an empty metal hybrid orbital (Figure 7.3a). These bonds are reinforced by synergic back-bonding from a metal d orbital or d hybrid orbital to the ligand. The π electron systems of conjugated ring systems can also form complexes with metals. Their structures are explained in a similar manner, the π electrons of the ligand being donated to the metal whilst the metal forms synergic back-bonds with the ligand (Figure 7.3.b). However, the precise nature of the sequence of energy levels used to explain the structure of the complex is more complex and in many cases is still a subject for much discussion.

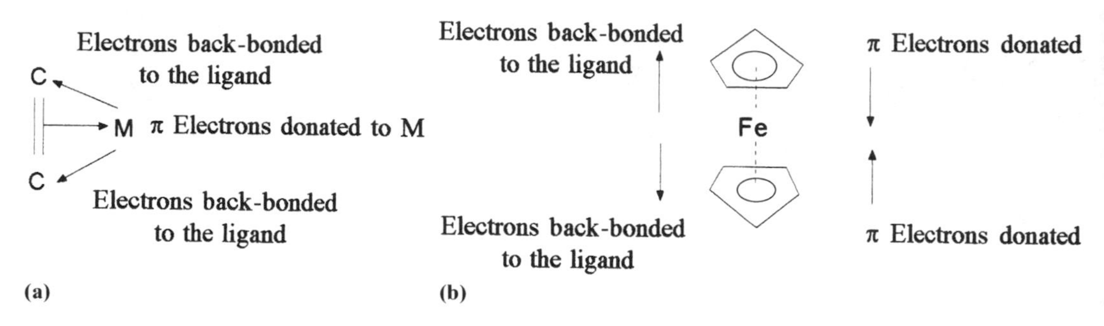

Figure 7.3. The bonding (a) alkene complexes and (b) cyclopentadienyl complexes.

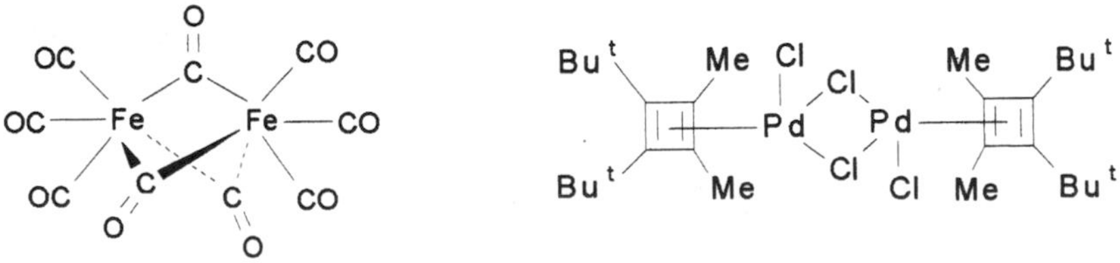

$Fe_2(\mu_2\text{-}CO)_3(CO)_6$ $Pd_2(\mu_2\text{-}Cl)_2$**(1,2-Dimethyl-3,4-*t*-butylbutadiene)**$_2$

Figure 7.4. Examples of complexes with bridging ligands.

7.2.2 Bridging Ligands

Ligands can form bridges between metal atoms. For example, in $Mn_2(\mu_2\text{-}Br)_2(CO)_8$ bromine acts as a bridge between the manganese atoms by using one of its lone pairs of electrons (dative covalent bond →) to bond to one manganese atom and its unpaired electron (covalent bond —) to bond to the other manganese atom.

$Mn_2(\mu_2\text{-}Br)_2(CO)_8$

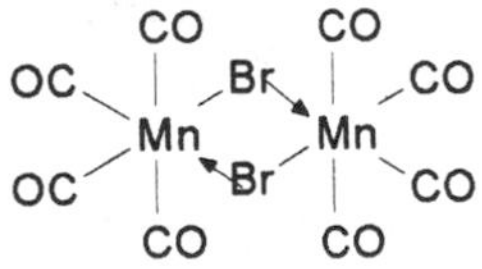

The symbol μ in the formula of the complex shows the presence of a ligand bridge in the structure of a complex. The subscript in both names and formulae shows the number of metal ions linked by that bridge (Figure 7.4).

7.2.3 Metal–Metal Bonds

The structures of complexes can contain metal–metal bonds. For example, both cobalt and manganese form carbonyl complexes whose structures are explained by the existence of a metal-to-metal bond (Figure 7.5).

Figure 7.5. The metal–metal bonds of (a) $Co_2(CO)_8$ and (b) $Mn_2(CO)_{10}$ give the metal atoms stable electronic configurations with 18 electrons in the outer shell.

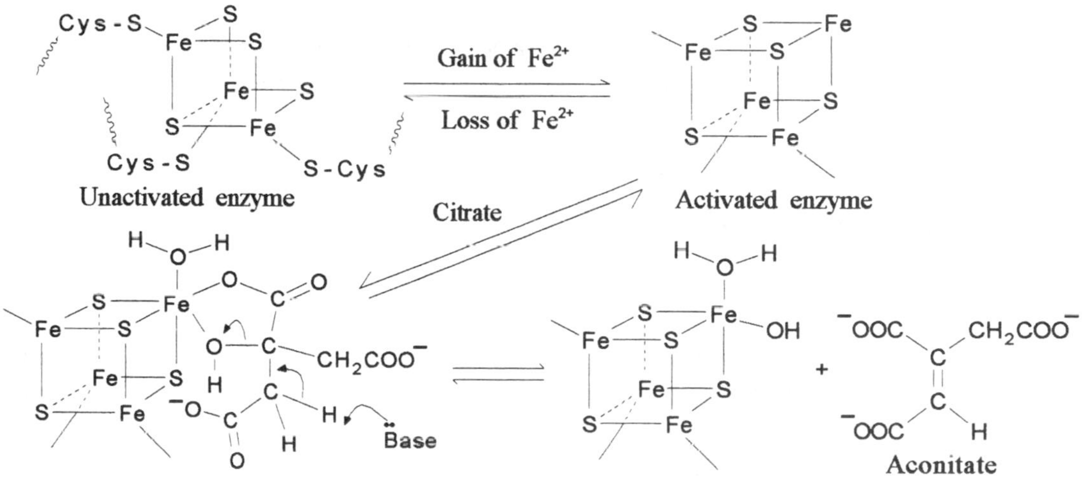

Figure 7.6. The structure of the Fe_3S_4 iron cluster in aconitase and its mode of action. Note that the sulphur atoms binding the cysteine to the cluster are not included in the formula used to represent the cluster and that in the interests of clarity the peptide chains are omitted after the first structure.

7.2.4 Metal Clusters

Metal–metal bonding and ligand bridge formation can result in complexes whose structures contain several metal atoms in relatively close proximity. These localised concentrations of metal atoms are commonly referred to as **metal clusters**.

Metal clusters are found in some proteins. They are often associated with the biological activity of the molecule and are commonly referred to by their molecular formulae. For example, the enzyme aconitase contains an Fe_3S_4 iron–sulphur cluster (Figure 7.6) with an approximately cubic shape, which catalyses the conversion of citrate to aconitate. The Fe_3S_4 is activated by the binding of a fourth Fe^{2+} ion to the vacant corner of the iron cluster. This ion forms a coordination compound with the citrate by covalently bonding to the carboxylate and accepting an electron pair from the hydroxy group. The latter weakens the C–OH bond, allowing the hydroxyl group to act as a leaving group.

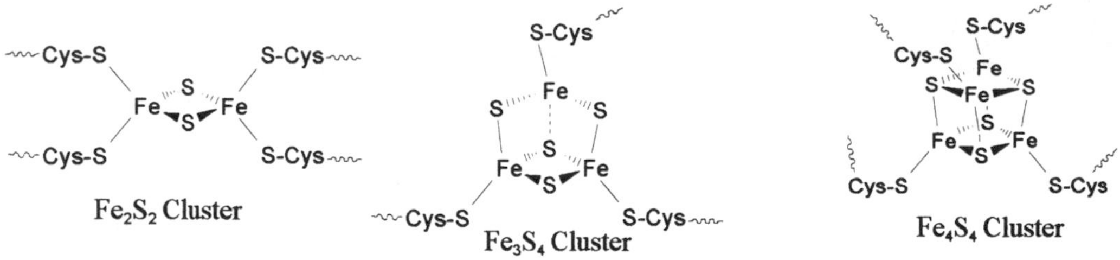

Figure 7.7. The structures of the three iron clusters in succinate dehydrogenase.

Different types of cluster may be found in the same enzyme, for example, succinate dehydrogenase contains three different iron–sulphur clusters (Figure 7.7).

7.3 Stability

The stability of complexes plays a major role in their biological and chemical activity. Metals exhibit a preference for particular ligands whereas ligands will preferentially bond to certain metals. In this context a 'new' ligand (L) may displace an 'old' ligand (Y) from a complex.

$$M(Y)_n + L \rightleftharpoons M(T)_{n-1}L + Y$$

This has important medicinal implications when one considers that most drugs contain groups that can act as ligands. Attempts to measure metal–ligand selectivity in terms of the relative strengths of metal–ligand bonds are based on two methods: (i) equilibrium constants and (ii) the concept of hard and soft acids and bases.

7.3.1 Stability and Equilibrium Constants

This method of stability assessment assumes that the formation of any complex is a dynamic equilibrium process. Consequently, it is possible to use the equilibrium constant for the formation of the complex by the direct reaction of the metal and ligand as a measure of its stability. Because complexes are mainly formed in aqueous solutions, the majority of quantitative determinations have been made in aqueous solution and they involve the displacement of hydrated water molecules. However, by convention, displaced water molecules are ignored because the concentration of water is so large by comparison to the concentration of the complex that it is effectively the same constant value for all complexes.

The formation of a complex by the direct reaction of a metal ion and a ligand can be regarded as taking place in a series of steps because it is improbable that all the ligands will bond to the

Stepwise reactions:

$$M + L \rightleftharpoons ML \qquad K_1 = \frac{[ML]}{[M][L]}$$

$$ML + L \rightleftharpoons ML_2 \qquad K_2 = \frac{[ML_2]}{[ML][L]}$$

$$ML_2 + L \rightleftharpoons ML_3 \qquad K_3 = \frac{[ML_3]}{[ML_2][L]}$$

$$\vdots$$

$$ML_{n-1} + L \rightleftharpoons ML_n \qquad K_n = \frac{[ML_n]}{[ML_{n-1}][L]}$$

Overall reaction:

$$M + nL \rightleftharpoons ML_n \qquad \beta_n = \frac{[ML_n]}{[M][L]^n}$$

Figure 7.8. The stepwise equations and overall formation stability equilibrium constants for the formation of ML_n.

metal at precisely the same time. Consider, for example, the synthesis of a complex ML_n from a metal ion M and a monodentate ligand L. The formation can be considered as a series of n steps where n is the appropriate coordination number of the metal. Each of these steps has its own equilibrium constant referred to as its **stepwise formation constant K_n** (Figure 7.8). The value of K_n for each of these steps is a measure of the affinity of the metal for the ligand: the larger the value, the greater the ligand–metal bond strength.

The overall stability formation constant (β_n) is related to the stepwise formation constants by the expression:

$$\beta_n = K_1 \cdot K_2 \cdot K_3 \cdot \ldots K_n \qquad (7.1)$$

Overall and stepwise stability constants are normally recorded for convenience as log values: the larger the value, the more stable the complex and the greater the affinity of the metal for the ligands.

The formation stability constant data accumulated since the 1960s allow a number of **generalisations** to be made with regard to the stabilities of complexes:

(i) Complexes formed by a metal and a specific ligand tend to be more stable when the metal is in the +3 as against the +2 oxidation state.

(ii) Metals of the first transition series in their +2 oxidation state will form complexes whose stability is usually in the order:

$$Hg > Cu > Ni \sim Pb > Co \sim Zn > Fe > Mn \quad (\text{Note: } Mn > Mg > Ca)$$

(iii) Where ligands contain the same donor atoms, a ligand that can form a chelated ring (L-L) will form more stable complexes than a ligand (L) that does not form a chelated ring, that is, $\beta_{L\text{-}L} > \beta_L$.

It is stressed that these statements are generalisations and that there are many exceptions. However, as a general rule, in systems where more than one complex can be formed the complex with the larger formation stability constant will be the most stable and therefore should be formed in the highest yield. However, all the possible compounds will be present in the system at equilibrium.

$$\underset{\text{Equilibrium lies to the left when } \beta_{MY} > \beta_{ML}}{MY + L} \rightleftharpoons \overset{\text{Equilibrium lies to the right when } \beta_{MY} < \beta_{ML}}{ML + Y}$$

7.3.2 Hard and Soft Acids and Bases (HASB)

This approach classifies the reactants as either a hard or soft acid or base. It regards the complex formation process as being a Lewis acid–base type of reaction. The species that donates the electrons is the base, whereas the species that is able to accept the electrons is the acid. In most cases the metal is the acid and the ligand is the base.

$$M \curvearrowleft :L \longrightarrow M:L$$

The terms hard and soft refer to the availability and mobility of the electrons possessed by the acid or base. Soft species have electrons that are easily removed (relatively mobile) whereas hard species have electrons that are firmly held (not very mobile). Softness is associated with a low density of charge whereas hardness is related to a high charge density. For example, a hard acid will have a high positive charge density and a small size whereas a soft acid would have a low positive charge density and be a large-sized species. A large number of species have been classified as hard and soft acids and bases using these definitions but, as expected with all such definitions, a number of borderline cases are also known (Table 7.4).

The concept of hard and soft acids and bases can be used to predict the relative strengths of bonds in complexes. A bond formed between two atoms with the same degree of electron mobility would be stable, having an almost even distribution of electrons. However, a bond formed between atoms with widely differing degrees of electron mobility would be less stable. Consequently, strong bonds are formed between hard acids and hard bases, soft acids and soft bases or borderline acids and borderline bases, whereas hard–soft bonding is generally much weaker. This is borne out in biological systems, where Ca^{2+} ions (hard acid) are frequently found coordinated with carboxylate (hard base), Fe^{3+} (hard acid) with either carboxylate or phenoxide (hard base), Cu^{2+} (borderline acid) with the nitrogens of the imidazole ring (borderline base) of histidine residues and cadmium (soft acid) with the sulphydryl groups (soft bases) of proteins. Metallothioneins, a group of small proteins whose structures contain about 30% cys-

Table 7.4. Examples of hard and soft acids and bases.

	Hard	Soft	Borderline
Acids	H^+, Li^+, Na^+, K^+, Be^{2+}, Mg^{2+}, Ca^{2+}, Sr^{2+}, Mn^{2+}, Al^{3+}, Cr^{3+}, Co^{3+}, Fe^{3+}	Cu^+, Ag^+, Au^+, Hg^+, Pd^{2+}, Cd^{2+}, Pt^{2+}, Hg^{2+}, Pt^{4+}, Te^{4+}	Fe^{2+}, Co^{2+}, Cu^{2+}, Zn^{2+}, Pb^{2+}, Sn^{2+}, Sb^{2+}, Bi^{3+}, NO^+, SO_2, R_3C^+, $C_6H_5^+$
	Si^{4+}, Ti^{4+}, Zr^{4+}, Sn^{4+}, Hf^{4+}, BF_3, AlH_3, $Al(Me)_3$, SO_3, RSO_2^+, RPO_2^+, HF, HCl	RS^+, I^+, Br^+, I_2, Br_2, O, Cl, Br, I	
Bases	H_2O, OH^-, F^-, Cl^-, RO^-, PO_4^{3-}, ClO_4^-, SO_4^{2-}, CO_3^{2-}, ROH, RO, ROR, RCOR, $RCOO^-$, NH_3, RNH_2, NH_2NH_2	R_2S, RSH, RS^-, I^-, SCN^-, CN^-, I^-, R^-, H^-, $S_2O_3^{2-}$, $CH_2{=}CH_2$, C_6H_6 CO, NO	Br^-, NO_3^-, SO_2^{2-}, N_2, $PhNH_2$, pyridine, N NH Imidazole

teine, are believed to protect cells by complexing with soft toxic metals such as mercury(II) and lead(II).

7.3.3 Medical Significance of Stability

The addition of a xenobiotic to a living organism could affect the balance of metal ions in that organism because the structures of most xenobiotic contain groups that are able to act as ligands. Consequently, the xenobiotic could form complexes that could either remove minerals and trace metals from the system by excretion or prevent essential minerals and trace element metals from carrying out their normal function or initiating a pathological response. These situations are likely to occur if the xenobiotic forms complexes that are more stable than those normally formed between the metal and the naturally occurring ligands in the system.

In view of the large numbers of different ligands found in biological systems it is not usually possible to predict accurately the effect of a xenobiotic on the metal ion balance of the system. Consequently, it is necessary in the development of a potential drug to investigate its effect on the metal ion balance of the body.

7.4 The General Roles of Metal Complexes in Biological Processes

Metal ions are found as part of the structures of many naturally occurring proteins These proteins are classified as **metalloproteins**. Metalloproteins that act as enzymes are subclassified as **metalloenzymes**. The metal ion in all types of metalloprotein may be bound either to an amino

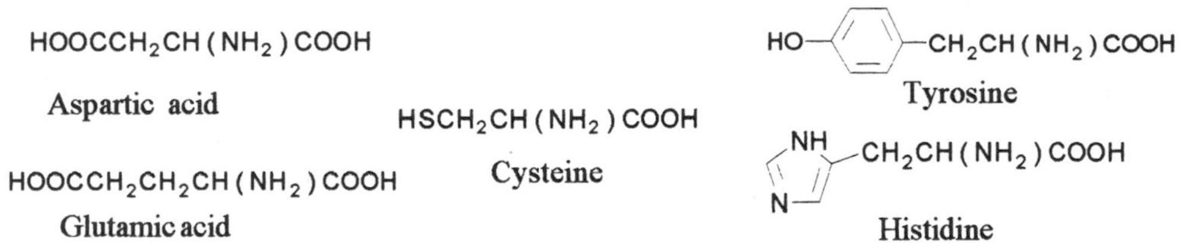

$HOOCCH_2CH(NH_2)COOH$

Aspartic acid

$HOOCCH_2CH_2CH(NH_2)COOH$

Glutamic acid

$HSCH_2CH(NH_2)COOH$

Cysteine

$HO-C_6H_4-CH_2CH(NH_2)COOH$

Tyrosine

$CH_2CH(NH_2)COOH$ (on imidazole ring: NH, N)

Histidine

Figure 7.9. Amino acids that are frequently found coordinated to metal ions in metalloproteins.

acid residue or to a prosthetic group in the protein. In the former case the metal ion is often coordinated to either one or more aspartic acid, cysteine, glutamine, histidine or tyrosine residues (Figure 7.9). Metal ions are also found as integral parts of the structures of other types of naturally occurring molecules, for example, iron(III) enterobactin (Figure 7.11a), which is involved in iron transport mechanisms.

The metals found in metalloproteins are often readily substituted by other metals. For example, Ca by Pb, Cd or Sr, Fe by Pu, K by Tl or Cs, Mg by Be or Al and Zn by Cd. These exchanges occur because the resultant complex is more stable than the original naturally occurring complex. It usually results in either the disruption of a biological process or the accumulation of the metal in the body. Both of these processes can lead to a pathological condition.

Naturally occurring metal ion complexes are involved in a wide variety of biological processes, including storage, transport, detoxification and catalysis, as well as in a structural role. A great deal of information is available concerning the storage, transfer and detoxification of iron but there is far less information available concerning the storage, transfer and detoxification of the other trace elements. This section sets out to give the medicinal chemist a better appreciation of the possible areas of impact of a new drug by presenting a general snapshot of the types of process in which coordination compounds make a significant contribution.

7.4.1 Storage

Metal ions that are not immediately required by the organism are usually stored in the organism in the form of complexes. For example, iron is mainly stored in mammals as ferritin, which is essentially a core of hydrated iron(III) oxide coated with a protein (Figure 7.10a). This protein coating consists of 24 pod-shaped protein subunits attached to the core by clusters of Fe(III) ions and Fe(II)–O–Fe(III) dimers attached to each other by oxo and hydroxo bridges. The core, which is about 8 nm in diameter, can contain up to 4500 Fe(III) ions. It has been suggested that the core is basically a hexagonal close-packed array of oxygen ions with Fe(III) occupying the octagonal sites. The close packing is interupted by a variety of other ions such as phosphate, hydroxide and sulphate, the composition varying according to the source of the ferritin. Ferritin is widely distributed in the organs of mammals, especially the liver, spleen and

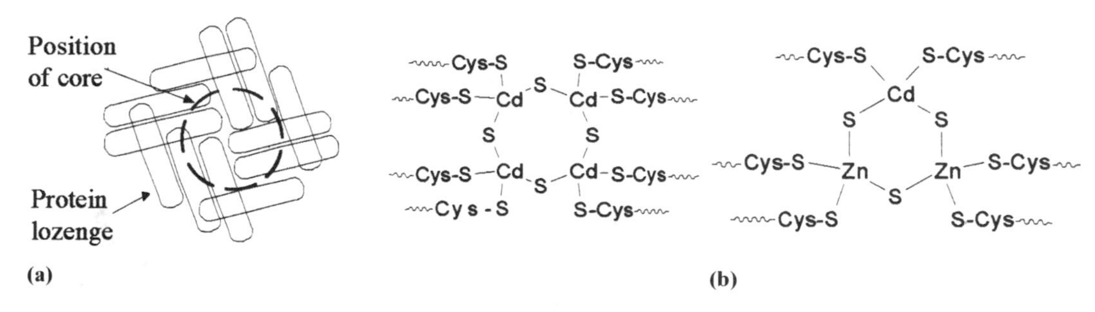

Figure 7.10. (a) Ferritin and (b) the cadmium and zinc–cadmium metal clusters of metallothioneins.

bone marrow. It also occurs in plants and bacteria. Factors that reduce the iron in a biological system will result in the liberation of replacement iron from ferritin stores. However, the mechanism by which the iron leaves the ferritin is not fully understood. Iron is also stored as haemosiderin, which is considered to be a degradation product of ferritin.

Little is known about the storage of other metals. However, it is believed that one of the functions of the small, cysteine-rich, multifunctional proteins called metallothioneins is the storage of ions such as copper(I) and zinc(II). In addition these proteins also readily bind toxic ions such as Cd^{2+} and Hg^{2+} (see section 7.4.3).

7.4.2 Transport

Naturally occurring metal complexes are involved in the transport of both metal ions and ligands. For example, in aerobic bacteria, iron is transported from its storage sites in the form of complexes known as siderophores whereas in mammals monomeric glycoproteins (M~80 000), known as transferrins, are utilised. Both siderophores and transferrins transport the iron as water-soluble Fe(III) complexes.

Two general types of siderophores are known, those based on hydroxamate ligands (ferrichromes and ferrioxamines, Figures 7.11b and 7.11c) and those having catecholate ligands (enterobactin, Figure 7.11a). In both cases there is hard-base to hard-acid complex bonding. In transferrins the Fe(III) ions are strongly ($\beta_n > 10^{20}$) but reversibly bound via the nitrogen and oxygen atoms of amino acid residues. Anions, with carbonate being preferred, are also found bound to the metal. In lactoferrin (Figure 7.11d), the transferrin present in milk, the iron X-ray crystallographic work has shown that the Fe(III) ion occupies a deep cleft in the structure between two protein domains. The Fe(III) ion is six-coordinated in the form of a distorted octahedron, four sites being occupied by amino acid residues and the remaining two by the carbonate or other anions. The function of the carbonate is not known but it is thought that the Fe(III) binds first to the domain where the carbonate is located and this is followed by the second protein domain closing like a hinge to lock the Fe(III) ion in place with a

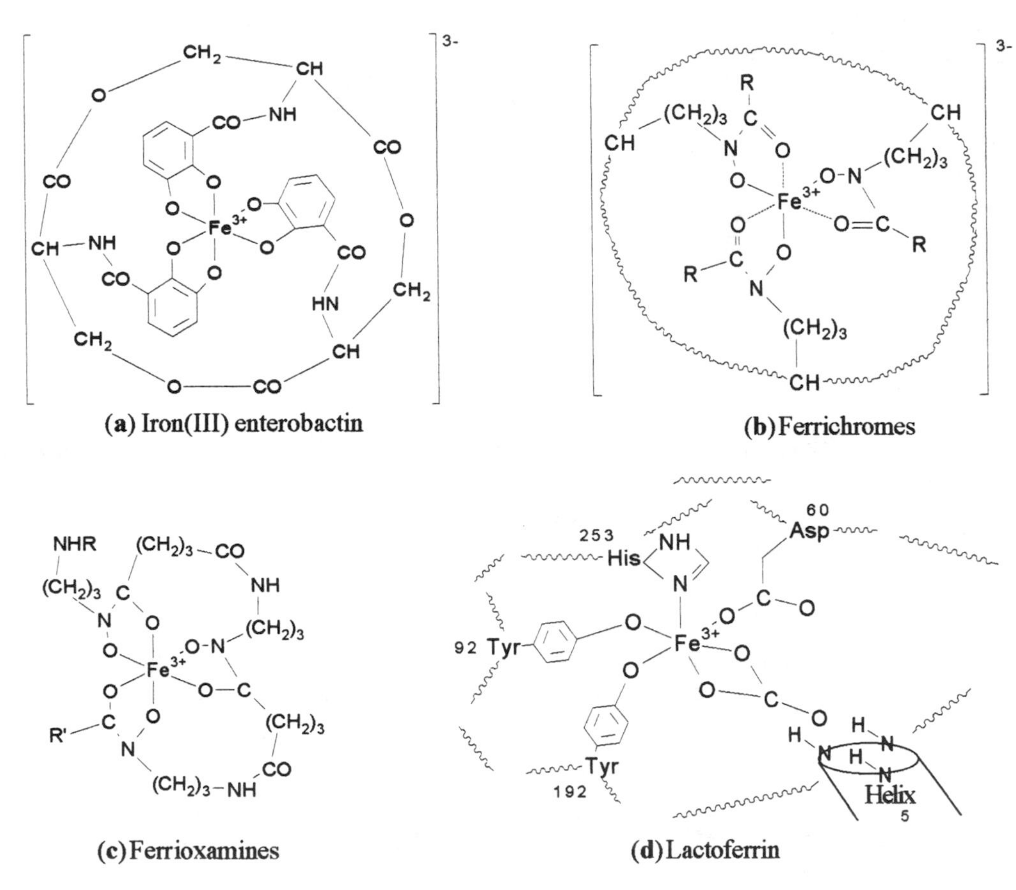

Figure 7.11. Structural formulae for typical siderophores and lactoferrin. ~~~ represents a polypeptide chain. These structures do not represent the three-dimensional shapes of these molecules.

relocation of the coordination. Furthermore, it has been suggested that the release mechanism is essentially the reversal of this process, changes in the conformations of the protein domains resulting in the molecule opening to release the iron.

The transport of other metal ions is also believed to involve complex formation. For example, the metallothioneins are believed to be involved in the transport as well as the storage of metals such as Cu(II), Zn(II) and Cd(II). Copper is also thought to be transported by the blue serum protein ceruloplasmin. However, apart from iron, the area of metal transport in biological systems is not well documented.

Metalloproteins are also involved in the transport of species other than metals. The best known system is the transport of oxygen, carbon dioxide and other compounds by haemoglobin (Hb). Haemoglobin transports oxygen from the lungs and delivers it to myoglobin (My) in the tissues.

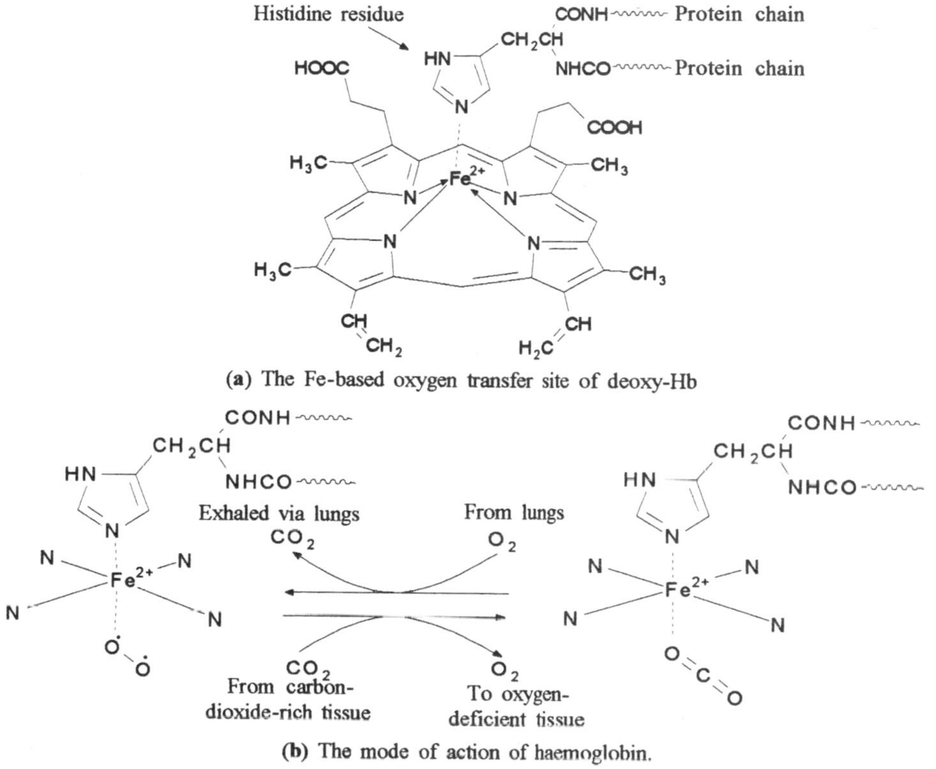

(a) The Fe-based oxygen transfer site of deoxy-Hb

(b) The mode of action of haemoglobin.

Figure 7.12. The structure of the oxygen transfer site and a general outline of the mode of action of haemoglobin. Reproduced from G. Thomas, *Chemistry for Pharmacy and the Life Sciences including Pharmacology and Biomedical Science*, 1996, by permission of Prentice Hall, a Pearson Education Company.

Haemoglobin is a metalloprotein comprising of four subunits, two with α- and two with β-polypeptide chains. Each of these polypeptide chains contains a porphyrin–iron(II) group, which is coordinated to the chain through the imidazole base of a histidine residue (Figure 7.12a). Myoglobin is effectively a monomer of haemoglogin consisting of a 153-residue polypeptide chain attached via a histidine residue to a porphyrin–iron(II) complex. The oxygen binding site in both haemoglobin and myoglobin is the Fe(II)–phorphyrin complex. When oxygen binds to this centre, the structure changes from a flattened five-coordinated pyramid in the deoxy state to a slightly distorted octahedral six-coordinated structure in the oxygenated state (Figure 7.12b). In both haemoglobin and myoglobin, the substituents of the porphyrin ring act like a picket fence, forming a hydrophobic pocket for the oxygen. Furthermore, there is another histidine residue situated in this pocket on the opposite side (distal side) of the oxygen from the Fe(II)-coordinated histidine that is able to hydrogen bond with the coordinated oxygen, thereby stabilising the complex. X-ray data suggest that this distal histidine residue also has the effect of sterically hindering the binding of carbon monoxide to haemo-

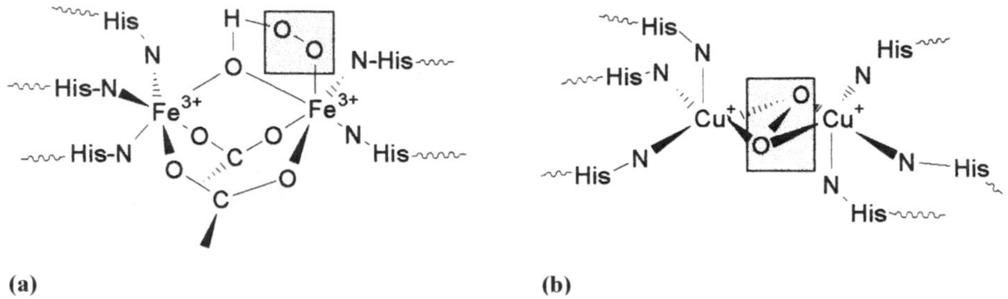

(a) **(b)**

Figure 7.13. The oxygen binding sites of (a) haemerythrins and (b) haemocyanins. The bound oxygen molecules are in the shaded areas.

globin. Because carbon monoxide usually forms more stable complexes with metal ions than other ligands, this appears to be nature's way of discouraging a toxic reaction.

The oxygen transport systems of other species also involve metalloproteins. For example, oxygen transfer systems in marine invertebrate involve haemerythrins (Hr) whereas that in molluscs and arthropods utilises haemocyanins (Hc). In spite of their names the structures of these complexes do not contain a metal–porphyrin group. Haemerythrin depends for its action on an Fe_2 site, the oxygen binding to a vacant coordination site on one of the iron atoms and a hydrogen bonded to the oxygen bridge (Figure 7.13a). In haemocyanin the active centre is a Cu_2 site, the oxygen forming a ligand bridge between the two copper(I) ions (Figure 7.13b).

7.4.3 Detoxification

Detoxification is the process by which a living organism converts unwanted species into harmless substances. Natural metal ion detoxification processes usually remove the metal ions from circulation by forming complexes that are stable at physiological pH but are readily excreted. For example, one of the possible roles of the multifunctional metallothioneins (see section 7.4.1) in mammals is as a natural detoxification agent, although this usage is not universally accepted. The soft sulphydryl groups of the cysteine residues of metallothioneins readily bind to soft metal ions such as Cd^{2+}, Ag^{+}, Hg^{+} and Hg^{2+}, the softer cadmium being able to displace both Zn^{2+} and Cu^{2+} from the protein. Furthermore, cadmium favours low coordination numbers of two and three, which makes it possible for 12–18 cadmium ions to bind to a metallothione molecule. Phytochelatins, a group of cysteine-rich glutathione-related peptides, appear to fulfil a similar detoxification function in plants and some fungi. Therapeutic metal detoxification is discussed in section 7.5.1.

Metalloproteins are also involved in the detoxification of unwanted ligands. For example, the detoxification of superoxide by Fe(III) SOD to oxygen and hydrogen peroxide (Figure 7.1) and the removal of carbon dioxide from tissue by haemoglobin (Figure 7.11).

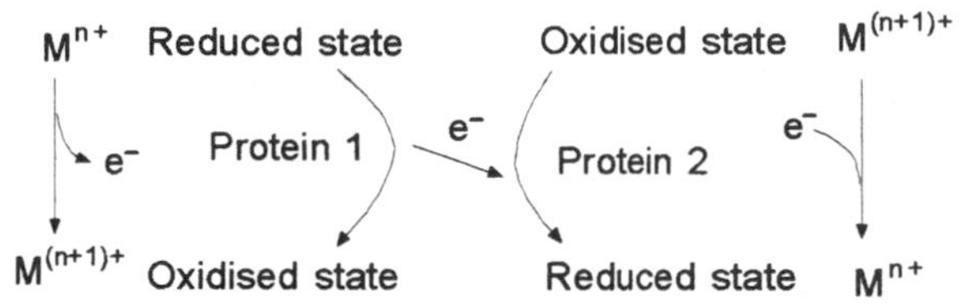

M^{n+} Reduced state
Oxidised state $M^{(n+1)+}$
e^-
Protein 1
e^-
Protein 2
e^-
$M^{(n+1)+}$ Oxidised state
Reduced state M^{n+}

Figure 7.14. A schematic representation of an intermolecular electron transfer process involving metal sites in different proteins.

7.4.4 Electron Transfer

Electron transfer reactions are, in this context, redox reactions in which electrons are either transferred from one protein to another (**intermolecular** transfer) or between two sites in the same protein (**intramolecular** transfer). The proteins undergo these reactions without catalysing changes in any substrate molecules. An electron transfer reaction can lead to a series of consecutive transfer reactions, ultimately activating a protein or enzyme that requires the stimulation of a redox reaction in order to perform its function. The reactions are membrane based, the proteins involved being embedded in a set sequence in close proximity in the membrane. This allows electrons to move from a site in one protein to a site in an adjacent protein. These sites can be up to 3 nm apart.

Metals with multiple oxidation states often constitute the active sites of the proteins. In these cases the process of electron transfer is essentially a redox process in which a change in the oxidation state of a metal in one protein is accompanied by the corresponding change of state in the next adjacent protein. For example, electrons generated by an oxidation process in one protein are transferred to the adjacent protein where they reduce the oxidised state of the metal (Figure 7.14). These oxidation state changes are often accompanied by a change in the geometry of the metallic site, which necessitates the change taking place via a suitable transition state. Iron clusters (Figure 7.7) are often involved in the electron transfer processes in mammals.

Reduction potentials are used as a measure of the system's ability to accept or donate electrons. The larger the positive value of a half-cell system, the greater the tendency to accept electrons, whereas the lower the negative value, the greater the tendency to donate electrons. In other words, a half-system with a high positive value will act as a strong oxidising agent whereas that with a low positive value will act as a strong reducing agent.

7.4.5 Metalloenzymes

One-third of the human body's enzyme systems are metalloenzymes. Their active sites are either coordinated single metal ions or, more commonly, metal clusters. Several active sites with different structures may be located within the same enzyme (see section 7.2.4). Metalloenzymes are classified according to their mode of action (see section 6.2). Each category includes metal clusters with different structures that are capable of bringing about the same substrate changes.

7.4.6 Interaction with Nucleic Acids

Metals are able to bond with DNA in a number of different ways. Their potential coordination sites are the oxygen and nitrogen atoms of the base residues, the hydroxyl groups of the sugar residues and the oxygen atoms of the phosphate residues. However, the main sites of coordination appear to be those found on the base and phosphate residues. Burger has suggested that the affinities of metals for the nitrogen atoms of the bases and nucleoside monophosphate residues follow the decreasing orders:

Bases

Order of affinity of metals for a base:
(dien)Pd > MeHg > Ag(I) > Cu(II) > Ni(II) > Co(II) > Zu(II)~La(III) > Mn(II) > Mg(II)
Order of affinity of bases for metals:
Uridine(N7-)~thymine(N3-)~guanosine(N1-)~inoisne(N1-) > cytidine(N3)~guanosine(N7) > inoisne(N7) > adenosine(N7) = adenosine(N1)

Nucleoside monophosphates

Order of affinity of metals for the monophosphate unit:
La(III) > Cu(II) > (dien)Pd(II) > Zn(II) > Ni(II) > Co(II)~Mn(II) > Mg(II)

The coordination of metal ions to DNA and RNA has been shown to have a number of important physiological and pharmacological actions:

(i) *Structure stabilisation*. Nucleic acid structures are stabilised by Na^+ and Mg^{2+} ions electrostatically bonding to adjacent phosphate residues. This stabilises the structure by preventing the mutual repulsion of the adjacent negatively charged phosphate residues. Potassium ions have been found to stabilise the structures of telomeres, the nucleic acid sequence at the tips of chromosomes that is essential to ensure appropriate DNA replication in cell division.

(ii) *Cofactors*. Divalent ions, such as Mg^{2+}, have been identified as being cofactors for activating catalytic RNA. For example, DNA polymerase is believed to act by Mg^{2+} ions binding the enzyme protein to the nucleoside triphosphate substrate. It is possible that metal ions are also involved in the selectivity of the system.

(iii) *Toxic effects*. The coordination of metals such as lead and mercury to nucleic acids can result in a toxic response.

(iv) *Drug action*. Coordination of metals to nucleic acids can result in a therapeutic effect, for example, antiviral or anticarcinogenic activity (see *cisplatin*, section 7.5.2).

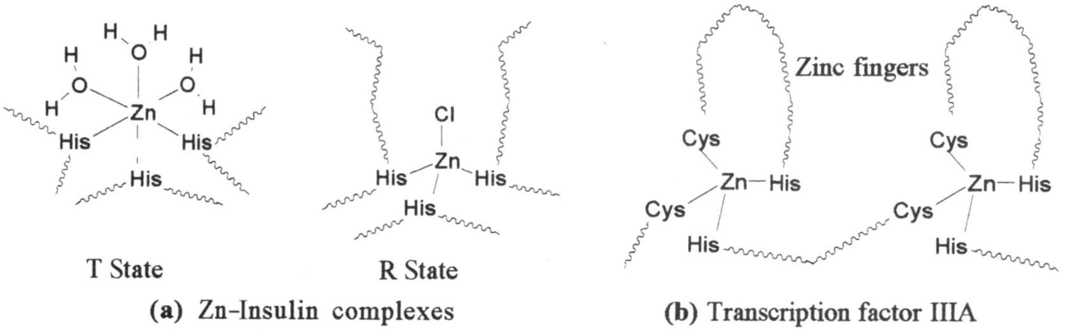

T State R State
(a) Zn-Insulin complexes
Zinc fingers
(b) Transcription factor IIIA

Figure 7.15. Examples of the influence of zinc on the structures of proteins.

7.4.7 Structural Role

All metal ions have a structural role in metalloproteins. Their coordination number determines the configuration of their area of the structure of the protein. For example, a metal with a coordination number of four is likely to have either an approximately tetrahedral or square planar arrangement for its ligands. In addition, metal ions influence the conformations of the adjacent peptide chains. Zinc ions are particularly interesting in that they appear to act as both intermolecular and intramolecular bridges in protein structures. For example, Zn^{2+} ions act as intermolecular bridges in the oligomeric storage forms of some hormones. The dimeric form of the human growth hormone (hGH) is due to the coordination of two zinc centres with histidine and glutamine residues. Similarly the peptide chains of the trimeric forms of insulin are connected by zinc ions. In the T and R insulin trimers, which have been identified as the storage forms of insulin in the pancreas, the zinc is bound to three histidine residues, one from each insulin molecule. However, in the T state the zinc is six-coordinated (approximately octahedral), the remaining ligands being water, whereas in the R state it is four-coordinated (approximately tetrahedral), the fourth ligand thought to be a chloride ion (Figure 7.15a). Intramolecular Zn^{2+} bridges have been found in the transcription factor IIIA (TFIIIA) from the African clawed toad *Xenopus* and proteins from other species. The zinc ion is responsible for the formation of the peptide chain of the protein into a loop of about 12–15 amino acid residues referred to as a zinc finger. The chain is coordinated to the zinc in an approximately tetrahedral configuration through cysteine and histidine residues (Figure 7.15b). This type of zinc finger is classified as a C_2H_2 type in order to distinguish them from the C_x type, where x is usually 4, 5 or 6, in which only cysteine residues are coordinated to the zinc. Zinc fingers can occur either singly or in tandem, with a number of fingers being found in one protein molecule.

7.5 Therapeutic Uses

Complexes and complexing agents are being used on an expanding scale to treat a variety of diseases. This aspect of medicinal chemistry is increasing in importance with advances in the knowledge of the role of metals in the physiological and pathological states of the body.

7.5.1 Metal Poisoning

The presence of excessive amounts of a metal in a living organism is responsible for a variety of syndromes. For example, Wilson's disease, symptoms of which are a malfunctioning liver, neurological damage and brown or green rings in the cornea of the eyes, is caused by a copper overload due to a genetically inherited metabolic defect. Excessive amounts of calcium results in calcification of tissue, cataracts, kidney stones and gallstones.

Metal poisoning occurs when the body's metal management system allows the concentration of the metal to reach toxic levels in sensitive areas of the body. The metal can enter the organism in a number of ways ranging from accidental ingestion, pollution of the food chain, skin absorption and breathed in as atmospheric pollutants. Treatment is based on the use of chelating agents that form stable complexes with the excess metal and are either easily excreted or deposited as harmless solids. However, it is possible that a second line of attack based on identifying the natural detoxification process for the metal will emerge.

The toxic effects of a metal are believed to be due to the formation of a stable complex between the metal and a species that is an essential component of a biological pathway in a living organism. The resulting complexes are unable to take part in the pathway and so its efficiency is reduced. Consequently, the organism develops the diseased state associated with toxic levels of the metal. However, these toxic levels can be reduced by the use of chelating agents to form stable complexes that are easily excreted (Figure 7.16). For this form of therapy to be effective the metal must form more stable complexes with the chelating agent than the naturally occurring ligands it coordinates with in the living organism. In other words, the stability formation constant (β_1) for the formation of the metal–chelate (L'M) must be greater than the stability constant (β_2) for the formation of the metal–natural ligand (LM). Furthermore, the chelating agent must also contain good water-solubilising groups (see section 3.7) as well as suitable ligands.

Chelating agents are used instead of monodentate ligands because their complexes tend to be more stable than those formed by monodentate ligands. For example, the stability formation constant $\log\beta$ value for $[Cu(NH_3)_4]^{2+}$ is 11.9 whereas that for the CuEDTA complex is 17.7. In addition, experimental evidence indicates that the stability of complexes usually increases with the number of bonds formed by the chelating agent to the metal. Furthermore, chelating agents that form five- and six-membered rings will form the most stable complexes. Consequently,

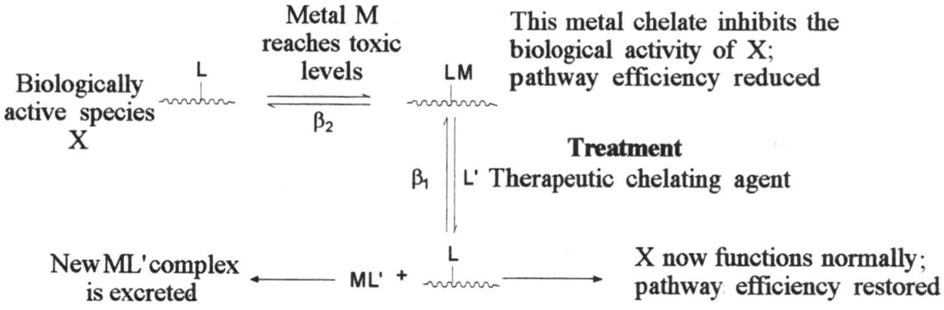

Metal M reaches toxic levels

Biologically active species X — L — β_2 ⇌ — LM — This metal chelate inhibits the biological activity of X; pathway efficiency reduced

β_1 L' Therapeutic chelating agent

Treatment

New ML' complex is excreted ← ML' + L → X now functions normally; pathway efficiency restored

Figure 7.16. A schematic representation of the general action of chelating agents in the treatment of metal poisoning. The addition of the chelating agent L' restores the pathway to full operation.

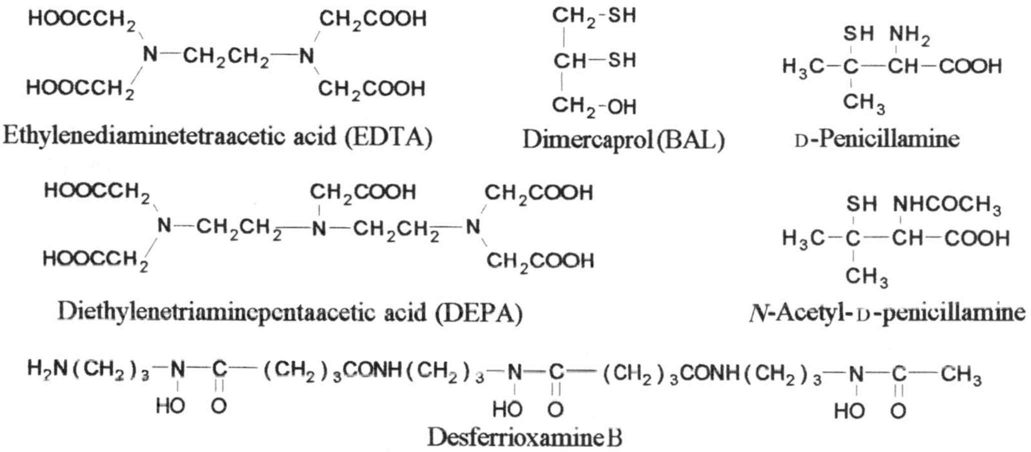

$(HOOCCH_2)_2N-CH_2CH_2-N(CH_2COOH)_2$

Ethylenediaminetetraacetic acid (EDTA)

$CH_2(SH)-CH(SH)-CH_2OH$

Dimercaprol (BAL)

$H_3C-C(SH)(CH_3)-CH(NH_2)-COOH$

D-Penicillamine

$(HOOCCH_2)_2N-CH_2CH_2-N(CH_2COOH)-CH_2CH_2-N(CH_2COOH)_2$

Diethylenetriaminepentaacetic acid (DEPA)

$H_3C-C(SH)(CH_3)-CH(NHCOCH_3)-COOH$

N-Acetyl-D-penicillamine

$H_2N(CH_2)_3-N(OH)-C(=O)-(CH_2)_3CONH(CH_2)_3-N(OH)-C(=O)-(CH_2)_3CONH(CH_2)_3-N(OH)-C(=O)-CH_3$

Desferrioxamine B

Figure 7.17. Examples of the chelation agents used to treat metal poisoning.

a compound that is to be used as a therapeutic chelating agent should have the following characteristics:

(i) ligand groups that are specific for the metal;
(ii) be a multiligand chelating compound;
(iii) form complexes that are more stable than the relevant naturally occurring ligands;
(iv) be easily excreted;
(v) have an LD_{50} greater than $400\,mg\,kg^{-1}$.

Chelating agents commonly used to treat cases of metal poisoning are ethylenediaminetetraacetic acid (EDTA), dimercaprol and penicillamine (Figure 7.17), the latter being a degradation product of penicillin. The structure of EDTA contains hard amine and carboxylate groups, which means that it readily coordinates with hard metals such as calcium and magnesium. Both of these elements are essential components of living organisms and their depletion would result in the malfunction of a number of biological processes as well as weakening of the bone structure of mammals. Consequently, EDTA is normally used in the form of its Na_2Ca(EDTA) salt in an attempt to reduce the Ca^{2+} depletion. Dimercaprol (British anti-Lewisite, BAL) has the soft sulphydryl groups and so coordinates soft metal ions such as Cd^{2+}, Hg^{2+} and Cu^{+}. However the discovery of a natural detoxification process in bacteria resistant to mercury poisoning may result in a new genetic method of treatment for mercury poisoning. Penicillamine has both hard and soft ligand groups. It is used in the treatment of Wilson's disease and in chronic cases of lead and mercury poisoning. The siderophore desferrioxamine, which contains hard ligand groups, has been used to treat iron overload.

A number of compounds with similar types of active structural areas have been developed from the led compounds described in the previous paragraph. For example, EDTA has led to the development of polyaminocarboxylic acids such as DEPA (diethylenetriaminepentaacetic acid, Figure 7.17), which has been shown to be effective in cases of plutonium poisoning.

The toxicity and difficulty of administration of BAL has resulted in the use of sodium 2,3-dimercaptosuccinate and unithiol (Figure 7.18). Complexes formed by these chelating agents are charged in solution. Consequently, these complexes are less likely to cross biological membranes (see section 4.3.3) and be distributed around the body, which reduces the possibility of unwanted side effects. However, these charged complexes will accumulate in the liver and kidneys, which enhances their chances of excretion.

The toxicity of D-penicillamine has resulted in the use of the less toxic *N*-acetyl-D-penicillamine. This drug is very effective in treating methylmercury(II) poisoning and will extract this compound from brain tissue.

$$\mathrm{NaOOC{-}CH(SH){-}CH(SH){-}COONa} \rightleftharpoons \mathrm{{}^{-}OOC{-}CH(SH){-}CH(SH){-}COO^{-}} + 2Na^{+}$$

Sodium 2,3-dimercaptosuccinate

$$\mathrm{NaO_3SCH_2{-}CH(SH){-}CH_2SH} \rightleftharpoons \mathrm{{}^{-}O_3SCH_2{-}CH(SH){-}CH_2SH} + Na^{+}$$

Unithiol

Figure 7.18. Sodium 2,3-dimercaptosuccinate and unithiol. The thiol groups have a high affinity for metals.

Chelating agents can cause worse problems than they are designed to cure. For example, both BAL and penicillamine form complexes with cadmium that are more toxic than the metal that they are being used to remove. Furthermore, some ligands will form complexes that prevent the metal being excreted, that is, they effectively trap the metal in the body. Consequently, it is important to consider this aspect of chelation when designing clinical trials of new chelating agents.

7.5.2 Anticancer Agents

One of the best known uses of complexes is that of cisplatin in cancer therapy (Figure 7.19). This drug was discovered as a consequence of an investigation into the effects of electric fields on the growth of *Escherichia coli* bacteria. Barnett, Rosenberg and collaborators observed that normal cell division was inhibited and cells grew into long filaments. Eventually it was discovered that the cause of this growth was due to *cis*-diamminedichloroplatinum(II) (cisplatin) and *cis*-diamminetetrachloroplatinum(IV) generated *in situ* from the platinum electrodes and ammonium chloride in the solution used in the original study. Cisplatin was a well-documented compound and animal testing showed that it was active against testicular, cervical, ovarian, lung and other cancers. Consequently, drug development has focused on this compound and its analogues. However, it is not active against all forms of cancer.

The mechanism by which cisplatin acts is not fully elucidated. In the extracellular fluid cisplatin undergoes little chloride–water interchange because of the fluid's high chloride concentration. However, once the drug penetrates the cell, the chloride concentration is low enough to allow a significant chloride–water interchange with the formation of species such as

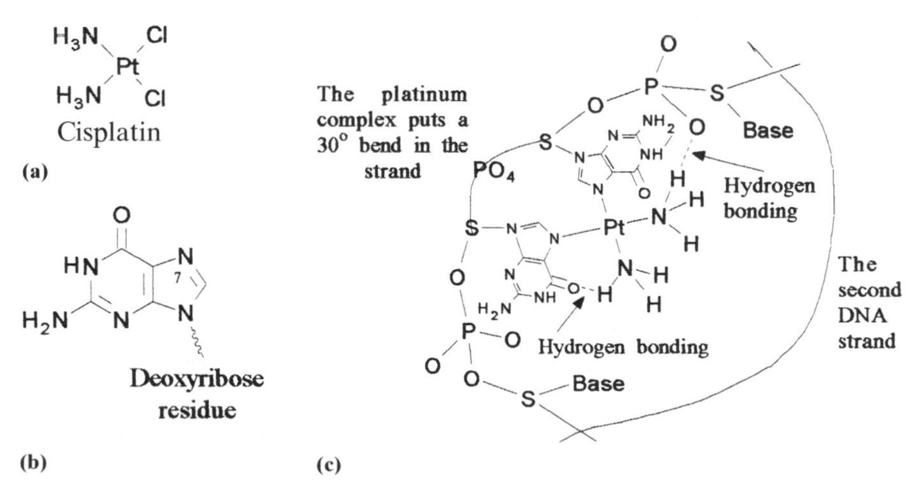

Figure 7.19. (a) Cisplatin, (b) a guanine residue and (c) the proposed structure of the cisplatin–DNA complex. An outline structure is given for guanine and S represents a sugar residue.

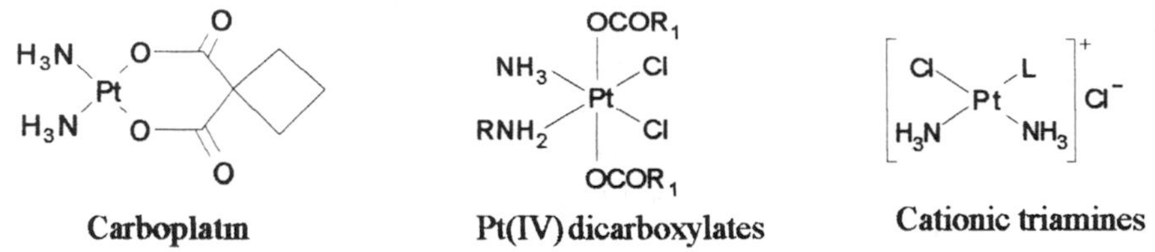

Carboplatin Pt(IV) dicarboxylates Cationic triamines

Figure 7.20. Examples of platinum complexes that are active against cancer. Carboplatin is less toxic than cisplatin and is now in clinical use. Pt(VI) dicarboxylate analogues are active against a number of ovarian carcinoma cell lines. Cationic triamines, where L is pyridine, a *para*-substituted pyridine, primidine or purine, are active against S 180 ascites and L 1210 tumours in mice.

$[Pt(NH_3)_2(H_2O)_2]^{2+}$ and $[Pt(NH_3)_2(OH)_2]$. Practical evidence suggests that these species coordinate with either the N-7 atoms of two guanine residues or the N-7 of a guanine residue with an adenine residue, to form intrastrand bridges linking two areas of the same strand (Figure 7.19).

This causes a distinct bend in the DNA at the point of platination, which leads to a supression of replication. The mechanism by which the formation of these complexes leads to the death or suppression of reproduction of the cancer cell is not fully understood but has been shown to involve proteins known as structure-specific recognition proteins (SSRPs). It is not certain how these proteins facilitate the anticancer activity of cisplatin but they appear to bind to the area of the DNA containing the platinum adduct.

Transplatin also binds to DNA but less is known about the nature of this binding. However, transplatin has also been shown to inhibit DNA replication but it has not been used clinically.

Cisplatin suffers from the serious disadvantage that it must be administered by intravenous infusion and it is very toxic. Its use can cause nausea, vomiting, renal dysfunction and leucopenia. In addition, some tumours become resistant to the drug and so researchers are actively seeking second-generation platinum(II) and platinum(IV) analogues with less severe side effects (Figure 7.20).

The importance of the discovery of cisplatin has stimulated the investigation of other metal complexes for antitumour activity. A number of compounds of main group (Ga, Ge, Sn and Zn) and transition metals (Ti, V, Cr, Mo, Mn, Fe, Cu and others) have been reported to have antitumour activity (Figure 7.21). However, few compounds have been studied in depth because investigations have been dominated by platinum coordination compounds.

Although metal complexes are being used as antitumour agents they can also cause cancers. For example, nickel carbonyl ($Ni(CO)_4$) is one of the most carcinogenic compounds known to man. Consequently, this aspect of the pharmacology of metal complexes must be borne in mind when designing clinical trials for complexes.

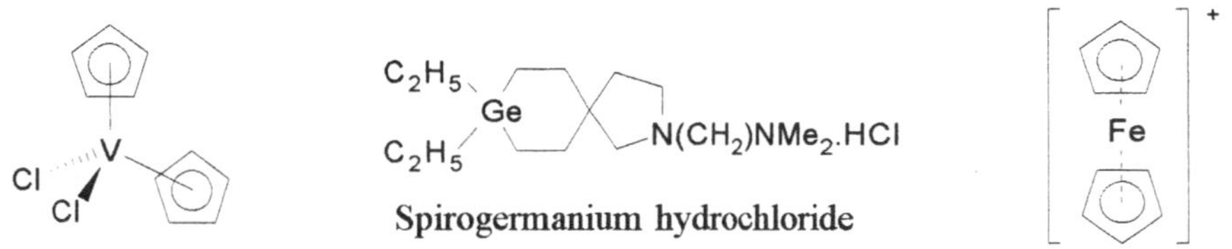

Vanadocene dichloride

Spirogermanium hydrochloride

Ferrocene salts

Figure 7.21. Examples of the complexes of other metals that have antitumour activity. Vanadocene dichloride and its titanium, molybdeum, tungsten, niobium and tantalum analogues exhibit antitumour activity. Ferrocene salts can exhibit activity against animal tumours.

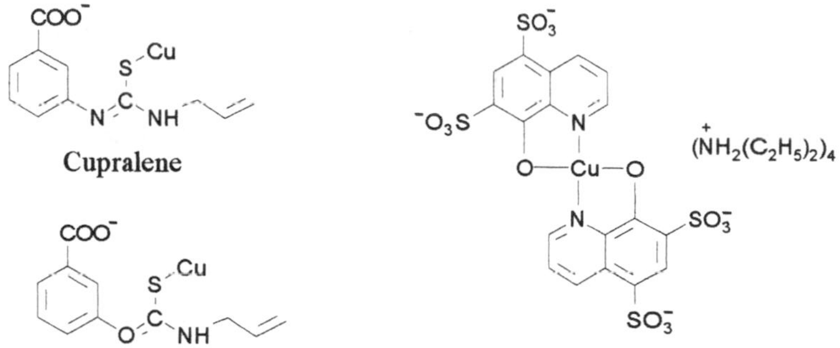

Cupralene

Alcuprin

Dicuprene

Figure 7.22. Examples of copper complexes with antiarthritic activity.

7.5.3 Antiarthritics

Rheumatoid arthritis affects over 5% of the population of the Western world. Its cause is unknown but it is thought that it results from a failure of the patient's immune system. Practical evidence also suggests that rheumatoid arthritis is related to a *local* imbalance in the concentration of copper. However, the relationship between the copper concentration and rheumatoid arthritis is still not clear.

A number of copper complexes have been found to be active against rheumatoid arthritis (Figure 7.22). This antiarthritic activity is believed to be due to their anti-inflammatory action. Penicillamine has also been used to treat rheumatoid arthritis even though it increases the rate of excretion of copper by urinary excretion (see section 7.5.1). It is thought that its antiarthritic activity is due to it mobilising the copper in the form of a complex that temporarily accumulates in the tissues, thereby reducing any inflammation.

Since 1940 a number of gold(I) thiolates have been used to treat rheumatoid arthritis (Figure 7.23), although little is known about their mode of action. However, they are slow to act and it can be several months before any beneficial effects are noticed. Furthermore, with the

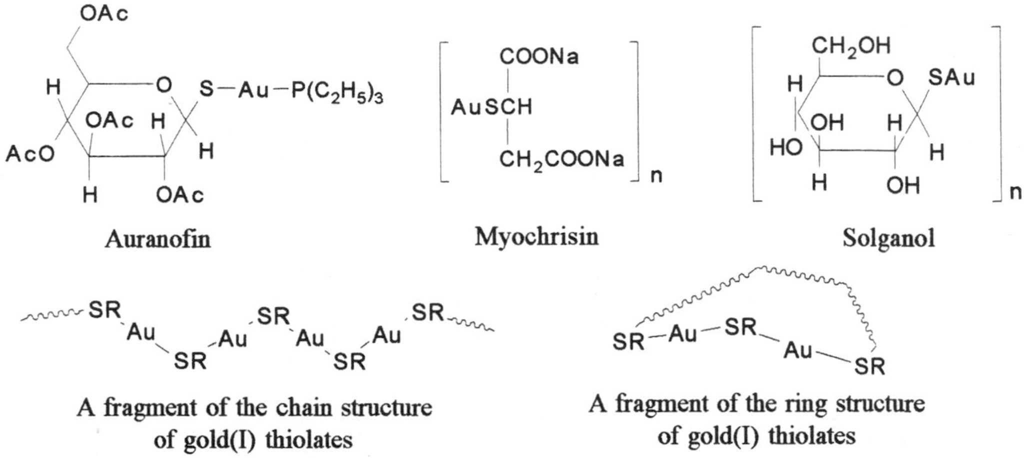

Auranofin

Myochrisin

Solganol

A fragment of the chain structure of gold(I) thiolates

A fragment of the ring structure of gold(I) thiolates

Figure 7.23. Gold(I) compounds used to treat rheumatoid arthritis. Experimental work has indicated that the 1:1 thiolates are small chain or ring polymers.

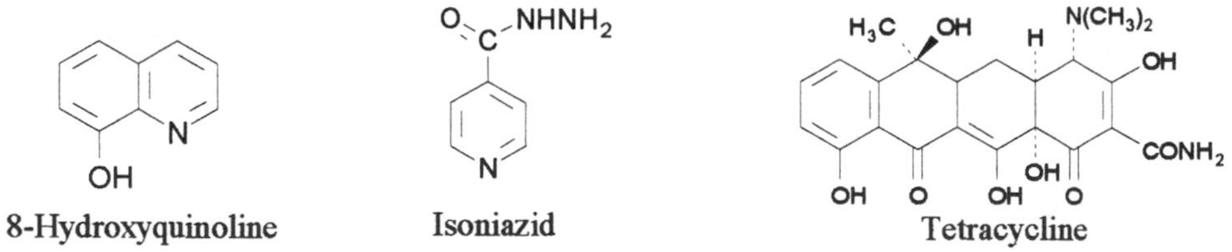

8-Hydroxyquinoline

Isoniazid

Tetracycline

Figure 7.24. Drugs that are believed to owe part of their action to chelation.

exception of auranofin, which is administered orally, they are given by a painful intramuscular injection.

7.6 Drug Action and Metal Chelation

The activities of a number of drugs can be related to their ability to chelate essential metals. For example, the action of 8-hydroxyquinoline (Figure 7.24) appears to be partly based on its ability to chelate iron. Analogues of 8-hydroxyquinoline that are unable to form chelates exhibit either a much reduced or zero activity. The activity of isoniazid is also attributed to its ability to chelate iron. However, the mode of action of the tetracyclines is possibly based on their chelation with magnesium. The magnesium is believed to form a bridge binding the tetracycline to the r-RNA in the bacteria. This is thought to halt the protein synthesising action of the bacterial ribosomes. Because most drugs contain potential ligand groups, it is possible that many compounds owe part of their pharmacological activity to complex formation.

7.7 Summary

Complexes are compounds with a metal ion to which are bonded species known as **ligands**. The total number of atoms forming bonds with the central metal ion is known as the **coordination number** of the metal ion. It is a characteristic property of the complex.

Ligands can be electrically neutral or charged species. They may be classified as:

(i) **bidentate, tridentate, tetradentate, etc.** depending on the number of atoms that can form with the metal ion;
(ii) **one, two, three, etc. electron donors** according to the number of electrons they supply to the metal ion;
(iii) **simple ligands**, which form a simple covalent bond to the metal, or **π-acceptor ligands**, which are also bonded by a synergic back-bond from the metal.

The number of atoms, n, of a **multidentate ligand** that donate electrons to the metal ion is shown in the **name and formula** of the ligand by η^n. **Chelating agents** are multidentate ligands that form ring structures with the central metal ion. Ligands can also form bridges linking two or more metal ions. The presence of a **ligand bridge** is indicated by the use of μ_n, where n is the number of ions linked by the bridge in either the formula or name of the complex.

Complexes may also contain **metal-to-metal bonds**. Structures in which several metal ions are found in relatively close proximity are known as **metal clusters.** Metal clusters may or may not contain metal-to-metal bonds.

The **stability of complexes** may be measured in terms of either the stepwise equilibrium constant (K_n), where n is the number of the steps, or the overall equilibrium constant (β) for their formation. **Formation stability constants** are recorded as the log of the relevant equilibrium constant: the larger the value, the more stable the complex. The formation stability constant gives a measure of the affinity of the ligand for the metal.

The affinity of ligands for metals and the relative strengths of the bond between the ligand and the metal may be predicted by the use of the concept of **hard and soft acids and bases**. In this approach, all complexes are formed by the acidic metal reacting with the basic ligand. Metals and ligands are classified as being either 'hard' or 'soft'. **Hard species** have electrons that are firmly held (lacking in mobility) whereas **soft species** have electrons that are easily donated (mobile). Strong bonds are formed between similar types of species, that is, hard with hard and soft with soft.

Metalloproteins are proteins containing metal ions as part of their structure whereas **metalloenzymes** are metalloproteins that function as enzymes. Both metalloproteins and metalloenzymes are involved in a wide variety of biological processes, such as metal storage, metal and ligand transport and enzyme action. The activity of metalloproteins is often linked to metal clusters in the structure.

Chelating agents are used to treat a variety of **diseases**, such as metal poisoning, cancer and rheumatoid arthritis. However, a xenobiotic whose structure contains ligand groups can upset the balance of metal ions in biological systems by forming complexes that are more stable than the naturally occurring complexes found in the biological system. This can lead to deficiency diseases.

7.8 Questions

(1) Outline, quoting relevant examples, the various uses of metallocomplexes in maintaining health in humans.

(2) Draw structural formulae for each of the following ligands. Classify the ligand in terms of its electron-donating power or the number of atoms bonding to the metal atom: (a) diene, (b) Cp, (c) EDTA and (d) ammine.

(3) Draw the structural formulae of each of the following compounds. Indicate on the formula the atom most likely to coordinate with a metal ion: (a) cysteine, (b) tyrosine, (c) 8-hydroxyquinoline and (d) dimercaprol.

(4) Explain, using suitable examples, the meaning of the terms (a) synergic bonding, (b) ligand bridge and (c) metal cluster in the context of the structures of complexes.

(5) Explain the meaning of the terms hard and soft acids and bases. Predict, using the concept of hard and soft acids and bases, whether it is possible for the following pairs of species to form a stable complex: (a) cysteine and Pt(II), (b) carbon monoxide and tin(IV), (c) magnesium(II) and benzene, (d) copper(I) and sodium ethanethiolate, (e) tin(II) and histidine, (f) cholesterol and iron(III), (g) cadmium and glutamic acid and (h) mercury(II) and acrylic acid. Indicate, in the case of stable complex formation, where the ligand coordinates with the metal ion.

(6) Calcium ions are known to initiate blood clotting. Suggest a compound that could be added to blood to prevent blood samples from clotting. Explain how the compound prevents blood clotting.

(7) List the general features that a compound should exhibit if it is to be suitable for use as a metal detoxification agent.

(8) The stability formation constants of a number of EDTA–metal complexes are given in Table 7.5.

   (a) What would be the effect of adding 0.01 mmol of EDTA to a solution containing 0.5 mmol of copper(II) and 0.01 mol of calcium(II) ions?

   (b) The Cr–EDTA complex is highly toxic. State whether EDTA is suitable to use as a heavy metal antidote for lead when a patient has accidentally ingested a solution containing chromium(II), mercury(II) and lead(II) ions.

Table 7.5. The log stability constants of some EDTA complexes.

Metal	Log β	Metal	Log β
Calcium(II)	10.6	Iron(III)	25.1
Chromium(II)	13.0	Lead(II)	17.0
Copper(II)	18.0	Mercury(II)	21.0
Iron(II)	14.3	Zinc(II)	16.5

(c) Explain the consequences of adding 0.05 mmol of EDTA to a solution containing a mixture of 0.05 mmol of iron(II) and 0.05 mmol of iron(III) ions. Why would the addition of excess zinc(II) ions to the solution have no effect on the concentration of iron ions in the mixture?

(9) Suggest a series of experiments to demonstrate that chelation with iron(III) may play a part in the action of a potential drug on *Staphylococcus aureus*. Practical details are not required.

8 Receptors and Messengers

8.1 Introduction

Receptors are specific areas of certain proteins and glycoproteins that are found either embedded in cellular membranes or in the nuclei of living cells. Any endogenous or exogenous chemical agent that binds to a receptor is known as a **ligand**. The general region on a receptor where a ligand binds is known as the **binding domain**. It should be remembered that the term domain is used to indicate an area of a biomacromolecule that is linked to a specific function of that molecule.

The binding of a ligand to a receptor may cause either a positive or negative biological response. For example, positive responses can result in either an immediate physiological response, such as the opening of an ion channel, or lead to a series of biochemical events that may result in the release of so-called secondary messengers (see section 8.4.2), such as cyclic adenosine monophosphate (cAMP). These secondary messengers promote a sequence of biochemical events that result in an appropriate physiological response (Figure 8.1). Alternatively, the binding of the ligand to the receptor may prevent a physiological response by either initiating the inhibition of an associated series of biological events (see section 8.4.2) or simply preventing the normal endogenous ligand from binding to a receptor. For example, β-blockers such as propranolol act by blocking the β-receptors for adrenaline. In all relevant cases, the mechanism by which any message carried by the ligand is translated through the receptor system into a tissue response is known as **signal transduction**.

Cyclic adenosine monophosphate (cAMP)

Propranolol (the *S* isomer is 100 times more active than the *R* isomer)

The binding of a drug to a receptor either inhibits the action of the receptor or stimulates the receptor to give the physiological responses that are characteristic of the action of the drug. Drugs that bind to a receptor and give a similar response to that of the endogenous ligand are known as **agonists** whereas drugs that bind to a receptor but do not cause a response are termed **antagonists**. However, drugs are not the only xenobiotic that can bind to a receptor; viruses,

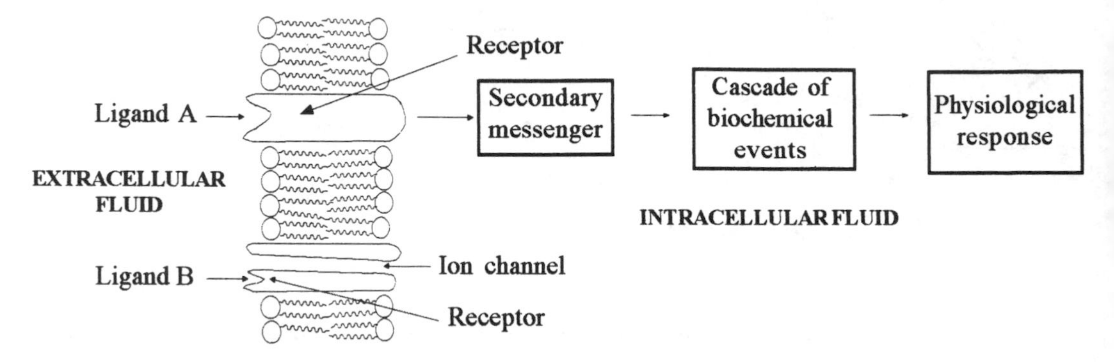

Figure 8.1. A schematic representation of some of the effects of ligands binding to receptors. The binding of ligand A to the extracellular surface of the receptor results in the activation of a secondary messenger, which is followed by a cascade of biochemical events and the appropriate physiological response. The binding of ligand B causes the ion channel to open.

bacteria and toxins may bind to the receptor sites of specific tissues, which could cause unwanted pharmacological effects to occur.

8.2 The Bonding of Ligands to Receptors

The forces binding ligands to receptors cover the full spectrum of chemical bonding, namely: covalent bonding, ionic bonding and dipole–dipole interactions of all types, including hydrogen bonding, charge-transfer bonding, hydrophobic bonding and van der Waals' forces (Figure 8.2). The bonds between the ligand and the receptor are assumed to be formed spontaneously when the ligand reaches the appropriate distance from its receptor for bond formation. To achieve this situation the ligand is transported by either diffusion or a transport protein to the receptor.

Covalent bonding is by far the strongest form of bond between a ligand and a receptor. It usually forms an irreversible link between the drug and the receptor. Consequently, except in cancer therapy and the inhibition of certain enzymes, covalent bond formation is seldom found in drug action. Ionic or electrostatic bonding is an important form of bonding between ligands and receptors because many of the functional groups on the receptor will be ionised at physiological pH. Ionic interactions are usually reversible. They are effective at distances that are considerably greater than those required by other types of bonding but their strength is inversely proportional to the square of the distance between the charges (the inverse square law). Electrostatic attractions in the form of ion–dipole attractive forces, dipole–dipole inter-

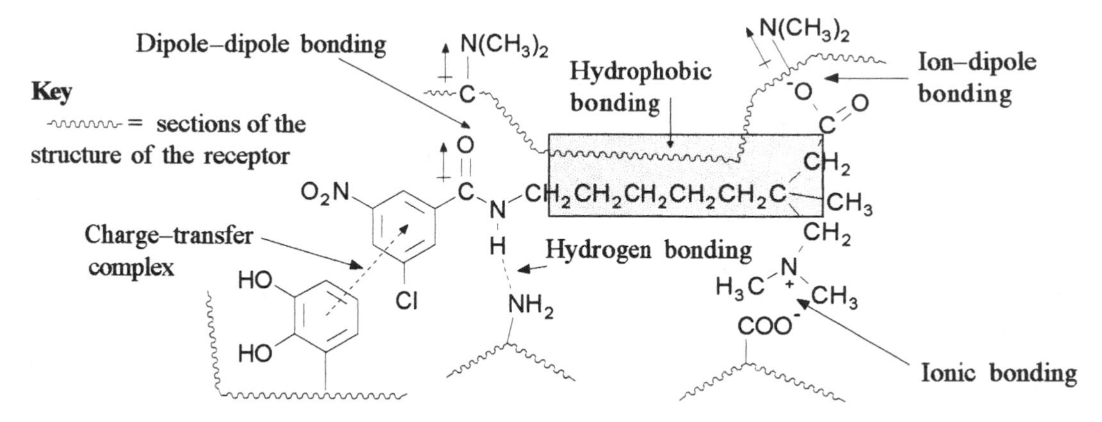

Figure 8.2. Theoretical examples of some of the common forms of bonding, excluding van der Waals' forces, found in drug–receptor interactions.

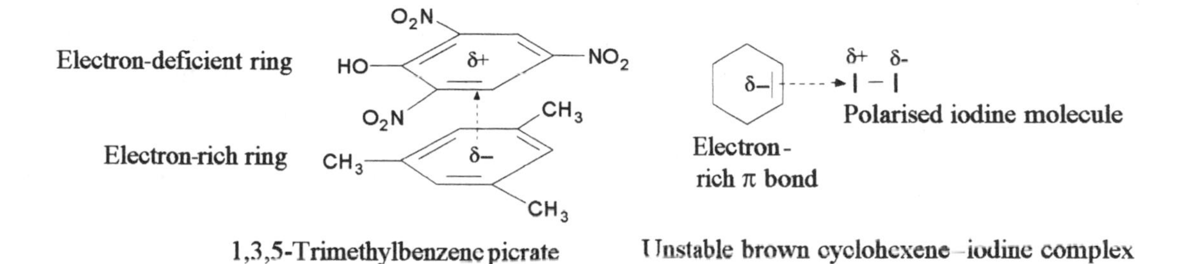

Figure 8.3. Examples of charge-transfer compounds.

actions and hydrogen bonding usually form weaker bonds than ionic bonds. However, they are important because of the large numbers of these bonds that are usually formed between the drug and the receptor.

Charge-transfer complexes may also be formed when an electron donor group is adjacent to an electron acceptor group. In this situation experimental evidence suggests that the donor may transfer a portion of its charge to the acceptor. As a result, one compound becomes partially positively charged with respect to the other and a weak electrostatic bond is formed (Figure 8.3). The precise nature of this type of bonding has not been fully elucidated but donors are usually π-electron-rich species and chemical moieties with lone pairs of electrons.

Hydrophobic bonding is a very weak form of bonding that occurs when non-polar sections of molecules are forced together by a lack of water solubility (Figure 8.4a). The precise nature of

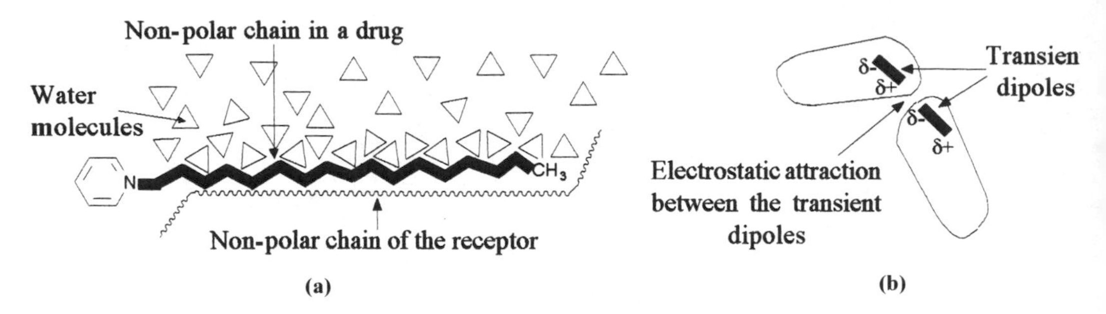

Figure 8.4. Hypothetical examples of (a) hydrophobic bonding and (b) London dispersive forces.

hydrophobic bonding is not known but the formation of hydrophobic bonds leads to a fall in the energy of the system and a more stable structure. Some workers do not believe that hydrophobic effects exist.

London dispersive forces may also contribute to the bonding between a drug and a receptor. These forces are very weak dipole–dipole interactions due to the formation of transient dipoles within a structure. Transient dipoles are time dependent. They arise because the electron distribution in a molecule varies with time, giving an uneven distribution of electrons that results in a temporary charge distribution and a transient dipole within the molecule (Figure 8.4b).

The binding of many drugs to their receptors is by weak reversible interactions:

$$\text{Drug} + \text{Receptor} \rightleftharpoons \text{Drug–Receptor} \quad (8.1)$$

This means that the binding of a drug to its receptor is concentration dependent. As the concentration of the ligand in the extracellular fluid increases, the equilibrium of Equation (8.1) moves to the right and the drug will bind to the receptor. However, when the concentration of the drug in the extracellular fluid falls, the equilibrium will move to the left and the drug–receptor complex will dissociate. Consequently, drugs and endogenous ligands become ineffective as soon as their concentrations fall below a certain limit, because an insufficient number of receptors are being activated by these ligands. Endogenous reduction of drug concentration is brought about by metabolism and excretion. Consequently, both of these processes will have a direct bearing on the duration of action of a drug.

Drugs that form strong bonds with their receptors do not readily dissociate from the receptor when their concentrations in the extracellular fluid fall. Consequently, drugs that act in this manner will often have a long duration of action. This is a particularly useful attribute for drugs used in the treatment of cancers where it is particularly desirable that the drug forms irreversible bonds to the receptors of tumour cells.

8.3 Structure and Classification of Receptors

The majority of the known receptors are proteins and glycoproteins, which are either embedded in the lipid membranes of cells or are found in the nuclei of living cells. The first receptors were isolated in the 1970s. Their isolation was difficult and usually involved disruption of the lipid membranes by detergents. This method of isolation inevitably resulted in conformational changes and this, together with the minute quantities isolated, yielded little structural information except for the primary sequence of amino acid residues. The minute amount of receptor proteins isolated from human cells was too small to allow the use of X-ray crystallography until it was discovered that many algae had receptors that were similar to the human receptors. These plants could be grown in large amounts, which enabled the isolation of a sufficient quantity of the receptor protein for X-ray crystallography studies. These studies resulted in the determination of the three-dimensional structures of many of these plant receptors and, by analogy, those of similar human receptors. Furthermore, recent advances in molecular biology coupled with a knowledge of the sequence of the amino acid residues of a receptor have now enabled microbiologists to produce receptor molecules by cloning techniques (see section 10.13.1). Investigation of these cloned receptors has yielded a great deal of information concerning the chemical structures of receptors.

Receptors are classified according to function into four so-called **superfamilies** of receptors (Table 8.1). The members of a superfamily will all have the same general structure and general mechanism of action. However, individual members of a superfamily tend to exhibit variations in the amino acid residue sequence in certain regions and also in the sizes of their extracellular and intracellular domains.

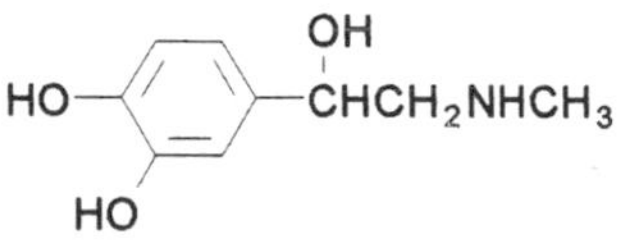

Adrenaline (Epinephrine)

Acetylcholine (ACh)

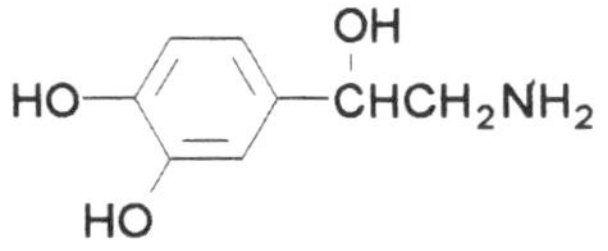

Noradrenaline (Norepinephrine)

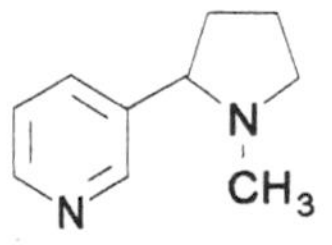

Nicotine

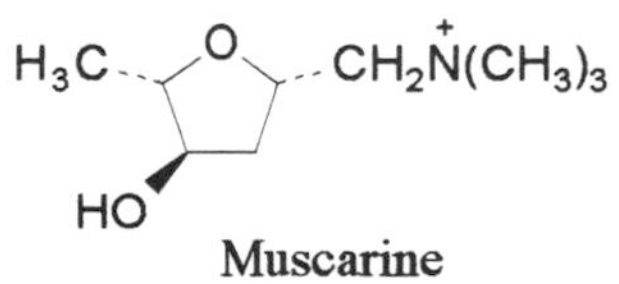

Muscarine

Each of the superfamilies is subdivided into a number of types of receptor whose members are usually defined by their endogenous ligand. For example, all receptors that bind acetylcholine (ACh) are of the cholinergic type and those that bind adrenaline and noradrenaline

Table 8.1. The four receptor superfamilies. The rectangles represent α-helices and the single lines represent polypeptide chains.

Family	Endogenous ligands	Examples	General structure
1.	Fast neurotransmitters	nAChR, $GABA_A$ receptor, glutamate receptor	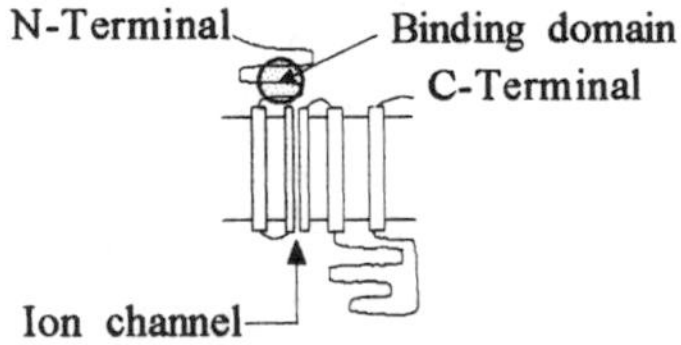 N-Terminal Binding domain C-Terminal Ion channel Receptors consist of four or five subunits of this type with a total of 16–20 membrane-spanning domains
2.	Hormones and slow transmitters. The receptor is coupled to the effector system by G-protein	mAChR and noradrenergic receptors	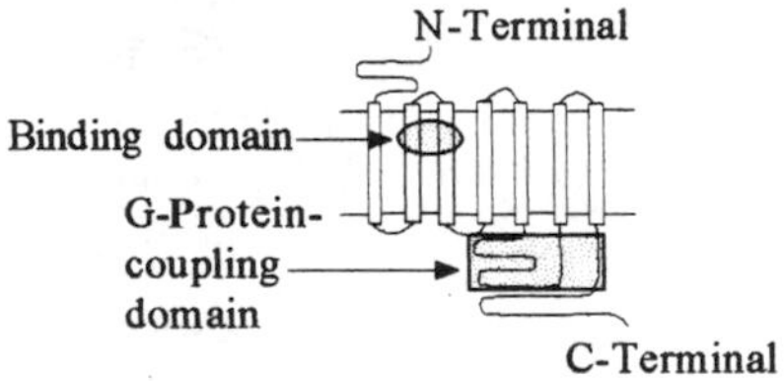 N-Terminal Binding domain G-Protein-coupling domain C-Terminal
3.	Insulin and growth factors. The receptor is linked to tyrosine kinase	Insulin receptors	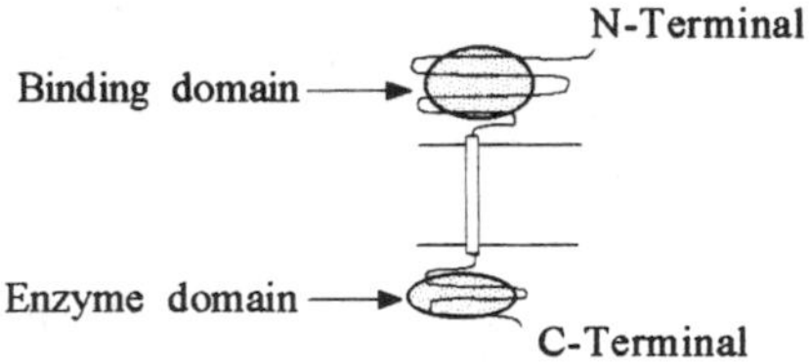 N-Terminal Binding domain Enzyme domain C-Terminal
4.	Steroid hormones, thyroid hormones and vitamins such as vitamin D and retinoic acid	Antidiuretic hormone (ADH) or vasopressin receptors	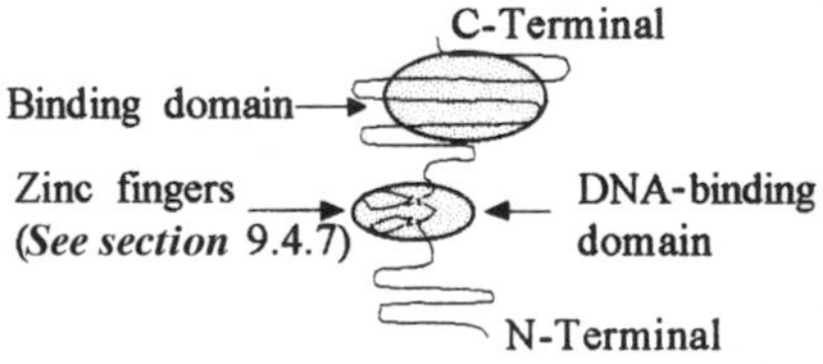 C-Terminal Binding domain Zinc fingers (*See section* 9.4.7) DNA-binding domain N-Terminal

are of the adrenergic type. These subtypes are further classified either according to the type of genetic code responsible for their structure or after the exogenous ligands that selectively bind to the receptor. For example, the endogenous ligand acetylcholine will bind to all cholinergic receptors but the exogenic ligand nicotine will only bind to nicotinic cholinergic receptors (nAChR). Similarly muscarine will only bind to muscarinic cholinergic receptors (mAChR). However, it is possible to differentiate between different types of receptor within a subtype. For example, three different types of muscarinic cholinergic receptors have been detected and a further two predicted from a study of the genes that code for this type of receptor (Figure 8.5). These five mAChR receptors are classified as m_n, where n is an integer between 1 and 5 inclusive.

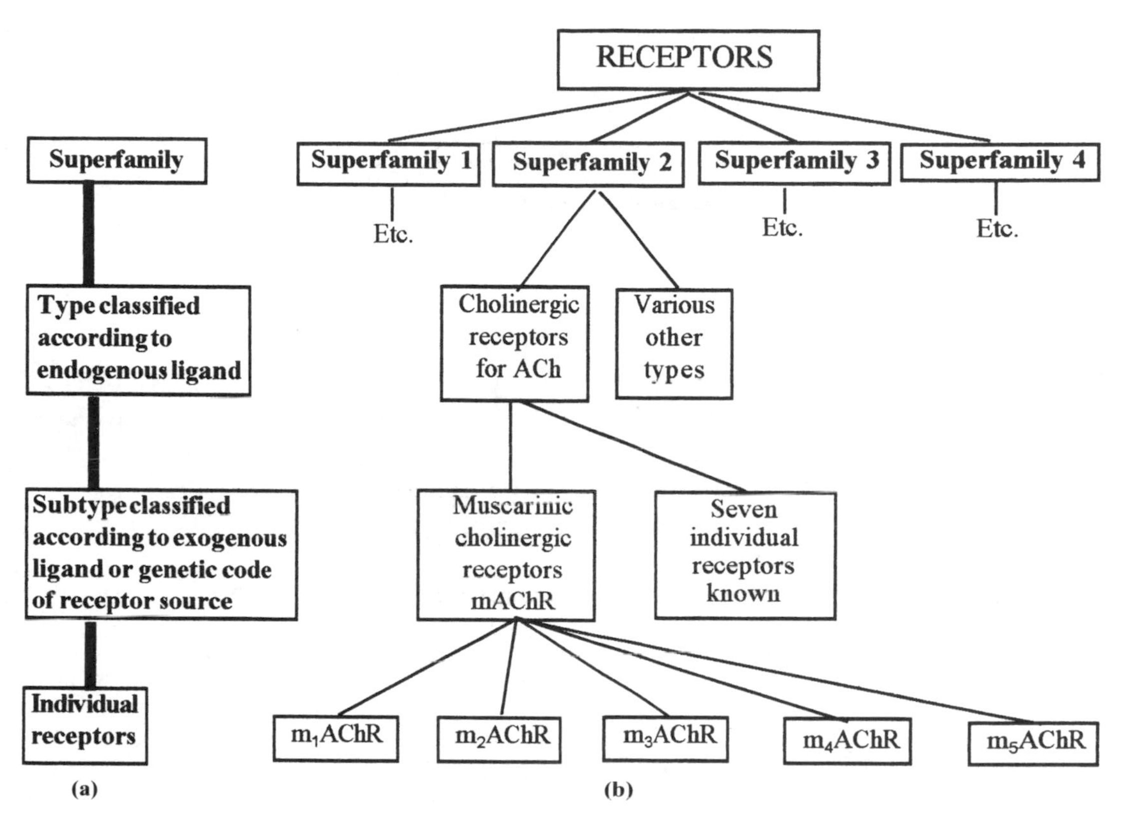

Figure 8.5. (a) A simplified outline of receptor classification. (b) The classification of muscarinic cholinergic receptors.

Similar classifications are used for other receptors. For example, adrenoceptors are classified as α and β subtypes. These subtypes are further classified according to the nature of their exogenous ligands. The various types are distinguished by the use of subscript numbers.

8.4 General Mode of Operation

The ligands that **activate** or **deactivate** (inhibit) a receptor are known as **primary** or **first** messengers. These first messengers may be hormones, neurotransmitters, other endogenous substances or xenobiotics including drugs, bacteria and viruses. Some receptors require one ligand for activation whereas other receptors require two ligands for activation.

Hormones are endogenous ligands (Figure 8.6) that are produced, stored and released by specialised organs either in response to a stimulus of the physical senses, such as light, odour, stress,

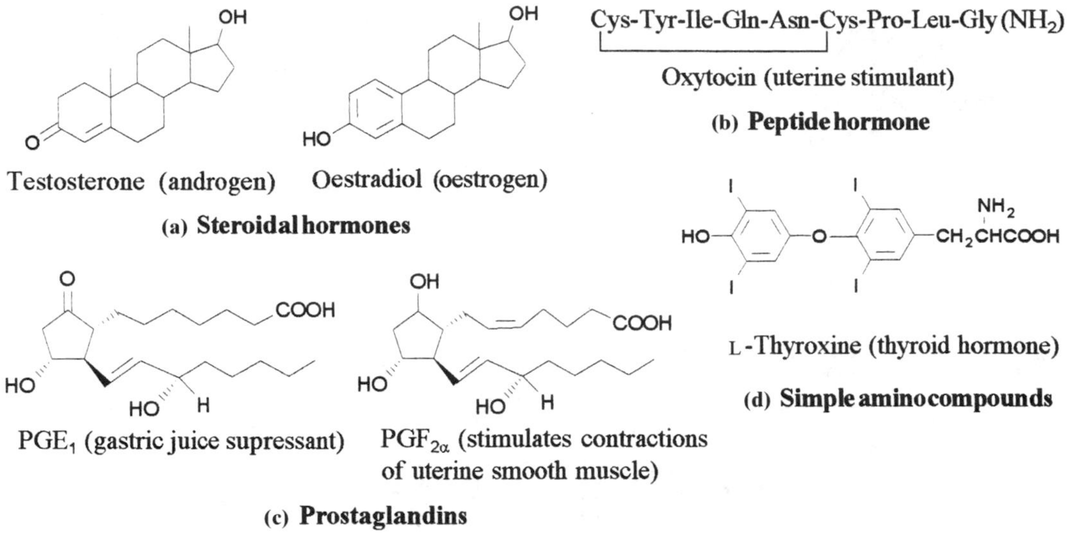

Testosterone (androgen) Oestradiol (oestrogen)

(a) Steroidal hormones

Cys-Tyr-Ile-Gln-Asn-Cys-Pro-Leu-Gly (NH_2)

Oxytocin (uterine stimulant)

(b) Peptide hormone

PGE_1 (gastric juice supressant) $PGF_{2\alpha}$ (stimulates contractions of uterine smooth muscle)

(c) Prostaglandins

L-Thyroxine (thyroid hormone)

(d) Simple amino compounds

Figure 8.6. Examples of four major classes of ligands that act as hormones in humans.

pain, or to a change in the concentration of a compound in a body fluid. For example, an increase in glucose concentration in blood stimulates the release of insulin. Many hormones are released into the bloodstream, where they may be transported distances of up to about 20 cm in order to reach their target receptor. Others, such as the prostaglandins, operate where they are formed. These hormones are referred to as local hormones or **autocoids**. Autocoids differ from ordinary hormones in that they are continuously synthesised and released into the circulation.

Neurotransmitters are the endogenous ligands (Figure 8.7) released by a neuron when it communicates with other neurons across a synaptic gap. The neurotransmitter is released from the presynaptic site and diffuses to its target on the postsynaptic neuron or an effector cell where it binds to the appropriate receptor. This activates the receptor, which in turn transmits the chemical message carried by the ligand to the cell (signal transduction). The mechanisms by which both hormones and neurotransmitters activate a receptor have been found to be remarkably similar.

The mechanisms by which chemical messages are received and delivered by a receptor are not fully understood. It is believed that some sections of the structure of the ligand are complementary in shape to sections of the receptor and so are able to bind to the receptor by appropriate bonds (Figure 8.8). However, the complete structure of the ligand does not exactly match that of the receptor site. This means that other sections of the ligand can only form very weak

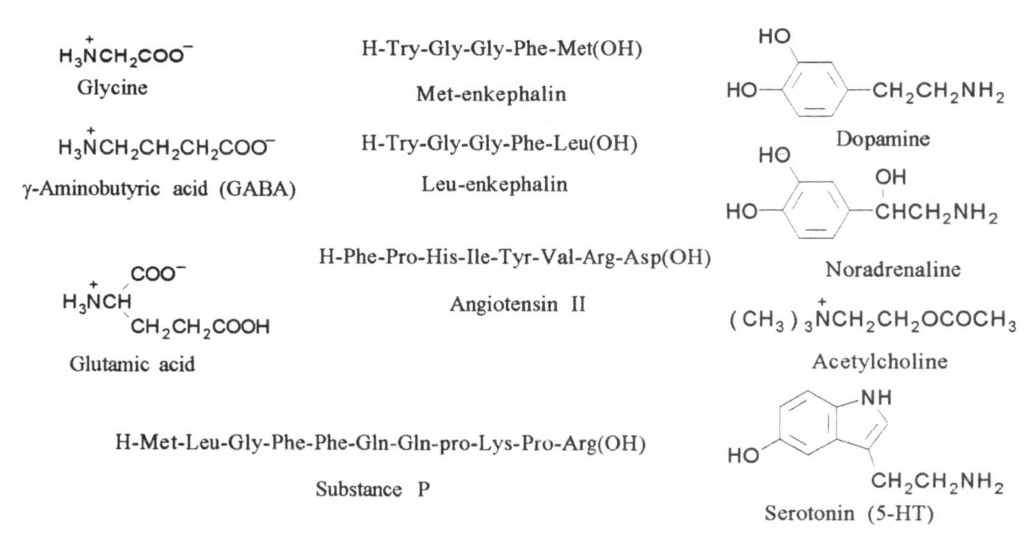

Figure 8.7. Examples of ligands that act as neurotransmitters in humans. These ligands are varied and include some amino acids, small peptides, acetylcholine, β-phenylethanolamines, catecholamines and their derivatives.

bonds with the receptor because they are not quite correctly aligned to maximise their interaction with the receptor. However, it is thought that the receptor then changes its conformation to maximise its interaction with these sections of the ligand. It is as though these sections of the ligand were acting as a magnet for the receptor.

The change in conformation caused by this maximising of the binding between receptor and ligand is thought to affect the whole of the receptor molecule and result in conformational changes in another domain of the receptor molecule. These subsequent changes may, for example, result in either the opening or closing of an ion channel (Figure 8.8a), the activation or deactivation of an enzyme, the activation of a G-protein (Figure 8.8b) or another biochemical event. The physiological significance of these events will depend on the nature of the process controlled by the receptor (see sections 8.4.1, 8.4.2, 8.4.3 and 8.4.4).

It is emphasised that it is not yet known whether the binding of the ligand occurs in a series of steps as described in the foregoing discussion or simultaneously in one step. However, although the binding must be strong enough to alter the conformation of the receptor, it must not be too strong to prevent the removal of the ligand from the receptor after it has delivered its message. This is particularly important for neurotransmitters. If they are not rapidly removed from their receptors the nerve receiving the message would be switched on or off for long periods of time, which could result in injury and death.

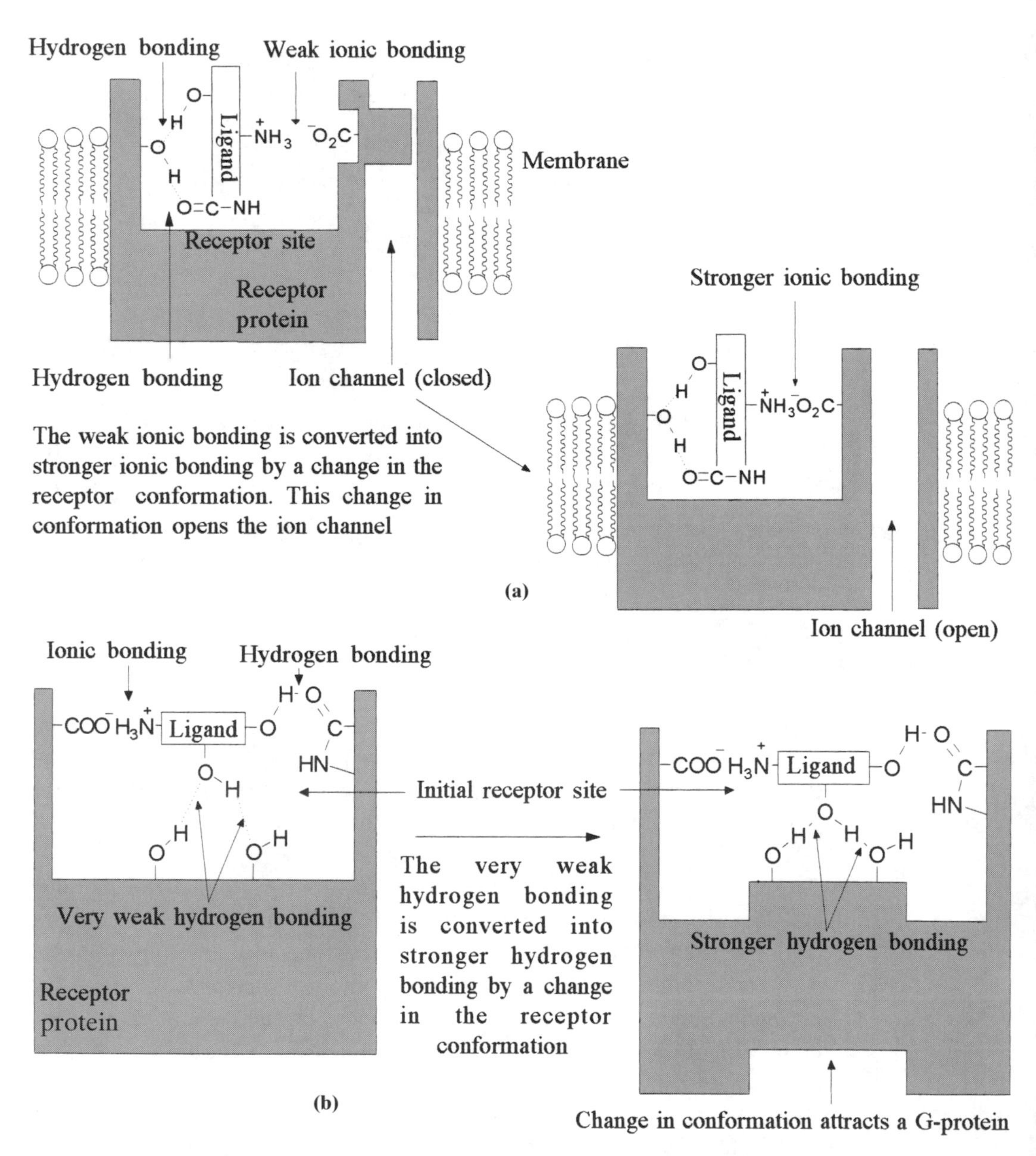

Figure 8.8. Diagrammatic representations of the general methods by which it is believed that a receptor transmits a chemical message when a ligand binds to the receptor site. The bonds shown are not the only types that can bind a ligand to a receptor. (a) The binding of the ligand causes a change in the conformation of the receptor molecule, which results in the opening of an ion channel (Superfamily type 1). The change in conformation that opens the channel may be some distance from the receptor site for the ligand and not as close as shown in the diagram. (b) The binding of a ligand results in a change of conformation that results in the receptor protein developing a strong affinity for a G-protein (Superfamily type 2).

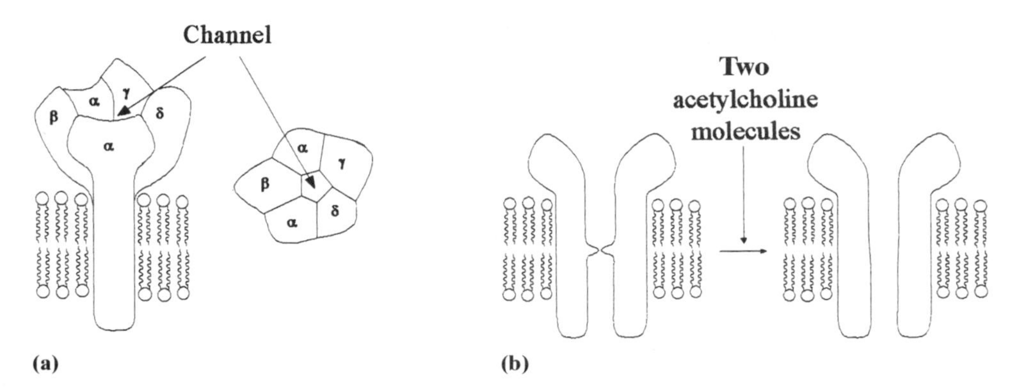

Figure 8.9 (a) The proposed general structure of the nACh receptor and (b) its proposed mode of action. Based on electron microscopy and other experimental data.

8.4.1 Superfamily Type 1

These receptors are membrane-embedded proteins that have both their C-terminal and N-terminals in the extracellular fluid. It usually has four to five membrane-spanning subunits surrounding a central pore. Each subunits consists of a sequence of 20–25 amino acid residues in the form of an α-helix. The different sequences are identified by a Greek letter and in some cases a suitable subscript. Sugar residues are attached to the extracellular N-terminal chain but these do not appear to take part in the receptor's biological function.

The ion channels of most ion channel receptors have two α-subunits. They normally require two ligand molecules to activate the receptor and open the ion channel. For example, the pore of the nicotinic acetylcholine (nACh) receptor, which consists of five subunits (two α- and one each of β-, γ- and δ-subunits), is only opened when an acetylcholine molecule binds to each of the α-subunits (Figure 8.9). In other words, the ion channel only opens when two acetylcholine molecules have bound to the receptor. The way in which the channel opens is not fully understood. However, Unwin showed in 1993 that about halfway through the membrane the five subunits protrude into the channel, effectively blocking it. He suggested that the binding of the actylcholine residues caused a conformational change that resulted in the removal of this protrudance.

The physiological response to the binding of the appropriate ligand to an ion channel receptor occurs in microseconds. The opening of the channel allows the passage of ions into or out of a cell, which leads to a variety of cellular effects. For example, the opening of ion channels in a nerve cell increases the flow of Na^+ ions into the neuron. This depolarises the membrane, which results in the rapid transmission of a nerve impulse along the nerve (see section 4.4.3). The opening of the Ca^{2+} ion channels of endothelial cells results in a flow of Ca^{2+} ions into the cell, which, with the calcium-binding protein calmodulin (CaM), activates the enzyme eNOS (see section 11.4).

8.4.2 Superfamily Type 2

Most of the receptors in this family have been found to consist of a single polypeptide chain containing 400–500 amino acid residues. The N-terminal of this chain lies in the extracellular fluid whereas the C-terminal is found in the intracellular fluid. The lengths and sequences of the terminal polypeptide chains vary considerably between members of this family. All the receptors have a group of seven transmembrane subunits that consist of 20–25 amino acid residues arranged in an α-helix. These transmembrane subunits are grouped around a central pocket that is believed to contain the receptor site.

The binding of a ligand to the receptor site results in a conformational change in the large intracellular polypeptide loop and C-terminal chain. These changes attract a protein known as **G-protein** because of its close association with guanine nucleotides. G-Proteins are a family of unattached proteins that are able to diffuse through the cytoplasm. They consist of three polypeptide subunits known as α-, β- and γ-subunits. In its resting state the G-protein has guanine diphosphate (GDP) bound to its α-subunit. Experimental evidence suggests that when the G-protein reaches the receptor the GDP is exchanged for guanine triphosphate (GTP) and in the course of this exchange the α-GTP subunit becomes detached and migrates to either the receptor of an ion channel (an effector protein) or the active site of an enzyme (an effector protein). The coupling of the α-GTP subunit to the receptor of the ion channel opens or closes the channel whereas the binding of the α-GTP subunit to the enzyme either inhibits or activates the enzyme (Figure 8.10). The action of the α-GTP subunit is terminated when the GTP is hydrolysed to GDP by the catalytic action of the α-subunit. When the GTP is converted to GDP the α-unit loses its affinity for the effector protein and migrates back to its β- and γ-subunits to complete the cycle. It should be noted that some researchers have suggested that ion channels are controlled by the β–γ-subunit and not the α-GTP subunit.

The type of activity initiated by a G-protein is thought to depend on the nature of the G-protein involved. It is now believed that there are three major groups of G-proteins, namely: G_s, which stimulates adenylate cyclases (AC); G_i, which inhibits adenylate cyclases; and G_o, which mediates neurotransmission in the brain by a pathway that is not yet fully understood. It is believed that these G-proteins act through specific receptors activated by the appropriate agonist. For example, an enzyme would be stimulated by one receptor but inhibited by a different receptor (Figure 8.11). It is interesting to note that cholera toxin bonds (conjugates) to a G_s-coupled receptor, resulting in the permanent activation of adenylyl cyclase. This induces excessive secretion of fluid from the gastrointestinal epithelium, which leads to excessive diarrhoea and dehydration, the classic symptoms of cholera.

G_s-Proteins activate two major classes of enzymes – the adenylate cylases (AC) and the phospholipases – to produce secondary messengers. The adenylate cyclases are peripheral proteins embedded in the interior surface of the cell membrane. They catalyse the conversion of ATP to the secondary messenger cAMP. Adenylate cyclase acts as an amplifier, one activated AC molecule catalysing the conversion of about 10 000 ATP molecules to cAMP. cAMP is able to

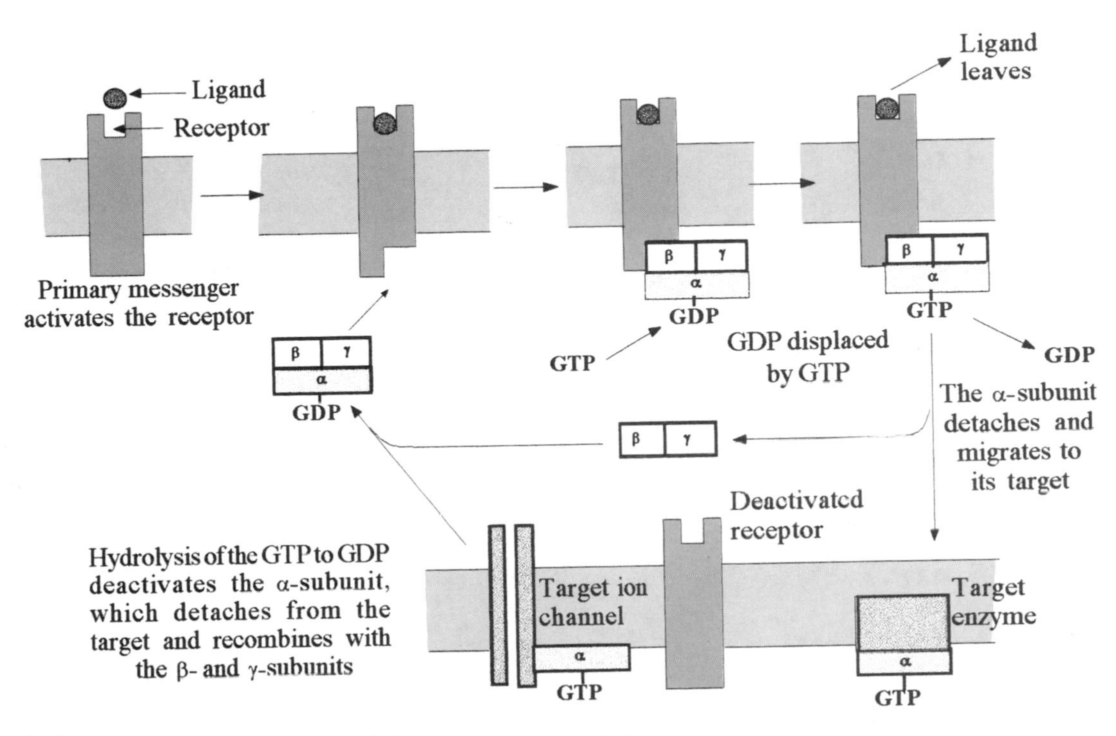

Figure 8.10. A diagrammatic representation of the general action of G-proteins.

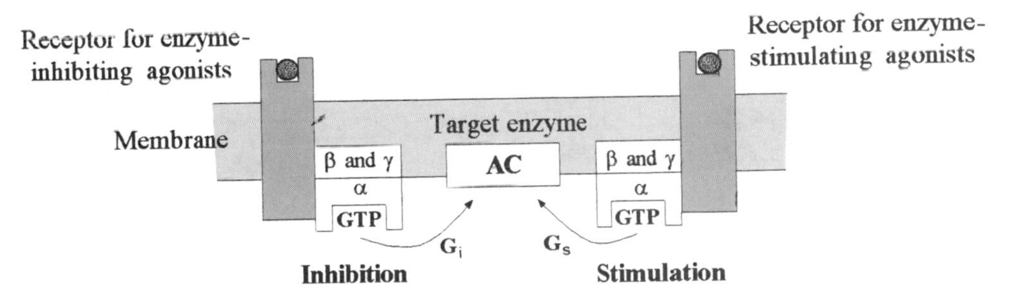

Figure 8.11. A diagrammatic illustration of the specific nature of receptors and G-proteins.

activate a number of different protein kinases (Figure 8.12). These enzymes control a wide variety of cell functions, such as energy metabolism and cell division. It is deactivated by phosphodiesterase. This enzyme catalyses its conversion to AMP, which is reconverted by a series of steps to ATP. It is interesting to note that both caffeine and theophylline inhibit phosphodiesterase.

Other members of the G-protein family activate phospholipase C. These enzymes catalyse the hydrolysis of phosphatidylinositol bisphosphate (PIP_2) to two secondary messengers

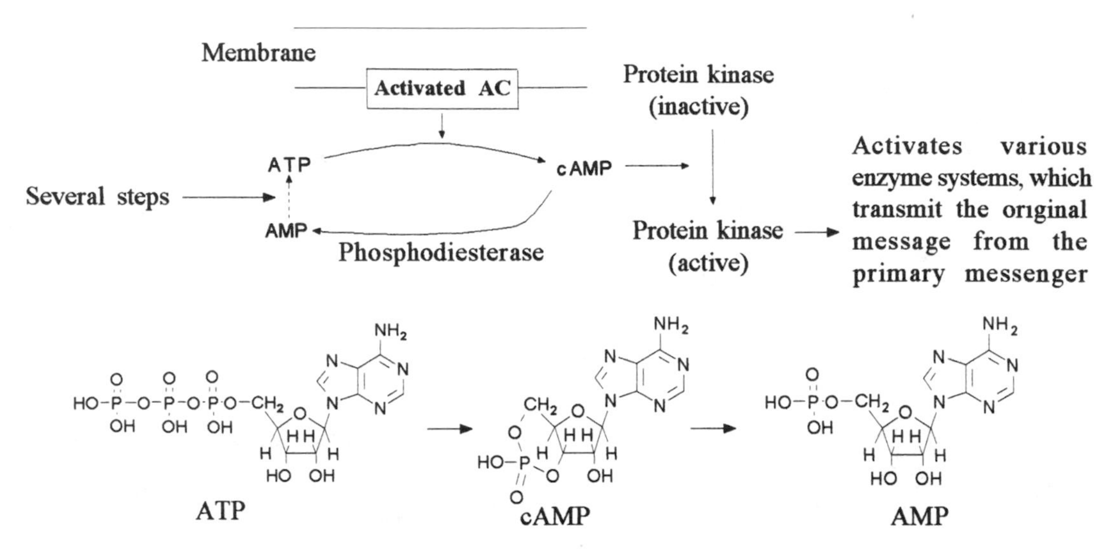

Figure 8.12. A schematic representation of the formation and action of the secondary messenger cAMP.

diacylglycerol (DAG) and inositol-1,4,5-triphosphate (IP_3). DAG activates membrane-bound protein kinase C (PKC), which in turn initiates various processes within the cell. IP_3 initiates the rapid release of Ca^{2+} ions from intracellular storage, which itself acts as a secondary messenger initiating a range of cellular responses. Both DAG and IP_3 are converted back to PIP_2, although this takes some time because the enzymes involved in this process are found in the cytoplasm (Figure 8.13).

Malfunctions of the inositol system have been linked to a number of illnesses such as manic depression and cancer. The former usually responds to treatment with lithium carbonate because lithium ions have been found to block the recycling pathway for IP_3. Cancer formation is believed to occur because protein kinase is involved in cell division.

8.4.3 Superfamily Type 3

Tyrosine-kinase-linked receptors have a single helical transmembrane subunit. This subunit is attached to large extracellular and intracellular domains. The extracellular domains are varied and their nature depends on that of their endogenous ligand. However, the intracellular domains of all these receptors contain a tyrosine-kinase residue as an integral part of their structures. In addition, it is believed that all these intracellular domains have an ATP binding site near the surface of the membrane whereas the substrate site is nearer to the end of the intracellular domain.

It is thought that the binding of the ligand to type 3 receptors normally causes the receptors to dimerise (Figure 8.14). This is believed to result in conformational changes that trigger the

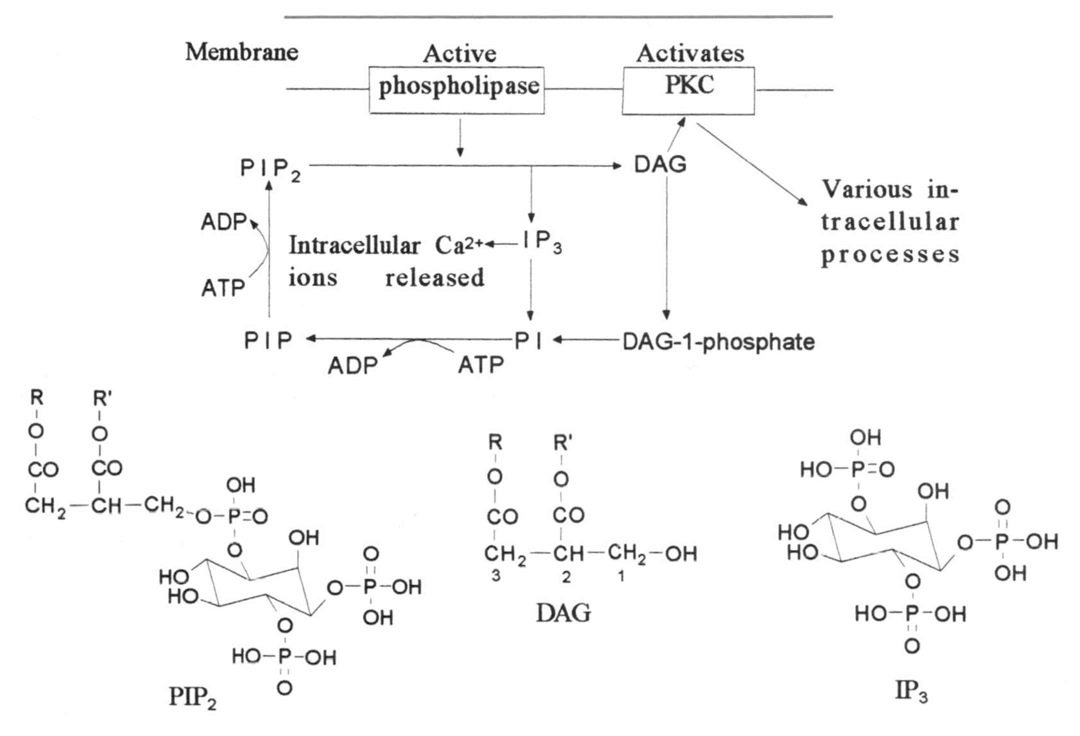

Figure 8.13. A schematic representation of the formation and action of the secondary messengers DAG and IP_3.

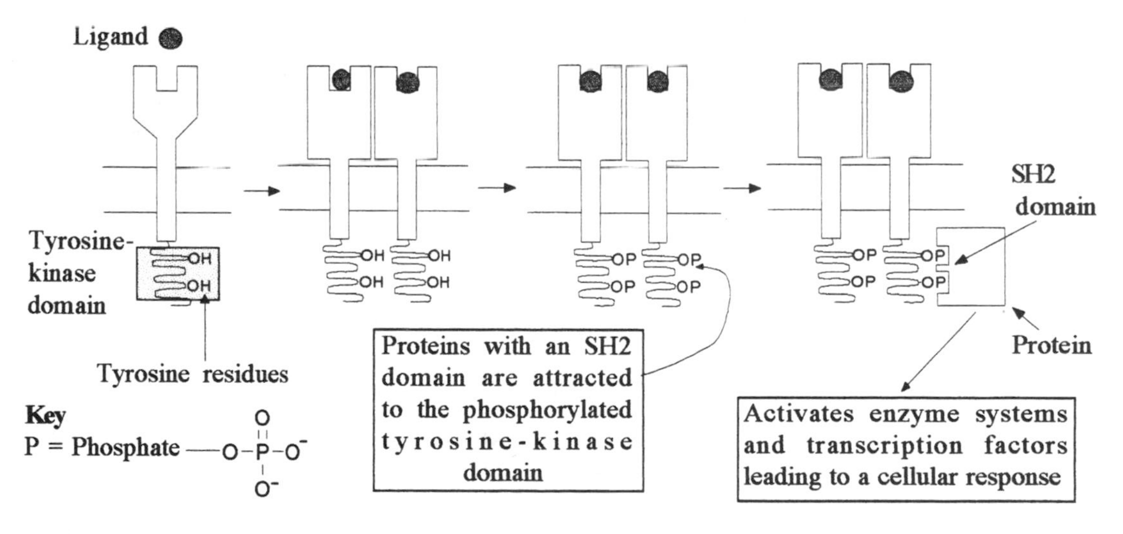

Figure 8.14. A diagrammatic representation of the mode of action of tyrosine-kinase-linked receptors. A number of different proteins with SH2 domains exist in the intracellular fluid. These proteins control a wide variety of cell functions.

autophosphorylation of various tyrosine resides in the intracellular domain. These tyrosine phosphorylated residues appear to act as high-affinity binding sites for specific intracellular proteins. The structures of these proteins appear to have in common a conserved sequence of about 100 amino acids referred to as the SH2 domain, which binds to specific regions of the phosphorylated tyrosine-kinase domain. This binding is thought to trigger either enzyme activity or the binding of further functional proteins to the SH2-containing protein, which leads to a wide variety of biological activities.

Many cytokine receptors appear to act in a similar way to tyrosine-kinase receptors even though they do not have a tyrosine-kinase domain. It is believed that the dimerisation of the cytokine receptors attracts a tyrosine-kinase unit in the cytoplasm to the intracellular domain, where it is autophosphorylated. This is followed by the attraction and binding of a specific SH2 protein, which ultimately leads to a cellular response.

The discovery of SH2 domains opens the way for the design of new drugs that will have structures that are either complementary to or mimic the action of the SH2 domain.

8.4.4 Superfamily Type 4

The members of this family of intracellular proteins are located within the nucleus of living cells, although they were first isolated from the cytoplasm of homogenised cells. This led to the erroneous conclusion that these receptors existed in the cytoplasm. They are activated by steroid and thyroid hormones, retinoic acid and vitamin D (Figure 8.15).

The members of the family are large proteins containing 400–1000 amino acid residues. Their structures have similar central sections of about 60 residue sequences that contain two loops of about 15 residues. These loops are known as zinc fingers because each loop originates from a group of four cysteine residues coordinated to a zinc atom (see section 7.4.7). The hormone

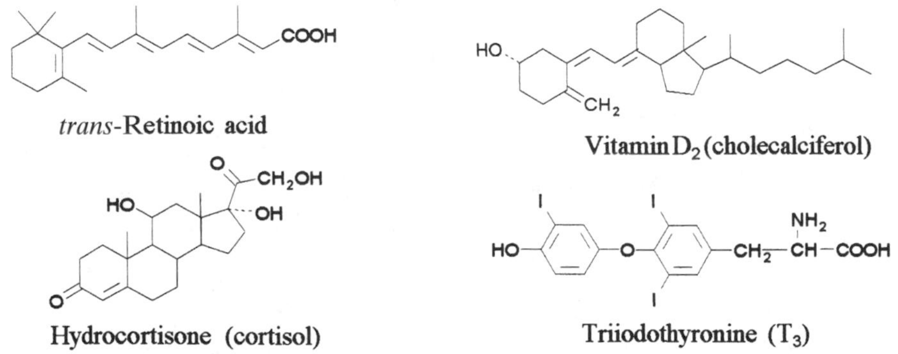

COOH

trans-Retinoic acid

HO

CH_2

Vitamin D_2 (cholecalciferol)

O

CH_2OH

HO

OH

O

Hydrocortisone (cortisol)

I I

NH_2

HO—O—CH_2—CH—COOH

I

Triiodothyronine (T_3)

Figure 8.15. Examples of the agonists of the members of Superfamily Type 4.

receptor lies on the C-terminal side of this central region whereas located on the N-terminal side is a highly variable region that controls gene transcription. The conformational changes caused by the binding of the hormone to the receptor are believed to expose the DNA-binding domain, which is normally hidden within the structure of the protein. This allows the DNA to bind to the protein. This is accompanied by an increase in RNA polymerase activity and, within minutes, the production of a specific mRNA. This mRNA controls the synthesis of a specific protein that produces the cellular response. However, this response can take from hours to days to develop. For example, glucocorticoids are believed to increase the concentration of lipocortin, which acts as an anti-inflammatory by inhibiting the activity of phospholipase A_2 (PLA_2).

8.5 Ligand–Response Relationships

The binding of a ligand (L) to a receptor (R) results in a loss of energy. This loss is broadly described as the **affinity** of the ligand for the receptor. The greater the affinity of the ligand for the receptor, the more easily it binds to that receptor. Clark in the 1920s envisaged this binding process as a dynamic equilibrium. He postulated, in his **occupancy theory**, that the binding of a ligand to a number of identical independent receptor sites activated those sites and that the magnitude of the corresponding response was proportional to the number of receptors occupied at equilibrium. Clark assumed that a maximum response would occur when all the receptors were occupied and that the response to the ligand would cease when the ligand dissociated from the receptor, that is:

$$\underset{\text{(Ligand–receptor complex)}}{\text{L–R}} \rightleftharpoons \underset{\text{(Ligand)}}{\text{L}} + \underset{\text{(Receptor)}}{\text{R}} \tag{8.2}$$

and:

$$K_D = \frac{[L][R]}{[L\text{-}R]} \tag{8.3}$$

where K_D is the dissociation constant for the process. It follows that K_D (= K_A, the affinity constant) will be a measure of the affinity of the ligand for the receptor: the smaller the value of K_D, the greater the affinity of the ligand for the receptor. Most agonists have K_D values of the order of 10^{-10}–$10^{-6}\,\text{mol}\,\text{dm}^{-3}$. These values are usually recorded as pD_2 values, where:

$$pD_2 = -\log K_D = -\log EC_{50} \tag{8.4}$$

because from Equation (8.20) (see section 8.6) $K_D = EC_{50}$, where EC_{50} is the molar concentration that produces half the maximum biological response observed when a ligand binds to a receptor. According to Clark's theory the EC_{50} corresponds to half the receptors being occupied by the ligand. The EC_{50} of a drug is normally determined *in vitro* and may be used to estimate the value of K_D, but the correlation is not always reliable. A low EC_{50} value will indicate a high affinity for the receptor. Both EC_{50} and pD_2 are used to compare the relative effects of the members of a series of analogues in the preclinical development of a new drug.

The affinity of a ligand may also be measured in terms of the tendency of a ligand to associate with the receptor, that is:

$$L + R \rightleftharpoons L\text{-}R \tag{8.5}$$

and:

$$K_a = \frac{[L\text{-}R]}{[L][R]} \tag{8.6}$$

where K_a is the association constant and is the reciprocal of K_D. High-affinity ligands have large K_a values.

A high proportion of the amount of a ligand in general circulation will bind to receptor sites for which it has a strong affinity. However, not all the ligand binds to these sites; some will bind to secondary sites for which the ligand has a lower affinity. Furthermore, binding of the ligand to these secondary sites could give rise to unwanted side effects, although this is not the only source of unwanted side effects.

8.5.1 Experimental Concentration–Response Curves

Quantitative information concerning the dose–response relationship of a drug is usually obtained from a study of the effect of the drug on an isolated tissue, such as guinea-pig ileum and rabbit jejunum. These studies are normally carried out using an organ bath (Figure 8.16). The lower end of the tissue is attached to the air line whereas the top is connected to an isometric transducer that detects changes in the tension in the tissue. A dose of the drug is administered and the response of the tissue is detected, amplified and recorded as a trace on a chart.

The appearance of the trace will depend on the nature of the study (Figure 8.17). It will depend on both the drug and the tissue. For example, prenalterol acts as a full agonist on thyroxine-treated guinea-pig atria, as a partial agonist on rat atria and as an agonist in canine coronary artery, whereas isoprenaline acts as a full agonist in all three tissues.

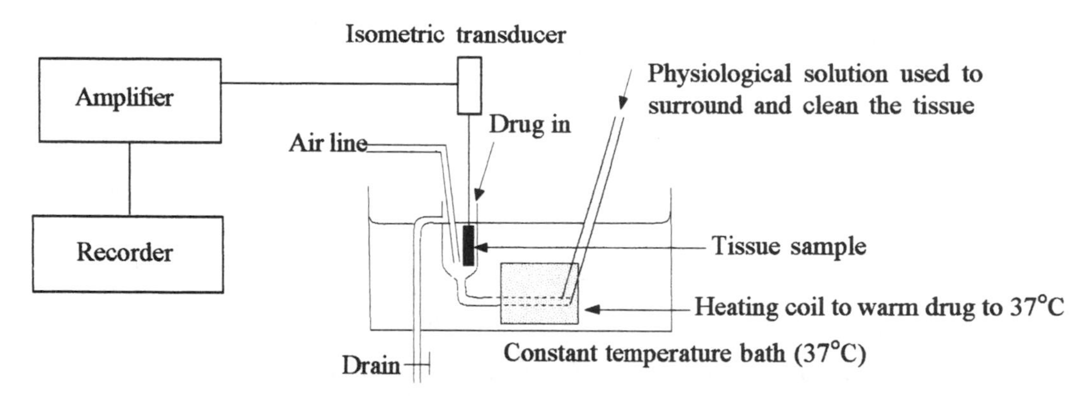

Figure 8.16. A schematic representation of a typical organ bath experiment. The drug is injected directly onto the tissue.

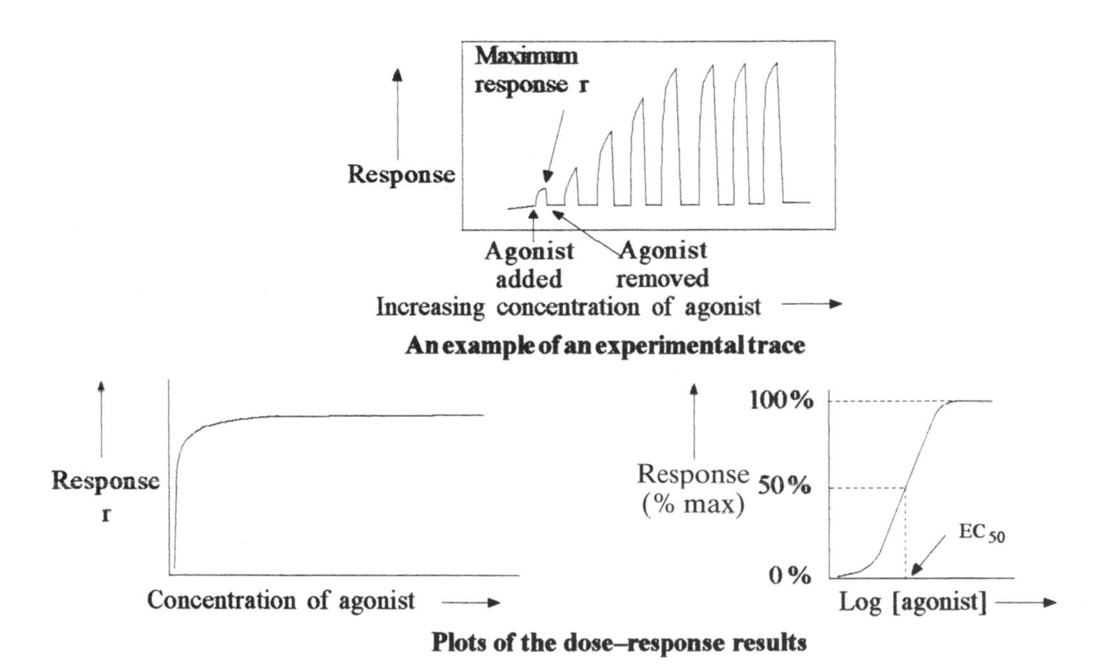

Figure 8.17. A simulation of the results of a typical molar concentration–response investigation for a full agonist. If the amount of the drug is measured in terms of the dose administered, the plots will have the same general appearance but the EC_{50} will now be recorded as the ED_{50}. The appearances of the trace will vary; the example shown is for an ideal set of experimental results.

8.5.2 Agonist Concentration–Response Relationships

The response due to an agonist increases with increasing concentration of the agonist until it reaches a maximum (Figure 8.17). At this point, further increases in dose have no further effect on the response. It should be noted that the scale on the x-axis of response curves may be

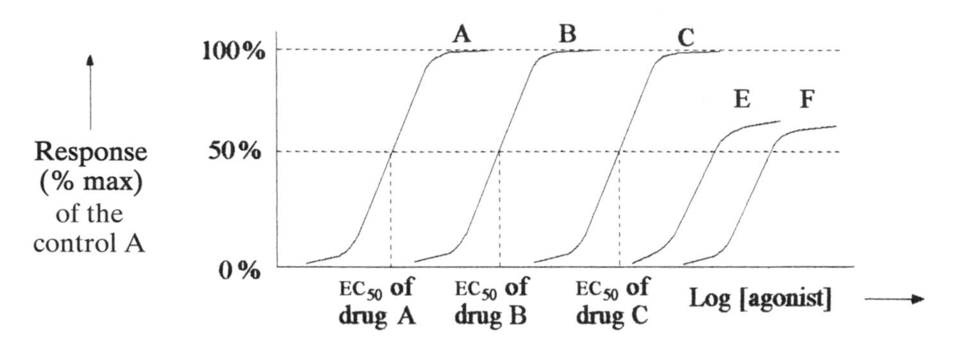

Figure 8.18. Ideal representations of the percentage response against log molar concentration plots for several agonists that cause the same response. Plot A is used as the control to calculate the percentage response. Plots B and C are for drugs acting on the same receptor site but with different EC_{50} values. The dose–response curves of these drugs are similar and their ED_{50} values can be read off the log dose axis of these plots. Plots E and F are for drugs in the same series that are partial agonists.

either in terms of the molar concentration or the dose (total amount used) of the agonist. The latter is referred to as a **dose–response** curve. Agonists with similar structures acting on the same receptor will usually exhibit similar log molar concentration–response plots (Figure 8.18). However, the slopes of these plots are not always similar (Figure 8.19). Furthermore, some agonists in the same structural series may show a lower maximum response value than the other members of the series. These compounds are referred to as partial agonists (see section 8.5.4) whereas those that show the maximum response are known as **full agonists**. Most drugs act as partial agonists.

Agonist-response plots may be used during preclinical drug development to compare the potency of the different analogues of a lead compound. These comparisons may be made in terms of the EC_{50} and ED_{50} values (see Figure 8.17) of the potential drugs. However, the values of both the EC_{50} and ED_{50} of a drug will depend on the biological effect being studied. For example, the EC_{50} value for aspirin used as an analgesic is significantly different from the EC_{50} value of aspirin when it is used as an anticoagulant.

8.5.3 Antagonist Concentration–Response Relationships

Antagonists inhibit the action of an agonist (Figure 8.20). They may be classified as either **competitive** or **non-competitive**. Competitive antagonists are more common than non-competitive antagonists.

A competitive antagonist binds to the same receptor molecule as an agonist but does not cause a response (Figure 8.20). As the concentration of the antagonist increases, the response due to

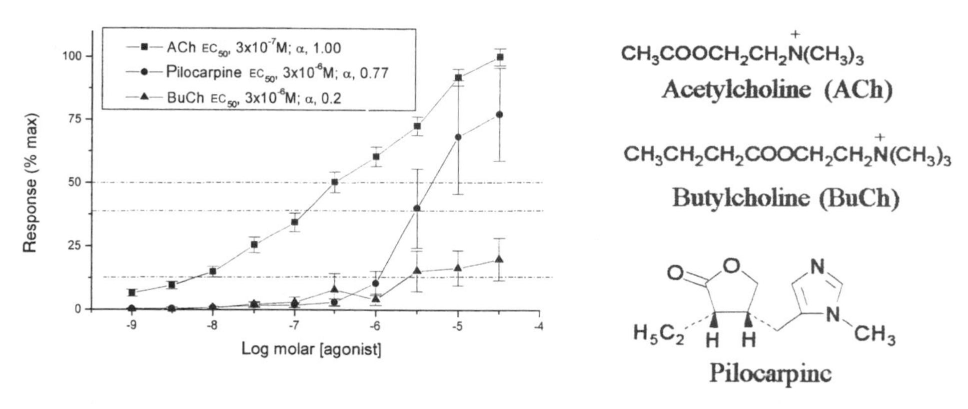

Figure 8.19. Experimental results showing the effect of different agonists on guinea-pig ilium. (Courtesy of Dr S. Arkle and the students of the 1998 cohort of the ABMS, BMS and MPharm degree courses.)

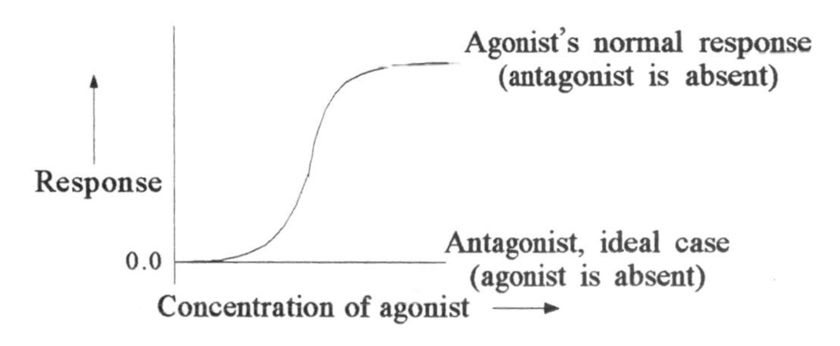

Figure 8.20. The effect of an ideal antagonist on the response of a receptor.

the agonist decreases. However, for competitive antagonists this decrease can be reversed by increasing the concentration of the agonist (Figure 8.21a).

This behaviour can be explained if the binding of both the agonist and competitive antagonist to the receptor are in dynamic equilibria (Figure 8.22). When the concentration of the competitive antagonist is increased, the position of the equilibrium K_1 moves to the left. However, as the concentration of agonist increases, the antagonist is displaced and the equilibrium moves to the right. This behaviour means that in the presence of a competitive antagonist a higher concentration of the agonist is required in order to obtain the same degree of response as that observed in the absence of the antagonist (Figure 8.22). In other words, the effect of a competitive antagonist will depend on the relative concentrations of the agonist and antagonist.

Originally, competitive antagonists were thought to compete for the same receptor site as the agonist. However, in most cases the structures of an agonist and its antagonists are not terribly similar (Figure 8.21). Consequently, it is now thought that the antagonist can also bind to a receptor at a different site close to the agonist receptor site. This site may be close enough

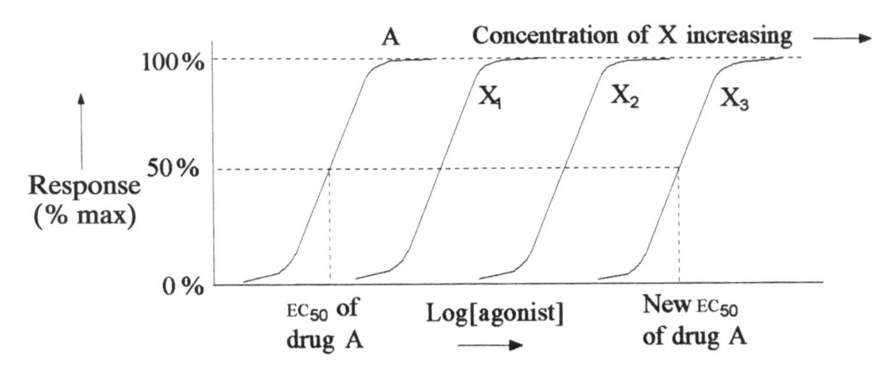

Figure 8.21. The effect of an ideal competitive antagonist X on the dose–response curves for an agonist A. Plot A is the dose–response curve for the agonist A in the absence of the antagonist X. Plots X_1 to X_3 are the dose–response curves for the agonist A in the presence of three different constant concentrations (X_1, X_2 and X_3) of the antagonist X. The value of the EC_{50} of A will depend on the concentration of the antagonist.

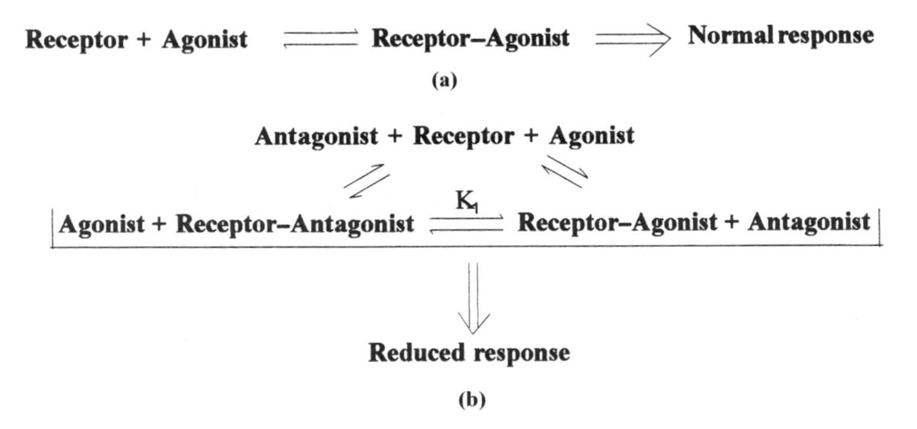

Figure 8.22. An outline of the mode of action of a competitive antagonist: (a) with no antagonist and (b) with an antagonist.

for the bound antagonist to overlap the agonist receptor site and so hinder the binding of the agonist to its receptor site. Alternatively, the binding of the antagonist at a site near the agonist receptor site may change the conformation of the agonist receptor site, which would also prevent the agonist from binding to its site.

The action of non-competitive antagonists is not dependent on the concentration of the agonist. Increasing the concentration of the agonist does not restore the degree of response (Figure 8.23a). It is believed that non-competitive antagonists bind irreversibly by strong bonds such as covalent bonds to allosteric sites on the receptor. This changes the conformation of the receptor site, which prevents the binding of the agonist to the receptor (Figure 8.23b).

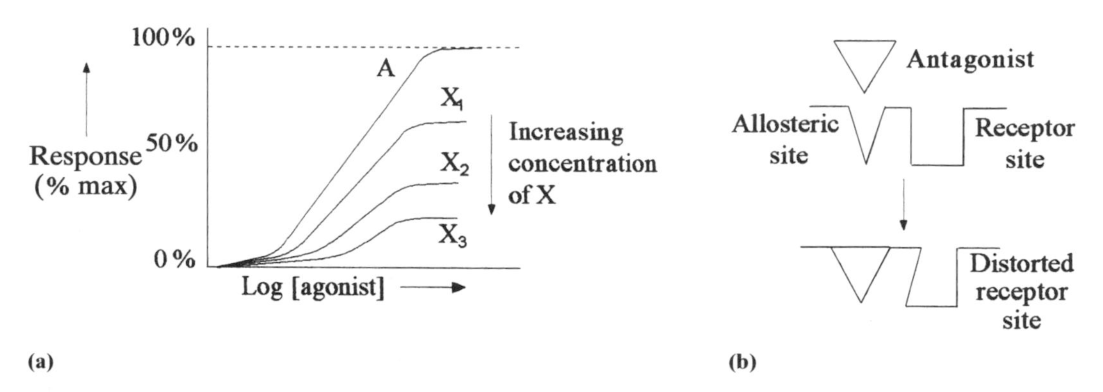

Figure 8.23. **(a)** The effect of a non-competitive antagonist on the dose–response curves for a drug A. Plot A is the dose–response curve for the agonist A in the absence of the antagonist X. Plot X_1 is the dose–response curve for the agonist A in the presence of a constant concentration (X_1) of the antagonist X. The effect of increasing the constant concentration of X is shown in the succeeding plots. (b) A representation of the general mode of action of a non-competitive antagonist.

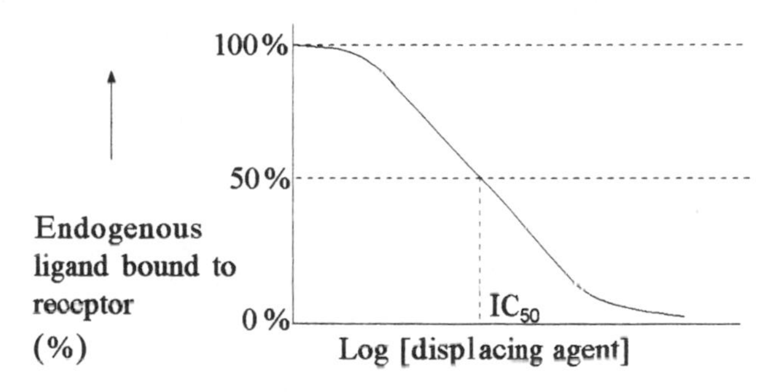

Figure 8.24. A displacement curve for the displacement of a ligand antagonist or agonist from a receptor. The IC_{50} is for the displacing agent.

The magnitude of the response to an agonist in the presence of an antagonist will depend on the relative affinities of the receptor for the antagonist and the agonist. The concentration at which an antagonist exerts half its maximum effect is known as its IC_{50} (Figure 8.24). This measurement may either refer to an antagonist displacing an agonist or an agonist displacing an antagonist. In both cases, the IC_{50} is a measure of the affinity of the drug for the receptor under the appropriate conditions. The smaller the value of the IC_{50}, the stronger the affinity of the displacing drug for the receptor.

IC_{50} values may be used in drug development to compare the potencies/affinities of a series of drugs that bind to the same receptor molecule and inhibit the same biological response. The values may be determined *in vitro* or *in vivo* and so can also be of use in estimating the dose of a drug required in preclinical trials.

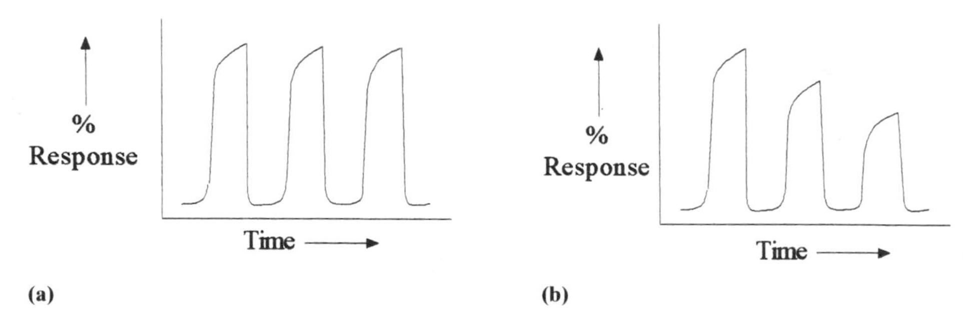

Figure 8.25. (a) No desensitisation. (b) Desensitisation. The start of each peak corresponds to the administration of an identical dose of the agonist.

8.5.4 Partial Agonists

Partial agonists are compounds that act as both agonists and antagonists. They act as antagonists in the usual manner by preventing endogenous ligands from binding to their receptors. However, the binding of the partial agonist to the receptor also results in a weak activation of the receptor, that is, a weak signal is transmitted to the appropriate domain of the receptor. The net result of these opposing effects is that the dose of agonist required to obtain a maximum response is higher and the maximum response is less than that of pure agonists with similar structures.

Partial agonism has been explained in two ways:

(i) The structure of the partial agonist is such that it can bind in two different ways to the receptor. In one orientation it acts as an agonist whereas in the alternative orientation it acts as an antagonist. As a result, the degree of activity of the receptor would depend on the relative numbers of molecules binding in each orientation.
(ii) An alternative explanation is that the partial agonist is a reasonable but not perfect fit to the receptor site and so its binding does not cause a great enough change in the receptor molecule's conformation to allow a full transmission of the signal. However, the conformational change is large enough to allow a weak transmission of the signal.

8.5.5 Desensitisation

The repeated exposure of a receptor to identical doses of a drug can in some cases result in a reduction of the response (Figure 8.25). For example, smooth and voluntary muscle will become insensitive when repeatedly exposed to a depolarising agent. It appears that the drug starts by acting as a full agonist but its repeated use results in partial agonistic action. This phenomenon is known as **desensitisation** or **tachyphylaxis**. At present there is no universally accepted explanation for desensitisation, but rate theory does account for the phenomenon (see section 8.6.3).

8.6 Ligand–Receptor Theories

A number of theories have been advanced to explain the action of ligands on receptors. They attempt to explain desensitisation, the relative potencies of different drugs acting on the same receptor and why a drug may act as an agonist in one tissue, a partial agonist in another and an antagonist in a third tissue, even though it is acting on the same receptor.

8.6.1 Clark's Occupancy Theory

Clark visualised the drug–receptor interaction as being a bimolecular dynamic equilibrium with the drug molecules continuously binding to and leaving the receptor (see also section 8.5), that is:

$$\textbf{Drug (D)} + \textbf{Receptor (R)} \rightleftharpoons \textbf{Drug–receptor complex (DR)} \Rightarrow \textbf{Response} \qquad (8.7)$$

Clark stated that the intensity of the response at any time was proportional to the number of receptors occupied by the drug: the greater the number occupied, the greater the pharmacological effect, that is:

$$\text{Response effect E} \propto [\text{DR}] \qquad (8.8)$$

According to Clark, a maximum response would be obtained when all the receptors were occupied, that is:

$$\text{Maximum response effect } E_{max} \propto [R_T] \qquad (8.9)$$

where R_T is the total number of receptors. It follows from Equations (8.8) and (8.9) that for a given dose of a drug the fraction of the maximum response is given by:

$$\text{Fraction of the maximum response } \frac{E}{E_{max}} = \frac{[\text{DR}]}{[R_T]} \qquad (8.10)$$

The dissociation of the drug–receptor complex may be represented as:

$$\text{D-R} \rightleftharpoons \text{D} + \text{R} \qquad (8.11)$$

applying the law of mass action:

$$K_D = \frac{[\text{D}][\text{R}]}{[\text{DR}]} \qquad (8.12)$$

where K_D is the dissociation constant for the drug receptor complex. But the total receptor concentration is:

$$[R_T]=[R]+[DR] \tag{8.13}$$

Substituting Equation (8.13) in Equation (8.12) gives:

$$K_D=\frac{[D]([R_T]-[DR])}{[DR]} \tag{8.14}$$

Rearranging Equation (8.14) gives:

$$K_D=\frac{[D][R_T]}{[DR]}-\frac{[D][DR]}{[DR]} \tag{8.15}$$

therefore:

$$K_D=\frac{[D][R_T]}{[DR]}-[D] \tag{8.16}$$

and:

$$K_D+[D]=\frac{[D][R_T]}{[DR]} \tag{8.17}$$

$$\frac{K_D+[D]}{[D]}=\frac{[R_T]}{[DR]} \tag{8.18}$$

Substituting Equation (8.18) in Equation (8.10) gives:

$$\frac{E}{E_{max}}=\frac{[DR]}{[R_T]}=\frac{[D]}{K_D+[D]} \tag{8.19}$$

Equation (8.19) shows that the relationship between E and molar drug concentration [D] is in the form of a rectangular hyperbola whereas that between E and log[D] is sigmoidal. These theoretical relationships derived using Clarks' theory are often in good agreement with the experimental results (Figure 8.26) obtained in a number of investigations. Furthermore, substituting the value of E/E_{max} = 1/2 in Equation (8.15) gives the relationship:

$$K_D = EC_{50} \tag{8.20}$$

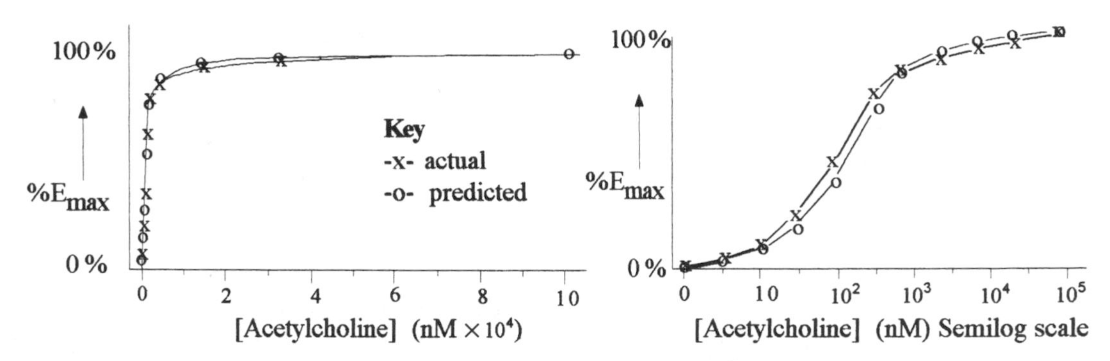

Figure 8.26. The correlation of experimental results and those predicted using Clark's theory for the stimulated contraction of guinea-pig ileum by acetylcholine.

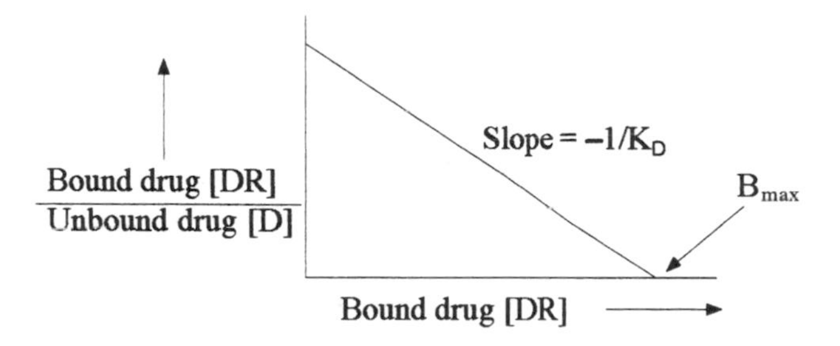

Figure 8.27. Scatchard plots for an ideal ligand binding to only one type of receptor. The steeper the slope of the line, the higher the affinity of the drug for the receptor. B_{max}, the intercept on the x-axis, gives an estimate of the maximum number of receptors to which the drug can bind. Any units that measure quantity may be used but the units for the x-axis are usually femto- or picomoles per milligram of protein.

However, in practice, this theoretical relationship does appear to be the exception rather than the rule.

The value of the dissociation constant K_D is a measure of the affinity of the drug for the receptor. Drugs with small K_D values have a large affinity for the receptor whereas those with high values have a low affinity. As a result, K_D values are used to compare the activities of a series of analogues during drug development. The value of K_D may be determined experimentally from tissue binding experiments using a radioactive form of the drug. The data obtained from this type of experimental work may be analysed using a Scatchard plot of the ratio of bound to free ligand against bound drug (Figure 8.27). This gives a straight line with a slope of $-1/K_D$ provided that the drug binds to only one type of receptor. In industry the data are now analysed by the use of a computerised method of least squares.

Although Clark's occupancy theory is still a cornerstone of pharmacodynamics, a number of its assumptions have now been shown to be incorrect. It is now known that:

(i) The formation of many drug-receptor complexes is not reversible.
(ii) The receptor sites are not always independent.
(iii) The formation of the complex may not be bimolecular. For example, two acetylcholine molecules bind to nACh receptors of ion channels (see section 8.4.1).
(iv) A maximum response may be obtained before all the receptors are occupied.
(v) The response is not linearly related to the proportion of receptors occupied, especially in the case of partial agonists.

In the 1950s Ariens and Stephenson separately modified Clark's theory to account for the existence of agonists, partial agonists and antagonists. They based their modifications on a proposal by Langley in 1905 that visualised the action of a receptor as taking place in two stages. The first stage was the binding of the ligand to the receptor, which was controlled by the ligand's affinity for the receptor. The second stage was the initiation of the biological response. Ariens said that this second step was governed by the ability of the ligand–receptor complex to initiate a response. Ariens called this ability the **intrinsic activity** (α) whereas Stephenson referred to it as the **efficacy** (e) of the ligand–receptor complex.

Intrinsic activity may be defined as:

$$\alpha = \frac{E_{max} \text{ of a drug}}{E_{max} \text{ of the most active agonist in the same structural series}} \tag{8.21}$$

Using the concept of intrinsic activity Clark's equation (Equation 8.19) becomes:

$$\frac{E}{E_{max}} = \frac{\alpha D}{K_D + D} \tag{8.22}$$

When $\alpha = 1$ for ligands with identical affinities for a receptor, Equation (8.22) reverts to Equation 8.19. This means that a normal response curve is obtained and the drug acts as a full agonist. However, when $\alpha = 0$ the percentage response is zero and the drug is a full antagonist. Intermediate values between 1 and 0 for α indicate a partial agonist (Figure 8.28).

In the 1950s Stephenson discovered that a maximum response was obtained when only a proportion of the available receptors were occupied. This discovery was in direct conflict with Clark's occupancy theory and led Stephenson independently to propose a two-stage route for receptor action. Independently of Ariens he proposed that the binding of a ligand to a receptor produced a stimulus (S) that was related to tissue response. The magnitude of the stimulus depends on both the affinity of the ligand for the receptor and its efficacy (e).

$$D + R \xrightarrow{\text{Affinity}} DR \xrightarrow{\text{Efficacy}} \text{Tissue response}$$

As a result, Clark's equation (Equation 6.19) was modified to:

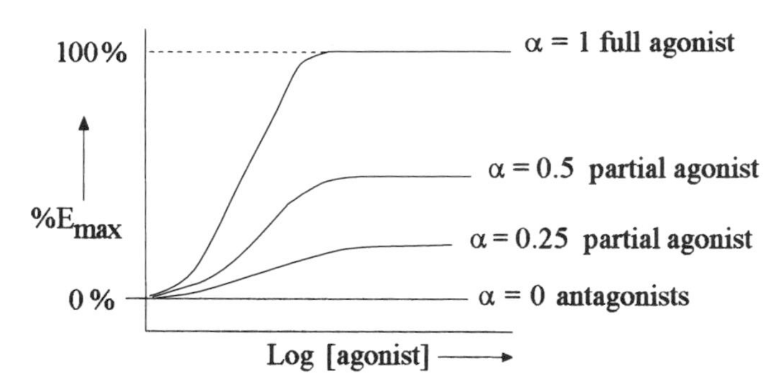

Figure 8.28. A pictorial representation of the variation of dose–response curves with the value of α. Values for α between 1 and 0 correspond to drugs that act as partial agonists, the degree of partial agonism depending on the value of α.

$$S = \frac{E}{E_{max}} = \frac{e[D]}{K_D + [D]} \tag{8.23}$$

Equation (8.23) shows that ligands with an e value of zero will have no biological response. Consequently, full antagonists will have an e value of zero. To obtain a positive response e must have a positive value and so agonists and partial agonists will have positive e values. Moreover, the higher the positive value, the greater the response (Figure 8.29) and the lower the dose of agonist required to achieve the maximum response. This means that agonists with a high efficacy will produce a maximum response even though they do not occupy all of the available receptor sites. Unoccupied receptors are known as **spare receptors**. Their presence increases the sensitivity of a receptor to other ligands.

It is now known that cells can contain several thousand receptors of a particular type. This number can increase (**upregulation**) or decrease (**downregulation**). These changes may be brought about by both pathological and physiological cell stimuli. They can affect drug response. For example, an increase (upregulation) in the number of receptors (R_T) moves the drug response curve to a lower concentration whereas a decrease will move it to a higher concentration (Figure 8.29).

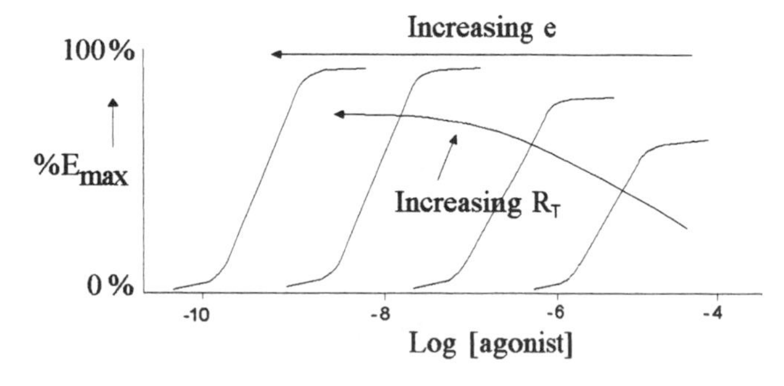

Figure 8.29. The effects on the dose–response curves of increasing e and R_T.

Stephenson observed that the magnitude of the response was not linearly related to the stimulus. This lead to a further modification of Clark's equation to:

$$S = \frac{E}{E_{max}} = f\left(\frac{e[D]}{K_D + [D]}\right) \tag{8.24}$$

where f is a function known as the **transducer function**. The transducer function represents the properties of the signal transducer mechanism that link the signal from the ligand to the tissue response, and is a characteristic of the responding tissue. As a result, the same ligand could have different transducer functions when it is bound to different tissues. This difference would explain why a ligand may act as an agonist in one tissue but as a partial agonist in a different tissue, even though it is acting on the same receptor. Furthermore, differences in the transducer functions of different ligands acting on the same receptor in the same tissue would also explain why their relative potencies may be different.

Potency depends on both the ligand–receptor complex and its efficacy. It is used to compare the relative effectiveness of different drugs and is defined as:

$$\text{Potency} = \varepsilon / K_D \tag{8.25}$$

where ε is the **intrinsic efficacy**, which is the efficacy per unit of receptors, that is:

$$\varepsilon = e / R_T \tag{8.26}$$

Because intrinsic efficacy is independent of the total number of available receptors, a drug with the same value for ε in different tissues is likely to be acting on the same receptor in those tissues. Conversely, if the values are different, the drug is likely to be acting on different receptors in the different tissues.

8.6.2 The Rate Theory

This theory was proposed by Paton in 1961 as an alternative to the occupancy theory. Paton proposed that the stimulus was produced only when the ligand first occupied the receptor site. Stimulation does not continue even though the site was still occupied. This is because the receptor undergoes a second conformational change that results in its inactivation. It is also likely that the new conformation enhances the binding of the agonist to that site, resulting in a more stable drug–receptor complex. Consequently, as long as the ligand is bound to the receptor, the receptor is unable to produce a further stimulus. As soon as the ligand disengages from the receptor, it returns to its original conformation. As a result, further stimulation of the receptor can now occur. Stimulation, according to the rate theory, is analogous to playing a piano:

the note (stimulation) only occurs when the key is struck. Holding the key down does not produce a continuous note. In contrast, the occupancy theory is like playing an organ: the note (stimulation) is maintained when the key is held down.

The rate theory suggests that the general type of activity exhibited by a drug is independent of the number of receptors, which explains the existence of spare receptors. Instead activity is a function of the rates at which the drug binds to and is released from the receptor. Full agonists rapidly bind and even more rapidly dissociate from the receptor. Partial agonists dissociate more slowly and antagonists will dissociate very slowly from the receptor. This theory, like Clark's theory, results in the establishment of a dynamic equilibrium between the ligand and the ligand–receptor complex. Consequently, the mathematics of rate theory is similar to that of the occupancy theory at equilibrium. According to rate theory, efficacy is related to the association and dissociation rate constants. However, correlation tends to be poor.

The rate theory also offers an explanation for desensitisation (see section 8.5.4). Ligands with a strong affinity for a receptor will remain bound when the second conformational change occurs. This second inactive conformation will be maintained for as long as the agonist occupies the receptor site. Consequently, because these receptors cannot respond to any further dose of the drug, the degree of response is smaller because the overall number of potentially active receptors is reduced. However, once the agonist leaves the receptor its structure reverts to its original conformation and the receptor is able to function normally. Obviously the rate of dissociation of the agonist will influence the degree of desensitisation. Furthermore, intrinsic efficacy becomes a measure of a ligand's ability to stabilise the ligand–receptor complex after binding.

8.6.3 The Two-state Model

In its simplest form, this model is based on the concept that receptors can exist in either an active or an inactive state. The active state is known as the **relaxed** or **R state** and the inactive state is referred to as the **tensed** or **T state**. Receptors in the R state can provide a stimulus but those in the T state are unable to produce a stimulus.

The two-state model postulates that in the absence of any ligands a population of receptors of the same type will consist of an equilibrium mixture of receptors in the R and T states:

$$\mathrm{T} \underset{k_{-1}}{\overset{k_1}{\rightleftharpoons}} \mathrm{R} \qquad (8.27)$$

where k_1 and k_{-1} are the rate constants for the forward and reverse processes, respectively. In the absence of ligands there will be no receptor stimulation, even though some of the receptors are in the R state. This situation may be regarded as being the *at rest* form of the receptor system. However, when the equilibrium of Equation (8.27) is to the right, by the interaction of a suitable ligand with the receptor the number of receptors in the R state will increase. This increase results in stimulation, which is followed by a tissue response. Conversely, ligands that cause the equilibrium to move to the left will inhibit stimulation and tissue response. This model offers a simple explanation of the general mechanism of action of agonists, antagonists and partial agonists.

Full agonists cause maximum receptor stimulation and so will have a strong affinity for the R state. This affinity will move the position of equilibrium to the right with a subsequent stimulation. For a drug to act as an agonist, $k_1 > k_{-1}$. The larger the ratio k_1/k_{-1}, the greater the efficacy of the drug. Antagonists do not cause any stimulation and so have a strong affinity for the T state. This affinity shifts the position of equilibrium to the left, resulting in no stimulation and no subsequent tissue response. For a drug to act as an antagonist, $k_1 < k_{-1}$; the smaller the ratio k_1/k_{-1}, the greater the efficacy of that drug. Partial agonists have an affinity for both states of the receptor, the degree of partial agonism exhibited depending on the relative values of k_1 and k_{-1}. This picture is a simplification and more detailed models, which are beyond the scope of this text, are discussed in a review by Kenakin (1987).

8.7 Drug Action and Design

Drugs can act at any stage of the signal transduction process. However, it is convenient to consider the drugs acting on receptors separately to those acting at other stages in the transduction process.

8.7.1 Agonists

Agonists often have structures that are similar to that of the endogenous ligand (Figure 8.30). Consequently, the normal starting point for the design of new agonists is usually the structure of the endogenous ligand or the structure of those parts of the ligand (its **pharmacophore**) that interact with the receptor. This information is normally obtained from a study of the binding of the endogenous ligand to the receptor using X-ray crystallography, nuclear magnetic resonance (NMR) and computerised molecular modelling techniques. However, it is emphasised that many agonists have structures that are not directly similar to those of their endogenous ligands.

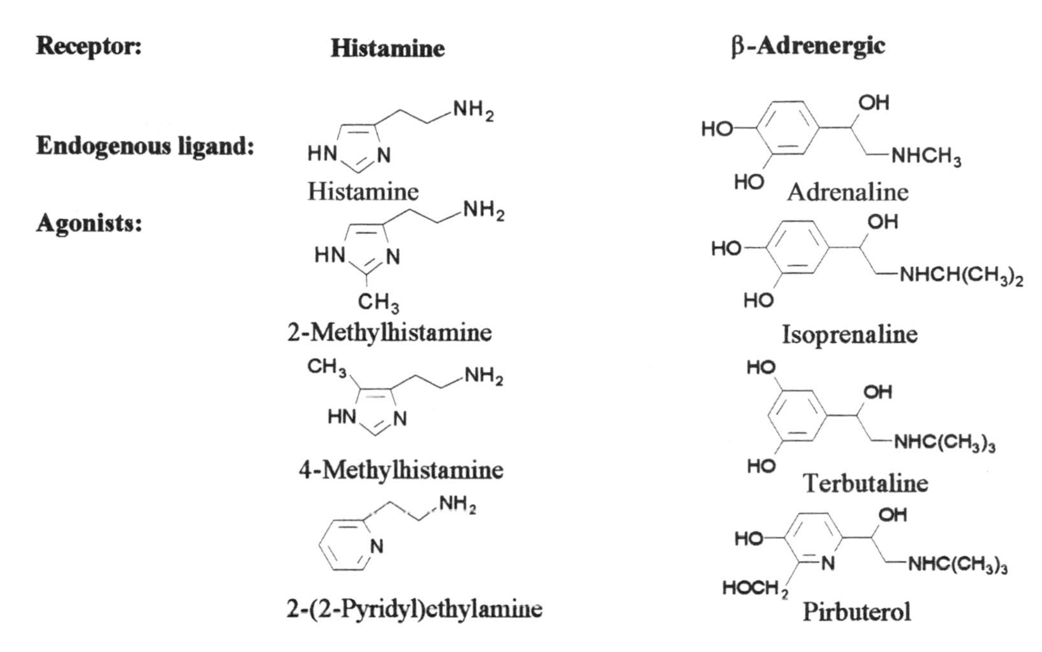

Figure 8.30. Examples of the structures of the agonists of some common receptors.

The usual approach to designing a new drug is to synthesise and investigate the activity of a series of compounds with similar structures to that of either the endogenous ligand or its pharmacophore (Table 8.2). This approach is based on the assumption that a new agonist is more likely to be effective if its structure contains the same binding groups and bears some resemblance to the endogenous ligand. It should be remembered that the groups involved in binding the endogenous ligand to the receptor are responsible for controlling both the binding of the endogenous ligand to the receptor and the ease with which it disengages from the receptor after its message has been received by the receptor. In other words, the nature of the binding groups controls the length of time a ligand remains bound to a receptor. Consequently, using mainly the same binding groups in the drug ensures that it has a good chance of fulfilling this requirement.

The nature of the binding groups is not the only structural factor that influences the activity of an agonist. The drug must also be of the correct size and shape to bind to and activate the receptor. Once again the initial approach is to use the structure of the endogenous ligand as a model. In the first instance the size of the pharmacophore of the potential drug should approximate to that of the pharmacophore of the endogenous ligand. However, it may be difficult to determine what section of the structure of the drug acts as its pharmacophore.

Information concerning the best shape for a new agonist may be obtained from a study of the conformations and configurations of a number of active analogues of the endogenous ligand.

Table 8.2. A series of compounds typical of that used in a search for a new agonist for the neurotransmitter acetylcholine. The activity is given in terms of the molar ratio needed to give the same degree of potency as acetylcholine.

	Structure	Activity	
		Cat blood pressure	Frog heart
Acetylcholine	$CH_3COOCH_2CH_2\overset{+}{N}(CH_3)_3$	1	1
	$CH_3COOCH_2CH_2\overset{+}{N}H(CH_3)_2$	50	50
	$CH_3COOCH_2CH_2\overset{+}{N}H_2CH_3$	500	500
	$CH_3COOCH_2CH_2\overset{+}{N}H_3$	2000	40000
	$CH_3COOCH_2CH_2\overset{+}{N}(C_2H_5)_3$	2000	10000
	$CH_3COOCH_2CH_2\overset{+}{P}(CH_3)_3$	13	12
	$CH_3COOCH_2CH_2\overset{+}{S}(CH_3)_2$	50	96

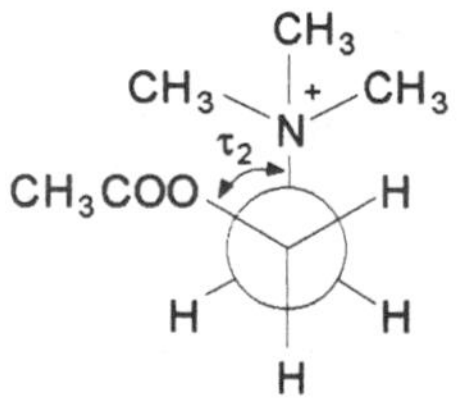

Acetylcholine

Structure	Torsion angle τ_2
$CH_3COOCH_2CH_2\overset{+}{N}(CH_3)_3\ Cl^-$ **Acetylcholine chloride**	+85
$CH_3COOCH(CH_3)CH_2\overset{+}{N}(CH_3)_3\ I^-$ **(+)Acetyl-2*S*-methylcholine iodide**	+85
$CH_3COOCH_2CH(CH_3)\overset{+}{N}(CH_3)_3\ I^-$ **(−)Acetyl-1*R*-methylcholine iodide**	+89
$CH_3COOCH(CH_3)CH(CH_3)\overset{+}{N}(CH_3)_3\ I^-$ **Acetyl-1*R*,2*S*-dimethylcholine iodide**	+76

Figure 8.31. The conformers of some acetylcholine receptor agonists. A torsion angle is positive when the bond of the specified front group of the drawn structure is rotated to the right, to eclipse the bond of the specified rear group. The angle is negative when the front group is rotated to the left.

For example, the torsion angle τ_2 of the endogenous mACh receptor agonist acetylcholine chloride (Figure 8.31) is +85°. Experimental work has shown that many agonist drugs acting on mACh receptors have τ_2 angles between +68° and +89°. Consequently, a proposed new drug will be more likely to exhibit agonistic cholinergic activity if its pharmacophore has a τ_2 angle within this range.

The configuration of a potential new agonist will also affect its shape and so should also be taken into account. For example, the effect of the cholinergic agonist (−)acetyl-2*S*-methylcholine chloride on guinea-pig ileum is about 24 times greater than its *R*(+) analogue, whereas (–)2*S*,3*R*,5*S*-muscarine iodide has about a 400 times greater effect than (+)2*R*,3*S*,5*R*-muscarine iodide on guinea-pig ileum.

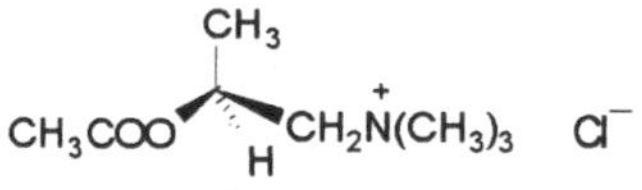

(–)Acetyl-2*S*-methylcholine chloride

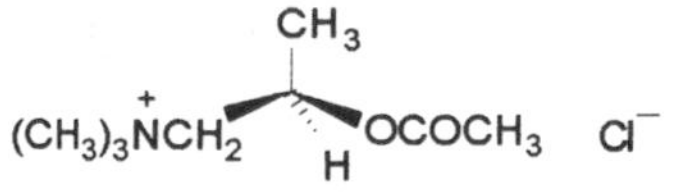

(+)Acetyl-2*R*-methylcholine chloride

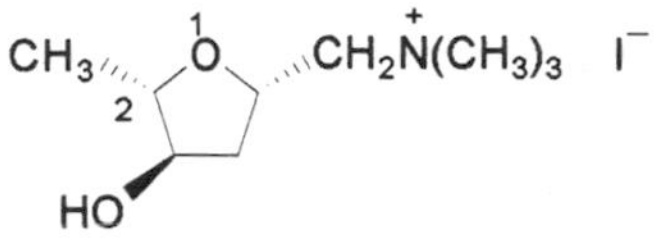

2*S*,3*R*,5*S*-Muscarine iodide

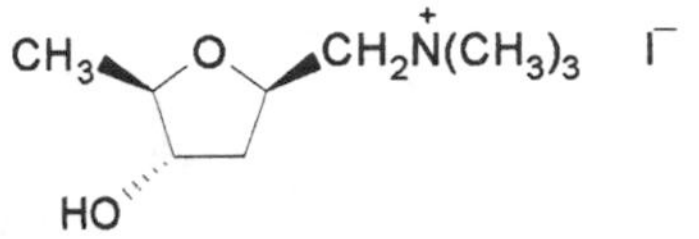

2*R*,3*S*,5*R*-Muscarine iodide

8.7.2 Antagonists

Antagonists act by inhibiting a receptor and so are used to reduce the effect of an endogenous ligand. Their action is classified as being either competitive or non-competitive (see section 8.5.2), depending on the nature of their binding to the receptor.

The structures of the antagonists acting on a receptor usually have little similarity to those of the endogenous ligands for the receptor (Figure 8.32). Consequently, the structure of the endogenous ligand cannot be taken as a good starting point for the design of a new antagonist. The ideal starting point for the design of a new antagonist would be the structure of the receptor. However, it is often difficult to identify the receptor and also obtain the required structural and stereochemical information. Consequently, although it is not the ideal starting point, many developments start with the structure and stereochemistry of either the endogenous ligand or any other known agonists and antagonists for the receptor. Because antagonists exert a stronger affinity for the receptor than its natural agonist, the binding groups selected for the new drug are often groups that could form stronger bonds with the receptor. As in the case of agonist drug design (see section 8.6.1), the relative sizes and shapes, that is, the conformations and configuration of the new antagonist and its binding site, should also be taken into account because this can have a significant effect on activity. For example, the antihistamine *S*(+)-dexchlorpheniramine is 200 times more potent than its *R*(–) isomer. If the structure of the receptor site is known in sufficient detail, molecular modelling (see section 2.5) can be of some use in the design of new antagonists.

8.8 Summary

Positive and negative **biological responses** are believed to be caused by the binding of a **ligand** to a receptor. Ligands are the endogenous and exogenous chemical agents that bind

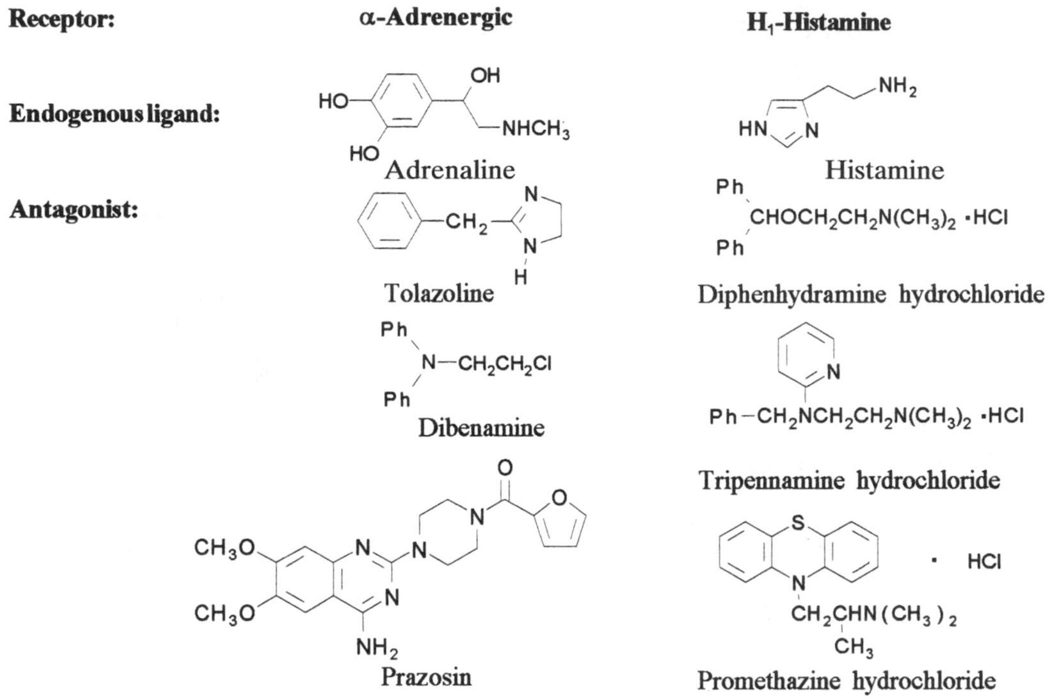

Receptor:	**α-Adrenergic**	**H_1-Histamine**
Endogenous ligand:	Adrenaline	Histamine
Antagonist:	Tolazoline	Diphenhydramine hydrochloride
	Dibenamine	Tripennamine hydrochloride
	Prazosin	Promethazine hydrochloride

Figure 8.32. Examples of some of the antagonists of some common receptors.

to a receptor. The endogenous ligands that bind to and activate a receptor are known as **first messengers**. First messengers are classified as neurotransmitters, hormones and autocoids.

Receptors are glycoproteins that are found either embedded in the membranes or inside the nuclei of cells. They are classified into **four superfamilies** according to their **general structures** and **modes** of action. The mechanism by which the chemical message carried by the ligand is converted by the receptor and its associated biochemical system is known as **signal transduction**. Signal transduction releases small molecules, known as **secondary messengers**, which target specific proteins within the cell. This stimulation results in the initiation of a series of biochemical events leading to a biological response. **Agonists** are drugs that give the same response as the endogenous ligand for a specific receptor. **Antagonists** are drugs that inhibit the response of a receptor. The binding of bacteria, viruses and toxins to a receptor may result in unwanted pharmacological effects.

Ligands bind to receptors by covalent bonding, ionic bonding (electrostatic bonding), dipole–dipole interactions (including hydrogen bonding), charge-transfer bonding, hydrophobic bonding and van der Waals' forces. Covalent bonding is the strongest form of ligand–receptor bond and it is usually irreversible. Ionic bonding between ligands and receptors is strong and normally reversible. It is effective over larger distances than the other forms of bonding. Hydrogen bonding between ligands and receptors is weak but stronger than the other forms of dipole–dipole interactions and hydrophobic bonding. However, although these types of bonding are weak they make a significant contribution to the binding of a ligand to a receptor because numerous bonds are formed between the ligand and the receptor. All the dipole–dipole bonds and hydrophobic bonding are reversible.

The binding of a **first messenger** or drug to a receptor is believed to cause a change in the conformation of that receptor. This change stimulates further biochemical changes that lead to a cellular response. The receptors of **Superfamily type 1** often require the binding of two endogenous ligand molecules for activation. These receptors control the opening and closing of ion channels. The receptors of **Superfamily type 2** have seven membrane-spanning subunits. These subunits form a pocket that is believed to be the binding site for endogenous ligands. The binding of a ligand to this type of receptor results in the attraction of a **G-protein** to an intracellular domain of the receptor. G-Proteins are proteins consisting of α-, β- and γ-subunits. They can freely diffuse through the cytoplasm. The binding of a G-protein to the receptor is followed by a series of biochemical changes (see Figure 8.7), which either results in the activation/deactivation of an enzyme system or the opening/closing of an ion channel. **Superfamily type 3** receptors have only one transmembrane subunit. Built into their extracellular structure is a tyrosine-kinase enzyme system. The binding of a ligand to a receptor is believed to result in dimerisation of the receptors and also phosphorylation of the tyrosine residues in the tyrosine-kinase domain. Proteins with an SH2 domain are attracted to and bind via their SH2 domain to the phosphorylated tyrosine-kinase domain, which triggers a wide variety of biological activities. **Superfamily type 4** receptors are found in the nuclei of living cells. These receptors are large proteins with conserved central sections containing two loops of about 12–15 amino acid residues held in position by coordination with zinc ions (zinc fingers). The binding of a ligand to the hormone receptor site, which lies on the C-terminal side of the central region, is believed to expose a DNA-binding domain, which ultimately leads to the production of specific mRNA that controls the production of the protein required for a specific cell response.

The **affinity** of a ligand is a measure of the ease with which a ligand binds to a receptor. **Clark's occupancy theory**, which envisaged the ligand–receptor system as being in the form of a dynamic equilibrium (see Equation 8.2), uses the **dissociation constant K_D** for the ligand–receptor complex as a measure of the affinity of a ligand for a receptor: the lower the value of K_D, the higher the affinity of the ligand for the receptor. K_D values are usually recorded as pD_2 values (see Equation 8.3). EC_{50} and ED_{50} (see Figure 8.17) are also used as a measure of the affinity of a ligand for a receptor. pD_2, EC_{50} and ED_{50} are used to compare the relative effects of the analogues of a lead compound in drug development.

Ligand dose–response relationships are normally determined *in vitro* using isolated tissue. Plots of response against molar concentration are hyperbolic in shape whereas those of percentage maximum response against log molar concentration are usually sigmoid in shape. **Agonists** with similar structures acting on the same receptor will often exhibit similar parallel plots of percentage maximum response against log molar concentration. In a series of compounds, agonists that exhibit the highest response are known as **full agonists** and those that exhibit a lower response are known as **partial agonists** in that series. Most drugs are partial agonists. **Antagonists** do not trigger a response when they bind to a receptor. Their presence inhibits the action of an agonist.

Antagonists are classified as **competitive** and **non-competitive** antagonists. The binding of a **competitive** antagonist to a receptor is reversible. Its action is dependent on the concentration of the agonist: the higher the concentration of the antagonist, the higher the concentration of a competitive agonist needed to obtain a maximum response (see Figure 8.21). The binding of a **non-competitive** antagonist to a receptor is irreversible. Its action is independent of the concentration of agonist: the higher the concentration of the non-competitive agonist, the lower the maximum response of the agonist (see Figure 8.23). The concentration at which an antagonist displaces half of the endogenous ligand from a receptor is known as the IC_{50} (see section 8.24) for that system. It is a measure of the affinity of the agonist for the receptor and has some use in drug development. Clark's original theory has been modified by **Ariens** and **Stephenson**. to account for the existence of agonists, partial agonists and antagonists. Independently, they proposed that the action of a receptor was a two-stage process. The first was the binding of the ligand to the receptor and the second was the initiation of the biological response. This second stage is governed by the ability of the ligand–receptor complex to produce a response. Ariens defined this ability to produce a response by the concepts of intrinsic activity (α), whereas Stephenson used efficacy (e). **Ariens** modified Clark's equation to the form shown in Equation (8.22): for full agonists $\alpha = 1$; for partial agonists α has a value between 0 and 1; for antagonists $\alpha = 0$ (see Figure 8.28). **Stephenson** introduced the concept that the binding of a ligand to the receptor produced a stimulus (S) that was related to the response. The stimulus was related to both the affinity of the ligand for the receptor as well as its efficacy. Stephenson modified Clark's equation to the form shown in Equation (8.23) to account for the fact that a maximum response may be obtained when only a proportion of the available receptors were occupied. For full antagonists e = 0, whereas for agonists and partial agonists e has a positive value. The higher the value of e, the higher the response and the lower the dose required to obtain a maximum response. Stephenson also introduced the concept of a transducer function to account for the fact that the magnitude of the response is not linearly related to the stimulus.

The **rate theory** proposed by **Paton** states that the stimulus only occurs when the ligand first occupies the receptor site. Stimulation does not continue even though the receptor site is still occupied. Further stimulation of the receptor can only occur when the ligand has disengaged from the receptor. The rate theory states that the rate of disengagement of the drug from the receptor governs its general type of activity. The order of disengagement rates is:

Agonists (very rapid) > Partial agonists > Antagonists (very slow).

Rate theory offers an explanation of **desensitisation** (**tachyphylaxis**). Desensitisation occurs when repeated exposure of a receptor to identical doses of a drug results in a reduction of the response to the drug (see section 8.5.4). The rate theory states that it occurs because the rate of dissociation of an agonist is so slow that some of the receptors are still occupied and therefore unable to be stimulated when the next dose of the agonist is given.

The **two-state model** of receptor action postulates that receptors exist in two forms, the relaxed state (R) and the tensed state (T). Receptors in the R state can provide a stimulus but those in the T state are unable to produce a stimulus. In the absence of any ligands a population of a group of receptors of the same type exist as a dynamic equilibrium mixture of receptors in the R and T states. Stimulation occurs when a ligand binds to a receptor in the R state, but no stimulation occurs when ligands bind to receptors in the T state. **Full agonists** have a strong affinity for the R state. When they bind to a receptor in its R state the equilibrium moves to increase the number of receptors in the R state, which results in stimulation. **Antagonists** have a strong affinity for the T state. When they bind to a receptor in its T state the equilibrium moves to increase the number of receptors in the T state, which inhibits stimulation. **Partial agonists** have an affinity for both states of the receptor, the degree of partial agonism depending on the position of equilibrium.

Agonists often have structures that are similar to the corresponding ligand (see section 8.7.1) but the structures of **antagonists** can be quite different to that of the endogenous ligand for the receptor (see section 8.7.2).

8.9 Questions

(1) Explain the meaning of the terms (a) ligand, (b) endogenous molecule, (c) agonist, (d) antagonist, (e) first messenger and (f) secondary messenger in the context of medicinal chemistry.

(2) Explain the meaning of the abbreviation m_2AChR in the context of the classification of receptors.

(3)

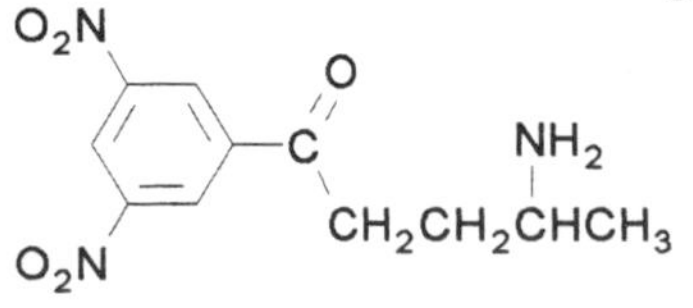

O_2N O_2N C O NH_2 $CH_2CH_2CHCH_3$

(A)

Peptide chain CONH $CH_2CH_2CHCONH$ H N CH_3 CH_3 CH_3

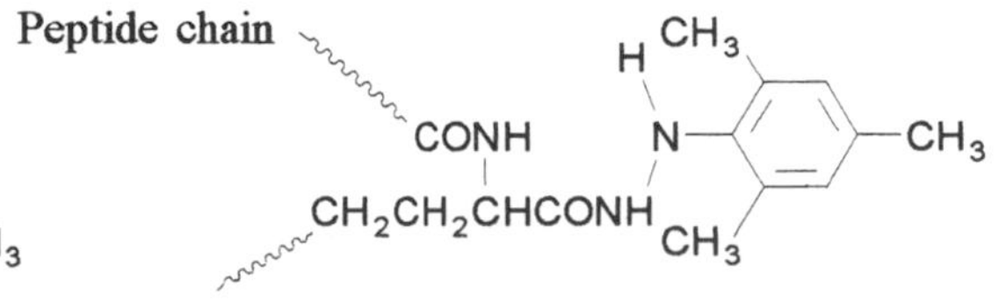

The structure of the receptor site B

Suggest where the compound A would form (a) strong bonds and (b) weak bonds with the receptor site B. The receptor B and compound A are drawn to the same scale.

(4) Distinguish between the terms hormone, neurotransmitter and autocoid.

(5) Explain the meaning of the term **affinity** in the context of ligand receptor relationships. What is the relationship between K_A, K_a and K_D.

(6) Outline how cAMP is generated within a cell by the action of a G-protein.

(7) Draw structural formulae for (a) DAG and (b) IP_3. What is the cellular function of these compounds?

(8) Distinguish between competitive and non-competitive antagonists.

(9) Outline the main features of the rate theory for explaining the relationships between ligands and receptors. How does this theory explain desensitisation?

(10) Outline a strategy for designing a new agonist for histamine receptors. The answer should include both the structures of possible analogues and an outline of the experimental work that would be needed to support the drug development.

9 Drug Metabolism

9.1 Introduction

Drug metabolism or biotransformations are the chemical reactions that are responsible for the conversion of drugs into other products within the body before and after they have reached their sites of action. Almost all of these reactions are enzyme catalysed and so they will exhibit the general characteristics of enzyme-controlled processes, that is, they will tend to be stereospecific, many will exhibit Michaelis–Menten kinetics (see section 6.8) and all will be affected by substrate concentration, pH and temperature (see section 8.7).

The amount of a drug reaching a receptor will depend on the quantity taken up by other tissues and how much of the drug is metabolised before it reaches its site of action. Consequently, a knowledge of the concepts of drug metabolism is useful in both the design of new drugs and the improvement of existing drugs. Furthermore, drug metabolism usually occurs by more than one route (Figure 9.1). Each route normally consists of a series of reactions that result in the formation of compounds (metabolites) that may also be pharmacologically active (see section 9.2). Consequently, in the development of a new drug it is important to document the behaviour of the metabolic products of a drug as well as that of their parent drug in the body.

9.1.1 Phase I and Phase II Metabolic Reactions

The reactions involved in drug metabolism are classified for convenience as **Phase I** (the so-called 'asynthetic route') and **Phase II** (synthetic route) reactions. Both of these types of reaction normally produce metabolites that are more water soluble and hence more easily excreted than the original drug.

In Phase I reactions the increase in water solubility is usually brought about by either introducing polar water-solubilising groups such as -OH, -COOH and -SO_3H or unmasking polar water-solubilising groups. The latter includes reactions such as hydrolysis. For example, procaine is metabolised in the bloodstream by hydrolysis catalysed by esterases to yield compounds with more polar acid and alcoholic groups.

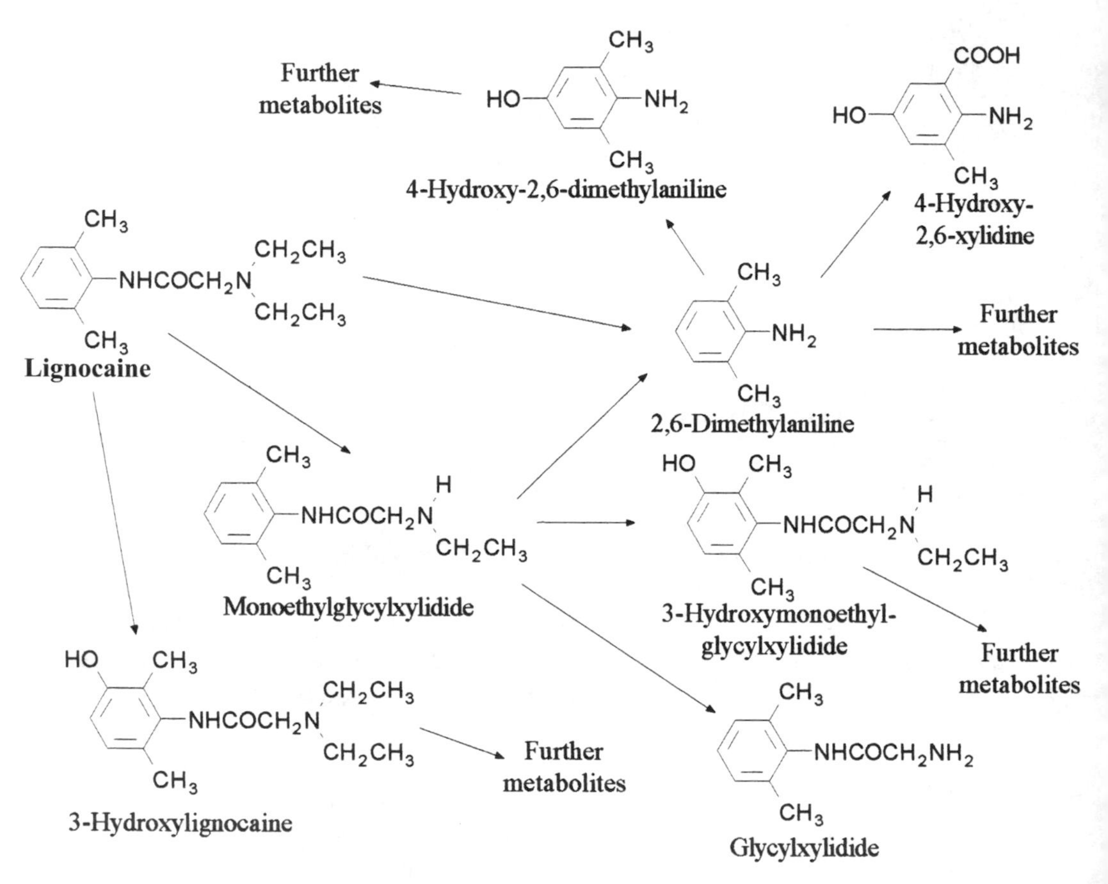

Figure 9.1. An outline of the known metabolic pathways of the local anaesthetic lignocaine.

Procaine (local anaesthetic) $H_2N-C_6H_4-COOCH_2CH_2N(CH_2CH_3)_2$

Esterases

4-Aminobenzoic acid $H_2N-C_6H_4-COOH$ + $HOCH_2CH_2N(CH_2CH_3)_2$ 2-Diethylaminoethanol

Phase II reactions, which are also known as **conjugation reactions**, increase water solubility by combining a water-solubilising reactant with the drug to form a highly polar derivative sometimes referred to as a **conjugate**. For example, the metabolism of aromatic carboxylic acids involves the conjugation of the acid with glycine via an acetyl coenzyme-A (CoA) intermediate. The hippuric acid conjugate is very water soluble and is readily excreted through the kidneys.

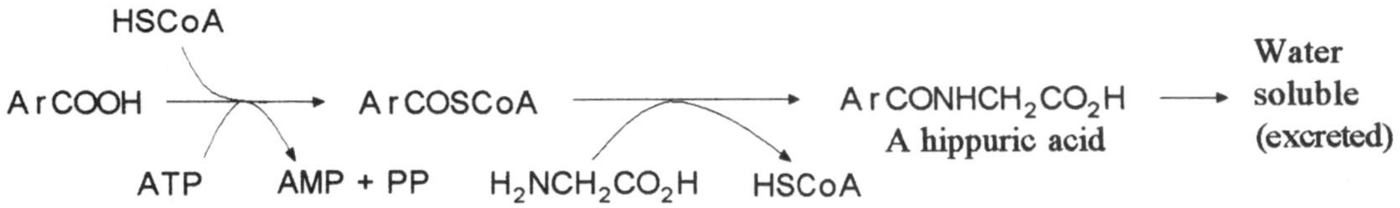

$$ArCOOH \xrightarrow[ATP \;\to\; AMP + PP]{HSCoA} ArCOSCoA \xrightarrow[H_2NCH_2CO_2H \;\to\; HSCoA]{} ArCONHCH_2CO_2H \longrightarrow \text{Water soluble (excreted)}$$

A hippuric acid

It is interesting to note that the administration of sodium benzoate and the measurement of the concentration of its hippuric acid metabolite in the urine was the basis of a test for checking the detoxification function of the liver. A reduced concentration of hippuric acid in the urine indicated the possibility of liver damage. However, this test is no longer used.

9.1.2 The Stereochemistry of Drug Metabolism

The body contains a number of non-specific enzymes that form part of its defence against unwanted xenobiotics. Drugs are metabolised both by these enzymes and by the more specific enzymes that are found in the body. The latter enzymes usually catalyse the metabolism of drugs that have structures related to those of the normal substrates of the enzyme and so are to a certain extent stereospecific (see section 6.5). The stereospecific nature of some enzymes means that enantiomers may be metabolised by different routes, in which case they could produce different metabolites (Figure 9.2).

In some cases an inert enantiomer is metabolised into its active enantiomer. For example, *R*-ibuprofen is inactive but is believed to be metabolised to the active analgesic *S*-ibuprofen.

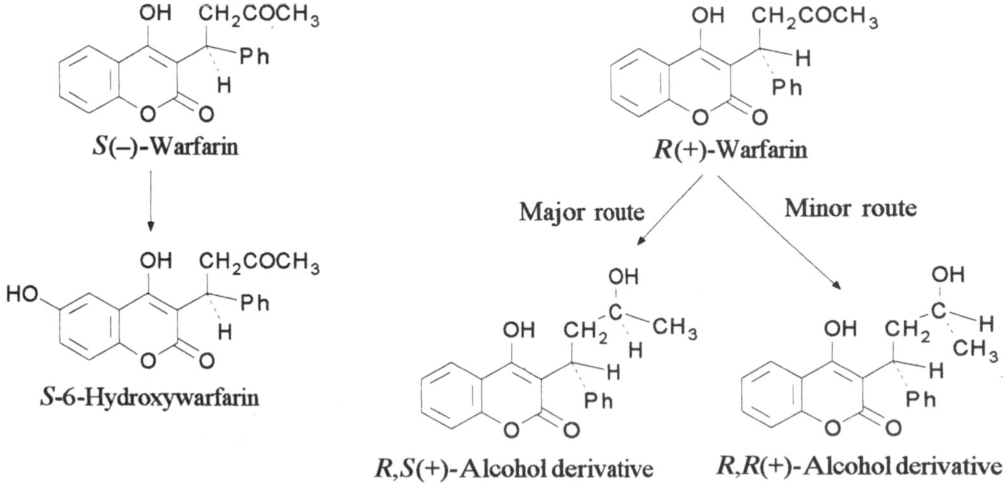

S(–)-Warfarin → *S*-6-Hydroxywarfarin

R(+)-Warfarin → Major route: *R,S*(+)-Alcohol derivative; Minor route: *R,R*(+)-Alcohol derivative

Figure 9.2. The different metabolic routes of *S*(–)-warfarin and *R*(+)-warfarin in humans.

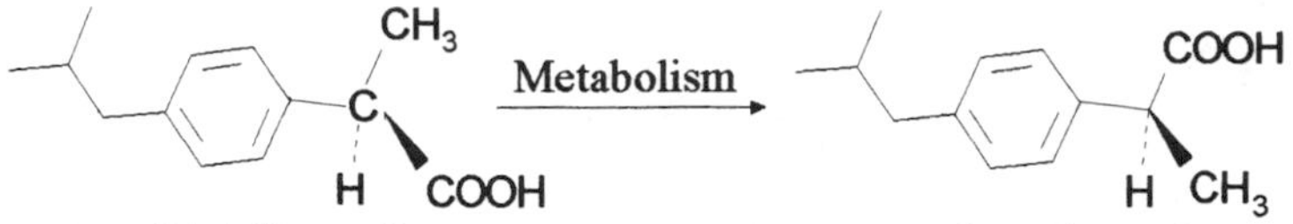

CH_3 Metabolism COOH

H COOH H CH_3

R(−)-Ibuprofen *S*(+)-Ibuprofen

A direct consequence of the stereospecific nature of many metabolic processes is that racemic modifications must be treated as though they contain two different drugs, each with its own pharmacokinetic and pharmacodynamic properties. Investigation of these properties must include an investigation of the metabolites of each of the enantiomers of the drug. Furthermore, if a drug is going to be administered in the form of a racemic modification, the metabolism of the racemic modification must also be determined because this could be different from that observed when the pure enantiomers are administered separately.

9.1.3 Biological Factors Affecting Metabolism

Metabolic differences are found within a species. The fundamental cause of these differences in healthy humans is believed to be due to variations in **age**, **gender** and **genetics**. Diseases can also affect drug metabolism. In particular, diseases that affect the liver, which is the major site for metabolism, will have a large effect on drug metabolism. Diseases of organs such as the kidneys and lungs, which are less important centres for metabolism, will also affect the excretion of the resulting metabolic products. Consequently, when testing new drugs, it is essential to design trials to cover all these aspects of metabolism.

(i) *Age.* The ability to metabolise drugs is lower in the very young (under 5 years) and the elderly (over 60 years). However, it is emphasised that the quoted ages are only approximate and the actual changes will vary according to the individual and his/her life style. In the foetus and the very young (neonates) many metabolic routes are not fully developed. This is because the enzymes required by metabolic processes are not produced in sufficient quantities until several months after birth. For example, when chloramphenicol was used to treat bacterial infections in premature babies it was found to have a high mortality rate, which fell considerably when the babies were 30 days old. This high mortality rate was attributed to the premature babies having too little glucuronate transferase, the enzyme that catalyses the conversion of the drug to its readily excreted water-soluble glucuronate. Consequently, the concentration of the drug built up to fatal levels in the babies. It is now known that the body synthesises glucuronate transferase over the first 30 days of life and as a result the mortality rate falls.

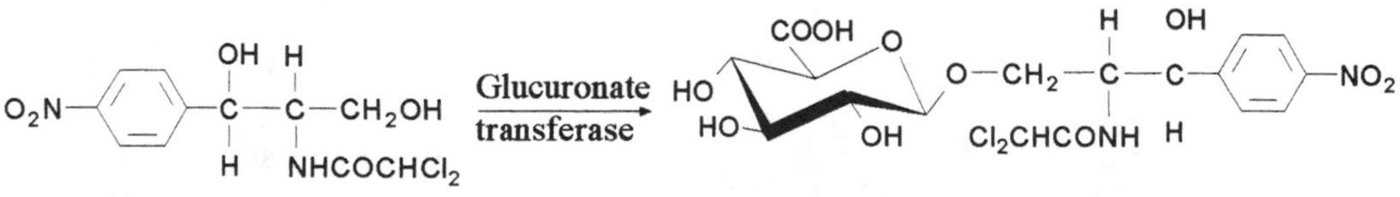

OH H

O_2N C C CH_2OH Glucuronate transferase

H $NHCOCHCl_2$

COOH O H OH

HO O CH_2 C C NO_2

HO OH $Cl_2CHCONH$ H

Chloramphenicol Chloramphenicol glucuronide derivative

Some xenobiotics are able to cross the placenta from the mother to the foetus and so any drugs used by the mother are also likely to pass into the foetus. Because a number of the enzyme systems that are required by the body for Phase II drug metabolism reactions have been found to be present in negligible to low concentrations in the foetus, these drugs can affect biological processes in the developing foetus, resulting in teratogenic and other undesirable effects. Furthermore, where drug metabolism occurs the more water-soluble metabolites of the drug are likely to accumulate on the foetal side of the placenta. Obviously, if these metabolic products are pharmacologically active they could also be detrimental to the development of the foetus. However, not all drugs cause damage to the foetus, administration of betamethasone to the mother several days before birth has been shown to aid the development of the surfactants in the lungs of the foetus.

Betamethasone

Children (above 5 years) and teenagers usually have the same metabolic routes as adults. However, their smaller body volume means that smaller doses are required to achieve the desired therapeutic effect. It is well documented that a person's ability to metabolise drugs decreases with age. This decline in a person's ability to metabolise drugs has been attributed to the physiological changes that accompany aging. It does not pose too many problems between the approximate ages of 20 and 60 years but can become more significant in the elderly. In general, after sixty the body gradually loses its capacity to metabolise and eliminate the drug and its metabolites (Figure 9.3). This leads to higher blood concentrations of the drug and the possibility of an increase in the adverse effects of the drug. This is normally countered by giving reduced doses of relevant drugs to the patient.

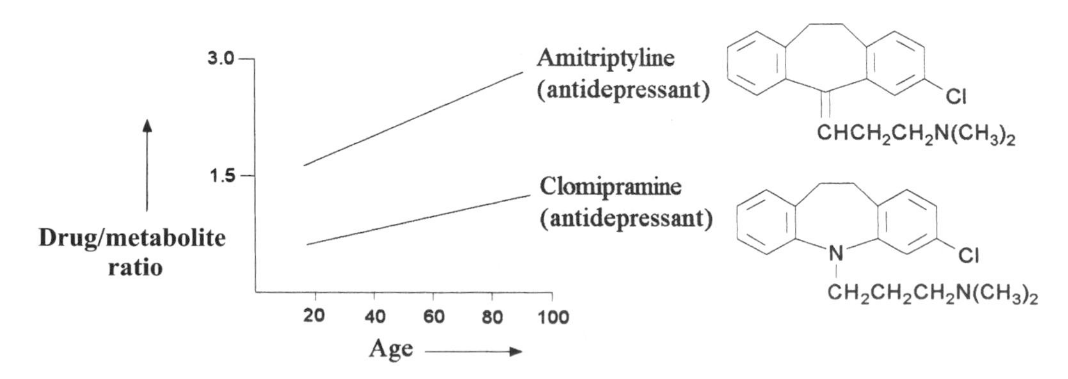

Figure 9.3. The variation of the drug/metabolite ratio of some drugs with age.

(ii) *Gender.* The metabolic pathway followed by a drug is normally the same for both males and females. However, some gender-related differences in the metabolism of anxiolytics, hypnotics and a number of other drugs have been observed. For example, Wilson has found that diazepam has an average half-life of 41.9 h in females but only 32.5 h in males. Such differences have been attributed in some instances to significant differences in enzyme concentrations. For example, women have a significantly lower concentration of alcohol dehydrogenase and so do not metabolise alcohol so rapidly as men.

Pregnant women will also exhibit changes in the rate of metabolism of some drugs. For example, the metabolism of both the analgesic pethidine and the antipsychotic chlorpromazine is reduced during pregnancy.

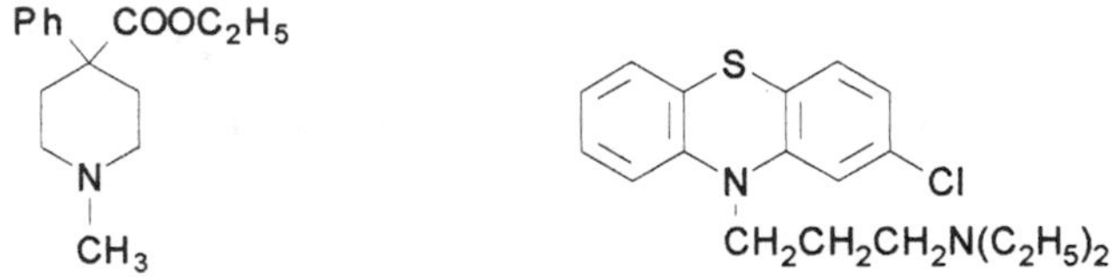

Ph $COOC_2H_5$ N CH_3

Pethidine

S N Cl $CH_2CH_2CH_2N(C_2H_5)_2$

Chlorpromazine

(iii) *Genetic variations.* Variations in the genetic codes of individuals can result in the absence of enzymes, low concentrations of enzymes or the formation of enzymes with reduced activity. These differences in enzyme concentration and activity result in individuals exhibiting different metabolic rates and in some cases different pharmacological responses for the same drug. For example, the short-acting neuromuscular blocking agent suxamethonium chloride owes its short duration of action to rapid metabolism by Phase I hydrolysis. However, some individuals cannot metabolise the drug by this route because they do not possess the necessary enzymes. Consequently, in these cases the duration of drug action is prolonged, with often fatal consequences.

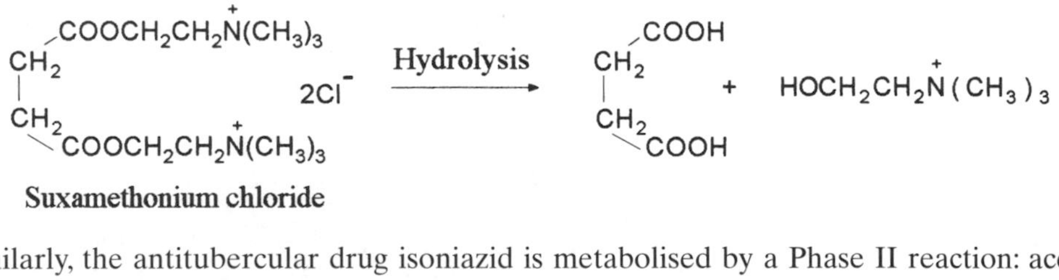

$$CH_2(COOCH_2CH_2\overset{+}{N}(CH_3)_3)-CH_2(COOCH_2CH_2\overset{+}{N}(CH_3)_3)\ 2Cl^- \xrightarrow{\text{Hydrolysis}} CH_2(COOH)-CH_2(COOH) + HOCH_2CH_2\overset{+}{N}(CH_3)_3$$

Suxamethonium chloride

Similarly, the antitubercular drug isoniazid is metabolised by a Phase II reaction: acetylation. In some patients this acetylation is fast and in others slow. Slow acetylation is found in 75% of Caucasians and Negroes but in only 10% of Japanese and Eskimos. Patients are often classified as fast or slow acetylators.

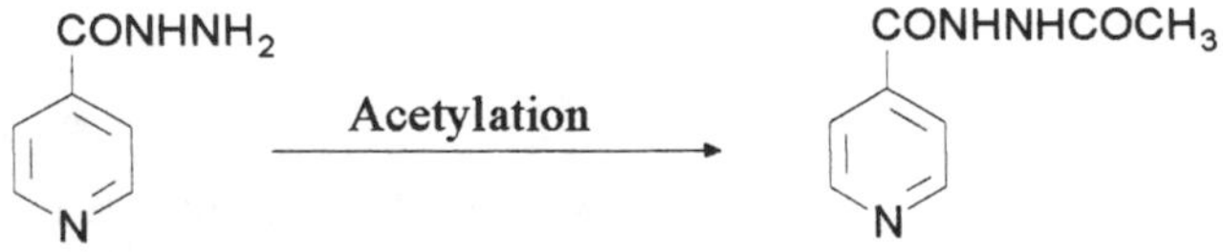

$CONHNH_2$ N —Acetylation→ $CONHNHCOCH_3$ N

Isoniazid

An individual's inability to metabolise a drug could result in that drug accumulating in the body. This could give rise to unwanted effects.

9.1.4 Environmental Factors Affecting Metabolism

The metabolism of a drug is also affected by life style. For instance, poor diet, drinking, smoking, self-medication and drug abuse may all have an influence on the rate of metabolism. A drug, for example, often competes for the same active site as another drug. Consequently, the use of over-the-counter self-medicaments can affect the rate of metabolism of an endogenous ligand or a prescribed drug. Because the use of over-the-counter medicaments is widespread, it can be difficult to assess the results of some large-scale clinical trials.

9.1.5 Species and Metabolism

Different species often respond differently to a drug. For example, a dose of $50\,mg\,kg^{-1}$ body mass of hexobarbitone will anaesthetise humans for several hours but the same dose will only anaesthetise mice for a few minutes.

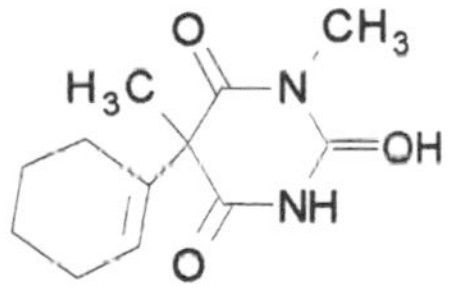

Hexobarbitone

The main reasons for the different reponses to a drug by members of different species are believed to be due to differences in their metabolism. These metabolic differences may take either the form of different metabolic pathways for the same compound or different rates of metabolism when the pathway is the same. Both deviations are thought to be due to enzyme deficiencies or sufficiencies.

9.2 Secondary Pharmacological Implications of Metabolism

Metabolites may be either pharmacologically inactive or active. Active metabolites may exhibit a similar activity to the drug, a different activity or be toxic.

9.2.1 Inactive Metabolites

Routes that result in inactive metabolites are classified as **detoxification** processes. For example, the detoxification of phenol results in the formation of phenyl hydrogen sulphate, which is pharmacologically inactive. This compound is very water soluble and so is readily excreted through the kidney.

Phenol —(Phenol sulphokinase; 3'-Phosphoadenosine-5'-phosphosulphate (PAPS))→ Phenyl hydrogen sulphate

9.2.2 Metabolites with a Similar Activity to the Drug

In this situation the metabolite can exhibit a different potency or duration of action, or both, with respect to the original drug. For example, the anxiolytic diazepam, which has a sustained action, is metabolised to the anxiolytic temazepam, which has a short duration of action. This in turn is further metabolised by demethylation to the anxiolytic oxazepam, which also has a short duration of action.

Diazepam —(Hydroxylation)→ Temazepam —(N-Demethylation)→ Oxazepam

9.2.3 Metabolites with a Dissimilar Activity to the Drug

In these cases the activity of a metabolite has no relationship to that of its parent drug. For example, the antidepressant iproniazid is metabolised by dealkylation to the antitubercular drug isoniazid.

Iproniazid —(N-Dealkylation)→ Isoniazid

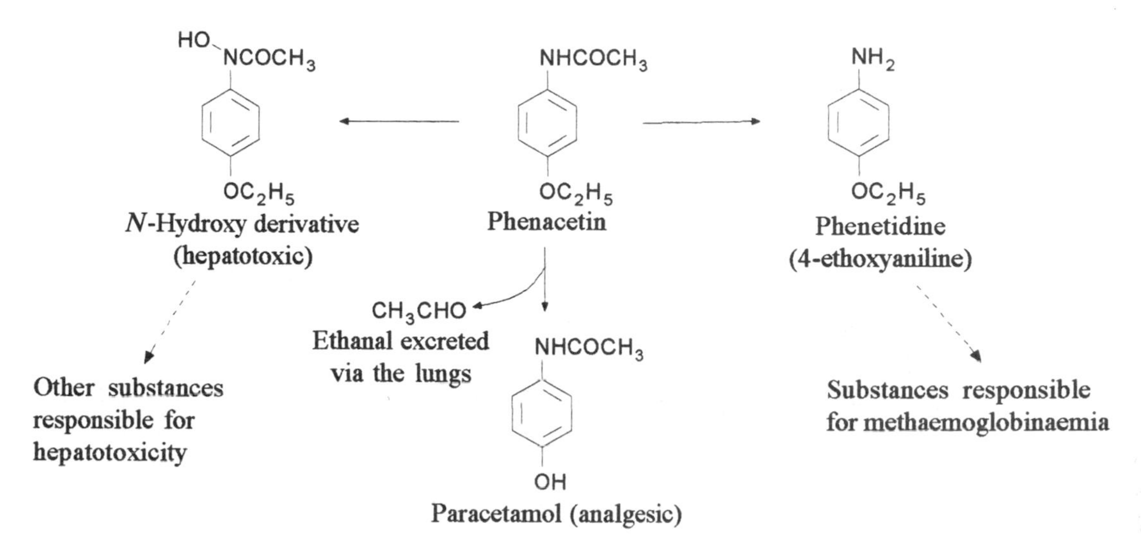

Figure 9.4. An outline of part of the metabolism of phenacetin.

9.2.4 Toxic Metabolites

The toxic action usually arises because the metabolite either activates an alternative receptor or acts as a precursor for other toxic compounds. For example, deacylation of the analgesic phenacetin yields *p*-phenetidine, which is believed to act as the precursor of substances that cause the condition methaemoglobinaemia. This condition, which causes headaches, shortness of breath, cyanosis, sickness and fatigue, is caused by the presence of methaemoglobin, a modification of haemoglobin in the blood (Figure 9.4). Phenacetin is also metabolised via its *N*-hydroxy derivative, which is believed to cause liver damage.

9.3 Sites of Action

Drug metabolism can occur in all tissues and most biological fluids. However, the widest range of metabolic reactions occurs in the liver. A more substrate-selective range of metabolic processes takes place in the kidney, lungs, brain, placenta and other tissues. These processes are often limited to a particular type of reaction, such as glucuronidisation.

Oral administration is the most popular route for the administration of drugs. Drugs administered in this manner can be metabolised as soon as they are ingested. However, the first region where a significant degree of drug metabolism occurs is usually in the gastrointestinal tract and within the intestinal wall. For example, a number of enzymes are responsible for the hydrolysis of the ester and amide groups of drugs in the GI tract whereas the bronchodilator isoprenaline undergoes considerable 3-O-methylation and sulphate conjugation within the intestinal wall.

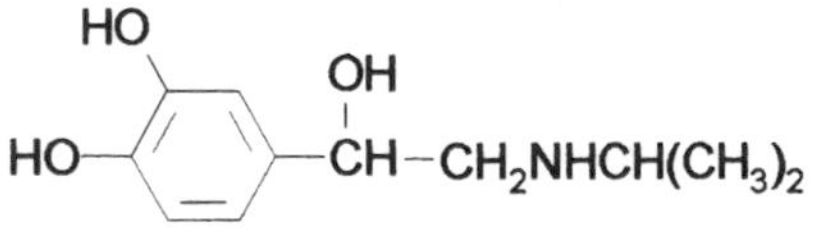

Isoprenaline

Once absorbed from the GI tract, many potential and existing drugs are extensively metabolised by first pass metabolism (see section 5.5). For example, the first-pass metabolism of some drugs like lignocaine is so complete that they cannot be administered orally. The bioavailability of other drugs, such as nitroglycerine (vasodilator), pentazocine (narcotic analgesic), propranolol (antihypertensive) and pethidine (narcotic analgesic), is significantly reduced by their first-pass metabolism.

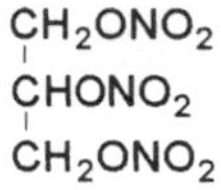

Nitroglycerine

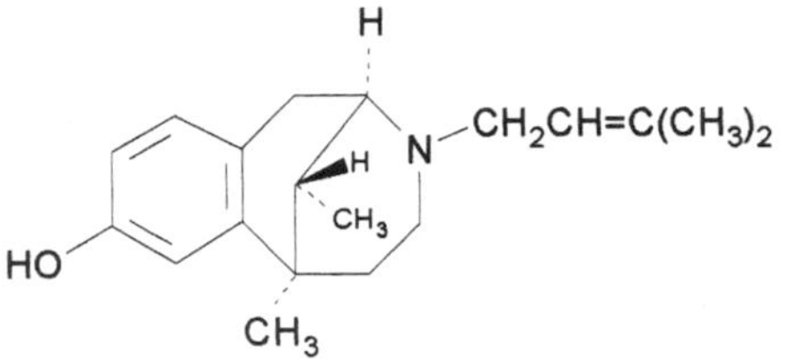

Pentazocine

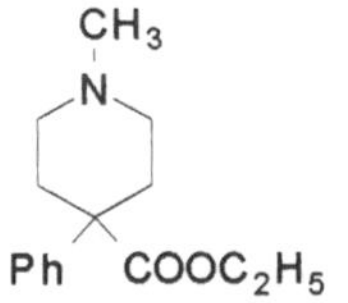

Pethidine

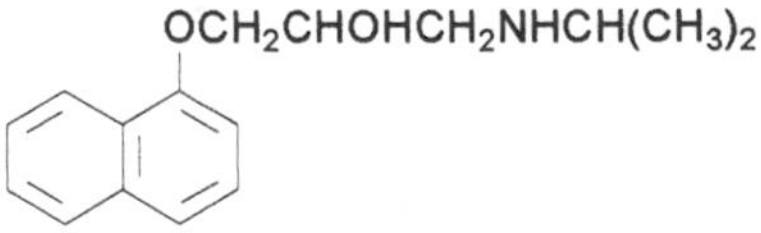

Propranolol

9.4 Phase I Metabolic Reactions

The main Phase I metabolic reactions are oxidation, reduction and hydrolysis. Within each of these areas the reactions undergone will largely depend on the nature of the available enzymes, the hydrocarbon skeleton of the drug and the functional groups present. A knowledge of the structural features of a molecule makes it possible to predict its most likely metabolic reactions and products. However, the complex nature of biological systems makes a comprehensive prediction difficult. Consequently, in practice the identification of the metabolites of a drug and their significance is normally carried out during its preclinical and Phase I trials. However, prediction of the possible products can be of some help in these identifications, although it should not be allowed to obscure the possible existence of unpredicted metabolites.

The identity of metabolites is obtained by a variety of investigative methods. A commonly used method is to incorporate radioactive tracers, such as C-14 and tritium (H-3 or T), into the drug. After ingestion by the test animal any radioactive compounds are isolated from the relevant organs, urine or faeces and identified by an appropriate analytical method. C-14 is preferred to tritium because the latter is prone to exchange reactions and the C—T bond can be broken in the course of the enzyme-catalysed reaction. Consequently, when using tritium, it must be incorporated into the structure of the drug in such a way that it cannot be replaced by other non-radioactive hydrogen isotopes or exchanged with the active hydrogen groups of compounds that occur naturally in the test animal.

9.4.1 Oxidation

Oxidation is by far the most important Phase I metabolic reaction. The main enzymes involved in the oxidation of xenobiotics appear to be the so-called **mixed-function oxidases** or **monooxygenases**, which are found mainly in the smooth endoplasmic reticulum of the liver but also occur, to a lesser extent, in other tissues. The mechanism by which these mixed-function oxidases operate may involve a series of steps, each step being controlled by an appropriate enzyme system. These steps can be either oxidative or reductive in nature. For example, oxidation of aliphatic C—H bonds involving the cytochrome P-450 family is believed to take place by a series of interrelated steps. These steps are catalysed by cytochrome P-450, cytochrome P-450 reductase and cytochrome b_5 reductase, with either NADPH or NADH as coenzymes (Figure 9.5).

Cytochrome P-450 (CYP-450) is a haem–protein molecule in which the haem residue is protoporphyrin IX containing coordinated iron. When the system is at rest the iron is in the fully oxidised Fe^{3+} state. It has been proposed that the substrate initially binds to the active site. This is followed by a reduction of the Fe^{3+} of the cytochrome P-450 to the Fe^{2+} state by a one-electron transfer catalysed by P-450 reductase. At this point molecular oxygen binds to the Fe^{2+} and is converted by a series of steps to an activated oxygen species. This species reacts to form the appropriate product.

The complete system is non-specific, catalysing the oxidation of a wide variety of substrates. This is probably due to both the non-selectivity of the enzyme and also the existence of numerous isozymes. Furthermore, lipid-soluble xenobiotics are good substrates for the cytochrome P-450 monooxygenases because high concentrations of the enzyme system are found in lipoidal tissue.

Flavin monooxygenases (FMO) are also an important family of non-selective mixed-function oxidases. These enzymes, which have a FAD (flavin adenine dinucleoide) prosthetic group, require either NADH or NADPH as coenzymes. They catalyse the oxidation of nucleophilic groups such as aromatic rings, amines, thiols and sulphides but will not metabolise substrates with anionic groups. Like cytochrome P-450, the mechanism of FMO-catalysed oxidation is

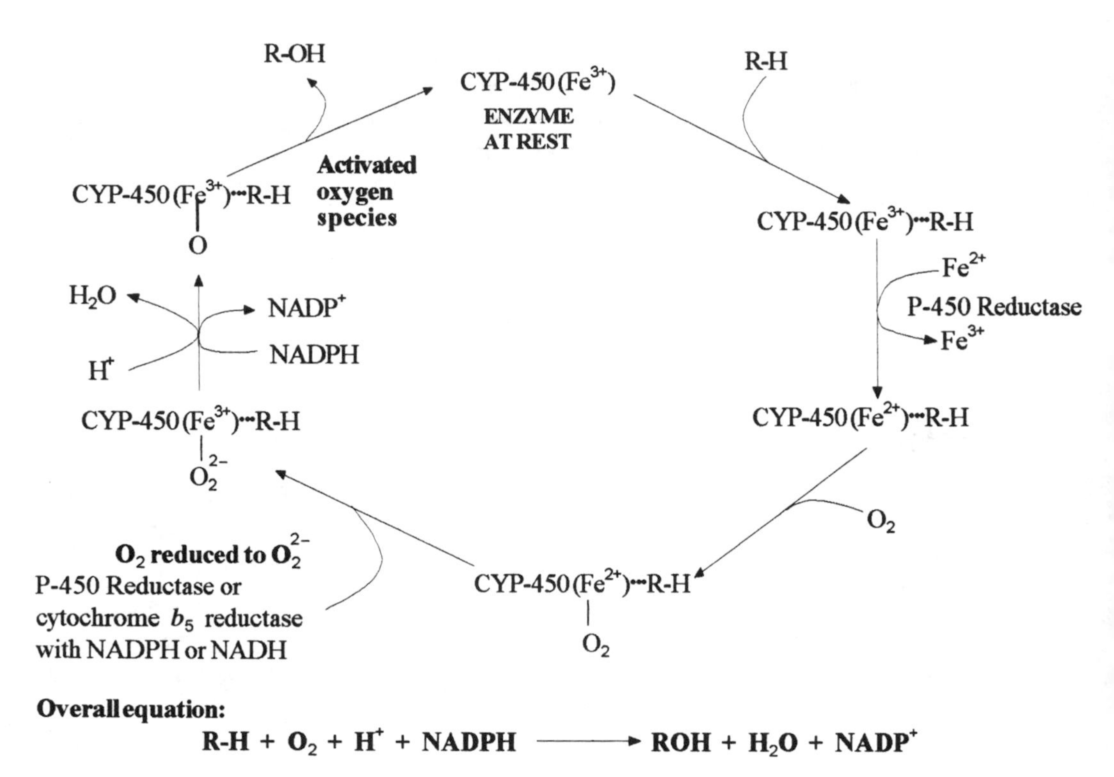

Figure 9.5. The proposed mechanism of action of cytochrome P-450. The abbreviation of CYP-450 is used for this system because the P-450(Fe^{2+})•••R-H complex forms a derivative with carbon monoxide that has a λ_{max} at 450 nm.

thought to involve a number of oxidative and reductive steps. The flavin residue is believed to play a major part in the oxidation, which is thought to occur by the attack of the nucleophile on the terminal oxygen of the hydroxyperoxide residue formed on this residue (Figure 9.6).

A number of other enzymes, such as monoamine oxidase, alcohol dehydrogenase and xanthine oxidase, are also involved in drug metabolism. These enzymes tend to be more specific, oxidising xenobiotics related to the normal substrate for the enzyme.

9.4.2 Reduction

Reduction is an important reaction for the metabolism of compounds that contain reducible groups, such as aldehydes, ketones, alkenes, nitro groups, azo groups and sulphoxides. The products of many of these reductions are functional groups that subsequently could undergo further Phase II reactions (see section 9.6) to form derivatives that are more water soluble than the original drug. Reduction of some functional groups results in the formation of stereoisomers.

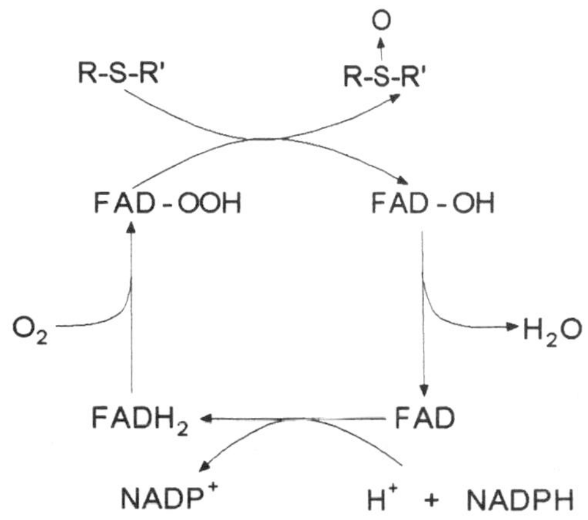

Figure 9.6. A simplification of the proposed mechanism of action of FMO in the metabolism of a sulphide.

Although this means that two metabolic routes may be necessary to deal with the products of the reduction, only one product usually predominates. For example, *R*(+)-warfarin is reduced to a mixture of the corresponding *R*,*S*(+) and *R*,*R*(+) diastereoisomers, the *R*,*S*(+) isomer being the major product.

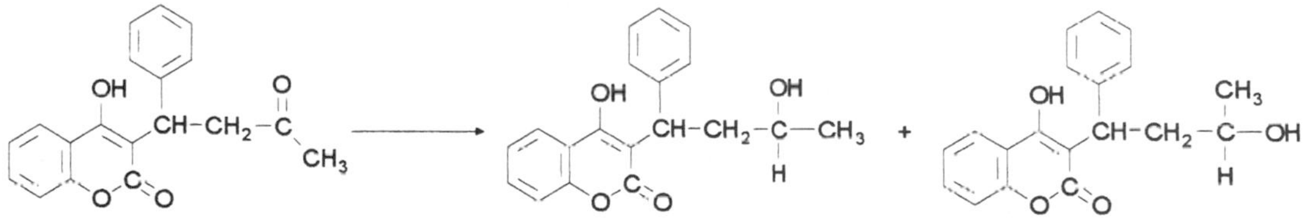

***R*(+)-Warfarin** → **1*R*,3*S*(+)-Alcohol (major product)** + **1*R*,3*S*(+)-Alcohol (minor product)**

The enzymes used to catalyse metabolic reductions are usually specific in their action. For example, aldehydes and ketones are reduced by soluble aldo–keto reductases, nitro groups by soluble nitro reductases and azo groups by multicomponent microsomal reductases. All these enzymes are found mainly in the liver but also occur in the kidneys and other tissues. Many of these enzymes require NADPH as a coenzyme.

9.4.3 Hydrolysis

Hydrolysis is an important metabolic reaction for drugs whose structures contain ester and amide groups. All types of esters and amides can be metabolised by this route. Ester hydrolysis is often catalysed by non-specific esterases in the liver, kidney and other tissues as well as pseudocholinesterases in the plasma. Amide hydrolysis is also catalysed by non-specific esterases as well as by non-specific amidases, carboxypeptidases, amino peptidases and decyclases. These enzymes are found in various tissues in the body. More specific enzyme systems are able to hydrolyse sulphate and glucuronate conjugates as well as hydrate epoxides, glycosides and other moieties. Hydration in the context of metabolism is the cleavage of a bond by the incorporation of water into the structure. In all these various reactions the products formed may be converted subsequently by Phase II reactions into inactive and/or more water-soluble conjugates that are more easily excreted.

The hydrolysis of esters is usually rapid whereas that of amides is often much slower. This makes esters suitable as prodrugs (see section 9.9) and amides a potential source for slow-release drugs.

9.5 Examples of Phase I Metabolic Reactions

The reactions in this section are classified according to functional group. Each of the functional groups in the drug is a potential starting point for a metabolic pathway. Consequently, it is emphasised that the reactions used in this section to illustrate a general process represent only one of the possible reactions by which the drug being used as the example could be metabolised. Moreover, it is again emphasised that most drugs are metabolised by several routes, each of which normally involves a series of reactions. In other words, the metabolite of one reaction becomes the starting point of a succeeding metabolic reaction, this process being repeated until the drug has been converted to a compound that can either be excreted or is pharmacologically inert. Because the rates at which compounds are metabolised vary, some metabolites may accumulate in the body (see section 9.7). In addition, some metabolites accumulate in the body because there is no mechanism for their excretion. These compounds may have an adverse effect on health.

The existence of this sequence of metabolites increases the possibility of a toxic reaction and so a new drug should be designed to be metabolised and excreted by as short a pathway as possible once it has acted.

9.5.1 Alkanes

Saturated aliphatic C—H bonds are normally metabolised by hydroxylation. Hydroxylation normally occurs at either the ω or ω-1 C—H bond of a saturated hydrocarbon chain or the α-C—H bond next to an electron-withdrawing group, such as a benzene ring, chloro, aliphatic amino, keto or ester group. The latter oxidation is preferred to the ω or ω-1 hydroxylation. For example, the methyl group of tolbutamide is hydroxylated in preference to its butyl side chain.

General reactions:

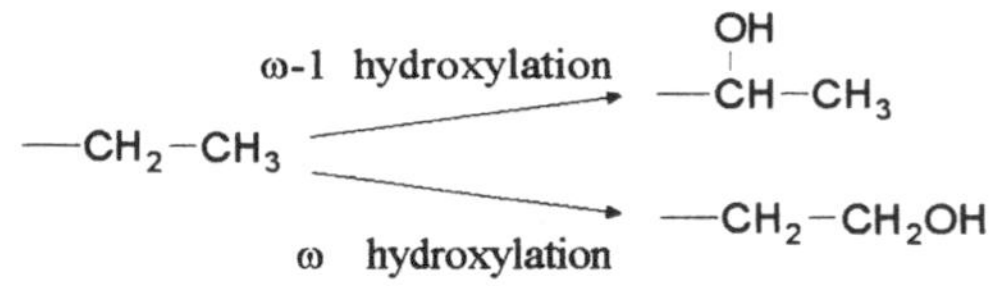

Examples:

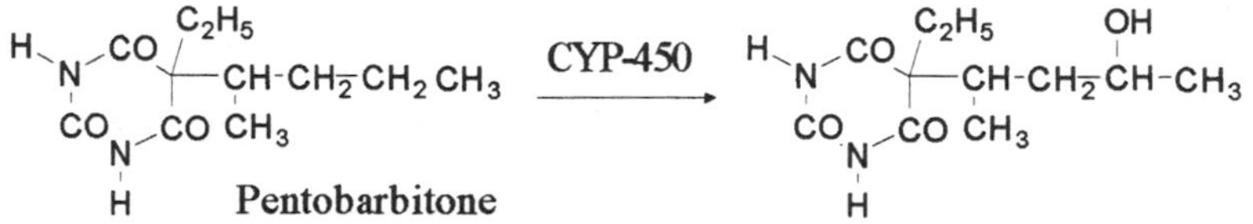

Pentobarbitone (sedative, hypnotic)

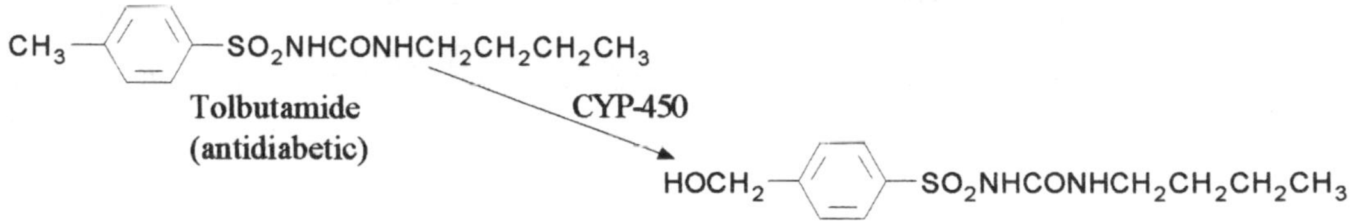

Tolbutamide (antidiabetic)

9.5.2 Alkenes

The main route for the metabolism of alkene C=C bonds is metabolic oxidation to the corresponding epoxide (oxirane). These oxiranes are relatively stable but are thought to undergo further reaction with nucleophiles (see section 9.5.13). For example, the anticonvulsant carbamazepine has been shown to form the stable 10,11-epoxide, which is believed to undergo subsequent hydration to the 10,11-diol that has been isolated from the urine of patients taking the drug.

General reaction:

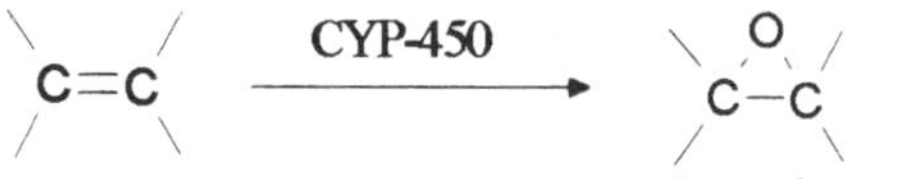

An oxirane

Examples:

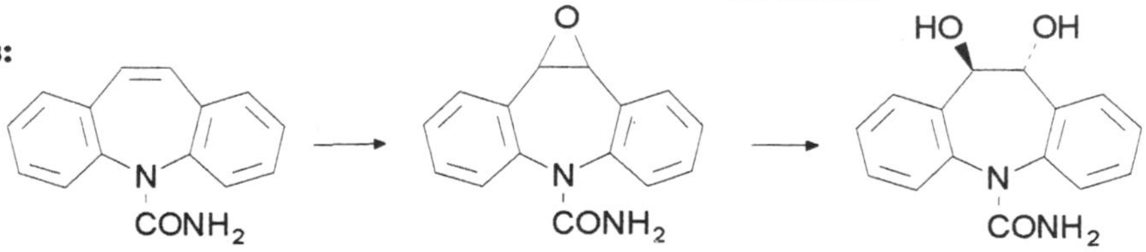

Carbamazepine (anticonvulsant)

Carbamazepine-10,11-epoxide (a possible anticonvulsant)

Carbamazepine-10,11-diol (excreted)

Reduction of alkenes is rare but some α,β-unsaturated ketones are reduced. For example, the hormone progesterone is reduced to a mixture of the 5α- and 5β-pregnane-3,20-diones.

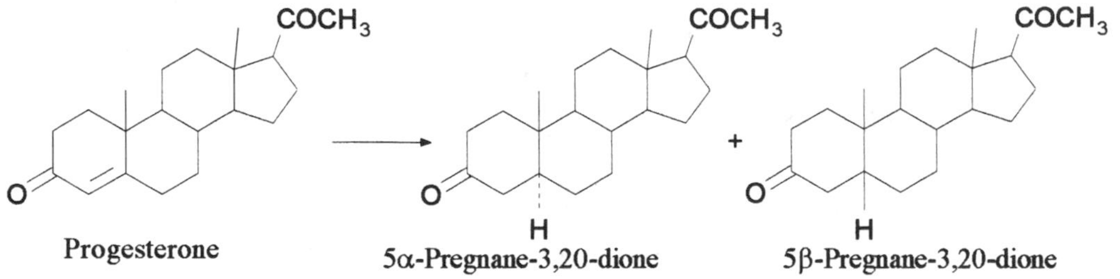

$COCH_3$ $COCH_3$ $COCH_3$

Progesterone → 5α-Pregnane-3,20-dione + 5β-Pregnane-3,20-dione

9.5.3 Alkynes

Some alkyne groups are thought to be oxidised to very reactive oxirenes, which undergo further reaction in a number of ways.

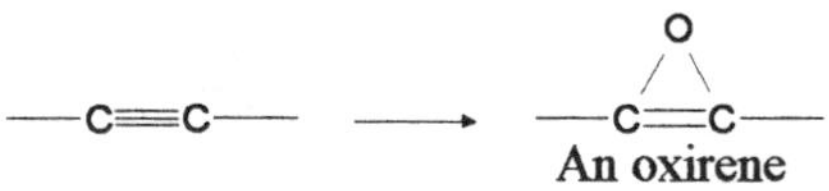

—C≡C— → An oxirene

Most alkyne groups appear to be stable to metabolism and compounds containing these groups are metabolised by reactions involving other groups. For example, the progestogen norgestrel is mainly metabolised by reduction of the ketone group.

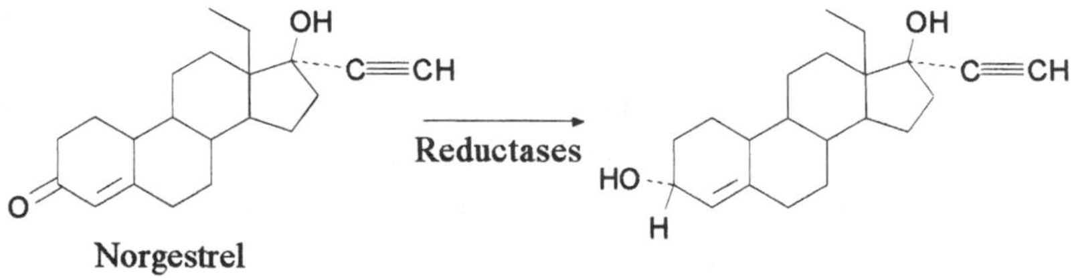

Norgestrel —Reductases→

9.5.4 Aromatic C—H

The main metabolic oxidation of aromatic C—H bonds is hydroxylation, often in the *para* position. Hydroxylation is believed to proceed by an epoxide intermediate known as an arene oxide. The phenolic metabolites of these hydroxylation are often metabolised further to the corresponding water-soluble sulphates and glucuronates that are excreted in the urine.

General reaction:

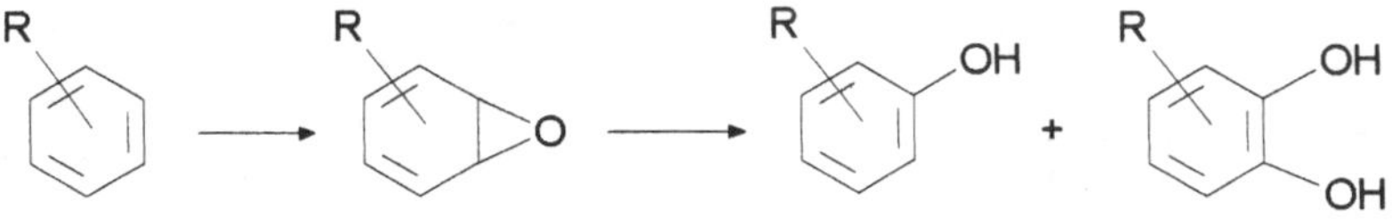

Example:

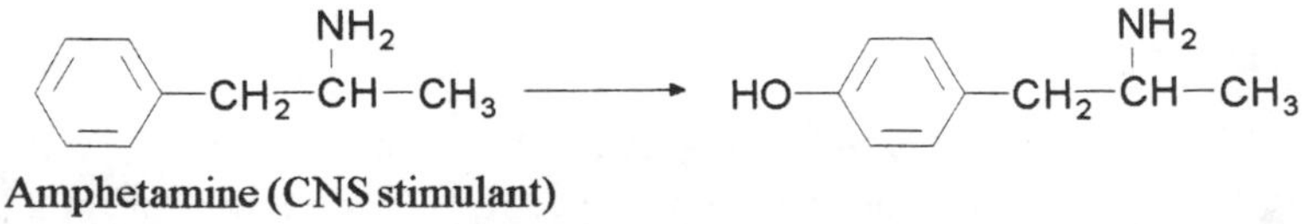

Amphetamine (CNS stimulant)

The reactive electrophilic nature of the epoxide ring means that the arene oxide intermediates are likely to react spontaneously with the nucleophilic functional groups of any adjacent endogenous biological molecules, such as glutathione, nucleic acids and proteins. These reactions have been shown, in some cases, to result in the formation of compounds that could affect the normal biological function of the original biological molecule.

Key:
GSH Glutathione
M-Nu Biologically active molecule with a nucleophilic group

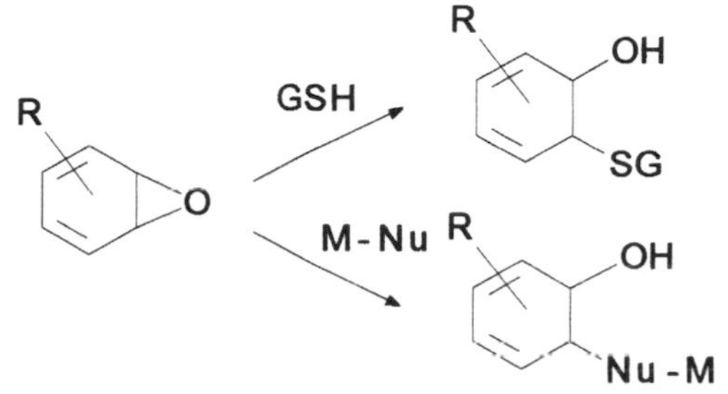

R, O, GSH, R, OH, SG, M-Nu, R, OH, Nu-M

9.5.5 Heterocyclic Rings

Metabolism of heterocyclic rings usually involves either the hetero atom or hydroxylation of an α-carbon atom. The products of this hydroxylation are usually unstable, often decomposing by reactions involving cleavage of the carbon–heteroatom bond (see section 9.5.9). Sulphur heteroatoms are usually oxidised to the corresponding sulphoxide or sulphone. Similarly, unsaturated nitrogen heteroatoms may be metabolised to the corresponding oxide. However, this reaction is usually a minor metabolic pathway for drugs. Hydroxylation of aromatic heterocyclic rings is also rare. However, heterocyclic rings with a secondary amino hetero-group may be converted to the corresponding lactam.

General reactions:

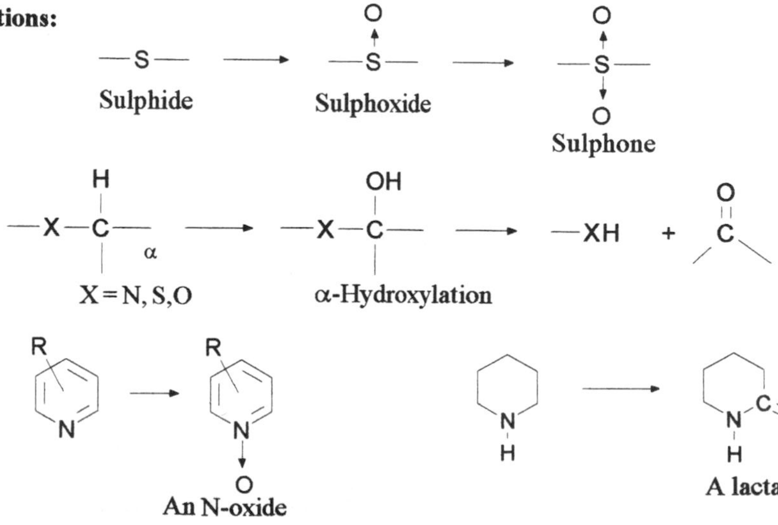

—S— → —S(→O)— → —S(→O)(→O)—

Sulphide, Sulphoxide, Sulphone

—X—C(H)— → —X—C(OH)— → —XH + C=O

α, X=N, S,O, α-Hydroxylation

R, N → R, N→O

An N-oxide

N-H → N-H, C=O

A lactam

Examples:

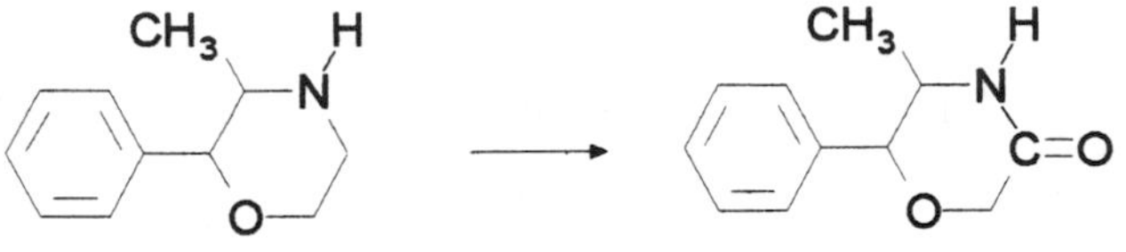

Phenmetrazine (anorectic)

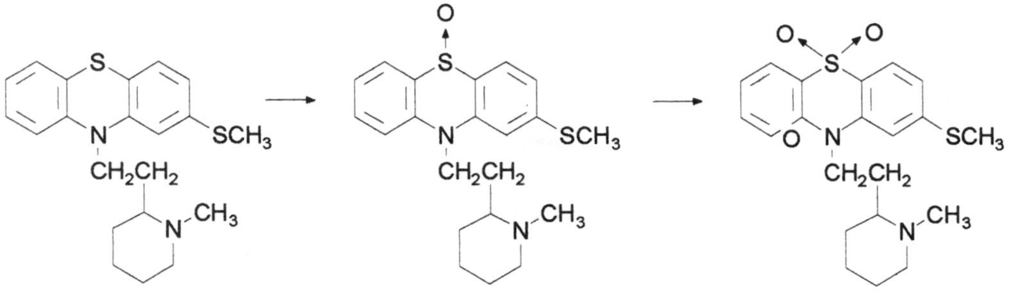

Thioridazine (antipsychotic)

9.5.6 Alkyl Halides

Oxidative dehydrohalogenation is an important route for many alkyl halide groups in drugs and xenobiotics. It is believed that this reaction involves an initial CYP-450-mediated α-hydroxylation followed by a spontaneous elimination of a hydrogen halide to yield carbonyl derivatives such as aldehydes, ketones and acyl halides. *gem*-Trihalides usually react more readily than *gem*-dihalides and monohalides.

Key:
Hal = F, Cl or Br

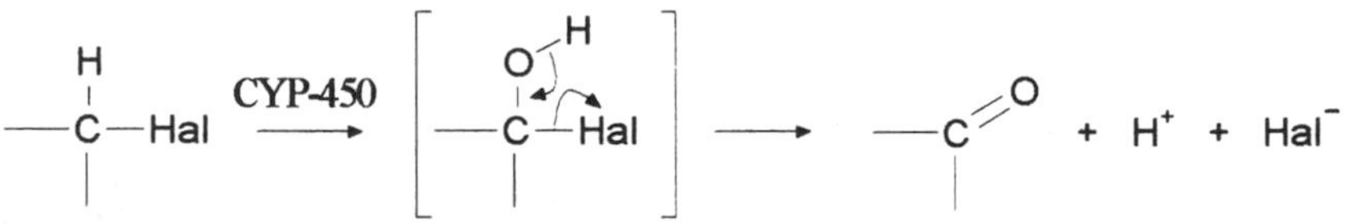

CYP-450

+ H^+ + Hal^-

The reactive acyl and carbonyl compounds produced by the oxidative dehydrohalogenation may react further with water to form less toxic carboxylic acids. However, these intermediates may also react with the nucleophilic groups of the many naturally occurring biologically active molecules present, such as DNA, proteins, carbohydrates and lipids. These subsequent reactions could explain the toxicity of some drugs because these reactions may prevent the normal function of the biological molecules. For example, it is believed that the toxicity of chloramphenicol could be due to this type of interaction:

Chloramphenicol (antibiotic) → Oxamyl chloride derivative + HCl

Oxamyl chloride derivative + H_2O → Chloramphenicol oxamic acid derivative + HCl

Oxamyl chloride derivative + Nu—**Biological molecule** → Nu-**Biological molecule** adduct + HCl

Reductive metabolism of polyhalogenated compounds can also occur by a free radical mechanism. Normally this type of metabolic process results in the formation of halides with fewer halogen atoms, alkenes and reactive carbenes or produces peroxide free radicals by reaction with oxygen. The reactive carbenes and peroxide free radicals are potential sources of toxic products in the body.

General reactions:

One-electron reduction (e, −Hal⁻); Oxygen; (hydrogen abstraction); A peroxide free radical; A carbene formed by a one-electron reduction

Examples:

Dichlorodiphenyltrichloroethane (DDT, an insecticide) → (−HCl) Dichlorodiphenyldichloroethene (DDD); → (−Cl⁻) Dichlorodiphenyldichloroethane (DDE)

The use of DDT is now banned in many countries because it causes convulsions and respiratory paralysis in mammals. Like DDT, both DDE and DDD accumulate in the fatty tissues of the body.

9.5.7 Aryl Halides

Aryl halides are usually too stable to take part in metabolic reactions. Most aryl halides utilise a metabolic route that is based on either hydroxylation of the aromatic ring or one of the other functional groups in the drug or xenobiotic. If the structure of the drug does not contain a group that can be utilised by the body's metabolic pathways the drug is likely to accumulate in the fatty tissues of the body, which may result in a toxic reaction.

9.5.8 Alcohols

Primary alcohols are normally metabolised by oxidation to the corresponding aldehyde at a relatively rapid rate. Subsequently, these aldehydes are frequently metabolised further to the corresponding carboxylic acid. Secondary alcohols are metabolised more slowly to the appropriate ketone. This reaction is not a very important metabolic route because the ketones produced are often readily reduced back to the alcohol. The C—OH group of tertiary alcohols is not usually metabolised by oxidation.

General reactions:

Primary alcohol: $RCH_2OH \xrightarrow[NAD^+ \rightarrow NADH]{\text{Alcohol dehydrogenase}} RCHO \longrightarrow RCOOH$

Secondary alcohol: $RR'CHOH \xrightarrow[NAD^+ \rightarrow NADH]{\text{Alcohol dehydrogenase}} RR'C{=}O$

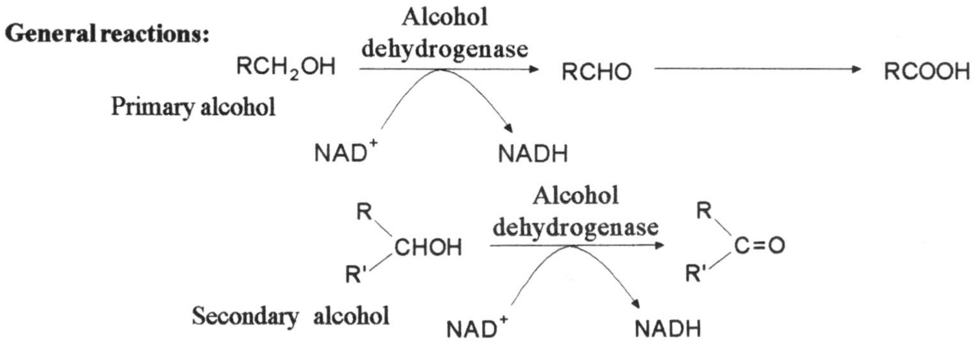

9.5.9 Amines

All types of aliphatic amine can be oxidised by CYP-450 and FMO. However, each class of amine yields different types of compound.

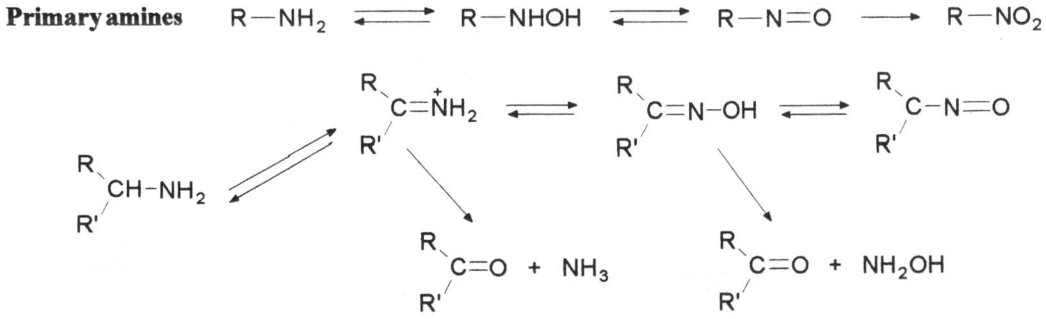

General reactions:

Primary amines $R{-}NH_2 \rightleftharpoons R{-}NHOH \rightleftharpoons R{-}N{=}O \longrightarrow R{-}NO_2$

$RR'CH{-}NH_2 \rightleftharpoons RR'C{=}\overset{+}{N}H_2 \rightleftharpoons RR'C{=}N{-}OH \rightleftharpoons RR'C{-}N{=}O$

$RR'C{=}\overset{+}{N}H_2 \longrightarrow RR'C{=}O + NH_3$

$RR'C{=}N{-}OH \longrightarrow RR'C{=}O + NH_2OH$

Secondary amines $R(R')NH \rightleftharpoons R(R')N{-}OH$ **Tertiary amines** $R(R')(R'')N \rightleftharpoons R(R')(R'')N \rightarrow O$

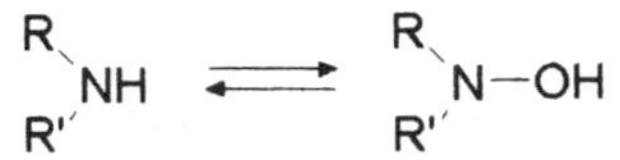

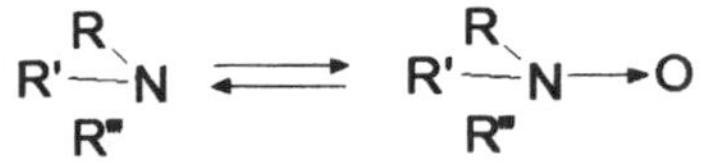

Examples:

$C_6H_5{-}CH_2{-}C(CH_3)_2{-}NH_2 \longrightarrow C_6H_5{-}CH_2{-}C(CH_3)_2{-}NHOH$

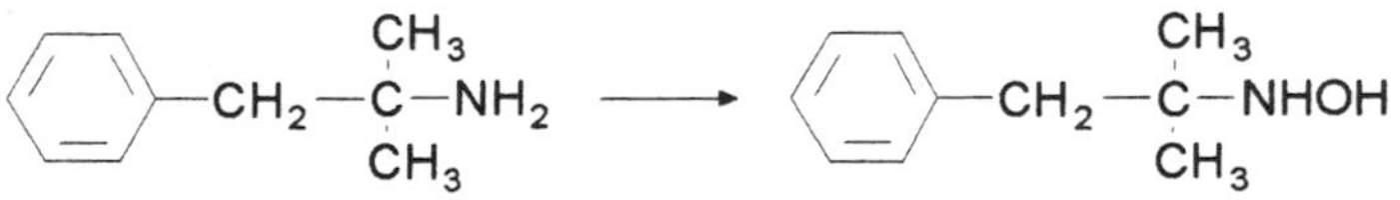

Phentermine
(anorectigenic agent)

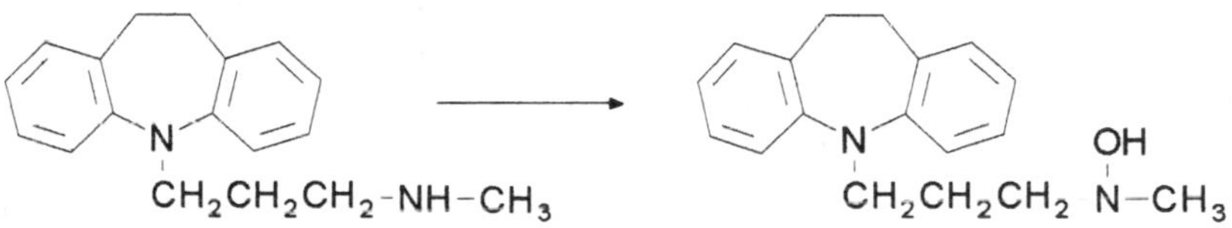

Desmethylimipramine ($N{-}CH_2CH_2CH_2{-}NH{-}CH_3$) $\longrightarrow$ $N{-}CH_2CH_2CH_2\ N(OH){-}CH_3$

Desmethylimipramine
(antipsychotic)

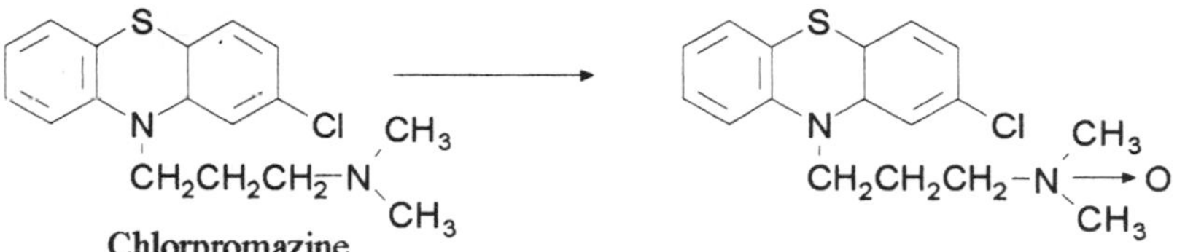

Chlorpromazine ($N{-}CH_2CH_2CH_2{-}N(CH_3)_2$, Cl) $\longrightarrow$ $N{-}CH_2CH_2CH_2{-}N(CH_3)_2 \rightarrow O$

Chlorpromazine
(antipsychotic)

Chlorpromazine oxide (34%)

Primary amines attached to a methylene group may also be converted to the corresponding aldehyde by monoamine oxidase (MAO). These aldehydes may be oxidised further to the carboxylic acid.

General reactions: $RCH_2NH_2 \xrightarrow{\text{MAO}} RCHO \xrightarrow[\text{oxidation}]{\text{Further}} RCOOH$

Examples:

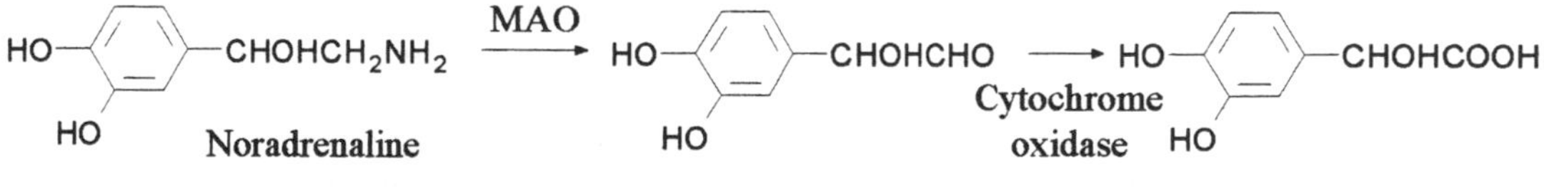

(HO)(HO)$C_6H_3{-}CHOHCH_2NH_2 \xrightarrow{\text{MAO}}$ (HO)(HO)$C_6H_3{-}CHOHCHO \xrightarrow[\text{Cytochrome oxidase}]{}$ (HO)(HO)$C_6H_3{-}CHOHCOOH$

Noradrenaline

All types of amine with α-C—H bonds may undergo α-hydroxylation, catalysed by CYP-450, to an unstable carbinolamine intermediate. This intermediate spontaneously decomposes with cleavage of the C—N bond to the corresponding carbonyl and nitrogen compounds.

General reaction:

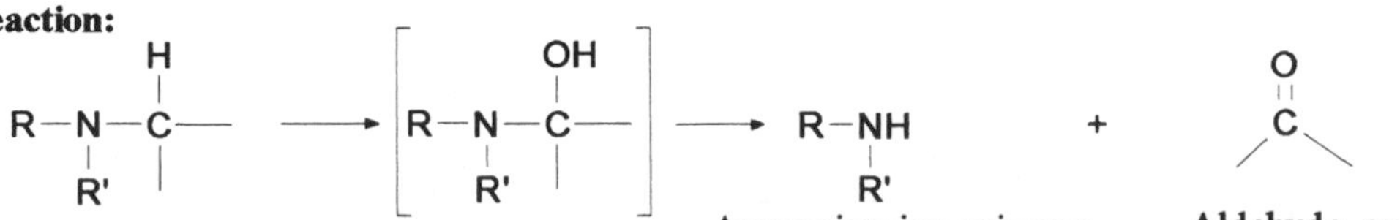

$R{-}N(R'){-}C(H) \longrightarrow [R{-}N(R'){-}C(OH)] \longrightarrow R{-}NH{-}R' + C{=}O$

Ammonium ion, primary or secondary amine

Aldehyde or ketone

Key: R and R' = H or ≠ H

Primary amines attached to a methinine group are oxidised by CYP-450 to the corresponding ketone with the liberation of ammonium ions. Aliphatic secondary and tertiary amines are metabolised by this route to amines, aldehydes and ketones. The process is referred to as **N-dealkylation**. With tertiary amines the first alkyl group is relatively easy to remove but removal of the second is often more difficult and does not always occur to any great extent. However, small alkyl groups, such as methyl, ethyl and isopropyl, are rapidly removed as methanal, ethanal and propanone, respectively. These metabolites are often excreted via the lungs. With secondary amines small alkyl groups also may be removed rapidly. However, tertiary and secondary amines with α-C—H bonds on both sides of the amino group may be metabolised by all the possible hydroxylation routes. For example, metabolism of *N*-benzylamphetamine yields 1-phenylpropanone, benzaldehyde, phenylmethylamine and 1-phenyl-2-aminopropane.

Examples:

Primary amines

Amphetamine (CNS stimulant anorectic) —CYP-450→ 2-Phenylpropanone + $\overset{+}{N}H_4$

Secondary amines

N-Benzylamphetamine → 1-Phenyl-2-aminopropane + Benzaldehyde

N-Benzylamphetamine → 1-Phenylpropanone + Phenylmethylamine

Tertiary amines

Imipramine (antidepressant) → (HCHO) → Desmethylimipramine (antipsychotic) → (HCHO) → Bisdesmethylimipramine

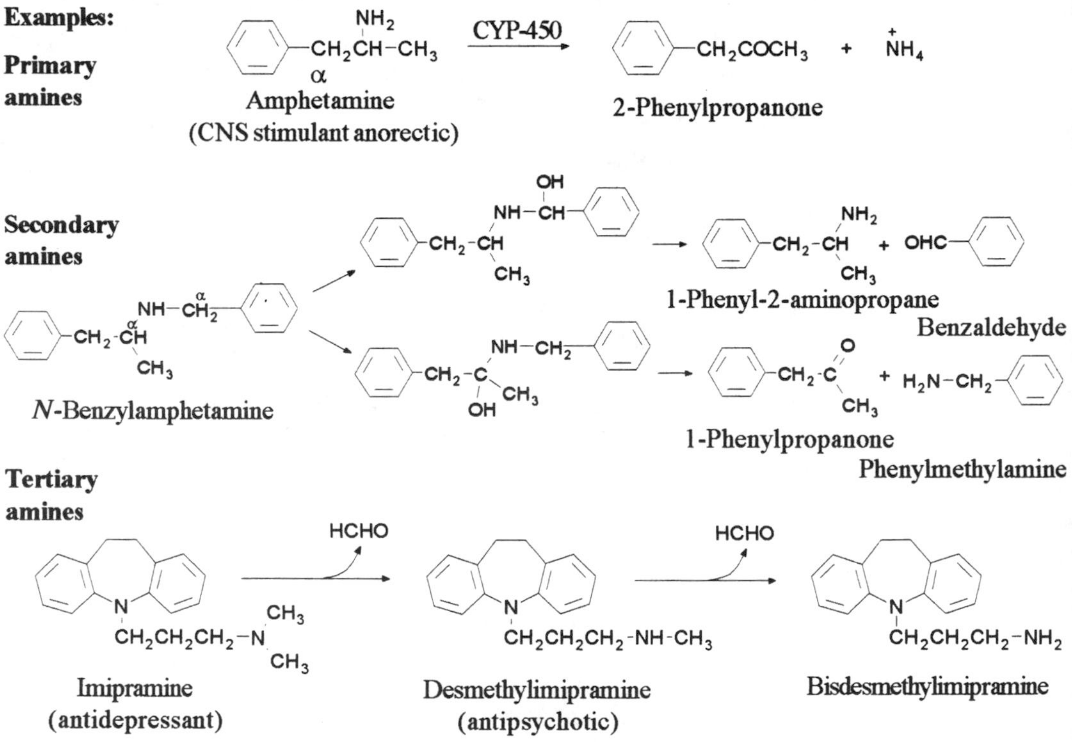

9.5.10 Amides

Amide bonds, which include peptide links and imides, may be metabolised by hydrolysis. However, initially only one of the amide bonds of an imide is cleaved. In most cases, amide hydrolysis is comparatively slow compared to ester hydrolysis and a high proportion of an amide drug is often excreted unchanged. For example, about 60% of a dose of procainamide may be excreted in the urine.

General reactions: $$RCONHR' \xrightarrow{H_2O} RCOOH + R'NH_2$$

Examples:

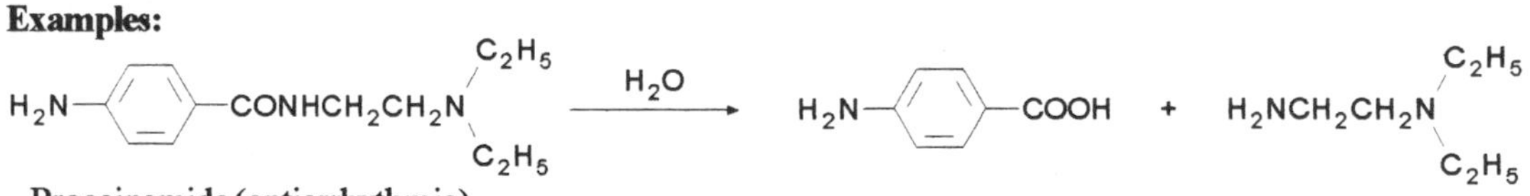

$$H_2N{-}C_6H_4{-}CONHCH_2CH_2N(C_2H_5)_2 \xrightarrow{H_2O} H_2N{-}C_6H_4{-}COOH + H_2NCH_2CH_2N(C_2H_5)_2$$

Procainamide (antiarrhythmic)

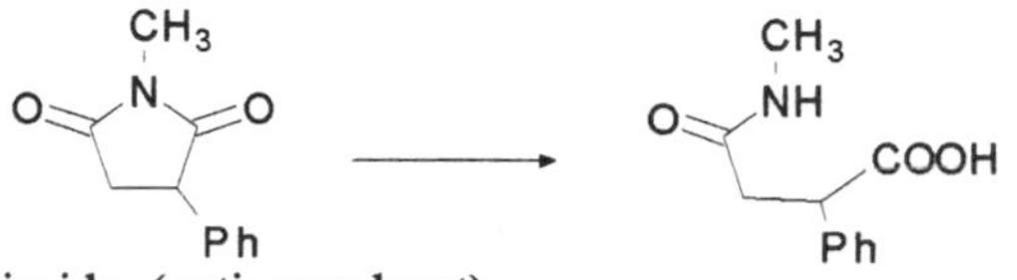

CH_3, N, O, O, Ph → CH_3, NH, O, COOH, Ph

Phensuximide (anticonvulsant)

Amides may also be metabolised by oxidative α-hydroxylation and dealkylation or N-hydroxylation. Oxidative α-hydroxylation and dealkylation occur at the α-C—H bond of the amine residue. It is believed to follow a similar mechanism to that for the dealkylation of amines. Amides with small alkyl groups may undergo extensive metabolic dealkylation.

General reactions:

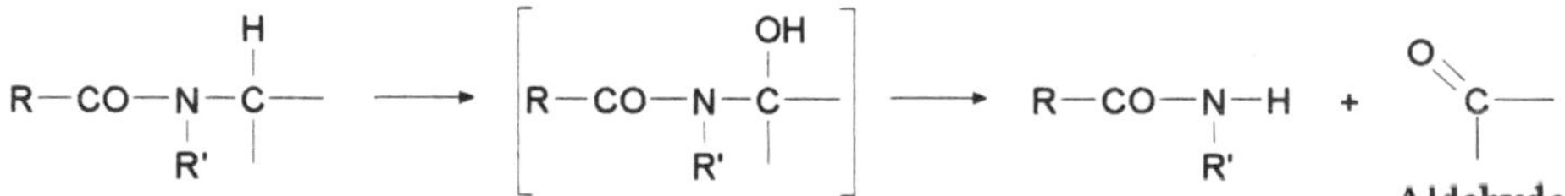

$$R{-}CO{-}N(R'){-}C(H) \longrightarrow [R{-}CO{-}N(R'){-}C(OH)] \longrightarrow R{-}CO{-}N(R'){-}H + O{=}C$$

Aldehyde or ketone

Example:

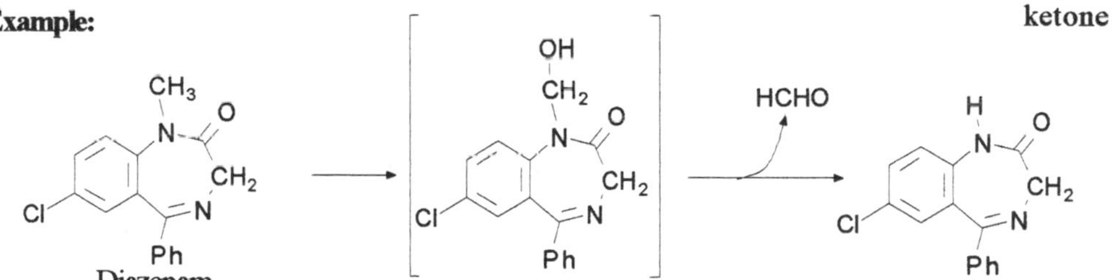

CH_3, N, O, CH_2, N, Cl, Ph → [OH, CH_2, N, O, CH_2, N, Cl, Ph] →(HCHO) H, N, O, CH_2, N, Cl, Ph

Diazepam (anxiolytic)

Demethyldiazepam (anxiolytic)

N-Hydroxylation is usually a minor route for amide metabolism. However, the hydroxylated metabolites produced by this reaction are often the precursors of highly toxic compounds. For example, the carcinogen 2-acetylaminofluorene is believed to owe its carcinogenic activity to a metabolic route that starts with the formation of an *N*-hydroxyl intermediate.

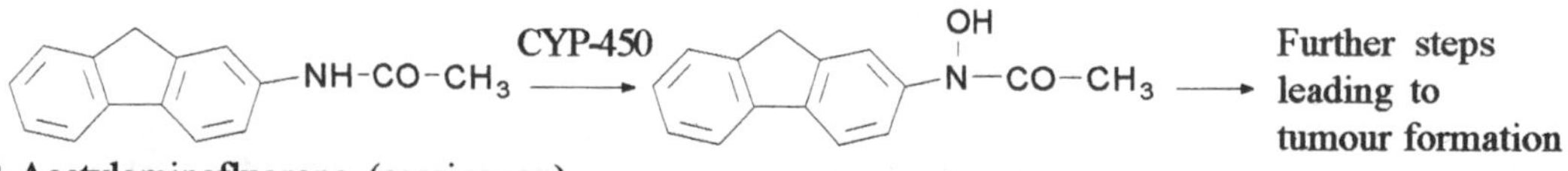

$NH{-}CO{-}CH_3 \xrightarrow{CYP\text{-}450} N(OH){-}CO{-}CH_3 \longrightarrow$ Further steps leading to tumour formation

2-Acetylaminofluorene (carcinogen)

9.5.11 Aldehydes and Ketones

Aldehydes may be oxidised or reduced. However, they are so easily oxidised to carboxylic acids that they do not normally undergo metabolic reduction unless the aldehyde is resistant to oxidation.

General reactions:

$$\underset{\text{Carboxylic acid}}{RCOOH} \xleftarrow{\text{Oxidation}} \underset{\text{Aldehyde}}{RCHO} \xrightarrow{\text{Reduction}} \underset{\text{Primary alcohol}}{RCH_2OH}$$

Examples:

$$\underset{\text{Glyoxal}}{CHO{-}CHO} \xrightarrow[\text{(LAD)}]{\text{Liver aldehyde dehy rogenase}} \underset{\text{Oxalic acid}}{COOH{-}COOH}$$

$$\underset{\text{Chloral (hypnotic)}}{CCl_3CHO} \longrightarrow \underset{\text{Trichloroethanol}}{CCl_3CH_2OH}$$

Ketones are resistant to oxidation and so are normally metabolised by reduction. Reduction of unsymmetrical ketones may give a mixture of isomers. The secondary alcohols formed are frequently metabolised further by appropriate Phase II reactions.

General reaction:

$$\underset{\text{Ketone}}{RR'C{=}O} \longrightarrow \underset{\text{Secondary alcohol}}{RR'CH{-}OH}$$

Example:

R(+)-Warfarin → 1*R*,3*S*(+)-Alcohol (major product) + 1*R*,3*R*(+)-Alcohol (minor product)

9.5.12 Carboxylic Acids

Aliphatic acids can be metabolised by oxidation at their α- and β-carbon atoms provided that there are methylene groups adjacent to the carboxyl group. β-Oxidation is the most common, proceeding via an acetyl-CoA thioester intermediate. It always involves the loss of two carbons and is terminated when a branch in the chain is reached. α-Oxidation can occur when β-oxidation is not possible. It results in the loss of a single carbon atom.

β-Oxidation:

$$Ph{-}CH_2CH_2\overset{\beta}{C}H_2CH_2COOH \xrightarrow{HS\overline{CoA}} Ph{-}CH_2CH_2CH_2CH_2COS\overline{CoA}$$

$$\downarrow$$

$$Ph{-}COOH \xleftarrow{\text{Process is repeated}} Ph{-}\overset{\beta}{C}H_2CH_2COOH + CH_3COS\overline{CoA}$$

α-Oxidation: $RCH_2CH_2\underset{\alpha}{\overset{CH_3}{C}}HCH_2COOH \longrightarrow RCH_2CH_2\overset{CH_3}{C}HCOOH$

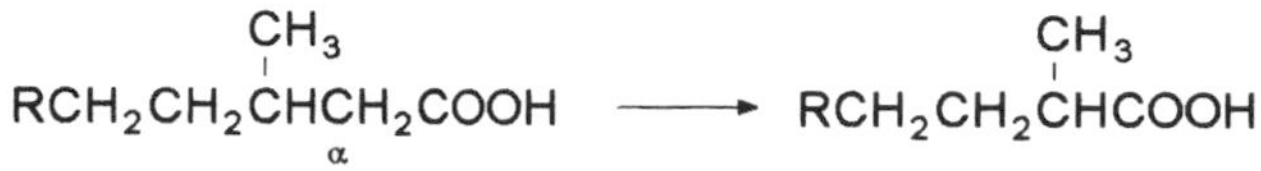

9.5.13 Epoxides

The carbon atoms of epoxide rings are electrophilic in nature due to the electron-withdrawing effect of the oxygen atom. Consequently, epoxides are readily hydrated by epoxide hydrases to the corresponding diols (Figure 9.7). However, due to their electrophilic nature they can also react with the nucleophilic groups of a variety of biological molecules. These latter reactions are possibly responsible for the toxic effects of some drugs.

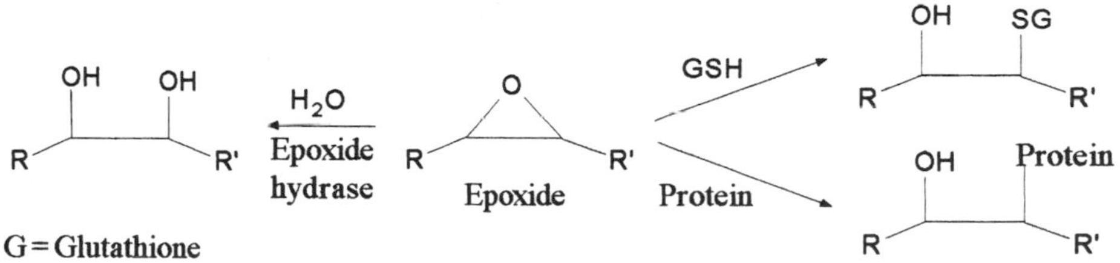

Figure 9.7. Examples of some of the metabolic routes for epoxides.

9.5.14 Ethers

The main route for ether metabolism appears to be oxidative dealkylation. It is catalysed by mixed-function oxidases and is believed to occur by a similar mechanism to that for the dealkylation of amines, producing alcohols, phenols, aldehydes and ketones. Small alkyl groups are often removed in preference to larger groups. The alcoholic and phenolic metabolites produced have often been found to undergo further Phase II conjugation reactions.

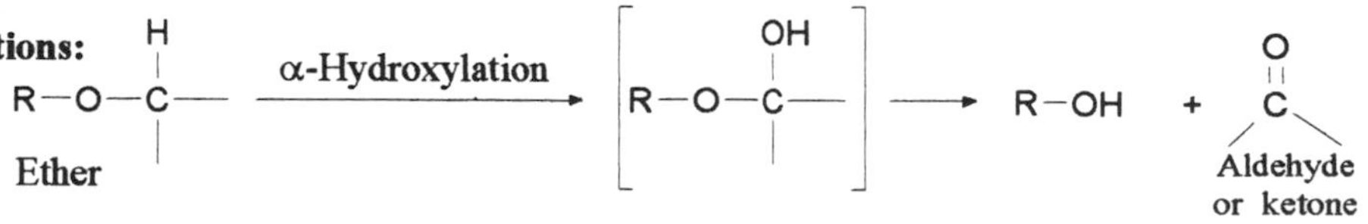

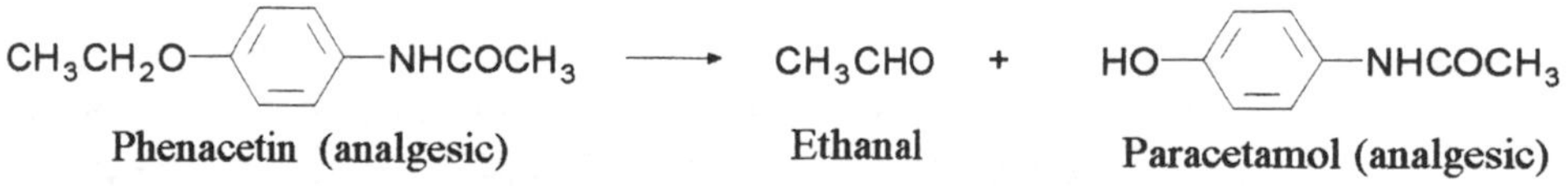

9.5.15 Esters

The main metabolic pathway for esters is hydrolysis. Hydrolysis is often rapid provided that the ester is not sterically hindered. The reaction is catalysed by esterases and other hydrolytic enzymes present in the plasma and various tissues. The alcohols, phenols and acids produced often take part in further Phase II conjugation reactions to form derivatives that are readily excreted.

General reactions:

$$\underset{\text{Ester}}{RCOOR'} \xrightarrow{H_2O} RCOOH + R'OH$$

Example:

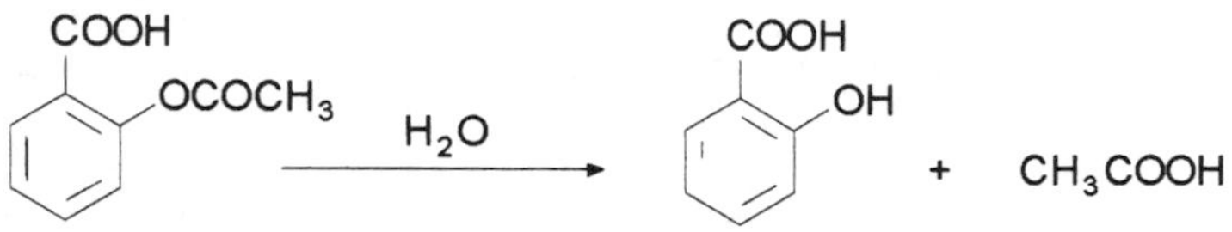

Aspirin (analgesic) $\xrightarrow{H_2O}$ Salicylic acid + CH_3COOH

Carbamates are also metabolised by hydrolysis to the carbamic acid, which decarboxylates spontaneously to the amine.

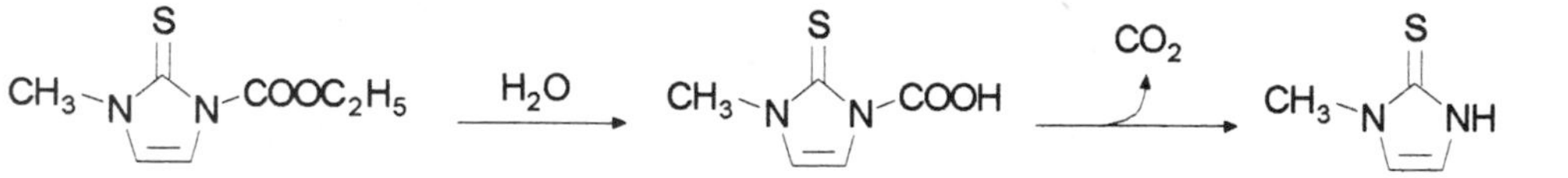

Carbimazole (thyroid inhibitor) $\xrightarrow{H_2O}$ carbamic acid $\xrightarrow{-CO_2}$ Methimazole (antihyperthyroid)

9.5.16 Hydrazines

Hydrazines are usually metabolised by reduction to the corresponding amines. However, some *gem*-substituted nitrogen hydrazines are oxidised by FMO enzymes to the corresponding oxide.

$$RNHNHR' \longrightarrow RNH_2 + R'NH_2$$

$$RR'NNH_2 \xrightarrow{FMO} RR'N(\rightarrow O)NH_2$$

9.5.17 Nitriles

Aliphatic nitriles are metabolised by oxidation to the corresponding aldehyde with the liberation of highly toxic cyanide ions. The reaction is believed to proceed by an α-hydroxylation mechanism.

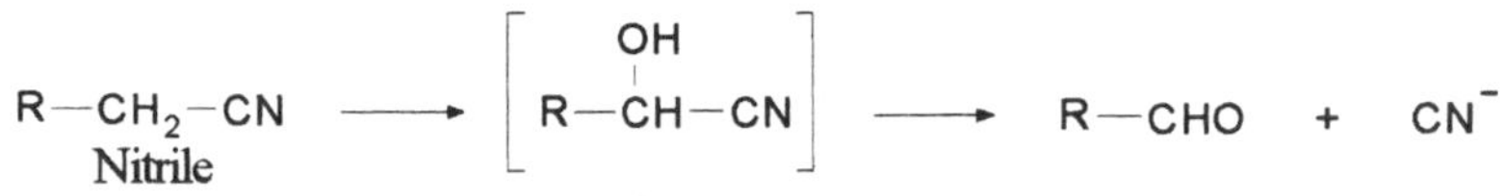

$$R{-}CH_2{-}CN \longrightarrow \left[R{-}\underset{}{\overset{OH}{CH}}{-}CN \right] \longrightarrow R{-}CHO + CN^-$$

Nitrile

Aromatic nitriles are usually metabolised by hydroxylation of the aromatic ring. Reduction to the corresponding aliphatic amine is slow and normally a minor route.

$$Ar{-}CN \rightarrow Ar{-}CH_2{-}NH_2$$

9.5.18 Nitro

Nitro groups are usually reduced to the corresponding amine by nitroreductases. It is believed that the reaction proceeds through nitroso and hydroxylamine intermediates.

General reactions:

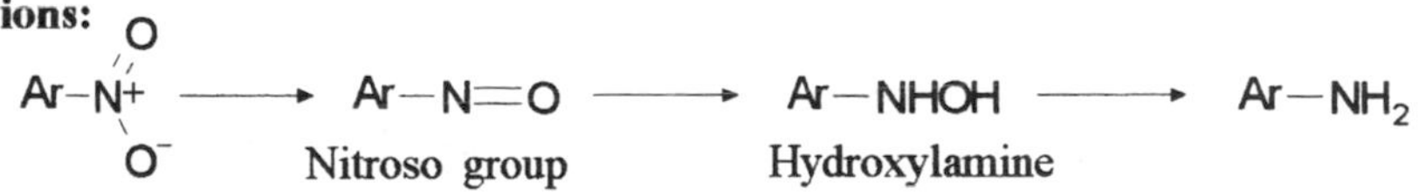

$$Ar{-}N^+(=O){-}O^- \longrightarrow Ar{-}N{=}O \longrightarrow Ar{-}NHOH \longrightarrow Ar{-}NH_2$$

Nitroso group — Hydroxylamine

Example:

Nitrofurantoin (antibacterial) → Aminofurantoin (antibacterial)

9.5.19 Sulphides and Sulphoxides

The most important route for the metabolism of sulphides is oxidation catalysed by FMO to the sulphoxide. This sulphoxide may undergo further oxidation catalysed by FMO to a sulphone. However, sulphoxides may also be reduced metabolically to sulphides, but sulphones are not usually reduced to sulphoxides.

General reactions:

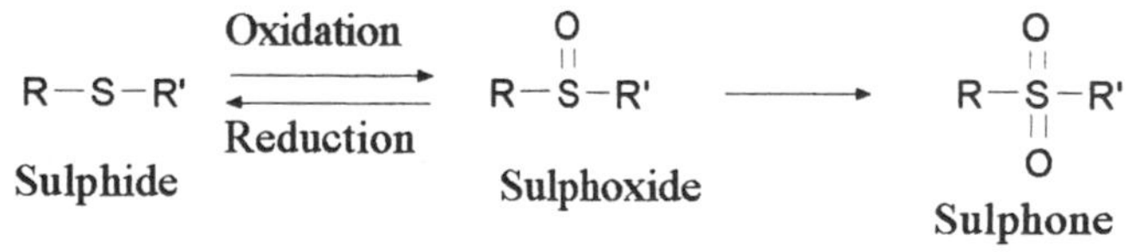

$$R{-}S{-}R' \underset{\text{Reduction}}{\overset{\text{Oxidation}}{\rightleftharpoons}} R{-}S(=O){-}R' \longrightarrow R{-}S(=O)_2{-}R'$$

Sulphide — Sulphoxide — Sulphone

Example:

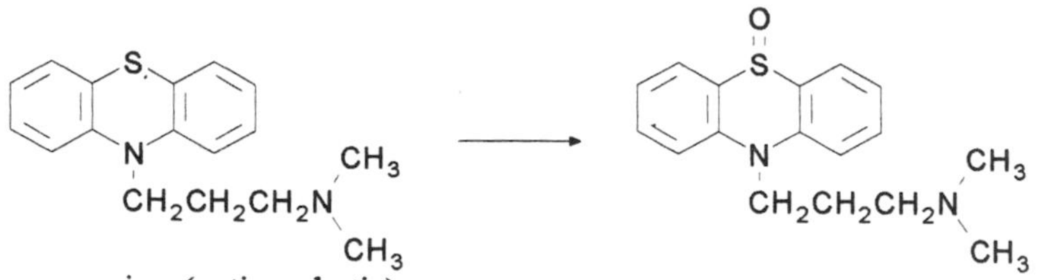

Chlorpromazine (antipsychotic) → Chlorpromazine-*S*-oxide

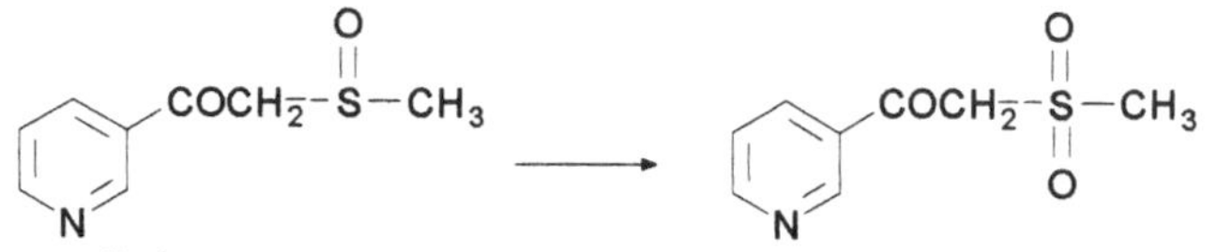

Oxisuran →

Sulphides are also metabolised by dealkylation. This is believed to proceed by a similar α-hydroxylation mechanism to that found for amines and ethers.

General reaction:

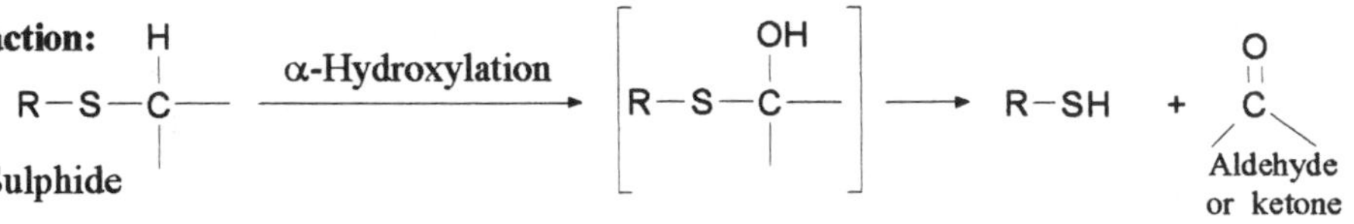

Sulphide —α-Hydroxylation→ [R—S—C(OH)] → R—SH + C=O (Aldehyde or ketone)

Example:

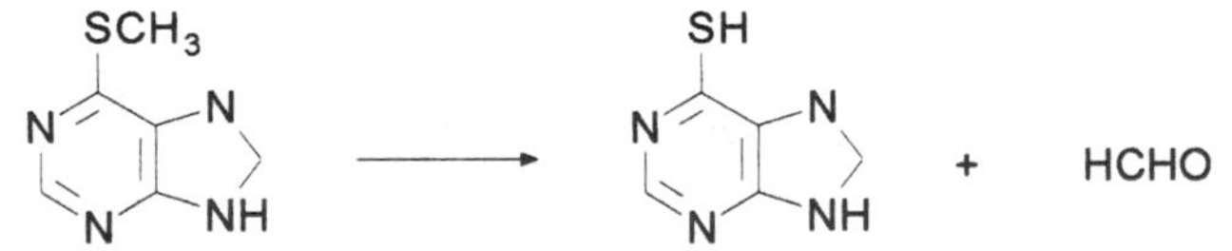

6-Methylthiopurine → 6-Mercaptopurine (anticancer) + HCHO

Disulphides are usually reduced to the corresponding thiols.

General reaction: R—S—S—R' → RSH + R'SH

Example:

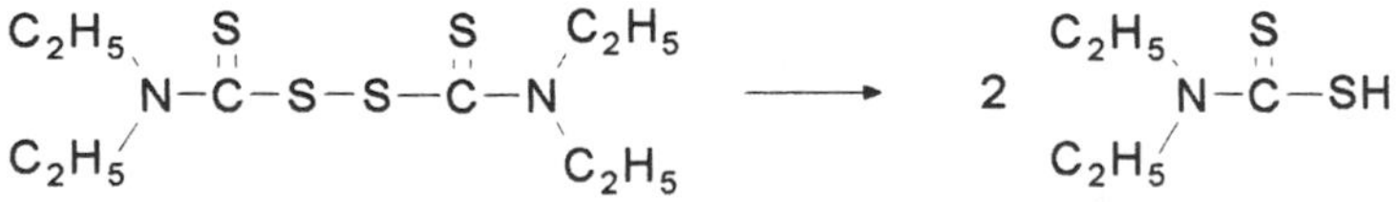

Disulphiram (alcohol deterrent) → 2 $(C_2H_5)_2N{-}C(=S){-}SH$

9.5.20 Thiols

Thiols may be oxidised by FMO to the corresponding disulphide.

$$RSH \rightarrow \underset{\text{Disulphide}}{R{-}S{-}S{-}R}$$

9.6 Phase II Metabolic Routes

Phase II conjugation reactions may occur at any point in the metabolism of a drug or xenobiotic. However, they often represent the final step in the metabolic pathway before excretion. The conjugates formed are usually pharmacologically inactive although there are some notable exceptions. They are excreted in the urine and/or bile.

The reactions commonly involved in Phase II conjugation are acylation, sulphate formation and conjugation with amino acids, glucuronic acid glutathione and mercapturic acid. Methylation is also regarded as a Phase II reaction although it is normally a minor metabolic route. However, it can be a major route for phenolic hydroxy groups. In all cases, the reaction is usually catalysed by a specific transferase.

9.6.1 Acylation

N-Acetylation is an important route for primary aromatic amines (Ar—NH_2), simple sulphonamides (—SO_2NH_2), hydrazines (—$NHNH_2$) and hydrazides (—$CONHNH_2$). The reaction is catalysed by *N*-acetyltransferases with acetyl-coenzyme A supplying the acetyl group. The water solubility of the products of these reactions is not greatly enhanced when compared to that of the parent compound but they are usually pharmacologically inactive. However, it has been reported that *N*-acetylisoniazid is more toxic than isoniazid.

General reaction: —NH_2 $\xrightarrow{N\text{-Acetyltransferase}}$ —$NHCOCH_3$ ($CH_3COSCoA$ → HSCoA)

Example:

SO_2NH_2 / NH_2 (Sulphanilamide (antibacterial)) → $SO_2NHCOCH_3$ / NH_2 and SO_2NH_2 / $NHCOCH_3$ → $SO_2NHCOCH_3$ / $NHCOCH_3$

Sulphanilamide
(antibacterial)

9.6.2 Sulphate Formation

Sulphate conjugation is an alternative metabolic route for phenols and occasionally some alcohols and amines. The sulphate originates from the body's sulphate pool. It is believed to be activated by conjugation with ATP to form 3′-phosphoadenosine-5′-phosphosulphate (PAPS),

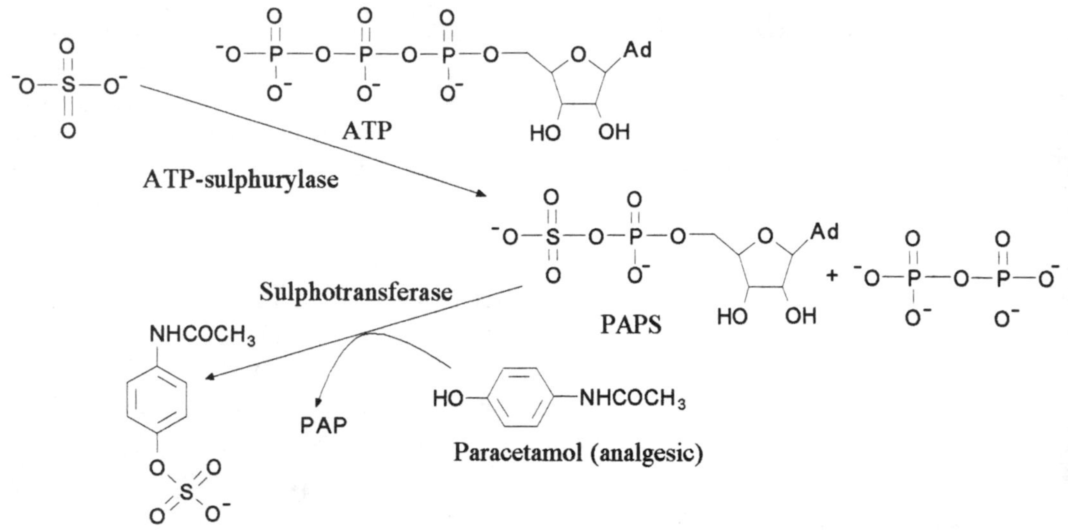

Figure 9.8. Sulphate conjugation of paracetamol.

the process being catalysed by Mg^{2+} ions and ATP-sulphurylase. The sulphate is transferred from PAPS to the xenobiotic to form the sulphate conjugate, this reaction being catalysed by soluble sulphotransferases (Figure 9.8). The sulphates formed are water soluble and, with the exception of steroids, are excreted in the urine. Steroids commonly undergo biliary excretion. However, sulphate conjugation can occur in the GI tract, which can cause a significant decrease in a drug's bioavailability.

Sulphate conjugates are usually pharmacologically inactive. However, the sulphate conjugates of *N*-hydroxy compounds are often toxic. For example, it has been suggested that the hepatotoxicity and nephrotoxicity of phenacetin could be due to the formation of the *O*-sulphate esters, which have been shown to bind to microsomal proteins.

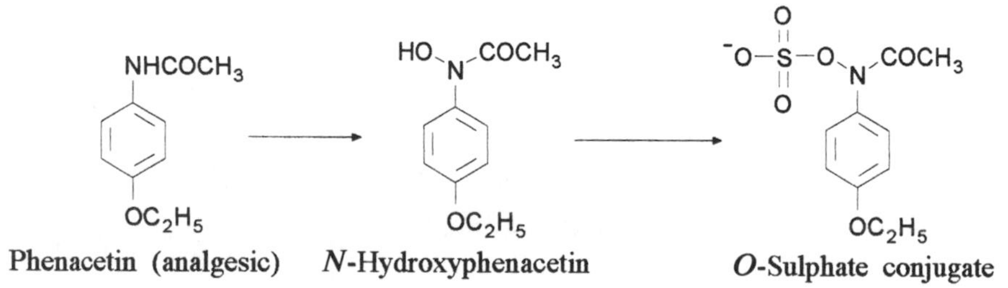

Phenacetin (analgesic) *N*-Hydroxyphenacetin *O*-Sulphate conjugate

9.6.3 Conjugation with Amino Acids

Conjugation with amino acids is an important route for the metabolism and elimination of carboxylic acids. It is thought that the reaction proceeds in a series of steps. Initially the carboxylic acid reacts with ATP to form a monoadenosine phosphate (AMP) intermediate. This intermediate is converted to an active acyl-coenzyme A intermediate by the action of coenzyme A. The coenzyme A intermediate undergoes nucleophilic displacement with the amino acid. All these reactions are catalysed by various *N*-acyl transferases. The resulting amino acid conjugates are more water soluble than the parent acid and are mainly excreted in the urine. However, some biliary excretion also occurs. The quantity of amino acids available for conjugation is limited and so usually few amino acid conjugates are found in the urine.

COOH
Benzoic acid
ATP
Pyrophosphate
C(=O)AMP
AMP
HS̄CoA
C(=O)SCoA
H_2NCH_2COOH
Glycine
HS̄CoA
C(=O)$NHCH_2COOH$
Glycine conjugate

COOH
Cinnamic acid
NH_2
$H_2NCOCH_2CH_2CHCOOH$
Glutamine
COOH
$CONHCHCH_2CH_2CONH_2$
Glutamine conjugate

In humans, aromatic carboxylic acids are often excreted as their glycine conjugates. Glutamine conjugates is not common and have been detected for only a few aryl-substituted aliphatic carboxylic acids. Other species can utilise different amino acids, for example, ornithine in birds and alanine in hamsters and mice.

9.6.4 Conjugation with Glucuronic Acid

Conjugation with glucuronic acid is possibly the most important of all the Phase II reactions. This is probably because there is a good supply of glucuronic acid in the body. Numerous alcohols, phenols, amines, thiols and some carboxylic acids have been found to be metabolised by this route. The xenobiotic reacts with the activated form of glucuronic acid, uridine diphosphate glucuronic acid (UDPGA), to form a highly water-soluble glucuronide conjugate (Figure 9.9). This reaction is catalysed by microsomal uridine diphosphate glucuronyl transferases (UDPG transferases).

A number of drugs, such as phenylbutazone and sulphinpyrazone, have been shown to form glucuronides in which a carbon atom is directly attached to the glucuronide residue.

UDPGA

Key: R =
$CH_2CH_2CH_2CH_3$ **Phenylbutazone (anti-inflammatory)**
CH_2CH_2SOPh **Sulphinpyrazone (uricosuric)**

Glucuronides are excreted in both the urine and the bile. In both cases acylglucuronides may be hydrolysed and reabsorbed. Furthermore, the acyl group can either migrate to other positions on the glucuronide ring or to another cellular constituent. The latter rearrangement is believed to be responsible for the toxic behaviour of some drugs.

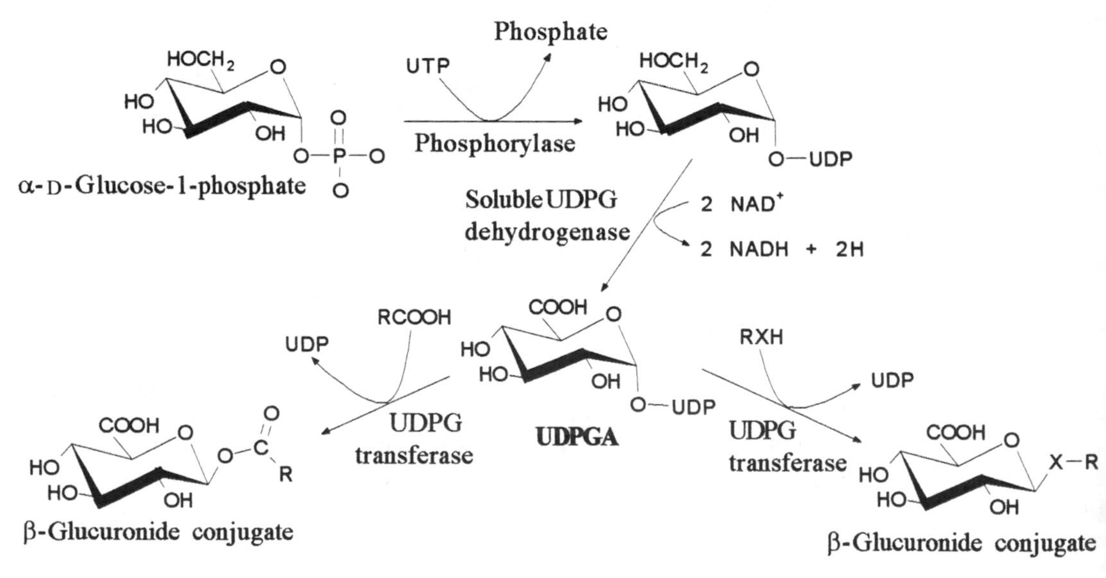

Figure 9.9. An outline of the mechanism for the formation of glucuronate conjugates. UTP is uridine triphosphate and X is O (alcohols and phenols), S (thiols) or NH (amines).

9.6.5 Conjugation with Glutathione

Conjugation with the nucleophilic thiol group of glutathione (GSH, Figure 9.11) is the first step in the elimination of compounds whose structures contain an electrophilic centre. These electrophilic centres are caused by the presence of electron-withdrawing groups, such as halide, nitro, sulphonate, nitrate, epoxide and organophosphate groups in the structure (Figure 9.10). Reaction occurs by nucleophilic substitution, displacement or addition, the reactions being catalysed by a family of enzymes known as glutathione *S*-transferases, which are found in most tissues. These enzymes are able to catalyse the conjugation of glutathione with a wide variety of compounds possessing electrophilic centres.

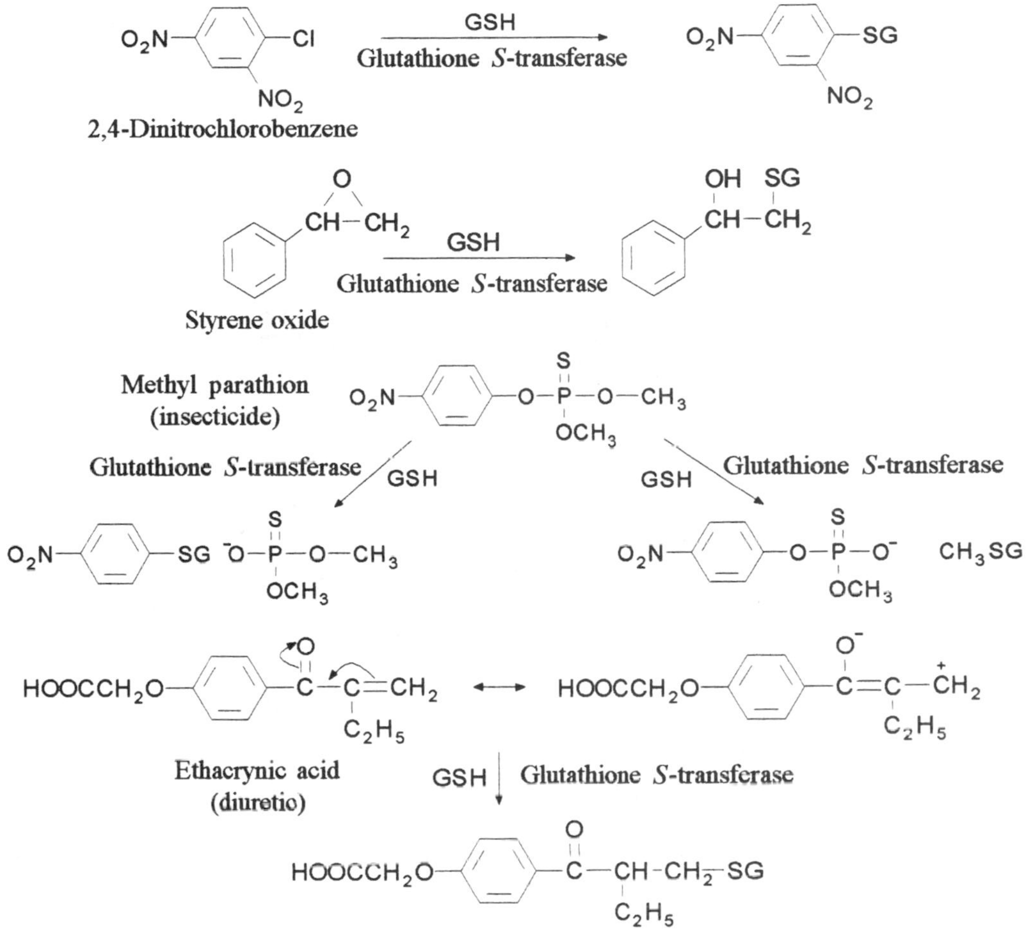

Figure 9.10. Examples of glutathione conjugation.

Once formed, many of the conjugates are converted to mercapturic acid derivatives, which are excreted (Figure 9.11). However, some organic nitrate conjugates are converted to the corresponding alcohol derivatives with the formation of glutathione disulphide.

The widespread occurrence of glutathione, coupled with its reactivity, has resulted in it being regarded as a detoxification agent protecting other essential cellular molecules from attack by electrophilic xenobiotics.

9.6.6 Methylation

Methylation has been found to occur with some xenobiotics whose structures contain phenol, thiol and amino groups. Some *N*-heterocyclics have also been found to be metabolised by this

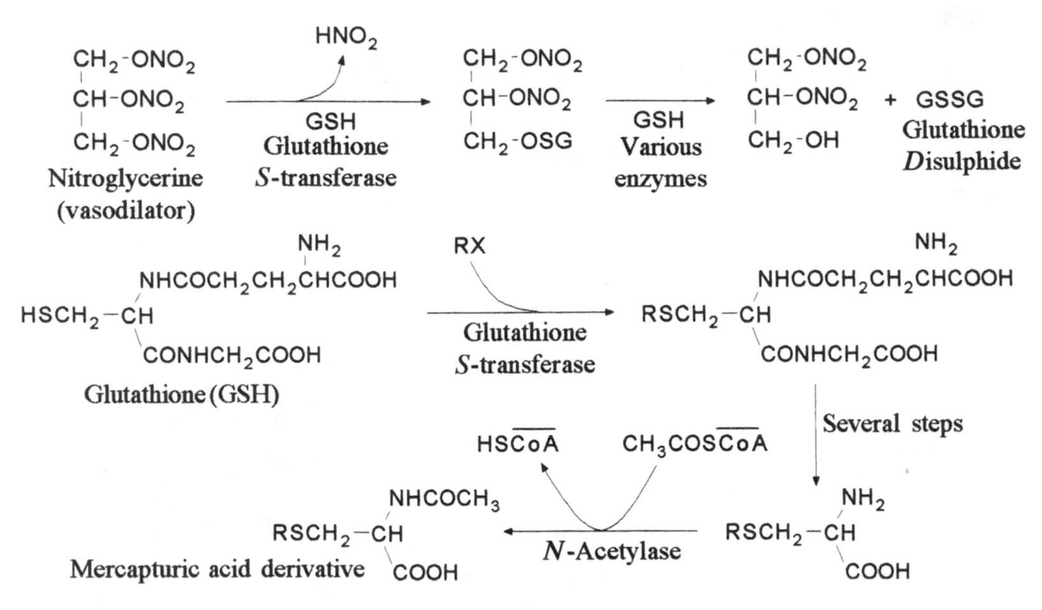

Figure 9.11. The formation of mercapturic acid derivatives.

route. Methylation is catalysed by specific methyltransferases. In each case, it is thought that *S*-adenosylmethionine (SAM) is the coenzyme. Methylation, with the exception of amines, which can form very water-soluble quaternary salts, does not lead to metabolites that are more water soluble than their immediate precursors. This lack of improved water solubility means that most of these methylated xenobiotics exhibit a reduced ease of excretion. However, this does not normally pose a problem because these derivatives are usually pharmacologically inactive.

General reaction:

$HOOCCH(NH_2)CH_2CH_2\overset{+}{S}(CH_3)Ad$ (*S*-Adenosylmethionine (SAM)) → $HOOCCH(NH_2)CH_2CH_2SAd$

$RXH \xrightarrow{\text{X-Methyltransferase}} RXCH_3$

Key: Ad = 5-Adenosyl residue
X = O, S or NH

Examples:

Dopamine (neurotransmitter) $\xrightarrow{SAM}$ 3-Methoxytyramine

$CH_2\text{-}SH$, $CH\text{-}SH$, $CH_2\text{-}OH$ $\xrightarrow{SAM}$ $CH_2\text{-}SCH_3$, $CH\text{-}SCH_3$, $CH_2\text{-}OH$

Dimercaprol (heavy metal poisoning antidote)

Nicotine (neurotransmitter) $\xrightarrow{SAM}$ Quaternary ammonium salt

9.7 Pharmacokinetics of Metabolites

The activity of a metabolite will have a direct bearing on the safe use and dose of a drug administered to a patient (see section 9.2). A build-up in the concentration of a metabolite may have a serious effect on a patient. Consequently, when investigating the pharmacokinetics of a drug it is also necessary to obtain data concerning the elimination of its metabolites.

Information concerning a metabolite's kinetics in humans is usually obtained by administering the drug and measuring the change in concentration with time of the resulting metabolite in the plasma. The metabolite may be administered separately where independent data concerning its activity and pharmacokinetics are required. However, observations made from metabolite administration can be suspect due to bioavailability problems.

The total administered dose (A) of a drug is excreted partly unchanged and partly metabolised (Figure 9.12). Because metabolites are produced in the appropriate body compartment, a metabolite may be partly or fully metabolised before it reaches the plasma. In these cases the amount of metabolite found by analysis of plasma samples is only a fraction of the amount of the metabolite produced by the body. For simplicity, the discussions in this text will assume that all the metabolite produced reaches the plasma.

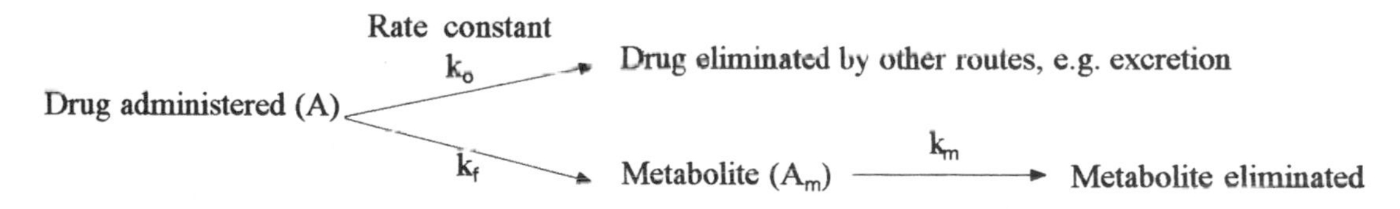

Figure 9.12. A schematic representation of the general metabolic routes for a drug in the body.

The rate of change of concentration of a metabolite (dM/dt) in the plasma is given by:

$$dM/dt = \text{Rate of formation} - \text{Rate of elimination} \tag{9.1}$$

Because most biological processes exhibit first-order kinetics, Equation (9.1) becomes:

$$dM/dt = k_f A - k_m A_m \tag{9.2}$$

where k_f and k_m are the rate constants for the metabolite formation and elimination processes, respectively. Furthermore if the rate of drug elimination by all other pathways is first order then:

$$\text{Rate of drug elimination} = kA \tag{9.3}$$

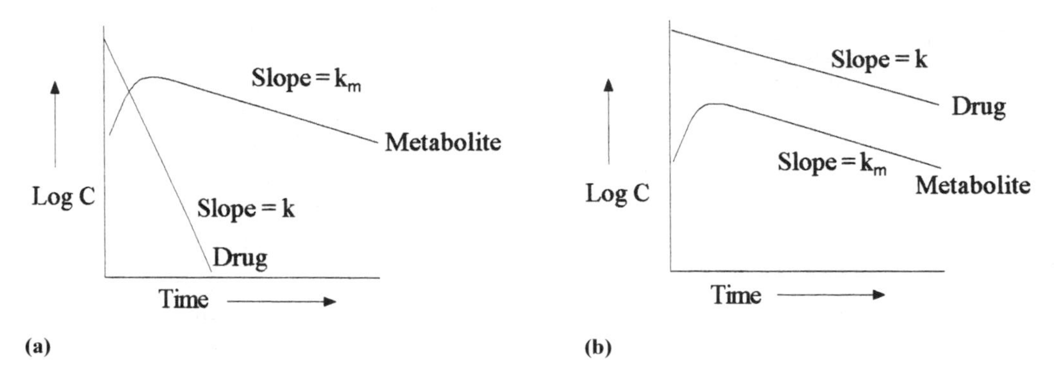

Figure 9.13. Representations of typical log concentration–time plots for a drug and metabolite exhibiting first-order kinetics showing the general changes when (a) $k > k_m$ and (b) $k < k_m$.

where $k = k_f + k_o$. Consequently, if $k > k_m$ the metabolite will accumulate in the plasma because the clearance of the metabolite is slower than that of the drug responsible for forming the metabolite. This accumulation could pose a problem if the metabolite is pharmacologically active. However, if $k < k_m$ the metabolite will not accumulate in the body because the metabolite is eliminated faster than it is formed. The values of k and k_m can be determined experimentally from log plots of plasma measurements of the drug and metabolite (Figure 9.13).

Most metabolic pathways consist of a series of steps. The importance of this series is not the number of steps but whether the pathway has a rate-determining step. In other words, is there a metabolite bottleneck where the rate of elimination of a metabolite is far slower than the rate of clearance of the drug ($k > k_m$)? At such a point the concentration of the metabolite would increase to significant amounts, which could lead to potential clinical problems if the metabolite was pharmaceutically active. Consequently, to avoid problems of this nature a metabolite should be eliminated faster than the drug.

9.8 Drug Metabolism and Drug Design

A knowledge of the metabolic pathway of a drug may be used to design analogues with a different metabolism to that of the lead. This difference may either make the analogue more stable or increase its ease of metabolism relative to the lead. This is achieved by modifying the structure of the drug. However, these modifications must not have a detrimental effect on the pharmacological activity of the drug.

Increasing the metabolic stability and hence the duration of action of a drug is usually achieved by replacing a reactive group by a less reactive one. For example, N-dealkylation can be prevented by replacing an *N*-methyl group by an *N*-*t*-butyl group. Reactive ester groups are replaced by less reactive amide groups. Oxidation of aromatic rings may be reduced by introducing strong electron acceptor substituents such as chloro (—Cl), quaternary amine (—NR_3), carboxylic acid (—COOH), sulphonate (—SO_3R) and sulphonamide (—SO_2NHR) groups. For example, replacement of the arylmethyl substituent of the antidiabetic tolbutamide by chlorine yielded the antidiabetic chlorpropamide; a compound with a considerably longer half-life.

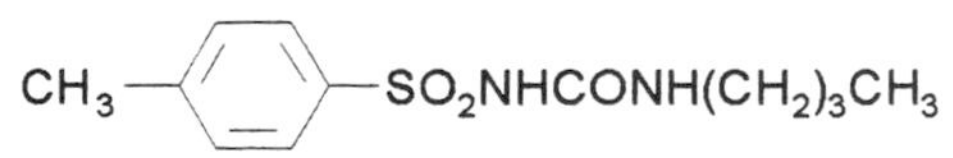

$CH_3-C_6H_4-SO_2NHCONH(CH_2)_3CH_3$

Tolbutamide (mean $t_{1/2}$ = 7 h)

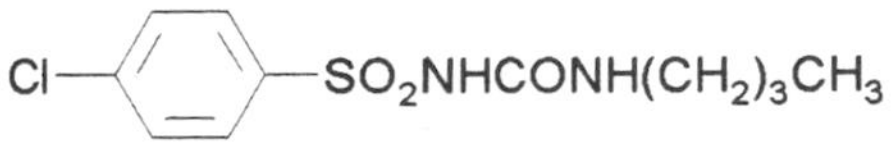

$Cl-C_6H_4-SO_2NHCONH(CH_2)_3CH_3$

Chlorpropamide (mean $t_{1/2}$ = 35 h)

However, these modifications can result in a change of pharmacological activity. For example, the replacement of the ester group in the local anaesthetic procaine by an amide group produced procainamide, which acts as an antiarrhythmic.

$H_2N-C_6H_4-COOCH_2CH_2N(C_2H_5)_2$

Procaine

$H_2N-C_6H_4-CONHCH_2CH_2N(C_2H_5)_2$

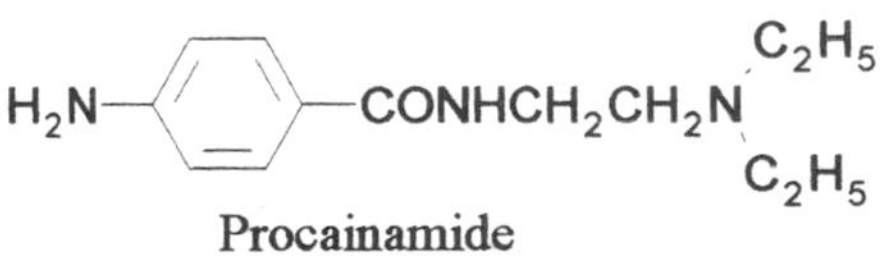

Procainamide

The ease of metabolism of a drug may be increased by incorporating a metabolically labile group, such as an ester, in the structure of the drug. This type of approach is the basis of prodrug design (see section 9.9). It has also led to the development of so-called **soft drugs**. These are biologically active compounds that are rapidly metabolised by a predictable route to pharmacologically non-toxic compounds. The advantage of this type of drug is that its half-life is so short that the possibility of the patient receiving a fatal overdose is considerably reduced. For example, the neuromuscular blocking agent succinylcholine is almost fully metabolised by hydrolysis in about 10 min, which considerably reduces the chances of a fatal overdose. However, this drug can still be fatal because some people do not have the esterase necessary for hydrolysis.

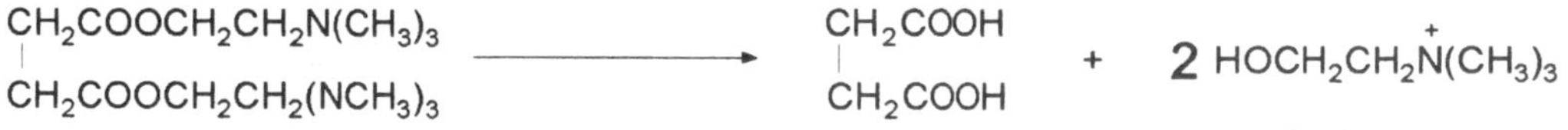

$$\begin{matrix} CH_2COOCH_2CH_2N(CH_3)_3 \\ | \\ CH_2COOCH_2CH_2(NCH_3)_3 \end{matrix} \longrightarrow \begin{matrix} CH_2COOH \\ | \\ CH_2COOH \end{matrix} + 2\ HOCH_2CH_2\overset{+}{N}(CH_3)_3$$

Succinylcholine **Succinic acid** **Choline**

However, incorporation of labile groups into a structure can also change the nature of a drug's biological action.

Changing the metabolic pathway of a drug may also be used to develop analogues that do not have the undesirable side effects of the lead compound. For example, the local anaesthetic lignocaine is also used as an antiarrhythmic. In this respect, its undesirable convulsant and emetic side effects are caused by its metabolism in the liver by dealkylation to the mono-*N*-ethyl derivative. The removal of the *N*-ethyl substituents and their replacement by an α-methyl group gives the antiarrhythmic tocainide. Tocainide cannot be metabolised by the same pathway as lignocaine and does not exhibit convulsant and emetic side effects.

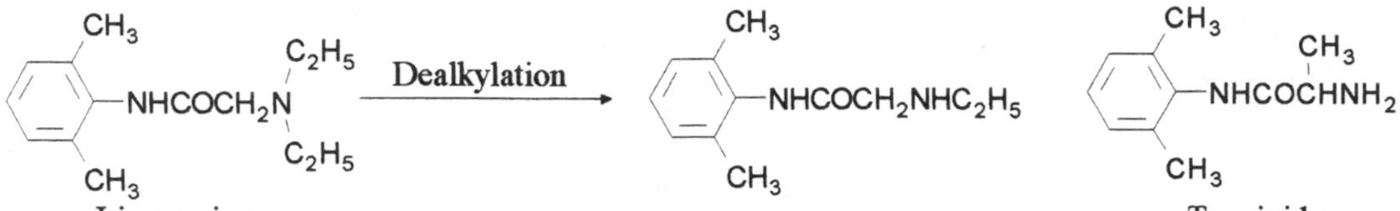

CH_3 / C_2H_5 / $-NHCOCH_2N$ / C_2H_5 / CH_3 —Dealkylation→ CH_3 / $-NHCOCH_2NHC_2H_5$ / CH_3 ; CH_3 / CH_3 / $-NHCOCHNH_2$ / CH_3

Lignocaine Tocainide

9.9 Prodrugs

Prodrugs are inactive compounds that yield an active compound on metabolism in the body. This metabolism may or may not be enzyme controlled.

Prodrugs may be broadly classified into two groups, namely: **bioprecursor** and **carrier** prodrugs. They may also be subclassified according to the nature of their action.

9.9.1 Bioprecursor Prodrugs

Bioprecursor prodrugs are compounds that already contain the embryo of the active species within their structure. They produce the active species on metabolism. For example, the first prodrug discovered, the antibacterial prontosil, is metabolised *in vivo* to its active metabolite sulphanilamide.

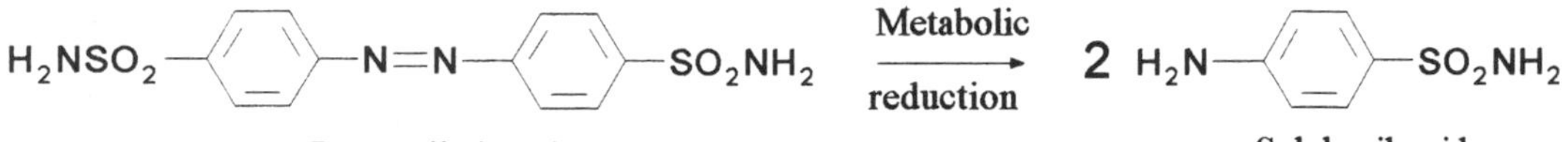

$H_2NSO_2-C_6H_4-N{=}N-C_6H_4-SO_2NH_2$ —Metabolic reduction→ $2\ H_2N-C_6H_4-SO_2NH_2$

Prontosil (inactive) Sulphanilamide

Similarly, the inactive prodrug cyclophosphamide is metabolised in the liver to the active phosphoramidate mustard.

H / N O / P CH_2CH_2Cl / O N / CH_2CH_2Cl —Liver CYP-450→ H_2N O / P CH_2CH_2Cl / HO N / CH_2CH_2Cl

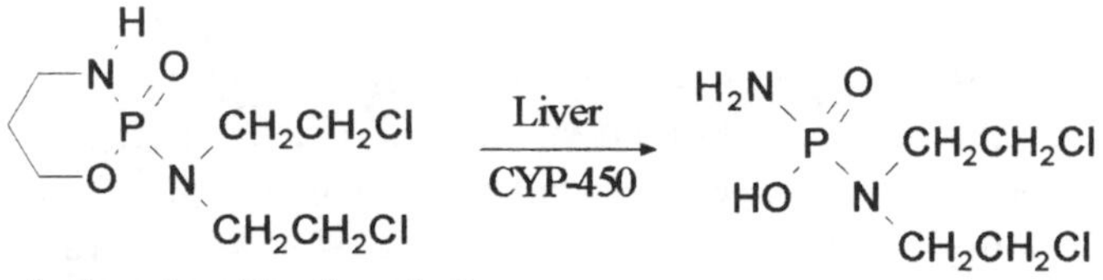

Cyclophosphamide (inactive)
Phosphoramidate mustard (antineoplastic)

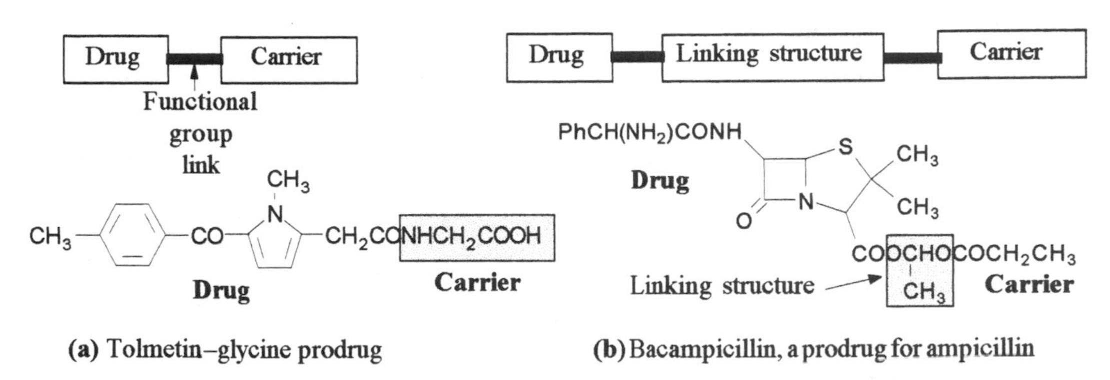

Figure 9.14. Examples of (a) bipartate and (b) tripartate prodrug systems.

9.9.2 Carrier Prodrugs

Carrier prodrugs are formed by combining an active drug with a carrier species to form a compound with the desired chemical and biological characteristics: for example, a lipophilic moiety to improve transport through membranes. The link between carrier and active species must be a group such as an ester or amide, which can be easily metabolised once absorption has occurred or the drug has been delivered to the required body compartment. The overall process may be summarised by:

$$\text{Carrier} + \text{Active species} \xrightarrow{\text{Synthesis}} \text{Carrier prodrug} \underset{}{\overset{\text{Metabolism}}{\rightleftharpoons}} \text{Carrier} + \text{active species}$$

Carrier prodrugs that consist of the drug linked by a functional group to the carrier are known as **bipartate prodrugs** (Figure 9.14). **Tripartate prodrugs** are those in which the carrier is linked to the drug by a link consisting of a separate structure. In these systems, the carrier is removed by an enzyme-controlled metabolic process and the linking structure by either an enzyme system or a chemical reaction.

Ideally, all types of carrier prodrug should meet the following criteria:

(i) the prodrug should be less toxic than the drug;
(ii) the prodrug should be inactive or significantly less active than the parent drug;
(iii) the rate of formation of the drug from the prodrug should be rapid enough to maintain the drug's concentration within its therapeutic window.
(iv) the metabolites from the carrier should be non-toxic or have a low degree of toxicity;
(v) the prodrug should have an improved bioavailability if administered orally;
(vi) the prodrug should be site specific.

Carrier prodrugs seldom meet all these criteria. However, a good example of a carrier prodrug meeting most of these criteria is bacampicillin, one of a number of prodrugs for the antibiotic

Table 9.1. Examples of the functional groups used to link carriers with drugs.

Drug group (D—X)	Type of group	Examples of R groups
Alcohol, phenol (D—OH)	Ester: D—COOR	Alkyl, phenyl, —$(CH_2)_2NR_2$, —$(CH_2)_n$CONR′R″, —$(CH_2)_n$NHCOR, —CH_2OCOR′
Amines (all types), imides and amides (>NH)	Amide: >NCOR	Alkyl, phenyl, —CH_2NHCOAr, —CH_2OCOR″
	Carbamate: >NCOR	—OCHR′OCOR″, —$OCH_2OPO_2H_3$
	Imine: >N = CHR	Aryl
Aldehydes and ketones (>C = O)	Acetals: >C$(OR)_2$	Alkyl
	Imine: >C = NR	Aryl, —OR
Carboxylic acids (D—COOH)	Ester: D—COOR	Alkyl, aryl, —$(CH_2)_n$NR′R″, —$(CH_2)_n$CONR′R″, —$(CH_2)_n$NHCOR′R″, —CH(R)OCOR, —CH(R)OCONR′R″

ampicillin. Only about 40% of an orally administered dose of ampicillin is absorbed. Relatively large quantities of the drug have to be administered in order to reach its therapeutic window. This means that significant amounts of the drug remain in the GI tract, damaging the intestinal flora. The prodrug becampicillin is about 98–99% absorbed, which reduces its toxic action in the GI tract. Furthermore, a considerably smaller dose is required in order to maintain the plasma concentration within the drug's therapeutic window. This further reduces the risk of toxic responses. Moreover, the metabolism of the prodrug yields carbon dioxide, ethanal and ethanol, all of which are natural metabolites in the body.

Becampicillin → + CO_2 + C_2H_5OH → **Ampicillin** + CH_3CHO

The choice of functional group used as a metabolic link depends both on the functional groups occurring in the drug molecule (Table 9.1) and the need for the prodrug to be metabolised in the appropriate body compartment.

The precise nature of the structure of the carrier used to form a carrier prodrug will depend on the intended outcome (see section 9.9.3).

9.9.3 The Design of Prodrug Systems for Specific Purposes

The nature of the carrier used will largely control a drug's bioavailability. Consequently, the selection of a suitable carrier enables the medicinal chemist to change the biological properties of a drug. Careful selection of a carrier usually improves a drug's performance and in some cases has been used to direct the drug to specific areas.

9.9.3.1 Improving Absorption

The transport of a drug through a membrane depends largely on its relative solubilities in water and lipids. If the drug is too water soluble it will not enter the membrane, but if it is too lipid soluble it will enter but not leave the membrane. Good absorption requires that a drug's hydrophilic–lipophilic nature is in balance. The selection of a suitable carrier can be used to fine-tune this balance and consequently improve absorption of the drug.

Lipophilic carriers are used to increase the lipophilic nature and hence the absorption of the drug. This is usually achieved by combining the carrier with a polar group(s) on the drug (Table 9.2). However, the carrier must be selected so that the new compound is able to act as a prodrug and release the active drug in the body. For example, adrenaline, when used to treat glaucoma, is poorly absorbed through the cornea. However, converting it to the less polar prodrug dipivaloyladrenaline by forming the di-trimethylethanoate derivative masks the polar phenolic hydroxy groups, which makes the molecule more lipophilic and results in a better absorption. Adrenaline is liberated by the action of the esterases found in the cornea and aqueous humour.

$(CH_3)_3CCOO$, $(CH_3)_3CCOO$, OH, $NHCH_3$ — Esterases → — HO, HO, OH, $NHCH_3$

Dipivaloyladrenaline — Adrenaline

It is difficult to select a lipophilic carrier that will provide the degree of lipophilic character required. If the carrier is too lipophilic the prodrug will remain in the membrane.

Table 9.2. Examples of the derivatives used to improve the lipophilic nature of drugs.

Functional group	Derivative
Acids	An appropriate ester
Alcohols and phenols	An appropriate ester
Aldehydes and ketones	Acetal
Amines	Quaternary ammonium derivatives, amino acid peptides and imines

The absorption of a drug will also depend on its water solubility. A drug must have a suitable water solubility if it is to be transported through a membrane by passive diffusion (see section 4.3.3). Carriers with water-solubilising groups have been used to produce prodrugs with a better water solubility than the active drug. For example, amino acid carriers have been used to prepare water-soluble derivatives of the local anaesthetic benzocaine and the sodium succinate derivative has been used for the glucocorticoid methylprenisolone.

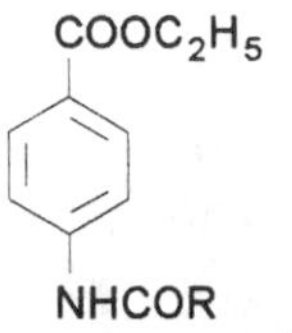

Benzocaine prodrugs (R = glycine, alanine, valine and leucine residues)

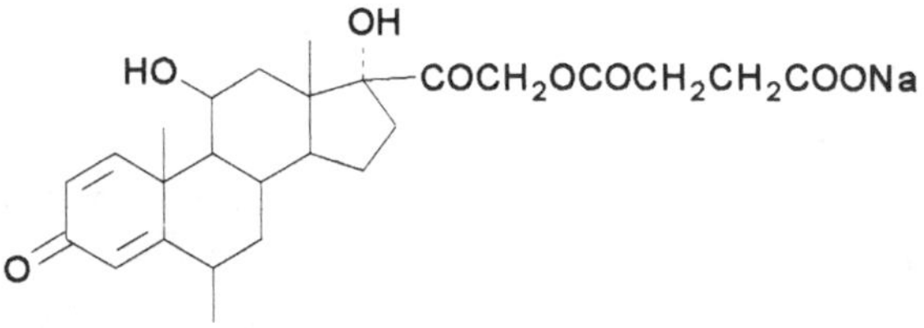

Methylprenisolone sodium succinate

Water-solubilising carriers should have ionisable groups that can form salts, groups that can hydrogen bond with water or both (see sections 3.5 and 3.6).

9.9.3.2 Improving Patient Acceptance

Odour and taste are important aspects of drug administration. A drug with a poor odour or too bitter a taste will be rejected by patients, especially children. Furthermore, a drug that causes pain when administered by injection can have a detrimental effect on a patient. The formation of a carrier prodrug can sometimes alleviate some of these problems. For example, palmitic acid and other long-chain fatty acids are often used as carriers because they usually form prodrugs with a bland taste. For example, the antibiotic clindamycin has a very bitter taste that makes it unsuitable for use with children. Furthermore it causes considerable pain on injection. However, it was found that the palmitate ester was not bitter and the phosphate caused less pain than the parent drug when injected. In both cases the drug is released by enzyme action. The use of fatty acid esters to improve patient acceptance does, however, reduce the water solubility and increase the lipid solubility of the drug. This could affect the drug's bioavailability.

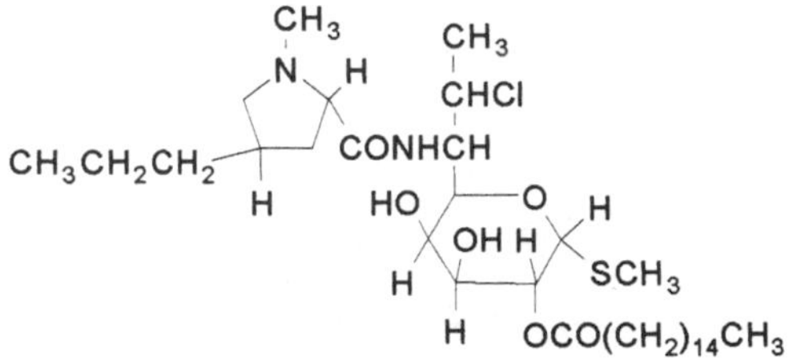

Clindamycin-2-palmitate

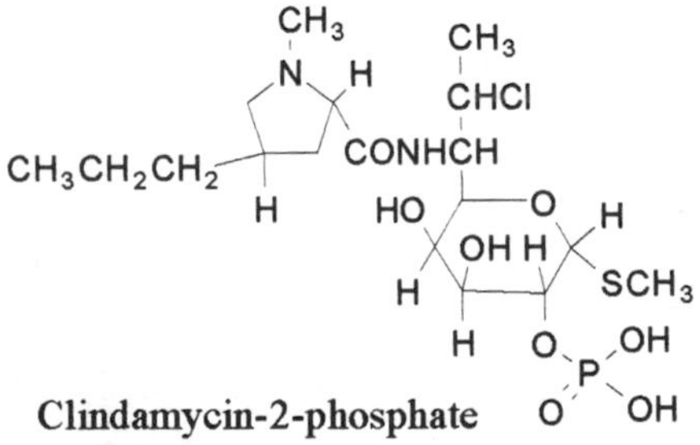

Clindamycin-2-phosphate

9.9.3.3 Slow Release

Prodrugs may be used to prolong the duration of action by providing a slow-release mechanism for the drug. Slow, prolonged release is particularly important for drugs that are used to treat psychoses where the patient requires medication that is effective over a long period of time.

Slow release and subsequent extension of action is often provided by the slow hydrolysis of amide- and ester-linked fatty acid carriers. Hydrolysis of these groups can release the drug over a period of time that can vary from several hours to weeks. For example, the use of glycine as a carrier for the anti-inflammatory tolmetin sodium results in the duration of its peak concentration being increased from about 1 h to 9 h.

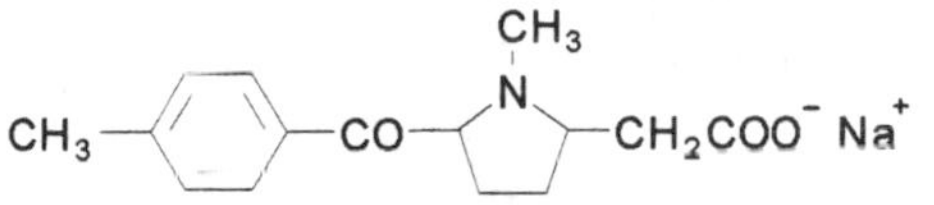

CH_3–C_6H_4–CO–(N-CH_3 pyrrole)–$CH_2COO^- Na^+$

Tolmetin sodium

CH_3–C_6H_4–CO–(N-CH_3 pyrrole)–$CH_2CONHCH_2COOH$

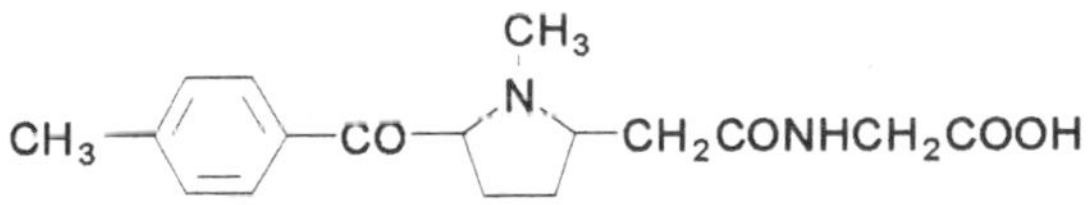

Tolmetin-glycine prodrug

Slow-release carrier prodrugs are also used as the basis of depot preparations, which are administered by intramuscular injection. For example, a single dose of the almost insoluble carrier prodrug cycloguanil embonate will slowly release the antimalarial drug cycloguanil at therapeutic levels over a period of several months.

Cycloguanil embonate

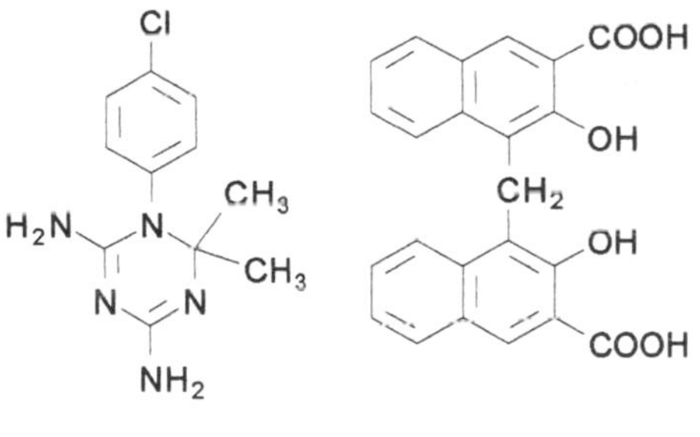

Cl, H_2N, N, CH_3, CH_3, N, N, NH_2, COOH, OH, CH_2, OH, COOH

9.9.3.4 Site Specificity

When a drug is absorbed into the body it is not transported to just its site of action but is rapidly distributed through all the available body compartments. This means that one must use a higher concentration than that which is necessary to achieve a favourable therapeutic result. An undesirable consequence of using higher concentrations is the increased possibility of the unwanted pharmacological effects of the drug becoming significant. Carrier prodrug systems offer a possible solution to this problem. In theory, it should be possible to design a carrier prodrug that would only release the drug in the vicinity of its site of action. Furthermore, once released, the drug should remain mainly in the target area and only slowly migrate to other areas. In addition the carrier should be metabolised to non-toxic metabolites. Unfortunately, these requirements have only been achieved in a number of cases.

The criteria outlined in the previous paragraph for site-specific prodrug systems have been found to be difficult to meet. However, the site-specific carrier prodrug approach has been successfully used to design drugs capable of crossing the blood–brain barrier. This barrier will only allow the passage of very lipophilic molecules unless there is an active transport mechanism available for the compound. A method developed by Bodor and other workers involved the combination of a hydrophilic drug with a suitable lipophilic carrier, which, after crossing the blood–brain barrier, would be rapidly metabolised to the drug and carrier. Once released, the hydrophilic drug is unable to recross the blood–brain barrier. The selected carrier must also be metabolised to yield non-toxic metabolites. Carriers based on the dihydropyridine ring system have been found to be particularly useful in this respect. This ring system has been found to have the required lipophilic character for crossing not only the blood–brain barrier but also other membrane barriers. The dihydropyridine system is particularly useful because it is possible to vary the functional groups attached to the dihydropyridine ring so that the carrier can be designed to link to a specific drug. Once the dihydropyridine prodrug has crossed the blood–brain barrier it is easily oxidised by the oxidases found in the brain to the hydrophilic quaternary ammonium salt, which cannot return across the barrier, and relatively non-toxic pyridine derivatives in the vicinity of its site of action.

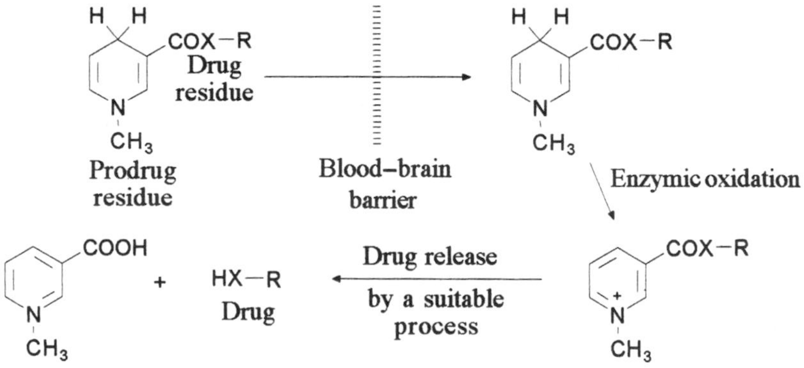

COX–R Drug residue
Prodrug residue
Blood–brain barrier
Enzymic oxidation
Drug release by a suitable process
HX–R Drug

Site-specific prodrugs are usually activated by an appropriate enzyme system. Consequently, a method of approach followed by some workers is to design prodrugs that are activated by enzymes that are found mainly at the target site. This strategy has been used to design antitumour drugs because tumours contain higher proportions of phosphatases and peptidases than normal tissues. For example, diethylstilboestrol diphosphate (Fosfestrol) has been used to deliver the oestrogen agonist diethylstilboestrol to prostatic carcinomas.

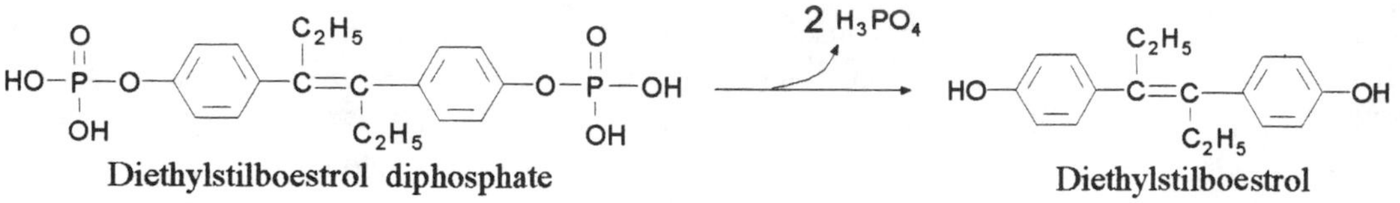

$2\ H_3PO_4$

Diethylstilboestrol diphosphate → Diethylstilboestrol

Unfortunately this approach has not been very successful for producing site-specific antitumour drugs. However, site-specific prodrugs have been developed to deliver drugs to a number of other sites (Table 19.3).

Table 9.3. Examples of site-specific prodrug delivery systems. Carrier residues are shaded.

Example	Prodrug	Target	Enzyme that releases the drug
Pilocarpine (antiglaucoma)	C_2H_5; CH_3; N; N; (1); (2); $CH_2OCO(CH_2)_{12}CH_3$; O	Eye	(1) Esterases (topical application) (2) Non-enzymic dealkylation
Adrenaline (antiglaucoma)	$(CH_3)_3CCOO$; $(CH_3)_3CCOO$; OH; $NHCH_3$	Eye	Esterases (topical application)
Sulphamethoxazole (antibacterial)	O–N; CH_3; $NHSO_2$; (2); $NHCOCH_2CH_2CH$; COOH; CH_3CONH; (1)	Kidney	(1) *N*-Acylamino acid deacylase (2) γ-Glutamyl transpeptidase
Dopamine (vasodilator)	HO; HO; (1); NH_2; $CH_2CHNHCOCH_2CH_2CHCOOH$; (2); COOH	Liver	(1) γ-Glutamyl transpeptidase (2) Dopa decarboxylase
Dexamethasone (anti-inflammatory)	$HOCH_2$; H; O; OH H; HO; H; H; OH; $O–CH_2–CO$; HO; OH; CH_3; F; O	Bowel	β D-Glucosidase (from colonic microflora)
β-Lactams (antibiotic)	RCONH; S; CH_3; CH_3; N; O; $COOCH_2OCO$; CH_3; N	Brain	Esterases
γ-Aminobutamide (anticonvulsant)	F; OH; $C=NCH_2CH_2CH_2CONH_2$; Cl	Brain	Chemical fission

9.9.3.5 Minimising Side Effects

Prodrug formation may be used to minimise toxic side effects. For example, salicylic acid is one of the oldest analgesics known. However, its use can cause gastric irritation and bleeding. The conversion of salicylic acid to its prodrug aspirin by acetylation of the phenolic hydroxy group of salicylic acid reduces the degree of stomach irritation because aspirin is mainly converted to salicylic acid by esterases after absorption from the GI tract. This reduces the amount of salicylic acid in contact with the gut wall lining.

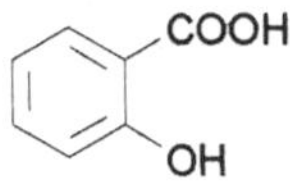

COOH, OH — Salicylic acid

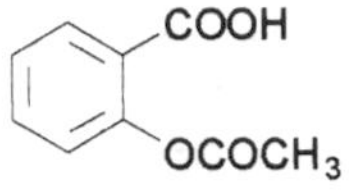

COOH, $OCOCH_3$ — Aspirin

9.10 Summary

Drug metabolism (biotransformation) is the transformation of a drug to other products. A drug is often metabolised by several different routes. The ultimate products of these metabolic pathways are pharmacologically inert compounds that are usually more water soluble than the original drug. This aids their excretion via the kidney although not all metabolites are excreted by this or other routes; some may accumulate in the body. The reactions involved in drug metabolism are classified as Phase I and Phase II reactions. **Phase I metabolic reactions** usually produce metabolites by either introducing or unmasking polar groups. **Phase II reactions** (conjugation reactions) usually increase water solubility by combining the drug with a water-solubilising compound to form so-called conjugates. The **Phase I** and **Phase II reactions** by which drug metabolism occurs are very varied. A selection is listed in sections 9.5 and 9.6, respectively.

Metabolism is affected by **genetic variations** within a species, **age** and **gender**. Ethnic genetic variations can lead to considerable differences in drug metabolism. Age can also have a considerable effect on metabolism. Its effects are particularly noticeable in the very young and the elderly. However, drug metabolism shows little difference between gender in humans, although pregnant women exhibit changes in the metabolism of some drugs. Drug metabolism is affected by **life style**. **Different species** can have different metabolic pathways for the same drug.

Metabolites may or may not exhibit pharmacological activity. Active metabolites may have either a similar, but reduced or enhanced, pharmacological activity or a different pharmacological action to the drug. The latter may be detrimental to the health of the patient.

Drug metabolism can occur in all biological tissues and fluids. However, its main centre metabolism is the liver. The GI tract, kidneys, lungs and plasma are also important centres of drug metabolism. The metabolism that occurs when a drug passes around the circulatory system for the first time is known as **first-pass metabolism**. First-pass metabolism usually occurs mainly in the liver. The main first-pass metabolic routes are **oxidation**, **reduction** and **hydrolysis**. **Oxidation** is the most important type of metabolic reaction. The main enzymes catalysing oxidative processes are the mixed-function oxidase systems and the cytochrome P-450 (CYP-450) and flavin monooxidase (FMO) families. These enzymes are not specific in their action. **Reduction** is usually catalysed by enzyme systems that are specific for a particular type of substrate. **Hydrolysis** is an important metabolic reaction for xenobiotics that contain ester and amide functional groups.

The **rate of elimination of a metabolite** from the body is an important pharmacokinetic property of the drug as well as the metabolite. If the rate of elimination of a drug is greater than the rate of elimination of the metabolite, the metabolite will accumulate in the body. However, if the rate of elimination of a drug is less than the rate of elimination of the metabolite, the metabolite will not accumulate in the body. The accumulation of a metabolite in the body may result in toxic side effects. Drugs may be designed to be more metabolically stable by replacing a metabolically active group by a less reactive group. Similarly, drugs may be designed to be less metabolically stable by incorporating a metabolically labile group in its structure.

Prodrugs are compounds that are biologically inactive but are metabolised to an active metabolite that is responsible for the drug's action. They are classified as either **bioprecursor** or **carrier prodrugs**. **Bioprecursor prodrugs** already contain the active species within their structures. **Carrier prodrugs** are formed by combining an existing drug with a compound known as a **carrier**, which gives the desired chemical and biological properties to the resulting compound. The link between the drug and carrier is a functional group that can be metabolised in the body. The nature of the carrier will depend on the available functional groups on the drug and the properties required for the prodrug. Prodrugs may be designed to improve absorption, improve patient acceptance, reduce toxicity and also for the slow release of drugs in the body. A number of prodrugs have been designed to be site specific (see Table 9.3).

9.11 Questions

(1) Explain the significance of each of the members of following pairs of terms: (a) Phase I and Phase II reactions; (b) conjugated structure and conjugation reaction; (c) carrier and bioprecursor prodrugs; (d) soft drugs and prodrugs.
(2) Explain why hippuric acids ($ArCONHCH_2COOH$) are likely to be water soluble.
(3) List the main biological factors that could influence drug metabolism. Outline their main effects.
(4) Outline the types of pharmacological activities that a metabolite could exhibit.

(5) Explain, by means of half electronic equations, how cytochrome P-450 is reduced by cytochrome P-450 reductase when a C–H bond is oxidised by cytochrome P-450 in a metabolic pathway.

(6) Outline, by means of general equations, how conjugation with glycine is used to metabolise aromatic acids. Suggest a chemical reason for the product of this process being readily excreted by the kidney.

(7) Suggest, by means of chemical equations and/or notes, feasible initial steps for the metabolism of each of the following compounds: (a) pethidine and (b) 4-aminoazobenzene.

(8) Explain why glucuronic conjugates are highly water soluble. Suggest reasons for glucuronate conjugation being a major Phase II metabolic route.

(9) (a) What is the desired objective of the drug metabolism? How is this normally achieved?
 (b) Suggest a series of metabolic reactions that could form a feasible metabolic pathway for *N,N*-diethylaminobenzene.

(10) The following scheme represents the hypothetical metabolic pathway of a drug. The values in parentheses are the rate constants for the appropriate step.

$$\text{Drug} \xrightarrow{(0.04)} \text{A} \xrightarrow{(0.30)} \text{B} \xrightarrow{(4.67)} \text{C} \xrightarrow{(0.004)} \text{D} \xrightarrow{(2.49)} \text{E} \longrightarrow \text{Excreted}$$

$$\text{B} \xrightarrow{(0.04)} \text{F} \longrightarrow \text{Excreted}$$

 (a) Explain the significance of the rate constants for the metabolism of the drug to stage B.
 (b) What is the significance of the rate constants for the metabolism of B to F and C, respectively.
 (c) Where is the rate-determining step of the series? What is its significance?

(11) Why is it necessary, in some instances, to design drugs with a very rapid rate of metabolism?

(12) Design a prodrug that could be used to transport the diethanoate ester of dopamine (A) across the blood–brain barrier. Show by means of notes and equations how this prodrug would function.

10 Nucleic Acids

10.1 Introduction

The nucleic acids are the compounds that are responsible for the storage and transmission of the genetic information that controls the growth, function and reproduction of all types of cells. They are classified into two general types: the **deoxyribonucleic acids (DNA)**, whose structures contain the sugar residue β-D-deoxyribose; and the **ribonucleic acids (RNA)**, whose structures contain the sugar residue β-D-ribose (Figure 10.1).

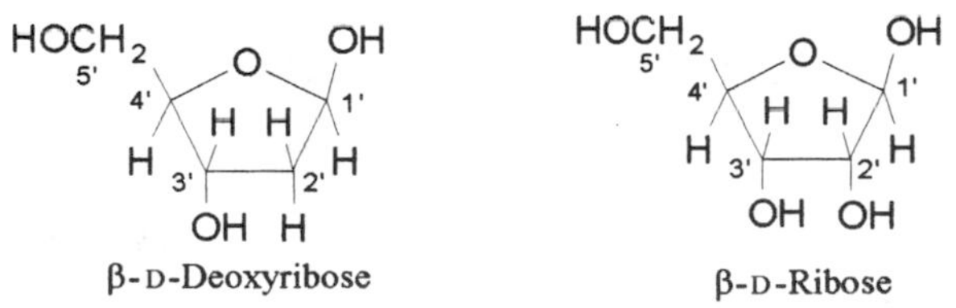

β-D-Deoxyribose β-D-Ribose

Figure 10.1. The structures of β-D-deoxyribose and β-D-ribose.

Both types of nucleic acids are polymers based on a repeating structural unit known as a **nucleotide** (Figure 10.2). These nucleotides form long single-chain polymer molecules in both DNA and RNA.

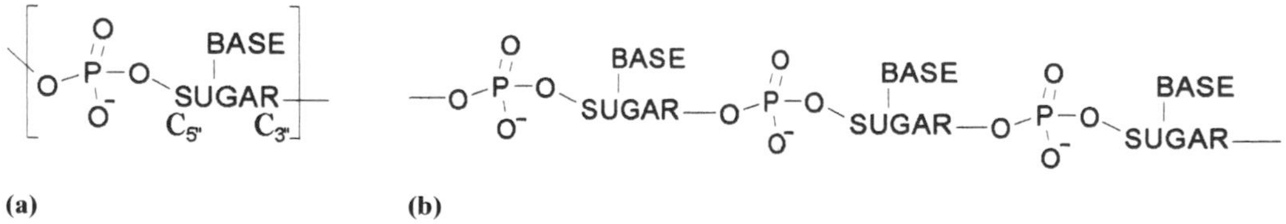

(a) (b)

Figure 10.2. The general structures of (a) nucleotides and (b) a schematic representation of a section of a nucleic acid chain.

Each nucleotide consists of a purine or pyrimidine base bonded to the 1′ carbon atom of a sugar residue by a β-*N*-glycosidic link (Figure 10.3). These base–sugar subunits, which are known as **nucleosides**, are linked through the 3′ and 5′ carbons of their sugar residues by phosphate units to form the polymer chain.

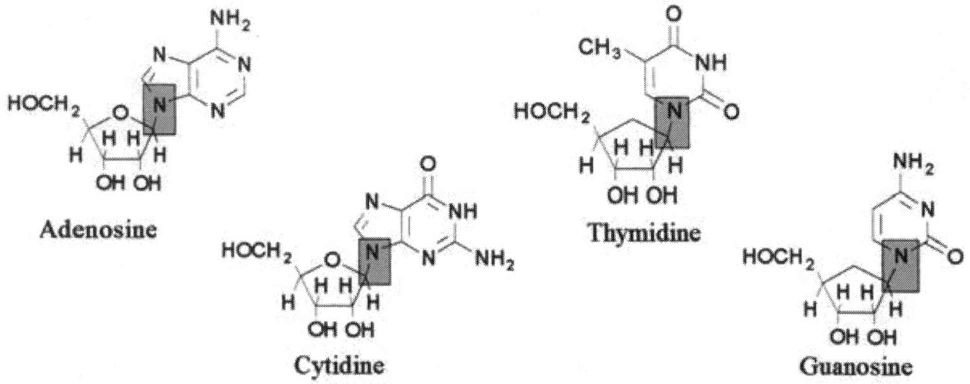

Figure 10.3. Examples of the structures of some of the nucleosides found in RNA. The β-*N*-glycosidic link is shaded. The corresponding nucleosides in DNA are based on deoxyribose and use the same name but with the prefix deoxy.

10.2 Deoxyribonucleic Acids (DNA)

DNA occurs in the nuclei of cells in the form of a very compact DNA–protein complex called **chromatin**. The protein in chromatin consists mainly of **histones**, a family of relatively small positively charged proteins. The DNA is coiled twice around an octomer of histone molecules with a ninth histone molecule attached to the exterior of these minicoils to form a structure like a row of beads spaced along a string (Figure 10.4). This 'string of beads' is coiled and twisted into compact structures known as miniband units, which form the basis of the structures of **chromosomes**. Chromosomes are the structures that form duplicates during cell division in order to transfer the genetic information of the old cell to the two new cells.

10.2.1 Structure

DNA molecules are large with relative molecular masses up to one trillion (10^{12}). The principal bases found in their structures are adenine (A), thymine (T), guanine (G) and

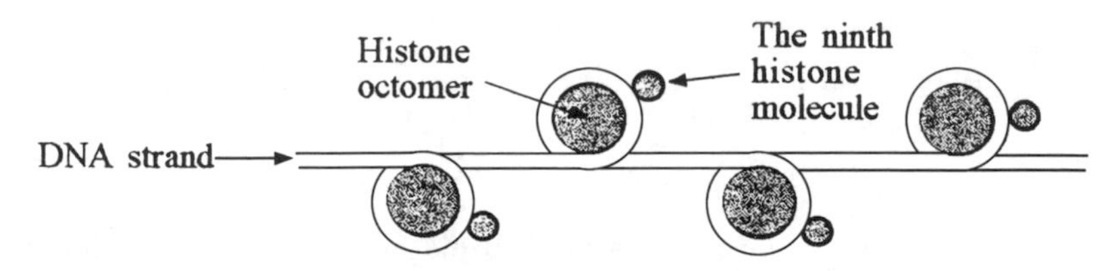

Figure 10.4. The 'string of beads' structure of chromatin. The DNA strand is wound twice around each histone octomer. A ninth histone molecule is bound to the exterior surface of the coil.

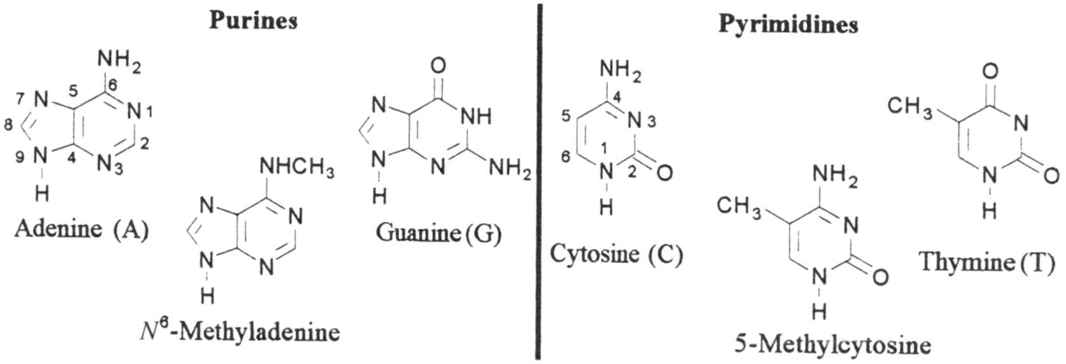

Purines

Adenine (A)

N^6-Methyladenine

Guanine (G)

Pyrimidines

Cytosine (C)

5-Methylcytosine

Thymine (T)

Figure 10.5. The purine and pyrimidine bases found in DNA. The numbering is the same for each type of ring system.

cytosine (C), although derivatives of these bases are found in some DNA molecules (Figure 10.5). Those bases with an oxygen function have been shown to exist in their keto form.

Chargaff showed that the molar ratios of adenine to thymine and guanine to cytosine are always approximately 1:1 in any DNA structure although the ratio of adenine to guanine varies according to the species from which the DNA is obtained. This and other experimental observations lead Crick and Watson in 1953 to propose that the three-dimensional structure of DNA consisted of two single molecule polymer chains held together in the form of a double helix by hydrogen bonding between the same pairs of bases, namely: the adenine–thymine and cytosine–guanine base pairs (Figure 10.6). These pairs of bases, which are referred to as **complementary base pairs**, form the internal structure of the helix. They are hydrogen bonded in such a manner that their flat structures lie parallel to one another across the inside of the helix. The two polymer chains forming the helix are aligned in opposite directions. In other words, at the ends of the structure one chain has a free 3′-OH group and the other chain has a free 5′-OH group. X-ray diffraction studies have since confirmed that this is the basic three-dimensional shape of the polymer chains of the B-DNA, the natural form of DNA. This form of DNA has about ten bases per turn of the helix. Its outer surface has two grooves known as the minor and major grooves, respectively, which act as the binding sites for many ligands. Two other forms of DNA, the A and Z forms, have also been identified but it is not certain if these forms occur naturally in living cells.

Electron microscopy has shown that the double helical chain of DNA is folded, twisted and coiled into quite compact shapes. A number of DNA structures are cyclic and these compounds are also coiled and twisted into specific shapes. These shapes are referred to as supercoils, supertwists and superhelices, as appropriate.

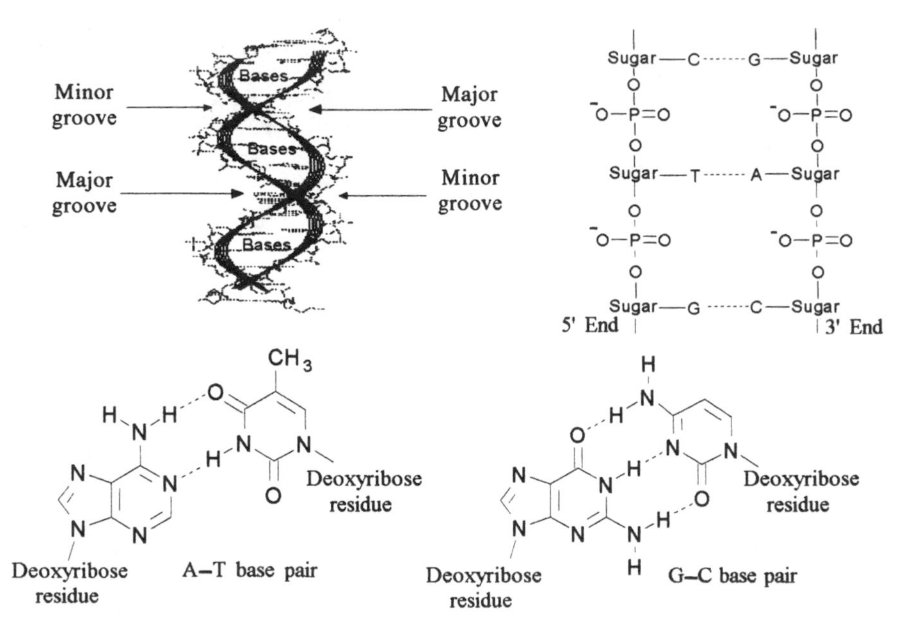

Figure 10.6. The double helical structure of B-DNA. Interchanging of either the bases of a base pair and/or base pair with base pair does not affect the geometry of this structure. Reproduced from G. Thomas, *Chemistry for Pharmacy and the Life Sciences including Pharmacology and Biomedical Science*, 1996, by permission of Prentice Hall, a Pearson Education Company.

10.3 The General Functions of DNA

The DNA found in the nuclei of cells has three functions:

(i) to act as a repository for the genetic information required by a cell to reproduce that cell;
(ii) to reproduce itself in order to maintain the genetic pool when cells divide;
(ii) to supply the information that the cell requires to manufacture specific proteins.

Genetic information is stored in a form known as **genes** by the DNA found in the nucleus of a cell (see section 10.4).

The duplication of DNA is known as **replication.** It results in the formation of two identical DNA molecules that carry the same genetic information from the original cell to the two new cells that are formed when a cell divides (see section 10.5).

The function of DNA in protein synthesis is to act as a template for the production of the various RNA molecules necessary to produce a specific protein (see section 10.6).

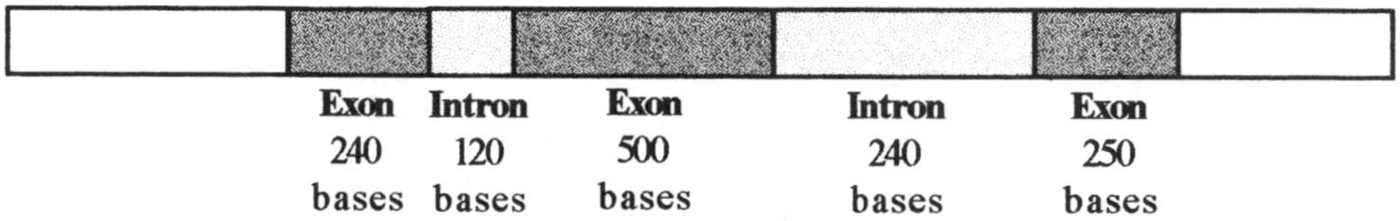

Exon 240 bases | Intron 120 bases | Exon 500 bases | Intron 240 bases | Exon 250 bases

Figure 10.7. A schematic representation of the gene for the β-subunit of haemoglobin.

10.4 Genes

Each species has its own internal and external characteristics. These characteristics are determined by the information stored and supplied by the DNA in the nuclei of its cells. This information is carried in the form of a code based on the consecutive sequences of bases found in sections of the DNA structure (see section 10.7). This code controls the production of the peptides and proteins required by the body. The sequence of bases that act as the code for the production of one specific peptide or protein molecule is known as a gene.

Genes can normally contain from several hundred to 2000 bases. Changing the sequence of the bases in a gene by adding, subtracting or changing one or more bases may cause a change in the structure of the protein whose production is controlled by the gene. This may have a subsequent knock-on effect on the external or internal characteristics of an individual. For example, an individual may have brown instead of blue eyes or their insulin production may be inhibited, which could result in that individual suffering from diabetes. A number of medical conditions have been attributed to either the absence of a gene or the presence of a degenerate or faulty gene in which one or more of the bases in the sequence have been changed.

In simple organisms, such as bacteria, genetic information is usually stored in a continuous sequence of DNA bases. However, in higher organisms the bases forming a particular gene may occur in a number of separate sections known as **exons**, separated by sections of DNA that do not appear to be a code for any process. These non-coding sections are referred to as **introns**. For example, the gene responsible for the β-subunit of haemoglobin consists of 990 bases. These bases occur as three exons separated by two introns (Figure 10.7).

The complete set of genes that contain all the hereditary information of a particular species is called a **genome**. The Human Genome Project, initiated in 1990, sets out to identify all the genes that occur in human chromosomes and also the sequence of bases in these genes. This will create an index that can be used to locate the genes responsible for particular medical conditions. For example, the gene in region q31 of chromosome 7 is responsible for the protein that controls the flow of chloride ions through the membranes in the lungs. The changing of about three bases in exon number 10 gives a degenerate gene that is known to be responsible for causing cystic fibrosis in a large number of cases.

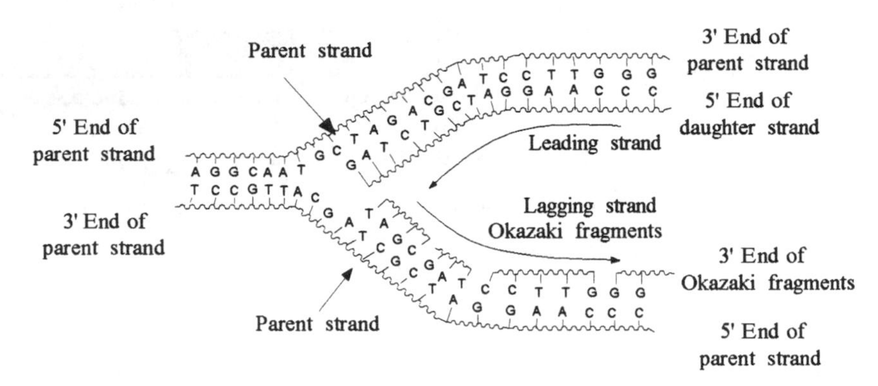

Figure 10.8. A schematic representation of the replication of DNA. The arrows show the direction of growth of the leading and lagging strands. Reproduced from G. Thomas, *Chemistry for Pharmacy and the Life Sciences including Pharmacology and Biomedical Science*, 1996, by permission of Prentice Hall, a Pearson Education Company.

10.5 Replication

Replication is believed to start with the unwinding of a section of the double helix (Figure 10.8). Unwinding may start at the end or more commonly in a central section of the DNA helix. It is initiated by the binding of the DNA to specific receptor proteins that have been activated by the appropriate first messenger (see section 8.4). The separated strands of the DNA act as templates for the formation of a new daughter strand. Individual nucleotides, which are synthesised in the cell by a complex route, bind by hydrogen bonding between the bases to the complementary parent nucleotides. This hydrogen bonding is specific: only the complementary base pairs can hydrogen bond. In other words, the hydrogen bonding can only be between either thymine and adenine or cytosine and guanine. This means that the new daughter strand is an exact replica of the original DNA strand bound to the parent strand. Consequently, replication will produce two identical DNA molecules.

As the nucleotides hydrogen bond to the parent strand they are linked to the adjacent nucleotide, which is already hydrogen bonded to the parent strand, by the action of enzymes known as DNA polymerases. As the daughter strands grow, the DNA helix continues to unwind. However, *both* daughter strands are formed at the same time in the 5′ to the 3′ direction. This means that the growth of the daughter strand that starts at the 3′ end of the parent strand can continue smoothly as the DNA helix continues to unwind. This strand is known as the **leading strand**. However, this smooth growth is not possible for the daughter strand that started from the 5′ of the parent strand. This strand, known as the **lagging strand**, is formed in a series of sections, each of which still grows in the 5′ to 3′ direction. These sections, which are known as Okazaki fragments after their discoverer, are joined together by the enzyme DNA ligase to form the second daughter strand.

Replication, which starts at the end of a DNA helix, continues until the entire structure has been duplicated. The same result is obtained when replication starts at the centre of a DNA helix. In this case, unwinding continues in both directions until the complete molecule is duplicated. This latter situation is more common.

DNA replication occurs when cell division is imminent. At the same time, new histones are synthesised. This results in a thickening of the chromatin filaments into chromosomes (see section 10.2). These rod-like structures can be stained and are large enough to be seen under a microscope.

10.6 Ribonucleic Acids (RNA)

Ribonucleic acids are found in both the nucleus and the cytoplasm. In the cytoplasm RNA is located mainly in small spherical organelles known as **ribosomes**. These consist of about 65% RNA and 35% protein. Ribonucleic acids are classified according to their general role in protein synthesis as: messenger RNA (mRNA); transfer RNA (tRNA); and ribosomal RNA (rRNA). Messenger RNA informs the ribosome as to what amino acids are required and their order in the protein, that is, they carry the genetic information necessary to produce a specific protein. This type of RNA is synthesised as required and once its message has been delivered it is decomposed. Transfer RNA transports the required amino acids in the correct order to the ribosome, where ribosomal RNA (rRNA) controls the synthesis of the required protein (see section 10.8).

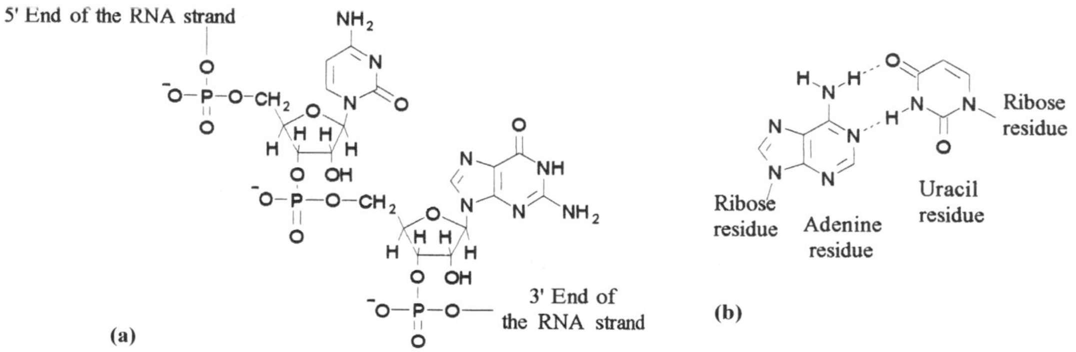

Figure 10.9. (a) The general structure of a section of an RNA polymer chain. (b) The hydrogen bonding between uracil and adenine. Reproduced from G. Thomas, *Chemistry for Pharmacy and the Life Sciences including Pharmacology and Biomedical Science*, 1996, by permission of Prentice Hall, a Pearson Education Company.

The structures of RNA molecules consist of a single polymer chain of nucleotides with the same bases as DNA, with the exception of thymine, which is replaced by uracil (Figure 10.9).

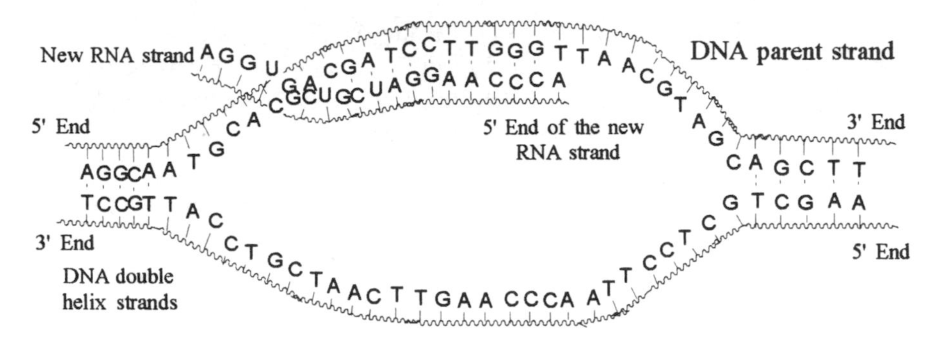

Figure 10.10. A schematic representation of a transcription process. Reproduced from G. Thomas, *Chemistry for Pharmacy and the Life Sciences including Pharmacology and Biomedical Science*, 1996, by permission of Prentice Hall, a Pearson Education Company.

These chains often contain single-stranded loops separated by short sections of a distorted double helix (Figure 10.11). These structures are known as **hairpin loops**.

All types of RNA are formed from DNA by a process known as transcription. It is thought that the DNA unwinds and the RNA molecule is formed in the 5′ to 3′ direction. It proceeds smoothly, with the 3′ end of the new strand bonding to the 5′ end of the next nucleotide (Figure 10.10). This bonding is catalysed by enzymes known as RNA polymerases. The sequence of bases in the new RNA strand is controlled by the sequence of bases in the parent DNA strand. In this way DNA controls the genetic information being transcribed into the RNA molecule. The strands of DNA also contain start and stop signals, which control the size of the RNA molecule produced. These signals are in the form of specific sequences of bases. It is believed that the enzyme **rho factor** could be involved in the termination of the synthesis and the release of some RNA molecules from the parent DNA strand. However, in many cases there is no evidence that this enzyme is involved in the release of the RNA molecule.

The RNA produced within the nucleus by transcription is known as **heterogeneous nuclear RNA** (hnRNA), premessenger RNA (pre-mRNA) or primary transcript RNA (ptRNA). Because the DNA gene from which it is produced contains both exons and introns, the hnRNA will also contain its genetic information in the form of a series of exons and introns complementary to those of its parent gene.

10.7 Messenger RNA (mRNA)

mRNA carries the genetic message from the DNA in the nucleus to a ribosome. This message instructs the ribosome to synthesise a specific protein. mRNA is believed to be produced in the nucleus from hnRNA by removal of the introns and the splicing together of the remain-

Table 10.1. The genetic code. Some codons act as start and stop signals in protein synthesis (see sections 10.10 and 10.11). Codons are written left to right, 5′ to 3′.

Code	Amino acid	Code	Amino acid	Code	Amino acid	Code	Amino acid
UUU	Phe	CUU	Leu	AUU	Ile	GUU	Val
UUC	Phe	CUC	Leu	AUC	Ile	GUC	Val
UUA	Leu	CUA	Leu	AUA	Ile	GUA	Val
UUG	Leu	CUG	Leu	AUG	Met	GUG	Val
UCU	Ser	CCU	Pro	ACU	Thr	GCU	Ala
UCC	Ser	CCC	Pro	ACC	Thr	GCC	Ala
UCA	Ser	CCA	Pro	ACA	Thr	GCA	Ala
UCG	Ser	CCG	Pro	ACG	Thr	GCG	Ala
UAU	Tyr	CAU	His	AAU	Asn	GAU	Asp
UAC	Tyr	CAC	His	AAC	Asn	GAC	Asp
UAA	**Stop**	CAA	Gln	AAA	Lys	GAA	Glu
UAG	**Stop**	CAG	Gln	AAG	Lys	GAG	Glu
UGU	Cys	CGU	Arg	AGU	Ser	GGU	Gly
UGC	Cys	CGC	Arg	AGC	Ser	GGC	Gly
UGA	**Stop**	CGA	Arg	AGA	Arg	GGA	Gly
UGG	Trp	CGG	Arg	AGG	Arg	GGG	Gly

ing exons into a continuous genetic message, the process being catalysed by specialised enzymes. The net result is a smaller mRNA molecule with a continuous sequence of bases that are complementary to the gene's exons. This mRNA now leaves the nucleus and carries its message in the form of a code to a ribosome.

The code carried by mRNA was broken in the 1960s by Nirenberg and other workers. These workers demonstrated that each naturally occurring amino acid had a DNA code that consisted of a sequence of three consecutive bases known as a **codon** and that an amino acid could have several different codons (Table 10.1). In addition, three of the codons are stop signals, which instruct the ribosome to stop protein synthesis. Furthermore, the codon that initiates the synthesis is always AUG, which is also the codon for methionine. Consequently, all protein synthesis starts with methionine. However, few completed proteins have a terminal methionine because this residue is normally removed before the peptide chain is complete. Moreover, methionine can still be incorporated in a peptide chain because there are two different tRNAs that transfer methionine to the ribosome (see section 10.8). One is specific for the transfer of the initial methionine whereas the other will only deliver methionine to the developing peptide chain. By convention, the three letters of codon triplets are normally written with their 5′ ends on the left and their 3′ ends on the right.

The mRNA's codon code is known as the **genetic code**. Its use is universal, all living matter using the same genetic code for protein synthesis. This suggests that all living matter must have originated from the same source and is strong evidence for Darwin's theory of evolution.

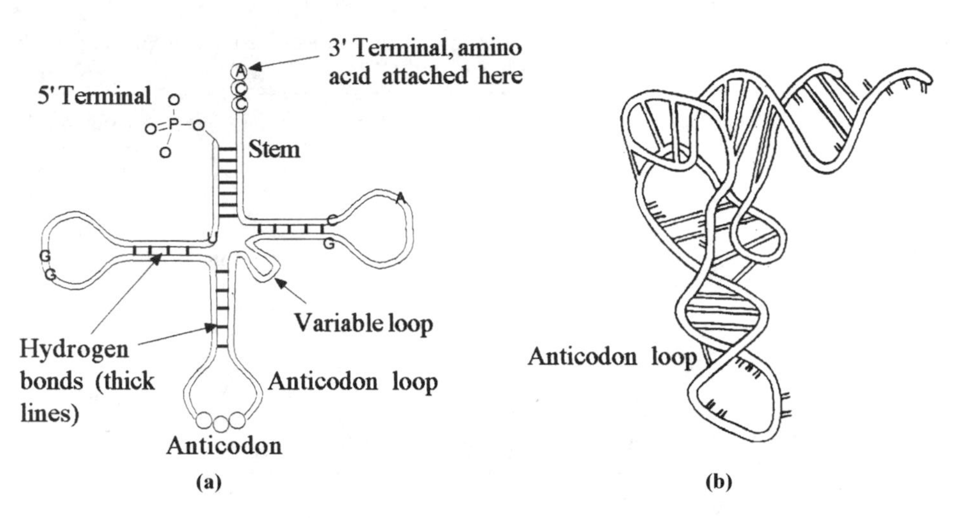

Figure 10.11. The general structures of tRNA. (a) The two-dimensional cloverleaf representation showing some of the invariable nucleotides that occur in the same positions in most tRNA molecules and (b) the three-dimensional L shape (From CHEMISTRY, by Linus Pauling and Peter Pauling. Copyright © 1975 by Linus Pauling and Peter Pauling. Used with permission of W. H. Freeman and Company)

10.8 Transfer RNA (tRNA)

2'-*O*-Methylguanosine

Inosine

tRNAs are also believed to be formed in the nucleus from the hnRNA. They are relatively small molecules that usually contain from 73 to 94 nucleotides in a single strand. Some of these nucleotides may contain derivatives of the principal bases, such as 2′-*O*-methylguanosine (OMG) and inosine (I). The strand of tRNA is usually folded into a three-dimensional L shape. This structure, which consists of several loops, is held in this shape by hydrogen bonding between complementary base pairs in the stem sections of these loops and also by hydrogen bonding between bases in different loops. This results in the formation of sections of double helical structures. However, the structures of most tRNAs are represented in two dimensions as a **cloverleaf** (Figure 10.11).

tRNA molecules carry amino acid residues from the cell's amino acid pool to the mRNA attached to the ribosome. The amino acid residue is attached through an ester linkage to the ribose residue at the 3′ terminal of the tRNA strand, which almost invariably has the sequence CCA. This sequence plus a fourth nucleotide projects beyond the double helix of the stem. Each type of amino acid can only be transported by its own specific tRNA molecule. In other words a tRNA that carries serine residues will not transport alanine residues. However, some amino acids can be carried by several different tRNA molecules.

The tRNA recognises the point on the mRNA where it has to deliver its amino acid through the use of a group of three bases known as an **anticodon.** This anticodon is a sequence of three bases found on one of the loops of the tRNA (Figure 10.11). The anticodon can only form base pairs with the complementary codon in the mRNA. Consequently, the tRNA will only hydrogen bond to the region of the mRNA that has the correct codon, which means its amino acid can only be delivered to a specific point on the mRNA. For example, a tRNA molecule with the anticodon CGA will only transport its alanine residue to a GCU codon on the mRNA. Furthermore, this mechanism will also control the order in which amino acid residues are added to the growing protein.

10.9 Ribosomal RNA (rRNA)

Ribosomes contain about 35% protein and 65% rRNA. Their structures are complex and have not yet been fully elucidated. However, they have been found to consist of two sections that are referred to as the **large** and **small** subunits. Each of these subunits contains protein and rRNA. In *Escherichia coli* the small subunit has been shown to contain a 1542-nucleotide rRNA molecule whereas the large contains two rRNA molecules of 120 (Figure 10.12) and 2094 nucleotides, respectively.

Experimental evidence suggests that rRNA molecules have structures that consist of a single strand of nucleotides whose sequence varies considerably from species to species. The strand is folded and twisted to form a series of single-stranded loops separated by sections of double helix (Figure 10.12). The double helical segments are believed to be formed by hydrogen bonding between complementary base pairs. The general pattern of loops and helixes is very similar between species even though the sequence of nucleotides are different. However, little is known about the three-dimensional structures of rRNA molecules and their interactions with the proteins found in the ribosome.

10.10 Protein Synthesis

Protein synthesis starts from the N-terminal of the protein. It proceeds in the 5′ to 3′ direction along the mRNA and may be divided into four major stages, namely: activation; initiation;

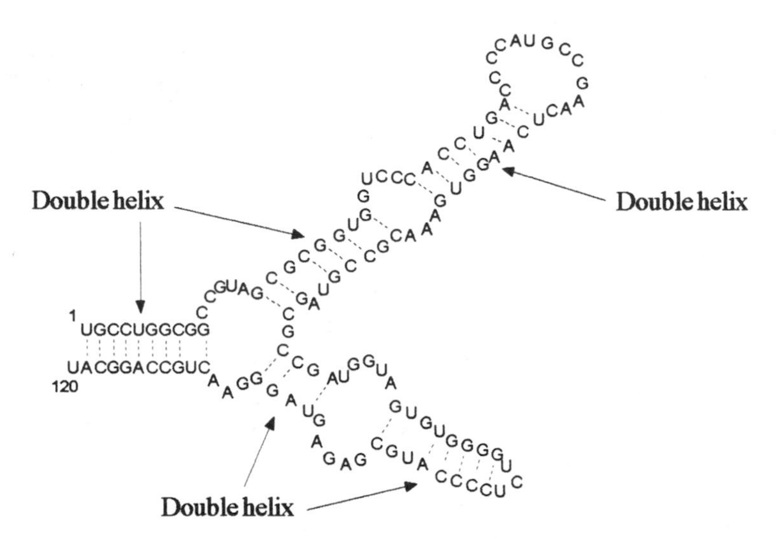

Figure 10.12. The proposed sequence of nucleotides in the 120-nucleotide subunit found in *Escherichia coli* ribosomes showing the single-stranded loops and the double helical structures. (Reprinted, with permission, from the *Annual Review of Biochemistry*, volume 53 © 1984 by Annual Reviews, www.AnnualReviews.org).

elongation; and termination. Activation is the formation of the tRNA–amino acid complex. Initiation is the binding of the mRNA to the ribosome and the activation of the ribosome. Elongation is the formation of the protein. Termination is the ending of the protein synthesis and its release from the ribosome. All these processes normally require the participation of protein catalysts, known as **factors**, as well as other proteins whose function is not always known. GTP and sometimes ATP act as sources of energy for the processes.

10.10.1 Activation

It is believed that the amino acids from the cellular pool react with ATP to form an active amino acid–AMP complex. This complex reacts with the specific tRNA for the amino acid, the reaction being catalysed by a synthase that is specific for that amino acid.

$$\underset{\text{Amino acid}}{\mathrm{RCH(NH_2)COOH}} + \underset{\text{ATP}}{{}^{-}\mathrm{O{-}P(=O)(O^-){-}O{-}P(=O)(O^-){-}O{-}P(=O)(O^-){-}O{-}A}} \longrightarrow \underset{\text{Amino acid–AMP complex}}{\mathrm{RCH(NH_2)CO{-}O{-}P(=O)(O^-){-}O{-}A}} + \underset{\text{Pyrophosphate}}{{}^{-}\mathrm{O{-}P(=O)(O^-){-}O{-}P(=O)(O^-){-}O^-}}$$

$$\mathrm{RCH(NH_2)CO{-}O{-}P(=O)(O^-){-}O{-}A} + \underset{\text{tRNA (3' Terminal, 5' Terminal)}}{\text{tRNA}} \longrightarrow \underset{\text{Aminoacyl–tRNA complex}}{\mathrm{RCH(NH_2)CO{-}O{-}tRNA}} + \underset{\text{AMP}}{{}^{-}\mathrm{O{-}P(=O)(O^-){-}O{-}A}}$$

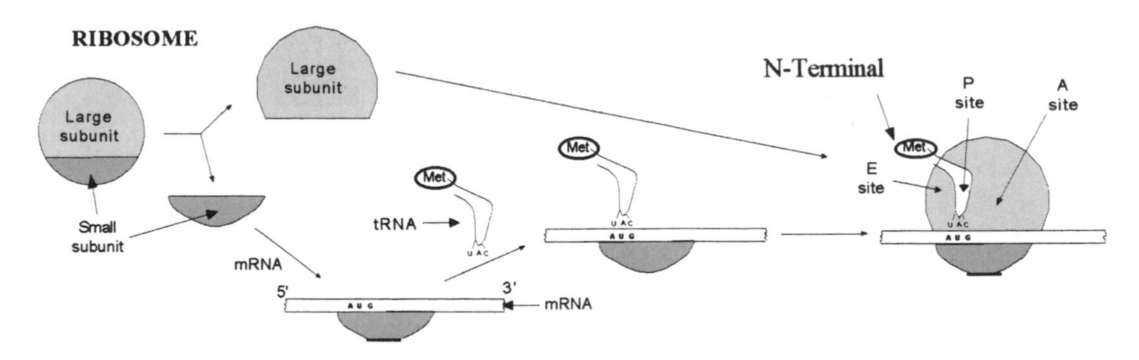

Figure 10.13. A schematic representation of the initiation of protein synthesis.

10.10.2 Initiation

The general mechanism of initiation is well documented but the finer details are still not known. It is thought that it starts with the two subunits of the ribosome separating and the binding of the mRNA to the smaller subunit. Protein synthesis then starts by the attachment of a methionine–tRNA complex to the mRNA so that it forms the N-terminal of the new protein. Methionine is always the first amino acid in all protein synthesis because its tRNA anticodon is also the signal for the ribosome system to start protein synthesis. Because the anticodon for methionine–tRNA is UAC, this synthesis will start at the AUG codon of the mRNA. This codon is usually found within the first 30 nucleotides of the mRNA. However, few proteins have an N-terminal methionine because once protein synthesis has started the methionine is usually removed by hydrolysis. As soon as the methionine–tRNA has bound to the mRNA the larger ribosomal subunit is believed to bind to the smaller subunit so that the mRNA is sandwiched between the two subunits (Figure 10.13). This large subunit is believed to have three binding sites called the **P (peptidyl), A (acceptor)** and **E (exit)** sites. It attaches itself to the smaller subunit so that its P site is aligned with the methionine–tRNA complex bound to the mRNA. This P site is where the growing protein will be bound to the ribosome. The A site, which is thought to be adjacent to the P site, is where the next amino acid–tRNA complex binds to the ribosome so that its amino acid can be attached to the peptide chain. The E site is where the discharged tRNA is transiently bound before it leaves the ribosome.

10.10.3 Elongation

Elongation is the formation of the peptide chain of the protein by a stepwise repetitive process. A great deal is known about the nature of this process but its exact mechanism is still not fully understood.

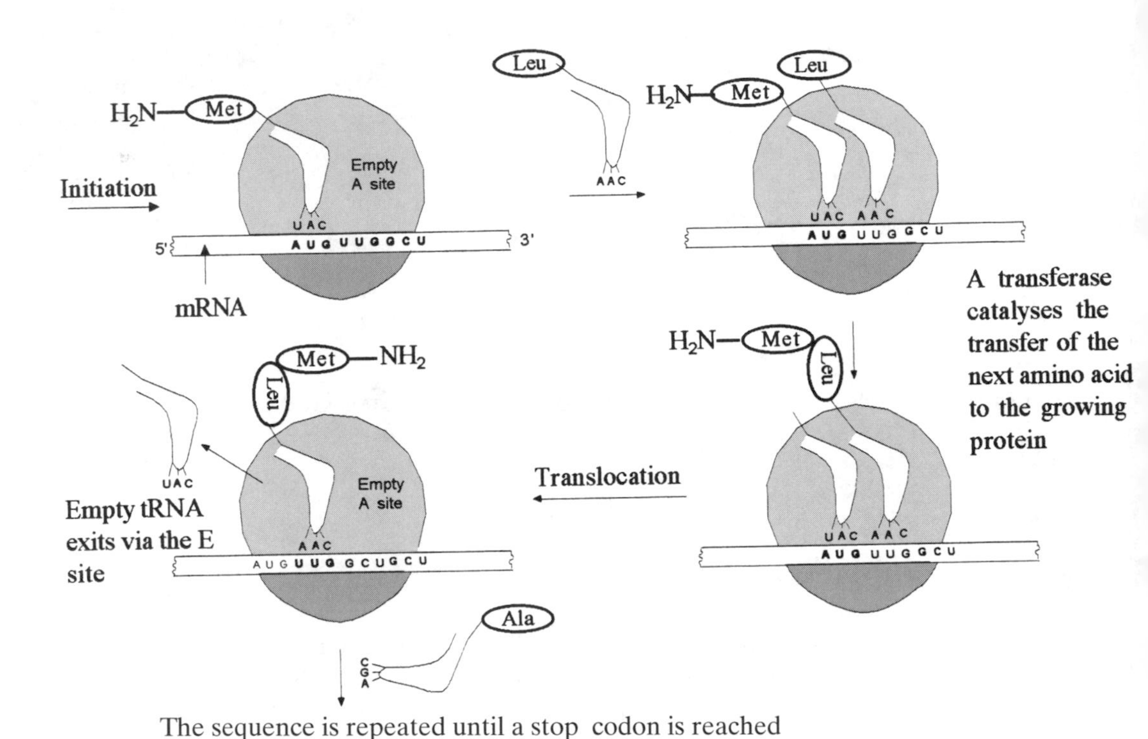

Figure 10.14. A diagrammatic representation of the process of elongation in protein synthesis.

The process of elongation is best explained by the use of a hypothetical example. Suppose that the sequence of codons, including the start codon, is AUGUUGGCUGGA . . . etc. The elongation process starts with the methionine–tRNA bound to the AUG codon of the mRNA (Figure 10.14). Because the second codon is UUG the second amino acid in the polypeptide chain will be leucine. This amino acid is transported by a tRNA molecule with the anticodon AAC because this is the only anticodon that matches the UUG codon on the mRNA strand. The leucine–tRNA complex '*docks*' on the UUG codon of the mRNA and binds to the A site. This docking and binding is believed to involve ribosome proteins, referred to as elongation factors, and energy supplied by the hydrolysis of guanosine triphosphate (GTP) to guanosine diphosphate (GDP). Once the leucine–tRNA has occupied the A site the methionine is linked to the leucine by means of a peptide link whose carbonyl group originates from the methionine. This reaction is catalysed by the appropriate transferase. It leaves the tRNA on the P site empty and produces an (NH_2)-Met-Leu-tRNA complex at the A site. The empty tRNA is discharged through the E site and at the same time the complete ribosome moves along the mRNA in the 5′ to 3′ direction so that the dipeptide–tRNA complex moves from the A site to the P site. This process is known as **translocation**. It is poorly understood but it leaves the A site empty and able to receive the next amino acid–tRNA complex. The whole process is then repeated in order to add the next amino acid residue to the peptide chain. Because the next mRNA codon in our hypothetical example is GCU this amino acid will be alanine (see Table 10.1). Subsequent amino acids are added in a similar way, the sequence of amino acid residues in the chain being controlled by the order of the codons in the mRNA.

10.10.4 Termination

The elongation process continues until a stop codon is reached. This codon cannot accept an amino acid–tRNA complex and so the synthesis stops. At this point the peptide–tRNA chain occupies a P site and the A site is empty. The stop codon of the mRNA is recognised by proteins known as release factors, which promote the release of the protein from the ribosome. The mechanism by which this happens is not fully understood but they are believed to convert the transferase responsible for peptide synthesis into a hydrolase, which catalyses the hydrolysis of the ester group linking the polypeptide to its tRNA. Once released, the protein is folded into its characteristic shape, often under the direction of molecular **chaperone** proteins.

10.11 Protein Synthesis in Prokaryotic and Eukaryotic Cells

The general sequence of events in protein synthesis is similar for both eukaryotic and prokaryotic cells. In both cases the hydrolysis of GTP to GDP is the source of energy for many of the processes involved. However, the structures of prokaryotic and eukaryotic ribosomes are different. For example, the ribosomes of prokaryotic cells of bacteria are made up of 50S (see Appendix 3) and 30S rRNA subunits whereas the ribosomes of mammalian eukaryotic cells consist of 60S and 40S rRNA subunits. The differences between the ribosomes of prokaryotic and eukaryotic ribosomes are the basis of the selective action of some antibiotics (see section 10.12).

10.11.1 Prokaryotic Cells

The first step in protein synthesis is the correct alignment of mRNA on the small subunit of the ribosome. In prokaryotic cells this alignment is believed to be due to binding by base pairing between bases at the 3′ end of the rRNA of the ribosome and bases at the 5′ end of the mRNA. This ensures the correct alignment of the AUG anticodon of the mRNA with the P site of the ribosome. The mRNA sequence of bases responsible for this binding occurs as part of the '*upstream*' (5′ terminal end) section of the strand before the start codon. This sequence is often

ϕX174 phage A protein	-AAU-CUU-**GGA-GG**C-UUU-UUU-**AUG**-GUU-CGU-
Ribosomal protein S12	-AAA-ACC-**AGG-AG**C-UAU-UUA-**AUG**-GCA-ACA-
trpL. **leader**	-GUA-AA**A-AGG-G**UA-UCG-ACA-**AUG**-AAA-GCA-
araB	-UUU-GGA-U**GG-AG**U-GAA-ACG-**AUG**-GCG-AUU-

Figure 10.15. Examples of Shine–Dalgarno sequences (bold larger type) of mRNA recognised by *Escherichia coli* ribosomes. These sequences lie about 10 nucleotides upstream of the AUG start codon for the specified protein.

known as the **Shine–Dalgarno** sequence after its discovers. Shine–Dalgarno sequences vary in length and base sequence (Figure 10.15).

The initiating tRNA in prokaryotic cells is a specific methionine–tRNA known as $tRNA_f^{Met}$, which is able to read the start codon AUG but not when it is part of the elongation sequence. $tRNA_f^{Met}$ is unique in that the methionine it carries is usually in the form of its *N*-formyl derivative.

***N*-Formyl-methionine–$tRNA_f^{Met}$**

$$CH_3SCH_2CH_2CH(NH{-}CO{-}H)COO\,tRNA$$

When AUG is part of the elongation sequence methionine is added to the growing protein by a different transfer RNA known as $tRNA_m^{Met}$, which also has the anticodon UAC. However, $tRNA_m^{Met}$ cannot initiate protein synthesis. Elongation follows the general mechanism for protein synthesis (see section 10.9). It requires a group of proteins known as elongation factors and energy supplied by the hydrolysis of GTP to GDP. Termination normally involves three release factors.

Experimental work has shown that an mRNA strand actively synthesising proteins will have several ribosomes attached to it at different places along its length. These multiple ribosome structures are referred to as **polyribosomes** or **polysomes**. The polysomes of prokaryotic cells can contain up to 10 ribosomes at any one time. Each of these ribosomes will be simultaneously producing the same polypeptide or protein; the further the ribosome has moved along the mRNA, the longer the polypeptide chain. The process resembles the assembly line in a factory. Each mRNA strand can in its lifetime produce up to 300 protein molecules.

In prokaryotic but not eukaryotic cells (see section 4.1), ribosomes are found in association with DNA. This is believed to be due to the ribosome binding to the mRNA as it is produced by transcription from the DNA. Furthermore, these ribosomes have been shown to start producing the polypeptide chain of their designated protein before transcription is complete. This means that in bacteria protein synthesis can be very rapid and in some cases faster than transcription. It has been reported that in some bacteria an average of 10 amino acid residues are added to the peptide chain every second.

10.11.2 Eukaryotic Cells

The initiation of protein synthesis in eukaryotic cells follows a different route from that found in prokaryotic cells although it still uses a methionine–tRNA to start the process. Eukaryotic mRNAs have no Shine–Dalgarno sequences but are characterised by a 7-methyl GTP unit at the 5′ end of the mRNA strand and a polyadenosine nucleotide tail at the 3′ end of the strand (Figure 10.16).

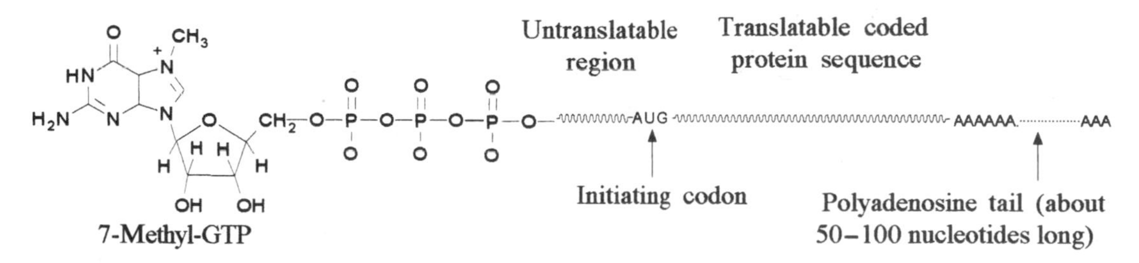

Figure 10.16. The general structure of eukaryotic mRNA molecules where ~~~~ represents the mRNA strand.

In eukaryotic cells, the initiating tRNA is a unique form of the activated methionine–tRNA ($tRNA_i^{Met}$). However, unlike in the case of prokaryotic cells, the methionine residue it carries is not formylated. The initiating process is started by this $tRNA_i^{Met}$ binding to the 40S subunit of the ribosome to form the so-called **preinitiation complex**, the process requiring the formation of a complex between $tRNA_i^{Met}$, various eukaryotic initiation factors (eIFs) and GTP. At this point the mRNA binds to the 40S preinitiation complex. This binding process is believed to involve a number of eukaryotic initiation factors and energy supplied by the conversions of GTP to GDP and ATP to ADP. Once the mRNA has bound to the preinitiation complex the 60S subunit recombines with the 40S unit to form the initiation complex (Figure 10.17).

The absence of the Shine–Dalgarno sequence means that an alternative mechanism must be available to align the first AUG codon of the mRNA with the P site of the ribosome. This mechanism is believed to direct the preinitiation complex to the first AUG codon of the mRNA.

Elongation in eukaryotic ribosomes follows the general mechanism for protein synthesis (see section 10.10.3) but involves different factors and proteins from those utilised by prokaryotic ribosomes. Termination only requires one release factor, unlike in prokaryotic ribosomes where three release factors are usually required.

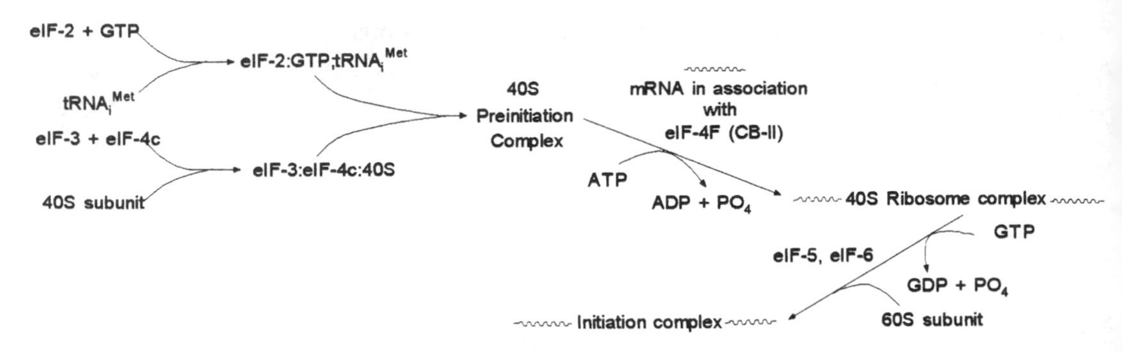

Figure 10.17. An outline of the formation of the protein synthesis initiation complex by the ribosomes of eukaryotic cells.

10.12 Bacterial Protein Synthesis Inhibitors (Antimicrobials)

Many protein inhibitors inhibit protein synthesis in both prokaryotic and eukaryotic cells (Table 10.2). This inhibition can take place at any stage in protein synthesis. However, some inhibitors have a specific action in that they inhibit protein synthesis in prokaryotic cells but not in eukaryotic cells, or vice versa. Consequently, a number of useful drugs have been discovered that will inhibit protein synthesis in bacteria but either have no effect or a very much reduced effect on protein synthesis in mammals.

The structures and activities of the drugs that inhibit protein synthesis are quite diverse. Consequently, only a few of the more commonly used drugs and structurally related compounds will be discussed in greater detail in this section.

Table 10.2. Examples of drugs that inhibit protein synthesis.

Drug	Action
Chloramphenicol	Blocks the enzymes that catalyse the transfer of the new amino acid residue to the peptide chain, that is, prevents elongation in **prokaryotic** cells
Cyclohexidine	Inhibits translocation of mRNA in **eukaryotic** ribosomes
Erythromycin	Blocks the enzymes that catalyse the transfer of the new amino acid residue to the peptide chain, that is, prevents elongation in **prokaryotic** cells.
Fusidic acid	Inhibits dissociation of the protein from the ribosome in both **prokaryotic** and **eukaryotic** cells
Streptomycin (see section 10.12.1)	Inhibits initiation of prokaryotic protein synthesis. It also causes the misreading of codons, which results in the insertion of incorrect amino acid residues into the structure of the protein formed in **prokaryotic** cells
Tetracycline	Inhibits the binding of aminoacyl–tRNA to the A site of **prokaryotic** ribosomes

10.12.1 Aminoglycosides

Streptomycin (Figure 10.18) is a member of a group of compounds known as aminoglycosides. These compounds have structures in which aminosugar residues in the form of mono- or polysaccharides are attached to a substituted 1,3-diaminocyclohexane ring by modified glycosidic-type linkages. This ring is either streptidine (streptomycin) or deoxystreptamine (kanamycin, neomycin, gentamicin and tobramycin).

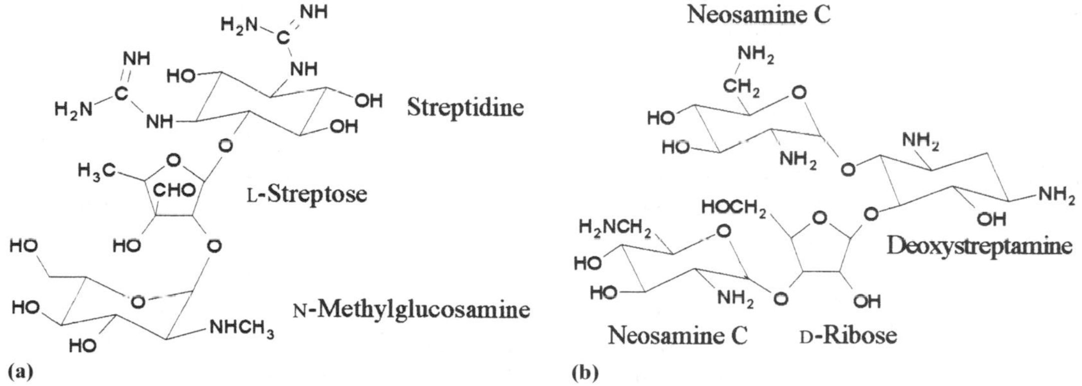

Streptidine
L-Streptose
N-Methylglucosamine
(a)

Neosamine C
Deoxystreptamine
Neosamine C
D-Ribose
(b)

Figure 10.18. The structures of (a) streptomycin and (b) neomycin C.

Streptomycin was the first aminoglycoside discovered (Schatz and co-workers, 1944) from cultures of the soil Actinomycete *Streptomyces griseus*. It acts by interfering with the initiation of protein synthesis in bacteria. The binding of streptomycin to the 30S ribosome inhibits initiation and also causes some amino acid–tRNA complexes to misread the mRNA codons. This results in the insertion of incorrect amino acid residues into the protein chain, which usually leads to the death of the bacteria. The mode of action of the other aminoglycosides has been assumed to follow the same pattern even though most of the investigations into the mechanism of the antibacterial action of the aminoglycosides have been carried out on streptomycin.

The clinically used aminoglycosides have structures closely related to that of streptomycin. They are essentially broad-spectrum antibiotics although they are normally used to treat serious Gram-negative bacterial infections (see section 4.2.5.1). Aminoglycosidic drugs are very water soluble. They are usually administered as their water-soluble inorganic salts but their polar nature means that they are poorly absorbed when administered orally. Once in the body they are easily distributed into most body fluids. However, their polar nature means that they do not easily penetrate the central nervous system (CNS), bone, fatty and connective tissue. Moreover, aminoglycosides tend to concentrate in the kidney where they are excreted by glomerular filtration.

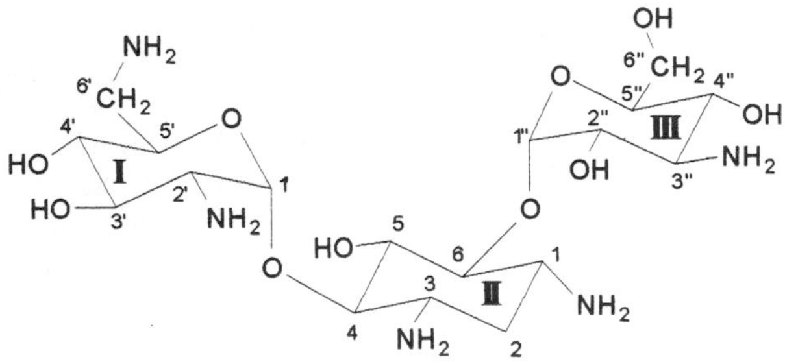

Figure 10.19. Kanamycin B.

Aminoglycoside-drug-resistant strains of bacteria are now recognised as a serious medical problem. They arise because dominant bacteria strains have emerged that possess enzymes that effectively inactivate the drug. These enzymes act by catalysing the acylation, phosphorylation and adenylation of the drug (see section 6.13). This results in the formation of inactive drug derivatives.

The activity of the aminoglycosides is related to the nature of their ring substituents. Consequently, it is convenient to discuss this activity in relation to the changes in the substituents of individual rings but, in view of the diversity of the structures of aminoglycosides, it is difficult to identify common trends. As a result, this discussion will be largely limited to kanamycin (Figure 10.19). However, the same trends are often true for other aminoglycosides whose structures consist of three rings, including a central deoxystreptamine residue.

Changing the nature of the amino substituents at positions 2′ and 6′ of ring I has the greatest effect on activity. For example, kanamycin A, which has a hydroxy group at position 2′, and kanamycin C, which has a hydroxy group at position 6′, are both less active than kanamycin B, which has amino groups at the 2′ and 6′ positions. However, the removal of one or both of the hydroxy groups at positions 3′ and 4′ does not have any effect on the potency of the kanamycins.

Modifications to ring II (the deoxystreptamine ring) greatly reduce the potency of the kanamycins. However, N-acylation and alkylation of the amino group at position 1 can give compounds with some activity. For example, acylation of kanamycin A gives 1-*N*-(L(−)-4-amino-2-hydroxybutyryl) kanamycin A (amikacin), which has a potency of about 50% of that of kanamycin A (Figure 10.20). In spite of this, amikacin is a useful drug for treating some strains of Gram-negative bacteria because it is resistant to deactivation by bacterial enzymes. Similarly, 1-*N*-ethylsisomicin (netilmicin) is as potent as its parent aminoglycoside sisomicin.

Changing the substituents of ring III does not usually have such a great effect on the potency of the drug as similar changes in rings I and II. For example, removal of the 2″ hydroxy group

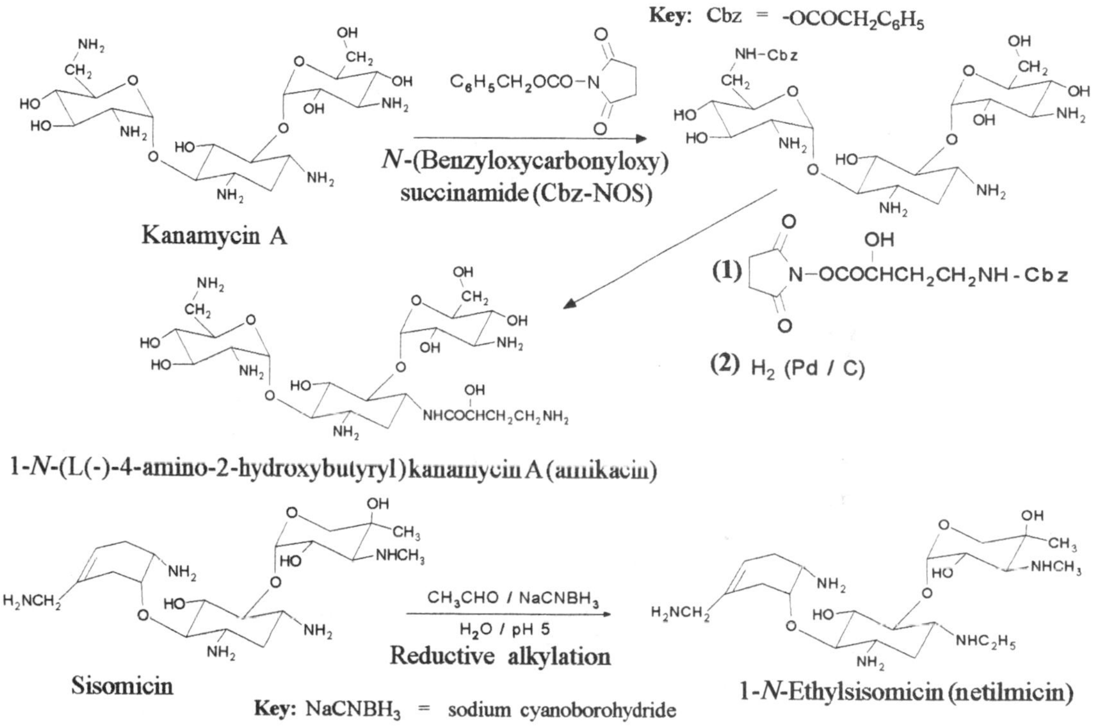
Key: Cbz = -OCOCH2C6H5
C6H5CH2OCO-N
N-(Benzyloxycarbonyloxy) succinamide (Cbz-NOS)
Kanamycin A
(1) N-OCOCHCH2CH2NH-Cbz
(2) H2 (Pd / C)
NHCOCHCH2CH2NH2
1-N-(L(-)-4-amino-2-hydroxybutyryl)kanamycin A (amikacin)
CH3CHO / NaCNBH3
H2O / pH 5
Reductive alkylation
Sisomicin
Key: NaCNBH3 = sodium cyanoborohydride
1-N-Ethylsisomicin (netilmicin)

Figure 10.20. An outline of the chemistry involved in the synthesis of the antibiotics amikacin and netilmicin. Cbz is frequently used as a protecting group for amines because it is easily removed by hydrogenation.

of gentamicin results in a significant drop in activity. However, replacement of the 2″ hydroxy group of gentamicin (Figure 10.21) by amino groups gives the highly active seldomycins.

10.12.2 Chloramphenicol

Chloramphenicol was first isolated from the microorganism *Streptomyces venezuelae* by Ehrlich and co-workers in 1947. It is a broad-spectrum antibiotic whose structure contains two asymmetric centres. However, only the D(−)-*threo* form is active.

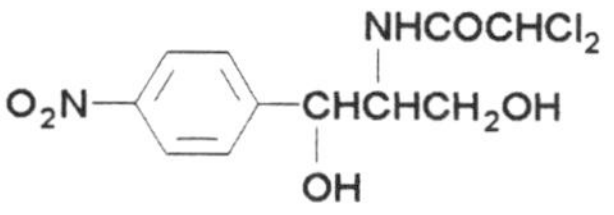
NHCOCHCl2
O2N
CHCHCH2OH
OH

D(−)-*threo*-Chloramphenicol

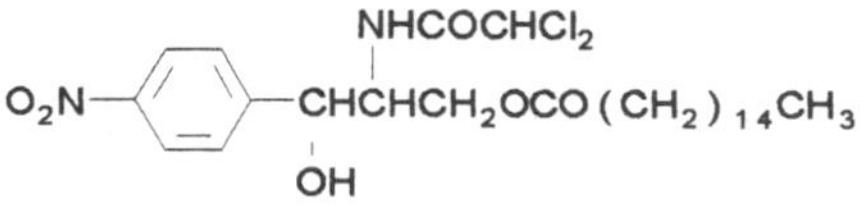
NHCOCHCl2
O2N
CHCHCH2OCO(CH2)14CH3
OH

D(−)-*threo*-Chloramphenicol palmitate

Chloramphenicol can cause serious side effects and so it is recommended that it is only used for specific infections. It is often administered as its palmitate in order to mask its bitter taste. The free drug is liberated from this ester by hydrolysis in the duodenum. Chloramphenicol has

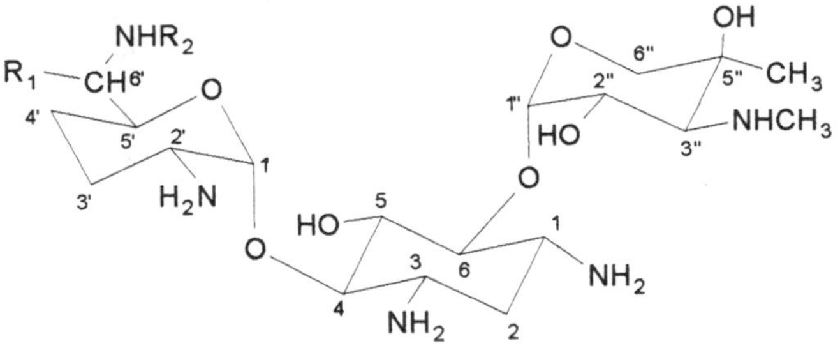

Key:
$R_1 = R_2 = CH_3$; gentamicin C_1
$R_1 = CH_3, R_2 = H$; gentamicin C_2
$R_1 = R_2 = H$; gentamicin C_2

Figure 10.21. The structures of gentamicin.

a poor water solubility (2.5 g dm^{-3}) and so it is sometimes administered in the form of its sodium hemisuccinate salt (see section 3.7.4.2), which acts as a prodrug.

Chloramphenicol acts by inhibiting the elongation stage in protein synthesis in prokaryotic cells. It binds reversibly to the 50S ribosome subunit and is thought to prevent the binding of the aminoacyl–tRNA complex to the ribosome. However, its precise mode of action is not understood.

Investigation of the activity of analogues of chloramphenicol showed that activity requires a *para*-electron-withdrawing group. However, substituting the nitro group with other electron-withdrawing groups gave compounds with a reduced activity. Furthermore, modification of the side chain, with the exception of the difluoro derivative, gave compounds that had a lower activity than chloramphenicol (Table 10.3). These observations suggest that D(−)-*threo*-chloramphenicol has the optimum structure of those tested for activity.

10.12.3 Tetracyclines

Tetracyclines are a family of natural and semisynthetic antibiotics isolated from various *Streptomyces* species, the first member of the group chlortetracycline being obtained in 1948 by Duggar from *Streptomyces aureofaciens*. A number of highly active semisynthetic analogues have also been prepared from the naturally occurring compounds (Table 10.4). The tetracyclines are a broad-spectrum group of antibiotics active against many Gram-positive and Gram-negative bacteria, rickettsiae, mycoplasmas, chlamydiae and some protozoa that cause malaria. A number of the natural and semisynthetic compounds are in current medical use.

The structures of the tetracyclines are based on an in-line fused four-ring system. Their structures are complicated by the presence of **up to** six chiral carbons in the fused-ring system. These normally occur at positions 4, 4a, 5, 5a, 6 and 12a, depending on the symmetry of the structure. The configurations of these centres in the active compounds have been determined by X-ray

Table 10.3. The activity against *Escheichla coli* of some analogues of chloramphenicol relative to chloramphenicol.

Analogue	Analogue chloramphenicol activity
$NHCOCHBr_2$ O_2N– –$CHCHCH_2OH$ OH	About 0.8
$NHCOCH_2Cl$ O_2N– –$CHCHCH_2OH$ OH	About 0.4
$NHCOCH_3$ O_2N– –$CHCHCH_2OH$ OH	Almost inactive
$NHCOCHCl_2$ O_2N– –$CHCHCH_2OH$ O	Almost inactive
$NHCOCF_3$ O_2N– –$CHCHCH_2OH$ OH	**1.7**

Table 10.4. The structures of the tetracyclines.

R_1 R_2 R_3 R_4 $N(CH_3)_2$ pK_{a3} (of the ammonium salt)
H H OH
7 6 5 4 3
8 5a 4a
9 11 12a 2
pK_{a2} 10 11a 12 OH 1 $CONH_2$ pK_{a1}
OH O OH O

Tetracycline	R_1	R_2	R_3	R_4
Chlortetracycline	Cl	CH_3	OH	H
6-Demethyl-7-chlorotetracycline (Demeclocycline)	Cl	H	OH	H
6-Deoxy-5-oxytetracycline (Doxycycline)	H	H	CH_3	OH
6-Demethyl-6-deoxy-5-hydroxy-6-methylenetetracycline (Methacycline)	H	CH_2		OH
Minocycline	$N(CH_3)_2$	H	H	H
Oxytetracycline	H	CH_3	OH	OH
Tetracycline	H	CH_3	OH	H

crystallography (Table 10.4). This technique has also confirmed that C_1 to C_3 and C_{11} to C_{12} were conjugated structures.

Tetracyclines are amphoteric, forming salts with acids and bases. They normally exhibit three pK_a ranges of 2.8–3.4 (pK_{a1}), 7.2–7.8 (pK_{a2}) and 9.1–9.7 (pK_{a3}), the last being the range for the corresponding ammonium salts. These values have been assigned by Leeson and co-workers to the structures shown in Table 10.4. These assignments have been supported by the work of Rigler and collegues. However, the assignments for pK_{a2} and pK_{a3} are opposite to those suggested by Stephens and collegues. Tetracyclines also have a strong affinity for metal ions, forming stable chelates with calcium, magnesium and iron ions. These chelates are usually insoluble in water, which accounts for the poor absorption of tetracyclines in the presence of drugs and foods that contain these metal ions. However, this affinity for metals appears to play an essential role in the action of tetracyclines.

Tetracyclines are transported into the bacterial cell by passive diffusion and active transport. Active transport requires the presence of Mg^{2+} ions and ATP possibly as an energy source. Once in the bacteria, tetracyclines act by preventing protein elongation by inhibiting the binding of the aminoacyl–tRNA to the 30S subunit of the prokaryotic ribosome. This binding has also been shown to require magnesium ions.

Tetracyclines also penetrate mammalian cells and bind to eukaryotic ribosomes. However, their affinity for eukaryotic ribosomes is lower than that for prokaryotic ribosomes and so they do not achieve a high enough concentration to disrupt eukaryotic protein synthesis. Unfortunately, bacterial resistance to tetracyclines is common. It is believed to involve three distinct mechanisms, namely: active transport of the drug out of the bacteria by membrane spanning proteins; enzymic oxidation of the drug; and ribosome protection by chromosomal protein determinants.

The structure–activity relationships of tetracyclines have been extensively investigated and reported. Consequently, the following paragraphs give only a synopsis of these relationships. This synopsis only considers general changes to both the general structure of the tetracyclines (Figure 10.22) and the substitution patterns of their individual rings.

Activity in the tetracyclines requires four rings with a *cis* A/B ring fusion. Derivatives with three rings are usually inactive or almost inactive. In general, modification to any of the substituent groups in the positions C-10, C-11, C-11a, C-12, C-12a, C-1, C-2, C-3 and C-4 results in a significant loss in activity. For example, replacement of the 2-carboxamide group by aldehyde and nitrile groups results in a significant loss in activity. N-Alkylation of the 2-amido group usually reduces activity, the reduction in activity increasing with increasing size of the group. Changes to the 4-dimethylamino group also usually reduce activity. This group must have an α-configuration and partial conversion of this group to its β-epimer under acidic conditions at room temperature significantly reduces activity.

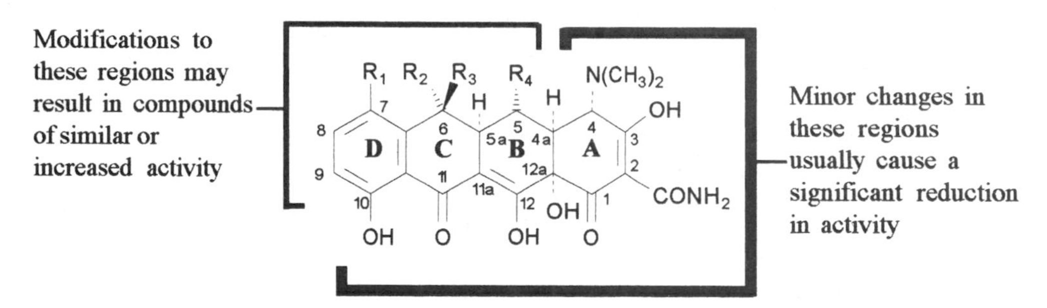

Figure 10.22. General structure–activity relationships in the tetracyclines.

α-Epimer of the 4-diethylamino group $\xrightleftharpoons{H^+}$ β-Epimer of the 4-diethylamino group

In addition, either removal of the α-dimethylamino group at position 4 or replacement of one or more of its methyl groups by larger alkyl groups also reduces activity. Ester formation at C-12a gives inactive esters, with the exception of the formyl ester, which hydrolyses in aqueous solution to the parent tetracycline. Alkylation of C-11a also gives rise to a loss of activity.

Modification of the substituents at positions 5, 5a, 6, 7, 8 and 9 may lead to similar or increased activity. Minor changes to the substituents at these positions tend to change the pharmacokinetic properties rather than activity (Table 10.5). A number of active derivatives have been synthesised by electrophilic substitution of C-7 and C-9 but the effect of introducing substituents at C-8 has not been studied because this position is difficult to substitute.

10.13 Drugs that Target Nucleic Acids

Drugs that target DNA and RNA either inhibit their synthesis or act on existing nucleic acid molecules. Those that inhibit the synthesis of nucleic acids usually act as either antimetabolites or enzyme inhibitors. The drugs that target existing nucleic acid molecules can, for convenience, be broadly classified into intercalating agents, alkylating agents and chain-cleaving agents.

Table 10.5. The pharmacokinetic properties of tetracycline hydrochlorides. The values given are representative values only because variations between individuals can be quite large.

Tetracycline	R_1	R_2	R_3	R_4	$P_{o/w}$ at pH 7.5	V_D ($dm^3 kg^{-1}$)	Cl_T ($dm^3 h^{-1}$)	$t_{1/2}$ (h)	Protein binding %
Chlortetracycline	Cl	CH_3	OH	H	0.12	1.2		6	55
Demeclocycline	Cl	H	OH	H	0.05	1.8		12	70
Doxycycline	H	H	CH_3	OH	0.63	0.7	1.7	16	90
Methacycline	H	CH_2		OH	0.40				
Minocycline	$N(CH_3)_2$	H	H	H		1.5	5	15	70
Oxytetracycline	H	CH_3	OH	OH	0.03	1.5		9	30
Tetracycline	H	CH_3	OH	H	0.04	2	15	6	45

However, it should be realised that these classifications are not rigid: drugs may act by more than one mechanism. Those drugs acting on existing DNA usually inhibit transcription whereas those acting on RNA normally inhibit translation. In both cases the net result is the prevention or slowing down of cell growth and division. Consequently, the discovery of new drugs that target existing DNA and RNA is a major consideration when developing new drugs for the treatment of cancer (see Appendix 4) and bacterial and other infections due to microorganisms.

10.13.1 Antimetabolites

Antimetabolites are compounds that block the normal metabolic pathways operating in cells. They act either by replacing an endogenous compound in the pathway by a compound whose incorporation into the system results in a product that can no longer play any further part in the pathway or by inhibiting an enzyme in the metabolic pathway in the cell. Both these types of intervention inhibit the targeted metabolic pathway to a level that hopefully has a significant effect on the health of the patient.

The structures of antimetabolites are usually very similar to those of the normal metabolites used by the cell. Those used to prevent the formation of DNA may be classified as antifolates, pyrimidine antimetabolites and purine antimetabolites. However, because of the difficulty of classifying biologically active substances (see section 1.6), antimetabolites that inhibit enzyme action are also classified as enzyme inhibitors.

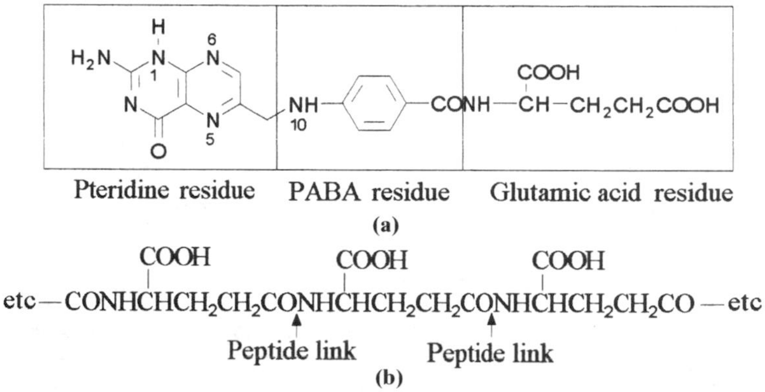

(a)

(b)

Figure 10.23. (a) The structure of folic acid. In blood, folic acids usually have one glutamate residue. However, in the cell they are converted to polyglutamates. (b) A fragment of a polyglutamate chain.

10.13.1.1 Antifolates

Folic acid (Figure 10.23) is usually regarded as the parent of a family of naturally occurring compounds known as folates. These folates are widely distributed in food. They differ from folic acid in such ways as the state of reduction of the pteridine ring and having carbon units attached to either or both of the N5 and N10 atoms.

In the body, folates are converted by a two-step process into tetrahydrofolates (FH_4) by the action of the enzyme dihydrofolate reductase (DHFR). Tetrahydrofolic acid is an essential cofactor in the biosynthesis of purines and thymine, which are required for DNA synthesis.

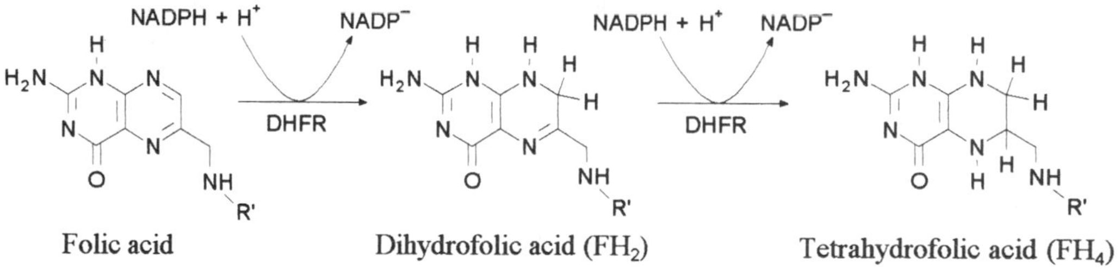

Folic acid antimetabolites have structures that resemble folic acid (Figure 10.24). They have a stronger affinity for DHFR than folic acid and act by inhibiting this enzyme at both stages in the conversion of folic acid to FH_4. This has the effect of inhibiting the formation of purines and thymine required for DNA synthesis. This inhibits cell growth, which prevents replication and ultimately leads to cell death.

Methotrexate is the only folate antimetabolite in clinical use. It is distributed to most body fluids but has a low lipid solubility, which means that it does not readily cross the blood–brain

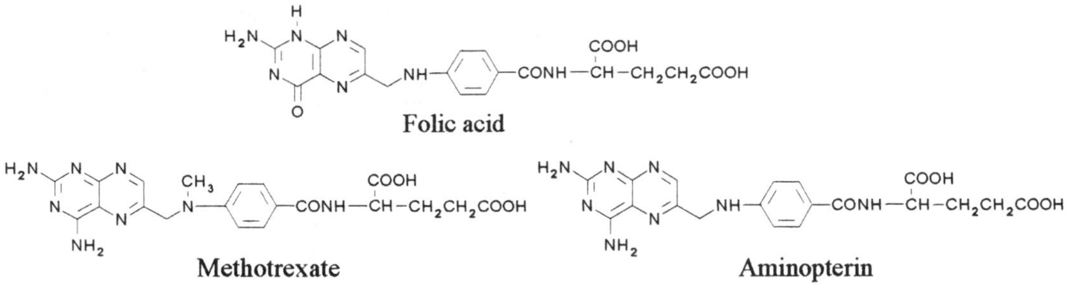

Folic acid

Methotrexate

Aminopterin

Figure 10.24. A comparison of the structures of folic acid antimetabolites with folic acid.

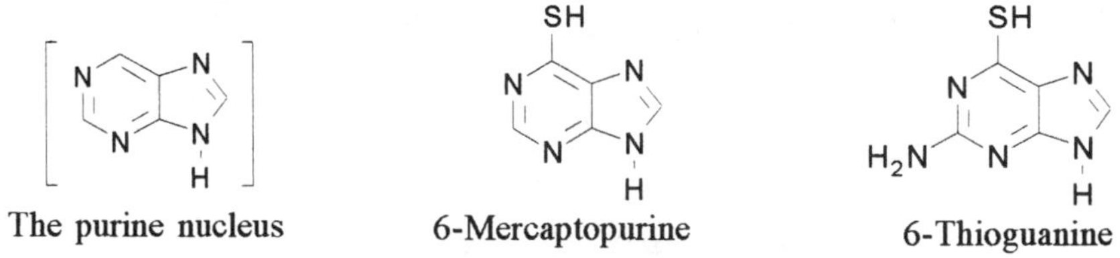

The purine nucleus

6-Mercaptopurine

6-Thioguanine

Figure 10.25. Examples of purine antimetabolites. The purine nucleus, on which the structures of the antimetabolites and the endogenous compounds they replace are based, is shown in square brackets.

barrier. It is transported into cells by the folate transport system and at high blood levels an additional second transport mechanism comes into operation. Once in the cell it is metabolised to the polyglutamate, which is retained in the cell for considerable periods of time. This is probably due to the polar nature of the polymer. Methotrexate is used to treat a variety of cancers, including head and neck tumours, and, in low doses, rheumatoid arthritis. It can cause vomiting, nausea, oral and gastric ulceration and depression of bone marrow, as well as other unwanted side effects.

10.13.1.2 Purine Antimetabolites

Purine antimetabolites are exogenous compounds, such as 6-mercaptopurine and 6-thioguanine, with structures based on the purine nucleus (Figure 10.25). They inhibit the synthesis of DNA and in some cases RNA by a number of different mechanisms. For example, 6-mercaptopurine is metabolised to the ribonucleotide 6-thioguanosine-5′-phosphate. This exogenous nucleotide inhibits several pathways for the biosynthesis of endogenous purine nucleotides. In contrast, 6-thioguanine is converted in the cell to the ribonucleotide 6-thioinosine-5′-phosphate. This ribonucleotide disrupts DNA synthesis by being incorporated into the structure of DNA as a false nucleic acid. Resistance to these two drugs arises because of a loss of the 5-phosphoribosyl transferase required for the formation of their ribonucleotides.

10.13.1.3 Pyrimidine Antimetabolites

These are antimetabolites whose structures closely resemble those of the endogenous pyrimidine bases (Figure 10.26a). They usually act by inhibiting one or more of the enzymes that are required for DNA synthesis. For example, fluorouracil is metabolised by the same metabolic pathway as uracil to 5-fluoro-2′-deoxyuridylic acid (FUdRP). FUdRP inhibits the enzyme thymidylate synthetase, which in its normal role is responsible for the transfer of a methyl group from the coenzyme methylenetetrahydrofolic acid ($MeFH_4$) to the C5 atom of deoxyuridylic acid (UdRP). The presence of the unreactive C5–F bond in FUdRP blocks this methylation, which prevents the formation of deoxythymidylic acid (TdRP) and its subsequent incorporation into DNA (Figure 10.26b). Fluorine was chosen to replace hydrogen at the C5 position of uracil because it is of a similar size to hydrogen (atomic radii: F, 0.13 nm; H, 0.12 nm). It was thought that this similarity in size would give a drug that would cause little steric disturbance to the biosynthetic pathway as well as being chemically inert. Analogues containing larger halogen atoms do not have any appreciable activity.

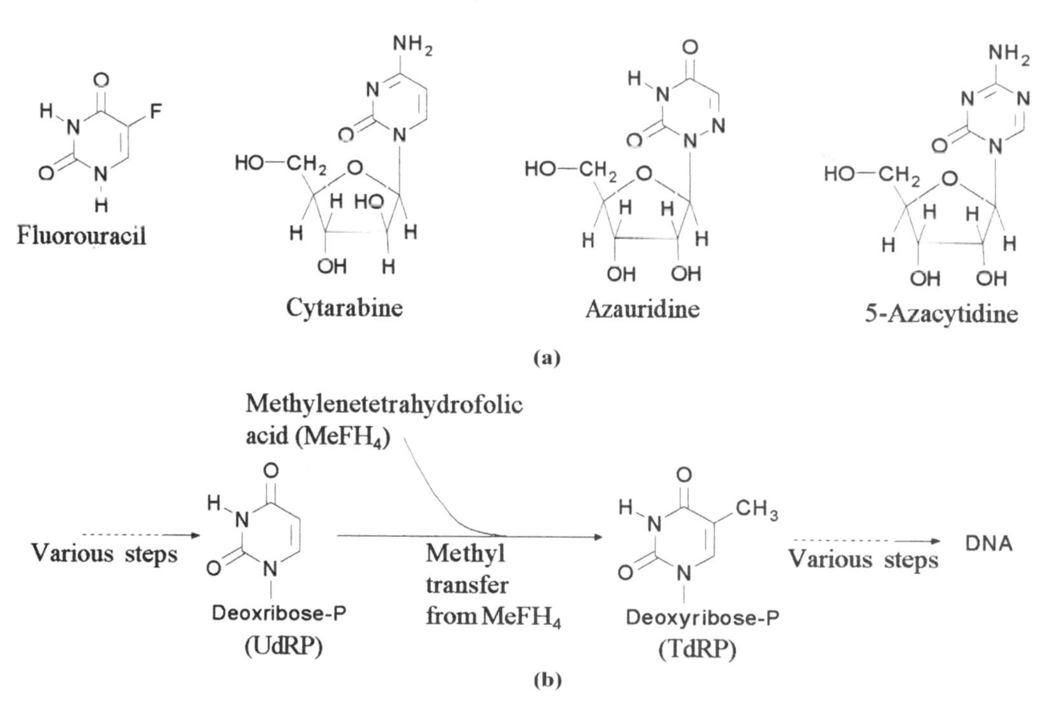

Figure 10.26. (a) Examples of pyrimidines that act as antimetabolites. It should be noted that cytarabine only differs from cytidine by the stereochemistry of the 2′ carbon. (b) The intervention of fluorouracil in pyrimidine biosynthesis.

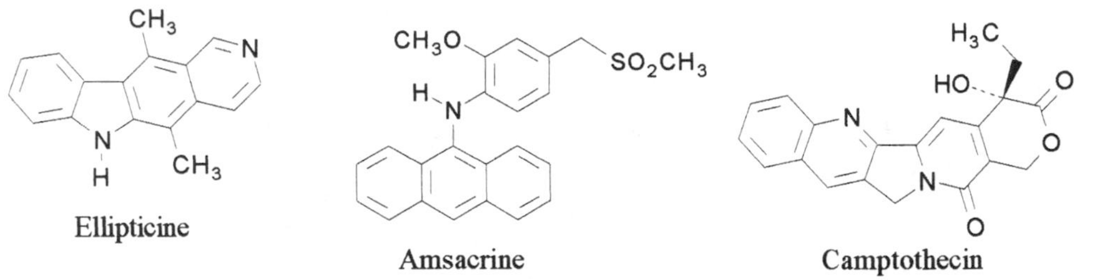

Ellipticine Amsacrine Camptothecin

Figure 10.27. Examples of topoisomerase inhibitors. Ellipticene acts by intercalation and inhibition of topoisomerase II enzymes. It is active against nasopharyngeal carcinomas. Amsacrine is used to treat ovarian carcinomas, lymphomas and myelogenous leukaemias. Camptothecin is an antitumour agent.

10.13.2 Enzyme Inhibitors

Enzyme inhibitors may be classified for convenience as those that inhibit the enzymes directly responsible for the formation of nucleic acids or the variety of enzymes that catalyse the various stages in the formation of the pyrimidine and purine bases required for the formation of nucleic acids.

10.13.2.1 Topoisomerases

Topoisomerases are a group of enzymes that are responsible for the supercoiling, the cleavage and rejoining of DNA. Their inhibition has the effect of preventing transcription. A number of compounds (Figure 10.27) are believed to act by inhibiting these enzymes. It is thought that some intercalators act in this manner although it is not clear whether the drug binds to the topoisomerase prior to or after the enzyme has formed a DNA–enzyme complex.

10.13.2.2 Enzyme Inhibitors for Purine and Pyrimidine Precursor Systems

A wide range of compounds are active against a number of the enzyme systems that are involved in the biosynthesis of purines and pyrimidines in bacteria. For example, sulphonamides inhibit dihydropteroate synthetase (see section 6.12.1), which prevents the formation of folic acid, whereas trimethoprim inhibits dihydrofolate reductase, which prevents the conversion of folic acid to tetrahydrofolate (see section 10.13.1.1). In both of these examples the overall effect is the inhibition of purine and pyrimidine synthesis, which results in the inhibition of the synthesis of DNA. This restricts the growth of the bacteria and ultimately prevents it from replicating, which gives the body's natural defences time to destroy the bacteria. Because sulphonamides and trimethoprim inhibit different stages in the same metabolic pathway, they are often used in conjunction (Figure 10.28). This allows the clinician to use lower and therefore safer doses.

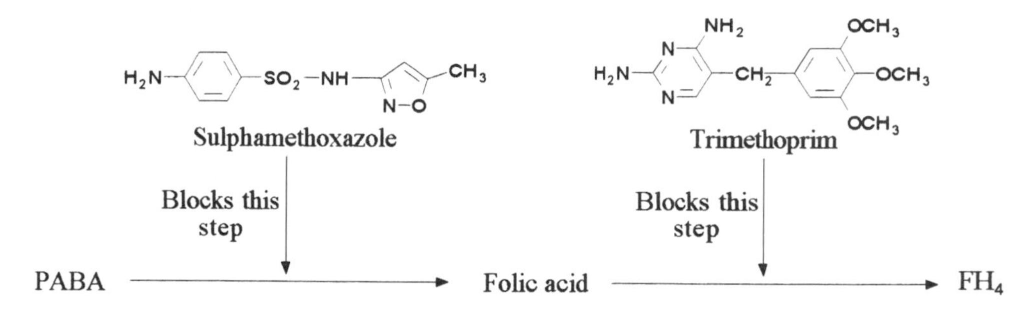

Figure 10.28. Sequential blocking using sulphamethoxazole and trimethoprim.

10.13.3 Intercalating Agents

Intercalating agents are compounds that insert themselves between the bases of the DNA helix (Figure 10.29). This insertion causes the DNA helix to partially unwind at the site of the intercalated molecule. This inhibits transcription, which blocks the replication process of the cell containing the DNA. However, it is not known how the partial unwinding prevents transcription but some workers think that it inhibits topoisomerases (see section 10.12.2.1). Inhibition of cell replication can lead to cell death, which reduces the size of a tumour, the number of '*free*' cancer cells or the degree of infection, all of which will contribute to improving the health of the patient.

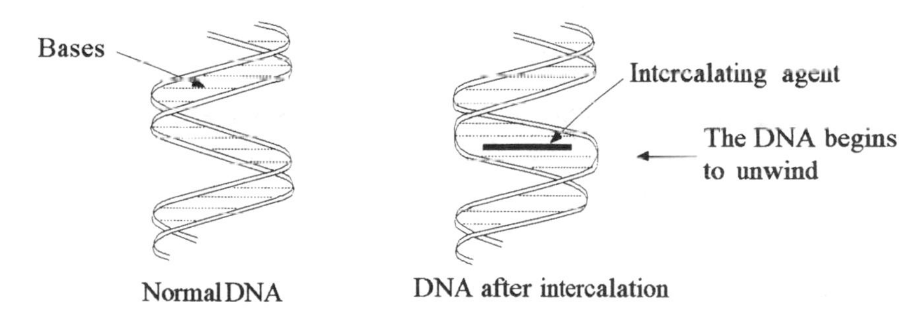

Figure 10.29. A schematic representation of the distortion of the DNA helix by intercalating agents. The horizontal lines represent the hydrogen-bonded bases. The rings of these bases and intercalating agent are edge on to the reader.

The insertion of an intercalation agent appears to occur via either the minor or major grooves of DNA. Compounds that act as intercalating agents must have structures that contain a flat fused aromatic or heteroaromatic ring section that can fit between the flat structures of the bases of the DNA. It is believed that these aromatic structures are held in place by hydrogen bonds, van der Waals' forces and charge-transfer bonds (see section 8.2).

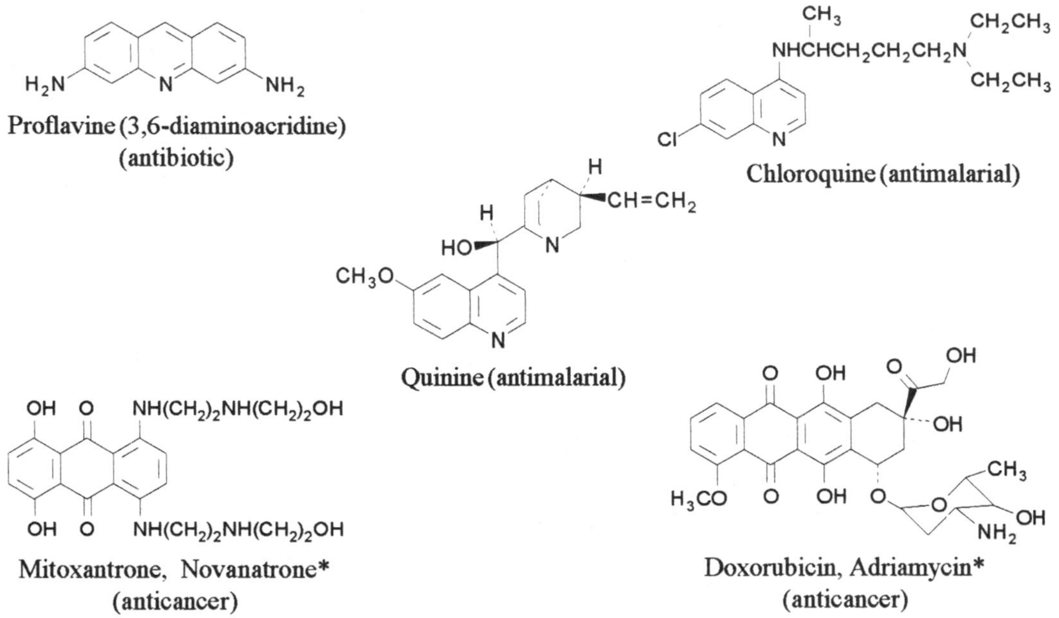

Proflavine (3,6-diaminoacridine)
(antibiotic)

Chloroquine (antimalarial)

Quinine (antimalarial)

Mitoxantrone, Novanatrone*
(anticancer)

Doxorubicin, Adriamycin*
(anticancer)

Figure 10.30. Examples of intercalating agents. *Trade name.

Drugs whose mode of action includes intercalation are the antimalarials quinine and chloroquine, the anticancer agents mitoxantrone and doxorubicin, and the antibiotic proflavine (Figure 10.30). In each of these compounds it is the flat aromatic ring system that is responsible for the intercalation. However, other groups in the structures may also contribute to the binding of a drug to the DNA. For example, the amino group of the sugar residue of doxorubicin forms an ionic bond with the negatively charged oxygens of the phosphate groups of the DNA chain, which effectively locks the drug into place. A number of other drugs appear to have groups that act in a similar manner.

Some intercalating agents exhibit a preference for certain combinations of bases in DNA. For example, mitoxantrone appears to prefer to intercalate with cytosine–guanosine-rich sequences. This type of behaviour does open out the possibility of selective action in some cases.

10.13.4 Alkylating Agents

Alkylating agents are believed to bond to the nucleic acid chains in either the major or minor grooves. In DNA the alkylating agent frequently forms either intrastrand or interstrand cross-links. Intrastrand cross-linking agents form a bridge between two parts of the same chain (Figure 10.31). This has the effect of distorting the strand, which inhibits transcription.

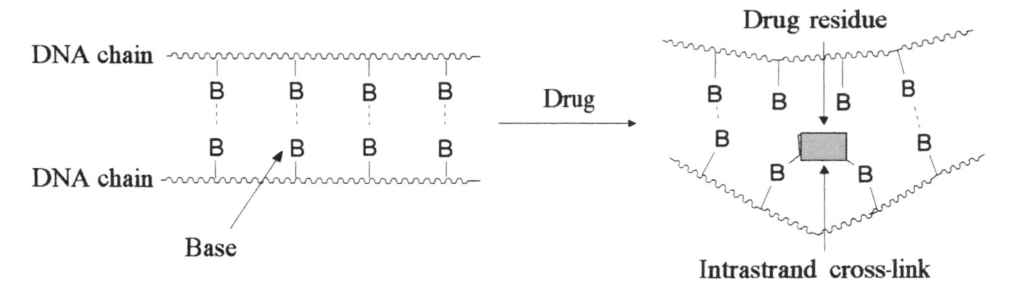

Figure 10.31. A schematic representation of the intrastrand cross-linking.

Interstrand cross-links are formed between the two separate chains of the DNA, which has the effect of locking them together (Figure 10.32). This also inhibits transcription. In RNA only intrastrand cross-links are possible. However, irrespective of whether or not it forms a bridge, the bonding of an alkylating agent to a nucleic acid inhibits replication of that nucleic acid. In the case of bacteria this prevents an increase in the size of the infection and so buys the body time for its immune system to destroy the existing bacteria. However, in the case of cancer it may lead to cell death and a beneficial reduction in tumour size.

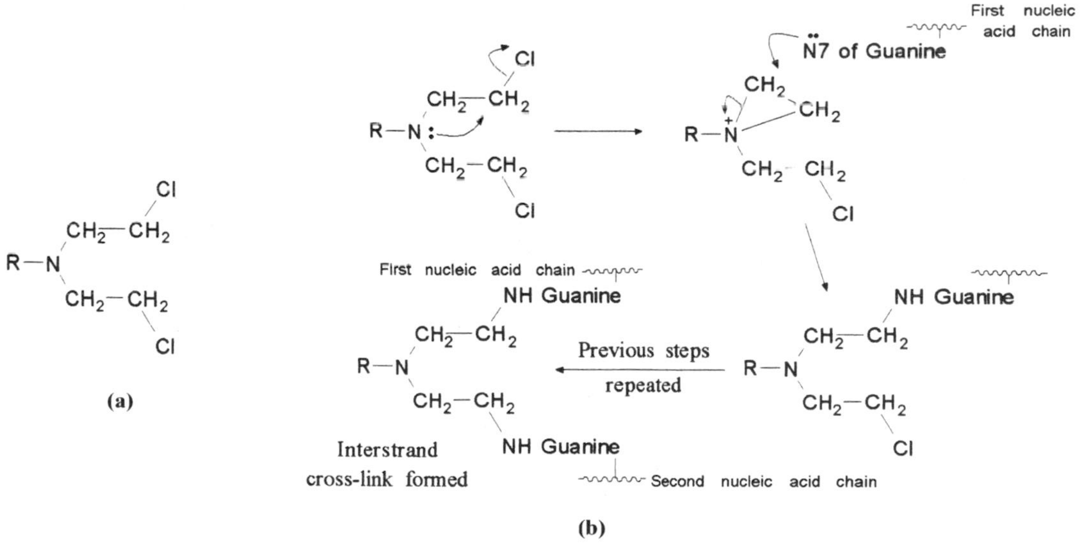

First nucleic acid chain
N7 of Guanine
First nucleic acid chain
NH Guanine
Previous steps repeated
Interstrand cross-link formed
NH Guanine
Second nucleic acid chain

(a)

(b)

Figure 10.32. (a) The general structure of nitrogen mustards. (b) The proposed mechanism for forming interstrand cross-links by the action of aliphatic nitrogen mustards.

The nucleophilic nature of the nucleic acids means that alkylating agents are usually electrophiles or give rise to electrophiles. For example, it is believed that a weakly electrophilic β-carbon atom of an aliphatic nitrogen mustard alkylating agent such as mechlorethamine

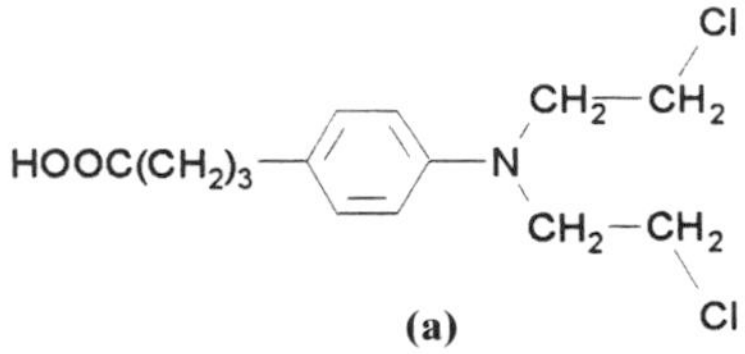

Cl
CH_2-CH_2
$HOOC(CH_2)_3$ N
CH_2-CH_2
Cl
(a)

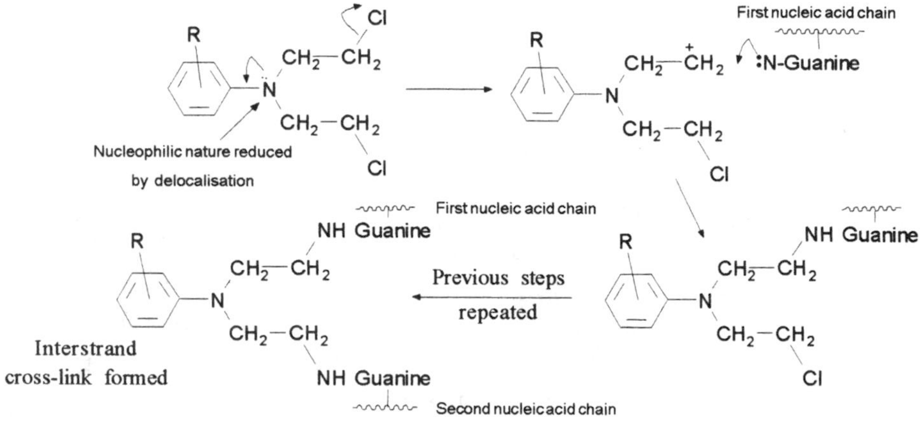

R
Cl
CH_2-CH_2
N
CH_2-CH_2
Cl
Nucleophilic nature reduced by delocalisation
R
First nucleic acid chain
$CH_2-\overset{+}{C}H_2$:N-Guanine
N
CH_2-CH_2
Cl
First nucleic acid chain
NH Guanine
R
CH_2-CH_2
N
CH_2-CH_2
NH Guanine
Second nucleic acid chain
Interstrand cross-link formed
Previous steps repeated
R
NH Guanine
CH_2-CH_2
N
CH_2-CH_2
Cl
(b)

Figure 10.33. (a) The structure of chlorambucil and (b) a proposed mode of action for some aromatic nitrogen mustards.

(Mustine) is converted to the more highly electrophilic aziridine ion by an internal nucleophilic substitution of a β-chlorine atom. This is thought to be followed by the nucleophilic attack of the N7 of a guanine residue on this ion by what appears to be an S_N2 type of mechanism. Because these drugs have two hydrocarbon chains with β-chloro groups, each of these chloro groups is believed to react with a guanine residue in a different chain of the DNA strand to form a cross-link between the two nucleic acid chains (Figure 10.32).

The electrophilic nature of alkylating agents means that they can also react with a wide variety of other nucleophilic biomacromolecules. This accounts for many of the unwanted toxic effects that are frequently observed with the use of these drugs. In the case of the nitrogen mustards, attempts to reduce these side effects have centred on reducing their reactivity by discouraging the formation of the aziridine ion before the drug reaches its site of action. The approach adopted has been to reduce the nucleophilic character of the nitrogen atom by attaching it to an electron-withdrawing aromatic ring. This produced analogues that would only react with strong nucleophiles and resulted in the development of chlorambucil. This drug is one of the least toxic nitrogen mustards, being active against malignant lymphomas, carcinomas of the breast and ovary and lymphocytic leukaemia. It has been suggested that because of the reduction in the nucleophilicity of the nitrogen atom these aromatic nitrogen mustards do not form an aziridine ion. Instead they react by direct substitution of the β-chlorine atoms by guanine, which is a strong nucleophile, by an S_N1 type of mechanism (Figure 10.33).

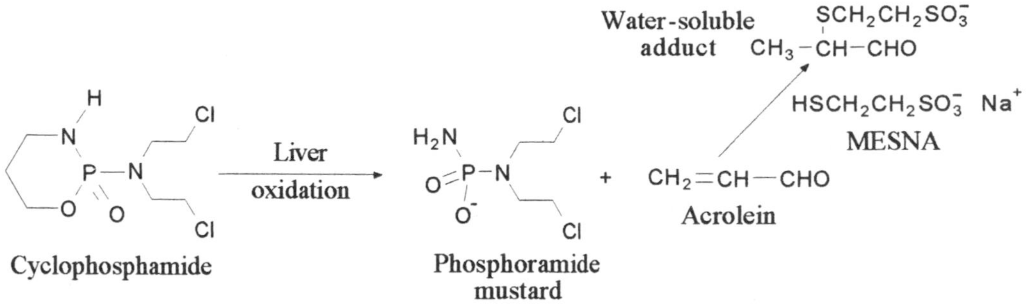

Figure 10.34. Cyclophosphamide and the formation of phosphoramide mustard, the active form of this drug.

Further attempts to reduce the toxicity of nitrogen mustards were based on making the drug more selective. Two approaches have yielded useful drugs. The first was based on the fact that the rapid synthesis of proteins that occurs in tumour cells requires a large supply of amino acid raw material from outside the cell. Consequently, it was thought that the presence of an amino acid residue in the structure of a nitrogen mustard might lead to an increased uptake of that compound. This approach resulted in the synthesis of the phenylalanine mustard melphalan (Table 10.6). The L-form of this drug is more active than the D-form and so it has been suggested that the L-form may be transported into the cell by means of an L-phenylalanine active transport system.

The second approach was based on the fact that some tumours were thought to contain a high concentration of phosphoramidases. This resulted in the synthesis of nitrogen mustard analogues whose structures contained phosphorus functional groups that could be attacked by this enzyme. It led to the development of the cyclophosphamide (Figure 10.34), which has a wide spectrum of activity. However, the action of this prodrug has now been shown to be due to phosphoramide mustard formed by oxidation by microsomal enzymes in the liver rather than hydrolysis by tumour phosphoramidases. The acrolein produced in this process is be-lieved to be the source of myelosuppression and haemorrhagic cystitis associated with the use of cyclophosphamide. However, co-administration of the drug with sodium 2-mercaptoethanesulphonate (MESNA) can relieve some of these symptoms. MESNA forms a water-soluble adduct with the acrolein, which is then excreted in the urine.

Some alkylating agents act by decomposing to produce an electrophile that bonds to a nucleophilic group of a base in the nucleic acid. For example, temozolomide (Table 10.6) enters the major groove of DNA where it reacts with water to from nitrogen, carbon dioxide, an aminomidazole and a methyl carbonium ion ($\overset{+}{C}H_3$). This methyl carbonium ion then methylates the strongly nucleophilic N7 of the guanine bases in the major groove.

A range of different classes of compound can act as nucleic acid alkylating agents (Table 10.6). Within these classes a number of compounds have been found to be useful drugs. In many

Table 10.6. Some examples of the classes and compounds of anticancer agents that act by alkylation of nucleic acids. It is emphasised that this table only lists some of the classes of alkylating compound that are active against cancers.

Class (based on)	Examples	Active against
Nitrogen mustard 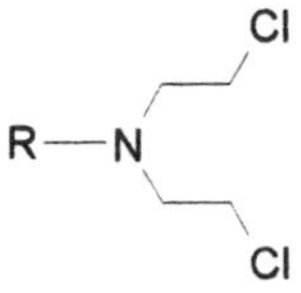	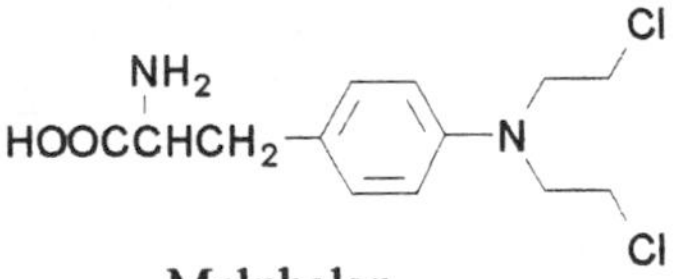 **Melphalan**	Multiple myeloma, ovarian and breast carcinoma
Triazeneimidazoles 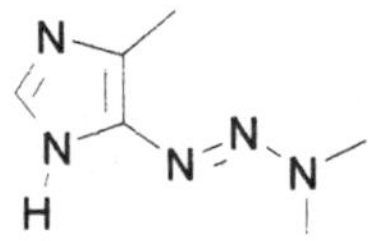	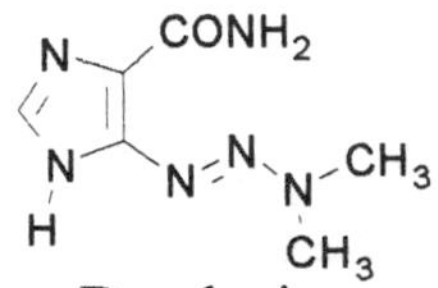 **Dacarbazine**	Wide range including malignant lymphomas, malanomas and sarcomas
Alkyl dimethanesulphonates $RO_2SOCH_2CH_2CH_2CH_2OSO_2R$	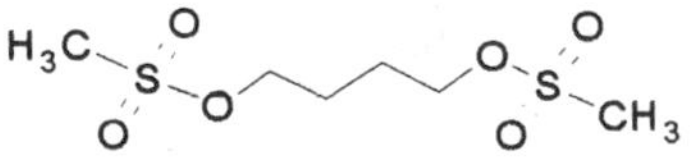 **Busulphan**	Granulocytic leukaemia
Nitrosoureas —NHCON(NO)—	 $ClCH_2CH_2NHCON(NO)CH_2CH_2Cl$ **Carmustine**	Cancers of the brain and cerebrospinal fluid
Imidazotetrazinones 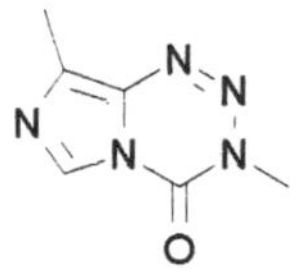	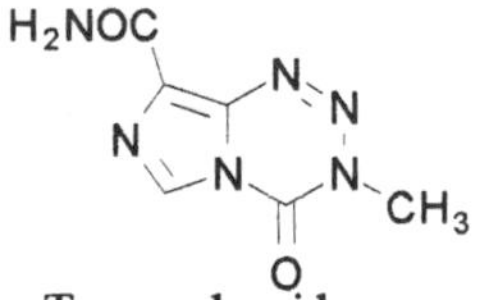 **Temozolomide**	Melanoma and brain tumours
Platinum complexes 	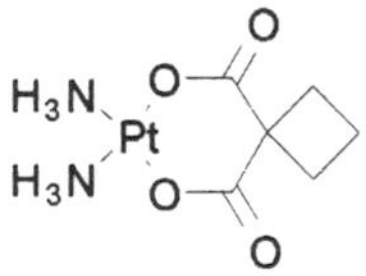 **Carboplatin**	Testicular and ovarian cancers
Carbinolamines —NR—CHOH–	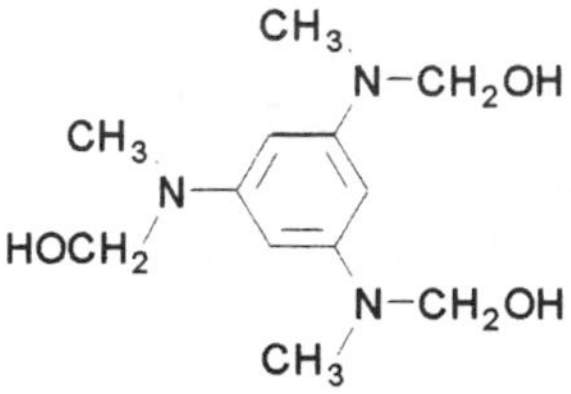 **Trimelamol**	Ovarian cancer

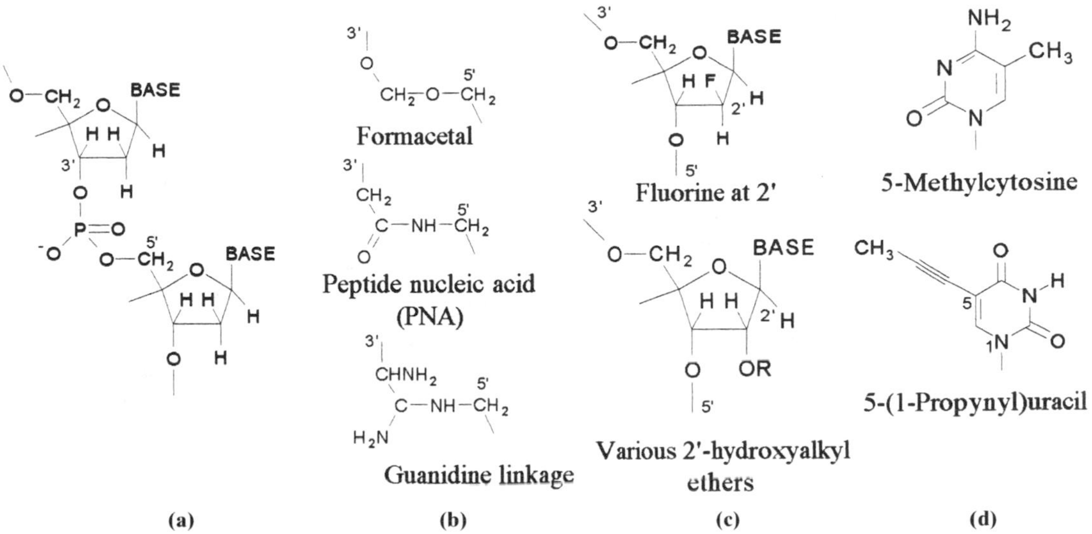

Figure 10.35. Development routes for antisense drugs. Examples of: (a) a section of the backbone of a deoxyribonucleic chain; (b) backbone modifications; (c) sugar residue modifications; and (d) base modifications.

cases their effectiveness is improved by the use of combinations of drugs. Their modes of action are usually not fully understood but a large amount of information is available concerning their structure-action relationships.

10.13.5 Antisense Drugs

The concept of antisense compounds or sequence-defined oligonucleotides (ONs) offers a new specific approach to designing drugs that target nucleic acids. The idea underlying this approach is that the antisense compound contains the sequence of complementary bases to those found in a short section of the target nucleic acid. This section is usually part of the genetic message being carried by an mRNA molecule. The antisense compound binds to this section by hydrogen bonding between the complementary base pairs. This inhibits translation of the message carried by the mRNA, which inhibits the production of a specific protein responsible for a disease state in a patient.

Antisense compounds were originally short lengths of nucleic acid chains that had base sequences that were complementary to those found in their target RNA. These short lengths of nucleic acid antisense compounds were found to be unsuitable as drugs because of poor binding to the target site and short half-lives due to enzyme action. However, they provided lead compounds for further development (Figure 10.35). Development is currently taking three basic routes:

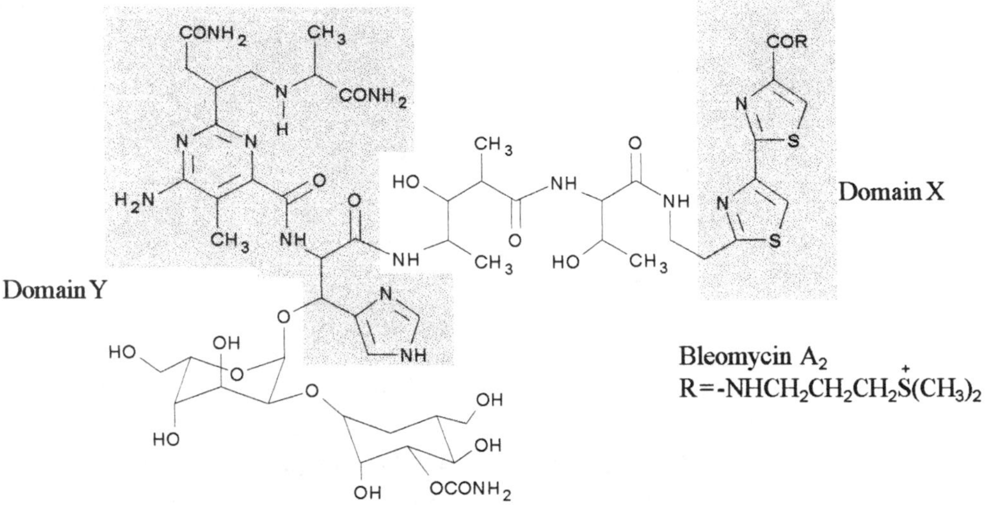

Domain Y

Domain X

Bleomycin A_2
$R = -NHCH_2CH_2CH_2\overset{+}{S}(CH_3)_2$

Figure 10.36. The bleomycins. The drug bleomycin sulphate is a mixture of a number of bleomycins.

(i) modification of the backbone linking the bases to increase resistance to enzymic hydrolysis;
(ii) changing the nature of the sugar residue by either replacing some of the free hydroxy groups by other substituents or forming derivatives of these groups;
(iii) modifying the nature of the substituent groups of the bases.

Antisense compounds are able to bind to both RNA and DNA. In the latter case they form a triple helix. At present, antisense drugs are still in the early stages of their development but the concept has aroused considerable interest in the pharmaceutical industry.

10.13.6 Chain-cleaving Agents

The interaction of chain-cleaving agents with DNA results in the breaking of the nucleic acid into fragments. Currently, the main cleaving agents are the bleomycins (Figure 10.36) and their analogues. However, other classes of drug are in the development stage.

The bleomycins are a group of naturally occurring glycoproteins that exhibit antitumour activity. When administered to patients they tend to accumulate in the squamous cells and so are useful for treating cancers of the head, neck and genitalia. However, the bleomycins cause pain and ulceration of areas of skin that contain a high concentration of keratin, as well as other unwanted side effects.

The action of the bleomycins is not fully understood. It is believed that the bithiazole moiety (domain X in Figure 10.36) intercalates with the DNA. In bleomycin A_2 the resulting adduct is thought to be stabilised by ionic bonding between the phosphate units of the DNA and the sulphonium ion of the side chain. The intercalation is believed to be followed by domain Y forming a complex with Fe(II). This complex reacts with oxygen to form free radicals, which react further to cause chain fragmentation. The chemical nature of this fragmentation is such that the nucleic acid sections cannot be rejoined by DNA ligases.

10.14 Viruses

Viruses are infective agents that are considerably smaller than bacteria. They are essentially packages, known as **virions**, of chemicals that invade host cells. However, viruses are not independent and can only penetrate a host cell that can satisfy the specific needs of that virus. The mode of penetration varies considerably from virus to virus. Once inside the host cell viruses take over the metabolic machinery of the host and use it to produce more viruses. Replication is often lethal to the host cell, which may undergo lysis to release the progeny of the virus. However, in some cases the virus may integrate into the host chromosome and become dormant. The ability of viruses to reproduce means that they can be regarded as being on the borderline of being living organisms.

10.14.1 Structure

Viruses consist of a core of either DNA or, as in the majority of cases, RNA fully or partially covered by a protein coating known as the **capsid** (Figure 10.37). The capsid consists of a number of polypeptide molecules known as **capsomers.** The capsid that surrounds most viruses consists of a number of different capsomers although some viruses will have capsids that only contain one type of capsomer. It is the arrangement of the capsomers around the nucleic acid that determines the overall shape of the virion. In the majority of viruses, the capsomers form a layer or several layers that completely surround the nucleic acids. However, there are some viruses in which the capsomers form an open-ended tube that holds the nucleic acids.

In many viruses the capsid is coated with a protein-containing lipid bilayer membrane. These are known as **enveloped viruses**. Their lipid bilayers are often derived from the plasma membrane of the host cell and are formed when the virus leaves the host cell by a process known as **budding**. Budding is a mechanism by which a virus leaves a host cell without killing that cell. It provides the virus with a membrane whose lipid components are identical to those of the host (Figure 10.37). This allows the virus to penetrate new host cells without activating the host's immune system.

Viruses bind to host cells at specific receptor sites on the host's cell envelope. The binding sites on the virus are polypeptides in its capsid or lipoprotein envelope. Once the virus has bound

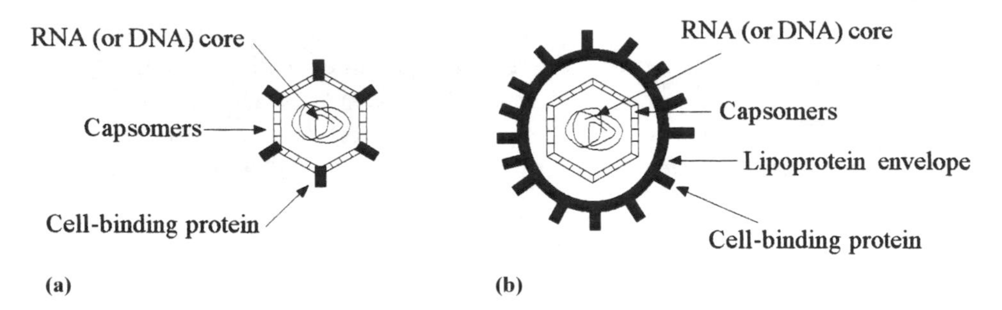

Figure 10.37. (a) Schematic representations of the structure of a virus (a) without a lipoprotein envelope (naked virus) and (b) with a lipoprotein envelope.

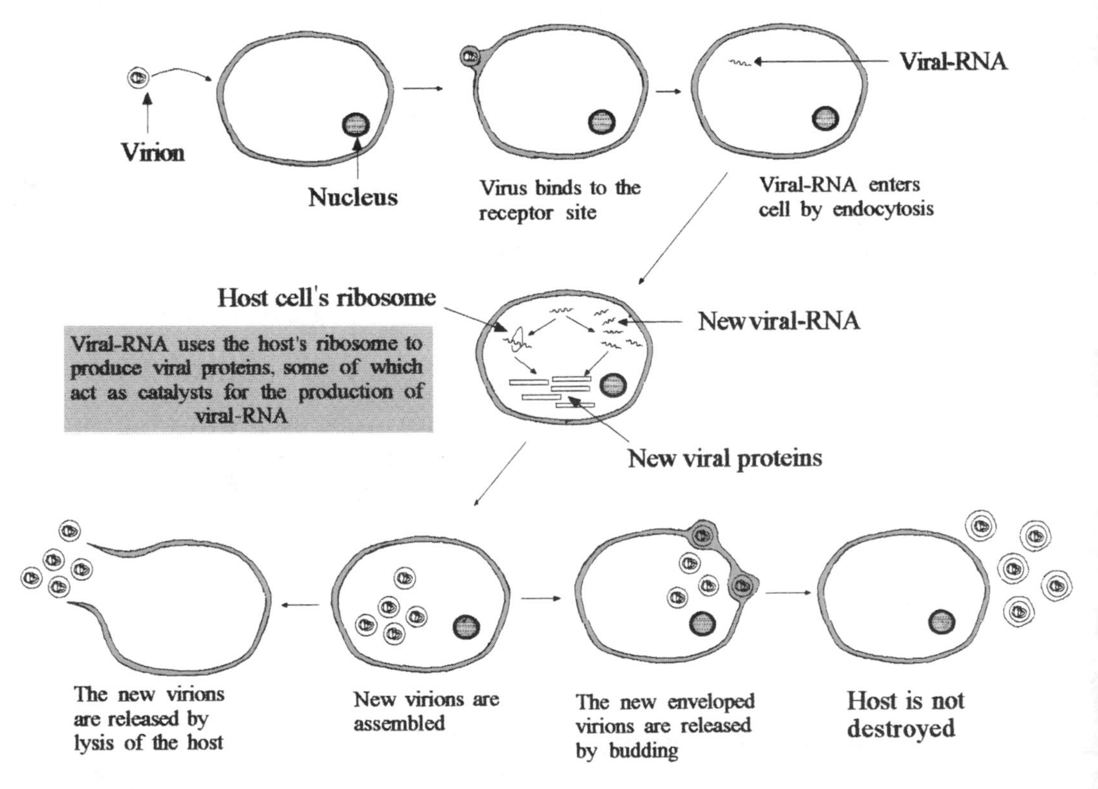

Figure 10.38. A schematic representation of the replication of RNA-virus.

to the receptor of the host cell the virus–receptor complex is transported into the cell by receptor-mediated endocytosis (see section 4.3.6). In the course of this process the protein capsid and any lipoprotein envelopes may be removed. Once it has entered the host cell the viral nucleic acid is able to use the host's cellular machinery to synthesise the nucleic acids and proteins required to produce a number of new viruses (Figure 10.38).

10.14.2 Replication

The replication of a virus requires that virus to reproduce its DNA or RNA and the proteins for its capsid coat. RNA-viruses can be broadly divided into two general types, namely: **RNA-viruses** and **RNA-retroviruses**. In both cases the virion has to enter the cell before replication can occur. Once in the host the two types of RNA-virus produce viral-mRNA by two different general routes (see sections 10.14.2.1. and 10.14.2.2). A great deal of information is available concerning the details of the mechanism of virus replication but this section will only outline the main points. For greater detail the reader is referred to specialist texts on virology.

10.14.2.1 RNA-viruses

RNA-virus replication usually occurs entirely in the cytoplasm. The viral-mRNA either forms part of the RNA carried by the virion or is synthesised by an enzyme already present in the virion. This viral-mRNA is used to produce the necessary viral proteins by translation using the host cell's ribosomes and enzyme systems. Some of the viral proteins are enzymes that are used to catalyse the reproduction of more viral-mRNA. The new viral-RNA and viral proteins are assembled into a number of new virions that are ultimately released from the host cell by either lysis or budding (Figure 10.38).

10.14.2.2 RNA-retroviruses

Retroviruses synthesise viral-DNA using their viral-RNA as a template. This process is catalysed by enzyme systems known as **reverse transcriptases** that form part of the virion. The viral-DNA is incorporated into the host genome to form a so-called **provirus**. Transcription of the provirus produces new viral-RNA and viral-mRNA. The viral-mRNAs are used to produce viral proteins, which, together with the viral-RNA, are again assembled into new virions. These virions are released by budding, which in many cases does not kill the host cell. Retroviruses are responsible for some forms of cancer and AIDS (see Appendix 5).

10.14.2.3 DNA-viruses

Most **DNA-viruses** enter the host cell's nucleus where formation of viral-mRNAs by transcription from the viral DNA is brought about by the host cell's polymerases. These viral mRNAs are used to produce viral proteins by translation using the host cell's ribosomes and enzyme systems. Some of these proteins will be enzymes that can catalyse the synthesis of more viral DNA. This DNA and the viral proteins synthesised in the host cell are assembled into a number of new virions that are ultimately released from the host by either cell lysis or budding.

Table 10.7. Examples of some of the groups of virus that cause disease.

Virus group	Disease	Characteristics	
		RNA/DNA	Envelope/naked
Parvovirus	Gastroenteritis	DNA	Naked
Herpes	Cold sores	DNA	Enveloped
Picornavirus	Polio and hepatitis A	RNA	Naked
Retrovirus	AIDS and leukaemia	RNA	Enveloped
Paramyxovirus	Measles, mumps and parainfluenza	RNA	Enveloped
Rhabdovirus	Rabies	RNA	Enveloped

10.14.3 Viral Diseases

Viral infections of host cells are common occurrences. Most of the time such infection does not result in illness because the body's immune system can usually deal with this viral invasion. When illness occurs it is often short lived and leads to long-term immunity. However, a number of viral infections can lead to serious medical conditions (Table 10.7). Some viruses, like HIV (human immunodeficiency virus), the aetiological agent of AIDS, are able to remain dormant in the host for a number of years before becoming active whereas others, such as herpes zoster (shingles), can give rise to recurrent bouts of the illness. Both chemotherapy and preventative vaccination are used to treat patients. However, the main clinical approach has been, and still is, vaccination because it has been difficult to design drugs that only target the virus.

10.14.4 Antiviral Drugs

It has been found that viruses utilise a number of virus-specific enzymes during replication. These enzymes and the processes they control are significantly different from those of the host cell to make a useful target for medicinal chemists. Consequently, antiviral drugs normally act by inhibiting viral nucleic acid synthesis, inhibiting attachment to and penetration of the host cell or inhibiting viral protein synthesis.

10.14.4.1 Nucleic Acid Synthesis Inhibitors

These drugs usually act by inhibiting the polymerases or reverse transcriptases required for nucleic acid synthesis. They are usually analogues of the purine and pyrimidine bases found in the nucleic acids (Table 10.8). Their general mode of action often involves conversion to the corresponding triphosphate by the host cell's cellular kinases. This conversion may also involve specific viral enzymes in the initial monophosphorylation step. These triphosphate drug derivatives are incorporated into the nucleic acid chain where they terminate its formation. Termination occurs because the drug residues do not have a 3′ hydroxy group necessary for the

Table 10.8. Examples of antiviral agents that inhibit nucleic acid synthesis.

Drug	Structure	Outline of action
Acyclovir	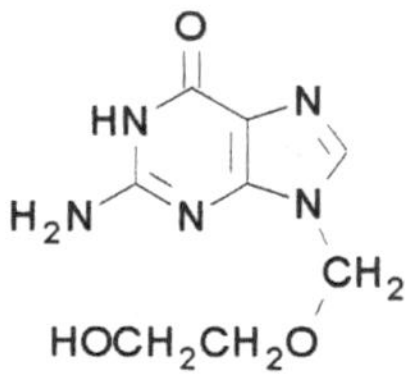	Drug —(Viral thymidine kinase)→ monophosphate —(Cellular kinases)→ triphosphate. The triphosphate inhibits DNA-polymerases and terminates the nucleic acid chain
Vidarabine	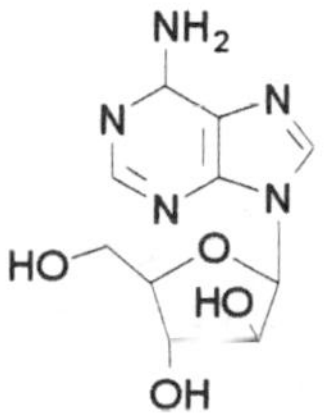	Drug —(Cellular kinases)→ triphosphate. Vidarabine triphosphate inhibits DNA-polymerases and ribonucleotide reductases. May also be incorporated into the nucleic acid chain
Tribavirin (ribavirin)	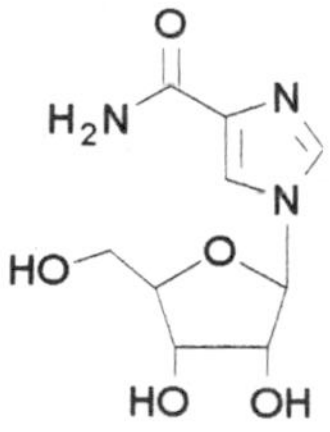	The drug inhibits viral RNA-polymerase and inosine monophosphate dehydrogenase
Zidovudine (AZT)	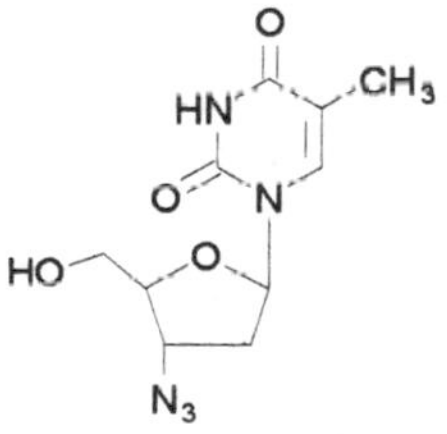	Drug —(Cellular kinases)→ triphosphate. The triphosphate inhibits reverse transcriptase in retrovirus. The incorporation of AZT triphosphate into the growing DNA strand terminates the formation of the strand
Didanosine	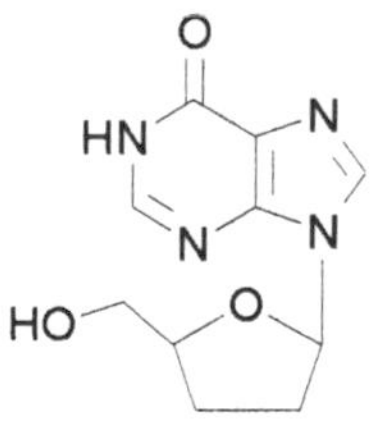	Drug —(Viral and cellular kinases)→ monophosphate → triphosphate. The drug inhibits reverse transcriptase in retrovirus by acting as a chain terminator

phosphate ester formation required for further growth of the nucleic acid chain. This effectively inhibits the polymerases and transcriptases that catalyse the growth of the nucleic acid. In addition, these drugs also inhibit other enzymes involved in the nucleic acid chain formation. It is not possible to list all the known antiviral agents in this text so only a representative selection are discussed.

Acyclovir. Acyclovir (Table 10.8) was the first effective antiviral drug. It is effective against a number of herpes viruses, notably simplex, varicella-zoster (shingles), varicella (chickenpox) and Epstein–Barr virus (glandular fever). The action of acyclovir is more effective in virus-infected host cells because the viral thymidine kinase is a more efficient catalyst for the monophosphorylation of acyclovir than the thymidine kinases of the host cell. This leads to an increase in the concentration of acyclovir triphosphate, which has 100-fold greater affinity for viral DNA-polymerase than human DNA-polymerase. However, resistance has been reported due to changes in the viral mRNA responsible for the production of the viral thymidine kinase. Acyclovir may be administered orally and by intravenous injection as well as topically. Orally administered doses have a low bioavailability.

Vidarabine. Vidarabine is active against herpes simplex and herpes varicella-zoster. However, the drug does give rise to nausea, vomiting, tremors, dizziness and seizures. In addition it has been reported to be mutagenic, teratogenic and carcinogenic in animal studies. Vidarabine is administered by intravenous infusion and topical application. It has a half-life of about 1 h, the drug being rapidly deaminated to arabinofuranosyl hypoxanthine (Ara-HX) by adenosine deaminase. This enzyme is found in the serum and red blood cells. Ara-HX also exhibits a weak antiviral action and has a half-life of about 3.5 h.

Deamination

Vidarabine → Arabinofuranosyl hypoxanthine (Ara-HX)

Tribavirin (ribavirin). Tribavirin is effectively a guanosine analogue. It is active against a wide variety of DNA- and RNA-virus. It is mainly used in aerosol form to treat influenza and other respiratory viral infections. Intravenous administration in the first 6 days of onset has been effective in reducing deaths from Lassa fever to 9%. Tribavirin has also been shown to delay the onset of full-blown AIDS in patients with early symptoms of HIV infection. However, administration of the drug has been reported to give rise to nausea, vomiting, diarrhoea, deterioration of respiratory function, anaemia, headaches and abdominal pain.

Zidovudine (AZT). Zidovudine is active against the retroviruses (see section 10.14.2.2) that cause AIDS and certain types of leukaemia. It also inhibits cellular α-DNA-polymerase but only at concentrations in excess of 100-fold greater than those needed to treat the viral infection. The drug may be administered orally or by intravenous infusion. The bioavailability from oral administration is good, the drug being distributed into most body fluids and tissues. However, when used to treat AIDS it has given rise to gastrointestinal disorders, skin rashes, insomnia, anaemia, fever, headaches, depression and other unwanted effects. Resistance increases with time. This is known to be due to the virus developing mutations that result in changes in the amino acid sequences in the reverse transcriptase.

Didanosine. Didanosine is used to treat some AZT-resistant strains of HIV. It is also used in combination with AZT to treat HIV. Didanosine is administered orally in dosage forms that contain antacid buffers to prevent conversion by the stomach acids to hypoxanthine. However, in spite of the use of buffers the bioavailability from oral administration is low. The drug can cause nausea, abdominal pain and peripheral neuropathy, amongst other symptoms. Drug resistance occurs after prolonged use.

10.14.4.2 Host Cell Penetration Inhibitors

The principal drugs that act in this manner are amantadine and rimantadine (Figure 10.39).

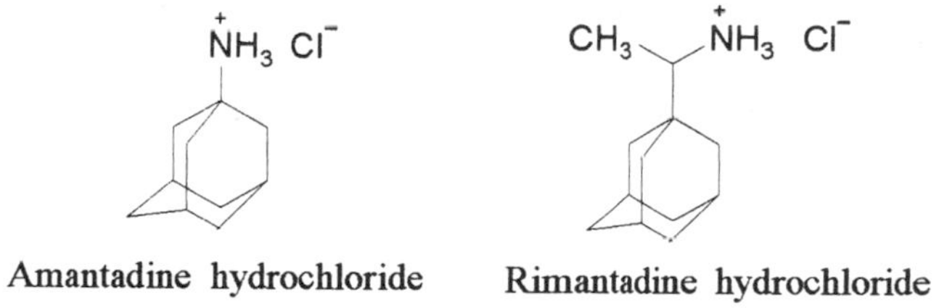

$\overset{+}{N}H_3$ Cl^- — Amantadine hydrochloride

CH_3 $\overset{+}{N}H_3$ Cl^- — Rimantadine hydrochloride

Figure 10.39. Examples of host cell penetration inhibitors.

Amantadine hydrochloride. Amantadine hydrochloride is effective against influenza A virus but is not effective against the influenza B virus.

When used as a prophylactic, amantadine hydrochloride is believed to give up to 80% protection against influenza A virus infections. The drug acts by blocking an ion channel in the virus membrane formed by the viral protein M_2. This is believed to inhibit the disassembly of the core of the virion and its penetration of the host (see section 10.14.1).

Amantadine hydrochloride has a good bioavailability on oral administration, being readily absorbed and distributed to most body fluids and tissues. Its elimination time is 12–18h. However, its use can result in depression, dizziness, insomnia and gastrointestinal disturbances, amongst other unwanted side effects.

Rimantadine hydrochloride. Rimantadine hydrochloride is an analogue of amantadine hydrochloride. It is more effective against influenza A virus than amantadine. Its mode of action is probably similar to that of amantadine. The drug is readily absorbed when administered orally but undergoes extensive first-pass metabolism. However, in spite of this its elimination half-life is double that of amantadine. Furthermore, CNS side effects are significantly reduced.

10.14.4.3 Inhibitors of Viral Protein Synthesis

The principal compounds that act as inhibitors of protein synthesis are the **interferons**. These compounds are members of a naturally occurring family of glycoprotein hormones (relative moleculer mass 20000–160000) that are produced by nearly all types of eukaryotic cell. Three general classes of interferons are known to occur naturally in mammals, namely: α-interferons produced by leucocytes, β-interferons produced by fibroblasts and γ-interferons produced by T lymphocytes. At least twelve α-, one β-and two γ-interferons have been identified.

Interferons form part of the human immune system. It is believed that the presence of virions and pathogens in the body switches on the mRNA that controls the production and release of interferon. This release stimulates other cells to produce and release more interferon. Interferons act by initiating the production in the cell of proteins that protect the cells from viral attack. The main action of these proteins takes the form of inhibiting the synthesis of viral-mRNA and viral-protein synthesis. α-Interferons also enhance the activity of killer T cells associated with the immune system (see section 11.5.3).

A number of α-interferons have been manufactured and proved to be reasonably effective against a number of viruses and cancers. Interferons are usually given by intravenous, intramuscular or subcutaneous injection. However, their administration can cause adverse effects, such as headaches, fevers and bone marrow depression, that are dose related.

The formation and release of interferon by viral and other pathological stimulation has resulted in a search for chemical inducers of endogenous interferon. Administration of a wide range of compounds has resulted in induction of interferon production. However, no clinically useful compounds have been found for humans, although tilorone is effective in inducing interferon in mice.

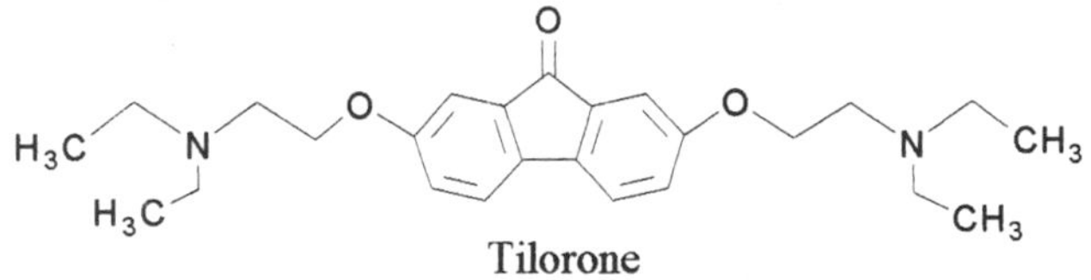

Tilorone

10.15 Recombinant DNA (Genetic Engineering)

The body requires a constant supply of certain peptides and proteins if it is to remain healthy and function normally. Many of these peptides and proteins are only produced in very small quantities. They will be produced only if the correct genes are present in the cell. Consequently, if a gene is missing or defective an essential protein will not be produced, which can lead to a diseased state. For example, cystic fibrosis is caused by a defective gene. This faulty gene produces a defective membrane protein (CFTR) that will not allow the free passage of chloride ions through the membrane. The passage of chloride ions through a normal membrane into the lungs is usually accompanied by a flow of water molecules in the same direction. In membranes that contain CFTR the transport of water through the membrane into the lungs is reduced. As a result, the mucus in the lung thickens. This viscous mucus clogs the lungs and makes breathing difficult, a classic symptom of cystic fibrosis. It also provides a breeding ground for bacteria that cause pneumonia.

Several thousand hereditary diseases found in humans are known to be caused by faulty genes. Recombinant DNA technology (genetic engineering) offers a new way of combating these hereditary diseases by either replacing the faulty genes or producing the missing peptides and proteins so that they can be given as a medicine.

10.15.1 Gene Cloning

Bacteria normally use the same genetic code as humans to make peptides and proteins. However, in bacteria the mechanism for peptide and protein formation is somewhat different. It is not restricted to the chromosomes but can also occur in extranuclear particles called **plasmids**. Plasmids are large circular supercoiled DNA molecules whose structure contains at least one gene and a start site for replication. However, the number of genes found in a plasmid is fairly limited, although a bacterium will contain a number of identical copies of the same plasmid.

It is possible to isolate the plasmids of bacterial cells. The isolated DNA molecules can be broken open by cleaving the phosphate bonds between specific pairs of bases by the action of enzymes known as either **restriction enzymes** or **endonucleases**. Each of these enzymes, of which over 500 are known, will only cleave the bonds between specific nucleosides. For example, EcoRI cleaves the phosphate link between guanosidine and adenosidine whereas XhoI cuts the chain between cytidine and thymine nucleosides. Cutting the strand can result in either **blunt ends**, where the endonuclease cuts across both chains of the DNA at the same points, or **cohesive ends** (sticky ends) where the cut is staggered from one chain to the other (Figure 10.40). The new non-cyclic structure of the plasmid is known as *linearised* DNA in order to distinguish it from the new *insert* or *foreign* DNA. This foreign DNA must contain the required gene, a second gene system that confers resistance to a specific antibiotic and any other necessary information. It should be remembered that a eukaryotic gene is made up of exons separated by introns (sequences that have no apparent use).

Mixing the foreign DNA and the linearised DNA in a suitable medium results in the formation of extended plasmid loops when their ends come into contact (Figure 10.41). This contact is converted into a permanent bond by the catalytic action of an enzyme called DNA ligase. When the chains are cohesive the exposed single chains of new DNA must contain a com-

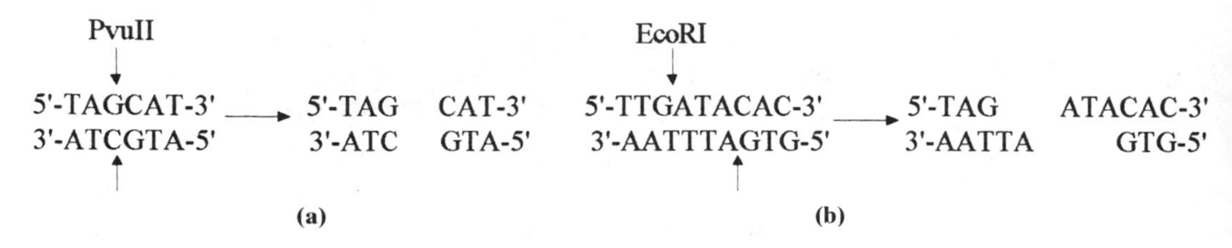

Figure 10.40. (a) Blunt and (b) cohesive cuts with compatible adhesive cuts.

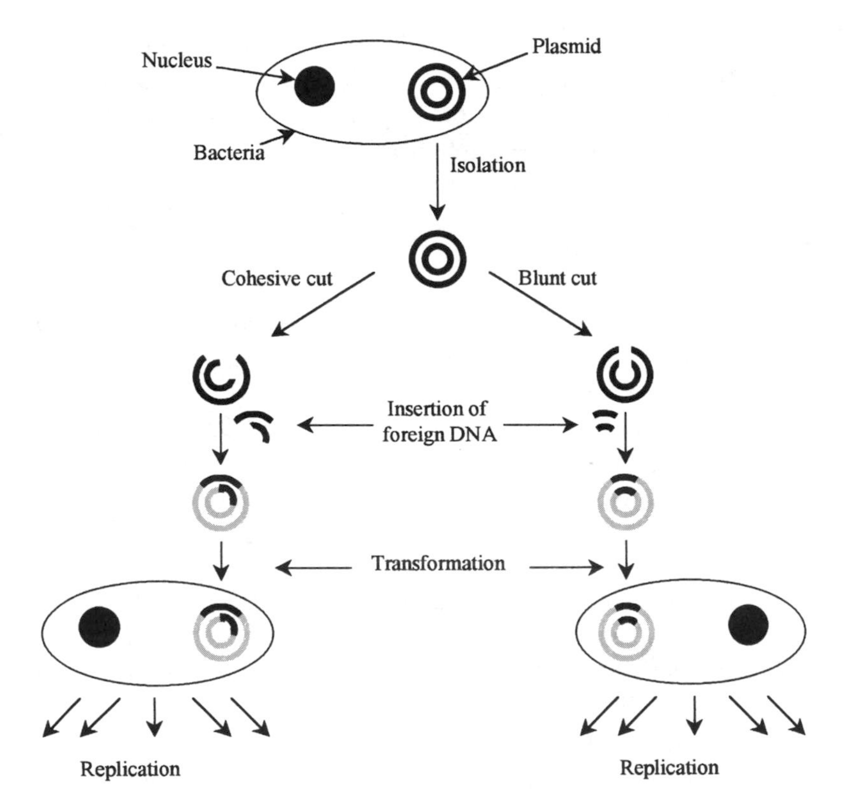

Figure 10.41. A representation of the main steps in the insertion of a gene into a plasmid.

plementary base sequence to the exposed ends of the linearised DNA. The hydrogen bonding between these complementary base pairs tends to bind the chains together prior to the action of the DNA ligase, hence the name 'sticky ends'. The new DNA of the modified plasmid is known as **recombinant DNA**. However, the random nature of the techniques used to form the modified plasmids means that some of the linearised DNA reforms the plasmid without incorporating the foreign DNA, but there are ways to select the recombinant plasmid.

The new plasmids are reinserted into the bacteria by a process known as **transformation**. Bacteria are mixed with the new plasmids in a medium containing calcium chloride. This medium makes the bacterial membrane permeable to the plasmid. However, not all bacteria will take up the modified plasmids. These bacteria can easily be destroyed by specific antibiotic action because they do not contain plasmids with the appropriate protecting gene. This makes isolation of the bacteria with the modified plasmids relatively simple. These modified bacteria are allowed to replicate and in doing so produce many copies of the modified plasmid. Under favourable conditions one modified bacterial cell can produce over 200 copies of the new plasmid. The gene in these modified plasmids will use the bacteria's internal machinery to automatically produce the appropriate peptide or protein. Because many bacteria replicate at a very rapid rate this technique offers a relatively quick way of producing large quantities of essential naturally occurring compounds that cannot be produced by other means.

Plasmids are not the only vehicles that can be used to transport DNA into a bacterial host cell. Foreign DNA can also be inserted into **bacteriophages** and **cosmids** by similar techniques. Bacteriophages (phage) are viruses that specifically infect bacteria whereas a cosmid is a hybrid between a phage and a plasmid that has been specially synthesised for use in gene cloning. Plasmids can be used to insert fragments containing up to 10 kilobase pairs (kbp), phages up to 20 kbp and cosmids 50 or more kbp. All vehicles used in gene cloning are referred to as **vectors**.

It is not always necessary to use a vector to place the recombinant DNA in a cell. If the cell is large enough, the recombinant DNA may be placed in the cell by using a micropipette whose overall tip diameter is less than 1 μm. Only a small amount of the recombinant DNA inserted in this fashion is taken up by the cells chromosomes. However, this small fraction will increase to a significant level as the cell replicates.

10.15.2 Medical Applications

The main uses of gene cloning in the medical field are:

(i) to correct genetic faults and absences;
(ii) to manufacture rare essential natural compounds.

10.15.2.1 Gene Therapy

A wide range of undesirable medical conditions are due to the presence of either defective genes that contain an incorrect base sequence or the absence of genes. For example, a defective gene is responsible for the substitution of a glutamic acid residue by a valine residue in haemoglobin (Hb). This results in the formation of haemoglobin S (HbS), which is responsible for sickle cell anaemia (Appendix 6), a disease commonly found in central and west Africa. The absence of genes producing a growth hormone leads to stunted growth in children.

The use of gene cloning in medicine is known as **gene therapy**. There are two fundamental approaches: **germline** and **somatic**. In germline therapy a fertilised egg is removed from the mother and the required gene inserted by cloning techniques. The egg is replanted in the mother and if the process has been successful the gene will be present in all the cells of the new individual. This removes the source of the disease and in theory could be used to prevent all inherited disorders.

The somatic approach uses cells that are not involved in the reproduction of the organism and so any alterations will not be transmitted to any future generations of cells and people. The technique involves removing cells from the body, infection with the required gene and the return of the cell to the body. This technique appears to offer a solution to inherited genetic disorders such as haemophilia and thalassaemia. It has been used to sucessfully treat children with adenosine deaminase (ADA) deficiency. This is a single gene deficiency disease that leads to an almost total lack of white blood cells and, as a result, almost no natural immunity.

10.15.2.2 Manufacture of Pharmaceuticals

The body produces peptides and proteins, often in extremely small quantities, that are essential for its well being. The absence of the necessary genes means that the body does not produce these essential compounds, resulting in a deficiency disease that is usually fatal. Treatment by supplying the patient with sufficient amounts of the missing compounds is normally successful. However, extraction from other natural sources is usually difficult and yields are often low. For example, it takes half a million sheep brains to produce 5 mg of somatostatin, a growth hormone that inhibits secretion of the pituitary growth hormone. Furthermore, unless the source of the required product is donated blood there is a limit to the number of cadavers available for the extraction of compounds suitable for use in humans. Moreover, there is also the danger that compounds obtained from human sources may be contaminated by viruses such as HIV, hepatitis and others that are difficult to detect. Animal sources have been used but only a few human protein deficiency disorders can be treated with animal proteins.

Gene cloning is used to obtain human recombinant proteins. However, some proteins will also need post-translational modification such as glycosylation and/or the modification of amino acid sequences. These modifications may require forming different sections of the peptide chain in the culture medium and chemically combining these sections *in vitro*. The genes required for these processes are synthesised using the required peptide as a blueprint. For example, human recombinant insulin may be produced in this manner (Figure 10.42). The genes for the A and B chains of insulin were synthesised separately. They were cloned separately, using suitable plasmids, into two different bacterial strains. One of these strains is used to produce the A chain whereas the other is used to produce the B strain. The chains are isolated and attached to each other by *in vitro* disulphide bond formation. This last step is inefficient and human recombinant insulin is also made by forming

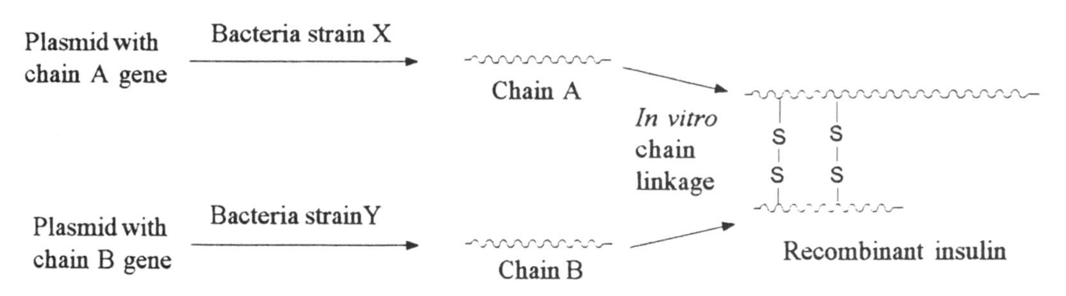

Figure 10.42. An outline of the synthesis of recombinant human insulin.

recombinant proinsulin by gene cloning. The proinsulin is converted to recombinant insulin by proteolytic cleavage.

10.16 Summary

Nucleic acids are the compounds in a cell that are responsible for the storage and transmission of genetic information that controls the growth, function and reproduction of the cell. All nucleic acid molecules consist of polymer-like structures made up from basic units called **nucleotides**. Nucleotides consist of a purine or pyrimidine base linked by a β-*N*-glycosidic link to a sugar residue with a phosphate residue at position 5′. They are linked through their phosphate residue to the 3′ position of the next nucleotide. **Nucleosides** are base–sugar residues.

Nucleic acids are classified into DNA and RNA, where the structure of DNA contains β-D-deoxyribose and that of RNA contains β-D-ribose. **DNA** occurs as the DNA–protein complex chromatin, which is found in **chromosomes**. Chromosomes occur in the nuclei of all types of cell. DNA is also found in plasmids and other structures found in bacterial cells. **RNA** is found in both the nucleus and the cytoplasm of cells. Both DNA and RNA are found in the structures of viruses.

The main **bases** found in **DNA** are adenine, cytosine, thymine and guanine. DNA consists of two long-chain nucleic acid molecules held in an α-helix by hydrogen bonding between specific pairs of bases known as complementary base pairs. A gene is the sequence of bases in DNA that acts as the code for the production of a specific protein by a cell. In eukaryotic cells the sequence of bases forming a gene usually occurs as a number of separate sections known as **exons**. Each exon is separated from its neighbour by **introns**. Introns are sequences with no apparent genetic function. The reproduction of DNA is known as **replication**. Replication starts with the unwinding of the double helix to form single-chain parent templates for the formation of a new DNA chain and double helix. The new daughter chains grow from their 5′ ends in the 3′ to 5′ direction of their parent chain.

RNA is a single-chain polymer of nucleosides with their sugar residues linked 3′ to 5′ by phosphate units. It contains the same bases as DNA with the exception of thymine, which is replaced by uracil. RNA is classified as messenger RNA (mRNA), transfer RNA (tRNA) and ribosomal (rRNA). All types of RNA are formed from heterogeneous nuclear RNA (hnRNA), which is produced from DNA by a process known as **transcription**. **mRNA** carries the **genetic code** that instructs the ribosomes to synthesise a specific protein. The genetic code carried by mRNA is in the form of a sequence of three consecutive bases known as a **codon**. A codon corresponds to a specific amino acid. Each amino acid can have several different codons. There are also codons incorporated into the mRNA that start or stop protein synthesis.

Most **tRNAs** have structures based on three loops joined by short sections of a double chain in the form of an α-helix, plus a loop of variable size. Each tRNA is responsible for transporting a specific amino acid from the cell's amino acid pool to the mRNA attached to a ribosome. The amino acid is attached through an ester link to the ribose residue at the 3′ end of the tRNA. The tRNA molecule has a sequence of three consecutive bases known as an **anticodon** on its anticodon loop. These bases are complementary to those found on a specific codon of mRNA. The tRNA binds to mRNA by hydrogen bonding between the complementary base pairs of the anticodon of the tRNA and the codon of the mRNA. tRNA uses the complementary anticodon–codon binding to deliver its amino acid residue to the correct point on the mRNA.

Ribosomes are organelles that are responsible for the synthesis of the proteins required by the cell. Ribosomes contain rRNA subunits and a number of proteins. **Protein synthesis** starts with the ribosome splitting into one large and one small subunit. The mRNA binds to the small subunit. Protein synthesis is divided into four general stages, namely: activation, initiation, elongation and termination. **Activation** is the formation of a reactive aminoacyl–tRNA complex prior to the start of protein synthesis. **Initiation** occurs when a methionine–tRNA complex binds to the AUG start codon of the mRNA bound to the small subunit of a ribosome. It is complete when the large subunit of the ribosome rebinds to the small ribosome subunit–mRNA–methionine–tRNA complex in such a way that the mRNA–methionine–tRNA complex is sandwiched between the two subunits at the P site of the large subunit of the ribosome. **Elongation** is effectively a two-step repetitive process that forms the peptide chain. In the first step an aminoacyl–tRNA complex binds to the A site of the ribosome that is adjacent to the P site. The growing peptide on the P site is transferred to and combines with the amino acid residue of the aminoacyl–tRNA complex bound to the A site. At this point, in the second step, the peptide is **translocated** to the P site by a movement of the ribosome along the mRNA strand, leaving the A site vacant for a new aminoacyl–tRNA complex. These two steps are repeated until the protein is formed. **Termination** of protein synthesis occurs when a stop codon is reached. This codon is recognised by **release factors** that promote the release of the protein by a mechanism that is not fully understood. mRNA strands actively synthesising proteins will have several ribosomes attached to it, all synthesising different molecules of the same protein.

Protein synthesis follows the same general routes in both **prokaryotic** and **eukaryotic cells** but differs in some of the details. In prokaryotic cells the correct alignment of the mRNA in the ribosome is ensured by the Shine–Dalgarno sequences of bases near the 5′ end of the mRNA chain. The mRNA of eukaryotic cells has no Shine–Dalgarno sequences but does have a 7-methyl-GTP at its 5′ end and a poly A unit at its 3′ end. In eukaryotic cells the mRNA is correctly aligned by the use of an intermediary preinitiation $tRNA^{met}$ complex. Termination in prokaryotic cell ribosomes usually requires three different release factors, as against only one in eukaryotic cell ribosomes.

Some **protein synthesis inhibitors** are selective for prokaryotic cells having little or no effect on eukaryotic cell protein synthesis. For example, the aminoglycosides (see section 10.12.1), chloramphenicol (see section 10.12.2) and the tetracyclines (see section 8.11.3) act mainly on prokaryotic protein synthesis. Bacterial resistance to these drugs arises in a variety of ways, ranging from active transport of the drug out of the bacteria by membrane-spanning proteins to deactivation of the drug by enzymes that catalyse acylation, phosphorylation and adenylation of amino and hydroxy groups in the drug.

Drugs that target nucleic acids either inhibit the synthesis of the nucleic acids or act on existing nucleic acid molecules. Compounds inhibiting the formation of the precursors of nucleic acids are referred to as **antimetabolites.** They block normal metabolic processes of the cell by either replacing the endogenous metabolite or inhibiting an enzyme. Compounds acting as antimetabolites include **antifolates** (see section 10.13.1.1), which have structures resembling that of folic acid, and **purine** (see section 10.13.1.2) and **pyrimidine** (see section 10.13.1.3) **antimetabolites**, which have structures closely resembling those of the endogenous purine and pyrimidine bases. A wide range of enzyme inhibitors (see section 6.12.1) inhibit the purine and pyrimidine precursors of nucleic acids.

Drugs acting on existing nucleic acids include intercalating agents (see section 10.13.3), alkylating agents (see section 10.13.4), antisense drugs (see section 10.13.5) and chain-cleaving agents (see section 10.13.6). **Intercalation agents** have structures that contain a flat structure that fits inside the α-helix of DNA, causing it to partially unwind. They are believed to act by inhibiting transcription. **Alkylating agents** usually prevent transcription of DNA by bonding to the bases in the nucleic acid chain. **Interstrand links** are formed between the two separate strands of DNA whereas **intrastrand links** are formed between two parts of the same strand of DNA. **Antisense drugs** have structures that mimic the complementary base sequence of its target, which is usually mRNA. They are believed to prevent mRNA from delivering its message to the ribosome by binding to a complementary base sequence of mRNA. **Chain-cleaving agents**, such as the bleomycins, cause the nucleic acid chain to break into two or more fragments.

Viruses are infective agents, smaller than bacteria, that are able to invade a host cell. They contain a central core of either RNA or DNA fully or partially surrounded by a protein coating called a **capsid**. **Capsomers** are the individual proteins that form the capsid. Enveloped viruses have an additional external lipoprotein envelope that surrounds the capsid. Viruses replicate by entering the host cell and then using the host cell's biological systems to produce multiple copies of its own nucleic acids and proteins. Viruses are broadly divided into **RNA-virus**, **RNA-retrovirus** and **DNA virus**, according to their mode of action and structure. They all use the host cell's biological systems to form multiple copies of their own proteins and nucleic acids. These new proteins and nucleic acids may directly cause a pathological addition and/or may be assembled into new virions in the host cell that are released from the host by either lysis or budding. In many cases budding does not result in the death of the host cell.

Antiviral drugs act by inhibiting viral nucleic acid synthesis, inhibiting attachment to and penetration of the host cell or inhibiting viral protein synthesis. **Viral nucleic acid synthesis inhibitors** (see section Table 8.8) usually inhibit either DNA-polymerases or reverse transcriptases. They are usually analogues of the purine and pyrimidine bases found in the nucleic acids. The main **attachment and penetration inhibitors** are amantadine hydrochloride and rimantadine hydrochloride (see section 10.14.4.2). **Interferons** act as inhibitors of viral protein synthesis. A number of compounds have been found to induce the production of interferons in animals.

Defective and missing genes are responsible for a number of diseased states. **Gene cloning** is the introduction of a DNA gene sequence into a living cell. In gene cloning the new gene is inserted into a so-called vector that is inserted into the living cell. **Vectors** are usually plasmids, bacteriophages and cosmids. **Plasmids** are large supercoiled DNA molecules that are found in the cytoplasm of bacteria. They may be cleaved at specific points by restriction enzymes, the new gene being inserted into the cut and joined to the original DNA by DNA ligases to form so-called recombinant DNA. The new plasmid is reinserted into the bacteria where bacterial reproduction results in the production of more recombinant DNA molecules Similar techniques to that used for plasmids are used to introduce DNA into bacteriophages and cosmids.

Gene cloning (gene therapy) has been used successfully in clinical trials to treat diseases that are caused by the absence of a gene or the presence of defective gene cells so that the progeny will also carry the new gene. It has also been used to manufacture many of the specialised peptides and proteins required by humans with deficiency diseases related to the absence of these peptides and proteins.

10.17 Questions

(1) Distinguish carefully between the members of the following pairs of terms:
  (a) nucleotide and nucleoside;
  (b) introns and exons;
  (c) codons and anticodons.

(2) Explain how cytosine and guanine form a complementary base pair.
(3) The sequence AATCCGTAGC appears on a DNA strand. What would be the sequence on (a) the complementary chain of this DNA and (b) a transcribed RNA chain?
(4) What are the two main functions of DNA?
(5) How does RNA differ from DNA? Outline the functions of the three principal types of RNA.
(6) What is the genetic code?
(7) Why is thymine never found in a human codon?
(8) Explain in outline the significance of the P, A and E sites of ribosomes.
(9) What is the sequence of amino acid residues in the peptide formed from the following mRNA:
UUCGUUACUUAGAUGCCCAGUGGUGGGUACUAAUGGCUCGAG
(10) How does protein synthesis in prokaryotic cells differ from that in eukaryotic cells?
(11) Explain why the structure of chloramphenicol is probably the optimum one for antibiotic activity.
(12) Outline a general strategy for discovering a more active antibiotic than tetracycline using tetracycline as the lead compound.
(13) Draw the general structure of nitrogen mustards. How do these drugs inhibit the transcription of DNA?
(14) Explain why the incorporation of didanosine (Table 10.6) into a strand of RNA stops the synthesis of the RNA.
(15) Describe the way in which the antibiotic proflavin disrupts the transcription of DNA. What part do the amino groups of proflavin play in the mode of action of this drug?
(16) What are antisense drugs? How do they inhibit mRNA?
(17) Explain the meaning of the classification 'antimetabolite'. Outline a strategy for designing a new antimetabolite for a biological process.
(18) Describe the essential differences between RNA-viruses, retroviruses and DNA-viruses.
(19) Outline, using suitable examples, the general mode of action of antiviral drugs.
(20) Explain how recombinant DNA is produced using plasmids.
(21) Describe the difference between germline and somatic gene therapy.

11 Nitric Oxide

11.1 Introduction

In the late 1980s and early 1990s it was confirmed by several groups of workers that nitric oxide was a chemical messenger released by the endothelium and other tissues in mammals. It was tentatively identified as being the endothelium-derived relaxing factor (EDRF) discovered earlier in the 1980s and has now been linked to a multitude of physiological and pathophysiological states in mammals. For example, it is now known to be involved in the control of blood pressure, neurotransmission and the immune defence system of the body. Excessive production has been linked to atherosclerosis, hypotension, Huntington's disease, Alzheimer's disease and AIDS dementia, whereas underproduction has been related to thrombosis, vasospasm and impotence. This diversity of action has resulted in there being a considerable interest in the relationship of nitric oxide to both healthy and diseased states.

Nitric oxide is a colourless paramagnetic gas (boiling point –151.7°C), sparingly soluble in water (2–3 mmol dm^{-3}). It is produced in mammals by the enzyme-catalysed interaction of molecular oxygen and arginine (see section 11.4). However, whether it is nitric oxide or a derivative of nitric oxide that is ultimately responsible for the observed physiological response to nitric oxide generation is still under dispute.

11.2 The Structure of Nitric Oxide

Nitric oxide is a relatively stable free radical with one unpaired electron (Figure 11.1). However, unlike many reactive free radicals, nitric oxide does not dimerise appreciably in the gas phase at room temperature and pressure although it appears to form N_2O_2 in the liquid state:

$$O{=}N\cdot \quad \cdot N{=}O \rightleftharpoons O{=}N{-}N{=}O$$

The simple molecular orbital (MO) picture for nitric oxide shows that the unpaired electron is in an antibonding MO (Figure 11.1). Consequently, this electron should be readily lost resulting in the formation of the nitrosonium ion (NO^+). This is in agreement with nitric oxide's low ionisation potential of 9.25 eV compared to a value of 15.56 eV for nitrogen gas.

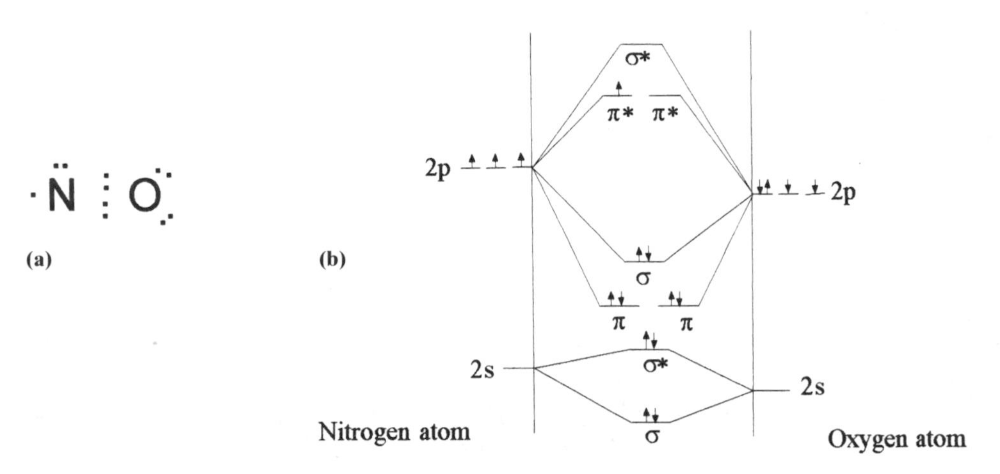

Figure 11.1. (a) The simple electronic structure of nitric oxide. (b) The MO energy diagram of the nitric oxide. Only the outer shell electrons are shown.

11.3 The Chemical Properties of Nitric Oxide

It is necessary to understand the chemistry of nitric oxide and related compounds in order to predict and understand their behaviour *in vivo.* The chemistry of nitric oxide is quite varied. It can act as a free radical, electrophile and oxidising agent. It gives rise to a series of salts and complexes with metals. Many of the chemical species identified in laboratory reactions have also been found in biological systems. It is believed that many of these species could be formed in biological systems by reactions similar to those found in the *in vitro* laboratory experiments although at present few of these reactions have been detected *in vivo*.

In biological systems, nitric oxide often appears to be closely associated with many other simple nitrogen–oxygen-containing species such as nitrogen dioxide, nitrogen trioxide, nitrogen tetroxide, nitrite, nitrate and peroxynitrite (Figure 11.2). Nitrogen dioxide and peroxynitrite are believed to cause tissue damage and so for this reason the relevant chemistry of these substances will also be included in this section.

11.3.1 Free Radical Reactions

Although nitric oxide is a relatively stable free radical it readily reacts with other free radicals. Two important free radicals that are widely distributed in mammals are oxygen and superoxide.

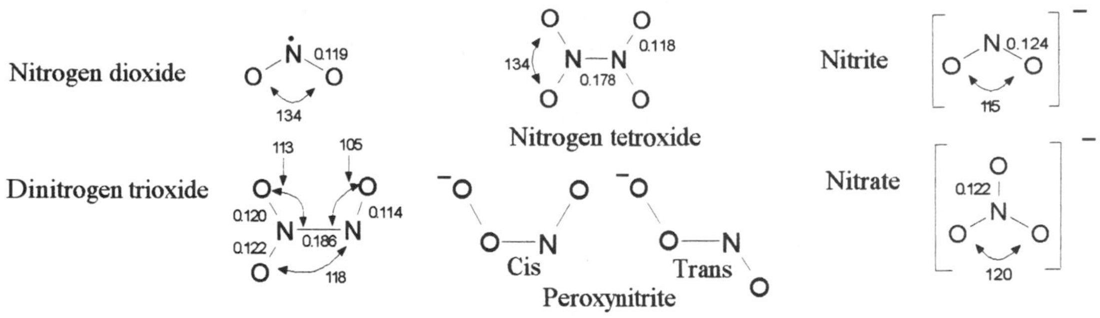

Figure 11.2. The structures of some compounds related to nitric oxide. Bond lengths are in nanometres.

Table 11.1. The $t_{1/2}$ values for various concentrations of nitric oxide in air.

Initial concentration of NO in air (ppm)	Half life
10	7 h
10 000	24 s

11.3.1.1 Reaction with Oxygen (Oxidation)

The molecular orbital picture of oxygen shows that it is a diradical and able to react with other free radicals. It reacts with nitric oxide in both the gaseous state and under aqueous aerobic conditions.

$$2NO + O_2 \rightarrow 2NO_2$$

In both the gaseous and aerobic aqueous states the reaction has been shown to be first order with respect to oxygen and second order with respect to nitric oxide.

$$-d[NO]/dt = k[NO]^2[O_2] \quad (11.1)$$

Because the overall reaction is third order, the half-life of nitric oxide is not constant but will depend on its initial concentration (Table 11.1). Furthermore, high concentrations of nitric oxide react more rapidly with oxygen than low concentrations. This means that once a low concentration of nitric oxide is achieved the nitric oxide could be present in the system for a considerable period of time. However, nitric oxide has a short biological life and so must be mainly metabolised by a route that does not involve reaction with oxygen.

The nitrogen dioxide formed in these reactions readily dimerises to nitrogen tetroxide, which reacts with water to form a mixture of nitrite and nitrate ions.

$$2NO_2 \rightleftharpoons N_2O_4 \xrightarrow{H_2O} NO_2^- + NO_3^- + 2H^+$$

This reaction does not play a significant part in the metabolism of nitric oxide because the *in vivo* rate of formation of nitrogen dioxide is slow. However, under physiological conditions the nitrogen dioxide formed combines with residual nitric oxide to form dinitrogen trioxide, which rapidly reacts with water to form nitrite ions.

$$NO + NO_2 \rightleftharpoons N_2O_3 \xrightarrow{H_2O} 2NO_2^- + 2H^+$$

This reaction is believed to be a minor *in vivo* route for the inactivation of nitric oxide.

The reaction of nitric oxide with oxygen is believed to proceed by the formation of nitrosyldioxyl radical, which offers a **possible route** for the formation of peroxynitrite in biological systems (see section 11.3.1.2).

$$O_2 + \cdot NO \rightleftharpoons \underset{\text{Nitrosyl radical}}{\cdot OONO} \xrightarrow{\cdot NO} ONOONO \longrightarrow 2NO_2$$

$$\cdot OONO \xrightarrow{1e} {}^-OONO \quad \text{(Possible route to peroxynitrite)}$$

Biological evidence suggests that nitrogen dioxide, nitrosyldioxyl and peroxynitrite are all involved in the physiological chemistry of nitric oxide although little is currently known about their chemical action in biological systems. Nitrogen dioxide is a strong oxidising agent, a good nitrosylating agent (adds NO to a structure), a good nitrating agent (adds NO_2 to a structure) and a free radical (Figure 11.2). All these reactions could have implications for the action of nitric oxide in biological systems.

It is interesting to note that both nitric oxide and nitrogen dioxide react with the high concentrations of hydrogen peroxide found in the pulmonary alveolar macrophages to form nitrite ions, nitrate ions and hydroxyl free radicals. Furthermore, laboratory experiments have demonstrated that the main reaction of high concentrations of nitrogen dioxide with C=C bonds is addition to form a nitro free radical. At low concentrations of the order of those normally found in biological systems nitrogen dioxide mainly acts as a free radical and abstracts an allyl hydrogen to form nitrous acid and an allyl free radical. In both cases, the free radicals produced have been shown to react to form a variety of products including nitro and, in the presence of molecular oxygen, hydroperoxyl derivatives whereas the nitrous acid would nitrosate amines that were added to the reaction mixture.

$$-CH{=}CH{-}CH_2- \xrightarrow{NO_2} -\underset{}{\overset{NO_2}{\overset{|}{C}H}}{-}\dot{C}H{-}CH_2-$$

$$-CH{=}CH{-}CH_2- \xrightarrow{NO_2} -CH{=}CH{-}\dot{C}H- + HNO_2$$

At biological concentrations, hydrogen abstraction appears to be the dominant reaction of nitrogen dioxide with unsaturated fatty acids. These laboratory reactions indicate that nitrogen dioxide formed in biological systems could initiate the auto-oxidation of fatty acids in lipid membranes, resulting in membrane damage. In addition, the nitrous acid could nitrosate naturally occurring amines such as DNA, which could impair their function.

11.3.1.2 Reaction with Superoxide (Oxidation)

Nitric oxide reacts with superoxide to form peroxynitrite. The reaction is very fast and is probably a major route for the metabolism of nitric oxide.

$$^{-}O\text{-}O\bullet + \bullet NO \rightarrow {}^{-}OONO$$

Peroxynitrite is a stable anion at alkaline pH (pK_a 6.8 at 37°C). It has been suggested that the stability of the peroxynitrite ion is due to its structure being held in a *cis* conformation by internal forces of attraction. When this conformer is protonated the *cis*-peroxynitrous acid formed can more easily isomerise to the more labile *trans* conformer.

cis-Peroxynitrite ⇌ (H^+) *cis*-Peroxynitrous acid → *trans*-Peroxynitrous acid

At neutral pH peroxynitrite is rapidly protonated to form the unstable peroxynitrous acid, which rapidly decomposes to nitrogen dioxide, hydroxyl radicals in about 20–30% yield and nitrate ions by two separate routes.

$$^{-}OONO + H^+ \longrightarrow HOONO \longrightarrow NO_2 + \dot{O}H$$

$$HOONO \longrightarrow NO_3^- + H^+$$

The stability of peroxynitrite allows it to diffuse considerable distances through biological systems as well as across membranes before it reacts. This reactivity is believed to take three main routes:

(i) at physiological pH in the presence of hydrogen ions peroxynitrite can result in the formation of an intermediate with hydroxyl free-radical-like reactivity;
(ii) peroxynitrite can react with metal ions and the metal centres of superoxide dismutase (SOD) to form a nitrating agent with similar reactivity to the nitronium ion ($\overset{+}{N}O_2$). This nitrating agent readily nitrates phenolic residues such as the tyrosine residues of lysozyme and histone;
(iii) peroxynitrite reacts with sulphydryl groups of proteins and other naturally occurring molecules (see section 11.3.4).

Many pathological conditions are thought to result in the tissues simultaneously producing nitric oxide and superoxide. Normally SOD would deactivate the superoxide but kinetic studies show that nitric oxide is produced in large enough quantities and fast enough to prevent this deactivation. Consequently, it has been suggested that in view of its reactivity peroxynitrite may be a major species involved in these conditions.

11.3.2 Salt Formation

Both ^{+}NO and ^{-}NO ions are known. Their existence may be explained by the low ionisation (9.5 eV) and reduction potentials (0.39 eV) of nitric oxide.

$$\underset{\text{Nitrosonium ion}}{e + {}^{+}NO} \xleftarrow[\text{Oxidation}]{} NO \xrightarrow[\text{Reduction}]{1e} \underset{\text{Nitroxyl ion}}{-NO}$$

11.3.2.1 Nitrosonium Salts

A number of well-characterised salts containing the nitrosonium ion are known. For example, nitrosonium hydrogen sulphate ($NOHSO_4$), nitrosonium tetrafluoroborate ($NOBF_4$), nitrosonium halides (NOX, where X = Cl or Br) and nitrosonium perchlorate ($NOClO_4$) have been prepared. Some of these salts exhibit considerable covalent character. For example, experimental data indicate that the nitrosonium halides are only about 50% ionic.

$$NOHSO_4 + H_2O \rightarrow HNO_2 + H_2SO_4$$

Nitrosonium salts are easily hydrolysed to form nitrous acid and the constituent acid. For example, nitrosonium hydrogen sulphate hydrolyses to a mixture of nitrous and sulphuric acids.

This ease of hydrolysis means that nitrosonium salts can only be prepared in the laboratory under anhydrous conditions. However, nitrosations involving nitrosonium compounds are known in both synthetic and biological chemistry (see section 11.3.5.1).

11.3.2.2 Nitroxyl Ion (NO^-)

This is a less-well-characterised species. However, it is believed to be formed in the reduction of NO by cuprous (Cu^+) superoxide dismutase (Cu^+SOD).

$$Cu^{+}SOD + NO \rightleftharpoons {}^{-}NO + Cu^{2+}SOD$$

Nitroxyl ions have been shown to react rapidly with molecular oxygen to form peroxynitrite, which would suggest that the nitroxyl ion would have a very short life in oxygenated tissue.

11.3.3 Reaction as an Electrophile

Nitric oxide can act as an electrophile because its electronic configuration is one electron short of a stable octet. It readily reacts with thiols, amines and other nucleophiles. For example, nitric oxide reacts in this manner with primary and secondary amines to form adducts known as NONO-ates. The mechanisms of these reactions are probably more complex than the equations suggest. However, it is believed that in the initial step of these reactions the nitric oxide appears to function as an electrophile, reacting with the amine to form an NO complex. This complex acts as a free radical and reacts with a second nitric oxide molecule to form the *N*-nitroso derivative of the amine salt, which in the presence of excess amine decomposes to the free base form of the NONO-ate adduct.

$$R_2N{-}H + {}^{\bullet}N{=}O \rightleftharpoons H{-}\overset{+}{N}R_2{-}{}^{\bullet}N{-}O^- \;+\; {}^{\bullet}N{=}O \xrightarrow{R_2NH} R_2N{-}N(O^-){-}N{=}O \cdot R_2\overset{+}{N}H_2$$

"NONO-ate adduct"

The NONO-ate adduct is unstable in aqueous solutions, yielding nitric oxide. The rate at which the nitric oxide is produced has been shown to depend on the pH, temperature and structure of the amine. Consequently, NONO-ate adducts may have possible use as drugs to treat cases where the production of endogenous nitric oxide is impaired (see section 11.6.2).

Nitric oxide reacts with proteins containing thiol groups, under physiological conditions, to form *S*-nitrosothiol derivatives. *In vitro* evidence suggests that nitric oxide is transported in the plasma in the form of stable *S*-nitrosothiols, about 80% of which are *S*-nitroso-serum albumins. These *S*-nitrosothiols are believed to act as a depot for nitric oxide, maintaining vascular tone. Furthermore, it is thought that *S*-nitrosothiols could be intermediates in the cellular action of nitric oxide (see section 11.4.1).

11.3.4 Reaction as an Oxidising Agent

Nitric oxide has been reported to oxidise thiols under basic conditions to disulphides and hyponitrous acid. The reaction is believed to proceed via the nitrosothiol.

$$RSH \xrightarrow{\text{Base}} RS^-$$

$$N{=}O + RS^- \rightarrow RS{-}\ddot{N}{-}O^-$$

$$2RS{-}\ddot{N}{-}O^- \xrightarrow{2H^+} \underset{\text{Hyponitrous acid}}{HO{-}N{=}N{-}OH} + \underset{\text{Disulphide}}{RSSR}$$

Nitrogen dioxide has also been reported to react with thiols in a similar manner provided that the thiol is present in excess. Excess nitrogen dioxide oxidises disulphides to other products.

$$2RSH + NO_2 \rightarrow RSSR + NO + H_2O$$

11.3.5 Complex Formation

Complexes containing nitric oxide can be prepared by a variety of methods. This section outlines those that could have biological analogues.

11.3.5.1 By Direct Reaction with Nitric Oxide

Nitric oxide will react with complexes that have unused coordination sites.

$$[CoCl_2L_2] + NO \rightarrow [CoCl_2L_2NO]$$

11.3.5.2 By Substitution of Another Ligand

Nitric oxide will displace another ligand such as carbon monoxide from a complex.

$$[Fe(CO)_5] + 2NO \rightarrow [Fe(CO)_2(NO)_2] + 3CO$$

11.3.5.3 By Substitution or Addition of the Nitrosonium Ion

This method uses either a nitrosonium salt (see section 11.3.2.1) or acidified nitrite or an organic nitrite as the source of the nitrosonium ion.

From a nitrosonium salt such as $NOBF_4$:

$$[Rh(CNR)_4]^+ + NO^+ \rightarrow [Rh(CNR)_4(NO)]^{2+}$$

Generated from nitrite by acidification with carbonic acid:

$$K[Fe(CO)_3(NO)] + KNO_2 + CO_2 + H_2O \rightarrow [Fe(CO)_2(NO)_2] + 2KHCO_2$$

Generated from an organic nitrite:

$$[Fe(CO)_3(PPh_3)_2 + RONO + H^+ \rightarrow [Fe(CO)_2(NO)(PPh_3)_2]^+ + CO + ROH$$

11.3.5.4 Structural Features of Nitric Oxide Complexes

Nitric oxide can be bonded to the metal in three distinct ways:

(i) Complexes in which the MNO bond is linear or almost linear, usually lying between 160° and 180°. In these complexes the nitric oxide donates its odd electron to the metal atom and binds to the metal through a two-electron bond (Figure 11.3a).
(ii) Complexes in which the MNO bond angle is bent and lies between 120° and 140°. In these structures the nitric oxide is considered to be a one-electron donor to the metal atom (Figure 11.3b).
(iii) Complexes in which the nitric oxide acts as a bridge (Figure 11.3c). More than one bridge may link the metal atoms in these complexes.

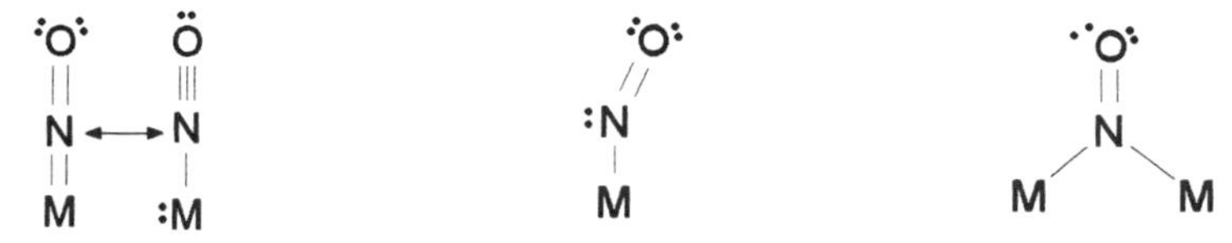

$Na_2[Fe(CN)_5NO]$
$[Mn(CO)_2(NO)(PPh_3)_2]$
$[Co(Cl)_2(NO)(PMePh_2)_2]$

(a) Linear

$[Co(NH_3)_5NO]^{2+}$
$[Rh(Cl)_2(NO)(PPh_3)_2]$
$[Ir(Cl)_2(NO)(PPh_3)_2]$

(b) Bent

$[\{Cr(\eta^5\text{-}C_5H_5)(NO)(\mu_2\text{-}NO)\}_2]$
$[Mn_3(\eta^5\text{-}C_5H_5)_3(\mu_2\text{-}NO)_3(\mu_3\text{-}NO)]$

(c) Bridge

Figure 11.3. Representations of the bonding in linear, bent and bridge NO complexes. Examples of linear, bent and bridge complexes are given below each structure. Complexes that contain a mixture of these MNO structures are also known.

Bent, linear and bridge nitric oxide ligands can be distinguished by their characteristic infrared and NMR spectra. The shapes of the resulting structures can be predicted with a reasonable degree of accuracy using the simple hybridisation rules assuming that the nitrogen in linear complexes is sp and it is sp^2 in bent and bridge complexes. However, for accurate work the structures used should be those determined by X-ray crystallography.

11.3.6 Nitric Oxide Complexes with Iron

Iron is widely distributed in mammalian cells, both as free ions and complexed with a wide variety of proteins. Nitric oxide readily forms complexes with both ferrous and ferric ions in the presence of other suitable ligands as well as reacting with the Fe^{II} and the Fe^{III} centres of iron containing naturally occurring molecules. The coordination state of iron in the resulting complexes is often 4, the nitric oxide taking up any vacant coordination sites.

11.3.6.1 Nitric Oxide–Iron–Protein Complexes

Electron paramagnetic resonance spectroscopy showed that nitric oxide reacts with cysteine, histidine and other amino acids in the presence of ferrous ions to form four-coordinate nitric oxide–iron–amino acid complexes (Figure 11.4). Spectroscopy also showed that nitric oxide reacted with proteins in the presence of ferrous ions to form complexes that were either associated with thiol groups or imidazole groups. Proteins whose structures contained a high proportion of thiol groups formed complexes involving the thiol groups in preference to the imidazole groups. Electron spin resonance has also shown that proteins whose structures did not contain iron bound to haem but included thiol groups formed complexes in the presence of free iron where one iron was complexed to two thiol groups and two nitric oxide molecules. These structures are probably similar to that proposed for the nitric oxide–iron–cysteine complex (Figure 11.4a). Complexes of this type are called **dinitrosylirondithiol** complexes.

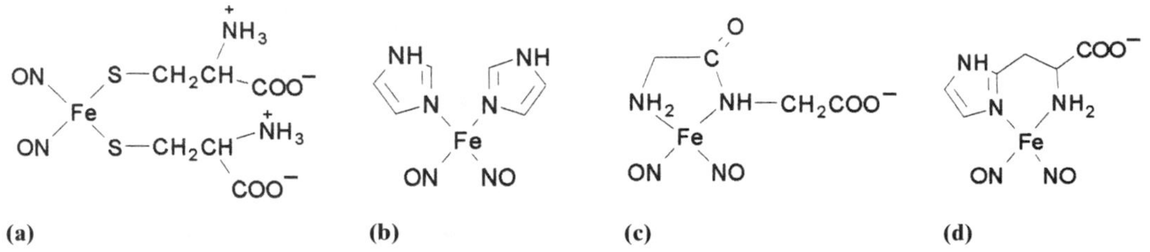

Figure 11.4. Proposed structures for the nitric oxide–iron complexes of (a) cysteine, (b) imidazole, (c) glycylglycine and (d) histidine.

The interaction of nitric oxide with cellular iron and proteins usually results in a loss of enzyme activity. However, it is not clear in the case of the proteins with thiol groups if the action of nitric oxide is the direct cause of the loss of enzyme activity. This is because the dinitrosylirondithiol protein complexes could also lose their activity by exchanging their protein-thiol ligands (RSH) with other protein-thiol ligands (R′SH).

$$(NO)_2Fe(RS)_2 + 2R'SH \rightarrow (NO)_2Fe(R'S)_2 + RSH$$

A further complication is that nitrogen dioxide can also react in the same manner as nitric oxide and to form dinitrosylirondithiols. Consequently, it could be that nitrogen dioxide produced from nitric oxide is the *in vivo* source of dinitrosylirondithiols, in which case it would be nitrogen dioxide that is responsible for the loss of enzyme activity. The problem is at present unresolved. However, the formation of dinitrosylirondithiols has been associated with a wide variety of types of tissue damage and it has also been found that cellular iron is a major target of nitric oxide.

11.3.6.2 Nitric Oxide–Iron–Haem–Protein Complexes

Experimental work shows that nitric oxide reacts with both the Fe^{II} and the Fe^{III} oxidation states of iron bound to haem in protein molecules. Nitric oxide undergoes a reversible reaction with proteins containing Fe^{III}–haem groups to form a nitrosyl–iron complex in which the iron–nitric oxide bond is best described by the canonical forms:

$$\text{Protein–haem–}Fe^{III}\text{–NO} \leftrightarrow \text{Protein–haem–}Fe^{II}\text{–}^{+}\text{NO}$$

The electron-deficient nitrogen explains why these complexes react with many nucleophiles (see section 11.3.6.1) to form the corresponding nitroso compounds and reducing the Fe^{III}–haem to Fe^{II}–haem.

$$\text{NO} + Fe^{III}\text{–haem–protein} \rightarrow \text{ON–}Fe^{III}\text{–haem–protein}$$

$$\text{ON–}Fe^{III}\text{–haem–protein} \xrightarrow{\text{Nucleophile(Nu)}} Fe^{II}\text{–haem–protein} + \text{ON–Nu}$$

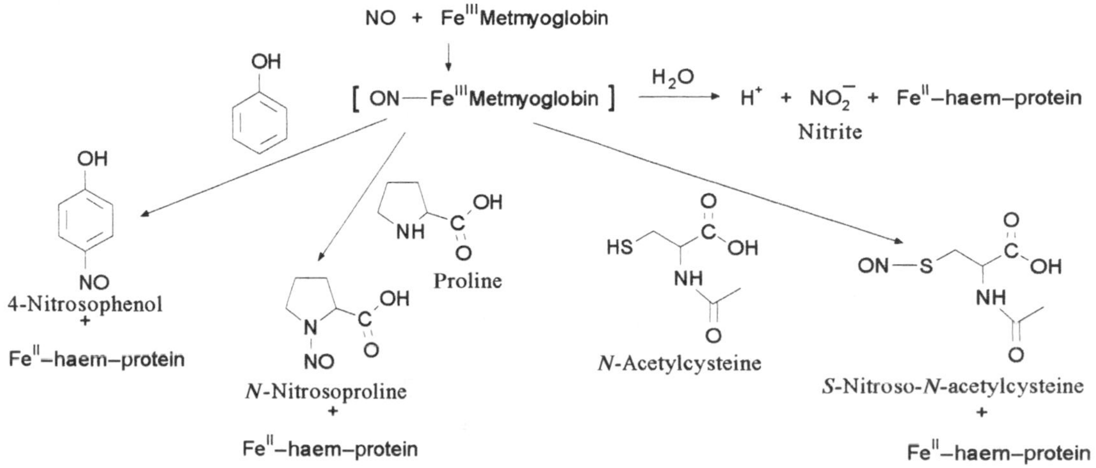

Figure 11.5. Examples of the reactions of nucleophiles with metmyoglobin–nitric oxide mixtures.

Castro and Wade have shown that nitric oxide reacts with metmyoglobin to form a nitric oxide–Fe^{III}–haem complex that nitrosates a wide variety of nucleophiles (Figure 10.5). The Fe^{II}–haem-protein complex formed in these reactions rapidly reacts with any nitric oxide present to yield an ON–Fe^{II}–haem-protein complex. Nitrite produced by the competitive reaction of water with this complex is also a product of many of these reactions. A number of other proteins ON-Fe-metmyoglobin groups, such as catalase, peroxidase and human haemoglobin, have also been shown to react under similar conditions with phenol to form 4-nitrosophenol, nitrite and Fe^{II}–haem–protein but the yield with human haemoglobin is very small. However, iron(III) cytochrome *c* and iron(III) cytochrome P-450 do not convert phenol to 4-nitrosophenols in the presence of nitric oxide. In the former case the iron(II) nitrosylcytochrome *c* and nitrite are formed. This suggests that it is the conformation of the protein about the Fe–NO site that controls nitrosation. It has been suggested on the basis of these reactions that nitric oxide could provide an *in vivo* route for the formation of the highly carcinogenic nitrosamines.

Nitric oxide reacts with proteins containing Fe^{II}–haem groups to form stable NO–Fe^{II}–haem complexes.

$$\mathrm{NO + Fe^{II}{-}haem{-}protein \rightarrow ON{-}Fe^{II}{-}haem{-}protein}$$

It is now well established that nitric oxide also targets haem–iron centres in cells to form nitrosyl complexes. This reaction, which is the main route of nitric oxide metabolism, is also linked to pathological conditions.

11.3.7 The Chemical Properties of Nitric Oxide Complexes

The differences in the structures of linear, bent and bridge nitric oxide complexes accounts for some of the differences in their chemical behaviour.

11.3.7.1 Electrophilic Behaviour

Some linear MNO complexes act as electrophiles because the nitrogen atom of the nitric oxide has donated three electrons to the metal. This leaves the nitrogen deficient in electrons and so open to attack by nucleophiles such as OH^-, RO^-, RS^- and RNH_2 (see section 11.3.6.2).

$$\text{Protein–haem–Fe}^{III}\text{–NO} \xrightarrow{RSH} \text{RS–N=O} + \text{Protein–haem–Fe}^{II}$$

S-Nitrosothiol

$$Na_2[(CN)_5FeNO]^{2-} \xrightarrow{OH^-} \left[(CN)_5Fe{-}N(OH){=}O\right]^{3-} \underset{}{\overset{OH^-}{\rightleftharpoons}} [(CN)_5FeNO_2]^{4-}$$

$$Na_2[(CN)_5FeNO]^{2-} \xrightarrow{RS^-} \left[(CN)_5Fe{-}N(SH){=}O\right]^{3-}$$

Sodium nitroprusside a vasodilator

In solution a red colouration is produced that is used as a test for thiols

S-Nitrosothiols slowly decompose, releasing nitric oxide, and so are of potential use as nitric oxide donors. However, the mechanism by which they release their nitric oxide is not clear. Furthermore, there is evidence to suggest that *S*-nitrosothiols are agents that act on soluble guanylyl cyclase.

11.3.7.2 Nucleophilic Behaviour

Both bent and bridge nitric oxide complexes undergo electrophilic attack by H^+ and other electrophiles. This is because the nitrogen of the nitric oxide has only donated one electron to the metal atom and so it has a lone pair of electrons that can react with electrophiles. In addition, the oxygen atom of the nitric oxide also has lone pairs, which can act as nucleophilic centres and undergo attack by electrophiles. For example, attack by H^+ can occur at both the nitrogen and oxygen nucleophilic centres, the initial attack being followed by further reaction in some cases.

$$(Ph_3P)_2(CO)ClOsNO \xrightarrow{HCl} (Ph_3P)_2(CO)Cl_2Os{-}N(H){=}O$$

$$(\eta_5\text{-}C_5H_4MeNO)_3Mn_3(\mu_3\text{-}NO) \xrightarrow[\text{from } HBF_4]{H^+} [(\eta_5\text{-}C_5H_4MeNO)_3Mn_3(\mu_3\text{-}NOH)]^+ \xrightarrow[\text{from } HBF_4]{H^+} [(\eta_5\text{-}C_5H_4MeNO)_3Mn_3(\mu_3\text{-}NH)]^+$$

11.3.7.3 Oxidation of the Nitric Oxide Ligand

It has been reported that some nitric oxide complexes undergo oxidation to the corresponding nitrogen dioxide complex.

$$L_nM(NO) \xrightarrow{\frac{1}{2}O_2} L_nM(NO_2)$$

Some complexes containing a nitrogen dioxide ligand have been shown to be involved in oxygen transfer reactions to alkenes, disulphides and other organic species.

$$\text{alkene} \xrightarrow{(CH_3CN)_2PdClNO_2} \text{ketone} + \text{cyclopropanone} + \text{allylic alcohol} + CH_3CN + (CH_3CN)PdClNO$$

This capacity of nitrogen dioxide complexes to transfer oxygen means that nitric oxide could also act through this route.

$$L_nM(NO) \xrightarrow{^1/_2O_2} L_nM(NO_2) \xrightarrow{\text{Oxygen transfer}} L_nM(NO)$$

11.3.7.4 Exchange Reactions

These are reactions in which the nitric oxide is exchanged with a ligand in another complex. For example:

$$RhNO(PPh_3)_3 + CoCl(PPh_3)_3 \rightleftharpoons RhCl(PPh_3)_3 + CoNO(PPh_3)_3$$

11.3.8 Reaction of Nitric Oxide with Complexes

Nitric oxide may either displace other ligands from a complex or react to form a complex that does not contain a simple nitric oxide unit.

11.3.8.1 Displacement Reactions

These are essentially exchange reactions in which the nitric oxide displaces other ligands from their complexes.

$$Co_2(CO)_8 + 2NO \rightarrow 2Co(NO)(CO)_3 + 2CO$$

This type of reaction is believed to be responsible for the activation of the enzyme guanylyl cyclase by nitric oxide in cells. The binding of nitric oxide to the iron atom of the haem nucleus of this enzyme releases a histidine residue. It has been suggested that this histidine residue acts as either a catalyst or a nucleophile, which increases the activity of the guanylyl cyclase. Other ligands that bind to the iron of haem do not liberate a histidine residue or activate guanylyl cyclase. Furthermore, removal of the nitric oxide deactivates the enzyme.

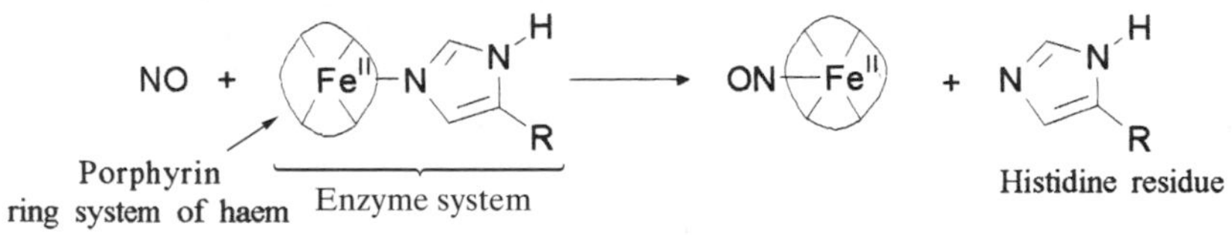

$$NO + Fe^{II}\text{–N(histidine, R)} \longrightarrow ON\text{–}Fe^{II} + \text{histidine residue}$$

Porphyrin ring system of haem; Enzyme system; Histidine residue

11.3.8.2 Other Reactions

The reaction of a complex with nitric oxide can result in a variety of new complexes whose structures do not contain a simple (NO) nitric oxide residue (Figure 11.6).

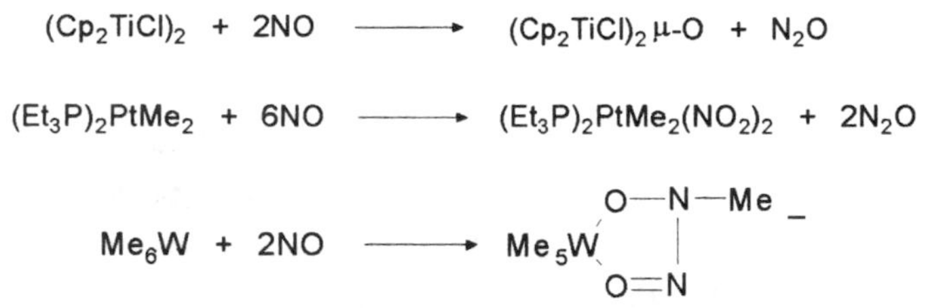

$$(Cp_2TiCl)_2 + 2NO \longrightarrow (Cp_2TiCl)_2\,\mu\text{-}O + N_2O$$

$$(Et_3P)_2PtMe_2 + 6NO \longrightarrow (Et_3P)_2PtMe_2(NO_2)_2 + 2N_2O$$

$$Me_6W + 2NO \longrightarrow Me_5W(O\text{–}N(Me)\text{–}N{=}O)$$

Figure 11.6. Examples of the reaction of nitric oxide with complexes.

Nitric oxide rapidly reacts with oxymyoglobin (O_2-My-Fe^{II}) to form nitrate and metmyoglobin (My-Fe^{III}).

$$O_2\text{-My-}Fe^{II} \xrightarrow{NO} \text{My-}Fe^{III} + NO_3^-$$

Oxyhaemoglobin has been found to react in a similar fashion to produce methaemoglobin. It has been proposed that both of these reactions involve the initial formation of peroxynitrite, which decomposes to nitrate.

$$O_2\text{–}Fe^{II}\text{–haemoprotein} + NO \rightarrow Fe^{III}\text{–haemoprotein} + ONOO^-$$
$$ONOO^- \rightarrow NO_3^-$$

The reaction with oxyhaemoglobin is thought to be a major route for the metabolism of nitric oxide in red blood cells.

11.3.9 The Chemistry of Related Compounds

It is still not certain that the physiological and pathological effects attributed to nitric oxide are directly due to that species. Many of the simple nitrogen compounds that can be formed from nitric oxide also react with the same compounds as nitric oxide. Consequently, the biochemistry of nitric oxide cannot be considered in isolation. The chemistry of compounds where nitric oxide may be a biological precursor must also be considered.

11.3.9.1 Nitrogen Dioxide

Nitrogen dioxide (NO_2) is a brown gas (b.p. 21°C) that is denser than air. It is a free radical (Figure 11.2) that exists at room temperature and pressure in equilibrium with its colourless dimer, nitrogen tetroxide (N_2O_4).

$$O_2N\cdot + \cdot NO_2 \rightleftharpoons O_2N{-}NO_2 \text{ Nitrogen tetroxide}$$

Nitrogen dioxide occurs as an atmospheric pollutant, is produced by bacteria and is found in tobacco smoke. It is well documented as a serious health hazard.

Nitrogen dioxide is an oxidising agent ($E^{\theta'}$ + 0.99 V). It has been reported that it initiates the auto-oxidation of unsaturated fatty acids in lipids. It is known to attack pulmonary lipids leading to membrane damage. *In vitro* experiments have shown that at high concentrations it forms nitro-substituted free radicals whereas at low concentration it forms allyl free radicals (Figure 11.7). These radicals react to form a variety of products by reactions that *in vivo* could lead to pathological effects. For example, the formation of nitrous acid by the action of nitrogen dioxide on unsaturated fatty acid residues could be another possible source of highly carcinogenic nitrosoamines.

At low concentrations nitrogen dioxide has been reported to react with thiols to form nitric oxide and thiol free radicals. The reaction is thought to proceed via a nitrosothiol intermediate although the reaction has not been observed in biological systems.

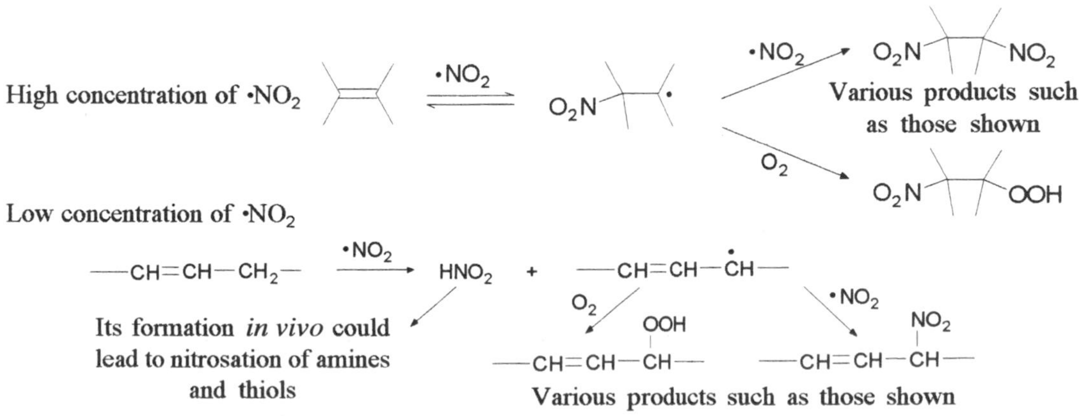

High concentration of $\cdot NO_2$

Various products such as those shown

Low concentration of $\cdot NO_2$

$—CH{=}CH—CH_2— \xrightarrow{\cdot NO_2} HNO_2 + —CH{=}CH—\dot{C}H—$

Its formation *in vivo* could lead to nitrosation of amines and thiols

Various products such as those shown

Figure 11.7. Reactions of nitrogen dioxide.

$$2RSH + NO_2 \rightarrow RS\bullet + \underset{\text{Nitrosothiol}}{RS{-}NO} + H_2O$$

$$RS{-}NO \rightarrow RS\bullet + \bullet NO$$

Nitrogen dioxide can also act as a nitrating agent. For example, nitrogen dioxide readily nitrates tyrosine residues in proteins to form nitrotyrosine and tyrosine biphenyl derivatives, the reaction rate increasing with pH. It has been proposed that this reaction occurs through a hydrogen abstraction, which would also account for the formation of tyrosine biphenyl derivatives.

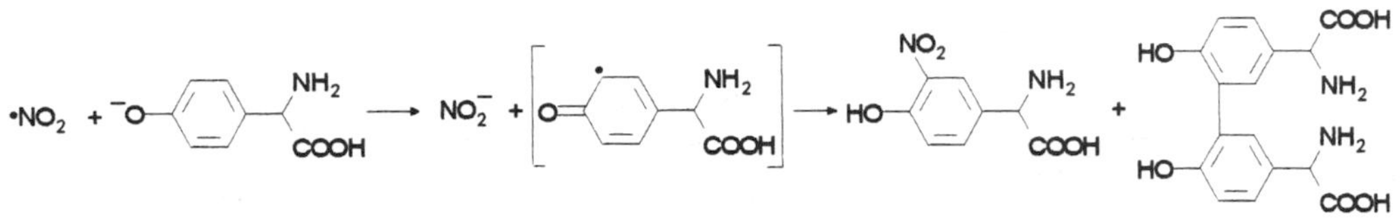

Nitration of tyrosine residues of proteins is known to change protein function. Consequently, nitration could be a major pathological route for tissue injury, especially as extensive nitration has been found in a number of pathological conditions.

11.3.9.2 Dinitrogen Trioxide (N_2O_3)

Dinitrogen trioxide (structure: see Figure 11.2) is an intense blue liquid formed by the reaction of nitric oxide with oxygen or dinitrogen tetroxide.

$$2NO + 2O_2 \rightarrow N_2O_3 \quad 2NO + N_2O_4 \rightarrow 2N_2O_3$$

It is not stable, decomposing to nitric oxide and nitrogen dioxide at room temperature and above.

$$N_2O_3 \rightarrow NO + NO_2$$

Dinitrogen trioxide is effectively the anhydride of nitrous acid, forming an unstable blue solution in water and non-polar solvents at room temperature.

$$2HNO_2 \rightleftharpoons N_2O_3 + H_2O$$

Dinitrogen trioxide is a powerful oxidising agent and a strong nitrosating agent. For example, it has been shown to rapidly oxidise oxymyoglobin (O_2-My-Fe^{II}) to metmyoglobin (My-Fe^{III}) and oxyhaemoglobin (O_2-Hb-Fe^{II}) to methaemoglobin (Hb-Fe^{III}).

$$O_2\text{-MY-Fe}^{II} \xrightarrow{N_2O_3} \text{My-Fe}^{III} + NO_3$$

$$O_2\text{-Hb-Fe}^{II} \xrightarrow{N_2O_3} \text{Hb-Fe}^{III} + NO_3$$

Dinitrogen trioxide in the form of nitrous acid also nitrosates amines and thiols.

11.3.9.3 Nitrite

Nitrite is ingested in food where it is used as a preservative, and is produced in mammals as metabolites of nitric oxide, nitrate and other substances. For example, nitrate is reduced in the oral cavity by bacteria and in the stomach by intestinal flora. Nitrite is also formed by the reduction of nitrogen dioxide (see section 11.3.1.1).

$$NO_3^- \xrightarrow[\text{Stomach}]{H^+/H_2O} NO_2^- + NH_3$$

$$\bullet NO_2 + 1e \xrightarrow{\text{Reduction}} NO_2^-$$

Nitrite is mildly toxic (tolerance limit ~100 mg kg^{-1} day^{-1}), autocatalysing the oxidation of oxyhaemoglobin to methaemoglobin, nitrogen dioxide and hydrogen peroxide. Methaemoglobin is formed in sufficient quantity to cause methaemoglobinaemia. Hydrogen peroxide is a very strong oxidising agent and is highly toxic when produced *in vivo*. It also forms weak complexes with methaemoglobin.

$$Fe^{II}{-}HbO_2 \xrightarrow{NO_2^-} Fe^{III}{-}Hb + NO_2 + H_2O_2$$

In acid conditions nitrite forms blue solutions of nitrous acid, which is a good nitrosating agent. Nitrous acid readily nitrosates amines, aromatic ring systems and many other organic species.

$$NO_2^- \xrightarrow{H^+} HNO_2$$

$$R_2NH \xrightarrow{HNO_2} R_2NNO + H_2O$$

$$\text{Phenol} \xrightarrow{HNO_2} \text{4-Nitrosophenol}$$

Phenol → 4-Nitrosophenol

Nitrous acid is unstable and in aqueous solution at room temperature and above it disproportionates to nitric oxide and nitrate.

$$3HNO_2 \rightarrow H_3^+O + NO_3^- + 2NO$$

Nitrous acid can also undergo one-electron reduction to nitric oxide in the presence of a suitable reducing agent

$$HNO_2 + H^+ + e \rightarrow NO + H_2O \quad E^\theta = 1.0\,V$$

An increase in the concentration of nitrate and nitrite ions in the urine is frequently used as an indication of the involvement of nitric oxide production in a biological system.

11.3.9.4 Peroxynitrite

Peroxynitrite is formed *in vivo* by the reaction of nitric oxide and superoxide (see section 10.3.1.2). Superoxide is normally scavenged by superoxide dismutase (SOD) but nitric oxide reacts so rapidly with superoxide that it outcompetes SOD to form peroxynitrite. Peroxynitrite is a strong oxidising agent. It reacts with SOD and other metalloproteins to form the nitronium ion ($\overset{+}{N}O_2$), which is a powerful nitrating agent, readily nitrating the tyrosine residues found in proteins. This nitration has been shown to lead to a loss of protein activity.

$$^{-}OONO \quad + \quad SOD \longrightarrow NO_2^{+} \xrightarrow[\text{Tyrosine residues}]{\text{—HNCH}_2\text{CO—}}$$

OH, NO_2, —$HNCH_2CO$—

It has been proposed that nitration initiated by peroxynitrite is a factor in amyotrophic lateral sclerosis (ALS). It is suggested that SOD mutants have a reduced superoxide scavenging effect, which results in over-production of peroxynitrite. The excess peroxynitrite nitrates tyrosine residues, which prevents the normal signal transduction by growth factors that support the survival of motor neurons. This would lead to a gradual loss of motor neurons and their associated activity, which is the principal feature of ALS.

11.4 The Cellular Production and Role of Nitric Oxide

Nitric oxide is produced *in vivo* by the catalytic oxidation of L-arginine by a family of enzymes known as nitric oxide synthases (NOS). The reaction requires nicotinamide adenosine diphosphate (NADPH) as a cofactor and produces nitric oxide and L-citrulline (Figure 11.8) in a 1:1 molar ratio. As a result, the concentration of citrulline is often used as an estimate of the *in vivo* concentration of nitric oxide.

Nitric oxide synthases have been broadly classified as constitutive NOS (cNOS) and inducible NOS (iNOS) enzymes (see Table 11.2). cNOS enzymes appear to be present at an approximately constant level in the host cell but only produce nitric oxide when activated by the Ca^{2+}-binding protein calmodulin (see section 6.4.2). Conversely, iNOS enzymes are not present in the cell but are produced in response to stimulants of host and bacterial origin. Activation of cNOS results in the production of a short burst of nitric oxide at a low concentration whilst activation of iNOS results in the continuous production of nitric oxide at a high concentration.

cNOS is present in endothelial and neural tissue. The cNOS enzymes in these tissues are not identical but have very similar properties. It has been shown that the endothelium and neural tissue production of nitric oxide is initiated by agonists such as acetylcholine, ADP, bradykinin

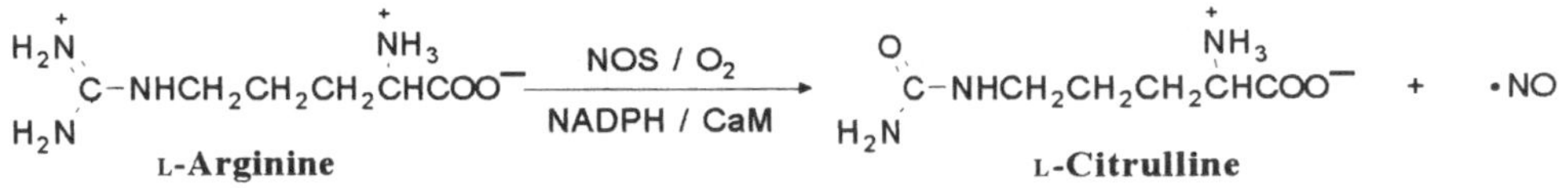

$$\mathrm{H_2\overset{+}{N}{=}C(NH_2){-}NHCH_2CH_2CH_2CH(\overset{+}{N}H_3)COO^-}\ \xrightarrow[\text{NADPH / CaM}]{\text{NOS / }O_2}\ \mathrm{O{=}C(NH_2){-}NHCH_2CH_2CH_2CH(\overset{+}{N}H_3)COO^-} + \bullet NO$$

L-Arginine → **L-Citrulline**

Figure 11.8. A schematic representation of the formation of nitric oxide. CaM = calmodulin.

Table 11.2. Some characteristic properties of cNOS and iNOS.

	cNOS		**iNOS**
	Neuronal NOS	**Endothelial NOS**	
Cellular location	Cytosolic (aqueous medium)	Particulate (membrane-bound)	Cytosolic (aqueous medium)
Ca^{2+}-Dependent	Yes	Yes	No

and glutamate. The binding of these agonists to appropriate receptors causes an increase in cellular calcium. This, together with the Ca^{2+}-binding protein calmodulin (CaM), activates the cNOS present in the cell to produce nitric oxide. It has also been observed that an increase in the shear stress in blood flow on the endothelium due to exercise can stimulate the synthesis of nitric oxide.

Experimental work has shown that the first step in the formation of nitric oxide is the synthesis of N^{ω}-hydroxy-L-citrulline as an enzyme-bound intermediate by a two-electron oxidation involving molecular oxygen, NADPH and CaM. This intermediate is converted to citrulline with the liberation of nitric oxide by an overall three-electron oxidation that also involves molecular oxygen, NADPH and CaM. The concentration of cellular L-arginine is maintained by the recycling of the L-citrulline to L-arginine. However, it is has been shown that a low local concentration of cellular L-arginine results in impairment of endothelium-dependent relaxation, which is alleviated by an infusion of L-arginine.

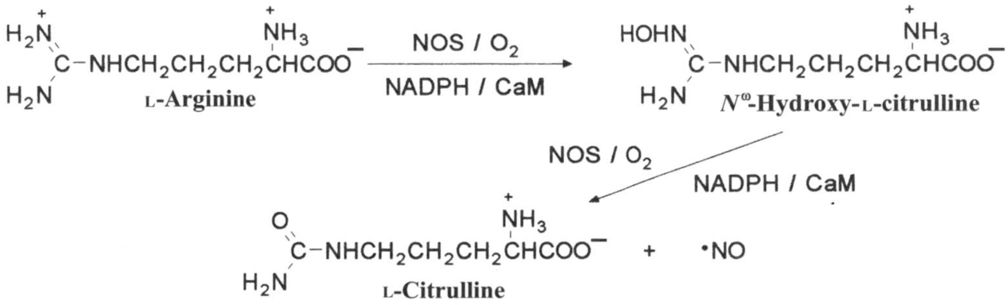

$$\mathrm{H_2\overset{+}{N}{=}C(NH_2){-}NHCH_2CH_2CH_2CH(\overset{+}{N}H_3)COO^-}\ \xrightarrow[\text{NADPH / CaM}]{\text{NOS / }O_2}\ \mathrm{HOHN{=}C(NH_2){-}NHCH_2CH_2CH_2CH(\overset{+}{N}H_3)COO^-}$$

L-Arginine → **N^{ω}-Hydroxy-L-citrulline**

$$\xrightarrow[\text{NADPH / CaM}]{\text{NOS / }O_2}\ \mathrm{O{=}C(NH_2){-}NHCH_2CH_2CH_2CH(\overset{+}{N}H_3)COO^-} + \bullet NO$$

L-Citrulline

iNOS is found in a wide variety of cells, such as mast cells, macrophages, Kupffer cells and neutrophils. It is also found in endotheial cells and vascular smooth muscle. Unlike cNOS it is not calcium dependent. However, it has been found that in macrophages iNOS calmodulin is tightly bound to the inducible enzyme and so probably plays a part in its action but possibly by a different mechanism to that found in other NOS enzymes. iNOS enzymes are activated by the presence of substances such as bacterial toxins, γ-interferon and interleukin-1β. Activation of iNOS results in the continuous production of a high concentration of nitric oxide.

These general forms of NOS reflect the two distinct general modes of action of nitric oxide. With cNOS the enzyme produces bursts of nitric oxide that transmit a message to the target cells without damaging those cells. With iNOS the enzyme is responsible for the continuous production of nitric oxide in sufficient concentration to damage and kill cells that may or may not be of benefit to the organism. For example, activated immune cells produce amounts of nitric oxide that are lethal to harmful target cells such as those found in cancers and invasive parasites but overproduction of nitric oxide has been linked to the death of pancreatic β-cells in insulin-dependent diabetes mellitus.

11.4.1 General Mode of Action

Nitric oxide has a short biological life undergoing rapid metabolism to nitrate, nitrite and other species (see sections 11.3.1 and 11.3.6). This short life means that nitric oxide is not able to diffuse any distance through the system before there is a significant decrease in its concentration. As a result, its targets must be close to its source and activated by the concentration of nitric oxide in their vicinity. For example, release of nitric oxide from the endothelium by the cNOS route is now known to cause a local relaxation of the underlying smooth muscle surrounding blood vessels, which results in a reduction of blood pressure.

The prevailing school of thought is that nitric oxide synthesised through the cNOS route is believed to act by binding to the iron in the haem unit (see section 11.3.6.1) that constitutes the active site of the enzyme, soluble guanylyl cyclase (soluble GC). This alters the conformation of the enzyme, which activates it to act as a catalyst. However, some workers believe that the nitric oxide is converted to an *S*-nitrosothiol and it is this compound that nitrosates the soluble GC (see section 11.5.1). Activation of the soluble GC enzyme, by either of these routes, results in the conversion of guanosine triphosphate (GTP) to cyclic guanosine monophosphate (cGMP) in the target cell. cGMP is known to interact with various proteins and in doing so changes their biological activity. For example, an increase in cGMP concentration has been shown to inhibit Na^+ channels of the kidney and to decrease the $[Ca^{2+}]$ in smooth muscle and platelets. Dissociation of the nitric oxide from the active ON–Fe^{II}–haem enzyme complex deactivates the guanylyl cyclase.

Biochemical effects
GTP → cGMP
NO + Soluble GC → ON - Fe - GC → Fe - GC + NO
Removed from the site of action
Endothelium → NO $\xrightarrow{RSH}$ RSNO

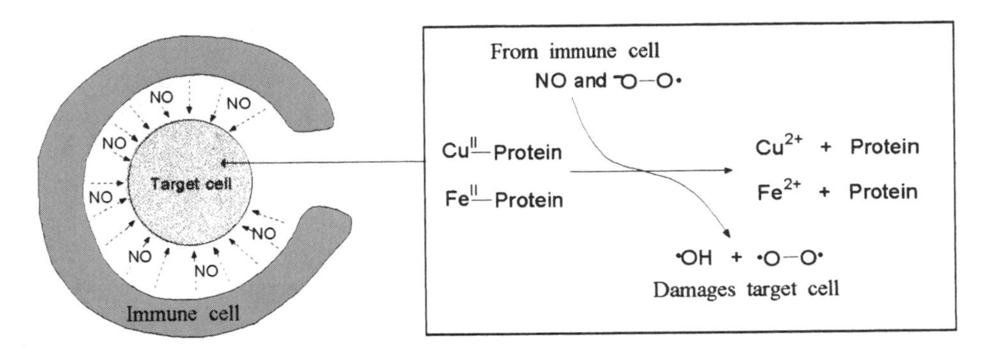

Figure 11.9. A schematic representation of the action of nitric oxide as a killer molecule.

The nitric oxide released by the cNOS route that does not bind to a haem target area may take part in nitrosation reactions (see section 11.3.5.1) or react with thiols (see section 11.3.4) to form nitrosothiols that decompose to release nitric oxide. It has been suggested that nitrosothiols, such as nitrosocysteine, may act as a depot for nitric oxide, thereby prolonging its action. Furthermore, it has also been suggested that nitric oxide binds to the thiol groups of mammalian albumin and is transported in the plasma.

Nitric oxide generated by the immune system through the iNOS pathway acts as a killer molecule. The high concentration of nitric oxide (nanomoles) produced by this process causes lethal oxidative injuries to the target cells, such as cancer and parasite cells. Little is known about this process but it is now thought that the nitric oxide reacts in conjunction with superoxide, which is also produced by activated immune system cells. The immune cells increase their surface area and fold around their target cells or microorganisms. Once in position they release nitric oxide, which attacks the copper and iron-complexed proteins in the target cell, liberating copper and iron ions from these proteins (Figure 11.9). This is accompanied by the formation of hydroxyl free radicals and molecular oxygen that cause massive oxidative injury to the target cell. It is thought that this may also be the mechanism of cellular injury in some pathological conditions.

11.4.2 Suitability of Nitric Oxide as a Chemical Messenger

Nitric oxide is a unique chemical messenger. All the other chemical messengers transmit information by means of their shape and ability to bind to a receptor (see sections 8.2 and 8.6). Nitric oxide does not depend on its shape to transmit information. Its action appears to be due to its redox reactivity. Furthermore, unlike other messenger molecules, nitric oxide is not stored *in situ* and released under a specific stimulation but it is synthesised as required. In other words, unlike classical messengers it is synthesised on demand and then diffuses to its target. Verma and collegues have suggested that carbon monoxide may also be a messenger of this type.

Nitric oxide is ideally suited as a locally acting chemical messenger in spite of its toxic nature because:

(i) it is soluble in both water (about 2 mmol dm^3 at 20°C) and lipids;
(ii) its small size enables it to cross cell membranes as easily as oxygen and carbon dioxide;
(iii) it diffuses faster in water than oxygen and carbon dioxide;
(iv) it is a short-lived species in biological systems, reacting with oxygen and other biological molecules;
(v) its short life in biological systems results in a decrease in nitric oxide concentration as distance from its source increases. This decreasing concentration gradient results in its 'message' being localised;
(vi) of its reactivity and the ease of formation of ON–Fe^{II}–haem complexes;
(vii) of the reversibility of ON–Fe^{II}–haem complex formation, which enables guanylyl cyclase to switch off after the nitric oxide is removed by the system.

11.4.3 Metabolism

The main metabolic route for nitric oxide is diffusion into the blood where it reacts with oxyhaemoglobin to form methaemoglobin and nitrate and small amounts of nitrite (see section 11.3.8.2). The methaemoglobin is converted back to haemoglobin by reductases whereas the nitrate is transferred to the serum and excreted in the urine. Small amounts of nitrate are thought to be reduced by bacterial action in the intestinal tract and be exhaled as ammonia and nitrogen.

A minor route for the metabolism of nitric oxide is oxidation to nitrogen dioxide (see section 11.3.1.1). A more important process is the reaction with superoxide to form peroxynitrite, which has been implicated in some pathological conditions (see section 11.3.1.2).

The higher concentration of nitric oxide produced by iNOS activity leads to a very much higher concentration of nitrate in the urine. However, this not a reliable indicator of the involvement of endogenous nitric oxide formation in the action of the immune system in man because nitrate is also produced from external sources. For example, nitric oxide is inhaled from the atmosphere and tobacco smoke and nitrate is also produced from nitrite used as a preservative in food.

11.5 The Role of Nitric Oxide in Physiological and Pathophysiological States

Nitric oxide is both a chemical messenger and a cyctotoxic agent, the former being initiated by cNOS and the latter by iNOS. In its chemical messenger mode it acts as an initiator for biological processes that are essential for a healthy organism. However, in its cytotoxic mode it can be both beneficial and harmful. Nitric oxide forms part of the immune system but it is also believed to be the toxic agent that initiates tissue damage involved in some pathological conditions. This section sets out to survey the role of nitric oxide in some of the many normal physiological and pathophysiological situations in which it is found.

11.5.1 The Role of Nitric Oxide in the Cardiovascular System

The cardiovascular system is the network of blood vessels, including the heart, through which blood flows to all parts of the body. In 1980 it was shown by Furchgott and Zawadzki that removal of the endothelium prevented the relaxation effect of acetylcholine on blood vessels. This lead to the discovery that stimulation of the endothelial cells resulted in the release of a substance that Furchgott called endothelium-derived relaxing factor (EDRF). This was followed by the discovery that many vasoactive substances were found to release EDRF from endothelial cells. The biological action and chemical properties of EDRF were found to resemble those of nitric oxide and so in 1987 it was proposed independently by Furchgott and Ignarro that EDRF was nitric oxide. Later work by Moncada and co-workers indicated that nitric oxide was responsible for the biological activity of EDRF. As a result, a number of workers have suggested that EDRF is a nitrosothiol (see section 11.4.1) produced by the action of nitric oxide on a suitable thiol. However, in spite of the controversy about EDRF it has become apparent that nitric oxide plays an important role in both the healthy and the unhealthy states of the cardiovascular system. Evidence suggests that the generation of too little nitric oxide can result in hypertension (high blood pressure), angina and impotence, whereas the synthesis of too high a concentration of nitric oxide is thought to be related to circulatory shock, inflammation and strokes.

Nitric oxide generated in the endothelium targets the underlying smooth muscle cells, causing these cells to relax. This dilates (widens) the blood vessel, allowing an increased flow rate that reduces blood pressure. It has also been suggested that nitric oxide reduces the adhesion and aggregation of platelets and leucocytes, which would also help to increase the flow of blood through the vessel. Therefore, it appears that the role of nitric oxide produced by cNOS synthesis in a healthy endothelium is to control and maintain a healthy blood flow. If this is so, damage to the endothelium, which results in endothelium dysfunction, could be responsible for some cardiovascular diseases.

Endothelium dysfunction has been shown to be caused by a genetic defect or acquired as a result of a poor diet, smoking or sedentary life style. It appears likely that endothelium dysfunction is due to an increase in superoxide production in the endothelium. This superoxide reacts with any available nitric oxide to form peroxynitrite (see section 11.3.1.2). The reaction reduces the amount of nitric oxide diffusing to the smooth muscle cells. Consequently, these cells are less likely to relax, which could lead to an increase in blood pressure and further endothelium damage. This effect is compounded by the recent discovery that endothelin, the endogenous vasoconstrictor produced by the endothelium to balance the vasodilating effect of nitric oxide, has the effect of increasing blood pressure. Furthermore, the formation of the highly toxic peroxynitrite ion could result in tissue-damaging reactions.

The formation of a large concentration of nitric oxide leads to circulatory shock (hypotension or reduced blood pressure) due to excessive vasodilation. Experimental evidence suggests that these large doses of nitric oxide are produced by the iNOS route and are also associated with the lowering of blood pressure that occurs with endotoxin, haemorrhagic and septic shock. Septic shock is also accompanied by inflammation. Consequently, iNOS inhibitors are a potential source of drugs to treat these conditions.

Experimental evidence indicates that nitric oxide may act as a chemical messenger in the function of the lungs, possibly acting as a regulator in pulmonary circulation by acting on pulmonary smooth muscle. Poor nitric oxide synthesis has been associated with hypoxic pulmonary vasoconstriction, and increased levels of nitric oxide are exhaled by asthmatics. Inhalation of nitric oxide has been used successfully in the experimental treatment of acute pulmonary hypertension. However, this treatment could cause a toxic reaction in cases where the patient has an active lung infection. The presence of such an infection could stimulate the natural production of nitric oxide (see section 11.4.1) and this, together with the inhaled dose, could result in the nitric oxide concentration reaching toxic levels.

11.5.2 The Role of Nitric Oxide in the Nervous System

A cNOS enzyme has been isolated from rat brain. It has been found to be very similar to endothelium cNOS in that it is calmodulin dependent (see section 11.4). The enzyme is mainly located in the neurons of the granule cell layer of the cerebellum. Little cNOS was found in the remainder of the brain's neurons. *In vitro* experimental evidence has indicated that this brain cNOS is activated by glutamate stimulation and that the nitric oxide produced appears to target Purkinje cells. Purkinje cells have a high concentration of guanylyl cyclase and stimulation of nitric oxide production by glutamate increases the concentration of cGMP in these cells. NOS has also been found throughout the central nervous system and the peripheral nervous system. It exists exclusively in neurons but, unlike its occurrence in the brain, no organised pattern of distribution has been observed and only a small percentage of the neurons contain NOS.

The function of nitric oxide in the brain is not clear. Evidence suggests it could act as a neurotransmitter with its action initiated by glutamate. It has been suggested that the glutamate is released from the presynaptic nerve terminal by exocytosis, triggered by a nerve signal. It diffuses across the synaptic gap to the postsynaptic nerve terminal where it interacts with *N*-methyl-D-aspartate receptors (NMDA receptors). These receptors are coupled to calcium channels and their activation allows calcium to flow into the postsynaptic nerve terminal. This calcium combines with calmodulin to activate cNOS, which converts L-arginine to L-citrulline

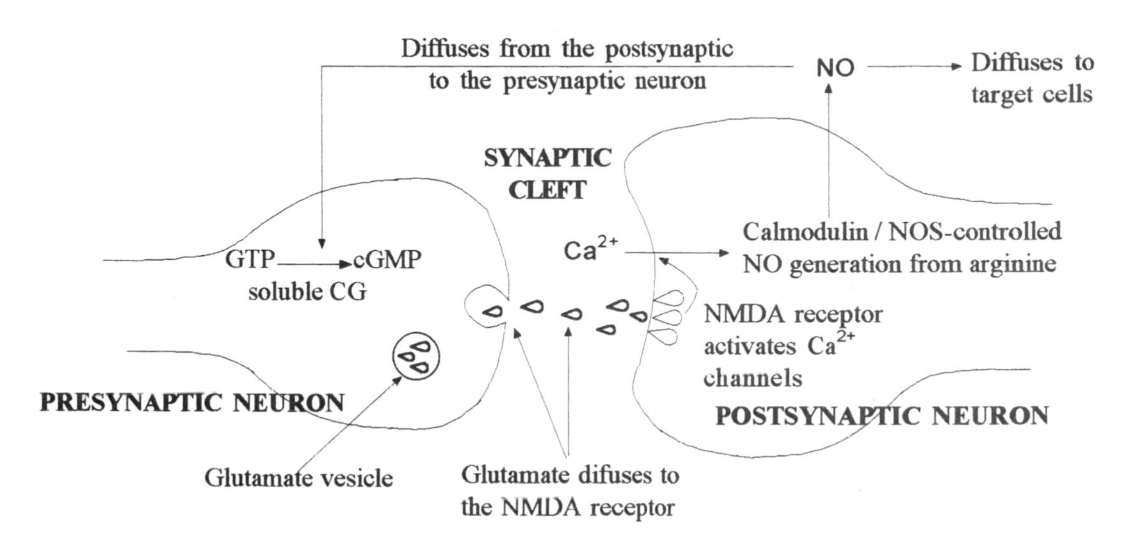

Figure 11.10. A schematic outline of the mechanism of the action of nitric oxide as a neurotransmitter. The cGMP produced in the presynaptic neuron by the action of NO initiates phosphorylation and other biochemical processes.

and nitric oxide (Figure 11.10). It is proposed that the nitric oxide formed either diffuses to nearby target cells or may react with specific thiols in the neuron to form *S*-nitrosothiols, which are stored in vesicles until released by a voltage-dependent mechanism. One target of the nitric oxide, irrespective of whether it is synthesised directly or released from an *S*-nitrosothiol, appears to be the presynaptic nerve, where it stimulates the conversion of GTP to cGMP. The purpose of this conversion is not known.

The sustained release of high concentrations of nitric oxide has been linked to strokes as well as possibly causing other forms of cell damage and death. In strokes, cells lose their ability to exclude Ca^{2+} ions. This means that the calmodulin in the cell will remain active (see section 6.4.2). Because cNOS is Ca^{2+} ion dependent, the presence of this Ca^{2+}-activated calmodulin will switch on cNOS, resulting in the production of large amounts of nitric oxide. It has been suggested that this nitric oxide combines with superoxide to form peroxynitrite and it is this powerful oxidising agent that destroys cell membranes, either by direct reaction with components of the membrane or by producing hydroxy and other free radicals that attack cell membranes.

Experimental evidence also suggests that nitric oxide can act as a neuromediator and that the controlled release of nitric oxide in low concentration is part of the normal function of the brain. It has been suggested that it could be a mediator for blood flow and neuronal activity and have an influence on the underlying mechanisms of long-term memory and depression. In addition, it has also been suggested that one role of the nitric oxide produced in the brain could be to protect some neurons against oxidative damage.

Nitric oxide acts as a transmitter in the peripheral nervous system of the urogenital and gastrointestinal tracts. It appears to play a major part in gastric dilation and maintaining the compartmentalisation of the gastrointestinal tract. Its absence has been linked to infantile pyloric stenosis and male impotence (see section 11.5.4).

11.5.3 Nitric Oxide and Diabetes

Destruction of pancreatic β-cells is known to occur in insulin-dependent diabetes mellitus (IDDM or type I diabetes) The loss of 80% or more of these cells results in insulin deficiency, which reduces the body's ability to control blood glucose. Cell death is believed to be caused by a variety of endogenous agents and also the autoimmune system. The mechanism by which these agents act is not known and is the subject of much controversy. However, it was thought that nitric oxide may be involved both in normal cell operation and IDDM. The main evidence for this deduction is summarised as follows:

(i) cNOS has been found in pancreatic β-cells.
(ii) Arginine stimulates the release of insulin from pancreatic β-cells in the presence of glucose. This process is inhibited by the nitric oxide inhibitor L-NMMA.
(iii) It had been shown that the nitric oxide inhibitors N^G-monomethyl-L-arginine (NMMA) and N^G-nitro-L-arginine methyl ester (NAME) reduced or prevented the inhibition of insulin secretion by interleukin 1 (IL-1), one of the agents that are known to prevent insulin secretion.
(iv) It has been demonstrated that IL-1 stimulates the formation of nitrite and cGMP: the former is a metabolite of nitric oxide whereas cGMP is produced when nitric oxide targets cells.
(v) IL-1β appears to induce the formation of iron–nitrosyl complexes. This would implicate nitric oxide involvement, especially as complex formation was prevented by NMMA.
(vi) Patients with IDDM have been shown to have impaired endothelium-dependent vasodilation.

In vitro experimental evidence has indicated that the overproduction of nitric oxide may destroy pancreatic β-cells during the development of IDDM. It has been proposed by Corbett and McDaniel that IL-1 released from macrophages binds to specific receptors on the pancreatic β-cells. This activates tyrosine kinase, which, through a series of intermediaries, activates the expression of iNOS. The high concentration of nitric oxide produced by this enzyme inhibits the activity of other essential enzymes by interaction with their iron–sulphur centres. This brings about cell dysfunction and ultimately cell distruction. Although this proposal has yet to be proved, specific iNOS inhibitors could be of value in preventing the development of IDDM.

11.5.4 Nitric Oxide and Impotence

NOS has been located in the adventitial layer of the penile arteries. Release of nitric oxide has been shown to cause a dose-dependent rapid relaxation of the human corpus cavernosium with a subsequent penile erection. NOS inhibitors have been shown to prevent this relaxation whereas nitric oxide sources mimic the NOS effect. Consequently, it appears that nitric oxide is an important mediator in penile erection. This erection is believed to be due to the action of cGMP formed from GTP by the action of guanylyl cyclase activated by nitric oxide (see section 11.4.1). Biological activity is terminated by the cGMP being converted to 5′-guanosine monophosphate (5′-GMP) by the action of phosphodiesterase type 5 (PDE5). Consequently, it is possible that impotence could be treated by either injections of nitric oxide donors into the corpus cavernosium or inhibition of PDE5. The latter approach lead to the discovery of Sildenafil (Viagra). Sildenafil inhibits the action of PDE5, which prevents the deactivation of cGMP and as a result the termination of the physiological response. The drug is effectively compensating for low nitric oxide production by allowing cGMP to accumulate.

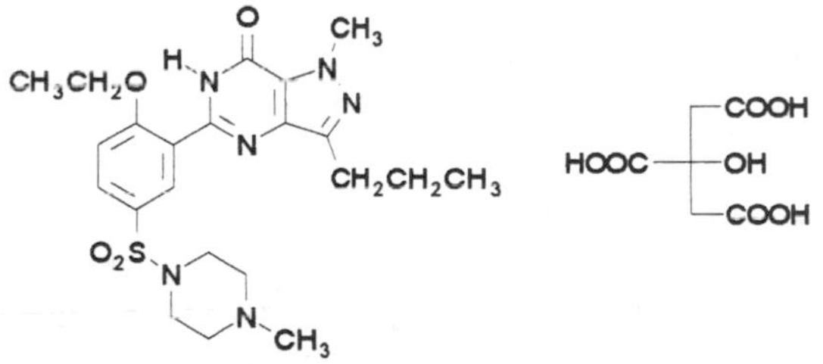

Sildenafil citrate

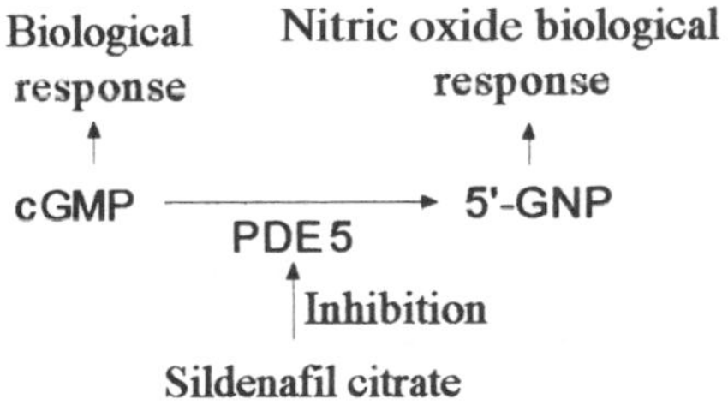

Biological response ← cGMP —PDE5→ 5'-GNP → Nitric oxide biological response

PDE5 ← Inhibition — Sildenafil citrate

It is likely that impotence in chronic diabetic men is due to a failure of nitric oxide synthesis.

11.5.5 Nitric Oxide and the Immune System

The immune system is an organism's natural defensive system against pathogens (microorganisms and viruses). In mammals immunity is due to the different members of a group of white cells known as lymphocytes. These cells are produced in the bone marrow and effectively police the body by being able to move through the intercellular spaces as well as the bloodstream. They operate (immune response) in two general ways:

(i) cellular immunity: the defensive mechanism is triggered by the presence of foreign antigens but without the immediate production of antibodies. It is mediated by T lymphocytes or T cells.
(ii) humoral immunity: B lymphocytes or B cells produce antibodies in response to a foreign antigen such as a foreign macromolecule, carbohydrate, nucleic acid or protein. These antibodies trigger a defensive mechanism that destroys the invading foreign antigen.

The cellular immune response is triggered when a macrophage envelopes and partly decomposes a foreign antigen. The decomposition products are displayed on the surface of the macrophage bound to surface proteins known as major histocompatibility complex (MHC) proteins. These structures are recognised by cytotoxic T cells that cause the macrophage to release IL-1, which stimulates the production of large numbers of cytotoxic T cells. These killer T cells bind specifically to the foreign antigen, releasing perforin, a protein that lyses the target cell.

Humoral immunity is triggered when a foreign antigen binds to an immunoglobin displayed on the surface of the B cell. The cell responds by engulfing and partly decomposing the foreign antigen. These partial decomposition products are displayed on its surface in the form of complexes with Class II MHC protein found on the surface of B cells. This stimulates mature helper T cells to bind to the B cell, causing that cell to divide and produce large numbers of specialist plasma cells. These cells secrete antibodies that are specific for the foreign antigen. The antibodies bind to the foreign antigen and effectively label that antigen for destruction by either ingestion by phagocytes (phagocytosis) or activation of the complement system.

In vitro studies in 1987 by Hibbs, Vavrin and Taintor showed that cytotoxic activated macrophages (CAM) required L-arginine for their activity. These workers also observed that the L-arginine was metabolised to L-citrulline, nitrite and nitrate. Furthermore, NOS inhibitors prevented the formation of L-citrulline, nitrite and nitrate as well as preventing the cytotoxic action of CAM. Electron paramagnetic resonance spectroscopy has shown that the action of CAMs is accompanied by the formation of nitrosyl–iron complexes. Furthermore, superoxide, which readily reacts with nitric oxide (see section 11.3.1.2), inhibits the cytotoxic action of CAMs. These observations indicated that nitric oxide is involved in the operation of the immune system. It is now believed that one cytotoxic mechanism employed by macrophages involves engulfing the target cell and flooding it with nitric oxide (see section 11.4.1).

11.6 Therapeutic Possibilities

Nitric oxide is involved in both physiological and pathophysiological conditions. The pathophysiological effects can be divided conveniently into two types: those due to a lack of nitric oxide and those due to an excess of nitric oxide. Consequently, the main approaches to drug design are based on producing compounds that prevent the overproduction of nitric oxide or act as a source of nitric oxide. Gene manipulation is also being investigated as a means of controlling nitric oxide production. This section outlines only some of the approaches being followed in these areas.

11.6.1 Compounds that Reduce Nitric Oxide Generation

Reduction of nitric oxide production can be achieved by blocking the action of NOS or its activating processes, such as calcium ingress. Because iNOS is frequently associated with pathophysiological states, inhibitors that are specific for iNOS are of particular interest.

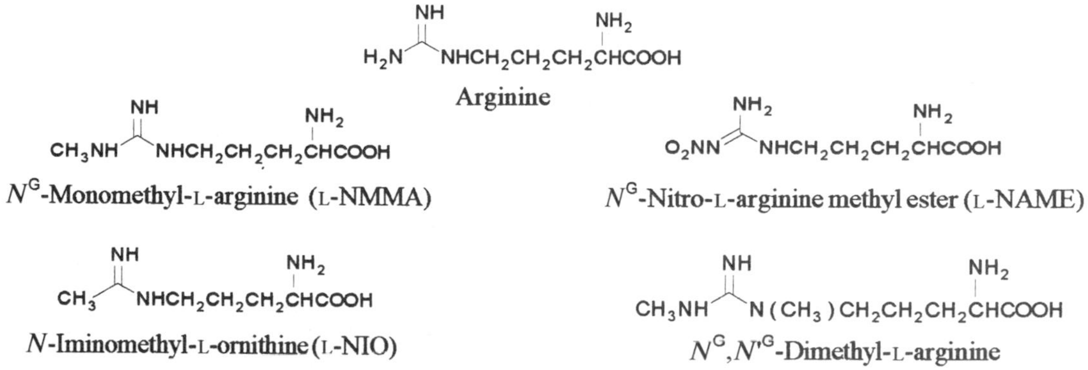

$H_2N(C{=}NH)NHCH_2CH_2CH_2CH(NH_2)COOH$

Arginine

$CH_3NH(C{=}NH)NHCH_2CH_2CH_2CH(NH_2)COOH$

***N*^{G}-Monomethyl-L-arginine (L-NMMA)**

$O_2NN{=}C(NH_2)NHCH_2CH_2CH_2CH(NH_2)COOH$

***N*^{G}-Nitro-L-arginine methyl ester (L-NAME)**

$CH_3(C{=}NH)NHCH_2CH_2CH_2CH(NH_2)COOH$

***N*-Iminomethyl-L-ornithine (L-NIO)**

$CH_3NH(C{=}NH)N(CH_3)CH_2CH_2CH_2CH(NH_2)COOH$

***N*^{G},*N*$'^{G}$-Dimethyl-L-arginine**

Figure 11.11. Examples of arginine analogues used to block NOS activity.

The obvious line of investigation is based on arginine and its analogues. A number of these analogues have been found to inhibit the formation of nitric oxide by acting as NOS blocking agents (Figure 11.11). These inhibitors have been used extensively to investigate the action of nitric oxide. *N*-Monomethyl-L-arginine (L-NMMA) has been found to increase blood pressure in man and other species whereas *N*-iminomethyl-L-ornithine (L-NIO) is an irreversible inhibitor of NOS in activated macrophages. However, none of these compounds are in general clinical use.

Several simple guanidino compounds (Figure 11.12) have also been found to inhibit nitric oxide synthesis. It is believed that these compounds inhibit nitric oxide synthesis by preventing the second stage of the oxidation of arginine (see section 11.4).

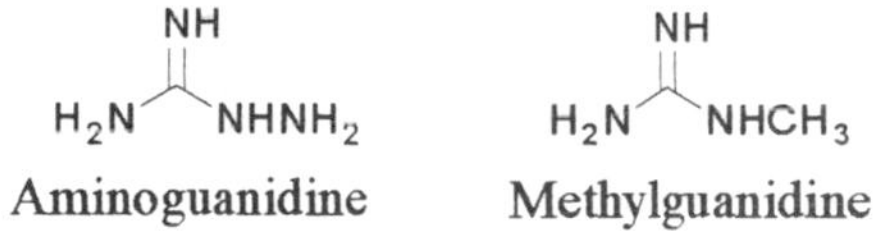

$H_2N(C{=}NH)NHNH_2$ — **Aminoguanidine**

$H_2N(C{=}NH)NHCH_3$ — **Methylguanidine**

Figure 11.12. Examples of guanidine derivatives used as NOS inhibitors.

Aminoguanidine has been shown to be a selective inhibitor for iNOS in animal models. It has a minimal effect on the cNOS that is required to maintain blood pressure. The selectivity of aminoguanidine is believed to be due to the presence of the hydrazine residue because replacement of this moiety by a methyl group which has similar overall shape and size, resulted in the loss of selectivity and a considerable loss of activity.

11.6.2 Compounds that Supply Nitric Oxide

Sodium nitroprusside and organic nitrates and nitrites (Figure 11.13) have been used for over 100 years to treat angina. Glycerol trinitrate has also been used to relieve impotence.

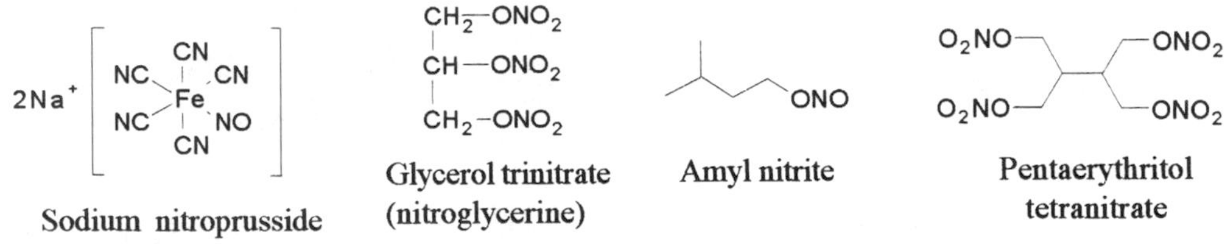

2Na$^+$ [Fe(CN)$_5$NO] — Sodium nitroprusside

CH_2-ONO_2 / $CH-ONO_2$ / CH_2-ONO_2 — Glycerol trinitrate (nitroglycerine)

ONO — Amyl nitrite

O_2NO, O_2NO, ONO_2, ONO_2 — Pentaerythritol tetranitrate

Figure 11.13. Examples of compounds used to treat cardiovascular diseases.

It has been demonstrated that sodium nitroprusside and organic nitrates and nitrites act by forming either nitric oxide or a nitric oxide adduct during their metabolism (Figure 11.14). The metabolic pathways of these drugs appear to be catalysed by enzymes that are specific for each drug. This specific nature would account for the wide diversity of pharmacological action of each of these drugs.

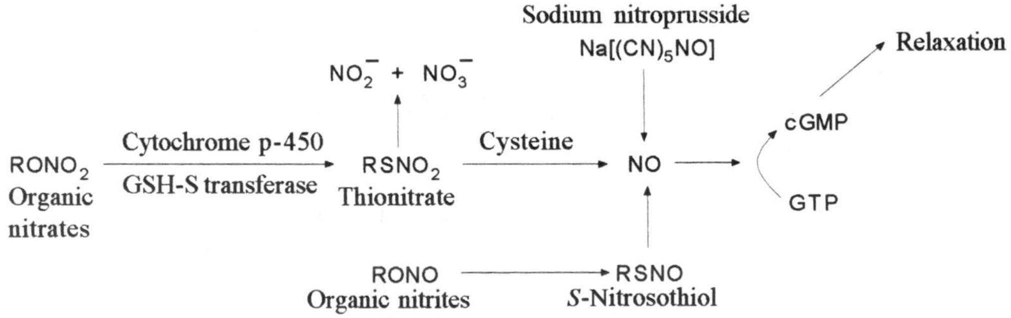

Sodium nitroprusside $Na[(CN)_5NO]$ → NO

$RONO_2$ Organic nitrates —(Cytochrome p-450, GSH-S transferase)→ $RSNO_2$ Thionitrate → $NO_2^- + NO_3^-$

$RSNO_2$ —(Cysteine)→ NO → (GTP → cGMP) → Relaxation

RONO Organic nitrites → RSNO *S*-Nitrosothiol → NO

Figure 11.14. A proposed biochemical pathway for various NO donors.

A knowledge of the chemistry and biochemical pathway of nitric oxide has resulted in several groups of compounds being investigated as leads to new drugs. For example, the suggestion that EDRF is an *S*-nitrosothiol has resulted in the investigation of a number of these compounds as potential drugs. *S*-Nitrosocaptopril, *S*-nitroso-*N*-acetylcysteine and *S*-nitroso-*N*-acetylpenicillamine have all been shown to have vasodilator properties in animals and may have some use in humans.

NONO-ates (see section 11.3.3) are also being investigated as potential sources of drugs. These compounds are prepared by the direct action of nitric oxide on a nucleophile.

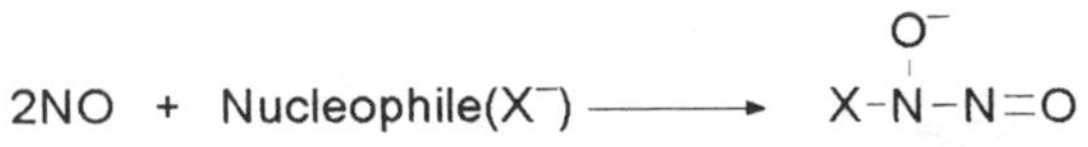

$$2NO + \text{Nucleophile}(X^-) \longrightarrow X{-}N(O^-){-}N{=}O$$

A NONO-ate

They are stable solids that spontaneously decompose in water. The rate of decomposition depends on the temperature, pH and the nature of the nucleophilic residue X. Because the rate of release of nitric oxide depends on the nature of X, NONO-ates could be useful as slow-release drugs. Furthermore, the spontaneous nature of the generation of nitric oxide means that the release of nitric oxide *in vivo* would depend on the chemical nature of the NONO-ate rather than the intervention of another biological process such as a redox system.

Sydnomines are a group of compounds used to treat angina. Molsidnomine is metabolised in the liver to 3-morpholino-sydnomine (SIN-1), which spontaneously releases nitric oxide under aerobic conditions. SIN-1 does not produce nitric oxide under anaerobic conditions, which suggests that the intervention of a redox system is necessary for nitric oxide release.

Liver esterases; ^{-}OH; O_2 → O_2^-; NO

Molsidnomine (SIN) (SIN-1) (SIN-1A)

11.6.3 The Genetic Approach

The cloning of a gene can enable researchers to understand the nature of the control of the expression of that gene (see section 10.15.1). This information is the starting point for the development of compounds that can either block or enhance the action of a specific gene. Each of the different isoforms of NOS is produced by a different gene. Consequently, control of the relevant gene would influence the production of the relevant NOS isoform and the subsequent generation of nitric oxide produced by that isoform. This should enable medicinal chemists to design specific NOS inhibitors and stimulants. A number of NOS enzymes have been cloned from a variety of sources (Table 11.3) but no compounds have yet been developed for clinical use.

11.7 Summary

Nitric oxide is a **chemical messenger** and **cytotoxic agent**. It is a stable free radical with one unpaired electron. Nitric oxide acts as a free radical and an electrophile, is oxidised and

Table 11.3. Sources of cloned NOS.

Type of NOS (human)	Sources (human)	Type of NOS	Sources (other)
Neuronal cNOS	Human cerebellum	Neuronal cNOS	Rat cerebellum
Endothelial cNOS	Human endothelial cells	Endothelial cNOS	Bovine endothelial cells
iNOS	Human hepatocytes	iNOS	Murine macrophages
		iNOS	Rat vascular smooth muscle

reduced, and forms salts and complexes. It reacts with **oxygen** to form nitrogen dioxide, which reacts with **water** under aerobic conditions to form mainly nitrite. Nitric oxide reacts with **superoxide** to form peroxynitrite, which is related to a number of pathological conditions. **Salts** containing the nitrosonium ion (NO^+) are well characterised but nitrosyl salts (NO^-) are less well documented. Nitric oxide acts as an **electrophile**, reacting with thiols, amines and other nucleophiles.

$$RSH + NO \rightarrow RSNO \text{ (}S\text{-nitrosothiols)}$$
$$R_2NH_2 + NO \rightarrow R_2NHNONO \text{ (NONO-ates)}$$

In basic conditions nitric oxide **oxidises** thiols to disulphides, nitrogen and nitrous oxide.

Nitric oxide forms **complexes** with metals in which the ligand can bond to the metal in three distinct ways:

(i) a linear complex with an M–N–O bond angle of approximately 160–180°;
(ii) bent complexes with an M–N–O bond angle of approximately 120–140°;
(iii) complexes in which the nitric oxide acts as a bridge between two metal ions.

These **complexes**, whose structures contain nitric oxide ligands, are formed by the reaction of nitric oxide, nitrosonium salts, acidified nitrite and organic nitrites with other metal complexes. The nitrogen atom in **linear nitric oxide complexes** can act as an **electrophile** because this atom is electron deficient. The nitrogen atom in **bent** and **bridge nitric oxide complexes** can act as a **nucleophile** because this atom has a lone pair of electrons that can act as a nucleophilic centre.

Nitric oxide forms complexes with both Fe(II) and Fe(III) bound to proteins and other biological molecules. It reacts with metmyoglobin to form a complex that nitrosates a wide variety of N, S and O nucleophiles with the formation of an $ON–Fe^{II}$–haem complex. Nitric oxide targets intercellular iron and will also displace ligands from iron–haem complexes. Simple chemical species, such as **nitrogen dioxide**, **dinitrogen trioxide**, **nitrite** and **peroxynitrite**, are also biologically active. Nitrogen dioxide acts as a free radical and as a nitrating and oxidising agent. Dinitrogen trioxide is unstable, decomposing to a mixture of nitric oxide and nitrogen dioxide. It is a strong oxidising and nitrosating agent. Nitrite is a good nitrosating agent and also reacts with oxyhaemoglobin to form methaemoglobin, nitrogen dioxide and hydrogen peroxide. Peroxynitrite is a strong oxidising agent that reacts with SOD to form the nitronium ion, which is a strong nitrating agent. Because all these chemical species exhibit chemical properties similar to those of nitric oxide, it can be difficult to identify the precise species responsible for a biological process.

Nitric oxide is produced *in vivo* by the catalytic oxidation of L-arginine to L-citrulline by a family of enzymes known as **nitric oxide synthases (NOS)**. The process requires CaM, NADPH and other cofactors. Two general types of NOS enzymes have been identified. **Constitutive**

nitric oxide synthase (cNOS), which is calcium dependent, produces nitric oxide in low concentration and short bursts. **Inducible nitric oxide synthase (iNOS)** is not calcium dependent. Activation of iNOS results in the continuous production of nitric oxide in a high concentration.

Nitric oxide has a short *in vivo* life and so it is believed that its targets must be close to its source. However, some workers think that it can be transported some distance from its source in the form of *S*-nitrosothiols. It is believed that nitric oxide activates **soluble guanylyl cyclase (soluble-GC)**, which catalyses the conversion of guanosine triphosphate (GTP) to cyclic guanosine monophosphate (cGMP), which triggers further biochemical effects that ultimately result in a physiological response. The activation of soluble guanylyl cyclase is also believed to occur by the action of *S*-nitrosothiols. Nitric oxide released into the blood is rapidly metabolised by reaction with oxyhaemoglobin to form methaemoglobin.

Nitric oxide usually acts as a **chemical messenger** when its production is initiated by cNOS. It normally initiates vasodilation in the cardiovascular system. However, the generation of large quantities of nitric oxide can result in circulatory and septic shock. Conversely, the production of low levels of nitric oxide has been linked to male impotence. Nitric oxide acts as a **cytotoxic agent** when it is generated by the action of iNOS. For example, the release of high concentrations of nitric oxide in the brain has been linked to strokes. In addition, it has been suggested that nitric oxide is responsible for the distruction of pancreatic β-cells in insulin-dependent diabetes mellitus. One of the routes by which cytotoxic activated macrophages of the immune system are believed to act is by engulfing the target cell and flooding it with nitric oxide.

The main approaches to the design of new drugs for the control of nitric oxide are:

(i) finding inhibitors for NOS, especially iNOS;
(ii) producing compounds that will spontaneously release nitric oxide;
(iii) finding substances that enhance or block the action of the various genes producing the NOS isoforms.

11.8 Questions

(1) Name each of the following compounds: (a) N_2O_3; (b) ONOO; (c) $ONSCH_2CH(NH_2)COOH$; and (d) $(C_2H_5)_2NNONO$.
(2) Predict how nitric oxide might be expected to react with each of the following: (a) oxygen; (b) nitrogen dioxide; (c) diethyl amine; (d) ethanethiol in the presence of ferrous ions; and (e) a mixture of metmyoglobin and glycine.
(3) Explain the meaning of the terms nitrosation and nitration. Illustrate the answer by reference to *in vivo* nitrosation and nitration of the tyrosine residues of proteins.

(4) Outline the biological significance of the reaction of nitric oxide with intercellular iron.
(5) Explain, outlining the necessary chemical evidence, why the major route for the deactivation of nitric oxide is not its reaction with oxygen.
(6) Describe the fundamental differences in the action of iNOS and cNOS with respect to nitric oxide production.
(7) What general types of biological action are associated with iNOS and cNOS.
(8) Describe, by means of equations and notes, how nitric oxide is generated in the endothelium. Give details of any intermediates and cofactors that are involved in the process.
(9) What enzyme does nitric oxide activate in endothelium smooth muscle cells? Show, by means of a chemical equation(s), the chemical process catalysed by this enzyme. Give one physiological result of this process.
(10) Suggest three general chemical approaches that could be used to deal with the pathological effects of nitric oxide. Illustrate the answer by reference to classes of compounds that are either used as drugs or have a potential drug use.
(11) List the evidence that suggests that nitric oxide is involved in insulin-dependent diabetes mellitus.
(12) Discuss the importance of *S*-nitrosothiols in the biological action of nitric oxide.

12 An Introduction to Organic Drug and Analogue Synthesis

12.1 Introduction

This chapter is intended to introduce some of the strategies used in the design of synthetic pathways. These pathways may be broadly classified as either **partial** or **full** synthetic routes. Partial synthetic routes are a combination of traditional organic synthesis and other methods. However, these routes tend to be more concerned with the large-scale production of proven drugs rather than the synthesis of lead compounds. They utilise processes such as fermentation and the extraction of starting materials from animals and plants. For example, fermentation is used to produce benzylpenicillin, which is used as the starting point for the manufacture of a number of penicillins (see section 12.3.3.1), whereas pig insulin is used as the starting material for the production of human insulin. These examples demonstrate that partial synthetic routes often utilise compounds with fairly complex structures as their starting point. In contrast, full organic syntheses start with readily available compounds, both synthetic and naturally occurring, but only utilise the standard methods of organic synthesis to produce the desired product.

12.2 Some General Considerations

12.2.1 Starting Materials

The choice of starting materials is important in any synthetic route. Commonsense dictates that they should be chosen on the basis of what will give the best chance of reaching the desired product. However, in all cases the starting materials should be cheap and readily available.

12.2.2 Practical Considerations

The chemical reactions selected for the proposed synthetic pathway will obviously depend on the structure of the target compound. However, a number of general considerations need to be borne in mind when selecting these reactions:

(i) The yields of reactions should be high. This is particularly important when the synthetic pathway involves a large number of steps.
(ii) The products should be relatively easy to isolate, purify and identify.
(iii) Reactions should be stereospecific because it is often difficult and expensive to separate enantiomers. However, the exclusive use of stereospecific reactions in a synthetic pathway is a condition that is often difficult to satisfy.
(iv) The reactions used in the research stage of the synthesis should be adaptable to large-scale production methods. The reactions used by research workers frequently use expensive exotic reagents and it is the job of pharmaceutical development chemists to find simpler cost-effective alternatives.

12.2.3 The Overall Design

All approaches are based on a knowledge of the chemistry of functional groups and their associated carbon skeletons. The design may result in either a **linear synthesis** where one step in the pathway is immediately followed by another:

$$A \rightarrow B \rightarrow C \rightarrow D \rightarrow \text{etc. to the target molecule}$$

or a **convergent synthesis** where two or more sections of the molecule are synthesised separately before being combined to form the target structure (see section 12.4.2). In both cases, the **disconnection** approach (see section 12.4.1) may be used to design the pathway and identify suitable starting materials. Alternative design strategies that can also be usefully employed to design a synthesis are:

(i) finding compounds with similar structures to the target molecule and modifying their synthetic routes, if known, to produce the target compound;
(ii) modifying natural products whose structures contain the main part of the target structure (see section 12.5).

An important aspect of all medicinal chemistry synthetic pathway design is divergency. Ideally, the chosen route should be such that it is relatively easy to modify the structure of the lead compound, either directly or during the course of its synthesis. This is an economic way of producing a greater range of analogues for testing and hence increasing the chance of discovering an active compound. Initially these modifications would normally take the form of changing the nature of side chains or introducing new substituents in previously unsubstituted positions. The synthetic pathway for the preparation of the lead compound should include stages where it is possible to introduce these new side chains and substituents. For example, the presence of an amino group in a structure opens out the possibility of introducing different side chains by N-acylation (Figure 12.1).

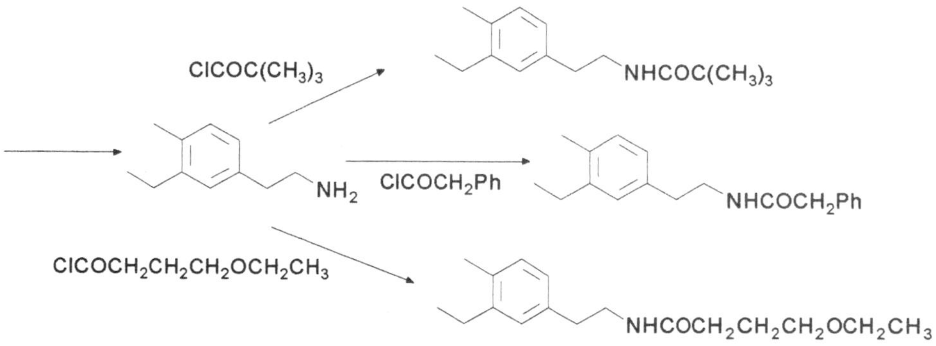

$ClCOC(CH_3)_3$ $NHCOC(CH_3)_3$

NH_2 $ClCOCH_2Ph$ $NHCOCH_2Ph$

$ClCOCH_2CH_2CH_2OCH_2CH_3$ $NHCOCH_2CH_2CH_2OCH_2CH_3$

Figure 12.1. A stage in a hypothetical drug design pathway illustrating some of the possibilities provided by the presence of an amino group in the structure of an intermediate. The products of the reactions illustrated could be either the final products of the design pathway or intermediates for the next stage(s) in the synthetic pathway.

12.2.4 The Use of Protecting Groups

The design of synthetic pathways often requires a reaction to be carried out at one centre in a molecule, the primary process, whilst preventing a second centre from either interfering with the primary process or undergoing a similar unwanted reaction. This objective may be achieved by careful choice of reagents and reaction conditions. However, an alternative is to combine the second centre with a so-called **protecting group** to form a structure that cannot react under the prevailing reaction conditions. A protecting group must be easy to attach to the relevant functional group, it must form a stable structure that is not affected by the reaction conditions and reagents being used to carry out the primary process and should be easily removed once it is no longer required (Table 12.1). However, in some circumstances, protecting groups may not be removed but converted into another structure as part of the synthesis.

12.3 Asymmetry in Syntheses

The presence of an asymmetric centre or centres in a target structure means that its synthesis requires either the use of **non-stereoselective reactions** and the separation of the resulting stereoisomers or the use of **stereoselective reactions** that mainly produce one of the possible enantiomers. This section introduces some of the general methods used to incorporate stereospecific centres into a target molecule. However, for a more comprehensive discussion the reader is referred to *Stereoselective Synthesis* by R. S. Atkinson, published by John Wiley and Sons Chichester (1995) and *Asymmetric Synthesis* by R. A. Aitken and S. N. Kilenyi, published by Blackie Academic and Professional, London (1994).

Table 12.1. Examples of protecting groups. The conditions used will vary and so only the principal reagents are shown. Other examples are given in Figures 2.23 and 2.33.

Functional group	Protecting group	Removal
Alcohols/phenols	**Benzyl ether group**	
	$-OH \xrightarrow[\text{Base}]{PhCH_2Cl} -O\text{-}CH_2Ph$ A benzyl ether	$-O\text{-}CH_2Ph \xrightarrow{\text{Hydrogenolysis}} -OH$ (catalytic hydrogenation)
Alcohols	**(i) Triphenylmethyl (trityl) ether group**	
	$-OH \xrightarrow[\text{Base}]{Ph_3CCl} -O\text{-}CPh_3$ A triphenylmethyl ether	$-O\text{-}CPh_3 \xrightarrow{CH_3COOH} -OH$ Acid conditions
	(ii) Acetate (also trifluoroacetate) ester group	
	$-OH \xrightarrow{CH_3COOCOCH_3} -O\text{-}CO\text{-}CH_3$ An acetate	$-O\text{-}CO\text{-}CH_3 \xrightarrow{\text{Aqueous base}} -OH$
	$-OH \xrightarrow{CF_3COOCOCF_3} -O\text{-}CO\text{-}CF_3$ A trifluoroacetate	$-O\text{-}CO\text{-}CF_3 \xrightarrow{\text{Aqueous base}} -OH$
Carboxylic acids	**(i) *t*-Butyl ester group**	
	$-COOH \xrightarrow[\text{Isobutene},\ H_2SO_4]{(CH_3)_2C{=}CH_2} -COO\text{-}C(CH_3)_3$ A *t*-butyl ester	$-COO\text{-}C(CH_3)_3 \xrightarrow[\text{conditions}]{\text{Dry acid}} -COOH$
	(ii) Trichloroethyl esters	
	$-COOH \xrightarrow[\text{Trichloroethanol}]{Cl_3CCH_2OH} -COO\text{-}CH_2CCl_3$ Trichloroethyl ester	$-COO\text{-}CH_2CCl_3 \xrightarrow[\text{Elimination}]{Zn} -COOH$
Amines	**(i) Ethanamide (acetamide) groups**	
	$-NH_2 \xrightarrow{CH_3COOCOCH_3} -NHCOCH_3$ An ethanamide	$-NHCOCH_3 \xrightarrow[\text{Hydrolysis}]{H^+/H_2O} -NH_2$
	(ii) Benzyloxycarbonamide groups	
	$-NH_2 \xrightarrow{PhCH_2OCOCl} -NHCOOCH_2Ph$; Benzyl chloromethanoate (benzyl chloroformate); An N-substituted urethane	$-NHCOOCH_2Ph \xrightarrow[\text{Hydrogenolysis}]{H_2/Pd} -NH_2$

Table 12.2. Examples of the pure enantiomers used to resolve racemic modifications by forming diastereoisomers. In all regeneration processes there is a danger of the racemic modification being reformed by racemisation.

Functional group	Enantiomers used (resolving agents)	Diastereoisomers	Regeneration
Carboxylic and other acids	A suitable base, e.g. (−)-Brucine (−)-Strychnine (−)-Morphine	Salts	Treatment with a suitable acid, e.g. HCl
Amines and other bases	A suitable acid, e.g. (+)-Tartaric acid (−)-Malic acid (+)-Camphor sulphonic acid	Salts	Treatment with a suitable base, e.g. NaOH
Alcohols	A suitable acid (see above)	Esters	Acid or base hydrolysis

12.3.1 The Use of Non-stereoselective Reactions to Produce Stereospecific Centres

Non-stereoselective reactions produce either a mixture of diastereoisomers or a racemic modification. Diastereoisomers exhibit different physical properties. Consequently, techniques utilising these differences may be used to separate the isomers. The most common methods of separation are fractional crystallisation and chromatography.

Racemic modification consisting of enantiomers (+)A and (−)A

Pure enantiomer e.g. (+)B isomer

Two isomers: [(+)A-(+)B] and [(−)A-(+)B]

Diastereoisomers

Separation by the use of physical methods

Pure (+)A-(+)B

Pure (−)A-(+)B

Regeneration of the enantiomer of A

(+)B

(+)B

Pure (+)A

Pure (−)A

Figure 12.2. A schematic representation of the use of diastereoisomers in the resolution of racemic modifications.

The separation (**resolution**) of a racemic modification into its constituent enantiomers is normally achieved by converting the enantiomers in the racemate into a pair of diastereoisomers by reaction with a pure enantiomer (Figure 12.2). Enantiomers of acids are used for racemates of bases whereas enantiomers of bases are used for racemates of acids (Table 12.2). Neutral compounds may sometimes be resolved by conversion to an acidic or basic derivative, which is suitable for diastereoisomer formation. The diastereoisomers are separated using methods based on the differences in their physical properties and the pure enantiomers are regenerated from the corresponding diastereoisomers by suitable reactions. For example,

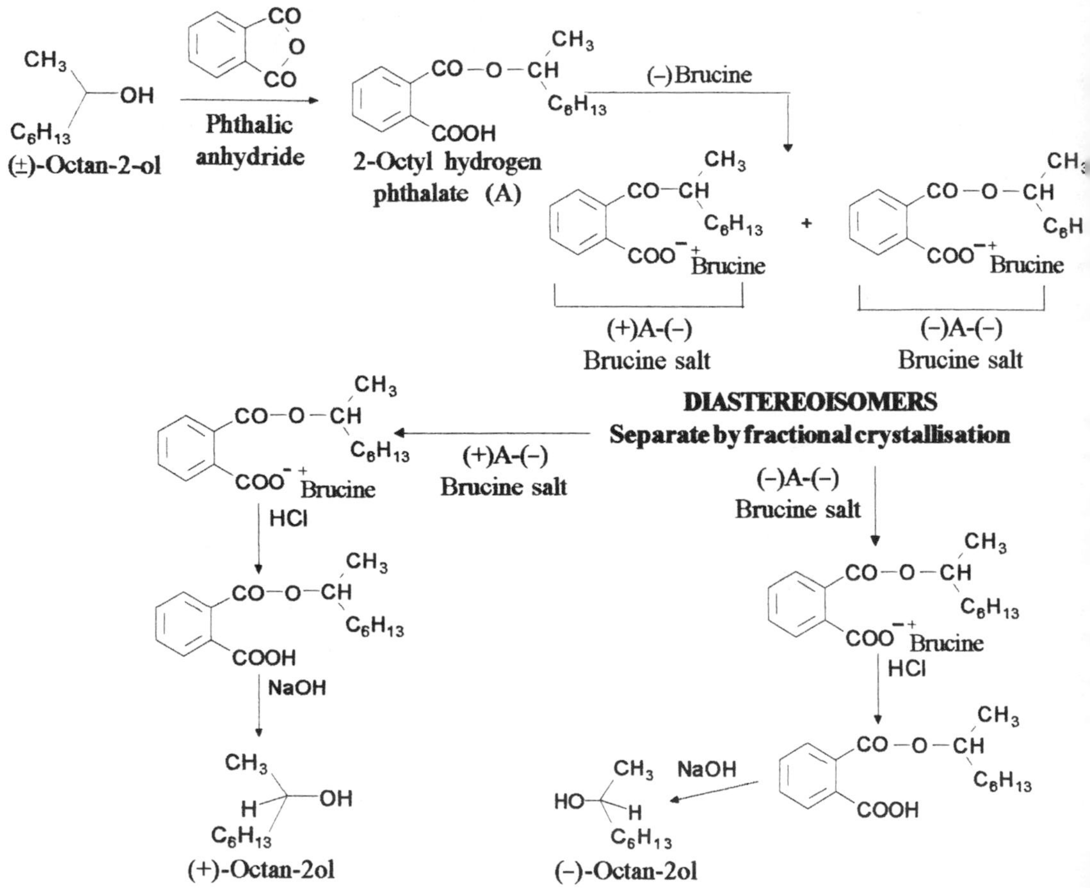

Figure 12.3. The reaction sequence used to resolve a racemic mixture of octan-2-ol.

(±)-octan-2-ol, a neutral compound, can be resolved into its separate enantiomers by conversion to the corresponding racemic hydrogen phthalate followed by treatment with (−)-brucine. The latter converts the phthalate esters in the racemic modification into a mixture of the corresponding brucine salts (Figure 12.3), which are diastereoisomers. These diastereoisomers are separated by fractional crystallisation. The phthalate esters are regenerated separately from their respective salts by treatment with hydrochloric acid and the octan-2-ol enantiomers liberated by treatment with sodium hydroxide. The pure enantiomers are isolated from their respective reaction mixtures by steam distillation.

The incorporation of the resolution of a racemic modification into a synthetic pathway considerably reduces the overall yield of the synthesis because the maximum theoretical yield of an enantiomer is 50%.

12.3.2 The Use of Stereoselective Reactions to Produce Stereospecific Centres

Stereoselective reactions are those that result in the selective production of one of the stereoisomers of the product. The extent of the selectivity may be recorded as the **enantiomeric excess** (e.e.) when the reaction produces a mixture of enantiomers and as the **diastereoisomeric excess** (d.e.) when it produces a mixture of diastereoisomers. Both of these parameters are defined as the difference between the yields of the stereoisomers expressed as a percentage of their total yield (Equation 12.1).

$$\text{e.e. or d.e.} = \frac{(\text{Yield of the major stereoisomer} - \text{Yield of the stereoisomer}) \times 100}{\text{Yield of the major stereoisomer} + \text{Yield of the manor stereoisomer}} \quad (12.1)$$

$$= \%\ \text{Major stereoisomer} - \%\ \text{Minor stereoisomer} \quad (12.2)$$

The values of e.e. and d.e. are obtained by quantitatively measuring the yields of the individual stereoisomers using suitable analytical methods. An e.e. or d.e. value of 0% means that the stereoisomers are produced in equal amounts. In the case of enantiomeric mixtures the product is likely to be in the form of a racemic modification. Conversely, an e.e. or d.e. value of 100% indicates that only one product is formed. This rarely occurs in practice; most reactions yield a mixture of isomers.

The principal factors that appear to influence the stereochemistry of a reaction are:

(i) the shape of the substrate about the reaction centre;
(ii) the nature of the reagent;
(iii) the mechanism of the reaction;
(iv) the catalyst used;
(v) the relative activation energies of the pathways used to produce the isomers.

These factors are interrelated and should not be considered in isolation when assessing the stereochemical potential of a reaction. However, it is more convenient to consider each of these factors separately in order to illustrate their influence on the stereoselectivity of a reaction.

(i) *The shape of the substrate molecule.* A reaction will yield a mixture of enantiomers when there is an equal possibility of the reagent approaching the reactive centre of the substrate from opposite directions (Figures 12.5 and 12.6). However, the same type of reaction will be stereospecific if steric hindrance reduces the chances of the reagent from attacking from more than one direction. For example, Davies and co-workers have developed a synthesis of the antihypertensive drug *S,S*-captopril (see section 6.12.2) that involves the introduction of the side chain chiral centre by alkylation of an enolate. This reaction is stereoselective because the iron–phosphorus–benzene ring complex of the sub-

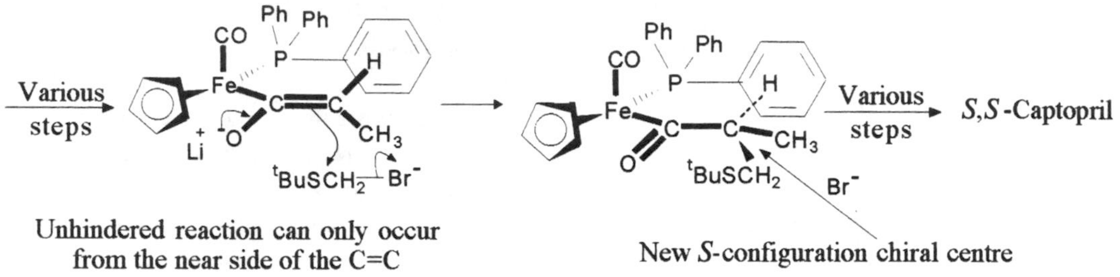

Various steps → S,S-Captopril

Unhindered reaction can only occur from the near side of the C=C

New *S*-configuration chiral centre

Figure 12.4. Stereoselective alkylation of an enolate in the synthesis of captopril. The heavier straight lines are bonds in the plane of the paper and the thin straight lines are bonds behind the plane of the paper. ^{t}Bu is a *t*-butyl group.

strate only allows the unhindered approach of the *t*-butylthiomethyl bromide from one side of the molecule. Consequently, reaction occurs mainly from that side and produces the *S*-configuration product (Figure 12.4).

(ii) *The nature of the reagent.* The nature of the reagent may affect the stereochemistry of the product of a reaction. Different reagents undergoing the same general type of reaction with the same reaction centre of a substrate may yield products that have different types of stereochemistry. For example, hydroxylation of the C=C bond of *E*-but-2-ene by osmium tetroxide yields a racemate (Figure 12.5a) but bromination of this compound produces the *meso*-dibromide (Figure 12.5b).

(iii) *The mechanism.* The mechanism by which a reaction proceeds could influence the stereochemistry of the product(s). For example, in theory nucleophilic substitution of a chiral alkyl halide by an S_N1 mechanism should result in the formation of a racemate because the nucleophile can attack the planar intermediate from either side (Figure 12.6a). However, the time taken for the halide ion to diffuse away from the carbonium ion means that for this period of time the attack of the nucleophile is restricted to one side of the intermediate. Consequently, reactions proceeding by an S_N1 mechanism normally yield a product that consists of a mixture of an enantiomer with the opposite configuration to the substrate and the racemic modification. However, nucleophilic substitution by an S_N2 mechanism at a chiral alkyl halide centre will produce one enantiomer because the attack of the nucleophile can only take place from the side opposite the halogen atom (Figure 12.6b).

(iv) *The catalyst used.* The action of enzymes catalysing reactions that produce asymmetric centres is usually stereospecific. Consequently, a number of enzymes have been used to bring about a number of stereoselective transformations (see section 12.3.3.1i). Furthermore, a number of non-enzyme catalysts have also been developed to catalyse the formation of chiral centres (see section 12.3.3.1ii).

(v) *The activation energy of the process.* The usefulness of a reaction in stereospecific synthesis will also depend on the activation energy of a process. Consider, for example, a reaction that produces a mixture of two stereoisomers. The activation energies for the formation of these stereoisomers will be the same if the stereoisomers are enantiomeric but *may be different* if the stereoisomers are diastereoisomeric (Figure 12.7). The relative

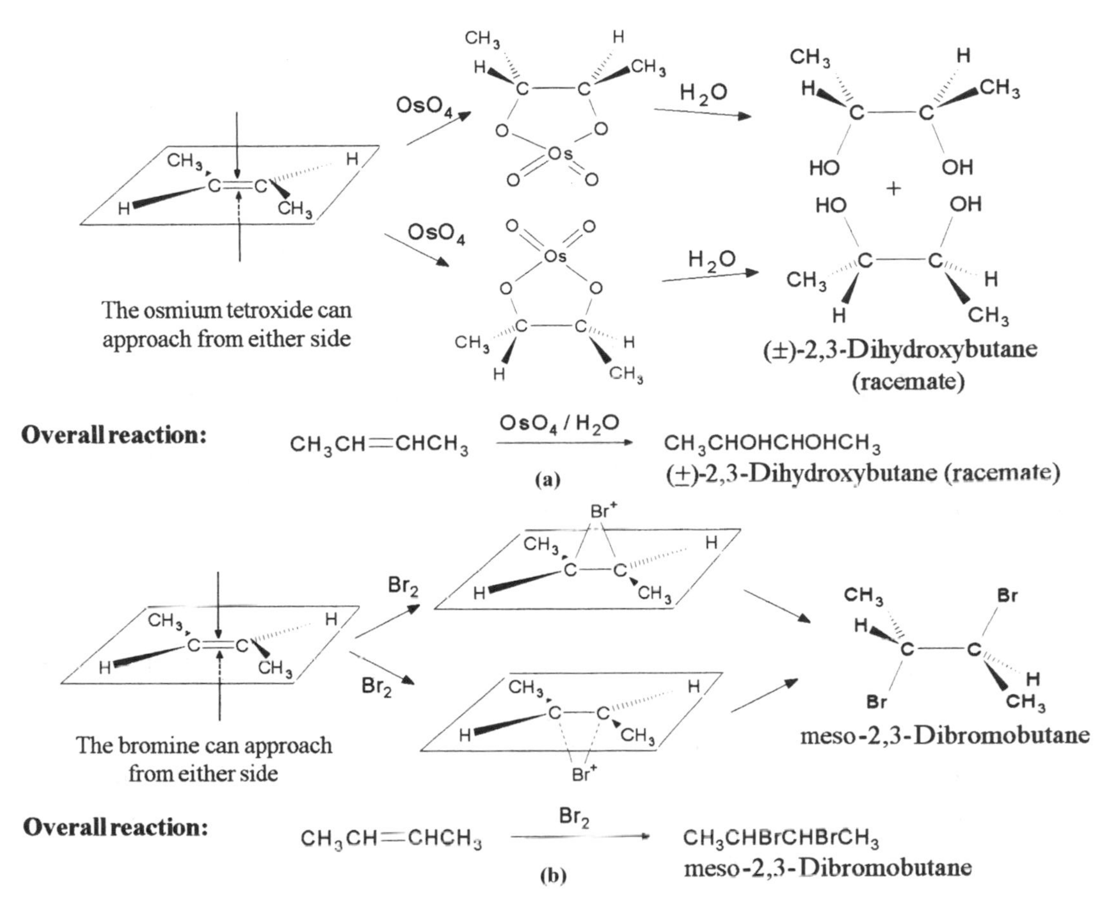

Figure 12.5. Examples of the effect of the nature of a reagent on the stereochemistry of a reaction. In both examples the reagent has an equal chance of attacking the C=C from either side.

proportions of the diastereoisomers produced in the reaction mixture will depend on the relative values of the activation energies of the pathways producing the stereoisomers. The greater this difference, the greater the chance that the reaction will be diastereoselective with respect to the product formed via the lowest activation energy pathway. This is because reactants will find it easier to acquire the energy necessary to overcome the lower activation energy barrier than the higher and so there is a greater chance of the reaction proceeding by the lower energy pathway. Furthermore, it can be shown that the relative rates of the two reactions, expressed in terms of their rate constants, are given by the equation:

$$k_1/k_2 = e^{[-\Delta G_1 + \Delta G_2]/RT} \qquad (12.3)$$

where k_1 is the rate constant for product 1, k_2 is the rate constant for product 2, R the ideal gas constant and T is the temperature in °K. Equation (12.3) shows that decreasing the

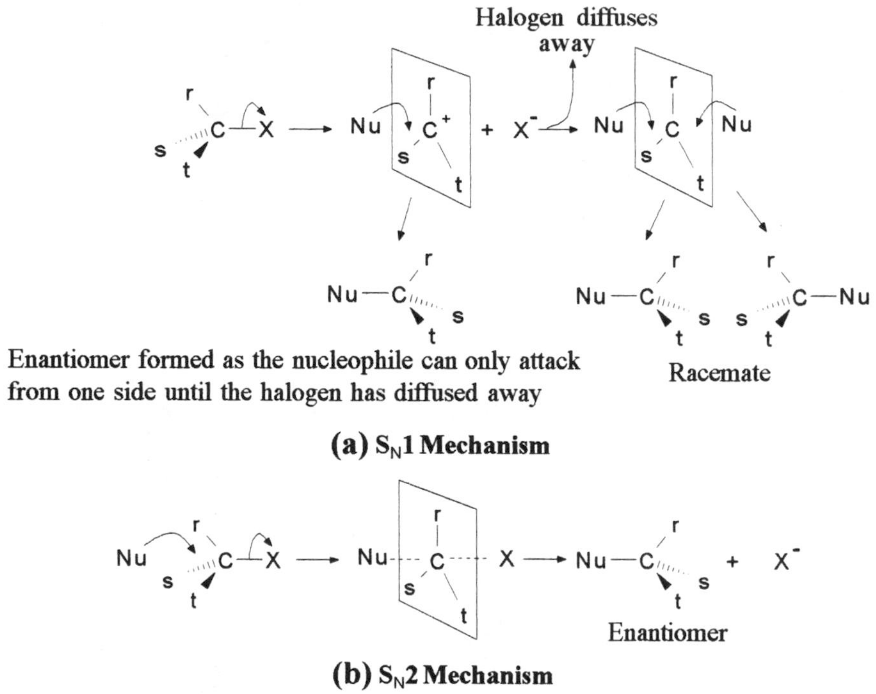

Halogen diffuses away
Nu
X
Nu
Nu
Enantiomer formed as the nucleophile can only attack from one side until the halogen has diffused away
Racemate
(a) S_N1 Mechanism
Nu
Enantiomer
(b) S_N2 Mechanism

Figure 12.6. The stereochemistry of nucleophilic substitutions at a chiral alkyl halide centre.

ΔG_2
ΔG_1
Energy of the system
Product 2
Product 1
Reaction ordinate

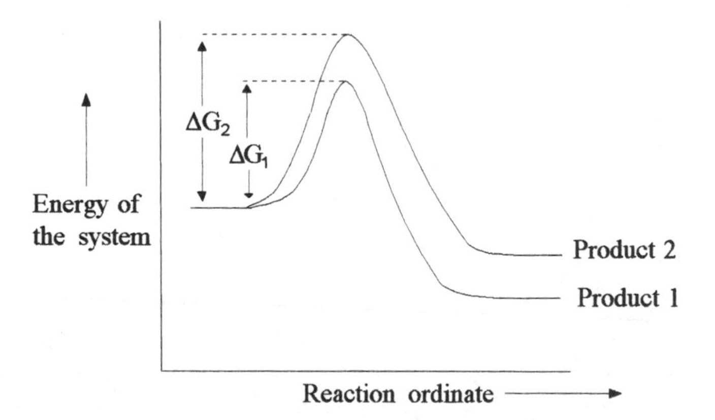

Figure 12.7. A schematic representation of the energy pathways of a reaction that produces two distereoisomers by processes that have different activation energies.

temperature increases the ratio of k_1 to k_2, which results in an increase in the yield of product 1. In other words, lowering the temperature of a diastereoselective reaction tends to make it more diastereoselective.

12.3.3 General Methods of Asymmetric Synthesis

There is no set method for designing an asymmetric synthesis. Each synthesis must be treated on its merits and in all cases success will depend on the skill and ingenuity of the research worker.

The range and scope of the reactions used in asymmetric synthesis are extremely large and consequently they are difficult to classify. In this text they are discussed under the broad headings of reactions that either require a catalyst or those that do not require a catalyst for their stereoselectivity. However, it is emphasised that this and the subdivisions used are a simplification and many reactions can fall into more than one category. Furthermore, it should be realised that several different general approaches may be used in the design of a synthetic pathway.

It is often difficult to visualise the three-dimensional structural changes that occur in stereoselective reactions. Consequently, it is recommended that the reader makes models of the relevant parts of the structures mentioned in order to appreciate how the three-dimensional reaction corresponds to the two-dimensional formulae on the page. A simple ball and spring tetrahedron with four different coloured balls is particularly useful in this context.

12.3.3.1 Methods that Depend on the Use of a Catalyst for their Stereoselectivity

(i) Methods using enzymes as catalysts. These methods can, in theory, use all types of substrates and reagents because the enzyme control will give the process stereoselectivity. These methods are economical in this use of chiral material but suffer from the disadvantage that they can require large quantities of the enzyme to produce significant quantities of the drug.

Enzyme-catalysed processes may be **single** or **interrelated multistep** processes. The latter usually use enzymes or microorganisms to produce asymmetric starting materials from basic raw materials. For example, the production of the various semisynthetic penicillins uses benzylpenicillin, phenoxymethylpenicillin and cephalosphorin C as starting materials because these can be produced from naturally occurring raw materials by fermentation. This section is only concerned with examples of the use of enzymes or microorganisms to catalyse a single transformation.

A wide variety of enzyme-controlled stereoselective transformations are known. These transformations include oxidations, reductions, reductive aminations, addition of ammonia, transaminations and hydrations (Figure 12.8). The microorganisms used as enzyme sources in these

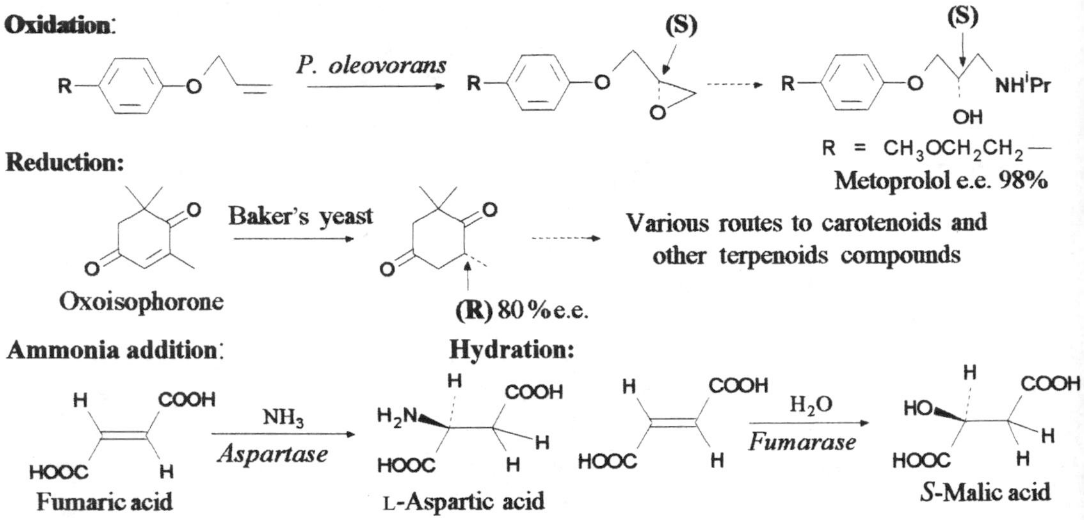

Figure 12.8. Examples of enzyme-controlled transformations.

transformations have either been produced in bulk after isolation from natural sources or produced from existing microorganisms by genetic engineering (see section 10.15). The configuration of the new asymmetric centre produced by a particular enzyme or organism will depend on the structure of the substrate. However, substrates whose reactive centres have similar structures will often produce asymmetric centres with the same configuration. (Table 12.3).

The enzyme preparations used in transformations may take the form of a solid isolated from its natural source. These preparations are referred to as **cell-free** enzymes and do not usually include the cofactors and coenzymes essential for enzyme action (see section 6.1). Consequently, the use of cell-free preparations often requires the addition of the appropriate cofactors and coenzymes. The use of complete microorganisms means that the necessary cofactors and coenzymes are already present but in order for the microorganism to be effective the substrate must be able to penetrate the cell envelope (see section 4.3).

Enzymes normally act in aqueous media, usually at room temperature and about pH 7 (see section 6.7). However, many of the substrates of interest to medicinal chemists are insoluble in water. Changing the solvent can have an effect on the structure of the active site of an enzyme and, as a consequence, its activity. Consequently, a cell-free enzyme preparation must be capable of acting in non-aqueous solvents. Lipases, a group of enzymes that catalyse ester hydrolysis and acetyl transfer reactions, satisfy this requirement in that they are active in cyclohexane and toluene amongst other hydrocarbon solvents. They have the advantages of being commercially available, many are inexpensive and they do not require a cofactor.

Table 12.3. Examples of the epoxides produced by *Pseudomonas oleovorans* and *Nocardia corallina*.

Enzyme system

Substrate	Epoxide	% e.e.
Epoxide formation by *P. oleovorans*		
Octene	2*R*-1,2-Epoxyoctane	70
Octa-1,7-diene	7*R*,8-Epoxyoct-1-ene	60
Butyl, prop-2-enyl ether	Butyl, 2*R*-glycidyl ether	85
Phenyl, prop-2-enyl ether	Phenyl, 2*R*-glycidyl ether	92
4-Fluorophenyl, prop-2-enyl ether	4-Fluorophenyl, 2*R*-glycidyl ether	99
Epoxide formation by *N. corallina B-276*		
Propene	1,2*R*-Epoxypropane	83
Heptadecene	1,2*R*-Epoxyheptadecene	81
3-Chloroethene	*S*-Epichlorohydrin	81
3,3,3,Trifluoropropene	3,3,3,Trifluoro-1,2*S*-epoxypropane	75
Pentyl, prop-2-enyl ether	Pentyl, 2*R*-glycidyl ether	84
Octyl, prop-2-enyl ether	Octyl, 2*R*-glycidyl ether	92

(ii) Methods using non-enzyme catalysts. A number of stereospecific non-enzyme catalysts have been developed that convert achiral substrates into chiral products. These catalysts are usually either complex non-organic (Figure 12.9) or organometallic compounds (Figure 12.10). The organometallic catalysts are usually optically active complexes whose structures contain one or more chiral ligands. An exception is the Sharpless–Katsuki epoxidation, which uses a mixture of an achiral titanium complex and an enantiomer of diethyl tartrate. The selection or development of a catalyst, reagent and reaction conditions for a transformation are normally made by considering similar stereoselective reactions in the literature.

Cl Cl O CH_3O H (1)

The chiral centre that gives rise to the racemate

CH_3Cl / $PhCH_3$ / NaOH / Catalyst

Cl Cl O CH_3O CH_3 92 % e.e.

Various steps → Indacrinone (a diuretic)

HO H H N^+ CF_3 Br^-

Figure 12.9. An example of a stereoselective transformation using a non-organic catalyst. The base deprotonates the racemic ketone (1) to form the enolate, which is alkylated (see Figure 12.4) under the influence of the catalyst to form a chiral centre with an *S*-configuration.

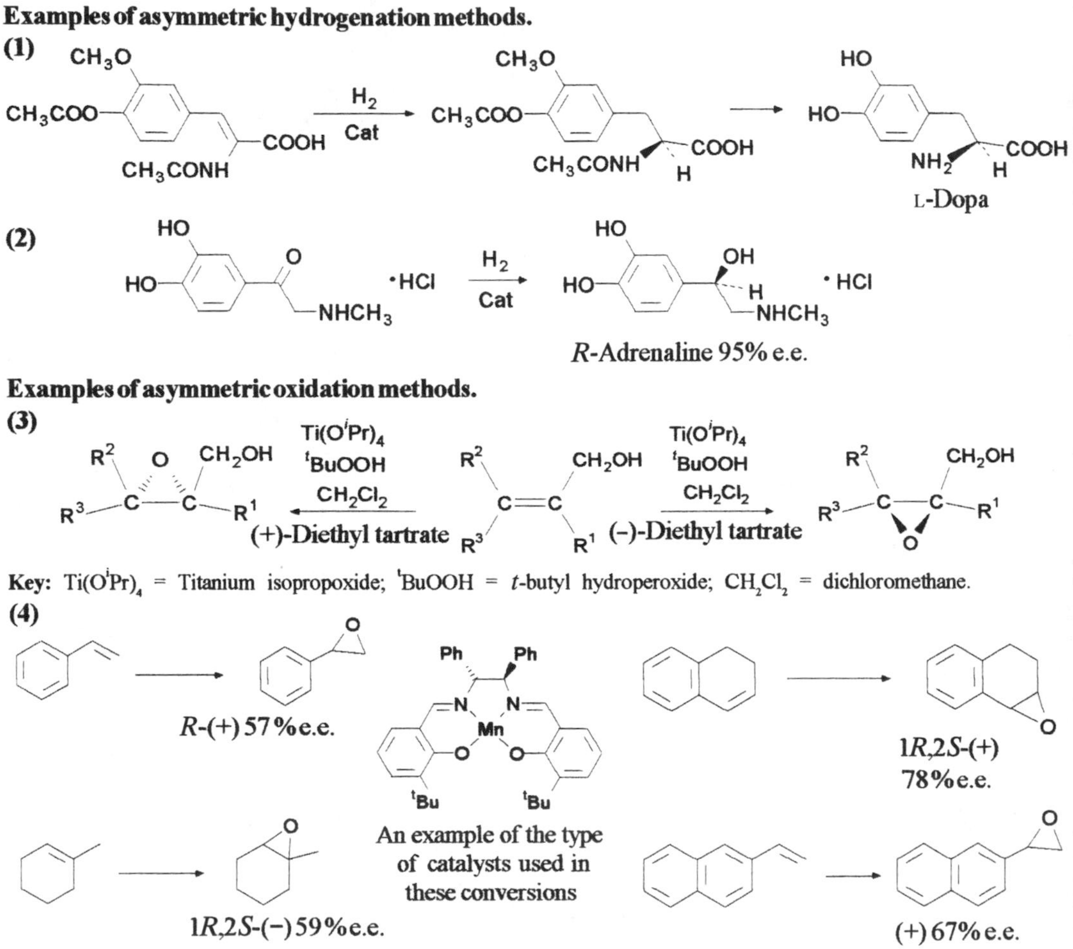

Examples of asymmetric hydrogenation methods.

(1) L-Dopa

(2) *R*-Adrenaline 95% e.e.

Examples of asymmetric oxidation methods.

(3) (+)-Diethyl tartrate; (−)-Diethyl tartrate

Key: $Ti(O^iPr)_4$ = Titanium isopropoxide; ^tBuOOH = *t*-butyl hydroperoxide; CH_2Cl_2 = dichloromethane.

(4) *R*-(+) 57% e.e.; 1*R*,2*S*-(+) 78% e.e.; 1*R*,2*S*-(−) 59% e.e.; (+) 67% e.e.

An example of the type of catalysts used in these conversions

Figure 12.10. Examples of the use of non-enzyme catalysts in stereoselective synthesis. (1) The stereoselective step in the process that Monsanto used to produce l-dopa. The catalyst is a rhodium complex with chiral phosphine ligands. (2) The synthesis of *R*-adrenaline using a rhodium–iron-based catalyst. (3) The Sharpless–Katsuki epoxidation. The stereochemistry of the product depends on which enantiomer of diethyl tartrate is used in the preparation. (4) Examples of epoxidations using manganese complexes of chiral Schiff bases, modified from W. Zhang, J. L. Loebach, S. R. Wilson and E. N. Jacobsen, *Journal of the American Chemical Society*, **112**, 2801 (1990).

12.3.3.2 Methods that Do Not Use Catalysts to Produce Stereoselectivity

These general approaches can be classified for convenience as:

(i) using chiral building blocks;
(ii) using a chiral auxiliary;
(iii) using achiral substrates and reagents.

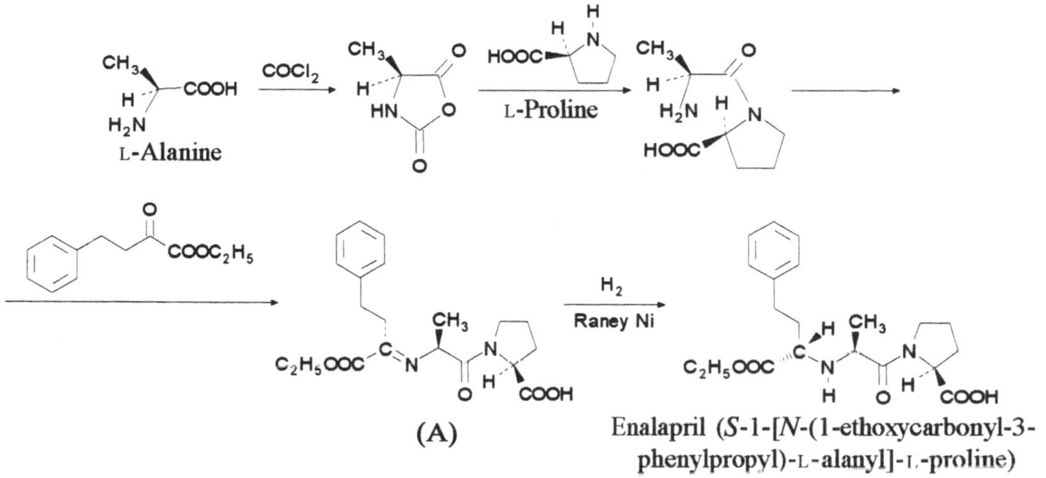

L-Alanine

L-Proline

(A)

Enalapril (*S*-1-[*N*-(1-ethoxycarbonyl-3-phenylpropyl)-L-alanyl]-L-proline)

Figure 12.11. A synthetic route for the preparation of the ACE inhibitor enalapril (see section 6.12.2). The configurations of L-alanine and L-proline (the reagent for stage 2) are retained in the final product. The reduction of the intermediate A is stereoselective, giving the *S*,*S*,*S*-isomer in 87% yield.

(i) Using chiral building blocks. These methods depend on the use of enantiomerically pure building blocks with the required configurations. A building block is treated with either chiral or achiral reagents to introduce the required asymmetric centre(s) into the product. In the latter case the stereochemistry of the substrate is used to make the reaction stereoselective (see Figure 12.12, stage 1 and also section 12.3.2). The products of these types of reaction range from a single enantiomer (Figure 12.11) to a mixture of diastereoisomers (Figure 12.12) that may be separated into their constituents (see section 12.3.1). In all cases, the reactions used in further stages of the synthesis should not affect the configurations of the chiral centres of the building blocks. However, in some instances reactions that cause an inversion of configuration may be used (Figure 12.13). The main sources of enantiomerically pure substrates and reagents are naturally occurring compounds, such as amino acids, amino alcohols, hydroxyacids, alkaloids, terpenes and carbohydrates. These materials are usually cheap and available in bulk.

(ii) Using a chiral auxiliary. This method is based on a three-step process. The achiral substrate is combined with a pure enantiomer known as a **chiral auxiliary** to form a chiral intermediate. Treatment of this intermediate with a suitable reagent produces the new asymmetric centre. The chiral auxiliary causes, by steric or other means (see section 12.3.2), the reaction to favour the production of one of the possible stereoisomers in preference to the others. Completion of the reaction is followed by removal of the chiral auxiliary, which may be recovered and recycled, thereby cutting down development costs (Figure 12.14).

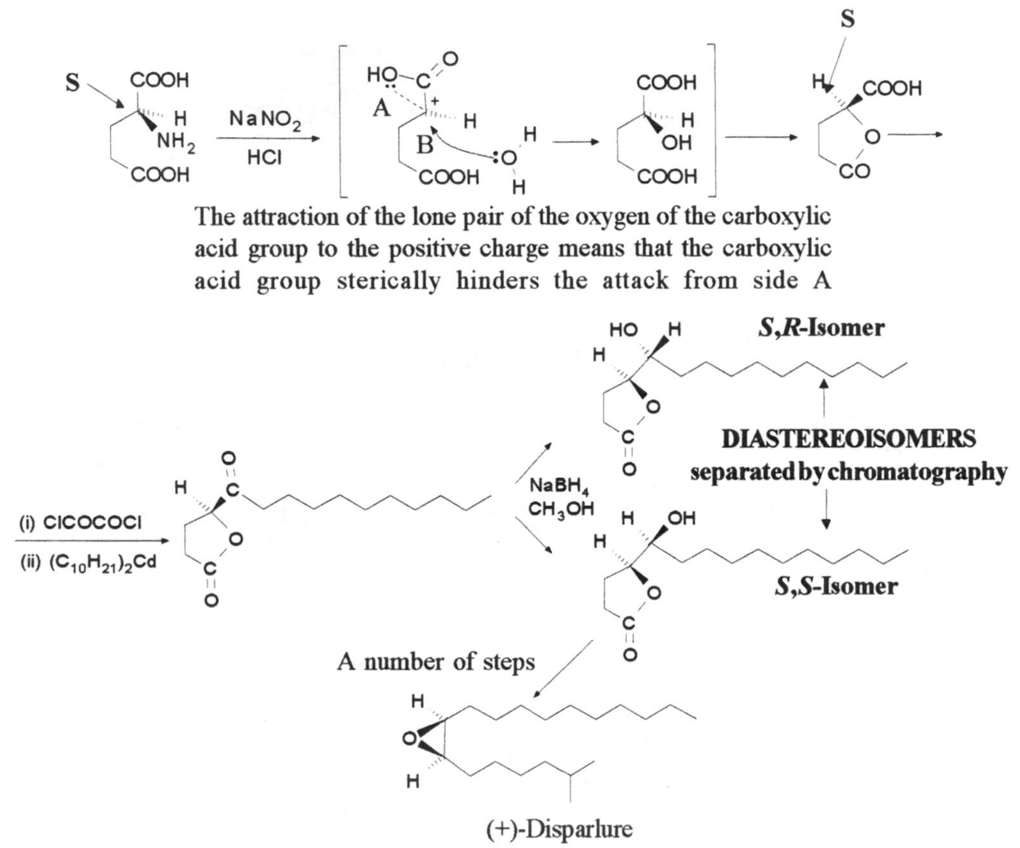

Figure 12.12. A scheme for the synthesis of (+)-disparlure, a pheromone produced by the female gypsy moth. This method starts with *S*(+)-glutamic acid, which contains one of the required asymmetric centres. In the first stage the presence of the adjacent carboxylic acid group prevents nucleophilic attack of water occurring from the A side (behind the paper) of the molecule. It only allows the nucleophile to attack from the B side (in front of the paper) and so the configuration of the glutamic acid residue is retained in the product. The second asymmetric centre is introduced by the non-selective reduction of a ketone intermediate to a mixture of hydroxylactone diastereoisomers that are separated by chromatography, and the *S,S*-isomer is converted by a series of seven steps into (+)-disparlure.

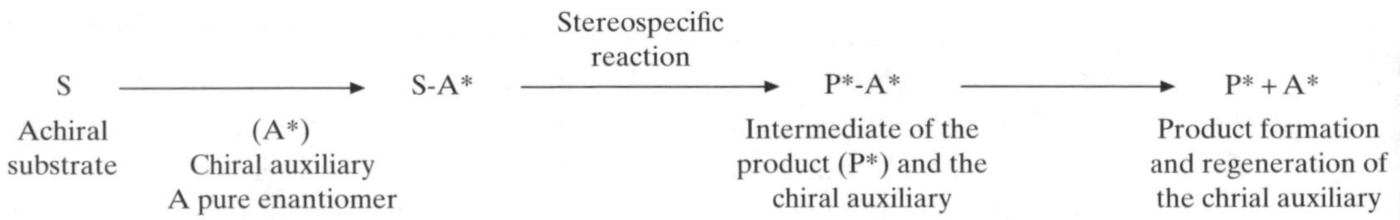

An advantage of this approach is that where the reaction used to produce the new asymmetric centre has a poor stereoselectivity, the two products of the reaction will be diastereoisomers because they contain two different asymmetric centres. These diastereoisomers may be separated by crystallisation or chromatography (see section 12.3.1) and the unwanted isomer discarded.

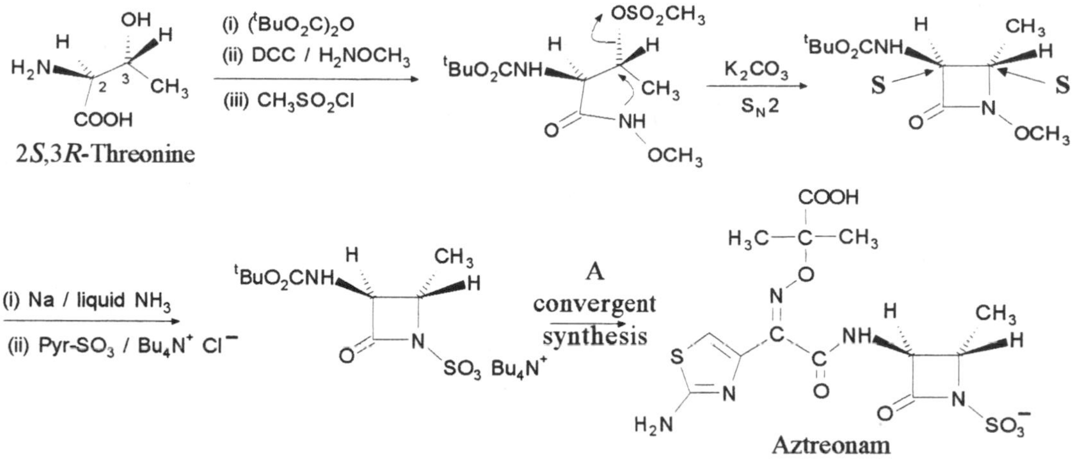

2*S*,3*R*-Threonine

Aztreonam

Figure 12.13. A scheme for the synthesis of the antibacterial aztreonam. The synthesis starts with the naturally occurring 2*S*,3*R*-threonine. The amino and hydroxy groups are protected before the acid is converted into the corresponding hydroxamate. An internal S_N2 closes the ring and also inverts the configuration of carbon 3 (see Figure 12.6a). The *N*-methoxy group is removed by reduction and the resulting β-lactam is sulphonated. The sulphonated β-lactam is converted to the free amine, which is coupled to the rest of the molecule (a convergent synthesis).

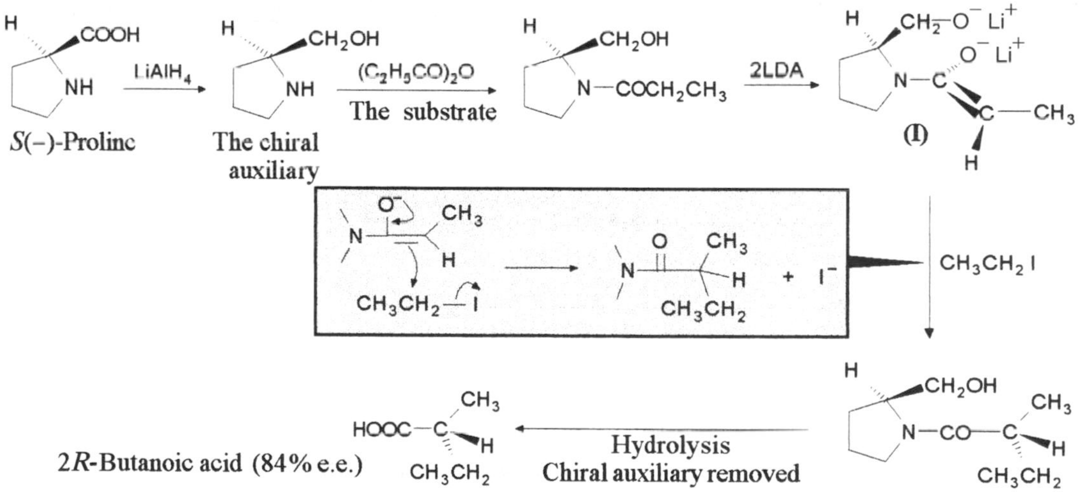

S(−)-Proline

The chiral auxiliary

The substrate

(I)

Hydrolysis
Chiral auxiliary removed

2*R*-Butanoic acid (84% e.e.)

Figure 12.14. The synthesis of 2*R*-methylbutanoic acid, illustrating the use of a chiral auxiliary. The chiral auxiliary is 2*S*-hydroxymethyltetrahydropyrrole, which is readily prepared from the naturally occurring amino acid proline. The chiral auxiliary is reacted with propanoic acid anhydride to form the corresponding amide. Treatment of the amide with lithium diisopropylamide (LDA) forms the corresponding enolate (I). The reaction almost exclusively forms the *Z*-isomer of the enolate in which the OLi units are well separated and possibly have the configuration shown. The approach of the ethyl iodide is sterically hindered from the top (by the OLi units or Hatoms) and so alkylation from the lower side of the molecule is preferred. Electrophilic addition to the appropriate enolate is a widely used method for producing the enantiomers of α-alkyl-substituted carboxylic acids.

(iii) Using achiral substrates and reagents. A wide variety of achiral substrates and reagents can give rise to asymmetric centres. However, the usefulness of these reactions in stereoselective synthesis will depend on their degree of stereoselectivity (see section 12.3.2). For example, electrophilic addition of hydrogen chloride to butene gives rise to a racemic mixture of the *R*- and *S*-isomers of 2-bromobutane because addition has an equal chance of occurring from either side of the C=C bond.

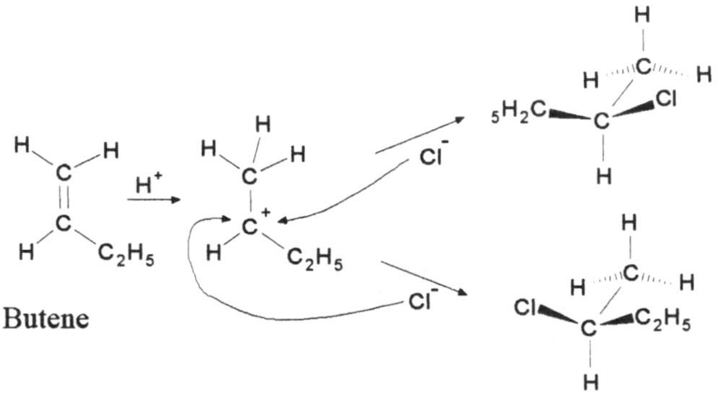

Butene

(±)-2-Bromobutane

12.4 Designing Organic Syntheses

The synthetic pathway for a drug or analogue must start with readily available materials and convert them by a series of inexpensive reactions into the target compound. There are no obvious routes because each compound will present a different challenge. The usual approach is to work back from the target structure in a series of steps until cheap commercially available materials are found. This approach is formalised by a method developed by Warren and known as either the **disconnection approach** or **retrosynthetic analysis**. In all cases the final pathway should contain a minimum of stages in order to keep costs to a minimum and overall yields to a maximum.

12.4.1 An Introduction to the Disconnection Approach

This approach starts with the target structure and then works **backwards** by artificially cutting the target structure into sections known as **synthons**. Each of these backward steps is represented by a double-shafted arrow (⇒) whereas ~~~~ is drawn through the disconnected bond of the target structure. Each of the possible synthons is converted on paper into a real compound known as a **reagent**, whose structure is similar to that of the synthon. All the possible disconnection routes must be considered. The disconnection selected for a step in the pathway is the one that gives rise to the best reagents for a reconnection reaction. This analysis is

Table 12.4. Examples of books and databases that catalogue chemical reactions.

Title	Author	Classification used
Books:		
Synthetic Organic Chemistry	Wagner and Zook	Lists reactions according to the functional group being produced
Organic Functional Group Preparations	Sandler and Karo	Lists reactions according to the functional group being produced
Reagents for Organic Synthesis	Fieser and Fieser	Lists reagents and their uses in alphabetical order (includes suppliers)
Carbanions in Synthesis	Ayres	Lists carbanion transformations
Oxidations in Organic Chemistry	Hudlicky	Lists and discusses transformations that can be brought about by oxidation
Reduction in Organic Chemistry	Hudlicky	Lists and discusses transformations that can be brought about by reduction
Databases:		
CASREACT	The Chemical Abstracts Research Service	Information from 1985. Covers single and multistep reactions. Includes CAS Registery numbers of reactants, products, rcagcnts, catalysts and solvcnts. It is structure-searchable
ISI Reaction Centre	Institute for Scientific Information	Data from 1840. Classified according to reaction type. Includes biological assays
Crossfire	Beilstein Information Service	Three main types of data: structural and properties; reactions including preparations; and chemical literature references

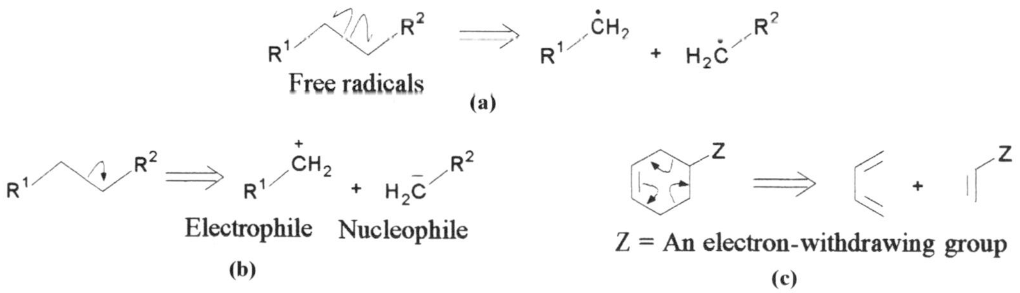

Free radicals

(a)

Electrophile Nucleophile

(b)

Z = An electron-withdrawing group

(c)

Figure 12.15. (a) Homolytic, (b) heterolytic and (c) pericyclic bond disconnections.

repeated with the reagents of each disconnection step until readily available starting materials are obtained. The selection of the reagents and the reactions for their reconnection may require extensive literature searches (Table 12.4).

In the disconnection approach, bonds are usually disconnected by either homolytic or heterolytic fission (Figure 12.15). However, some bonds may be disconnected by a reverse pericyclic mechanism (see Table 12.5, Diels–Alder).

Free radical disconnections are usually disregarded because it is difficult to predict the outcome of reconnection reactions that proceed through a free radical mechanism because these reactions tend to produce mixtures. Heterolytic disconnections result in the formation of electrophilic and nucleophilic species. The most useful heterolytic disconnections are those that either give rise to a stable species or occur by a feasible disconnection mechanism, such as hydrolysis, because these disconnections are more likely to have a corresponding reconnection reaction. Synthons are converted to a real reagent by converting them to a structurally similar compound that has the relevant electrophilic or nucleophilic centre. For example, a carbanion synthon with the structure RCH_2^- could correspond to a Grignard reagent RCH_2MgBr. Similarly, an electrophilic synthon $R\overset{+}{C}O$ could correspond to an acid halide RCOCl or ester RCOOR′. Where disconnection does not produce an obvious electrophilic or nucleophilic synthon, all possible structurally related reagents should be considered.

Disconnections are made by either disconnecting functional groups or the carbon skeleton. It is normal to first attempt to disconnect the sections that are held together by functional groups such as esters, amides and acetals because it is usually easier to find reconnection reactions for these functional groups. Consider, for example, the synthesis of the local anaesthetic benzocaine. The most appropriate disconnections are the ester and amine groups. At this point it is a matter of experience as to which disassembly route is followed. The normal strategy is to pick the synthons that give rise to reagents that can most easily be reformed into the product. Consequently, in this case the ester disconnection would appear to be the most profitable pathway because ester formation is relatively easy but it is not possible to directly introduce a nucleophilic amino group into a benzene ring.

Benzocaine

Key:

||| **Indicates the real compound (derived from the synthon) that is used in the reconnection reaction**

P-Aminobenzoic acid PABA

Ethanol is a readily available starting material but 4-aminobenzoic acid is not. Therefore, the next step is to consider the disconnection of the amino and carboxylic acid groups of 4-aminobenzoic acid. However, there are no simple inexpensive reactions for the reconnection of these groups. Consequently, the next step has to be a **functional group interconversion** (FGI). Disconnection arrows are usually used for FGIs but as no synthons are involved it is customary to use real structures in the relationships. FGIs are found by searching the literature for suitable reactions. This search, in the case of the current example, reveals that aromatic carboxylic acids may be produced by the oxidation of an aromatic methyl group, whereas an aromatic amine may be produced by reduction of the corresponding nitro group. Because amino groups are sensitive to oxidation, the disconnection via functional group interconversion follows the order:

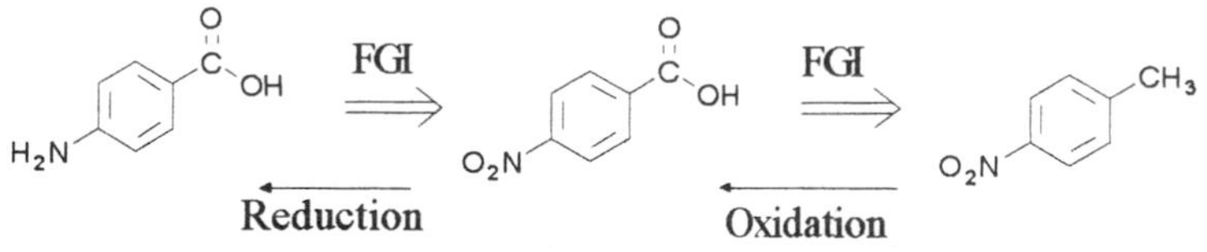

The final stage is to consider the disconnection of both the methyl and nitro groups.

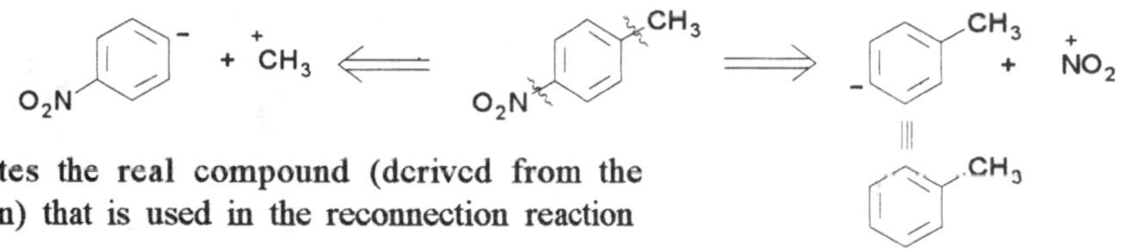

Toluene is a readily available compound so the best disconnection is the nitro group. This is also supported by the fact that it is easy to form 4-nitrotoluene by nitration of toluene. Consequently, the complete synthesis is:

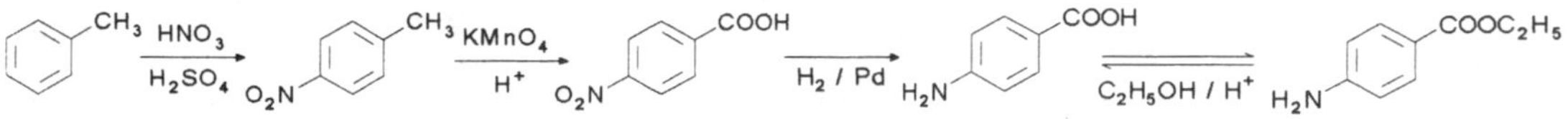

Carbon skeletons are usually disassembled at a point adjacent or near to a functional group or at a branch point in the structure. Consider, for example, the synthesis of 4-phenyl-1,4-butyrolactone (Figure 12.16). The scheme for the disconnection starts with the lactone. It is followed by FGI of the alcohol to the ketone. Although there are other possibilities, these are the most obvious logical disassembles for these stages. However, there are several possible disconnections at this point. Routes 1, 2 and 3 give synthons that do not give rise to readily available compounds. Further disconnections would be needed. However, route 4 gives access to synthons corresponding to readily available inexpensive starting materials that can undergo reconnection by a carbanion nucleophilic substitution. Consequently, this route, which is the shortest, becomes the preferred pathway for the synthesis. The design of a synthesis using the disconnection approach must take into account:

(i) the order of disconnection, which could influence the ease and direction of subsequent reactions;
(ii) the need to protect a reactive group in a compound by the use of a suitable protecting agent (see section 12.2.5 and Figure 12.15);
(iii) the need to incorporate the appropriate chiral centres into the structure (see section 12.3.3).

A wide range of disconnections linked to suitable reconnections are known (Table 12.5). However, a number of functional groups are usually best introduced by FGI (Table 12.6). Once

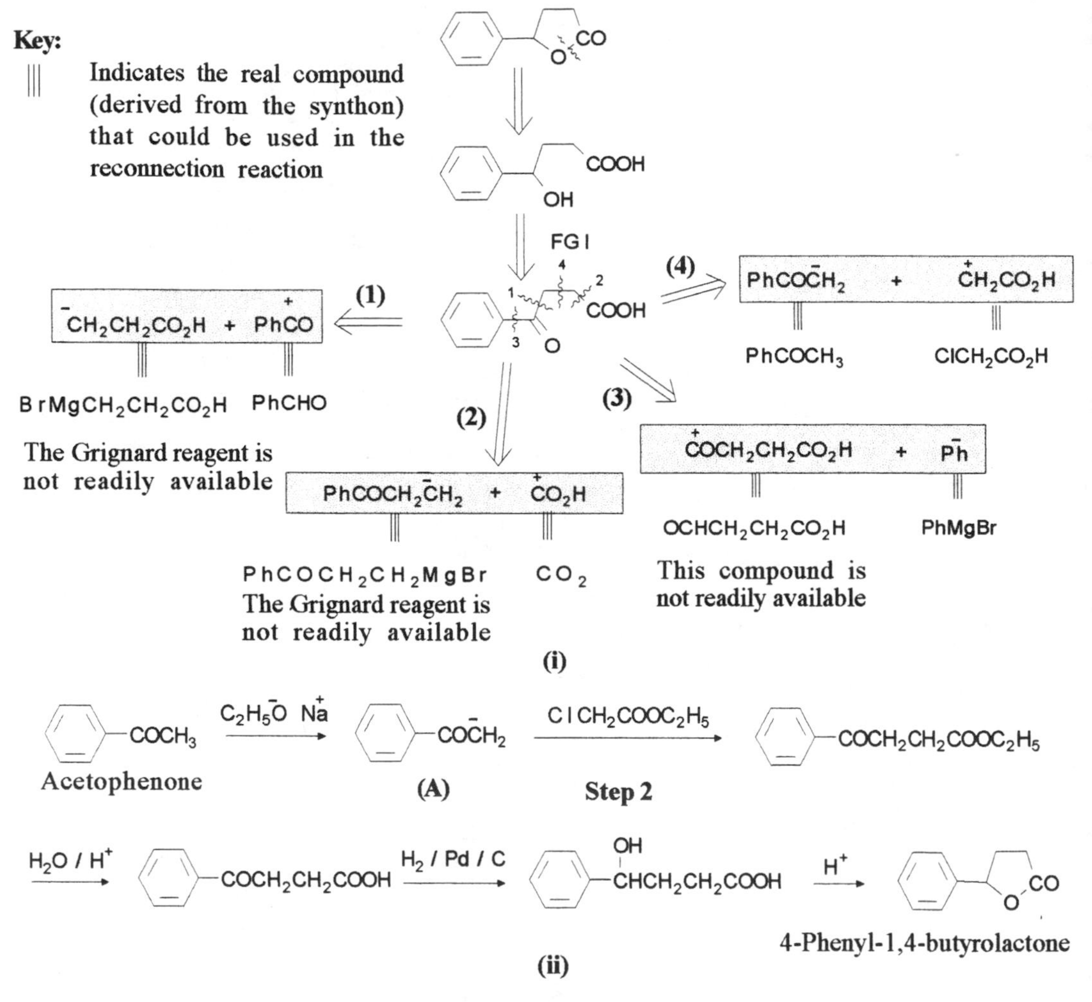

Figure 12.16. (i) The disconnection scheme for the synthesis of 4-phenyl-1,4-butyrolactone. The shaded structures are synthons. (ii) The complete synthesis is based on the disconnection flow chart. It is necessary to protect the acid at step 2 in the form of its ethyl ester in order to prevent it reacting with the carbanion A.

again it cannot be overemphasised that their use will depend on the experience of the designer, which only improves with usage and time.

12.4.2 Convergent Synthesis

The yields of the final product in a linear multistep synthesis (Figure 12.17a) may be small even though the yields of the individual steps are large. For example, a ten-step synthesis in which each of the steps has a yield of 90% will only have an overall yield of the target molecule of about 35%. A strategy known as **convergent synthesis** may be used to improve this overall

Table 12.5. Examples of the most useful disconnections and reconnection systems. The compounds used for the reconnection reactions are those normally associated with the synthons for the disconnection. These reactions are not the only reactions that could be used for the reconnection. Aolapted from S. Warren, *Organic Synthesis, the Disconnection Approach*, John Wiley and Sons, 1982.

Disconnections Examples of reconnection reactions

Functional group disconnections:

Ether:

$R^1{-}O{-}R^2 \Rightarrow R^1{-}O^- + {}^+R^2$

$R^1{-}OH \xrightarrow{Na} R^1{-}O^-\ Na^+ \xrightarrow{R^2-I} R^1{-}O{-}R^2$

$R^1{-}OH$: Source of the nucleophile; $R^2{-}I$: Source of the electrophile

For methyl ethers use dimethyl sulphate instead of R^2I

Amide:

$R^1NH{-}COR^2 \Rightarrow R^1{-}\bar{N}H + R^2{-}\overset{+}{C}{=}O$

$R^1{-}\bar{N}H$ ≡ All types of amines; $R^2{-}\overset{+}{C}{=}O$ ≡ Suitable acid derivatives

$R^1NH_2 + R^2COCl \longrightarrow R^1NH{-}COR^2$

R^1NH_2: Source of the nucleophile; R^2COCl: Source of the electrophile

Note: The acid anhydride will give the same compound but by a less vigorous reaction.

Ester:

$R^1CO{-}OR^2 \Rightarrow R^1{-}\overset{+}{C}{=}O + R^2{-}O^-$

$R^1{-}\overset{+}{C}{=}O$ ≡ Suitable acid derivatives; $R^2{-}O^-$ ≡ All types of alcohol and phenol

$R^1COCl + R^2OH \longrightarrow R^1CO{-}OR^2$

R^1COCl: Source of the electrophile; R^2OH: Source of the nucleophile

Note: The acid anhydride will give the same compound but by a less vigorous reaction.

Lactone:

Lactone (cyclic O–CO) $\Rightarrow$ O^- … $\overset{+}{C}{=}O$ (open chain)

Hydroxy acid (OH … COOH) $\xrightarrow{H^+}$ lactone (O … CO)

Source of both the nucleophile and electrophile

Acetal

$R^1R^2C(OR)_2 \Rightarrow R^1R^2C^+ + {}^-OR + {}^-OR$

$R^1R^2C{=}O \xrightarrow{ROH\ /\ H^+} R^1R^2C(OR)_2$

$R^1R^2C{=}O$: Source of the electrophile; ROH: Source of the nucleophile

Carbon–carbon disconnections:

Carbanion–electrophile disconnections: Disconnection should normally occur adjacent to an electron-withdrawing group. In each case, one of the compounds derived from the synthon should be able to form a carbanion and the structure of the other should contain an electrophilic centre. Reconnection is by means of a suitable carbanion substitution or condensation reaction.

$R{-}CH(Z){-}CH_2R^1 \Rightarrow R{-}\bar{C}H(Z) + {}^+CH_2R^1$

Z is an electron-withdrawing group, e.g.NO_2, COOR, SOR, SO_2R, CN

$RCO{-}CH_2{-}COOC_2H_5 \xrightarrow{Base} [RCO{-}\bar{C}H{-}COOC_2H_5]$ (Carbanion) $\xrightarrow{R^1CH_2-Br} RCO{-}CH(CH_2R^1){-}COOC_2H_5$

$R^1CH_2{-}Br$: Source of the electrophile

Friedel–Crafts disconnections:

$Ar{-}CO{-}R \Rightarrow Ar^- + {}^+C(=O){-}R$

Ar = an aromatic system.

$ArH \xrightarrow[AlCl_3]{RCOCl\ \text{or}\ RCOOCOR} ArCOR$

ArH: Source of the nucleophile

The position of substitution will depend on the nature of the substituents on the aromatic (Ar) ring system

Table 12.5. **Continued**, The reactions quoted are not the only reactions that could be used for the reconnection. Source: *Organic Synthesis, the disconnection approach,* S. Warren, John Wiley and Sons. 1982.

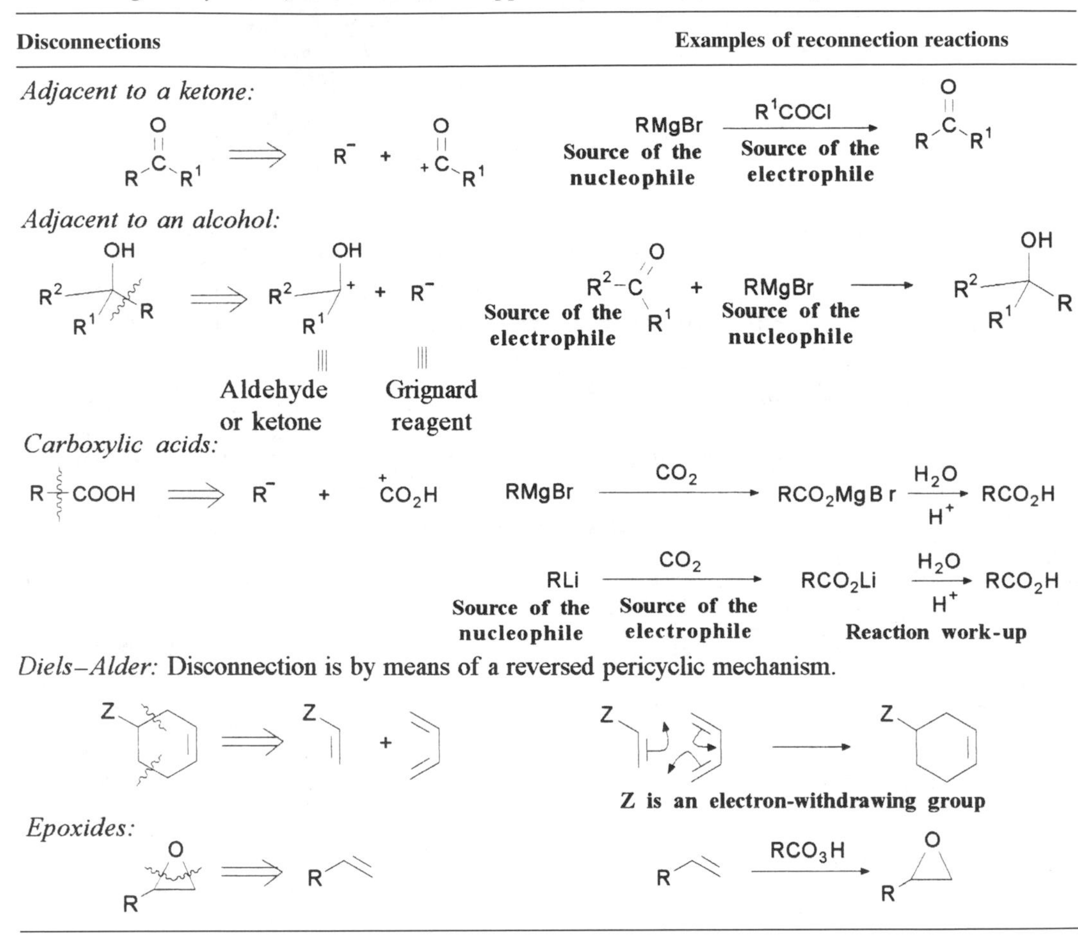

yield. In convergent syntheses, sections of the target molecule are prepared and joined together to form the target molecule (Figure 12.17b). A two-path convergent synthesis would have the effect of increasing the overall yield of the same ten-step synthesis to about 53% and also reducing the number of steps. It is not necessary to restrict convergence to two paths and the overall yield could be improved further by the use of several convergent pathways (Figure 12.17c).

The process of disconnection may be applied to target molecules by initially disconnecting large sections of the structure that could be suitable for individual linear synthesis. Consider, for example, the synthesis of the antihistamine *N,N*-dimethyl-2-[1-(4-chlorophenyl)-1-methyl-1-phenylmethoxy]ethylamine ({I} Figure 12.18). Disconnection of the ether group will

Table 12.6. Examples of reactions used to reverse FGIs, R can be both an aromatic and an aliphatic residue but Ar is only an aromatic residue.

Alkenes: By elimination of alcohols and halides.

HO–C–C–H $\xrightarrow{H^+}$ C=C   Br–C–C–H $\xrightarrow{OH^-}$ C=C

Aldehydes and ketones: By oxidation of the appropriate alcohol.

$RCH_2OH \xrightarrow{\text{Oxidation}} RCHO$ (Primary alcohol → Aldehyde) $R(R^1)CH\text{-}OH \xrightarrow{\text{Oxidation}} R(R^1)C{=}O$ (Secondary alcohol → Ketone)

Amines:

$R{-}NO_2 \xrightarrow{\text{Reduction}} R{-}NH_2$ $RCONHR^1 \xrightarrow{\text{Reduction}} RCH_2NHR^1$

$RNH_2 + O{=}C(R)(R^1) \xrightarrow{\text{Condensation}} RN{=}C(R)(R^1) \xrightarrow{\text{Reduction}} RNH{-}CH(R)(R^1)$

Useful one-carbon additions to aromatic ring systems:

$Ar{-}H \xrightarrow[\text{Chloromethylation}]{HCHO\ /\ HCl\ /\ ZnCl_2} Ar{-}CH_2Cl$ $Ar{-}H \xrightarrow[\text{Riemer–Tiemann}]{CHCl_3\ /\ OH} Ar{-}CHO$

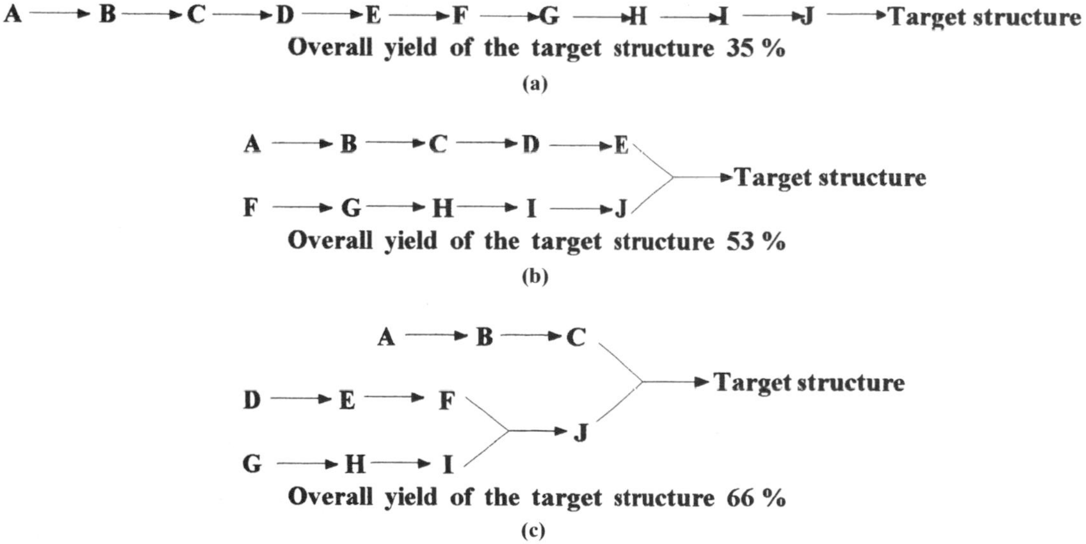

A → B → C → D → E → F → G → H → I → J → Target structure

Overall yield of the target structure 35 %

(a)

A → B → C → D → E ⟩
F → G → H → I → J ⟩ → Target structure

Overall yield of the target structure 53 %

(b)

A → B → C ⟩ → Target structure
D → E → F ⟩ → J
G → H → I ⟩

Overall yield of the target structure 66 %

(c)

Figure 12.17. A schematic representation of (a) linear and (b, c) convergent syntheses. The overall yields are based on a 90% yield for each step.

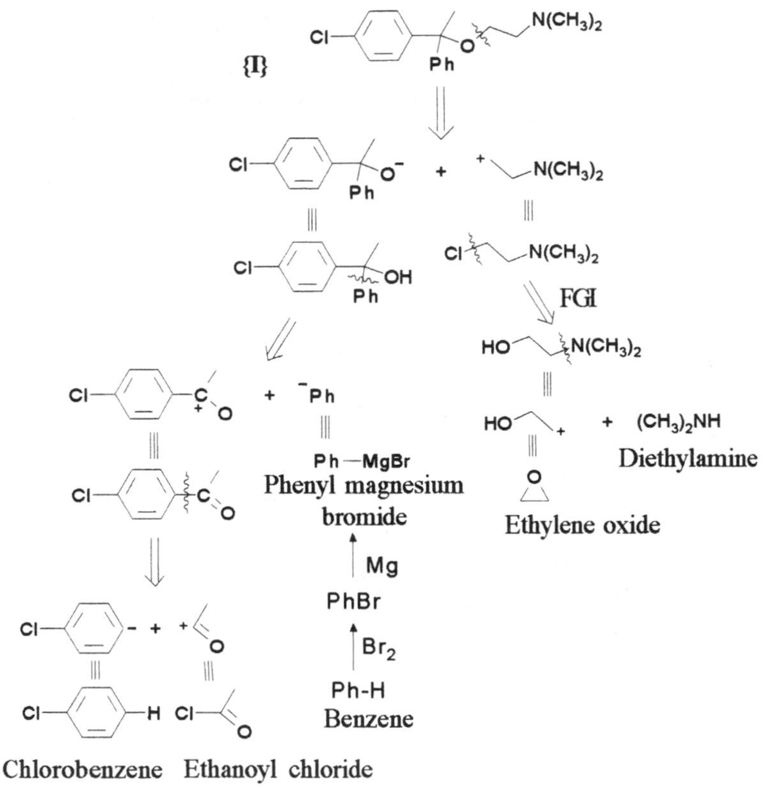

Figure 12.18. The disconnection scheme for the synthesis of the antihistamine *N,N*-dimethyl-2-[1-(4-chlorophenyl)-1-methyl-1-phenylmethoxy]ethylamine.

divide the target molecule into two reasonable-sized synthons. These synthons correspond to the nucleophilic and electrophilic precursors of the drug, namely the tertiary alcohol and the substituted alkyl halide. Because these compounds may be readily combined to form the drug it is now possible to design a convergent synthesis in which each of these compounds is prepared separately before being combined to form the antihistamine (Figure 12.18). First consider the synthesis of the tertiary alcohol. As tertiary alcohols are easily prepared using Grignard reagents, the obvious disconnections for simplifying the structure are the aromatic ring systems. The phenyl group is the better disconnection as it gives rise to the synthons of 4-chloroacetophenone and benzene. The latter can be reconnected to the 4-chloroacetophenone by a Grignard reaction using phenyl magnesium bromide. However, disconnection of the 4-chlorophenyl would have given the synthon of 1,4-dichlorobenzene, which would require the mono Grignard derivative for the reconnection step. Synthesis of this derivative would not be simple. The final step is the disconnection of the ketone from the 4-chloroacetophenone, which yields the readily available chlorobenzene and ethanoyl chloride.

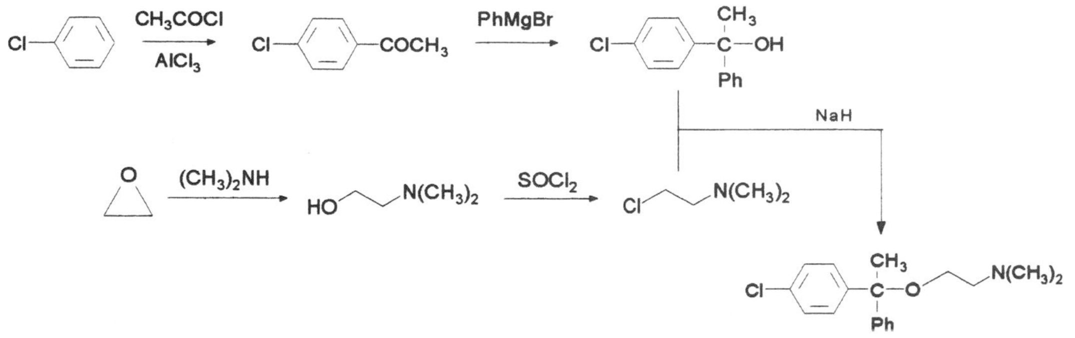

Figure 12.19. A scheme for the synthesis of *N,N*-dimethyl-2-[1-(4-chlorophenyl)-1-methyl-1-phenylmethoxy]ethylamine.

The synthesis of the substituted alkyl halide in the convergent synthesis starts with an FGI because halides are usually produced by either direct action of a halogen or substitution of an alcohol. Alcohols are produced from epoxides by nucleophilic substitution and so the disconnection of the secondary amino group leads to ethylene oxide and diethylamine as the readily available starting compounds for the synthesis (Figure 12.9).

12.5 Partial Organic Synthesis of Xenobiotics

Partial synthetic pathways use biochemical and other methods to produce the initial starting materials, and traditional organic synthesis to convert these compounds to the target structure. The principal methods are based on compounds produced by microbiological transformations, the use of enzymes and extraction from natural products. These methods are used to produce starting materials because it usually cuts down the cost of production and produces compounds whose structures have chiral centres with the required configurations. For example, the total synthesis of steroidal drugs is not feasible because of the many chiral centres found in their structures. Consequently, partial synthesis is the normal approach to producing new analogues and manufacturing steroidal drugs. For example, the starting material for the production of progesterone is diosgenin obtained from a number of *Dioscorea* species (a plant source). Diosgenin may be converted to pregnenolone acetate by a series of steps (Figure 12.20). This compound serves as the starting material for the synthesis of a number of steroidal drugs, including progesterone.

The disconnection approach may also be used to design the steps in the partial synthesis of the target molecules from these types of starting materials.

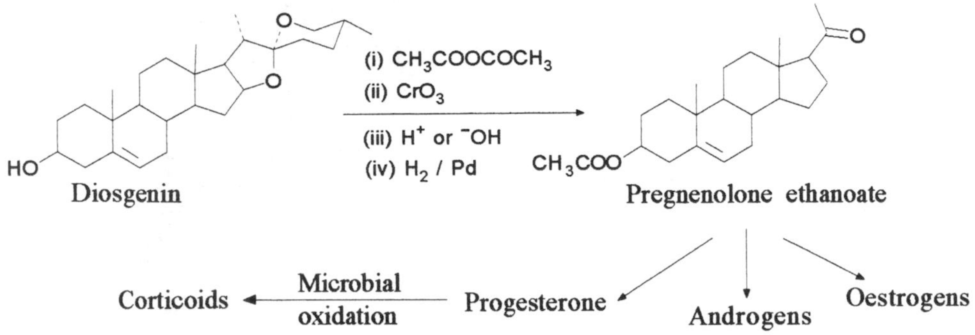

(i) $CH_3COOCOCH_3$
(ii) CrO_3
(iii) H^+ or ^-OH
(iv) H_2 / Pd

HO
Diosgenin
CH_3COO
Pregnenolone ethanoate

Corticoids ← Microbial oxidation — Progesterone
Androgens
Oestrogens

Figure 12.20. An outline of the synthesis of progesterone from diosgenin.

12.6 Summary

The **pathways** used to synthesise potential drugs may be classified as either **partial** or **full synthetic routes**. Partial routes tend to be a combination of organic synthesis and other methods whereas full synthetic routes only use organic synthesis methods. In both cases the starting materials should be cheap and readily available. Furthermore, the reactions used for synthetic pathways should be easy to carry out, the products should be easy to purify and identify, have high yields and be adaptable to large-scale production.

Synthetic pathways may be **linear**, where one step leads to another, until the target molecule is reached (see section 12.4). Alternatively they may be **convergent**, in which two or more sections of the target molecule are synthesised separately before being combined to form the target structure (see section 12.4.2). In both cases the disconnection approach may be used to design the synthetic pathway and identify suitable starting materials. However, all designs to produce a lead compound should also have features that enable them to be easily modified to produce a wide range of analogues.

The design of a synthetic pathway may require a reaction to be carried out at one centre in a molecule whilst preventing a second centre from reacting. This objective is usually achieved by the use of **protecting groups**. Protecting groups must produce stable structures under the prevailing reaction conditions but must be relatively easy to remove at an appropriate stage in the synthesis (see Table 12.1).

The presence of **chiral centres** in a target structure requires either the use of a **non-stereospecific reaction** and the subsequent resolution of the enantiomers or the use of

stereospecific reactions at a relevant stage in the pathway. Enantiomers are usually separated by forming **diastereoisomers** (see Figure 12.3). Diastereoisomers are usually separated by either fractional crystallisation or chromatography. **Stereospecific reactions** result in the selective production of an excess of one of the stereoisomers of the product. The extent of the excess is measured in terms of either the enantiomeric or diastereoisomeric excess. A number of interrelated factors influence stereospecificity, namely: the shape of the substrate about the reaction centre, the nature of the reagent, the mechanism of the reaction, the catalyst used and the relative activation energies of the pathways used to produce the isomers.

The **strategies used to introduce the chiral centres** into a molecule may be classified conveniently as **catalytic methods** or **non-catalytic methods**. **Enzyme-catalysed** reactions suffer from the disadvantage that they require large quantities of the enzyme or microorganism to produce significant quantities of the drug. However, a wide variety of enzyme-controlled stereospecific transformations are known. **Non-enzyme catalysts** may also be stereospecific. These catalysts are often complexes with chiral ligands.

Non-catalytic methods use chiral building blocks, a chiral auxiliary or achiral substrates and reagents. The **chiral building block method** depends on the use of reactions that do not affect the configuration of the asymmetric centre(s) of the building block. It may also require the use of enantiomerically pure substrates and reagents. The **chiral auxiliary method** combines an achiral substrate with a pure enantiomer known as the chiral auxiliary to form a chiral intermediate, which is treated with a suitable reagent to form the new asymmetric centre. This reaction is made stereospecific by the presence of the auxiliary, which is removed once the reaction is completed. The use of **achiral substrates and reagents** to form new asymmetric centres often results in the formation of a racemic modification that has to be resolved into its enantiomers.

The **disconnection approach** to the design of an organic synthesis starts with the target structure and works backwards by artificially cutting the structure into sections known as **synthons**. The most useful disconnections are made by heterolytic fission of the appropriate bond. Each synthon is converted into a related real compound known as a **reagent** (see Tables 12.5 and 12.6). These reagents act as either the electrophile or the nucleophile for a reconnection reaction for the disconnection. Disconnections are made by disassembling functional groups or carbon–carbon bonds. The disconnection analysis is continued for each reagent in the system until a set of readily available reagents is obtained.

The overall yield of a synthetic pathway may be increased by the use of **convergent and partial synthetic pathways**. **Partial synthetic pathways** are those that use biochemical and other methods to produce the initial starting materials for the synthesis. The disconnection method may also be used in the design of convergent and partial synthetic pathways.

12.7 Questions

(1) Explain the meaning of the terms (a) linear, (b) convergent and (c) partial synthetic pathways.
(2) Outline the practical considerations that need to be taken into account when selecting reactions for use in a synthetic pathway.
(3) What are the requirements for a good protecting group?
(4) Describe the factors that appear to influence the stereochemistry of a chemical reaction.
(5) Describe the use of catalysts in asymmetric synthesis.
(6) What is a chiral auxiliary? Suggest a feasible stereospecific synthesis for 2*R*-methylhexanoic acid starting from *S*(−)-proline and propanoic anhydride.
(7) Explain the meaning of the term 'synthon'. Draw the best synthons and their corresponding real compounds for each of the following compounds:

(a) CH_3 O O Ph (b) OC O (c) OH $COOC_2H_5$ $COOC_2H_5$ Cl (d) CO O CO

(8) What is the significance of the initials FGI in the disconnection approach to designing synthetic pathways? Suggest the best disconnection sequences for the synthesis of each of the following compounds:

(a) OH $NHCOCH_3$ (b) COOH

Selected Further Reading

Drug Safety

Glaxo Group Research, *Drug Safety, a Shared Responsibility*, Churchill Livingstone, Edinburgh, 1991.

General Chemistry

G. Thomas, *Chemistry for Pharmacy and the Life Sciences including Pharmacology and Biomedical Science*, Prentice Hall, London, 1996.

F. A. Cotton and G. Wilkinson, *Basic Inorganic Chemistry*, Fifth Edition, John Wiley and Sons, New York, 1988.

J. G. Morris, *A Biologist's Physical Chemistry*, Second Edition, Edward Arnold, London, 1974.

Synthetic Chemistry

S. Warren, *Organic Synthesis, the Disconnection Approach*, John Wiley and Sons, Chichester, 1982.

R. S. Atkinson, *Stereoselective Synthesis*, John Wiley and Sons, Chichester, 1995.

R. S. Ward, *Selectivity in Organic Synthesis*, John Wiley and Sons, Chichester, 1999.

A. N. Collins, G. N. Sheldrake and J. Crosby (Editors) *Chirality in Industry. The Commercial Manufacture and Applications of Optically Active Compounds*, John Wiley and Sons, Chichester, 1992.

Biochemistry

R. H. Garrett and C. M. Grisham, *Biochemistry*, Saunders College Publishing Harcourt Brace College Publishers, Fort Worth, TX, 1995.

R. H. Garrett and C. M. Grisham, *Molecular Aspects of Cell Biology*, Saunders College Publishing Harcourt Brace College Publishers, Fort Worth, TX, 1995.

D. Voet, J. G. Voet and C. W. Pratt, *Fundamentals of Biochemistry*, John Wiley and Sons, New York, 1999.

Pharmacology

H. P. Rang, M. M. Dale and J. M. Ritter, *Phrmacology*, Third Edition, Churchill Livingstone, Edinburgh, 1995.

A. Galbraith, S. Bullock, E. Manias, B. Hunt and A. Richards, *Fundamentals of Pharmacology*, Addison Wesley Longman Limited, Harlow, Essex, 1997.

Inorganic Medicinal Chemistry

D. E. Fenton, *Biocoordination Chemistry*, Oxford Chemistry Primers, Oxford, 1995.

D. M. Taylor and D. R. Williams, *Trace Elements in Medicine and Chelation Therapy*, The Royal Society of Chemistry, Cambridge, 1995.

S. J. Lippard and J. M. Berg, *Principles of Bioinorganic Chemistry*, University Science Books, Mill Valley, CA, 1994.

J. Lancaster (Editor), *Nitric Oxide, Principles and Actions*, Academic Press, San Diego, CA, 1996.

L. Ignaro and F. Murad, *Nitric Oxide, Biochemistry, Molecular Biology and Therapeutic Implications*, Academic Press, San Diego, CA, 1995.

Medicinal Chemistry

J. N. Delgado and W. A. Remers (Editors), *Wilson and Gisvold's Textbook of Organic Medicinal and Pharmaceutical Chemistry*, Tenth Edition, Lippincott-Raven, Philadelphia, PA, 1998.

H. J. Smith (Editor), *Smith and Williams' Introduction to the Principles of Drug Design and Action*, Third Edition, Harwood Academic Publishers, Amsterdam, 1998.

T. Nogrady, *Medicinal Chemistry*, Second Edition, Oxford University Press, New York, 1988.

F. D. King (Editor), *Medicinal Chemistry, Principles and Practice*, The Royal Society of Chemistry, Cambridge, 1994.

M. E. Wolf (Editor), *Burger's Medicinal Chemistry*, Fifth Edition, John Wiley and Sons, New York, 1997.

A. Gringauz, *Introduction to Medicinal Chemistry, How Drugs Act and Why*, Wiley-VCH Inc., New York, 1997.

D. Lednicer (Editor), *Chronicals of Drug Discovery*, Vols 1–3, John Wiley and Sons, New York, 1982–1993.

Combinatorial Chemistry

S. R. Wilson and A. W. Czarnick (Editors), *Combinatorial Chemistry, Synthesis and Application*, John Wiley and Sons, New York, 1997.

N. K. Terrett, *Combinatorial Chemistry*, Oxford University Press, Oxford, 1998.

Pharmacy and Related Subjects

M. E. Aulton (Editor), *Pharmaceutics, the Science of Dosage Form Design*, Churchill Livingstone, Edinburgh, 1988.

W. Sneader, *Drug Development from Laboratory to Clinic*, John Wiley and Sons, Chichester, 1986.

M. Rowland and T. N. Tozer, *Clinical Pharmacokinetics*, Third Edition, Williams and Wilkins, Baltimore, MD, 1995.

J. Merrills and J. Fisher, *Pharmacy Law and Practice*, Blackwell, Oxford, 1995.

Appendices

Appendix 1: Regression Analysis

In medicinal chemistry it is often desirable to obtain mathematical relationships in the form of equations between sets of data that have been obtained from experimental work or calculated using theoretical considerations. Regression analysis is a group of mathematical methods used to obtain such relationships. The data are fed into a suitable computer program, which, on execution, produces an equation that represents the line that is the best fit for those data. For example, an investigation indicated that the relationship between the activity and the partition coefficients of a number of related compounds appeared to be linear (Figure A.1). Consequently, these data could be represented mathematically in the form of the straight-line equation: $y = mx + c$. Regression analysis would calculate the values of m and c that gave the line of best fit to the data. When one is dealing with a linear relationship the analysis is usually carried out using the method of least squares.

Regression equations do not indicate the accuracy and spread the data. Consequently, they are normally accompanied by additional data that, as a minimum requirement, should include the number of observations used (n), the standard deviation of the observations (s) and the correlation coefficient (r).

The value of the correlation coefficient is a measure of how accurately the data match the equation. It varies from zero to one. A value of $r = 1$ indicates a perfect match. In medicinal chemistry r values greater than 0.9 are usually regarded as representing an acceptable degree of accuracy provided that they are obtained using a reasonable number of results with a suitable standard deviation.

The value of $100r^2$ is a measure of the percentage of the data that can be explained satisfactorily by the regression analysis. For example, a value of $r = 0.90$ indicates that 81% of the results can be explained satisfactorily by regression analysis using the parameters specified. It indicates that 19% of the data are not satisfactorily explained by these parameters and so indicates that the use of an additional parameter(s) might give a more acceptable account of the results. Suppose, for example, that regression analysis using an extra parameter gave a regression constant of 0.98. This shows that 96.04% of the data are now satisfactorily accounted for by the chosen parameters.

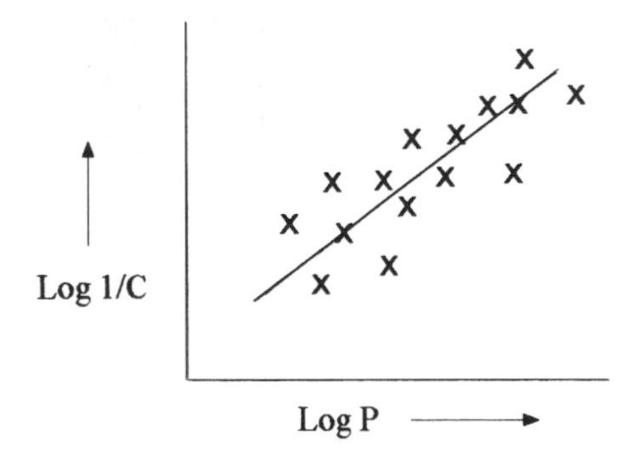

Figure A.1. A hypothetical plot of the activity (log 1/C) of a series of compounds against the logarithm of their partition coefficients (log P).

Appendix 2: Antibiotics

The site of action and the precise nature of the mechanism of action of many antibiotics are not known. However, the site of action of a number of commonly used antibiotics is known (Table A1). In all cases the action of antibiotics is classified as **bactericide** or **bacteriostatic**. The former destroys the bacteria whereas the latter prevents its reproduction. This difference can be an important consideration when deciding which antibiotic to use on a patient. The suffixes **-cide** and **-static** are in general use to indicate these types of action.

Table A.1. Examples of the sites and types of action of some antibiotics.

Drug	Site of action	Type of action
Polymixins	Plasma membrane	Bactericide
Lincomycins	Ribosome 50S subunit	Bacteriostatic
Rifampin	DNA or RNA	Bactericide
Chloramphenicol	Ribosome 50S subunit	Bacteriostatic
Erythromycin	Ribosome 50S subunit	Bacteriostatic
Actinomycin	DNA and RNA	Bactericide

Appendix 3: Svedberg Units (S)

Svedberg units (S) are a measure of the size of large molecules, especially nucleic acids. The unit is a measure of the rate of sedimentation of the molecule in an ultracentrifuge and is defined as:

$$S = 10^{-13}\,s$$

where s is the sedimentation coefficient of the molecule. The value of the sedimentation coefficient is usually corrected to the value that would be obtained at 20°C using a liquid that has the density and viscosity of water.

Appendix 4: Cancer

Cancer is a genetic disease that can occur in all types of body tissue. It is found in many forms, including solid tissue formations (tumours or neoplasms), leukaemias (blood cancer) and lymphomas (cancer of the lymphoid cells). Cancers are due to a reduction or loss of control of the growth of cells. This leads to a proliferation of cell growth. In its early stages the cells formed by this growth resemble the parent but as the cancer progresses they lose the appearance and function of the parent cell. This loss of function, if left unchecked, will become life threatening. For example, a growing tumour will obstruct, block and generally affect adjacent organs. If these are nerves it will cause pain. Furthermore, cancer cells are invasive. As the cancer grows the cells lose their adhesion and the malignant cells are carried in the blood to other parts of the body. These cells lodge in different parts of the body and grow into so-called secondary cancers.

Appendix 5: AIDS

AIDS is a disease that progressively destroys the human immune system. It is caused by the human immunodeficiency virus, which is a retrovirus. This virus enters and destroys human T4 lymphocyte cells. These cells are a vital part of the human immune system. Their destruction reduces the body's resistance to other infectious diseases, such as pneumonia, and some rare forms of cancer. Treatment is more effective when a mixture of antiviral agents is used.

Appendix 6: Sickle-cell Anaemia

Sickle-cell anaemia is caused by a defective gene. This gene results in the replacement of the glutamate at position 6 of the β-chains of haemoglobin (Hb) by a valine residue to produce sickle-cell haemoglobin (HbS). These HbS molecules aggregate into long polymer-like structures, which results in long sickle-shaped red blood cells instead of the normal round cells. These red blood cells are able to transport oxygen in the normal fashion but their shape causes them to lodge in capillaries, which causes tissue damage and impairs blood circulation. This results in headaches, dizziness and ultimately death.

Answers to Questions

Numerical answers may be slightly different from the ones you obtain due to differences in calculating methods and equipment. However, a correct answer is one that approximates to the given answer. Answers that require the writing of notes are answered by either an outline of the points that should be included in the answer or a reference to the appropriate section(s) of the text. Where questions have more than one correct answer only one answer is given.

Chapter 1

(1) (a) The *ortho* ethyl groups will sterically hinder the ester group. This will slow the rate of hydrolysis of the compound, which is likely to either increase its duration of action or inactivate it by preventing the compound from binding to its target site.

(b) The trimethylammonium group is permanently positively charged. This permanent charge will reduce the ease with which the molecule passes through biological membranes. In particular it would probably prevent the compound passing through the blood–brain barrier.

(c) The replacement of an ester group by an amide group can reduce the rate of metabolism of the compound by hydrolysis, which may prolong the action of a drug. It may also change the biological activity of the compound.

(2) The nature of the pathological target. Its site of action, nature of desired action, stability, ease of absorption and distribution, metabolism, dosage form and regimen.

(3) (a) See section 1.3, (b) see section 1.4, (c) see section 1.4, (d) see section 1.4, (e) see section 1.4.

(4) For definitions, see section 1.5. The factors affecting the pharmacokinetic phase are absorption, distribution, metabolism and elimination. The factor affecting the pharmacodynamic phase is the stereoelectronic structure of the compound.

(5) The reduction in pH reduces the negative charge of the albumin and so increases its electrophilic character. Therefore, as amphetamine molecules are nucleophilic in nature, their binding should improve with the decrease in pH. Part of this binding will involve salt formation between the amphetamine and the albumin. Amphetamine is more likely to form salts in which it acts as the positive ion as the electrophilic nature of the albumin increases.

(6) See section 1.5.3.

(7) Replace the ester (lactone) with a less easily hydrolysed amide group. Introduce bulky ethyl or propyl groups on either side of the lactone group to reduce the ease of hydrolysis. Use an enteric coating.
(8) (a) See section 1.2, (b) see sections 1.3 and 1.5, (c) see section 19.1, (d) see section 1.9.1.
(9) See section 1.9.

Chapter 2

(1) See section 2.1.
(2) See section 2.2.
   (a) Replacing a rigid structure by an equivalent-sized rigid structure gives a better chance of an active analogue. The shape of a rigid structure may give information about the size and shape of its receptor provided that the rigid part of the molecule is the part that binds to the receptor. If the part of a ligand that binds to the receptor is a rigid structure it can also indicate the best conformation for analogues to bind to the receptor.
   (b) Different conformations can result in different activities and potencies.
   (c) See Table 2.1.
(3) Structure Activity Relationships. SAR are the general relationships obtained from a study of the changes in activity with changes in the structure of a lead. These changes are used to find or predict the structure with the optimum activity. Example, see Figure 2.6.
(4) (a) See section 2.3.3.2. (b) CF_3. It is approximately the same size as a chlorine atom.
(5) (a) See section 2.3.3 and Table 2.3. (b) See section 2.3.
(6) (a) See section 2.3.2.5. (b) See section 2.3.2.1. (c) See section 2.3.2.6.
(7) See section 2.4.
(8) Lipophilicity; see section 2.4.1 (P and π) Shape; see section 2.4.3 (E_s and MR). Electronic effects; see section 2.4.2 (σ).
(9) (a) See section 2.4.4.
   (b) (i) n = the number of compounds used to derive the equation; s = the standard deviation for the equation; r = the regression constant, the nearer its value to 1 the better the fit of the data to the Hansch equation. Equations are normally said to have an acceptable degree of accuracy if r is greater than 0.9.
   (ii) See section 2.4.4, particularly Equation (2.18).
   (iii) A more polar substituent would have a negative π value (see Table 2.5) that could reduce the activity of the compound.
(10) (a) See section 2.4.5. (b) It is only possible to use it for analogues of a compound whose structure contains either a non-fused benzene residue and/or an aliphatic side chain substituent on a non-fused benzene residue.
(11) See sections 2.6.2 and 2.6.3.

(12) The objectives of the synthesis; do they require the formation of a library of separate compounds or mixtures?
The size of the library.
Solid or solution phase?
If solid phase is selected, use parallel synthesis or Furka's mix and split?
The nature of the building blocks and their ease of availability.
The suitability of the reactions used in the sequence.
The method of identification of the structures of final products.
The nature of the screening tests and procedure.
(13) See section 2.6.5.
(14) See section 2.6.5, particularly Figure 2.25.
(15) See sections 2.6.5.1 and 2.6.5.2.
(16) Adapt the scheme given in Figure 2.23 using the appropriate R groups.
(17) See sections 2.6.5.3, 2.6.5.4 and 2.6.5.5.
(18) See sections 2.5.6. and 2.5.7.

Chapter 3

(1) Compounds that are reasonably water soluble are more easily absorbed, transported to their site of action and eliminated from the body.
(2) See section 3.2.1.
(3) (a) (i) See section 3.3, (ii) see section 3.3, (b) see section 3.3.2.1.
(4) $2.99 \times 10^{-9}\,\mathrm{mol^3\,dm^{-9}}$.
(5) $3.844 \times 10^{-9}\,\mathrm{mol\,dm^{-3}}$.
(6) Convert mmHg to atmospheres and use Henry's law.
Solubility 0.89 mg per 100 g of water
(7) High polar group to carbon atom ratio (see section 3.5).
The presence of polar groups that can hydrogen bond to water molecules and ionise in water (see section 3.5).
(8) Form salts that would improve water solubility but would break down to yield the drug in the biological system (see section 3.6).
Introduce water-solubilising groups into a part of the structure that is not the pharmacophore of the drug (see section 3.7.3).
Formulate as a suitable dosage form (see sections 3.8 and 3.10).
(9) For general and specific methods, see section 3.7.4.
(a) Any method in sections 3.7.4.2–3.7.4.4 inclusive.
(b) Any method in section 3.7.4.5.
(c) Any method in section 3.7.4.6.
(10) (a) The degree of ionisation of codeine in the stomach is 99.99%. Most of the codeine is

in the form of the corresponding ions and so the drug will not be readily absorbed in the stomach.

(b) The degree of ionisation of codeine in the intestine is 98.44%. This is slightly less than the degree of ionisation in the stomach and so the absorption of codeine will be slightly better in the intestine than the stomach. However, a considerable concentration of the codeine is in the form of the corresponding ions and so the drug will still not be readily absorbed from the intestine.

(11) (a) P° from a graph of log P against log (1/C) is 133.4.

(b) The analogue has a different site of action to the local anaesthetics in the series used to determine P°.

(12) See section 3.11.

(13) (a) Olive oil/water, (b) *n*-octanol/ water, (c) chloroform /water.

(14) See sections 3.11.1.1, 3.11.1.2 and 3.11.1.3.

Chapter 4

(1) Consult index to find the appropriate pages.

(2) See section 4.2.

The integral proteins that have transmembrane spans normally have their C-terminals (COOH) in the intracellular fluid and their N-terminals in the extracellular fluid. These groups ionise in these fluids to form anionic (negative) and cationic (positive) ions. These ions are responsible for the charges on the surfaces of the membrane.

(3) See section 4.2.1 and Figure 4.3.

(4) See section 4.2.5.1.

(5) See Figure 4.11 and section 4.2.5.1 for the essential features.

(6) (a) See section 4.4.1.

(b) See sections 4.3.6 and 4.3.7.

(c) See section 4.1.

(d) See sections 4.3.4 and 4.3.5.

(7) The degrees of ionisation are calculated from the Henderson–Hasselbalch equation for bases (see Equation 3.5). The higher the degree of ionisation, the greater the amount of drug existing in the form of its cation. Because charged compounds are usually less easily absorbed than the electrically molecules, the activity of the drug will decrease with an increase in the degree of ionisation. Consequently, using this as the basis for the prediction, the activities of the drugs will be in the order:

Benzocaine (0.000126 : 1) > Cocaine (0.016 : 1) > Procaine (5.62 : 1)

where the valuer in parentheses are the ratios of ionised to unionised form of the drug.

(8) See section 4.3.5. Example: levodopa.

(9) See section 4.4.2. for definitions and mode of action. When the concentrations of the ions being transferred by the ionophore are the same on both sides of the membrane.

(10) See section 4.4.2.2. β-Lactam antibiotics have to penetrate the cell walls of the bacteria in order to inhibit the cross-linking of the cell wall during its regeneration. The thicker cell envelope of Gram-negative bacteria makes it more difficult for the drug to reach its site of action. Not all the prion channels in this envelope will transfer β-lactam drugs. β-Lactamases in the periplasmic space will also hydrolyse the drug.

(11) See Figure 4.37. Increase the distance between the rings by inserting one or two CH_2 groups. And/or change the amide group to an ester group.

(12) See Figure 4.31.

(13) (a) The carboxylic acid group makes the analogue more polar than B and so it is likely to be less easily absorbed than B.
(b) The amino group also makes the analogue more polar than B and so it is likely to be less easily absorbed than B.
(c) The ester group makes the molecule less polar than B and so it is likely to be more easily absorbed than B.

Chapter 5

(1) (a) See section 5.1 and Figure 5.2, (b) see section 5.1.2, (c) see section 5.4, (d) see section 5.4.1.4, (e) see section 5.5.

(2) Absolute bioavailability and half-life as a measure of the rate of elimination. These parameters would give an indication of the relative effectiveness of each of the compounds. Absolute bioavailabiliy would indicate the compound with the best absorption characteristics whereas half-life would show which compound was the most stable *in situ* and so have the best chance of being therapeutically effective. They would also indicate which compound would be effective using a minimum dose (the lower the dose, the lower the chances of unwanted side effects).

(3) See section 3.2.

(4) Frequent small doses but cyclosporin will be given more frequently than digoxin.

(5) Plot a graph of log C against t. The slope is equal to $k_{el}/2.303$. (a) $1.84\,h^{-1}$, (b) $4.12\,dm^3$, (c) $7.58\,dm^3\,h^{-1}$. The assumption made is that the elimination exhibits first-order kinetics.

(6) At 1 min, if the clearance rate is $5\,cm^3\,min^{-1}$ then $5\,cm^3$ will be clear of the drug, that is, 5/50 of the drug will have been removed leaving 45 mg of the drug in the compartment. In the next minute another 5/50th of the remaining amount of the drug will be removed because the clearance rate is constant but this will be removed from the 45 mg, leaving 40.5 mg, and so on. The figures corresponding to the times ar e:

Time lapse (min):	1	2	3	4	5	6	7	8	9	10
Drug remaining (mg):	45	40.5	36.5	32.8	29.5	26.5	23.8	21.4	19.3	17.4

A logarithmic plot using logarithm to **base 10** of the concentration against time is a straight line with a slope of 0.4584. Assuming a first-order elimination process, the value of k_{el} (see Figure 5.10) calculated from this slope is:

$$2.303 \times 0.4584 = 1.056$$

and the value of $t_{1/2}$ calculated from k_{el} using Equation (5.8) is 0.0656 min.

(7) Absolute bioavailability is defined by Equation (5.39). The i.v. data required are the answers to question 5, as follows:

(1) Calculate the AUC for the intravenously administered dose by substituting Equation (5.27) in Equation (5.17). This gives:

$$\text{AUC} = \frac{\text{Administered dose}}{V_d k_{el}} = \frac{30}{4.12 \times 1.84} = 3.96$$

(2) Substitute this value in Equation (5.39) to give the absolute bioavailability:

$$\text{Absolute bioavailability} = \frac{5.01/50}{3.96/30} = 0.76$$

This value indicates that the i.v. bolus administration gives a significantly better bioavailability than oral administration.

(8) The parameters that could be used to compare the biological activities of the analogues are:

Half-life (calculate using Equation 5.8), which would give a measure of the duration of the action. The longer the half-life, the longer the time the drug is available in the body.

Absolute bioavailability (calculate using Equation 5.39). The bigger the absolute bioavailability, the greater the chance of a favourable biological action.

Analogue:	A	B	C	D
Half-life (min):	5	25	15	40
Absolute bioavailability:	1.036	1.526	0.812	1.175

The best analogue is D because it has the longest half-life and a reasonable bioavailability.

(9) $C_{ss} = k_o/Cl_p$ and $Cl_p = V_d k_{el}$.
Calculate k_{el} from the value of $t_{1/2}$ (use Equation 5.8).
$t_{1/2} = 0.2772\,h^{-1}$.
Convert V_d to cm^3 and calculate the value of Cl_p. Substitute Cl_p in the expression for C_{ss}.
Answer: Rate of infusion $= 4.35\,\mu g\,cm^3\,h^{-1}$.

Chapter 6

(1) See section 6.8.2.
(2) See section 6.2.
(3) See section 6.4.
(4) See section 6.6.
(5) See section 6.9. Assume a single-substrate reaction and reversible inhibition.
(6) See section 6.9.
(7) See section 6.11. Compound A is a non-competitive inhibitor with a Lineweaver–Burk plot pattern the same as that shown in Figure 6.19. Compound B is a reversible inhibitor with a Lineweaver–Burk plot pattern the same as that shown in Figure 6.21. Irreversible inhibitors are more suitable because they cannot be reversed by a build-up of the substrate, which occurs when the enzyme is inhibited. However, they do tend to have more side effects.
(8) See section 6.4.3.2. (a) Must bind to enzyme's active site; contain a group that is converted to a group that can react with the enzyme's active site. (b) See Figure 6.23 for mechanism. Improvement of action: insert an electron acceptor group α to the C–Cl bond. This increases the electrophilic nature of the carbon, making it more susceptible to nucleophilic attack by a group from the active site of the enzyme.
(9) Various answers are possible but in general incorporate an electrophilic group in the structure of lactate, e.g. replace the OH by I. Reason: the active site will have nucleophilic groups.
(10) See section 6.10. The groups normally found at the active site of an enzyme are nucleophilic in nature. Pyrrole has an electron-deficient ring and so to increase binding use an electron acceptor substituent whereas to decrease binding use an electron donor substituent.
(11) See section 6.13.

Chapter 7

(1) This is an essay style question. A suggested approach is to consider the action of metallo complexes from two major points of view, namely: their role in maintaining normal biological functions and their use in drug therapy to treat certain pathological conditions. For examples, see section 7.1: active sites of enzymes, maintaining enzyme structure, treatment of heavy metal poisoning, cancer, rheumatoid arthritis.
(2) (a) Diethylenetriamine (structure, Table 7.1), tridentate, six-electron donor.
    (b) Cyclopentadienyl (structure, Table 7.2), pentadentate five-electron donor.
    (c) Ethylenediaminetetraacetic acid (structure, Figure 7.17), hexadentate, 12-electron donor.
    (d) Ammonia, NH_3, monodentate, two-electron donor.

(3) (a) See Figure 7.9. The S of the thiol is the most likely but the N of the amino and the O of the carboxylic acid groups could also coordinate.
(b) See Figure 7.9. The N of the amino and the O of the carboxylic acid groups.
(c) See Figure 7.24. The ring nitrogen and the O of the phenolic hydroxy group.
(d) See Figure 7.17. The S of the thiol groups.

(4) (a) See section 7.2.1.3. (b) See section 7.2.2. (c) See section 7.2.4.

(5) See section 7.3.2 for definitions.
(a) Yes (H to H), S of Cys to metal, (b) No, (c) No, (d) Yes (S to S), S of thiolate to metal, (e) Yes (B to B), = N- of the imidazole, (f) Yes, OH of cholesterol to metal, (g) No, (h) Yes (S to S), both alkene and COOH.

(6) EDTA. The EDTA removes Ca^{2+} ions from the blood by forming a complex. This inhibits the start of blood clotting.

(7) The general requirements for designing a compound for use in metal detoxification are:
(i) it must be able to act as a multidentate ligand;
(ii) the ligand groups in the structure should be specific for the metal;
(iii) preferably it should form water-soluble complexes that can be easily excreted;
(iv) the compound should be able to form five- or six-membered rings;
(v) form complexes that are charged in solution in order to prevent side effects.

(8) (a) Cu^{2+} forms a more stable complex with EDTA than Ca^{2+} ions. Consequently, the concentration of Cu^{2+} ions will be reduced more than the concentration of Ca^{2+} ions.
(b) Yes. It will chelate the lead and mercury in preference to the chromium because the log K values of the EDTA complexes of these metals are higher than that for the EDTA–chromium complex. However, some EDTA–chromium complex will be formed because all complexes are formed by dynamic equilibria so the patient must be carefully monitored during treatment.
(c) The concentration of Fe^{3+} ions will be reduced more than the concentration of Fe^{2+} ions because Fe^{3+} ions form a more stable complex with EDTA than Fe^{2+} ions. The addition of an excess of zinc ions has no effect because iron forms a more stable complex than zinc.

(9) Three types of experiment must be carried out:
(i) Determine the growth of the *Staphylococcus aureus* on treatment with the potential drug but in the absence of iron. This should show a normal growth pattern if the presence of iron is essential to the action of the drug.
(ii) Determine the growth of the *Staphylococcus aureus* in the presence of iron but in the absence of the drug. This would show a normal growth rate if the *Staphylococcus aureus* is not affected by the iron alone.
(iii) Determine the growth of the *Staphylococcus aureus* in the presence of both the drug and iron (III). The growth should be significantly reduced because both of the active factors are present.

Chapter 8

(1) The definitions for (a), (b), (c), (d) and (f) are to be found in section 6.1 and that for part (e) is given in 8.3.

(2) This is a type 2 muscarinic cholinergic receptor for acetylcholine. It is a member of the type 2 superfamily.

(3) A schematic representation of the possible bonds involved in the binding of the drug to the receptor. Hydrophobic bonding will occur between the carbon chains. This is not shown on the diagram.

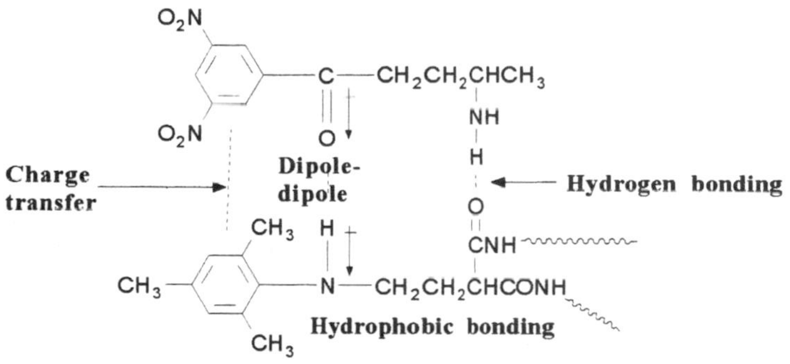

O_2N
C—$CH_2CH_2CHCH_3$
NH
O_2N
O
H
Charge transfer
Dipole-dipole
Hydrogen bonding
O
CH_3 H
CNH
CH_3
N—$CH_2CH_2CHCONH$
Hydrophobic bonding
CH_3

(4) See section 8.4.

(5) See section 8.5.

(6) See section 8.4.2.

(7) (a) See Figure 8.9. (b) DAG activates PKC, which controls a number of cellular processes. IP_3 initiates the rapid release of Ca^{2+} ions. These ions act as secondary messengers, initiating a range of cellular responses.

(8) Competitive antagonists compete for the same receptor as the agonist. However, competitive antagonists do necessarily bind to the same site as the agonist. They may bind to an alternative site close to the agonist,s receptor site. The binding of competitive antagonists is reversible. Consequently, the effect of a competitive antagonist may be reversed by increasing the concentration of the agonist. The increased concentration of the agonist displaces the antagonist from the receptor and restores the cellular response to the agonist Non-competitive antagonists also bind to or near to the same receptor site as the agonist. In this case the binding is not reversible and so increasing the concentration of the agonist does not restore the cellular response to the agonist.

(9) See section 8.6.2.

(10) See section 8.7.1.

Chapter 9

(1) (a) See section 9.1.1. (b, c) See section 9.9. (d) See section 9.8.

(2) A chemical species that can interact with water molecules will usually be more water soluble than one that does not interact with water molecules. The formation of hippuric acid increases the interaction of the molecule with water molecules and so is a more water-soluble conjugate. The carboxylic acid group will ionise in water to form an anion that could form ion–dipole bonds with water molecules. In addition the amide link can form hydrogen bonds with water molecules.

(3) See sections 9.1.3 and 9.1.5.

(4) See section 9.2.

(5) See Figure 9.4.

(6) See section 9.6.3. See the answer to question 3. The pH of the solution is correct for kidney excretion.

(7) **(a)**

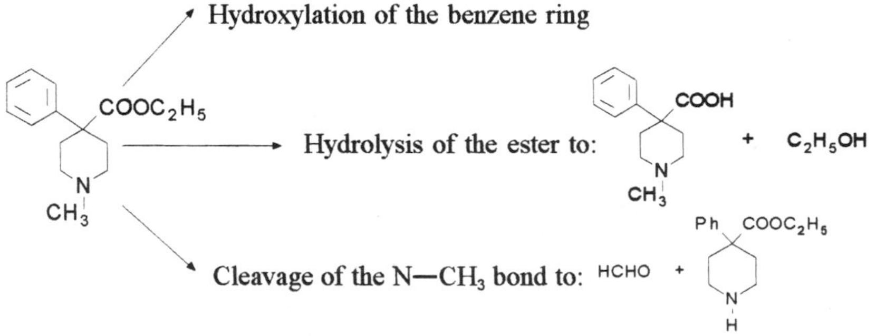

Hydroxylation of the benzene ring

$COOC_2H_5$ — N — CH_3 → **Hydrolysis of the ester to:** $COOH$, N, CH_3 + C_2H_5OH

Cleavage of the N—CH_3 bond to: HCHO + Ph $COOC_2H_5$ N H

(b) Hydroxylation of the benzene ring.

$-N{=}N-$ $-NH_2$ → Acylation → $-N{=}N-$ $-NHCOCH_3$

Reductive cleavage

$-NH_2$ H_2N- $-NH_2$

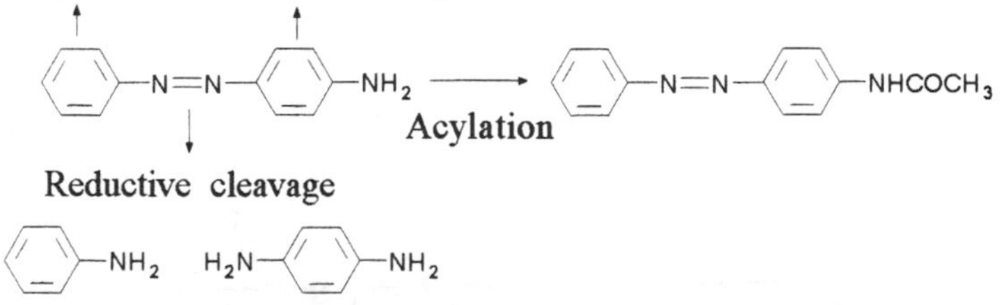

(8) A chemical species that can interact with water molecules will usually be more water soluble than one that does not interact with water molecules. There is a considerable interaction between glucuronic conjugates and water molecules because of the ionisable acid group and the numerous hydroxy groups that can hydrogen bond with water. There is a good supply of glucuronic acid in the body and supply and it is capable or reacting with many different functional groups on xenobiotics.

(9) (a) To inactive the drug. Metabolism to inert metabolites that are sufficiently water soluble to be readily excreted via the kidney.

(b) The α-carbon of an ethyl group *t*-amine is hydroxylated and cleaved to form ethanal and the *N*-ethylaminobenzene. Ethanal could be excreted via the lungs or be further metabolised to ethanoic acid. The *N*-ethylaminobenzene could be oxidised metabolically to the corresponding *N*-hydroxy compound or dealkylated to aminobenzene and ethanal.

(10) (a) A is metabolised faster than the drug so it does not accumulate in the body.
(b) B to C is the main metabolic route because this is a very much faster process than B to F.
(c) C to D. C will accumulate in the body. If C is pharmacologically active this could pose potential clinical problems for a patient.

(11) To avoid fatal overdoses.

(12) See section 9.9.3.4. Use *N*-methyldihydropyridine derivative as carrier. This carrier would require a substituent group that can bond to the dopamine diethanoate. The best group for this purpose is probably a carboxylic acid group because amides are slowly hydrolysed. This means that the prodrug has a good chance of reaching the blood – brain barrier in sufficient quantity to be effective. Once the prodrug has entered the brain it is easily oxidised to the quaternary salt, which, because of its charge, cannot return across the blood – brain barrier.

H H
$CONHCH_2CH_2$
OCOEt
OCOEt
N
CH_3

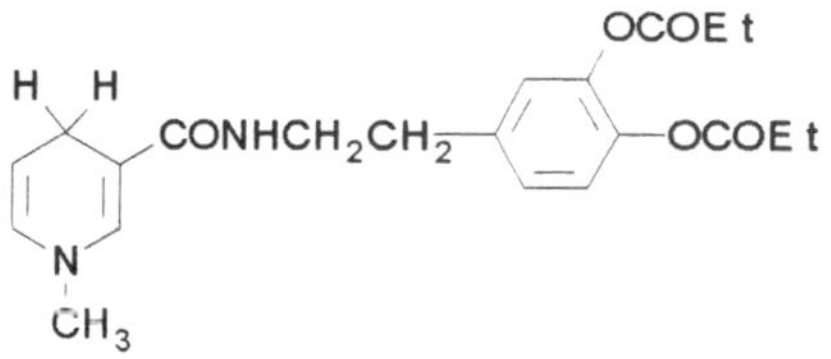

Chapter 10

(1) (a) See section 10.1. (b) See section 10.5. (c) See section 10.5.

(2) They are flat ring structures (fully conjugated structures). The positions of the appropriate functional groups when considering bond lengths and angles are complementary. The distance apart of these complementary functional groups is short enough to allow hydrogen bonding.

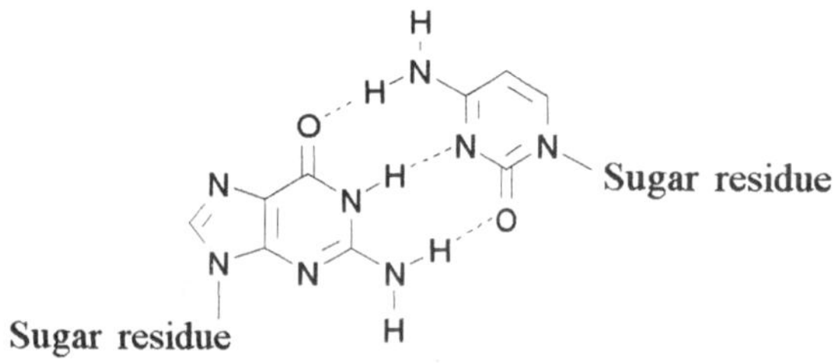

Sugar residue

Sugar residue

(3) (a) TTAGGCATCG. (b) AAUGGCUACG.

(4) Replication and acting as a source of genetic information for the synthesis of all the proteins in the body.

(5) See sections 10.1, 10.2 and 10.6. The main differences are:
   (i) DNA molecules usually have a very large relative molecular mass compared to RNA molecules;
   (ii) the structure of RNA contains the sugar residue ribose whereas that of DNA contains the sugar deoxyribose;
   (iii) RNA molecules consist of a single strand of nucleotides whereas DNA molecules consist of two nucleotide strands in the form of a supercoiled double helix.

(6) See section 10.7.

(7) The genetic code refers to mRNA and thymine does not occur in mRNA.

(8) The P site is the site where the growing peptide is attached to the ribosome. The A site is the site where the next amino acid–tRNA complex binds to the ribosome prior to the incorporation of the amino acid residue into the growing peptide.
   The E site is where the empty tRNA molecule exits from the ribosome.

(9) The first four codons are not involved in protein synthesis. Protein synthesis starts with AUG and stops with UAA. The peptide coded by the codons between the stop and start signals is:

Met-Pro-Arg-Gly-Gly-Try

(10) See sections 10.11.1 and 10.11.2.

(11) See Table 10.3 and section 10.12.2 for evidence. Changes to the substitution pattern of the aromatic ring and the side chain do not lead to analogues with an increased activity.

(12) Prepare a series of analogues based on the structure of tetracycline and test their effectiveness against suitable bacterial cultures. It is not easy to design a suitable such sequence of analogues and it will vary in detail from worker to worker. However, in this instance a list should include the following general considerations:
   (i) A series of compounds with one less ring than tetracycline but retaining the same substituent groups on the relevant rings. These substituent groups should also have the same stereochemistry as the original tetracycline molecule.
   (ii) The stereochemistry of the groups on tetracycline should be changed.
   (iii) Minor changes should be made to the positions of functional groups and ring types.
   (iv) Substituent groups should be changed for functional groups of similar size, shape and electron configuration (see section 2.3).
   (v) Substituents should be replaced by either larger or smaller substituents.
   Computer modelling is a useful technique for deciding what substituents should be used (see section 2.5). Analogues that are more active than tetracycline should be investigated in preference to any other analogues.

(13) See Figure 10.32 and section 10.13.4.

(14) See section 10.14.4.1. The tetrahydrofuran residue in the structure of the drug does not

have a 3′ hydroxy group, which prevents it forming a bond with a phosphate residue to continue the nucleic acid chain.

(15) Proflavin acts by intercalation (see section 10.13.3). The amino groups lock the drug in place by forming ionic bonds with the negative charges of the oxygens of the phosphate groups.

(16) See section 10.13.5.

(17) See section 10.13.1. Examine the chemistry of the biological pathway of the process. Decide which steps offer the best chance of intervention. Synthesise compounds with similar structures to the endogenous compounds being used in the pathway. Use fluorine and thiol groups to replace existing amino and methyl groups.

(18) See sections 10.14.2.1, 10.14.2.2 and 10.14.2.3.

(19) See sections 10.14.4, 10.14.4.1, 10.14.4.2 and 10.14.4.3.

(20) See section 10.15.1.

(21) See section 10.5.2.1.

Chapter 11

(1) (a) Dinitrogen trioxide, (b) peroxynitrite ion, (c) *S*-nitrosocysteine, (d) diethylaminoNONO-ate.

(2) (a) Nitrogen dioxide, (b) dinitrogen trioxide, (c) diethylaminoNONO-ate,
(d) CH_3CH_2S NO
Fe
CH_3CH_2S NO
(e) $ONNHCH_2COOH$

(3) Nitrosation, see section 11.3.6.2. Nitration, see sections 11.3.9.1 and 11.3.9.3.

(4) Haem iron: main metabolic pathway. Protein-bound iron: prevents the normal enzyme function and so is responsible for some pathological conditions.

(5) See section 11.3.1.1.

(6) iNOS is not Ca dependent; activation results in the continuous production of NO in a high concentration.
cNOS is Ca dependent and produces small amounts of NO in short bursts.

(7) iNOS is cytotoxic either in the physiological sense as part of the immune system or in the pathological sense in that it damages tissue.
cNOS: messenger NO for activating essential biological processes.

(8) See section 11.4.

(9) Soluble guanylyl cyclase. GTP $\rightarrow$ cGMP.
Relaxes smooth muscle leading to vasodilation and a reduction in blood pressure.

(10) For cases where there is an excess of nitric oxide, drugs that inhibit its production. For examples, see section 11.6.1.
For cases where there is a deficiency of nitric oxide, drugs that release nitric oxide. For examples, see section 11.6.2.

Genetic approach to control both types of problem. For examples, see section 11.6.3.

(11) See section 11.5.3.

(12) Use in (i) transport of NO in plasma (section 11.5) and (ii) potential NO prodrugs (section 11.6.2).

Chapter 12

(1) (a, b) See section 12.2.3. (c) See section 12.1.

(2) See section 12.2.

(3) See section 12.2.4.

(4) See section 12.3.2.

(5) See section 12.3.3.1.

(6) See Figure 12.14. Use propyl iodide instead of ethyl iodide in Figure 12.14.

(7) Synthon, see section 12.4.1.

(a)

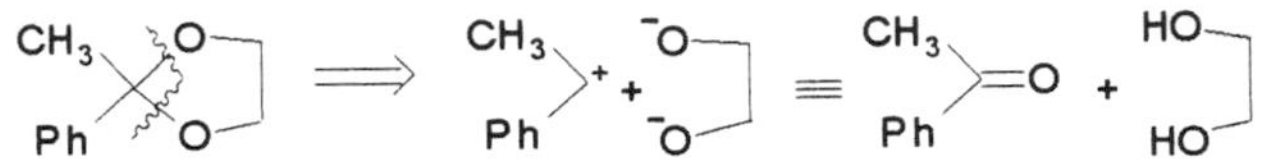

(b)

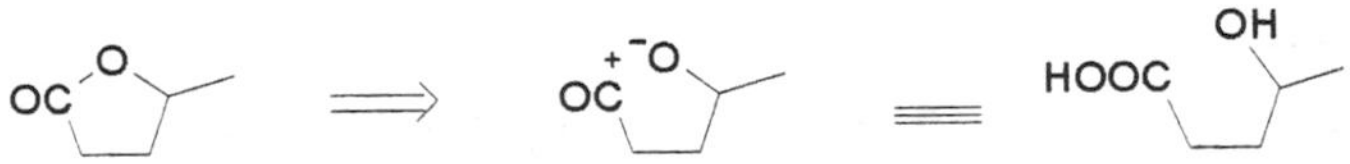

(c)

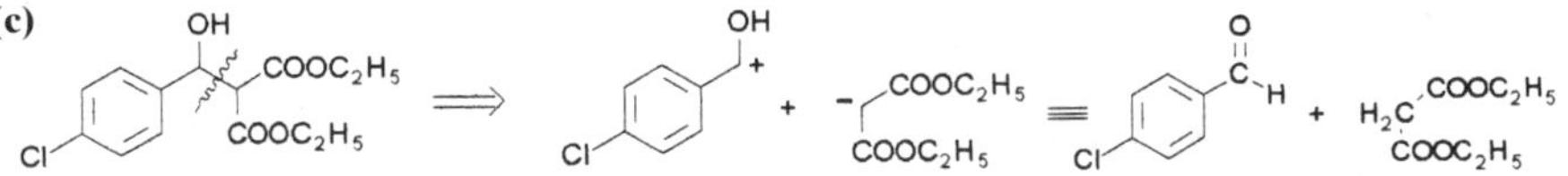

(d) A pericyclic disconnection.

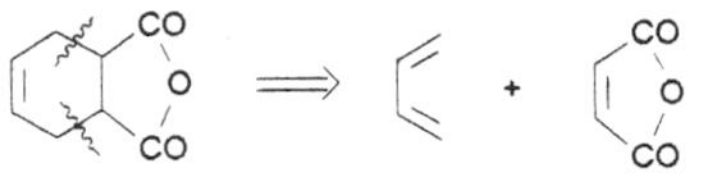

(8) FGI, see section 12.4.1.

(a)

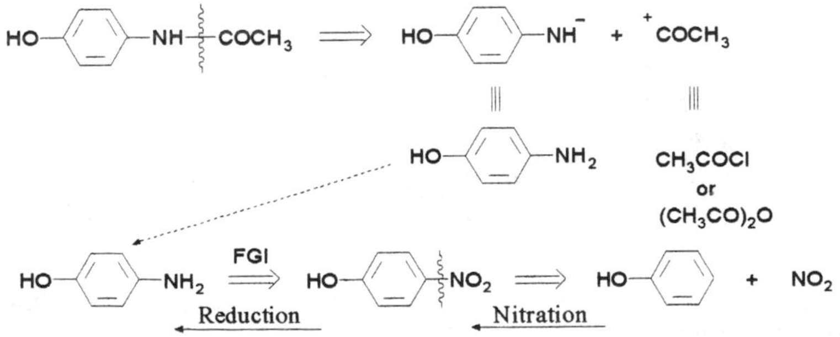

Reduction

Nitration

(b)

FGI ⟹ FGI ⟹

COOH Br OH

CN H_2O KCN

PBr_3

Reduction FGI

O O

COCl Benzene Friedel–Crafts + CH_3COCl Friedel–Crafts

Index